Recent Advances in Citrus Fruits

Sukhvinder Singh Purewal
Sneh Punia Bangar • Pinderpal Kaur
Editors

Recent Advances in Citrus Fruits

Springer

Editors
Sukhvinder Singh Purewal
University Centre for Research and Development
Chandigarh University
Mohali, Punjab, India

Sneh Punia Bangar
Department of Food, Nutrition and Packaging Sciences
Clemson University
Clemson, SC, USA

Pinderpal Kaur
Department of Food Science and Technology
Maharaja Ranjit Singh Punjab Technical University
Bathinda, Punjab, India

ISBN 978-3-031-37533-0 ISBN 978-3-031-37534-7 (eBook)
https://doi.org/10.1007/978-3-031-37534-7

This Springer imprint is published by the registered company Springer Nature Switzerland AG
The registered company address is: Gewerbestrasse 11, 6330 Cham, Switzerland

Paper in this product is recyclable.

Contents

About the Contributors

Rubai Ahmed Department of Physiology, University of Gour Banga, Malda, West Bengal, India

P. Arunkumar Department of Agricultural Entomology, Centre for Plant Protection Studies, Tamil Nadu Agricultural University, Coimbatore, Tamil Nadu, India

R. Arutselvan Regional Centre, ICAR-Central Tuber Crops Research Institute, Bhubaneswar, Odisha, India

Jhimli Banerjee Department of Physiology, University of Gour Banga, Malda, West Bengal, India

Renan Danielski Department of Biochemistry, Memorial University of Newfoundland, St. John's, NL, Canada

Sandeep Kumar Dash Department of Physiology, University of Gour Banga, Malda, West Bengal, India

Aria Dolatabadian Department of Plant Science, Faculty of Agricultural and Food Sciences, University of Manitoba, Winnipeg, MB, Canada

Isha Dudeja Department of Chemistry, Punjab Agricultural University, Ludhiana, Punjab, India

Sudip Kumar Dutta ICAR Research Complex for NEH Region, Sikkim Centre, Gangtok, Sikkim, India

K. Elango Department of Agricultural Entomology, Kumaraguru Institute of Agriculture, Tamil Nadu Agricultural University, Erode, Tamil Nadu, India

Nartunai Govindarajan Department of Pharmacognosy, Captain Srinivasa Murthy Central Ayurveda Research Institute, Central Council for Research in Ayurvedic Sciences, Chennai, Tamil Nadu, India

R. P. Jegadeeshkanth Department of Horticulture, Kumaraguru Institute of Agriculture, Tamil Nadu Agricultural University, Erode, Tamil Nadu, India

M. Kalaimani Department of Horticulture, MIT College of Agriculture and Technology, Musiri, Tamil Nadu, India

M. Kannan Department of Nano Science and Technology, Tamil Nadu Agricultural University, Coimbatore, India

Secil Karahuseyin Department of Pharmacognosy, Faculty of Pharmacy, Cukurova University, Adana, Türkiye

Avneet Kaur Department of Chemistry, University Institute of Science, Chandigarh University, Mohali, Punjab, India

Devinder Kaur Centre of Food Technology, University of Allahabad, Prayagraj, India

Norton Komora Department of Food Science and Nutrition, Faculty of Food Engineering, University of Campinas, Campinas, SP, Brazil

Bikash Kumar Department of Biosciences and Biomedical Engineering, Indian Institute of Technology, Indore, Madhya Pradesh, India

Mukesh Kumar Department of Biotechnology Engineering, Ambala College of Engineering & Applied Research Devsthali, Ambala, Haryana, India

Ravinder Kumar Department of Bio and Nano Technology, Guru Jambheshwar University of Science & Technology, Hisar, India

Sandeep Kumar Department of Physics, H.N.B. Government PG College, Prayagraj, India

Prem P. Kushwaha Department of Urology, Case Western Reserve University, Cleveland, Ohio, USA

Radha Kushwaha Centre of Food Technology, University of Allahabad, Prayagraj, India

Cláudia Maciel Universidade Católica Portuguesa, CBQF – Centro de Biotecnologia e Química Fina – Laboratório Associado, Escola Superior de Biotecnologia, Porto, Portugal

Monika Mahajan Punjab Agricultural University (PAU), Regional Research Station, Bathinda, Punjab, India

Ramandeep Kaur Mankoo Department of Chemistry, Punjab Agricultural University, Ludhiana, Punjab, India

Prathapa Reddy Maramreddy Department of Chemistry, Captain Srinivasa Murthy Central Ayurveda Research Institute, Central Council for Research in Ayurvedic Sciences, Chennai, Tamil Nadu, India

Gokul Marimuthu Department of Chemistry, Captain Srinivasa Murthy Central Ayurveda Research Institute, Central Council for Research in Ayurvedic Sciences, Chennai, Tamil Nadu, India

Rui Meneses Universidade Católica Portuguesa, CBQF – Centro de Biotecnologia e Química Fina – Laboratório Associado, Escola Superior de Biotecnologia, Porto, Portugal

Prabhat K. Nema Department of Food Engineering, National Institute of Food Technology Entrepreneurship & Management (NIFTEM), Kundli, Haryana, India

Merve Nenni Department of Analytical Chemistry, Faculty of Pharmacy, Cukurova University, Adana, Türkiye

Akansha Pandey Department of Botany, Captain Srinivasa Murthy Central Ayurveda Research Institute, Central Council for Research in Ayurvedic Sciences, Chennai, Tamil Nadu, India

Kalidas Pati Regional Centre, ICAR-Central Tuber Crops Research Institute, Bhubaneswar, Odisha, India

Pooja Department of Microbiology, Punjab Agricultural University, Ludhiana, Punjab, India

Chakra Siddappa Prashantkumar Department of Studies in Microbiology, Davangere University, Davangere, Karnataka, India

Chelladurai Premkumar Department of Pharmacognosy, Captain Srinivasa Murthy Central Ayurveda Research Institute, Central Council for Research in Ayurvedic Sciences, Chennai, Tamil Nadu, India

G. Rajiv Department of Horticulture, SRM College of Agricultural Sciences, Chengalpattu, Tamil Nadu, India

Palvi Rawal Department of Chemistry, Punjab Agricultural University, Ludhiana, Punjab, India

M. Prathapa Reddy Department of Chemistry, Captain Srinivasa Murthy Central Ayurveda Research Institute, Central Council for Research in Ayurvedic Sciences, Chennai, Tamil Nadu, India

Rohit Sadana Sant Longowal Institute of Engineering and Technology (SLIET), Longowal, Punjab, India

Sovan Samanta Department of Physiology, University of Gour Banga, Malda, West Bengal, India

Anupama Singh Department of Food Engineering, National Institute of Food Technology Entrepreneurship and Management, Sonipat, Haryana, India

Arashdeep Singh Department of Food Science & Technology, Punjab Agricultural University, Ludhiana, India

Vinti Singh Centre of Food Technology, University of Allahabad, Prayagraj, India

Sukhvinder Singh Purewal University Centre for Research & Development (UCRD), Chandigarh University, Mohali, Punjab, India

Kurra Sivadurga Department of Chemistry, Captain Srinivasa Murthy Central Ayurveda Research Institute, Central Council for Research in Ayurvedic Sciences, Chennai, Tamil Nadu, India

Sérgio Sousa Universidade Católica Portuguesa, CBQF – Centro de Biotecnologia e Química Fina – Laboratório Associado, Escola Superior de Biotecnologia, Porto, Portugal

Shweta Suri Department of Food Engineering, National Institute of Food Technology Entrepreneurship and Management, Sonipat, Haryana, India

Amity Institute of Food Technology, Amity University, Noida, Uttar Pradesh, India

Chidanandamurthy Thippeswamy Swamy Department of Microbiology, Captain Srinivasa Murthy Central Ayurveda Research Institute, Central Council for Research in Ayurvedic Sciences, Chennai, Tamil Nadu, India

Paula Teixeira Universidade Católica Portuguesa, CBQF – Centro de Biotecnologia e Química Fina – Laboratório Associado, Escola Superior de Biotecnologia, Porto, Portugal

About the Editors

Sukhvinder Singh Purewal, Ph.D., is presently working as Assistant Professor at the University Centre for Research and Development (UCRD), Chandigarh University, Mohali, Punjab, India. He also worked as Young Scientist at DST (SYST Scheme), Department of Food Science and Technology, Maharaja Ranjit Singh Punjab Technical University, Bathinda, during 2018–2022. His research interests include solid state fermentation, bioactive compounds from natural resources, antioxidants and fruits processing. He has presented his research in various national and international conferences and published more than 40 research papers/book chapters in national and international journals/books. Dr. Purewal has published one authored book entitled *Millets: Properties, Processing, and Health Benefits* and one edited book entitled *Maize: Nutritional Composition, Processing, and Industrial Uses* with CRC Press, Taylor & Francis Group. He is an active member of Association of Microbiologists of India (AMI), Mycological Society of India (MSI), Association of Food Scientists and Technologists (AFSTI), Mysore, India, and Indian Science Congress Association (ISCA). Dr. Purewal received his M.Sc., M.Phil. and Doctorate degrees in Biotechnology from Chaudhary Devi Lal University, Sirsa, Haryana (India). He has also worked on a UGC, Delhi sponsored Major Research Project during 2012–2015. He was awarded for best paper presentation in national as well as international conferences. Recently, Dr. Purewal has been awarded with prestigious "Young Scientist Award" by the "Association of Microbiologists of India (AMI)" for his work in the field of Food Microbiology.

Sneh Punia Bangar, Ph.D., is presently working as Post-doctoral Researcher at the Department of Food, Nutrition and Packaging Sciences, Clemson University, Clemson, USA. She also worked as Assistant Professor (C) at the Department of Food Science and Technology, Chaudhary Devi Lal University, Sirsa. Dr. Punia is involved in the mandated research activities of institution and has expertise on extraction and functional characterization of antioxidants, starch, their modifications and functional products. She has presented her research in various national and international conferences and published more than 150 research papers/book chapters in national and international journals/books. Dr. Punia has published more than 10 books with International publishers like Elsevier, Springer, CRC Press Taylor & Francis group and Royal Society of Chemistry. She also serves as the reviewer for various international journals. Dr. Punia received her M.Sc. and Doctorate degrees in Food Science and Technology from Chaudhary Devi Lal University, Sirsa, Haryana (India).

Pinderpal Kaur is a currently a Ph.D. research scholar at the Department of Food Science and Technology, Maharaja Ranjit Singh Punjab Technical University, Bathinda. Her research interests include antioxidant from natural resources, starch modifications and cereal technology. She has published more than 20 research papers in journals of international repute. She has actively participated in national and international conferences. She is an active life member of the Association of Food Scientists and Technologists (AFSTI), Mysore.

Abbreviations

2,4 D	2,4-Dichlorophenoxyacetic acid
5-HMF	5-Hydroxymethylfurfural
AA	Antioxidant activity
AAE	Ascorbic acid equivalent
ABTS	2,2′-Azinobis(3-ethylbenzothiazoline-6-sulfonic acid)
ACL	ATP citrate lyase
AcOEt	Ethanol ethyl acetate
Aco-GABA	Aconitase-γ-aminobutyric acid
AD	Atopic dermatitis
AD	Alzheimer's disease
AD	Anno Domini, Latin for "in the year of the Lord"
ADP	Adenosine diphosphate
AEDA	Aroma extract dilution analysis
AGE	Advanced glycation end-product
AKI	Acute kidney injury
ALS	Amyotrophic lateral sclerosis
AMPK	AMP-activated protein kinase
ANOVA	Analysis of variance
ASE	Accelerated solvent extraction
ATP	Adenosine triphosphate
BCO1	β-Carotene 15,15′-oxygenase 1 (beta-carotene 15, 15′-oxygenase 1)
BHA	Butylated hydroxyanisole
BMDMs	Bone marrow-derived macrophages
BP	Biphosphate
C. jabara	*Citrus jabara*
C. junos	*Citrus junos*
C. kinokuni	*Citrus kinokuni*
C. nobilis	*Citrus nobilis*
C. yuko	*Citrus yuko*
C/N	Carbon/nitrogen

Ca	Calcium
CA	Controlled atmosphere
CAA	Cellular antioxidant activity
CANTOS	Canakinumab Antiinflammatory Thrombosis Outcomes Study
CAPS	Cleaved amplified polymorphic sequences
CARD	Caspase recruitment domains
CCDC80	Coiled-coil domain containing 80
CDK	Cyclin-dependent kinase
CDP-ME2P	4-(Cytidine 5′-diphospho)-2-C-methyl-D-erythritol
CE	(+)-Catechin equivalent
CFB	Corrugated fiberboard
$CHCl_3$	Chloroform
CI	Chilling injury
CJ	*Citrus jabara*
CKD	Chronic kidney disease
CoA	Coenzyme A
COPD	Chronic obstructive pulmonary disease
COVID-19	Coronavirus disease 2019
COX	Cyclooxygenase
COX-2	Cyclooxygenase 2
CSE	Conventional solvent extraction
CUPRAC	Cupric ion reducing antioxidant capacity
CVDs	Cardiovascular diseases
Cyt-Aco	Cytosolic aconitase
DAS-ELISA	Double antibody sandwich enzyme-linked immunosorbent assay
DGE	Delayed gastric emptying
DHAP	Dihydroxyacetone phosphate
DMAPP	Dimethylallyl pyrophosphate
DMEM	Dulbecco's Modified Eagle Medium
DMPBQ	2,3-Dimethyl-6-phytyl-1,4-benzoquinol
DMSO	Dimethyl sulfoxide
DNA	Deoxyribonucleic acid
DOX	Doxorubicin
DPPH	2,2-Diphenyl-1-picrylhydrazyl
DW	Dry weight
DXP	1-Deoxy-D-xylulose 5-phosphate
EAE	Enzyme-assisted extraction
ELISA	Enzyme-linked immunosorbent assay
EMT	Epithelial-to-mesenchymal transition
EO	Essential oil
ERK	Extracellular signal-regulated kinase
EU	European Union
eV	Electronvolt
EW	Electrolyzed water

FAD	Flavin adenine dinucleotide
FAO	Food and Agriculture Organization
FBS	Fetal bovine serum
FD	Flavour dilution
FLG	Filaggrin protein gene
FOXO	Fork-head transcriptional factor
FPP	Farnesyl diphosphate
FRAP	Ferric reducing antioxidant potency
GABA	γ-Aminobutyric acid (gamma-aminobutyric acid)
GAD	Glutamate decarboxylase
GAE	Gallic acid equivalent
GBIF	Global Biodiversity Information Facility
GC	Gas chromatography
GC/FID	Gas chromatography with flame ionization detection
GC-MS	Gas Chromatography-Mass spectrometry
GC-O	Gas Chromatography–Olfactometry
GDP	Guanosine diphosphate
GEV	Grapefruit extracellular vesicle
GF	Glycosylated flavanones
GFR	Glomerular filtration rate
GGPP	Geranylgeranyl diphosphate
GLUT	Glucose transporters
GM-CSF	Granulocyte macrophage colony-stimulating factor
GOFA	4′-Geranyloxyferulic
GPDH	Glycerol-3-phosphate dehydrogenase
GPPS	Geranyl diphosphate synthase
GRAS	Generally Recognized as Safe
GSK3	Glycogen synthase kinase-3
GTP	Guanosine triphosphate
H1 NMR	Proton nuclear magnetic resonance
H_2O_2	Hydrogen peroxide
HAT	Hydrogen atom transfer
HDL	High-density lipoprotein
HGA	Homogentisate
$HgCl_2$	Mercuric chloride
HHP	High hydrostatic pressure
HMBPP	4-Hydroxy-3-methylbut-2-enyldiphosphate
HMDHP	6-Hydroxymethyldihydropterin
HMG	3-Hydroxy-3-methylglutaryl
HPLC	High-performance liquid chromatography
HRAR	Human recombinant aldose reductase
HRESIMS	High-resolution electrospray ionization mass spectrometry
HS-GC/MS	Headspace gas chromatography mass spectrometry
HS-SPME	Headspace solid phase micro-extraction

HS-SPME-GC-MS	Headspace solid phase micro-extraction gas chromatography mass spectrometry
HWD	Hot water dip
IBA	Indole-3-butyric acid
ICAM-1	Intercellular adhesion molecule 1
ID	Identification
IFN- γ	Interferon gamma
IFRA	The International Fragrance Association
IgE	Immunoglobulin E
IgG	Immunoglobulin G
IL	Interleukin
IMZ	Imazalil
iNOs	Inducible nitric oxide synthase
IPP	Isopentenyl diphosphate
IRS	Insulin receptor substrate
ISAAC	International Study of Asthma and Allergies in Childhood
JAK	Janus tyrosine kinase
JCP	Japanese cedar pollinosis
JDA	Japanese Dermatological Association
JNK	Jun N-terminal kinase
K	Potassium
$KMnO_4$	Potassium permanganate
LAB	Lactic acid bacteria
LDL	Low-density lipoprotein
LEKTI	Lymphoepithelial Kazal-type-related inhibitor
LGT	Limonoid glucosyltransferase
LOD	Limit of detection
LOQ	Limit of quantification
LPS	Lipopolysaccharide
m/z	Mass/charge ratio
MAE	Microwave-assisted extractions
MAP	Modified atmosphere packaging
MAPK	Mitogen-activated protein kinase
MDA	Malondialdehyde
MEcPP	2-C-methyl-derythritol 2,4-cyclodiphosphate
MEP	2C-methyl-d-erythritol-4-phosphate
MMD	Mineral metabolism disorder
MMT	Million metric tons
MPa	Mega Pascal
MPBQ	2-Methyl-6-phytyl-1,4-benzoquinol
MT	Metric ton
MT	Million metric tonnes
MUC5AC	Mucin-5AC
MVA	Mevalonic acid
NAA	Naphthalene acetic acid

NADP	Nicotinamide adenine dinucleotide
NB-UVB	Narrow-band UV B
NCBI	National Center for Biotechnology Information
NCD	Non-communicable diseases
NCI-H292	Human lung mucoepidermoid carcinoma cells
NET	Neutrophil extracellular trap
NFC	Not from concentrate
NF-κB	Nuclear factor kappa B
NGS	Next-generation sequencing
NK	Natural killer cell
NLEA	The Nutrition Labeling and Education Act of 1990
NO	Nitric oxide
Nrf2	Nuclear factor erythroid 2-related factor 2
OA	Osteoarthritis
OD	Optical density
OHCs	Oxygenated heterocyclic compounds
ORAC	Oxygen radical absorbance capacity
P5CS	Δ^1-Pyrroline-5-carboxylate synthetase
PAA	Peroxyacetic acid
pABA	*para*-Aminobenzoic acid
PAL	Phenylalanine ammonia lyase
PARP1	Poly(ADP-ribose) polymerase 1
Pb	Lead (in Latin-Plumbum)
PCR	Polymerase chain reaction
PD	Parkinson's disease
PE	Polyethylene
PEF	Pulsed electric field
PET	Polyethylene terephthalate
PFE	Pressurized fluid extractions
PGA	Phosphoglyceric acid
PHSE	Pressurized hot-solvent extraction
PHWE	Pressurized hot-water extraction
PI3K	Phosphoinositide 3-kinases
PLCγ	Phospholipase C-gamma
PLE	Pressurized liquid extractions
PMF	Polymethoxylated flavones
PON	Paraoxonase
PP	Pomelo peel
Ppm	Parts per million
PPP	Phytyl pyrophosphate
PVC	Polyvinyl chloride
QD	Quick decline
QE	Quercetin equivalent
qPCR	Quantitative-polymerase chain reaction

RANTES	Regulated upon activation, normal T cell expressed and secreted
RAPD	Random amplification of polymorphic DNA
RFLP	Restriction fragment length polymorphism
RH	Relative humidity
RNS	Reactive nitrogen species
ROS	Reactive oxygen species
RT-PCR	Reverse transcriptase-polymerase chain reaction
RTS	Ready to serve
SAFE	Solvent-assisted flavour evaporation
SARS-CoV	Severe acute respiratory syndrome coronavirus
$ScCO_2$	Supercritical CO_2
SCF	Supercritical fluid
sEH	Soluble epoxide hydrolase
SET	Single electron transfer
SFE	Supercritical fluid extractions
SHWE	Superheated water extraction
SOD	Superoxide dismutase
SOPP	Sodium-orthophenylphenate
SP	Stem pitting
SPINK-5	Serine protease inhibitor Kazal-type 5
SPS	Sucrose-phosphate synthase
STG	Shoot-tip grafting
SWE	Subcritical water extraction
TA	Titratable acidity
TAE	Tannic acid equivalent
TAL	Tyrosine ammonia lyase
TAW	Total aliphatic wax
TBARS	Thiobarbituric acid reactive substances
TBZ	Thiabendazole
TCA	Tricarboxylic acid
TCD	Thermal conductivity detector
TCS	Topical corticosteroids
TE	Trolox equivalent
TFC	Total flavonoid content
TGFB1	Transforming growth factor beta1
Th2	T Helper 2
TLC	Thin layer chromatography
TLR 2	Toll-like receptor 2
TLR	Toll-like receptor
TNF	Tumor necrosis factor
TNF-α	Tumor necrosis factor alpha
TPC	Total phenolic content
TSLP	Thymic stromal lymphopoietin
TSS	Total soluble solids

TTA	Total titratable acids
TTC	Total tannin content
TTS	Topical tacrolimus
TXB2	Thromboxane B2
UAE	Ultrasound-assisted extractions
UK	United Kingdom
UN	United Nations
UNEC	United Nations Economic Commission for Europe
UPLC	Ultra-performance liquid chromatography
US FDA	US Food and Drug Administration
US patent	United States patent
US	United States
USA	United States of America
USDA	United States Department of Agriculture
UTI	Urinary tract infection
UV B	Ultraviolet B
UV	Ultra violet
VCAM-1	Vascular cell adhesion molecule 1
VOC	Volatile organic compounds
WGS	Whole genome sequencing
WIPO	World Intellectual Property Organization
Wnt	Wingless/integrated
WP	Wettable powder
WST-8	Water-soluble tetrazolium salt

Chapter 1
Lemon and Lime

Avneet Kaur, Sukhvinder Singh Purewal, Arashdeep Singh Randhawa, Chidanandamurthy Thippeswamy Swamy, Bikash Kumar, Mukesh Kumar, and Ravinder Kumar

1.1 Introduction

Fruit plants with medicinal properties are gaining more consideration from researchers and pharmaceutical industries throughout the world. Fruits which possess high antioxidant properties are as important as medicines (Neri-Numa et al. 2020; Arruda

A. Kaur (✉)
Department of Chemistry, University Institute of Sciences, Chandigarh University, Mohali, Punjab, India
e-mail: avneet.e13480@cumail.in

S. Singh Purewal (✉)
University Centre for Research & Development (UCRD), Chandigarh University, Mohali, Punjab, India
e-mail: sukhvinder.e14148@cumail.in

A. S. Randhawa
Department of Food Science & Technology, Punjab Agricultural University, Ludhiana, India

C. T. Swamy
Department of Microbiology, Captain Srinivasa Murthy Central Ayurveda Research Institute, Central Council for Research in Ayurvedic Sciences, Arignar Anna Government Hospital Campus, Chennai, Tamil Nadu, India

B. Kumar
Department of Biosciences and Biomedical Engineering, Indian Institute of Technology, Indore, Madhya Pradesh, India

M. Kumar
Department of Biotechnology Engineering, Ambala College of Engineering & Applied Research Devsthali, Ambala Cantt, Haryana, India

R. Kumar
Department of Bio and Nano Technology, Guru Jambheshwar University of Science & Technology, Hisar, India

S. Singh Purewal et al. (eds.), *Recent Advances in Citrus Fruits*,
https://doi.org/10.1007/978-3-031-37534-7_1

Fig. 1.1 Botanical description of lemon and lime

et al. 2018; Pereira et al. 2018). Because their extracts, juices and other products based on them could be used as remedy to treat various diseased conditions and to improve the health of an individual. Secondary metabolites with health benefiting antioxidant properties are accumulated in various parts of plants ranging from roots, shoots, flowers, bark, leaves, fruits, seeds and even in fruit peels (Kaur et al. 2018; Dhull et al. 2016; Roesler et al. 2006; Malta et al. 2013; Pavan et al. 2014). Along with secondary metabolites fruits are also considered as rich source of vitamins and minerals which are usually involved in sustaining various biochemical processes within the body. India stands at top position in terms of lemon and lime productions (FAO 2020). Lemon belongs to *Rutaceae* family and possesses broad spectrum applications in food and pharmaceutical industries (Xi et al. 2017; Hamdan et al. 2011). Lime is another important dark green colored citrus fruit which is as important as lemon (Fig. 1.1). As compared to lemon the amount of sugar present in lime is higher. Juice extracted from lime is widely used in the preparation of limeade. Distinctive feature of lemon juice makes it an important substrate for preparation of lemon meringue pie and lemonade. Lemons and lime are good source of minerals like magnesium, calcium and potassium which are necessary for sustaining health life style. Presence of wide range of nutrients in lemon and lime makes them important from nutritional point of view. Chen et al. (2017) reported that citrus fruits are famous throughout the World for their health benefiting antioxidant potential. This chapter on lemon and lime provide in depth information about the nutritional composition, bioactive profile and health benefits of lemon and lime.

1.2 Lemons/Lime Production

Details of lemons and lime production in India and worldwide in terms of area and yield since last 10 years (2008–2018) are reported in Table 1.1 and Table 1.2 respectively. FAO (2020) data indicate that top five lemons and lime producers in the World were India (3,148,000 tonnes) followed by Mexico (2,547,834 tonnes); China, Mainland (2,482,884 tonnes); Argentina (1,989,400 tonnes) and Brazil (1,481,322 tonnes). In 2018, India was the top producer of lemons and lime with production (3,148,000 tonnes) (Fig. 1.2). Lemons and lime production in India was observed maximum during the 2018 (3,148,000 tonnes) and minimum during 2011

Table 1.1 Lemon and lime production detail of India since 2008–2018 (FAO 2020)

	2008	2009	2010	2011	2012	2013	2014	2015	2016	2017	2018
Area (ha)	302,800	316,050	295,600	219,000	234,000	255,200	286,410	268,000	245,000	248,000	286,000
Yield (hg/ha)	82,619	81,365	88,945	96,256	97,098	98,883	98,985	110,075	99,510	95,323	110,070
Production (tonnes)	2,501,700	2,571,530	2,629,200	2,108,000	2,272,100	2,523,500	2,835,020	2,950,000	2,438,000	2,364,000	3,148,000

Table 1.2 Worldwide production of Lemon and lime since 2008–2018 (FAO 2020)

	2008	2009	2010	2011	2012	2013	2014	2015	2016	2017	2018
Area (ha)	1,077,555	1,111,454	1,039,911	954,964	972,717	1,002,014	1,018,325	1,057,586	1,052,240	1,089,434	1,267,401
Yield (hg/ha)	156,373	147,842	141,594	157,695	154,481	154,035	156,129	159,874	163,787	160,234	152,823
Production (tonnes)	16,850,089	16,431,911	14,724,505	15,059,330	15,026,675	15,434,482	15,899,014	16,908,006	17,234,276	17,456,457	19,368,838

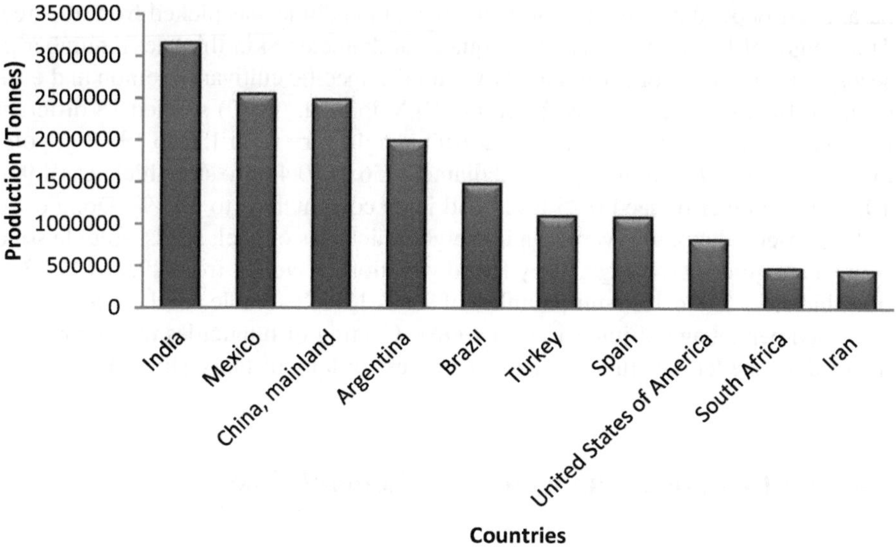

Fig. 1.2 Top producers of lemon and lime

(2,108,000 tonnes) (Table 1.1). Lemons and lime production (tonnes) at World level was observed maximum during 2018 (19,368,838 tonnes) and minimum during 2010 (14,724,505 tonnes) (Table 1.2).

1.3 Agrarian Conditions

The lemon tree generally requires constant sunlight for its maintenance and growth. The duration of sunlight significantly affects both qualities of fruit as well as quantity of specific nutrients in the fruits. Lemon tree which are grown under proper sunlight yields fruits with more sugar than those grown under shade. Temperature has significant effect on the lemon tree; the temperature below the required range damage flowers, foliage and fruits thereby affecting the consumer preference. The temperature below −1.66 °C may results in damage to fruits, woody stem and may result in leaf drop. Defoliation of lemon tree starts at temperature range −5.5 °C to −4.4 °C. Lemon tree is capable to grow in a variety of soil ranging from silty clam loam to sandy and even black soil. The pH requirement for the optimal growth of lemon tree is from 5.5 to 6.5. Lemon tree can tolerate heavy rainfall as well as humid conditions of environment. As compared to lemon, lime is more famous in India and widely cultivated in area which has low rainfall, dry climate having alluvial, loamy and black soil. In India the major lime producing area is Bihar, Rajasthan, Gujarat, Tamilnadu (T.N.) and Andhra Pradesh (A.P.). Consumer acceptability towards lemons varies with the size, color and juice content of lemon/lime. These parameters depend on the stage at which the lemons/lime was picked from the tree. The weight of fruit, polar diameter, equatorial diameter, skin thickness, number of seeds and percentage of juice may vary with the specific cultivar of lemon and even during different stages of growth/season. Di Vaio et al. (2010) studied 18 different lemon cultivars and they observed that fruit weight vary from 122 to 198.3 g; polar diameter 76.5 to 94.9 mm; equatorial diameter 56 to 70.4 mm; skin thickness 4.9 to 10.9 mm; number of seed 0.25 to 21 and juice content 16.2 to 30.3%. Dong et al. (2019) observed monthly variation in weight, thickness of peel, seeds, soluble solid content and juice percentage. They found variation on weight from 49.9 to 177.3 g; peel thickness 5.0 to 10.6 mm; number of seeds 12–32; soluble solid content 6.9 to 8.4% and percentage of juice 15.6 to 49.8%. Coating of fungicides may be recommended for the lemons/lime which is in process of long distance shipping.

1.4 Post Harvest Management of Lemon/Lime

Both lemon and lime comes under the category of non-climacteric fruits (which don't have capability to ripe after harvesting). Due to their non-climacteric nature, the shelf life is longer as compared to other fruits which make them suitable for marketing at long distances. After harvesting, lemon/lime has to go through sequential steps to enter in market for consumers use. Packing and transportation of fruits is one of the crucial phases for maintaining freshness in them. During transport, transpiration is one of the physical factors that affect their quality. During transpiration water starts evaporating from the fruit surface thereby resulting in textural changes and deterioration of quality (Ladaniya 2010). Due to improper handling and poor storage conditions the fruits suffer significant losses. Dirty bins and poor aeration during transportation and storage chamber result in fruit decaying. Prior to storage it is necessary to clean bins as well as fruits. Hot water dip is one of the physical treatments which have been applied as post harvest application for maintaining the fruit quality. During the treatment, lemon/lime has to be dipped in water (40 °C for 2–3 min) (Fallik 2004; Spadoni et al. 2013). The treatment is advantageous as it cleans the fruit and prevents the growth of surface microbes. Treatment with mild fungicide could also be useful to maintain the quality. Use of carbonate and bicarbonate salts (3% w/v) for fruits proved to more effective against attack of certain bacterial strains (Palou et al. 2001). To reduce the rate of fruit senescence, ozone gas is recommended in storage chamber as it has broad spectrum antimicrobial potential (Guzel-Sedym et al. 2004). Surface coating of both the fruits may be done to maintain the weight of fruit, reduce transpiration rate and chilling injuries during transportation. From consumer safety point of view edible coating is recommended to improve fruit appearance (Purewal and Sandhu 2020; Dhall 2012). Generally used coating materials include beeswax, shellac, carnauba and polyethylene (Purewal and Sandhu 2021).

1.5 Nutritional Profile

Both lemon and lime are rich source of specific nutrients including minerals/vitamins and bioactive compounds. Throughout the World, researchers are working on different fractions of lemon and lime fruit. The fruits are available in the market with different thickness of peel, coloration pattern, varying juice content and organoleptic properties. Internal juicy portion of lemon/lime is surrounded by an outer protection skin known as flavedo/exocarp (colored skin with oil secreting glands). Just next to flavedo another important part is Albedo (white spongy tissues) which varies in thickness with complex network of vascular bundles. Edible part of lemon/lime has small size densely packed juicy sacs which have juice and seeds. Czech et al. (2020) studied the mineral profile of different parts of lemon and lime and reported potassium (113–127 mg 100 g^{-1}); sodium (1.86–1.99 mg 100 g^{-1}); calcium (18–31.8 mg 100 g^{-1}); phosphorus (18–23.9 mg 100 g^{-1}); magnesium (8.4–11.5 mg 100 g^{-1}); iron (0.28–0.34 mg 100 g^{-1}); zinc (0.17–0.28 mg 100 g^{-1}); copper (0.04 mg 100 g^{-1}); manganese (0.05–0.06 mg 100 g^{-1}) and selenium content (1.78–4.13 mg 100 g^{-1}) (Table 1.3). Similarly for lime fruit the minerals present were: potassium (145–152 mg 100 g^{-1}); sodium (3.10–3.88 mg 100 g^{-1}); calcium (41.3–63.9 mg 100 g^{-1}); phosphorus (17.9–20.1 mg 100 g^{-1}); magnesium (11.6–13 mg 100 g^{-1}); iron (0.21–0.41 mg 100 g^{-1}); zinc (0.24–0.26 mg 100 g^{-1}); copper (0.04–0.06100 g^{-1}); manganese (0.01–0.04 mg 100 g^{-1}) and selenium (1.98–3.10 mg 100 g^{-1}). Fatin and

Table 1.3 Mineral profile of Lemon and lime (different parts)

Specific mineral	Lemon pulp	Lemon peel	Lemon (whole fruit)	Lime pulp	Lime peel	Lime (whole fruit)	References
Potassium (mg 100 g^{-1})	113	127	120	101–168	142–187	147	Czech et al. (2020) and Barros et al. (2012)
Sodium (mg 100 g^{-1})	1.89	1.99	1.86	1.6–3.10	3.88–54	3.14	
Calcium (mg 100 g^{-1})	18	31.8	25.9	41.3	63.9–214	51	
Phosphorus (mg 100 g^{-1})	18	23.9	21.8	17.9	20.1	19	
Magnesium (mg 100 g^{-1})	8.4	11.5	9.86	4.8–11.6	13–41	12.2	
Iron (mg 100 g^{-1})	0.28	0.34	0.31	0.10–0.21	0.41–0.94	0.31	
Zinc (mg 100 g^{-1})	0.17	0.28	0.22	0.093–0.24	0.15–0.26	0.24	
Copper (mg 100 g^{-1})	0.04	0.04	0.04	0.02–0.06	0.05–0.07	0.04	
Manganese (mg 100 g^{-1})	0.06	0.05	0.05	0.01–0.036	0.04–0.20	0.03	
Selenium (μg 100 g^{-1})	1.78	4.13	2.77	1.98	3.10	2.53	

Azrina (2017) reported the presence of ascorbic acid in lime, musk lime and kaffir lime. They observed that Kaffir lime posses' ascorbic acid of 37.2 mg 100 g^{-1} followed by lime (27.78 mg 100 g^{-1}) and musk lime (18.6 mg 100 g^{-1}). Chuku and Akani (2015) studied the proximate composition of lemon and lime. They observed ash (0.50%); lipids (1.15%); fiber (1.52%) and carbohydrates (10.61%). However, in lime they found ash content of 0.15% followed by lipid (0.05%); fiber (0.05%) and carbohydrate content (23%). Paul and Shaha (2004) studied the proximate composition and concentration of vitamins in lemon. They reported that lemon possess β-carotene (50 mg 100 g^{-1}); ascorbic acid (37 mg 100 g^{-1}); thiamine (0.02 mg 100 g^{-1}) and riboflavin (0.01 mg 100 g^{-1}). Further, results of proximate analysis indicate the presence of moisture (84.2%); fiber (1.6%); carbohydrates (10.8%) and protein (0.9%). Scientific studies demonstrated that oil extracted from both lemon and lime is an excellent source of specific fatty acids (Table 1.4).

1.6 Bioactive Profile

Oxidative stress inhibition and immunity enhancing properties in fruits is mainly due to presence of specific nutrients along with health benefiting bioactive compounds. Bioactive compounds are widely distributed in different fruit parts even the peel that are assumed as waste material could also be an important source of bioactive constituents (Singh et al. 2020). Fruits such as lemon and lime have long been used in traditional and skincare medicines as they are considered in category of natural resources which possess secondary metabolites of industrial importance (Wang et al. 2017; Hamdan et al. 2011; Pelletier et al. 2008). Bioactive compounds are group of specific metabolites that provide them free radical scavenging potential (Kaur et al. 2018; Dhull et al. 2016; Salar and Purewal 2016). Fruits that belong to family *Rutaceae* are well known for the presence of specific bioactive polyphenols. Earlier published reports demonstrated that as compared to other parts of fruits peel possess higher amount of bioactive constituents (Singh et al. 2020; Gorinstein et al. 2001). Phytochemical profile of citrus fruit peel indicates the presence of carbohydrates, alkaloids, tannins, fats, oils, cardiac glycosides, phytosterols, saponins, steroids, flavonoids and terpenes (John et al. 2017; Chauhan and Saxena 2019; Mathew et al. 2012) (Table 1.5). Guimaraes et al. (2010) studied crude juice, polar fraction of pulp and peel of lemon as well as lime. For lemon, they observed that the amount of phenolics present in crude juice was 11 mg GAE/g; juice polar fraction (8.4 mg GAE/g) and peel polar fraction (87.7 mg GAE/g). For lime, the phenolic in different fractions including crude juice was 9 mg GAE/g; juice polar fraction (7.5 mg GAE/g) and peel polar fraction (124.6 mg GAE/g) (Table 1.6). He et al. (2011) observed ferulic acid (14.4–97.8 μg/g); caffeic acid (4.5–30 μg/g) and chlorogenic acid (8.8–18.7 μg/g) in peels of citrus hybrids. Bioactive profile of lemon peel indicates the presence of ferulic acid (0.45 mg/g); p-coumaric acid (0.35 mg/g); caffeic acid (0.14 mg/g) and sinapic acid (0.42 mg/g) (Gorinstein et al. 2001). Casquete et al. (2015) studied lemon and lime peel for the detection of phenolic compounds

Table 1.4 Fatty acid profile and chemical composition of oil

Fruit part/experimental material	Nutrients	Amount	Reference
Citrus Seeds oil	Oil content (%)	34.92	Malacrida et al. (2012)
	Palmitic acid	21.03	
	Palmitoleic acid	0.65	
	Stearic acid	3.67	
	Oleic acid	20.8	
	Linoleic acid	44.31	
	Linolenic acid	8.96	
	Arachidic acid	0.31	
	Behenic acid	0.10	
	Lignoceric acid	0.17	
	Unsaturated fatty acids	74.72	
	Monounsaturated fatty acids	21.45	
	Polyunsaturated fatty acids	53.27	
	Essential fatty acids	53.27	
	Saturated fatty acids	25.28	
Lemon peel oil	α-thujene	0.82%	Baba et al. (2016)
	α-pinene	0.20%	
	Camphene	0.26%	
	β-pinene	8.81%	
	Myrecene	2.96%	
	α-phellandrene	0.19%	
	α-terpinene	0.73%	
	p-cymene	0.74%	
	Limonene	54.4%	
	β-cis-ocimene	0.20%	
	β-trans-ocimene	0.37%	
	γ-terpinene	12%	
	Terpinolene	2.08%	
	cis-p-mentha-2,8 dienol	0.29%	
	Linalool	0.76%	
	Fenchol	0.19%	
	Borneol	0.15%	
	Terpinene-4-ol	2.11%	
	α-terpineol	3.45%	
	cis-Geraniol	0.32%	
	β-citral	0.37%	
	trans-Geraniol	0.49%	
	Geranial	0.15%	
	Citronellol acetate	0.46%	
	Neryl acetate	1.81%	
	Geranyl acetate	1.60%	
	β-Caryophyllene	0.66%	
	α-trans-bergamotene	0.79%	
	(Z)-β-Farnesene	0.11%	
	Valencene	0.46%	
	α-selinene	0.16%	
	β-bisabolene	1.74%	

Table 1.5 Phytochemical profile of lemon/lime

Specific phytochemicals/tests	Methanol	Acetone	Chloroform	Petroleum ether	Ethyl acetate	Aqueous extract	Ethanol	References
Carbohydrates	+	−	+	+	+	+	+	John et al. (2017), Chauhan and Saxena (2019) and Mathew et al. (2012)
Alkaloids	+	−	+	+	+	+	−	
Tannins	−	−	+	−	+	+	+	
Fats and oils	+	−	+	+	+	+	+	
Reducing sugar	−	−	−	−	−	−	−	
Protein	−	−	−	−	−	+	+	
Cardiac glucosides	+	+	−	−	+	+	+	
Steroids	+	−	+	−	+	+	+	
Phytosterols	+	+	−	−	−	−	+	
Phenols	+	+	+	+	+	+	+	
Flavonoids	+	+	+	+	+	+	+	
Saponins	+	−	−	−	+	−	+	
Terpenes	+	+	+	−	+	−	−	

Table 1.6 Bioactive compounds/sugars in different fractions of lemon and lime

Specific compounds/sugars	Lemon zest	Lemon flesh	Lemon peel polar fraction	Lemon Juice polar fraction	Lemon crude juice	Lime peel polar fraction	Lime juice polar fraction	Lime crude juice	References
Phenolics (mg/g)	204.4	105.5	87.77	8.43	11.17	124.63	7.51	9.01	Guimaraes et al. (2010), Makni et al. (2018) and Bhat et al. (2011)
Flavonoids (mg/g)	27.5	56.1	15.96	1.43	0.22	13.61	2.36	0.26-0.43	
Ascorbic acid (µg/g)	–	–	938	348.76	417.44	1779.55	280.40	190.52	
Carotenoids (µg/g)	–	–	1.59	0.07	0.06	1.27	0.17	0.08	
Reducing sugars (mg/g)	–	–	291.26	6.88	22.43	38.34	6.83	23.81	
Flavonols (mg/g)	26.6	9.2	–	–	–	–	–	0.62	
Condensed tannins (mg/g)	138.3	26.6	–	–	–	–	–	–	

in them. They observed that phenolic acid was present in lemon peel (222.76 mg GAE/100 g) and lime peel (362.98 mg GAE/g). Londono et al. (2010) reported the presence of phenolic compounds in Tahiti lime peel (74.8 mg GAE/g). Gomez-Mejia et al. (2019) studied lemon peel and found phenolic compound of 5.9 mg GAE/g. Malacrida et al. (2012) studied the tocopherol and carotenoids composition of lemon seeds oil. Their observations show the presence of α-tocopherol (102.49 mg/kg); β-tocopherol (2.2 mg/kg) γ-tocopherol (1.33 mg/kg) and δ-tocopherol (18.98 mg/kg). Total carotenoid content was 4.36 mg/kg out of which major portion was for luteina 3.53 mg/kg followed by β-carotene (0.63 mg/kg) and β-cryptoxanthin (0.20 mg/kg).

1.7 Extraction of Bioactive Constituents

Bioactive compounds with broad spectrum health benefiting antioxidant properties have substantial potential to be used as raw substrate/material for the preparation of skin care cosmetic products, functional food products of industrial importance and pharmaceuticals. Scientific studies reveal the presence of antimicrobial, antimutagenicity, anti-cancerous and anti-ageing properties in bioactive compounds (Putnik et al. 2017; Bursac Kovacevic et al. 2016; Forbes-Hernandez et al. 2014). Substantial interest of researchers has been generated in citrus fruits owing to their health benefiting properties and economic importance. Sensitivity of target bioactive constituents makes the extraction techniques a difficult process for valorization of citrus fruits. Before performing extraction process it is most important to under the complexity of target bioactive compounds, temperature sensitivity, solubility and concentration of extraction phase along with required time duration (Purewal et al. 2020; Salar et al. 2016). Extraction process for the recovery of phenolic from natural resources could be performed via supercritical extraction (SE); microwave assisted extraction methods (MAEM); ultrasound assisted extraction process (UAEP) (Yaqoob et al. 2020; Mhiri et al. 2014; Fakhari et al. 2005). Al-Qassabi et al. (2018) studied the effect of extraction phases (hexane, chloroform, ethyl acetate, butanol, methanol and water) on phenolic compounds of lemons peel. Phenolic compounds in lemon peel was present in range from 5.07 to 25.24 mg/g. Highest amount of phenolics (25.24 mg/g) was observed in extract prepared using methanol whereas lowest amount (5.07 mg/g) was observed in extract prepared using butanol. Lopresto et al. (2014) compared the efficiency of High pressure-high temperature extraction (HPHTE) and conventional soxhlet extraction method for the recovery of D-limonene from lemon peel. The output of their investigation indicated that HPHTE method was superior over conventional extraction method as it save energy and process could be completed with a remarkable period of time with high yield. Further they noted that pre-drying of peel before extraction resulted in twofold increase in yield as compared to fresh one. Dong et al. (2019) studied the variation in phenolics and flavonoids content in methanolic extracts (80%) of lemon pulp and peel with respect to time. Phenolics in lemon pulp were in range from 1.5 to 3.5 mg GAE/g whereas lemon peel indicated the amount of phenolics in range from 4.2 to 7.9 mg GAE/g. Similarly total flavonoid content in lemon pulp and peel was 0.6 to 1.4 mg RE/g and 3.7 to 6.2 mg RE/g respectively. Wang et al. (2019) studied finger lime extract prepared in methanol (90%) and they found malic acid (1.3–4.8 mg/g); citric acid (8.2–73.4 mg/g) and quinic acid (0.5–5.4). Extraction detail to extract important bioactive constituents from lemon and lime is reported in Table 1.7.

1.8 Health Benefits of Specific Compounds

1.8.1 Hesperidin

Nowadays, benefits of bioactive constituents including flavonoids such as hesperidin are highly appreciated. Lemon is a rich source of hesperidin (1465 μg/g), the higher concentration being in pulp (622.8 μg/g) followed by juice (157.4 μg/g) (Xi et al. 2017). Saeidi et al. (2011) use HPLC method for the determination of hesperidin in lime juice (instrumental and hand squeezed extraction). They found that lime (instrumental extraction) collected from Jahrom, Minab and Rudan had hesperidin in the range from 13.2 to 19.8 μg/ml. However, the hand squeezed lime juice samples showed the compound in range between 10.4–15.1 μg/ml. Experimental studies reported that administration of flavonoids (hesperidin) may provide protection against certain chronic disorders, cancer and diabetes (Sugasawa et al. 2018; Xue et al. 2016; Homayouni et al. 2017). Wound healing is one of the challenging tasks especially in diabetes where the healing rate is very slow. Experimental observation shows the wound healing potential of hesperidin. Wessels et al. (2014) demonstrated that inclusion of hesperidin (0.05%) in medium for a period of 24 h results in significant wound healing/closing (39%). Li et al. (2018) observed that oral administration of hesperidin (100 mg kg^{-1}) for a duration of 21 days results in 97% success rate in wound healing.

1.8.2 Ferulic Acid

Ferulic acid is ubiquitous bioactive compound which arise from the metabolism of tyrosine and phenylalanine. It occurs in a variety of natural resources including fruits, vegetables and cereal grains. Presence of three distinct structural motifs makes ferulic acid active against free radicals/oxidative stress. Antioxidant properties in ferulic acid are mainly due to the presence of phenolic nucleus and unsaturated side chain which helps in forming phenoxy radical (resonance stabilized). Xu et al. (2008) studied lemon juice for the presence of specific bioactive constituents and they observed ferulic acid as major metabolite (35.77 mg GAE/L). Scientific studies confirm that as compared to other bioactive flavonoids, ferulic acid is readily

Table 1.7 Detailed extraction conditions and total phenolic compounds in lemon and lime

Source	Extraction phase	Concentration	Bioactive compounds	Specific compounds identified	References
Lemon (peel)	Methanol	80%	TPC 3.99 μg GAE/g	Gallic acid, chlorogenic, caffeic acid, eriocitrin, hesperidin, rutin, eriodictyol, hesperetin	Xi et al. (2017)
Lemon (pulp)	Methanol	80%	TPC 2.70 μg GAE/g	Chlorogenic, caffeic acid, hesperidin, rutin	
Lemon (juice)	Methanol	80%	TPC 0.52 μg GAE/g	Gallic, chlorogenic, caffeic acid, hesperidin, rutin, hesperetin	
Lemon juice	Methanol	80%	TPC 751.82 (mg GAE/L)	Caffeic p-Coumaric Ferulic Sinapic Protocatechuic p-Hydroxybenzoic Vanillic	Xu et al. (2008)
Peeled lemon	Ethanol	95%	TPC 164 mg/100 g	Ferulic acid Sinapic acid p-coumaric acod Caffeic acid ascorbic acid	Gorinstein et al. (2001)
Lemon peel	Ethanol	95%	TPC 190 mg/100 g	Ferulic acid Sinapic acid p-coumaric acod Caffeic acid ascorbic acid	
Lemon peel polar fraction	Water	Ultrapure distilled	TPC 87.7 mg GAE/g	–	Guimarães et al. (2010)
Lemon juice polar fraction			TPC 8.43 mg GAE/g		
Crude juice			TPC 11.17 mg GAE/g		
Lemon peel polar fraction	Water	Ultrapure distilled	TPC 124.63 mg GAE/g	–	
Lemon peel polar fraction			TPC 7.51 mg GAE/g		
Crude juice			TPC 9.01 mg GAE/g		

(continued)

Table 1.7 (continued)

Source	Extraction phase	Concentration	Bioactive compounds	Specific compounds identified	References
Lemon peel	Ethanol	40%	TPC 5.9 mg GAE/g	Coumaric acid, Ferulic acid, Rutin hesperidin	Gómez-Mejía et al. (2019)
Kaffir lime peel	Ethanol	52.9%	TPC 1291.8 mg GAE/g	–	Chan et al. (2009)
Lemon peel	Methanol		TPC 103 mg/g	Hesperidin Naringin Neohesperidin Narirutin	Huang and Ho (2010)
Lemon	Acetone	–	TPC 222.76 mgGAE/100 g	–	Casquete et al. (2015)
Lime			TPC 362.98 mgGAE/100 g		
Lemon	Methanol	–	TPC 223.2 mgGAE/100 g	–	Ghasemi et al. (2009)
Tahiti lime	Acetone	–	74.80 mg GAE/g	Hesperidin, Neohesperidin, Diosmin, Nobiletin and Tangeritin	Londoño-Londoño et al. (2010)
Lemon seed	Methanol	–	1196.71 mg GAE/kg	–	Malacrida et al. (2012)
Lemon peel	Acetone (microwave-assisted extraction)	40&50%	12.09 mg GAE/g	Gallic acid, Chlorogenic acid, Caffeic acid p-coumaric acid, Ferulic acid, Rutin, quercetin, Catechin	Nayak et al. (2015)
	Acetone (ultrasound-assisted extraction)		10.35 mg GAE/g		
	Acetone (accelerated solvent extraction)		6.26 mg GAE/g		
	Acetone (conventional solvent extraction)		10.21 mg GAE/g		
Lime leaves	Ethanol	–	96.5 μg GAE/mg	–	Namania et al. (2018)
Lemon	Methanol	80%	1882-2828 μg GAE/g	–	Ramful et al. (2010)

(continued)

Table 1.7 (continued)

Source	Extraction phase	Concentration	Bioactive compounds	Specific compounds identified	References
Lime peel	Ethanol (microwave-assisted extraction)	55%	53 mg GAE/g	–	Rodsamran and Sothornvit (2019)
	Ethanol (ultrasonic-assisted extraction)		54 mg GAE/g		
Kaffir lime	Water	–	23.65 mg GAE/g	–	Shafreen et al. (2018)
Key lime	Water	–	12.13 mg GAE/g	–	

available and remains in blood for a longer duration even compared to ascorbic acid (Srinivasan et al. 2007; Adam et al. 2002). Natural resources that are rich source of ferulic acid are gaining more important because of its broad range therapeutic effects. Ferulic acid has the capability to absorb ultraviolet radiations and whenever its combination with Vitamin-C and Vitamin-E is used its efficacy towards photo-protection is enhanced by 04–08-folds (Burke 2009; Saija et al. 2000; Lin et al. 2005). Studies demonstrated the inhibition potential of ferulic acid towards cholesterol synthesis by competitive inhibition of hydroxymethylglutaryl Co-A reductase activity (Ou and Kwok 2004; Kim et al. 2003; Ou et al. 2001). In sport foods, ferulic acid is used as an important ergogenic supplements for athletes (Headley and Massad 1999; Berning and Steen 1998). In food industries, ferulic acid is used to enhance the viscosity during gel formations from polysaccharides especially arabinoxylans and pectin (Oosterveld et al. 2001; Micard and Thibault 1999). Because of the presence of broad spectrum antimicrobial activity of ferulic acid, it is also used in preservation of food materials. Experimental studies confirm the anti-diabetic potential of ferulic acid as its supplementation results in successful decrease in plasma lipids as well as blood glucose (Barone et al. 2009; Ou et al. 2003; Sri Balasubashini et al. 2003). Use of ferulic acid rich resources is helpful during diabetic conditions as they may enhance the insulin secretion from beta cells (Adisakwattana et al. 2008). Supplementation of ferulic acid improves vascular functions and regulates blood pressure (Ardiansyah et al. 2008).

1.8.3 Sinapic Acid

Sinapine is a well known sinapic acid derivative which proved to have acetylcholine esterase inhibition potential (dose dependent) (He et al. 2008; Ferreres et al. 2009). Xu et al. (2008) reported 6.75 mg/L of sinapic acid in lime juice. Antibiotic resistant is one of the common problems faced by pharmaceutical industries and therefore need of new compounds with broad spectrum antimicrobial potential are continuously rising to combat drug resistant problem (Cowan 1999; Gibbons 2005; Saleem et al. 2009). Fruits, vegetables and other natural resources that possess significant amount of sinapic acid may be recommended as natural drug with broad spectrum antimicrobial potential [gram (+ve) and gram (−ve) bacterial strains] (Nowak et al. 1992). Kelly et al. (2008) reported the antifungal potential of syringaldehyde against *Candida guilliermondii*.

1.8.4 p-Coumaric Acid

Exposure of skin to ultraviolet radiations for longer duration results in early ageing problems and hyper-pigmentation (Fisher et al. 2002; Helfrich et al. 2008). Usually exposure to ultraviolet radiation may create an imbalance between the endogenous antioxidants and free radicals which ultimately affects the UV-protection shield of skin. An et al. (2010) reported that p-coumaric acid could protect the skin from damaging effects of radiations. As compared to cells (Human epidermal melanocytes) that were treated with p-coumaric acid, untreated cells showed higher rate of cell death during exposure to ultraviolet radiations. Exposure to radiations results in activation of peptidases (matrix metalloproteinases) which ultimately degrades the matrix protein (extracellular) and modulates intradermal tissues and forms wrinkles on skin (Brennan et al. 2003). Akdemir et al. (2017) studied Wistar rats for the evaluation of p-coumaric acid effects on Cisplatin (CIS)-induced hepatotoxicity and nephrotoxicity. After administration of CIS (24 h) the liver and kidneys were extracted out and assessed. As compared to untreated counterparts enhancement in malondialdehyde was observed in CIS treated group of rats along with decrease in superoxide dismutase activities and level of glutathione. Lemon juice possess significant amount of p-Coumaric acid (11.57 mg/L) (Xu et al. 2008).

1.8.5 Caffeic Acid

Bioactive compounds are those secondary metabolites which either inhibit or slow down the action triggered by reactive oxygen species/free radicals (Magnani et al. 2014). Antioxidant rich natural resources act as free radical scavengers/metal chelators. Demands for nutraceuticals as well as cosmaceuticals are increasing day by day because of the presence of health benefiting/skin caring potential in them. Xi et al. (2017) studied different parts of lemon fruit (peel, pulp and juice) and they observed concentration of caffeic acid as: peel (538.8 μg/g); pulp (233.5 μg/g) and juice (119.5 μg/g). Xu et al. (2008) reported 2.07 mg/L of caffeic acid in lemon juice. Wen et al. (2003) reported that as compared to other hydroxycinnamic acid, caffeic acid has the higher potential to inhibit Listeria. Canillac and Mourey (2004) demonstrated that the inhibition of microbial growth also depends of certain specific environmental factors, pH and sodium chloride conc. The whole mechanism of antimicrobial potential consist three major steps i) First step involve the change in cellular permeability and loss of nutrients ii) inactivation of enzymatic reactions occurring during synthetic pathways iii) destruction of genetic material (Branen 1993; Kim et al. 1995). Efficacy of caffeic acid towards reduction of mRNA expression (Interleucin-10) and inhibition of p38-MAPK activation has been reported in various studies (Yang et al. 2013; Staniforth et al. 2006).

1.8.6 Rutin

Rutin is one of the important bioactive metabolite that is present in fruits especially those belongs to Rutaceae family. Xi et al. (2017) reported the presence of rutin in lemon parts (peel, pulp and juice). Significant amount was observed in peel (31.55 μg/g); pulp (14.87 μg/g) and juice (3.82 μg/g). Rutin is necessary for the maintenance of blood circulation system as it provide required flexibility to blood vessels (capillaries and arteries) which play an important role in healthy life style. Experimental observations indicate the presence of anti-clotting potential of rutin which provide protection against the heart attacks and brain hemorrhage. Despite many health benefiting properties of rutin, it is always necessary to keep in mind the required concentration necessary for the healthy life. Consumption in amount greater than required may results in certain side effects (headache, stiffness in muscles and high WBC count).

1.8.7 Naringin

Citrus fruits are excellent source of naringin (flavonoid glycoside). The compound is bitter in nature and responsible for the presence of bitterness in juice extracted from citrus fruits. Yusof et al. (1990) studied skin, juice and seeds of lime for the presence of naringin in them. The found that lime skin is rich in naringin (517.2 μg/g) followed by lime juice (98.4 μg/g) and lime seeds (29.2 μg/g). Majority of population is suffering from different level of diseased condition such as obesity, diabetes, hypertension and cancer etc. Throughout the World, research is being carried out on plant derived natural secondary metabolites for the purpose to prepare specific drug that may have potential to combat metabolic disorders (Raja Kumar et al. 2019). Alam et al. (2013) reported that rats treated with naringin in concentration of 95 mg kg^{-1} Day^{-1} for a period of 56 days showed decreased rate of abdominal fat deposition. Pu et al. (2012) demonstrated that treatment of mice with naringin (200 mg kg^{-1}) for a period of 10 weeks results in reduced body weight and less visceral fat. Sharma et al. (2011) observed that diabetic rats when treated with naringin (50 and 100 mg kg^{-1}) for a period of 4 weeks showed regulated glucose utilization and insulin sensitivity. Administration of naringin (50 mg kg^{-1}) for a duration of 56 days results in decreased fasting sugar level with improved insulin in diabetic rats (Sinethemba, 2014). Treatment of rats with naringin (10 mg kg^{-1}) for duration of 46 days resulted in decreased malondialdehyde and increased activity of catalase and superoxide dismutase.

1.9 Summary

Experimental studies focusing on nutritional profile and health benefits of citrus fruits are gaining interest from food as well as pharmaceutical industries. Presence of broad spectrum health benefiting properties makes citrus fruits as important as medicines. Inclusion of citrus fruits in dietary chart (as per body requirements) could help to eradicated various disease causing factors from body. Bioactive compounds extracted from lemon and lime could be used in the preparation of novel functional food products and beverages.

References

Adam A, Crespy V, Levrat-Verny MA, Leenhardt F, Leuillet M, Demigne C, Remesy C (2002) The bioavailability of ferulic acid is governed primarily by the food matrix rather than its metabolism in intestine and liver in rats. J Nutr 132:1962–1968

Adisakwattana S, Moonsan P, Yibchok-Anun S (2008) Insulin-releasing properties of a series of cinnamic acid derivatives in vitro and in vivo. J Agric Food Chem 56:7838–7844

Akdemir FNE, Albayrak M, Calik M, Bayir Y, Gulcin I (2017) The protective effects of p-coumaric acid on acute liver and kidney damages induced by Cisplatin. Biomedicines 5:18

Alam MA, Kauter K, Brown L (2013) Naringin improves diet-induced cardiovascular dysfunction and obesity in high carbohydrate, high fat diet-fed rats. Nutrients 5:637–650

Al-Qassabi JSA, Weli AM, Hossain MA (2018) Comparison of total phenols content and antioxidant potential of peel extracts of local and imported lemons samples. Sustain Chem Pharm 8:71–75

An SM, Koh JS, Boo YC (2010) p-coumaric acid not only inhibits human tyrosinase activity in vitro but also melanogenesis in cells exposed to UVB. Phytother Res 24:1175–1180

Ardiansyah OY, Shirakawa H, Koseki T, Komai M (2008) Novel effects of a single administration of ferulic acid on the regulation of blood pressure and the hepatic lipid metabolic profile in stroke-prone spontaneously hypertensive rats. J Agric Food Chem 56:2825–2830

Arruda HS, Pereira GA, de Morais DR, Eberlin MN, Pastore GM (2018) Determination of free, esterified, glycosylated and insoluble-bound phenolics composition in the edible part of araticum fruit (*Annona crassiflora* Mart.) and its by-products by HPLC-ESI-MS/MS. Food Chem 245:738–749

Baba E, Acar U, Ontas C, Kesbic OS, Yilmaz S (2016) Evaluation of Citrus limon peels essential oil on growth performance, immune response of Mozambique tilapia Oreochromis mossambicus challenged with Edwardsiella tarda. Aquaculture 465:13–18

Barone E, Calabrese V, Mancuso C (2009) Ferulic acid and its therapeutic potential as a hormetin for age-related diseases. Biogerontology 10:97–108

Barros HRM, Ferreira TAPC, Genovese MI (2012) Antioxidant capacity and mineral content of pulp and peel from commercial cultivars of citrus from Brazil. Food Chem 134:1892–1898

Berning JR, Steen SN (1998) Nutrition for sport and exercise, 2nd edn. Aspen Publishers, Gaithersberg, p 130

Bhat R, Che-Kamaruddin NSB, Min-Tze L, Karim AA (2011) Sonication improves kasturi lime (Citrus microcarpa) juice quality. Ultrason Sonochem 18:1295–1300. https://doi.org/10.1016/j.ultsonch.2011.04.002

Branen AL (1993) Introduction to the use of antimicrobials. In: Davidson PP, Branen AL (eds) Antimicrobials in foods. Marcel Dekker, New York

Brennan M, Bhatti H, Nerusu KC, Bhagavathula N, Kang S, Fisher GJ, Varani J, Voorhees JJ (2003) Matrix metalloproteinase-1 is the major collagenolytic enzyme responsible for collagen damage in UV-irradiated human skin. Photochem Photobiol 78:43–48

Burke KE (2009) Prevention and treatment of aging skin with topical antioxidants. Skin Aging Handb:149–176. https://doi.org/10.1016/b978-0-8155-1584-5.50012-0

Bursac Kovacevic D, Putnik P, Dragovic-Uzelac V, Pedisic S, Rezek Jambra A, Herceg Z (2016) Effects of cold atmospheric gas phase plasma on anthocyanins and color in pomegranate juice. Food Chem 190:317–323

Canillac N, Mourey A (2004) Effects of several environmental factors on the anti-Listeria monocytogenes activity of an essential oil of Picea excels. Int J Food Microbiol 92:95–103

Casquete R, Castro SM, Martín A, Ruiz-Moyano S, Saraiva JA, Córdoba MG, Teixeira P (2015) Evaluation of the effect of high pressure on total phenolic content, antioxidant and antimicrobial activity of citrus peels. Innov Food Sci Emerg Technol 31:37–44

Chan SW, Lee CY, Yap CF, Wan Aida WM, Ho CW (2009) Optimisation of extraction conditions for phenolic compounds from limau purut (Citrus hystrix) peels. Int Food Res J 16:203–213

Chauhan N, Saxena J (2019) Phytochemical screening of yellow & green citrus limon peel extracts in different solvents. Int Res J Pharm 10:121–125

Chen XM, Tait AR, Kitts DD (2017) Flavonoid composition of orange peel and its association with antioxidant and anti-inflammatory activities. Food Chem 218:15–21

Chuku EC, Akani NP (2015) Determination of proximate composition and microbial contamination of fresh juice from three citrus species. J Biol Gen Res 1:1–8

Cowan MM (1999) Plant products as antimicrobial agents. Clin Microbiol Rev 12:564–582

Czech A, Zarycka E, Yanovych D, Zasadna Z, Grzegorczyk I, Klys S (2020) Mineral content of the pulp and peel of various citrus fruit cultivars. Biol Trace Elem Res 193:555–563

Dhall RK (2012) Advances in edible coatings for fresh fruits and vegetables: a review. Crit Rev Food Sci Nutr 53:435–450

Dhull SB, Kaur P, Purewal SS (2016) Phytochemical analysis, phenolic compounds, condensed tannin content and antioxidant potential in Marwa (*Origanum majorana*) seed extracts. Resour Eff Technol 2:168–174

Di Vaio C, Graziani G, Gaspari A, Scaglione G, Nocerino S, Ritieni A (2010) Essential oils content and antioxidant properties of peel ethanol extract in 18 lemon cultivars. Sci Hortic 126:50–55

Dong X, Hu Y, Li Y, Zhou Z (2019) The maturity degree, phenolic compounds and antioxidant activity of Eureka lemon [*Citrus limon* (L.) Burm. f.]: A negative correlation between total phenolic content, antioxidant capacity and soluble solid content. Sci Hortic 243:281–289

Fakhari AR, Salehi P, Heydari R, Ebrahimi SN, Haddad PR (2005) Hydrodistillation-headspace solvent microextraction, a new method for analysis of the essential oil components of Lavandula angustifolia mill. J Chromatogr A 1098:14–18

Fallik E (2004) Prestorage hot water treatment (immersion, rinsing and brushing). Postharvest Biol Technol 32:125–134

FAO (Food and Agricultural Organization of United Nations) (2020). https://www.fao.org/faostat/en/#data/QCL

Fatin NR, Azrina A (2017) Comparison of vitamin C content in citrus fruits by titration and high performance liquid chromatography (HPLC) methods. Int Food Res J 24:726–733

Ferreres F, Fernandes F, Sousa C, Valentao P, Pereira JA, Andrade PB (2009) Metabolic and bioactivity insights into Brassica oleracea var. acephala. J Agric Food Chem 57:8884–8892

Fisher GJ, Kang S, Varani J, Bata-Csorgo Z, Wan Y, Datta S, Voorhees JJ (2002) Mechanisms of photoaging and chronological skin aging. Arch Dermatol 138:1462–1470

Forbes-Hernandez TY, Giampieri F, Gasparrini M, Mazzoni L, Quiles JL, Alvarez-Suarez JM, Battino M (2014) The effects of bioactive compounds from plant foods on mitochondrial function: a focus on apoptotic mechanisms. Food Chem Toxicol 68:154–182

Ghasemi K, Ghasemi Y, Ebrahimzadeh MA (2009) Antioxidant activity, phenol and flavonoid contents of 13 citrus species peels and tissues. Pak J Pharm Sci 22:277–281

Gibbons S (2005) Plants as a source of bacterial resistance modulators and anti-infective agents. Phytochem Rev 4:63–78

Gómez-Mejía E, Rosales-Conrado N, Eugenia León-González M, Madrid Y (2019) Citrus peels waste as a source of value-added compounds: extraction and quantification of bioactive polyphenols. Food Chem 295:289–299. https://doi.org/10.1016/j.foodchem.2019.05.136

Gorinstein S, Martín-Belloso O, Park YS, Haruenkit R, Lojek A, Ĉíž M, Caspi A, Libman I, Trakhtenberg S (2001) Comparison of some biochemical characteristics of different citrus fruits. Food Chem 74:309–315

Guimaraes R, Barros L, Joao CM, Barriera M, Sousa J, Carvalho AM (2010) Targeting excessive free radicals with peels and juices of citrus fruits: grapefruit, lemon, lime and orange. Food Chem Toxicol 48:99–106

Guimarães R, Barros L, Barreira JCM, Sousa MJ, Carvalho AM, Ferreira ICFR (2010) Targeting excessive free radicals with peels and juices of citrus fruits: grapefruit, lemon, lime and orange. Food Chem Toxicol 48:99–106

Guzel-Sedym ZB, Greene AK, Seydim AC (2004) Use of ozone in the food industry. LWT 37:453–460

Hamdan D, El-Readi MZ, Tahrani A, Herrmann F, Kaufmann D, Farrag N, El-Shazly A, Wink M (2011) Secondary metabolites of ponderosa lemon (*Citrus pyriformis*) and their antioxidant, anti-inflammatory, and cytotoxic activities. Zeitschrift Für Naturforschung C 66:385–393

He L, Li HT, Guo SW, Liu LF, Qiu JB, Li F, Cai BC (2008) Inhibitory effects of sinapine on activity of acetylcholinesterase in cerebral homogenate and blood serum of rats. Zhongguo Zhongyao Zazhi 33:813–815

He D, Shan Y, Wu Y, Liu G, Chen B, Yao S (2011) Simultaneous determination of flavanones, hydroxycinnamic acids and alkaloids in citrus fruits by HPLC-DAD–ESI/MS. Food Chem127:880–885. https://doi.org/10.1016/j.foodchem.2010.12.109

Headley S, Massad S (1999) Nutritional supplements for athletes. National Association for Sport and Physical Education, Association of the American Alliance for Health, Physical Education, Recreation and Dance, pp 51–52

Helfrich YR, Sachs DL, Voorhees JJ (2008) Overview of skin aging and photoaging. Dermatol Nurs 20:177–183

Homayouni F, Haidari F, Hedayati M, Zakerkish M, Ahmadi K (2017) Hesperidin supplementation alleviates oxidative DNA damage and lipid peroxidation in type 2 diabetes: a randomized double-blind placebo-controlled clinical trial. Phytother Res 31:1539–1545

Huang YS, Ho SC (2010) Polymethoxy flavones are responsible for the anti-inflammatory activity of citrus fruit peel. Food Chem 119:868–873

John S, Monica SJ, Priyadarshini S, Arumugam P (2017) Investigation on phytochemical profile of Citrus limonum peel extract. Int J Food Sci Nutr 2:65–67

Kaur P, Dhull SB, Sandhu KS, Salar RK, Purewal SS (2018) Tulsi (*Ocimum tenuiflorum*) seeds: in vitro DNA damage protection, bioactive compounds and antioxidant potential. Food Measure 12:1530–1538

Kelly C, Jones O, Barnhart C, Lajoie C (2008) Effect of furfural, vanillin and syringaldehyde on *Candida guilliermondii* growth and xylitol biosynthesis. In: Adney WS, McMillan JD, Mielenz J, Klasson KT (eds) Biotechnology for fuels and chemicals. Humana Press, New York, pp 615–626

Kim JM, Marshall MR, Corneel JA, Preston JF, Wei CI (1995) Antibacterial activity of carvacrol, citral and geraniol against *Salmonella typhimurium* in culture medium and fish cubes. J Food Sci 60:1364–1374

Kim HK, Jeong TS, Lee MK, Park YB, Choi MS (2003) Lipidlowering efficacy of hesperetin metabolites in high-cholesterol fed rats. Clin Chim Acta Int J Clin Chem 327:129–137

Ladaniya M (2010) Fruit morphology, anatomy, and physiology. In: Citrus fruit: biology, technology and evaluation. Academic, Elsevier, pp 103–124
Li W, Kandhare AD, Mukherjee AA, Bodhankar SL (2018) Hesperidin, a plant favonoid accelerated the cutaneous wound healing in streptozotocin-induced diabetic rats: role of TGFB/SMADS and ANG-1/TIE-2 signaling pathways. EXCLI J 17:399–419
Lin FH, Lin JY, Gupta RD, Tournas JA, Burch JA, Selim MA, Monteiro-Riviere NA, Grichnik JM, Zielinski J, Pinnell SR (2005) Ferulic acid stabilizes a solution of vitamins C and E and doubles its photoprotection of skin. J Invest Dermatol 125:826–832
Londoño-Londoño J, de Lima VR, Lara O, Gil A, Pasa TBC, Arango GJ, Pineda JRR (2010) Clean recovery of antioxidant flavonoids from citrus peel: optimizing an aqueous ultrasound-assisted extraction method. Food Chem 119:81–87 https://doi.org/10.1016/j.foodchem.2009.05.075
Lopresto CG, Petrillo F, Casazza AA, Aliakbarian B, Perego P, Calabrò V (2014) A non-conventional method to extract D-limonene from waste lemon peels and comparison with traditional Soxhlet extraction. Sep Purif Technol 137:13–20
Magnani C, Isaac VLB, Correa MA, Salgado HRN (2014) Caffeic acid: a review of its potential use in medications and cosmetics. Anal Method 6:3203–3210
Makni M, Jemai R, Kriaa W, Chtourou Y, Fetoui H (2018) Citrus limon from Tunisia: phytochemical and physicochemical properties and biological activities. Biomed Res Int 2018:6251546
Malacrida CR, Kimura M, Jorge N (2012) Phytochemicals and antioxidant activity of citrus seed oils. Food Sci Technol Res 18:399–404
Malta LG, Tessaro EP, Eberlin M, Pastore GM, Liu RH (2013) Assessment of antioxidant and antiproliferative activities and the identification of phenolic compounds of exotic Brazilian fruits. Food Res Int 53:417–425
Mathew BB, Jatawa SK, Tiwari A (2012) Phytochemical analysis of citrus limonum pulp and peel. Int J Pharm Pharm Sci 4:369–371
Mhiri N, Ioannou I, Ghoul M, Boudhrioua NM (2014) Extraction methods of citrus Peel phenolic compounds. Food Rev Int 30:265–290
Micard V, Thibault JF (1999) Oxidative gelation of sugar-beet pectins: use of laccases and hydration properties of the crosslinked pectins. Carbohydr Polym 39:265–273
Namania JA, Baqira E, Ajwaa Al Abria AA, Hubaishia TA, Husainb A, Khana SA (2018) Phytochemical screening, phenolic content and antioxidant activity of Citrus aurantifolia l. leaves grown in two regions of Oman. Iran J Pharm Sci 14:27–34
Nayak B, Dahmoune F, Moussi K, Remini H, Dairi S, Aoun O, Khodir M (2015) Comparison of microwave, ultrasound and accelerated-assisted solvent extraction for recovery of polyphenols from Citrus sinensis peels. Food Chem 187:507–516
Neri-Numa IA, Pessoa MG, Arruda HS, Pereira GA, Paulino BN, Angolini CFF, Ruiz TG, Pastore GM (2020) Genipap (*Genipa americana* L.) fruit extract as a source of antioxidant and antiproliferative iridoids. Food Res Int 109252
Nowak H, Kujawa K, Zadernowski R, Roczniak B, KozŁowska H (1992) Antioxidative and bactericidal properties of phenolic compounds in rapeseeds. Eur J Lipid Sci Technol 94:149–152
Oosterveld A, Pol IE, Beldman G, Voragen AGJ (2001) Isolation of feruloylated arabinans and rhamnogalacturonans from sugar beet pulp and their gel forming ability by oxidative crosslinking. Carbohydr Polym 44:9–17
Ou S, Kwok KC (2004) Ferulic acid: pharmaceutical functions, preparation and applications in foods. J Sci Food Agric 84:1261–1269
Ou S, Bao H, Lan Z (2001) Advances on pharmacological study of ferulic acid and its derivatives. J Chin Med Mater 24:220–221
Ou L, Kong LY, Zhang XM, Niwa M (2003) Oxidation of ferulic acid by Momordica charantia peroxidase and related anti-inflammation activity changes. Biol Pharm Bull 26:1511–1516
Palou L, Smilanick JL, Usall J, Viñas I (2001) Control of postharvest blue and green molds of oranges by hot water sodium carbonate, and sodium bicarbonate. Plant Dis 85:371–376
Paul DK, Shaha RK (2004) Nutrients, vitamins and minerals content in common citrus fruits in the Northern region of Bangladesh. Pak J Biol Sci 7:238–242

Pavan V, Sancho RAS, Pastore GM (2014) The effect of in vitro digestion on the antioxidant activity of fruit extracts (*Carica papaya*, *Artocarpus heterophillus* and *Annona marcgravii*). LWT 59:1247–1251

Pelletier A, Lavoie JM, Chornet E (2008) Identification of secondary metabolites in citrus fruit using gas chromatography and mass spectroscopy. J Chem Educ 85:1555

Pereira GA, Arruda HS, de Morais DR, Eberlin MN, Pastore GM (2018) Carbohydrates, volatile and phenolic compounds composition, and antioxidant activity of calabura (Muntingia calabura L.) fruit. Food Res Int 108:264–273

Pu P, Gao DM, Mohamed S, Chen J, Zhang J, Zhou XY, Zhou NJ, Xie J, Jiang H (2012) Naringin ameliorates metabolic syndrome by activating AMP-activated protein kinase in mice fed a high-fat diet. Arch Biochem Biophy 518:61–70

Purewal SS, Sandhu KS (2020) Nutritional profile and health benefits of Kinnow: An updated review. Int J Fruit Sci 20:S1385–S1405. https://doi.org/10.1080/15538362.2020.1792390

Purewal SS, Sandhu KS (2021) Debittering of citrus juice by different processing methods: A novel approach for food industry and agro-industrial sector. Sci Hortic 276:109750. https://doi.org/10.1016/j.scienta.2020.109750

Purewal SS, Salar RK, Bhatti MS, Sandhu KS, Singh SK, Kaur P (2020) Solid-state fermentation of pearl millet with Aspergillus oryzae and Rhizopus azygosporus: effects on bioactiveprofile and DNA damage protection activity. Food Measure 14:150–162. https://doi.org/10.1007/s11694-019-00277-3

Putnik P, Bursac Kovacevic D, Rezek Jambrak A, Barba F, Cravotto G, Binello A, Lorenzo J, Shpigelman A (2017) Innovative "green" and novel strategies for the extraction of bioactive added value compounds from citrus wastes-a review. Molecules 22:680

Raja Kumar S, Mohd Ramli ES, Abdul Nasir NA, Ismail NHM, Mohd Fahami NA (2019) Preventive effect of Naringin on metabolic syndrome and its mechanism of action: a systematic review. Evid Based Complement Alternat Med 2019:9752826

Ramful D, Bahorun T, Bourdon E, Tarnus E, Auroma OI (2010) Bioactive phenolics and antioxidant propensity of flavedo extracts of Mauritian citrus fruits: potential prophylactic ingredients for functional foods application. Toxicology 278:75–87

Rodsamran P, Sothornvit R (2019) Extraction of phenolic compounds from lime peel waste using ultrasonic-assisted and microwave-assisted extractions. Food Biosci 28:66–73

Roesler R, Malta LG, Carrasco LC, Pastore G (2006) Evaluation of the antioxidant properties of the Brazilian Cerrado fruit *Annona crassiflora* (Araticum). J Food Sci 71:C102–C107

Saeidi I, Hadjmohammadi MR, Peyrovi M, Iranshahi M, Barfi B, Babaei AB, Dust AM (2011) HPLC determination of hesperidin, diosmin and eriocitrin in Iranian lime juice using polyamide as an adsorbent for solid phase extraction. J Pharm Biomed Anal 56:419–422

Saija A, Tomaino A, Tombetta D, Pasquale AD, Uccella N, Barbuzzi T, Paolino D, Bonina F (2000) In vitro and in vivo evaluations of caffeic and ferulic acids as topical photoprotective agents. Int J Pharm 199:39–47

Saleem M, Nazir M, Ali MS, Hussain H, Lee YS, Riaz N, Jabbar A (2009) Antimicrobial natural products: an update on future antibiotic drug candidates. Nat Prod Rep 27:238–254

Salar RK, Purewal SS (2016) Improvement of DNA damage protection and antioxidant activity of biotransformed pearl millet (Pennisetum glaucum) cultivar PUSA-415 using Aspergillusoryzae MTCC 3107. Biocatal Agric Biotechnol 8:221–227. https://doi.org/10.1016/j.bcab.2016.10.005

Salar RK, Purewal SS, Bhatti MS (2016) Optimization of extraction conditions and enhancement of phenolic content and antioxidant activity of pearl millet fermented with Aspergillusawamori MTCC-548. Resour Effic Technol 2:148–157. https://doi.org/10.1016/j.reffit.2016.08.002

Shafreen RB, Lubinska-Szczygeł M, Różańska A, Dymerski T, Namieśnik J, Katrich E, Gorinstein S (2018) Human serum interactions with phenolic and aroma substances of Kaffir (*Citrus hystrix*) and Key lime (*Citrus aurantifolia*) juices. J Lumin 201:115–122. https://doi.org/10.1016/j.jlumin.2018.04.010

Sharma AK, Bharti S, Ojha S, Bhatia J, Kumar N, Ray R, Kumari S, Arya DS (2011) Up-regulation of PPARγ, heat shock protein-27 and-72 by naringin attenuates insulin resistance, β-cell dysfunction, hepatic steatosis and kidney damage in a rat model of type 2 diabetes. Br J Nutr 106:1713–1723

Singh B, Singh JP, Kaur A, Singh N (2020) Phenolic composition, antioxidant potential and health benefits of citrus peel. Food Res Int 132:109114. https://doi.org/10.1016/j.foodres.2020.109114

Sinethemba NF (2014) Naringin reversal of some aspects of diabetic nephropathy in rats type 1 streptozotocin induced diabetes, pp 1–69

Spadoni A, Neri F, Bertolini P, Mari M (2013) Control of Monilinia rots on fruit naturally infected by hot water treatment in commercial trials. Postharvest Biol Technol 86:280–284

Sri Balasubashini M, Rukkumani R, Menon VP (2003) Protective effects of ferulic acid on hyperlipidemic diabetic rats. Acta Diabetol 40:118–122

Srinivasan M, Sudheer AR, Menon VP (2007) Ferulic acid: therapeutic potential through its antioxidant property. J Clin Biochem Nutr 40:92–100

Staniforth V, Chiu LT, Yang NS (2006) Caffeic acid suppresses UVB radiation- induced expression of interleukin-10 and activation of mitogen- activated protein kinases in mouse. Carcinogenesis 27:1803–1811

Sugasawa N, Katagi A, Kurobe H, Nakayama T, Nishio C, Takumi H, Higashiguchi F, Aihara K, Shimabukuro M, Sata M, Kitagawa T (2018) Inhibition of atherosclerotic plaque development by oral administration of α-glucosyl hesperidin and water-dispersible hesperetin in apolipoprotein E knockout mice. J Am Coll Nutr 38:1–8

Wang S, Yang C, Tu H, Zhou J, Liu X, Cheng Y, Luo J, Deng X, Zhang H, Xu J (2017) Characterization and metabolic diversity of flavonoids in citrus species. Sci Rep 7:10549

Wang Y, Ji S, Zang W, Wang N, Cao J, Li X, Sun C (2019) Identification of phenolic compounds from a unique citrus species, finger lime (*Citrus australasica*) and their inhibition of LPS-induced NO-releasing in BV-2 cell line. Food Chem Toxicol 129:54–63

Wen AM, Delaquis P, Stanich K, Toivonen P (2003) Antilisterial activity of selected phenolic acids. Food Microbiol 20:305–311

Wessels Q, Pretorius E, Smith CM, Nel H (2014) Te potential of a niacinamide dominated cosmeceutical formulation on fbroblast activity and wound healing in vitro. Int Wound J 11:152–158

Xi W, Lu J, Qun J, Jiao B (2017) Characterization of phenolic profile and antioxidant capacity of different fruit part from lemon (*Citrus limon* Burm.) cultivars. J Food Sci Technol 54:1108–1118

Xu G, Liu D, Chen J, Ye X, Ma Y, Shi J (2008) Juice components and antioxidant capacity of citrus varieties cultivated in China. Food Chem 106:545–551

Xue M, Weickert MO, Qureshi S, Kandala NB, Anwar A, Waldron M, Shafie A, Messenger D, Fowler M, Jenkins G, Rabbani N, Thornalley PJ (2016) Improved glycemic control and vascular function in overweight and obese subjects by glyoxalase 1 inducer formulation. Diabetes 65:2282–2294

Yaqoob M, Aggarwal P, Aslam R, Rehal J (2020) Chapter 15 – Extraction of bioactives from citrus. Green Sustainable Process for Chemical and Environmental Engineering and ScienceSupercritical Carbon Dioxide as Green Solvent. 357–377. Edited by Inamuddin, Asiri AM, Isloor AM https://doi.org/10.1016/B978-0-12-817388-6.00015-5 ISBN: 978-0-12-817388-6

Yang Y, Li Y, Wang K, Wang Y, Yin W (2013) P38/NF-kB/snail pathway is involved in caffeic acid-induced inhibition of cancer stem cells-like properties and migratory capacity in malignant human keratinocyte. PLoS One 8:1–8

Yusof S, Ghazali HM, King GS (1990) Naringin content in local citrus fruits. Food Chem 37:113–121

Chapter 2
Grapefruit

Avneet Kaur, Sukhvinder Singh Purewal, Arashdeep Singh Randhawa, and Chidanandamurthy Thippeswamy Swamy

2.1 Introduction

Fruits are multifunctional natural resources as they could be used as eatable material by both human as well as animals. The remaining waste of fruits after juice extraction could also be used as an important feed material. Further peel/waste of fruits could be industrially utilized as a source of oil/nutraceuticals (Czech et al. 2019; Rafiq et al. 2018). Grapefruit being a subtropical tree is well known for its quite sour to semi-sweet taste. Basically, grapefruit is a hybrid of sweet orange and pomelo which formed as a result of accidental cross (Fig. 2.1). Significant variability in grapefruit could be observed which ranges from its taste (sour, semi-sweet) to varying color (green; white, yellow and pink). Despite the nutritional profile, grapefruit is famous all over the world for its health benefitting bioactive phytochemicals which are present in juice and peels (Igual et al. 2019; Rechkemmer 2001). Grapefruit is an important eatable natural source which could fulfill the necessity of essential nutrients along with specific minerals/vitamins (Xu et al. 2008; Igual et al.

A. Kaur
Department of Chemistry, University Institute of Sciences, Chandigarh University, Mohali, Punjab, India

S. Singh Purewal (✉)
University Centre for Research & Development (UCRD), Chandigarh University, Mohali, Punjab, India
e-mail: sukhvinder.e14148@cumail.in

A. S. Randhawa
Department of Food Science & Technology, Punjab Agricultural University, Ludhiana, India

C. T. Swamy
Department of Microbiology, Captain Srinivasa Murthy Central Ayurveda Research Institute, Central Council for Research in Ayurvedic Sciences, Arignar Anna Government Hospital Campus, Chennai, Tamil Nadu, India

S. Singh Purewal et al. (eds.), *Recent Advances in Citrus Fruits*,
https://doi.org/10.1007/978-3-031-37534-7_2

Fig. 2.1 Grapefruit

2010). Presence of health benefitting compounds in fruits could helps to eradicate the chances of certain chronic disorders. From dietary point of view, inclusion of grapefruit in dietary chart is of utmost interest due to its low calorie and more health promising nutritional components (Pereira et al. 2018, 2020; Arruda et al. 2018). Nutritional profile of fruits often vary as the concentration of specific nutrients/bioactive constituents are under the influence of temperature, rainfall, pH and nutritional profile of soil and use of chemical as well as biofertilizer. Use of biofertilizer could helps in maintaining the growth and sustainability of crops (Kaur and Purewal 2019). However, the seasonal dependency and short life span limits its availability throughout the year (Cebadera-Miranda et al. 2019).

2.2 Grapefruit Production

Grapefruit detail in terms of area, production and yield since last 10 years (2008–2018) is reported in Tables 2.1 and 2.2. As per the detail collected from FAO (2020), top grapefruit producers in the World were China mainland (4,965,768 tonnes) followed by Viet Nam (657,660 tonnes); United States of America (558,830 tonnes); Mexico (459,610 tonnes); South Africa (445,385 tonnes) and India (257,750 tonnes). In year 2018, China Mainland was the top producer of grapefruit with production (4,965,768 tonnes). Worldwide, India stands at sixth place with grapefruit production (257,750 tonnes) (Fig. 2.2). Grapefruit production in India was observed maximum during the year 2016 (390,500 tonnes) and minimum during the year 2011 (196,000 tonnes) (Table 2.1). Grapefruit production (tonnes) at World level was observed maximum during the year 2018 (9,374,739 tonnes) and minimum during the year 2008 (7,496,457 tonnes) (Table 2.2).

Table 2.1 Grapefruit production detail of India since 2008–2018 (FAO 2020)

	2008	2009	2010	2011	2012	2013	2014	2015	2016	2017	2018
Area (ha)	8500	9000	10,994	8300	13,731	12,203	10,677	11,300	16,271	14,506	10,572
Yield (hg/ha)	235,294	240,000	237,038	236,143	234,864	233,790	232,736	230,531	239,998	242,660	243,799
Production (tonnes)	200,000	216,000	260,600	196,000	322,500	285,300	248,500	260,500	390,500	352,000	257,750

Table 2.2 Worldwide production of grapefruit since 2008–2018 (FAO 2020)

	2008	2009	2010	2011	2012	2013	2014	2015	2016	2017	2018
Area (ha)	310,677	317,350	320,255	324,518	321,518	320,906	319,105	354,133	359,950	332,690	373,735
Yield (hg/ha)	241,294	237,186	236,595	244,255	256,291	264,617	260,766	250,325	248,845	269,527	250,839
Production (tonnes)	7,496,457	7,527,086	7,577,065	7,926,536	8,240,213	8,491,714	8,321,169	8,864,859	8,957,158	8,966,891	9,374,739

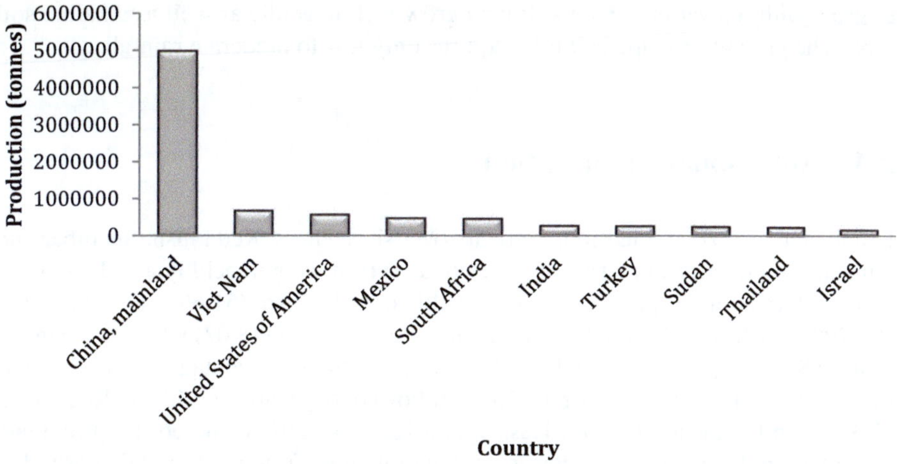

Fig. 2.2 Worldwide top ten producers of grapefruit in 2018

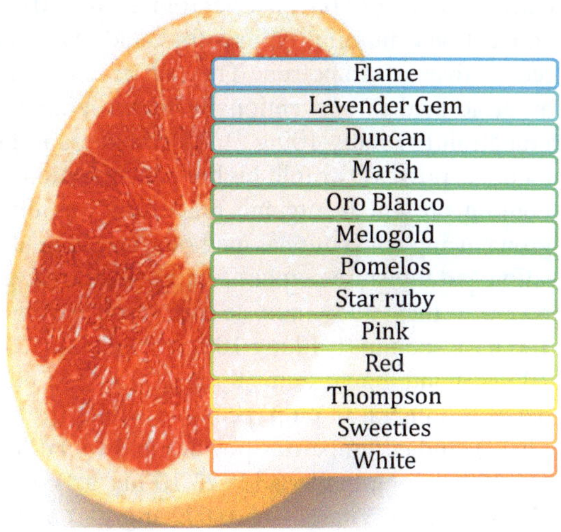

Fig. 2.3 Different types of grapefruit

2.3 Climatic Conditions

Usually the grapefruit tree grows up to 5–6 m however in some cases the height may reach 13–15 m. Leaves of grapefruit tree are dark green in color, thin and under the favorable conditions tree has white flowers (04 petals). Different cultivars of grapefruit are available in the market with varying color of segmented flesh. The name of different grapefruit types are reported in Fig. 2.3. Nutritional profile of different grapefruit types vary according to the conditions under which they are growing. The optimal temperature range required for the growth of grapefruit is 60–85 °F. Best storage temperature for the grapefruit is 10 °C. Grapefruit tree have the capability to grow within a variety of soils. It may grow well in acidic as well as alkaline soil type. The growth of grapefruit tree requires only low to moderate rainfall.

2.4 Nutritional Composition

Ishfaq et al. (2007) studied four cultivars (Marsh seedless, Red blush, Shamber and Duncan) of grapefruit in terms of juice content; fruit weight; acidity and TSS. They observed maximum juice content in cultivar red blush (51.6%) and Shamber (51.70%). TSS was observed maximum in cultivar Duncan (8.02) whereas the minimum TSS was present in Red blush (6.50). Maximal percentage of acidity was observed in cultivar Shamber (1.56%) followed by Marsh (1.48%); Red blush (1.38%) and Duncan (1.35%). Hassan and Hussein (2017) studied the proximate composition, lycopene and ascorbic acid in pink grapefruit and found that ascorbic acid was present in grapefruit in amount 52.2 mg 100 g^{-1}, lycopene (42.9 mg 100 g^{-1}) followed by ash (2.78%); protein (6.43%); fat (3.79%); fiber (8.22%) and carbohydrates (79.08%). Chebrolu et al. (2012) studied the effect of organic and conventional farming on ascorbic acid, carotenoids and limonoids content in Rio red cultivar of grapefruit. They observed significant difference in the amount of ascorbic acid in conventionally grown grapefruit (39.2 mg 100 g^{-1}) and organic grapefruit (41.8 mg 100 g^{-1}). As compared to grapefruit grown using conventional methods possess twofold higher concentration of lycopene. Whereas the amount of nomilin present in organic grapefruit was observed 77% higher than grapefruit grown using conventional methods. Czech et al. (2020) studied the mineral composition and moisture content of grapefruit (red, green and white) pulp and peel. The results of their study indicate the presence of moisture (92.3%); potassium (111 mg 100 g^{-1}); sodium (0.12 mg 100 g^{-1}); calcium (21.3 mg 100 g^{-1}); phosphorus (15 mg 100 g^{-1}) and magnesium (8 mg 100 g^{-1}) in pulp of red grapefruit. The individual detail of mineral in pulp, peel and whole fruit is reported in Table 2.3. Igual et al. (2019) studied grapefruit powder and found total fiber (255 mg g^{-1}); tartaric acid (16.45 mg g^{-1}); malic acid (23.10 mg g^{-1}); citric acid (29.43 mg g^{-1}); Vitamin-C (3.56 mg g^{-1}); Vitamin-A (0.077 mg g^{-1}) and Vitamin-E (0.017 mg g^{-1}). Kefford and Chandler (1970) reported the presence of terpenes in grapefruit peel oil as: d-limonene (93–95%) followed by myrcene (1.9%); α-pinene (1.6%); terpinene (0.5%) and cymene (0.4%).

2.4.1 *Bioactive Profile*

Bioactive phytochemical with antioxidant properties are the group of important secondary metabolites that are formed/secreted by natural resources under certain specific conditions (Dhull et al. 2020; Arruda et al. 2016; Salar et al. 2015; Siroha et al. 2016). Thousands of bioactive compounds were reported in scientific studies for

Table 2.3 Specific ingredients in grapefruit

	Red Grapefruit			Green Grapefruit			White Grapefruit			
	Whole fruit	Pulp	Peel	Whole fruit	Pulp	Peel	Whole fruit	Pulp	Peel	References
	Macronutrients									
Moisture (%)		92.3	80.4		91.9	80.2		90	79.9	Czech et al. (2020)
Potassium (mg 100 g^{-1})	120	111	132	126	123	133	124	117	129	
Sodium (mg 100 g^{-1})	0.16	0.12	0.25	0.26	0.21	0.31	0.21	0.16	0.25	
Calcium (mg 100 g^{-1})	27.4	21.3	36	30	24.5	38.9	26.2	22.6	34.8	
Phosphorus (mg 100 g^{-1})	21.2	15	20	22	19	22.5	17	17	19	
Magnesium (mg 100 g^{-1})	9.99	8	10	8.98	8	9.5	11.1	9	11	
	Micronutrients									
Iron (mg 100 g^{-1})	0.21	0.19	0.31	0.24	0.21	0.27	0.24	0.20	0.28	
Zinc (mg 100 g^{-1})	0.25	0.17	0.33	0.26	0.22	0.30	0.23	0.19	0.28	
Copper (mg 100 g^{-1})	0.06	0.03	0.08	0.07	0.05	0.08	0.07	0.05	0.07	
Manganese (mg 100 g^{-1})	0.06	0.01	0.10	0.06	0.01	0.10	0.04	0.02	0.07	
Selenium (µg 100 g^{-1})	1.48	1.28	1.68	2.79	2.05	3.55	2.13	1.79	2.28	

their health benefiting potential. Bioactive compounds formed within the natural resources either to participate during the signaling of hormones; germination of pollens; pigmentation or for protecting the plant from various pathogenic infections (Winkel-Shirley 2002; Treutter 2006; Agati et al. 2012). Phenolic compounds are also important from industrial point of view as they also contribute in enhancing the quality of food products, texture, taste, color, and health benefiting potential (Widelski et al. 2009). All over the world fruits with millions tons production represents the major immunity boosting part of daily dietary chart (Neri-Numa et al. 2020; De Paulo Farias et al. 2020; Roesler et al. 2006; Malta et al. 2013). Grapefruit juice, pomace and peel are rich in bioactive phytochemicals which are very important from nutritional point. Sicari et al. (2018) studied two varieties (marsh and Star ruby) of grapefruit for the detection of antioxidant properties in them. They observed the presence of quercetin, hesperetin, naringenin, rutin, naringin, poncirin and narirutin along with flavanones. The concentration of flavonoids they observed in studied varieties was ranged from 310 to 390 mg L^{-1}. Antioxidant potential in Star ruby and Marsh variety during the DPPH assay was 35.25% and 46%. Aludatt et al. (2017) reported that amount of free phenolic present in grapefruit was 2.4 mg GAE g^{-1} dried sample. Further they performed LC-MS/MS for the analysis of phenolics in grapefruit extracts. Analytical results indicate the presence of hesperidin (99.4%); vanillic acid (46.5%); rutin (40.6%); chlorogenic acid (5.9%); diosmin (4.3%); sinapic acid (2.2%) and p-coumaric acid (0.4%). Hassan and Hussein (2017) demonstrated the presence of bioactive compounds with antioxidant properties in pink grapefruit. They observed that phenolic compound was present in pink grapefruit in amount 10.78 mg GAE g^{-1} and flavonoids 1.73 mg CE g^{-1}. Agudelo et al. (2017) studied star ruby cultivar of grapefruit in terms of phytochemicals content and antioxidant potential. They observed that grapefruit possess phenolic compounds in amount 8.71 mg g^{-1} and flavonoids 8.56 mg g^{-1}. Aadil et al. (2017) studied bioactive profile of grapefruit and observed the amount of flavonoids (535.5 μg CE g^{-1}); flavonols (2.53 μg QE g^{-1}); antioxidant capacity (282.6 μg AAE g^{-1}) and ascorbic acid (28.6 mg 100 ml^{-1}). Castro-Vazquez et al. (2016) studied the bioactive profile of peel extracts (White and Pink grapefruit) and found phenolic compounds in amount 49.1 mg GAE g^{-1} dry weight (white grapefruit) and 27.9 mg GAE g^{-1} dry weight (Pink grapefruit). Antioxidant potential in peel extracts of white grapefruit during different antioxidant assays was as DPPH (32.4 mg trolox g^{-1}); FRAP (60.3 mg trolox g^{-1}) and ABTS (122.3 mg trolox g^{-1}). Pink grapefruit peel extracts showed activity as DPPH (25.1 mg trolox g^{-1}); FRAP (44.8 mg trolox g^{-1}) and ABTS (99.4 mg trolox g^{-1}). Xi et al. (2015) studied the phenolic composition and antioxidant properties of grapefruit parts. Antioxidant potential in flavedo during DPPH assay was 67.9% followed by ABTS 6.2 mM VC g^{-1} and FRAP (2.1 mM). Antioxidant potential in Albedo during DPPH assay was 20.5% followed by ABTS 2.39 mM VC g^{-1} and FRAP (1.1 mM). Antioxidant potential in segment membrane during DPPH assay was 35.6% followed by ABTS 5.8 mM VC g^{-1} and FRAP (2.7 mM). Antioxidant potential in juice vesicle during DPPH assay was 29.3% followed by ABTS 3.7 mM VC g^{-1} and FRAP (1.4 mM). Antioxidant potential in seeds during DPPH assay was 17.7% followed by ABTS 3.3 mM VC g^{-1} and FRAP (0.5 mM).

2.5 Health Benefits

2.5.1 Cancer

Scientific studies have their major focus on bioactive compounds of natural origin as they pose no risk to health and health benefiting in nature. Flavonoids are known for their potential to combat chronic disorders. Flavonoids showed their potential towards cancerous conditions using three different mechanisms (1) DNA damage protection (2) prevention of abnormal cellular growth (3) inhibition of cellular proliferation (Kaur et al. 2019; Marhuenda et al. 2019; Manthey et al. 2001). Leung et al. (2007) demonstrated that presence of aromatic ring in flavonoids makes them quite useful during the process of chemotherapy. The cancer preventive potential of flavonoids may vary with dose, cell type and cultural conditions of in-vitro models.

The mechanism behind the DNA damage protection activity in flavonoids might be due to their capability to absorb ultraviolet light. Flavonoids also play an important health benefiting role by preventing the promotion of abnormal cellular growth during the initial stage of carcinogenesis (Manach et al. 1996; Miller 2012). Miller et al. (2007) reported that as compared to tangeretin, hesperetin and nobiletin; naringin and naringenin give effective response towards inhibition of oral carcinogenesis. Hesperidin, an important bioactive compounds present in grapefruit showed its capability towards antiproliferative processes. Consumption of ascorbic acid rich fruits as per the daily dietary requirements could be helpful in preventing the occurrence of lungs and colorectal cancer (Turati et al. 2015; Kojo 2004). Bae and Kim (2016) reported that the persons consuming citrus fruits as per their daily dietary needs may be strong enough to prevent/delay the occurrence of gastric cancers.

2.5.2 Obesity

Balistreri et al. (2010) reported that obesity is kind of disease which is characterized by increase in body weight and enhancement in adipokines secretion which modulates body response. High level obesity increases the chance of insulin resistance, abnormal blood pressure and enhanced inflammatory response along with abnormal amount of blood lipids (Nakajima et al. 2014; Grundy et al. 2004). Consumption of citrus fruits rich phenolics and flavonoids could reduce the number of adipose cells and helps in maintaining healthy weight. Kanda et al. (2012) demonstrated that addition of flavonoids results in reduction in the rate of accumulation of lipids in cultured cells. Yoshida et al. (2010) observed that hesperetin and naringenin have the potential to inhibit the NFκB activation through TNF-α, with reduction in interleukin-6 (IL-6) secretion and inhibit extracellular signals regulated kinase (ERK) pathway which results in decreased activation of hormone sensitive lipase (HSL). Alam et al. (2013) studied the effect of naringin on reduction of weight gain in wistar rat. They observed that use of naringin results in reduced deposition of abdominal fat with better lipid profile and glucose tolerance. Grapefruit which have maximum bioactive compounds in them are reported to be involved in reduced fat deposition in hips and waist region. The consumption of grapefruit results in improvement in antioxidant level and superoxide dismutase with reduced oxidative stress and malondialdehyde.

2.5.3 Diabetes

Diabetes is one of leading problem in developing as well as developed nation. Majority of population in the world is suffering from the diabetes. Diabetes comes under a type of non-communicable disorders. As compared to normal person, a diabetic person is at high risk of mortality/morbidity (Marathe et al. 2017). One of the major features of metabolic syndrome is insulin resistance and hyperglycemic conditions. Insulin resistance is basically the decrease in response of peripheral tissues towards the action of insulin. Results of scientific studies showed that in diabetic persons concentration of IL-6 and TNF-α starts increasing with insulin resistance (Alam et al. 2014). Studies reported that use of fruits that are rich in naringin content markedly lower the glucose-6-phosphatase and PEP (phosphoenolpyruvate) carboxy kinase activity in mice (Jung et al. 2004). Scientific studies support the concept that hypoglycemic potential of naringin is mediated through glucose uptake in skeletal muscles (Zygmunt et al. 2010). The only treatment available for the diabetic persons is the use of pharmacological preperations, insulin and dietary as well as lifestyle changes. The use of natural resources with antidiabetic potential is highly favored by scientists/researchers as they help to understand the diabetes management process. The antidiabetic potential is present in the natural resources (fruits, vegetables, cereal and non-cereals) mainly due to the bioactive phytochemicals present in them. Fruits are used by majority of people for their tasty and delicious juice. In food industries, fruits that belong to citrus family are used in the preparation of specific beverages, squash, marmalades and jam, jelly etc. Citrus fruits possess broad range bioactivities ranging from antioxidant potential to anti-inflammatory, cardio-protective nature and anti-diabetic potential. Consumption of citrus fruits directly, in the form of juice and other food products provides many health benefits even in single dose (phenolics, flavonoids, alkaloids, carotenoids and essential minerals) (Lv et al. 2015).

2.5.4 Cardiovascular Diseases

One of the major causes of illness and death is cardiovascular disorder. The major reason behind the rise in cases with cardiovascular disorder is excess use of fried/junk foods, less exercise, smoking and sedentary life style. These factors contribute a lot during the plaque formation/cholesterol deposition in arteries which ultimately results even in sudden death. Other factors that contribute in boosting the person towards cardiovascular disorder are age of person, genetic disorder, gender and medical history (Marhuenda et al. 2019; Mendis et al. 2011). Epidemiological reports suggest that flavonoids and flavanones present in fruits of citrus family inversely affect the occurrence of cardiovascular disorder related symptoms. Inclusion of citrus fruits in dietary chart reduces the development of plaque formation in arteries including maintenance of BMI (body mass index) and blood pressure management (Mulvihill and Huff 2010). Naringenin rich fruit such as grapefruit play an important role during insulin resistance and dyslipidemia. In an experimental study, when mice fed the diet rich in plaque boosting components, hyperlipidemia results in atherosclerosis which was further controlled using naringenin treatment (Sung et al. 2012; Mulvihill and Huff 2010).

2.6 Summary

Philosophies that support fruits can be health promoting beyond its nutritional composition is gaining more acceptance within the consumers/scientific communities and industries as they help to solve the disease related problems in natural way. Studies confirmed the concept that grapefruit and extracts prepared from various parts of grapefruit could helpful as nutraceuticals and functional food products for the diabetes and cancer management. Selection of citrus fruits with specific nutrients and bioactive compounds could be used in the preparation of novel functional food products/prophylactic beverages.

References

Aadil RM, Zeng X, Jabbar S, Nazir A, Mann AA, Khan MKI, Ramzan AA (2017) Quality evaluation of grapefruit juice by thermal and high pressure processing treatment. Pak J Agric Res 30(3):249–257

Agudelo C, Barros L, Santos-Buelga C, Martínez-Navarrete N, Ferreira ICFR (2017) Phytochemical content and antioxidant activity of grapefruit (star ruby): a comparison between fresh freeze-dried fruits and different powder formulations. LWT 80:106–112

Alam MA, Kauter K, Brown L (2013) Naringin improves diet-induced cardiovascular dysfunction and obesity in high carbohydrate, high fat diet-fed rats. Nutrition 5(3):637–650

Alam MA, Subhan N, Rahman MM, Uddin SJ, Reza HM, Sarker SD (2014) Effect of citrus flavonoids, naringin and naringenin, on metabolic syndrome and their mechanisms of action. Adv Nutr 5(4):404–417

Aludatt MH, Rababah T, Alhamad MN, Al-Mahasneh MA, Ereifej K, Al-Karaki G, Al-Duais M, Andrade JE, Tranchant CC, Kubow S, Ghozlan KA (2017) Profile of free and bound phenolics extracted from citrus fruits and their roles in biological system: content, antioxidant, anti-diabetic and anti-hypertensive properties. Food Funct 8:3187–3197

Arruda HS, Pereira GA, Pastore GM (2016) Optimization of extraction parameters of total phenolics from *Annona crassiflora* Mart. (Araticum) fruits using response surface methodology. Food Anal Method 10(1):100–110

Arruda HS, Pereira GA, Pastore GM (2018) Brazilian Cerrado fruit araticum (*Annona crassiflora* Mart.) as a potential source of natural antioxidant compounds. Int Food Res J 25:2005–2012

Bae JM, Kim EH (2016) Dietary intakes of citrus fruit and risk of gastric cancer incidence: an adaptive meta-analysis of cohort studies. Epidemiol Health 38:e2016034

Balistreri CR, Caruso C, Candore G (2010) The role of adipose tissue and adipokines in obesity-related inflammatory diseases. Mediat Inflamm 2010:1–19

Castro-Vazquez L, Alañón ME, Rodríguez-Robledo V, Pérez-Coello MS, Hermosín-Gutierrez I, Díaz-Maroto MC, Jordan J, Galindo MF, Arroyo-Jiménez MM (2016) Bioactive flavonoids, antioxidant behaviour, and cytoprotective effects of dried grapefruit peels (*Citrus paradise* Macf.). Oxidative Med Cell Longev 2016:1–12

Cebadera-Miranda L, Domínguez L, Dias MI, Barros L, Ferreira ICFR, Igual M, Martínez-Navarrete N, Fernández-Ruiz V, Morales P, Cámara M (2019) Sanguinello and Tarocco (*Citrus Sinensis* [L.] Osbeck): bioactive compounds and color appearance of blood oranges. Food Chem 270:395–402

Chebrolu KK, Jayaprakasha GK, Jifon J, Patil BS (2012) Production system and storage temperature influence grapefruit vitamin C, limonoids, and carotenoids. J Agric Food Chem 60:7096–7103

Czech A, Zarycka E, Yanovych D, Zasadna Z, Grzegorczyk I, Klys S (2020) Mineral content of the pulp and peel of various citrus fruit cultivars. Biol Trace Elem Res 193:555–563

De Paulo Farias D, Neri-Numa IA, de Araújo FF, Pastore GM (2020) A critical review of some fruit trees from the *Myrtaceae* family as promising sources for food applications with functional claims. Food Chem 306:125630

Dhull SB, Punia S, Kidwai MK, Kaur M, Chawla P, Purewal SS, Sangwan M, Palthania S (2020) Solid-state fermentation of lentil (*Lens culinaris* L.) with *Aspergillus awamori*: effect on phenolic compounds, mineral content, and their bioavailability. Legume Sci:1–12

Grundy SM, Brewer HB Jr, Cleeman JI, Smith SC, Lenfant C (2004) Definition of metabolic syndrome: report of the national heart, lung, and blood Institute/American Heart Association conference on scientific issues related to definition. Circulation 109:433–438

Hassan MAM, Hussein SM (2017) Chemical and technological studies on pink grapefruit (citrus paradise l.) peels. 1-effect of storage conditions on gross chemical composition, phytochemical components and oil stability of pink grapefruit peels. World J Dairy Food Sci 12:115–123

Igual M, García-Martínez E, Camacho MM, Martínez-Navarrete N (2010) Effect of thermal treatment and storage on the stability of organic acids and the functional value of grapefruit juice. Food Chem 118:291–299

Igual M, Cebadera L, Cámara RM, Agudelo C, Martínez-Navarrete N, Cámara M (2019) Novel ingredients based on grapefruit freeze-dried formulations: nutritional and bioactive value. Foods 8(10):506

Ishfaq M, Ahmad S, Awan MZ, Nasir A (2007) Performance of grape fruit cultivars under agro-climatic conditions of Chakwal. Pak J Agri Sci 44(3):472–474

Jung UJ, Lee M-K, Jeong K-S, Choi MS (2004) The hypoglycemic effects of hesperidin and naringin are partly mediated by hepatic glucose-regulating enzymes in C57BL/KsJ-db/db Mice. J Nutr 134:2499–2503

Kanda K, Nishi K, Kadota A, Nishimoto S, Liu MC, Sugahara T (2012) Nobiletin suppresses adipocyte differentiation of 3T3-L1 cells by an insulin and IBMX mixture induction. Biochim Biophys Acta 1820:461–468

Kaur P, Purewal SS (2019) Biofertilizers and their role in sustainable agriculture. In: Giri B, Prasad R, Wu QS, Varma A (eds) Biofertilizers for sustainable agriculture and environment. Soil biology, vol 55. Springer, Cham, pp 285–300. https://doi.org/10.1007/978-3-030-18933-4_12

Kaur P, Purewal SS, Sandhu KS, Kaur M (2019) DNA damage protection: an excellent application of bioactive compounds. Bioresour Bioprocess 6:1–11

Kefford JF, Chandler BV (1970) The chemical constituents of citrus fruits. Academic, New York, p 246

Kojo S (2004) Vitamin C: basic metabolism and its function as an index of oxidative stress. Curr Med Chem 11:1041–1064

Leung HW, Lin CJ, Hour MJ, Yang WH, Wang MY, Lee HZ (2007) Kaempferol induces apoptosis in human lung non-small carcinoma cells accompanied by an induction of antioxidant enzymes. Food Chem Toxicol 45:2005–2013

Lv X, Zhao S, Ning Z, Zeng H, Shu Y, Tao O, Xiao C, Lu C, Liu Y (2015) Citrus fruits as a treasure trove of active natural metabolites that potentially provide benefits for human health. Chem Cent J 9:68

Malta LG, Tessaro EP, Eberlin M, Pastore GM, Liu RH (2013) Assessment of antioxidant and antiproliferative activities and the identification of phenolic compounds of exotic Brazilian fruits. Food Res Int 53(1):417–425

Manach C, Regerat F, Texier O, Agullo G, Demigne C, Remesy C (1996) Bioavailability, metabolism and physiological impact of 4-oxo-flavonoids. Nutr Res 16:517–544

Manthey JA, Guthrie N, Grohmann K (2001) Biological properties of citrus flavonoids pertaining to cancer and inflammation. Curr Med Chem 8:135–153

Marathe PH, Gao HX, Close KL (2017) American Diabetes Association standards of medical care in diabetes. J Diabetes 9:320–324

Marhuenda J, Cerdá B, Villaño D, Galindo A, Zafrilla P (2019) Citrus and health. Citrus – health benefits and production technology. https://doi.org/10.5772/intechopen.79283

Mendis S, Puska P, Norrving B (2011) World Health Organization. In: World Health Organization. Global atlas on cardiovascular disease prevention and control, Geneva. ISBN 92-4-456437-8

Miller E (2012) Neurodegenerative diseases. Adv Exp Med Biol 724:228–238

Miller EG, Peacock JJ, Bourland TC, Taylor SE, Wright JM, Patil BS, Miller EG (2007) Inhibition of oral carcinogenesis by citrus flavonoids. Nutr Cancer 60:69–74

Mulvihill EE, Huff MW (2010) Antiatherogenic properties of flavonoids: implications for cardiovascular health. Can J Cardiol 26:17A–21A

Nakajima VM, Macedo GA, Macedo JA (2014) Citrus bioactive phenolics: role in the obesity treatment. LWT 59(2):1205–1212

Neri-Numa IA, Torre AD, Oriani VB, Franch-Jr GC, Angolini CFF, Hubinger MD, Ruir ALTG, Pastore GM (2020) In vitro bioactivity approach of unripe genipap (*Genipa americana* L., Rubiaceae) fruit extract and its solid lipid microparticle. Food Res Int 127:108720

Pereira GA, Arruda HS, de Morais DR, Eberlin MN, Pastore GM (2018) Carbohydrates, volatile and phenolic compounds composition, and antioxidant activity of calabura (*Muntingia calabura* L.) fruit. Food Res Int 108:264–273

Pereira GA, Arruda HS, de-Morais DR, Macedo N, Araujo P, Pastore GM (2020) Mutamba (*Guazuma ulmifolia* Lam.) fruit as a novel source of dietary fibre and phenolic compounds. Food Chem 310:125857

Rafiq S, Kaul R, Sofi SA, Bashir N, Nazir F, Ahmad Nayik G (2018) Citrus peel as a source of functional ingredient: a review. J Saudi Soc Agric Sci 17:351–358

Rechkemmer G (2001) Funktionelle Lebensmittel-Zukunft der Ernahrung oder Marketing-Strategie? Molk Zeit Welt Milch 55:S-332

Roesler R, Malta LG, Carrasco LC, Pastore G (2006) Evaluation of the antioxidant properties of the brazilian cerrado fruit *Annona crassiflora* (Araticum). J Food Sci 71(2):C102–C107

Salar RK, Sharma P, Purewal SS (2015) In vitro antioxidant and free radical scavenging activities of stem extract of *Euphorbia trigona* Miller. TANG [Humanitas Medicine] 5:1–6

Sicari V, Pellicanò TM, Giuffrè AM, Zappia C, Capocasale M, Poiana M (2018) Physical chemical properties and antioxidant capacities of grapefruit juice (*Citrus paradisi*) extracted from two different varieties. Int Food Res J 25(5):1978–1984

Siroha AK, Sandhu KS, Kaur M (2016) Physicochemical, functional and antioxidant properties of flour from pearl millet varieties grown in India. Food Measure 10(2):311–318

Sung B, Prasad S, Gupta SC, Patchva S, Aggarwal BB (2012) Regulation of inflammation-mediated chronic diseases by botanicals. Adv Bot Res 62:57–132. Elsevier. ISBN 0065-2296

Turati F, Rossi M, Pelucchi C, Levi F, La Vecchia C (2015) Fruit and vegetables and cancer risk: a review of southern European studies. Br J Nutr 113:S102–S110

Xi W, Zhang G, Jiang D, Zhou Z (2015) Phenolic compositions and antioxidant activities of grapefruit (*Citrus paradise* Macfadyen) varieties cultivated in China. Int J Food Sci Nutr 66(8):858–866

Xu G, Liu D, Chen J, Ye X, Ma Y, Shi J (2008) Juice components and antioxidant capacity of citrus varieties cultivated in China. Food Chem 106:545–551

Yoshida H, Takamura N, Shuto T, Ogata K, Tokunaga J, Kawai K, Kai H (2010) The citrus flavonoids hesperetin and naringenin block the lipolytic actions of TNF-alpha in mouse adipocytes. Biochem Biophys Res Comms 394:728–732

Zygmunt K, Faubert B, MacNeil J, Tsiani E (2010) Naringenin, a citrus flavonoid, increases muscle cell glucose uptake via AMPK. Biochem Biophys Res Commun 398:178–183

Chapter 3
Mosambi

Vinti Singh, Radha Kushwaha, Devinder Kaur, and Sandeep Kumar

3.1 Introduction

3.1.1 Botanical Name, Common Name

Kingdom: Plantae
Genus: Citrus
Species: *C. limetta*
Family: Rutaceae
Botanical name: *Citrus lumia, Citrus limetta, and Citrus limettioides*
Common names: Persian lime, Sweet limetta, sweet lime, sweet lemon, and Mediterranean sweet lemon, Mosami, Mausam, Bergamot
Indian names: Mosambi

3.1.2 Natural and Cultural History

Mosambi (sweet lime) may be a cross between a Mexican lime and a sweet lemon or citron, which are native to the Mediterranean region and arrived in California from Mexico, but its exact origins are unknown. When mosambi's cultivar grew at the San Gabriel Mission in 1943, it was given the name "Millsweet." The Southeast Asian archipelago produces mosambi, or "sweet lime," which is a cross between a lemon and a citron (Inglese and Sortino 2019). The lime of Palestine, which is a

V. Singh · R. Kushwaha · D. Kaur
Centre of Food Technology, University of Allahabad, Prayagraj, India

S. Kumar (✉)
Department of Physics, H.N.B. Government PG College, Prayagraj, India

S. Singh Purewal et al. (eds.), *Recent Advances in Citrus Fruits*,
https://doi.org/10.1007/978-3-031-37534-7_3

hybrid of a citron and a sweet orange, is the most common of the sweet limes (*Citrus limettioides* Tan.).

3.1.3 Plant Description

A small tree with uneven branches and rather smooth, brownish-gray bark, *C. limetta* can grow to a height of 8 m (26 ft). It has several 1.5–7.5 cm (0.59–2.95 in) long thorns. The petioles range in length from 8 to 29 mm (0.31–1.14 in), and they have prominent, narrow wings. Compound leaves have acuminate leaflets that measure 5–17 cm (2.0–6.7 in) long and 2.8–8 cm (1.1–3.1 in) broad. White, 2–3 cm (0.79–1.18 in) broad flowers are produced. Fruits are oval, green, and ripen to yellow. The pulp is green. The white pith measures around 5 mm (0.20 in) thick. Despite having the label "sweet lime," the fruit looks more like a greenish orange. Tropical and subtropical conditions are ideal for *C. limetta* growth. At 5–7 years old, it starts to produce fruit; peak production occurs at 10–20 years. Seeds are used for its propagation (Bhaumik et al. 2018).

3.1.4 Origin and Distribution

Asian native Mosambi fruit, often referred to as sweet lime, is best grown in India, China, Malaysia, south Japan, Indonesia, Vietnam, and Thailand. Despite the fact that the Citrus genus has multiple origins, from China to Indonesia, a 2004 research in the "Agriculture Review" claims that the hills of Meghalaya and Nagaland are the true home of sweet limes (*Citrus tanaka*). The book "Fruits of Warm Climates," on the other hand, cites a wider range by adding India's central and northern regions. The fruit itself is a cross between a sweet citron or sweet lemon and a Mexican lime. Sweet limes are now grown throughout the Americas, Mediterranean, tropical Southeast Asia, Palestine, Egypt, and Syria. In some places of Florida and California, it's also a leisure fruit (Shivankar 2010).

3.1.5 Different Species and Cytogenetic

According to W.T. Swingle's classification (Nicolosi 2007), the sweet lime is a member of the Rutaceae family, the Auranciaceae subfamily, the limetta species, and the Citrus gender. There are 162 species in the Rutaceae family, which is grown worldwide in tropical and subtropical climates (Tanaka 1977).

Almost all cultivated varieties of Citrus, Fortunella, and Poncirus are diploid, and these genera have 18 chromosomes per diploid. *Citrus limettioides* (Sweet lime) is concentrated in Maharashtra, Uttaranchal, H.P. and NEH region, while*Citrus limited* (Sharbati lime) is found in the Baghmara area in Garo Hills, Meghalaya.

3.2 Production

Mosambi (sweet lime) is a widely grown crop in India, especially in the country's central region (Hodgson 1967). It is a native of Southeast Asia. Developing nations like China, India, Vietnam, Japan, Thailand, Malaysia, and Indonesia, are experiencing an increase in demand for output (Khan 2007). In India, the fruit is accessible twice in a year, from November to March and July to August. About 17.35 million tonnes of lime fruit are produced annually throughout the world. The biggest producers of sweet lime, accounting for 61.32% of global production, are China, Mexico, India, Argentina, and Brazil (FAOSTAT 2018). Over the past 5 years, the top three states of India growing sweet lime have been Andhra Pradesh (1800 tonnes), Maharashtra (500 tonnes), and Telangana (300 tonnes). India is gradually expanding the production of this fruitdue to itsincreasing demand.

3.3 Botanical Description

Mosambi, despite having a sweeter flavour, is a member of the citrus family. It is mildly seedy, round or oval in shape, and a little flat bottom. The skin is silky and starts out yellow with a little green tint before turning orange as it ripens. Due to its very low acid level, the flesh has a bland flavour despite being somewhat solid and juicy. White is the colour of flowers. They thrive in tropical and subtropical regions and can grow to a height of up to 25 ft (Bhaumik et al. 2018). The peak production period lasts between 10 and 20 years, and it is often propagated via seed. It is frequently consumed asbeverages, sorbets, pickles, jams, snacks and candies (Braddock 1995; Braddock and Cadwallader 1992) while most of the people eat it as a fruit.

3.4 Structure and Chemical Composition of Mosambi

The fruit's external cuticle, which is thinly covered by an epidermal layer (flavedo) holding numerous oil sacks packed with aromatic essential oil with significant commercial value, makes up the macroscopic structure of fruit (Miller et al. 1940). Flavedo, spongy layer parenchymatous cells, also called the albedo, which contains around 20% citrus pectin, is the next layer. The interior pulp or flesh (edible component) of the fruit is divided into segments (locules), each of which contains multiple juice sacs (vesicles) and seeds (Fig. 3.1). These segments are connected by a membrane made of thin epidermal tissue (Davis 1932). The term "axis" (or "core") refers to the white, spongy tissue in the centre of the fruit that resembles albedo.

The "rag" of the extracted juice is the collective term for the segment and core membranes (Matlack 1931). The primary factor in the popularity of citrus fruits is

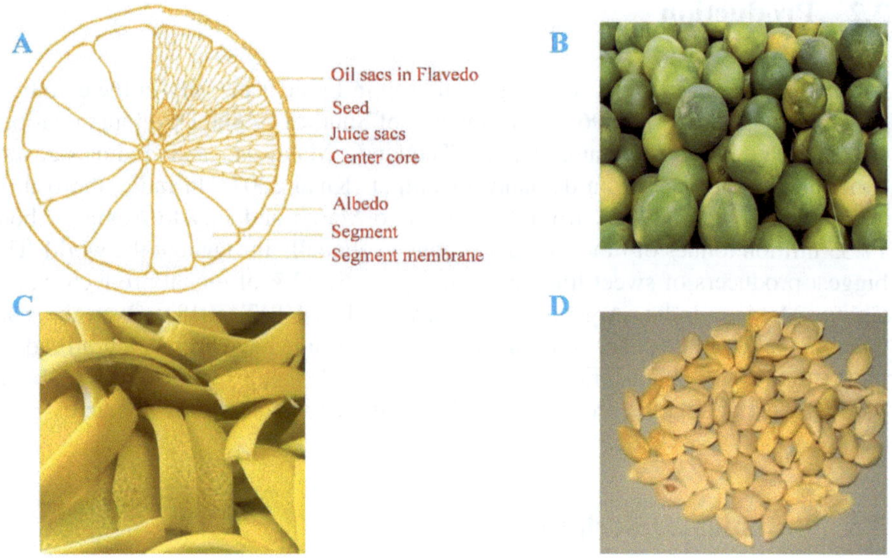

Fig. 3.1 (**a**) Schematic view of the cross section of an Mushambi, (**b**) A whole fruit, (**c**) Fruit peel, (**d**) Seed

their presence of significant organic acids, sugars, and amino acids. Dietary fruit is a good source of phytochemicals such as carotenoids and limonoids, as well as vitamins (C, A and B), minerals, fibre, and phytochemicals like phenolic compounds. The bioactivities of flavonoids, particularly poly-methoxy flavones (PMFs) present in citrus fruits, including their anti-oxidant (Chen et al. 2017) and anti-cancer properties (Ke et al. 2015), have gained increasing attention in recent years. Alarge group of researchers' (Bermejo and Cano 2012; Codoñer-Franch and Valls-Bellés 2010) study suggests that the fruit is positively connected with bone, cardiovascular health, and immune system. Fruit also has antioxidant and antimutagenic characteristics and has no fat, sodium or cholesterol.

3.5 Phytochemicals in Citrus Fruits Their Biological Activity of Mosambi

3.5.1 Flavonoids

Citrus fruit flavonoids have significant biological and pharmacological effects; they contain antibacterial, antifungal, anticancer, and antiviral properties (Burt 2004; Ortuño et al. 2006). A broad class of chemical compounds known as flavonoids are produced by plants. The chemical has a small molecular weight. Given their phenyl benzopyrene structure, flavonoids are polyphenolic chemicals that have two benzene rings (C6) connected by three carbon chains (C3) and a carbonyl group in the C position. Plants mostly include flavonoids as glycosides. Citrus has four different forms of flavonoids: flavanones, flavones, flavonols, and anthocyanins (Zhang 2007).

Citrus fruit flavonoids contain a variety of glycosides, such as hesperidin and naringin, as well as an extra group of O-methylated flavone aglycones, such as nobiletin and tangeretin (Wang et al. 2014). Citrus fruit peels are said to contain the highest concentrations of PMFs compared to other edible sections of the fruit (Wang et al. 2014). Citrus flavonoids have a wide range of health benefits, including anti-cancer, antiviral, anti-inflammatory, antidiabetic andanti-microbial actions. The biochemical analysis of the flavonoids found in orange peels shows that they increase the serum's antioxidant defences against lipid peroxidation and reduce oxidative stress in older people (Assini et al. 2013). Additionally, according to this biochemical investigation, flavonoids have beneficial effects against cancer, inflammation, and atherosclerosis (Romagnolo and Selmin 2012; Park and Pezzuto 2012; Mulvihill and Huff 2012).

These days, Chinese medicine practitioners frequently employ the dried form of tangerine peel (*Citri reticulatae*), specifically the pericarpium known as chen-pi, to cure a variety of ailments like dyspepsia, bronchial asthma and heart circulation (Chinese Pharmacopoeia Commission 2015). The most common flavonoid in tangerine peel, hesperidin, has been shown to have minor inhibitory efficacy against neuro inflammatory, but their combined effect is discovered to be considerable (Ho and Kuo 2014).

3.5.2 Carotenoids

Sweet oranges, mandarins, and grapefruits are excellent sources of carotenoids. The main function of carotenoids pigment components is to defend against the many diseases of the human body and control-health-related elements. Citrus fruits are rich in lutein and zeaxanthin, which are good for the body's immune system and eyes. A vivid source of carotenoids prevalent in Cara navel oranges and mandarin fruits. Comparing the substantial amounts of carotenoids found in citrus fruits reveals that Mandarins fruit has greater values than Grapefruit, then Orange fruits, while Lemon fruits don't contain any carotenoids.

3.5.3 Essential Oil

Citrus species are known to be abundant sources of aromatic compounds; in example, citrus fruit peels contain 400 volatile and nonvolatile chemicals. Essential oils have been utilized for medicinal and aromatic purposes since antiquity. Today, they are used in a variety of pharmacy, cosmetic, and allied industries due to their antispasmodic and antibacterial activities (Jung et al. 2004).

3.5.4 Mucilage

Citrus fruit seeds, peels, and pulp all contain mucilage, a fiber-like substance that when combined with water produces a gel-like structure. Citrus seeds contain psyllium, which helps to improve digestion and speed up the removal of cholesterol. Citrus fruits also have a higher ratio of soluble and insoluble fibre, which lowers cholesterol levels (Ono et al. 2011).

3.5.5 Tannin

In comparison to other citrus fruits, the grapefruit, lemon, and lime contain a larger quantity of tannins compounds; these compounds are crucial for stopping diarrhoea, reducing bleeding, and controlling other severe bodily discharges (Matsuda et al. 1991).

3.5.6 Polyphenols

The phenolic chemicals found in the peels of several fruits and vegetables have the power to scavenge free radicals, delaying the onset of degenerative illnesses. Lemon peel contains coumarins and tetrazene, the two compounds that can scavenge free radicals by hydrogen- or electron-donating processes, interrupt chain reactions, or eliminate ROS and RNS originators by quenching catalysts that start chain reactions (Guimarães et al. 2010). Some highly potent phytochemical compounds with names like 4′-Geranyloxyferulic (GOFA) and boropinic acid have emerged in the past decade from C. sinensis and kumquat (*Fortunella japonica*) for luxurious pharmaceutical drugs for neuroprotective, cancer chemo defensive, and anti-inflammatory activities (Genovese et al. 2014).

3.6 Antioxidant Properties

Mosambi naturally contains a lot of vitamin C and phytophenols (Gorinstein et al. 2004). The leaves, seeds, and floral part of it contain a significant amount of flavonoids, including polymethoxylated flavones and glycosylated flavanones (Imran et al. 2016). Due to their ability to suppress oxidation processes in food systems and their metal chelation activities andfree radical scavenging, phytophenolics are said to provide health advantages (Gundgaard et al. 2003; Taguri et al. 2004). The phytochemicals also aid in the prevention of diseases linked to DNA repair, pro-carcinogen inhibition, and the prevention of N-nitrosamine production (Shahidi 1997). Additionally, Mosambi supplies an adequate amount of potassium, pectin, and vitamin C. Mosambi contains potent anti-inflammatory, antibacterial, and antioxidant effects, and eating sweet limes is linked to a lower risk of developing cardiovascular diseases and several types of cancer. Additionally, mosambi's entire plant is employed in traditional medicine.

3.6.1 *Peels*

Peels from moambi trees are a good source of fibre and pectin. They have antiobesity and anti-diabetic qualities (Baker 1994; Riccardi et al. 2008). In the fruit juice industry, Mosambi peel is typically viewed as waste, yet it is actually the primary source of phenols, including flavonoids, pectin andessential oils. Jam and jelly can use pectin from mosambi peel as a source. Mosambi peel is a key ingredient in the creation of value-added meals with extra health benefits since it has strong antioxidant qualities and a high flavonoid content. Younis et al. (2015) described the impact of mosambi peel powder on the technical and functional properties of papaya jam. Additionally, mosambi peel oil can be utilised as a flavouring ingredient in a variety of meals and beverages and has a powerful aroma.

3.6.2 *Seed*

The citrus processing business generates a significant amount of waste from citrus seeds. From mosambi seed, the chemical industry extracts flavonoids and essential oils. Excellent sources of magnesium, sodium, potassium, iron, and calcium can be found in citrus seed oil (El-Adawy et al. 1999). The lipids obtained from mosambi seeds contain essential fatty acids as well as lipid-soluble bioactive compounds as carotenoids, polyphenols, and tocopherols (Malacrida et al. 2012). Mosambi seeds, which have antioxidant, anti-inflammatory, and antibacterial nutraceutical qualities, can alssed to extract aromatic oils. Numerous significant chemicals, including, myrcene, terpenes,-pinene, camphene, sabinene, −terpinene, linalool, p-cymene, andd-limonene, are found in citrus essential oils. The most common ingredient in citrus oil, d-limonene, has nutritional effects.

3.6.3 *Fruit Juice*

The mosambi juice is a cool summer beverage that is high in vitamin C and provides immediate energy (Arias and Ramón-Laca 2005). Water-soluble vitamins and minerals are abundant in mosambi juice which is less than half of the fruit's weight. Folic acid, which strengthens bones and joints, is another ingredient. There are many antioxidants, which strengthen the immune system and increase resistance to infections like the common cold. Fruit juice detoxifies the body and balances off unhealthy effects of fatty diets, stress, and pollution. Finally, the presence of limonenes makes it precious forour health.

3.6.4 Fruit

The fruit known as Mosambi is resourceful and has a sweet and sour flavour. Terpenes, oxygenated compounds, and nonvolatile chemicals including waxes and pigments make up the majority of the ingredients in citrus fruits (Kondo et al. 2000). Being the most prevalent terpene, D-limonene has antibacterial properties against gram-positive bacteria and also improves sodium benzoate's efficiency. D-limonene, a kind of limonene, has anti-inflammatory, anticancer, and antioxidant properties.

3.6.5 Leaves

It has been claimed that mosambi leaves have antihypertensive properties. Citrus limetta Risso leaf extract, according to Perez et al. (2010), causes angiotensin II to have a hypertensive impact.

3.6.6 Waste

The juice processing business generates a lot of by-product waste each year (Manthey and Grohmann 2001). The pulp, seeds, and peel, which together make up 40–60% of the raw material, are the primary byproducts of the citrus industry (Licandro and Odio 2002). The majority of these by-products are considered to be citrus wastes, which are rich in nutrients and phytochemicals that can be considered as bioactives. These bioactives can be employed effectively as food additives, enhancing agents during food processing, preservatives, and as parts of pharmaceutical medications (Middleton Jr and Kandaswami 1994). Citrus waste is used to produce phytochemicals that are used in the production of antifungal and antibacterial lotions and soaps, fragrances, and cosmetics. Citrus fruits also yield useful byproducts such beverages, pectin, bases, purees, molasses, peel seasoning, and marmalades. The rich flavonoids in mosambi peels have antioxidant qualities that stop oxidation in food systems (Taguri et al. 2004). Additionally, the majority of uses for citrus byproducts are as animal feed.

3.7 Antioxidantproperties of Its Products

Customers are paying increasing attention to functional foods since they offer health advantages above and beyond basic nutrition and lower the risk of diseases linked to diet. Functional foods high in antioxidants provide defence against a variety of diseases, such as atherosclerosis, diabetes mellitus, cancer, oxidative stress disease and cardiovascular disease. The products made from mosambi are rich in phytochemicals and antioxidants. In a study published by Imran et al. (2016), antioxidant extract was added to cookies to boost their nutritional value. Papaya jam also used mosambi peel powder, which improved its technological and fuctional attributes (Yonius et al. 2015). Additionally, phytophenols and antioxidants isolated from mosambi can be employed to boost the nutraceutical value of a variety of food formulations. As a result, the pharmaceutical and food industries can benefit from the isolation of bioactive chemicals from processing industry waste. Antioxidant properties of mosambi and its products are presented in Table 3.1.

3.8 Health Benefits

Mosambi is extraordinarinutritious. The examination of Mosambi's nutraceutical potential will confirm its potential for use in place of synthetic medicines in the future (Fig. 3.2). Very little fat is present. Below are a few of the health advantages:

3.8.1 Protect Against Scurvy

Scurvy is characterised by spongy, purple-colored gums, loose teeth, protruding eyes, dry, brownish skin, tongue and mouth ulcers, and cracked lips. It is brought on by a lack of vitamin C. Mosambi is the most abundant source of vitamin C and can be used as a primary preventative measure in the treatment of scurvy.

3.8.2 Improves Digestion

The plant chemical flavonoids, which have a pleasant scent, are abundant in Mosambi. By promoting the release of acids, bile, and digestive fluid, it quickens digestion. By neutralising the acidic digestive secretions, it eliminates the toxins from the body. As a result, it is frequently advised to those with digestive and gastrointestinal issues. Due to its high potassium content, sweet lime aids in reducing symptoms of dysentery, diarrhoea, nausea, vomiting, bloody amoebic

Table 3.1 Antioxidant properties of mosambi and its products

Fruit parts/ products	Properties	References
Fruit peel oils	7.21 mg/g TPC, 1.01 mg/g DPPH activity	Khan et al. (2012)
Peel extract	19.3% TPC, 44.5% DPPH, 65.6% antioxidant activity	Imran et al. (2016)
Fruit	26.7% antioxidant activity, 220 mg GAE/g of TPC, 50 mg QE/g of TFC	Sharma et al. (2017)
Peel	15.6–28.4 mg/100 g Vitamin C, 0.25–0.53 OD of ferric reducing antioxidant property	Singh et al. (2014)
Green tea infused with mosambi peel	31.38 mg TAE/g of TPC, 43.2 mg AAE/g of flavonoids, 9.83 mg AAE/g antioxidant activity	Kaur et al. (2021)
Juice	75.14 mg/100 g vitamin C	Tangirala et al. (2020)
Fruit	232 mg/L polyphenolic content	Paul and Bhattacharyya (2019)
Peel oil	77.12% antioxidant activity	Aruna et al. (2022)
Fresh peel	1.62 mg/g TPC, 1.0 mg/g Total antioxidant activity 10.0 µg IC50 of DPPH	Thakur et al. (2017)
Juice	60.47 mg/100 mL Vitamin C, 23.07 mg/100 mL Total phenolic content	Kumar et al. (2013)
Juice	Vitamin C (42.35 mg/100 g), total polyphenolic content (35.66 mg GAE/100 g), total flavonoid content (45.58 mg QE/100 g), antioxidant activity (DPPH 2.05 µmol TE/ml, CUPRAC 1.24 µmol TE/ml, FRAP 1.83 µmol TE/ml, ABTS 7.77 µmol TE/ml)	Nayak et al. (2020)
Peel powder	Eriocitrin (2.17 mg/100 g), Diosmin (15.73 mg/100 g), Hespertin (12.62 mg/100 g), Naringenin (3.48 mg/100 g), Tangeritin (13.88 mg/100 g), Rutin (5.39 mg/100 g), quercetin (4.62 mg/100 g), and about 72% antioxidant activity in terms of DPPH	Ali and Abass (2015)

dysentery and loose stools. Additionally, it has a lot of fibre, which controls constipation, healthy bowl movemen and digestion. Furthermore, d-limonene, a key ingredient in mosambi, has been shown to be successful in reducing gastroesophageal reflux disease symptoms (Wilkins Jr 2002).

3.8.3 Immunity Booster

Mosambi is an antioxidant powerhouse that controls the immune system and wards off sickness. Additionally, regular consumption of mosambi juice improves heart health by enhancing blood circulation. It also contains a lot of vitamin C which strengthens immunity (Bhaumik et al. 2018).

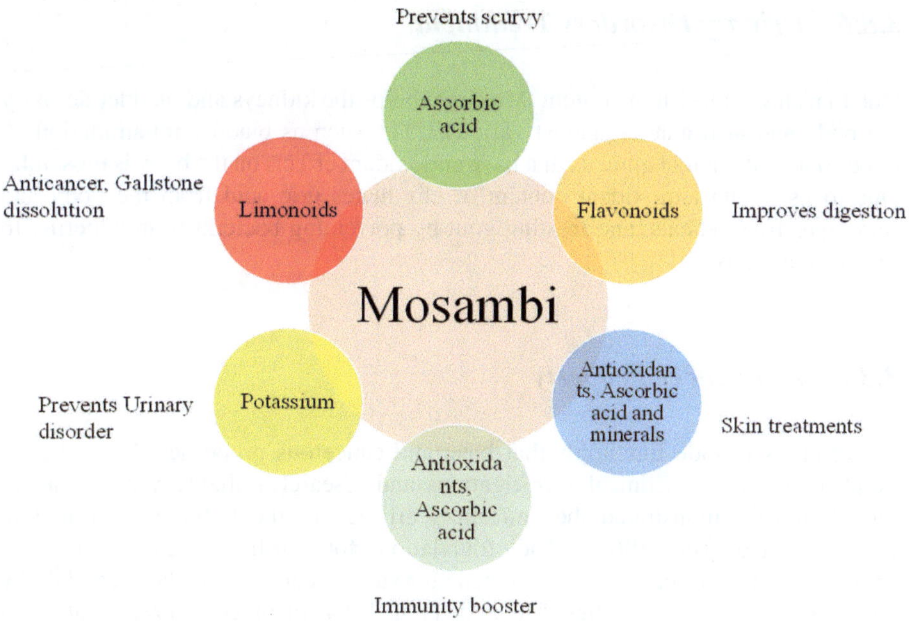

Fig. 3.2 Health benefits of Mosambi

3.8.4 Breathing Issues

Due to its anti-congestive qualities, vaporizers, inhalers, and balms all include mosambi. Juice from the mosambi tree can instantly relieve asthmatic cough.

3.8.5 Skin Treatments

Mosambi is a key ingredient in skincare and is a superior source of minerals, vitamin C, and antioxidants. Its juice is utilised as a vitamin and alternative medication. It has anti-inflammatory, antibacterial, and disinfecting effects. It is utilised in the production of numerous beauty items. Skin issues like pigmentation, spots, zit and blemishes are treated with it. Additionally, it helps to healthe chapped lips. Juice from mosambi has anti-aging qualities. It improves skin issues by purifying the blood. Its juice is frequently applied to dry, irritated skin to improve skin tone (Bhaumik et al. 2018).

3.8.6 *Urinary Disorders Treatment*

Due to its high potassium content, Mosambi helps the kidneys and bladder detoxify the body and also guards against various UTIs, such as bladder inflammation. A good source of phytochemicals that have antioxidant effects on the body is mosambi. Due to its strong antioxidant content, it eliminates uric acid from the body and scavenges free radicals, and treating gout by preventing bacteria from adhering to the bladder walls.

3.8.7 *Anticancer Property*

Sweet limes include limonoids that have anti-cancerous properties. It combats a number of cancers. Clinical investigations and research utilizing various cancer models have demonstrated the anticancer efficacy of the D-limonene found in mosambi (Hodgson 1967). The foundation for further research into the chemoprotective properties of d-limonene for various cancer types has been laid by animal studies. Several studies (Elegbede et al. 1984; Elson et al. 1988; Maltzman et al. 1989; Wattenberg 1983) suggested that orange peel oil or pure d-limonene administered to rodents inhibited chemically-induced mammary cancer in either the initiation or promotion phases, depending on the chemically-induced medium used (Crowell 1999; Uedo et al. 1999; Yano et al. 1999). D-limonene was shown by other experimental investigations to prevent the growth of stomach tumors, pulmonary adenoma, and for liver cancer (Dietrich and Swenberg 1991).

3.8.8 *Gallstone Dissolution*

Gallstones containing cholesterol can be broken up by D-limonene, according to clinical studies. D-limonene has been proven in an in vitro research to dissolve human gallstones within 2 h. In contrast, d-limonene injections into the gallbladder in animals dissolve and disintegrate gallstones, which are then eliminated through the common bile duct. Gallstones in patients were reported to dissolve after just three d-limonene infusions by Igimi et al. in 1976. D-limonene was found to be capable of dissolving human gallstones in 2 h in an in vitro research. Gallstones were broken up and eliminated through the common bile duct in animals after d-limonene was infused into the gallbladder. Infusions of 20 mL d-limonene given to patients after gallstone surgery disintegrated stones that were missed during surgery. Gallstone eradication occurs in some peopleafter only three infusions (Igimi et al. 1976).

3.8.9 Treatment of Dehydration and Sunstroke

Mosambi juice is a nutritious beverage that gives the body all the necessary vitamins. It is used as a substitute for carbonated beverages to slake thirst. Athletes would be better off choosing this. It lessens the chance of dehydration and cramping in the muscles. Furthermore, consuming fresh mosambi juice will help you avoid sunstroke. Sweet lime juice hydrates your body and makes you healthier thanks to its wealth of vitamins and minerals.

3.8.10 Antidepressant and Allergy-Fighting Properties

In the tissues of moshambi, apigenin exhibits depressive function, while certain flavonoids exhibit an antiallergic impact (Matsuda et al. 1991).

3.9 Traditional Uses of *Citrus limetta*

3.9.1 As Digestive Support

Citrus fruit (mosambi) juice's pleasant scent encourages the salivary glands to release saliva, which speeds up digestion. In addition, the flavonoids found in lime juice improve digestion by encouraging the release of bile, digestive fluids, and acids. Therefore, drinking mosambi juice frequently during the day can prevent stomach issues, indigestion, motion sickness, and dizziness. The mosambi juice's acids aid in the removal of toxins from the bowels, which relieves constipation. Constipation can be instantly relieved by drinking sweet mosambi juice with a dash of salt. Additionally, due to its high potassium content, it is effective in treating diarrhoea, loose stools, dysentery, and stomach problems. It prevents nausea and vomiting because of its delectable flavour. Bloody dysentery can also be treated with it, according to Gupta et al. (2021).

3.9.2 Anti-diabetic Benefits

Patients with diabetes can benefit from citrus fruit juice. It is advised to consume 2 teaspoons of citrus juice (mosambi), 4 teaspoons of amla juice, and 1 teaspoon of honey every morning on an empty stomach in order to manage diabetes.

3.9.3 In the Treatment of Scurvy

Scurvy is a disease brought on by a lack of vitamin C that is characterised by swollen gums, recurrent flu episodes, clod, and cracked lip corners. Citrus fruit is useful in treating scurvy since it is high in vitamin C.

3.9.4 Immunity Booster

Regular ingestion of citrus fruit juice enhances the heart's ability to control normal blood circulation. A significantly stronger immune system is the result.

3.9.5 Weight Reduction

Mosambi juice aids in weight loss because it is low in calories and fat. To burn more calories, you can sip on a mosambi juice and honey concoction.

3.9.6 Helpful During Pregnancy

The citrus fruit juice has a lot of calcium, which is good for the mother-to-be as well as the foetus.

3.9.7 Urinary Problems Treatment

Mosambi juice, which is high in potassium, aids in the treatment of urogenital diseases including cystitis. Cystitis, commonly referred to as a urinary tract infection, is an inflammation of the bladder (UTI). For prompt treatment from cystitis, mosambi juice cooked in water should be consumed shortly after cooling. Potassium helps the kidneys and bladder detoxify themselves, reducing the risk of a variety of UTIs (Bhaumik et al. 2018).

3.9.8 Common Cold Treatment

Citrus juice helps to prevent the common cold and strengthens the body's tolerance to cold weather because it is a high source of vitamin C.

3.9.9 Effects in Lowering Hyperlipidemia

Juice from mosambi trees improves blood pressure and decreases cholesterol.

3.9.10 Treatment for Various Skin Conditions

Since vitamin C is abundant, it naturally lightens the skin and is used in many cosmetic products as a vitamin and supplement replacement. The treatment of skin pigmentation, spots, and blemishes with citrus fruit is also beneficial.

3.9.11 Reduction of Swelling and Pain

Castor oil and citrus fruit (mosambi) juice applied to the affected area helps reduce discomfort and swelling (Gupta et al. 2021).

3.10 Postharvest Management, Processing and Storage

3.10.1 Propagation Methods and Planting Considerations

Cuttings from hardwood stems are used in commercial sweet lime propagation. It can also spread through nucellar seedlings on occasion, air layering, and budding. Sweet lime is quite polyembryony; 5.8 embryos on average have been recorded for each seed. The sweet lime is propagated from cuttings in India. The hardwood cuttings are made from chosen shoots of trees that consistently produce well. The cuttings root well in open areas during the monsoon season (Jauhari and Rahman 1959). Sweet lime cuttings that were 25 cm long and harvested in July were treated with IBA or NAA at rates ranging from 1500 to 6000 p.p.m. and then planted in sand-filled pots.

IBA at 1500 p.p.m. was closely followed by NAA at 1500 p.p.m. for optimal rooting, sprouting, and plant growth (Singh et al. 1986). The fast dip approach of applying IBA at 3000 ppm can improve the root development in cuttings. In ideal circumstances, rooting is finished in 4–6 weeks. However, the plants grown from stem cuttings are typically surface feeders with weak roots, which is a drawback (El-Sheikh 2005). Shield budding is a good method of propagation as long as it is done in March or April and only bud wood from the current growing season is used. Jatti khatti (rough lemon) is the appropriate rootstock in the Punjab, Karna Khatta in the state of Uttar Pradesh, and Jambheri in other regions of the nation(Arora et al. 2012).

3.10.2 Flowering and Fruiting

In the latter situation, leaves from the main axis may or may not be present. The flowers can be borne singly or in clusters. The production of leafy and leafless inflorescences was compared, and it was discovered that the former kind produced a higher proportion of hermaphrodite flowers and mature fruit (Singh and Dhuria 1960). On the same tree, sweet lime trees produce both staminate and harmaphrodite flowers, albeit the staminate blossom is more common. The low percentage of flawless flowers, not self-incompatibility as some workers claimed, is the cause of the shy bearing tendency of sweet lime plants. Average fruits may be enhanced to 139% by bending and ringing the branches and to 280% by spraying with IBA 100 ppm, as opposed to 109% fruits in the control, in order to increase the percentage of flawless blossoms (Motial 1964). On February 21, Planofix [NAA] or Ethrel [ethephon], both at 2,501,000 p.p.m. active component, were sprayed on seven-year-old sweet lime plants. The proportion of hermaphrodite flowers rose from 18.2% in the control to 34.1 and 32%, respectively, at the two compounds' maximum rates. Beginning in the fourth year, the sweet lime produces a weak and unprofitable crop; nevertheless, by the seventh year, a normal and profitable crop is produced. After flowering for around 6 months, the fruits are ready to eat. When PCPA is given via the leaves, it has been observed to promote fruit setting and provide the highest fruit output (336,345 fruit/tree) at concentrations of 75 and 100 ppm (68.5 and 70%, respectively) (Gangwar and Singh 1965).

3.10.3 Maturity, Harvesting, Yield

All citrus fruits are non-climacteric, which means they mature gradually over several weeks or months and take a while to fall off the tree. A poor measure of maturity, external colour variations during ripening are more a result of climate than ripeness. Internal measurements of °brix (sugar), acid content, and the °Brix/acid ratio are the best indicators of citrus fruit maturity. In the northern portions of India, from August to October, and in Assam, from September to November, the fruits are ready for harvesting. The primary harvesting season in the south is from August through September. Depending on the age of the plantation, the kind of soil, the temperature, and the ma3.nagement techniques used, the production ranges from 100 to 150 quintals per hectare.

3.10.4 Post Harvest Management and Storage

Application of ethephon @ 250 ppm coupled with 1% calcium acetate as foliar spray at maturity stage is advised to impart uniform yellow orange colour to the fruits. At room temperature, sweet limes have a lengthy shelf life and can stay fresh for up to 2 weeks. Sweet limes keep for 4–8 weeks in the refrigerator. Slices of the fruit can be frozen, though the limonin content of sweet limes may eventually make the pulp taste bitter. The fruit can be frozen in a "wet pack," which is created by drenching the slices in sweet syrup and placing them inside an airtight glass jar. The frozen juice can be stored for up to 6 months, but it's recommended to check on the fruit frequently to make sure that it doesn't turn stale.

3.10.5 Processing, Juice Extraction, Product Development

Sweet oranges and mandarins both have two primary crops. One is known as Ambiabahar (mango blossoming), which blooms in January (around the time of mango flowering, thus the name Ambia), and whose fruits are available from October to December. The other crop is mrigbahar (Monsoon bloom), which flowers in June and July and has fruit picked between February and April. Mandarins and sweet oranges typically reach maturity between 240 and 280 days after planting. Fruits that have reached the colour break stage are picked up every 2 or 3 days for 10–14 days. Lemons and limes mature around 150–160 days. There could be 2 or 3 crops of limes and lemons per year. It is advised to apply ethephon @ 250 ppm coupled with 1% calcium acetate as a foliar spray at the maturity stage to give the fruits a consistent yellow-orange colour. Ethylene gas can be used to de-green and develop colour in sweet oranges and mandarins. In a de-greening chamber, a temperature of 6–7 °C, 5–10 ppm of ethylene, and 90–95% RH can induce a colour shift in 48 h. There are available cold storage settings for keeping certain citrus fruits for an extended period of time. Citrus is pre-cooled using a forced air system. Sweet oranges can be stored for 4–8 weeks at 7–8 °C with 85–90% RH.

3.10.6 Diseases and their Management

3.10.6.1 Citrus Canker (*Xanthomonas citri*)

It is an extremely dangerous illness that affects citrus, especially moshambi. Disease manifests on fruits, twigs, and leaves. It shows as yellow dots on leaves. Which progressively get bigger, get rougher, browner, and develop elevated areas on both sides of the leaf. A yellow aura surrounds these specks. The lesions become tough and corky on the fruit's peel. Spraying 100 ppm streptocycline +2 g copper sulphate/L of water three times a year, in February, October, and December, is advised to control the disease.

3.10.6.2 Scab (*Elsinoe fawcctti*)

On the underside of the leaves, small, dark-brown, rough, irregularly elevated blemishes are most noticeable. Fruits and twigs can become diseased as well. It is suggested to use Bordeaux combination 2:2:250, 50% copper oxychloride at 4 g/L of water, Ziram 27 SC, or Dithane M45 at 2.5 g/L three times from June to August, each time at a 20-day interval.

3.10.6.3 Gummosis (*Phytophthora palmivora*)

Often known as foot rot, root rot, collar rot, or crown rot, is a disease that affects plants. The disease is typically more prevalent in orchards that have flood irrigation or inadequate soil drainage. Rootlets decay and diseased leaves drop as a result of fungus. The diseased plants produce lots of flowers.

The main sign of this disease is excessive gumming coming from the stem and branches. The following actions must be followed in order to control the disease: I. Gather and bury sick leaves or fruits deeply in the soil. II. Cleopatra can be used as a rootstock. Part III Avoid flood irrigation, IV. Use Bordeaux paint to cover the trunk sections up to a height of 20 cm. Spray the affected area of the trunk with the Bordeaux mixture 2:2:250 three times, i.e., in February before flowering, June, and late July; alternatively, apply two coats of Ridomil MZ as paint (2 g/100 ml of Linseed oil) or two applications of Aliette 80 WP (2.5 g/L of water) to the affected area of the trunk in April and September.

3.10.6.4 Wither Tip

Another name for it is die back (anthracnose). It may be due to physiological factors or the fungus Colletotrichum gloesporioides. Anthracnose symptoms show up on leaves, new shoots, and fruits. The necrotic patches on leaves resemble concentric rings of acervuli. Twig dead ends have a silvery grey look. Wither tip is known to cause leaves to shed and twigs to die back. In severe cases, immature fruit infection and the stem lead to fruit dropping and tree dieback.

3.11 Characterization of Waste and Byproducts

3.11.1 Animal Feed

The solid byproducts of citrus processing can be used as components of animal feed. Since it is the most accessible and cost-effective way of disposal, the majority of industries throughout the world dispose of their waste in fresh or dried form for the production of animal feed (Moreira et al. 2004). The pulp is dried (up to 90 g/100 g) and pelletized for selling after being twice pressed to lower the moisture content to 65–70% (w.b.) for dried silage production. Pectin, which is present in large quantities in citrus residue, helps to bond the water there. During drying, lime and calcium hydroxide or calcium oxide are added to pulp, theyraises pH 5.5–6.5, removes pectin's hydrophilic character, and makes it easier to separate the water and carbohydrates that are present in the pulp (Cocke and Muncie 1976; Teixeira 2001; Wing 2003). Bampidis and Robinson have thoroughly examined the use of citrus waste as ruminant feed and its effects on the animal's productivity (Bampidis and Robinson 2006). The appropriateness of the byproduct to the ruminant animal is related to its capacity to ferment high-fiber meals in the rumen (Grasser et al. 1995). During in vitro fermentation, citrus pulp is believed to create more lactic acid than sugar beet pulp, sorghum grain, or corn grain (Cullen et al. 1986). Therefore, adding pectin-rich citrus byproduct feed that is degradable and neutral detergent fibre to rations that support ruminant growth and lactation will not have a negative impact on the rumen ecology (Chavan et al. 2018).

3.11.2 Pectin Production

Citrus peel is a wonderful source of pectin, a type of soluble fibre that is utilised in the culinary industry, most notably in jam and jelly, as well as in candies, pharmaceuticals, and as a stabiliser in fruit juices (Sulieman et al. 2013). Citrus peel has an exceptionally high level of flavonoids and the extremely bitter chemical naringin, hence the peel must be removed before pectin extraction (Puri et al. 2008). Vacuum impregnation technology is utilised to remove these bitterness-forming chemicals. Vacuum-infusing diluted naringinase, sweeteners, colouring, and flouring agents into Albedo results in an 81% reduction in bitterness and an improvement in acceptance (Baker and Wicker 1996). The traditional method of extracting pectin involved heating apple or citrus peel with hot water (60–100 °C) while simultaneously acidifying with a mineral or organic acid. Although the standard procedure is straightforward, it takes more time to extract the pectin and results in pectin structural breakdown (Mesbahi et al. 2005; Levigne et al. 2002). Pectin acid or HCl sulfuric acid is used in industrial pectin extraction processes, both of which produce significant amounts of toxic waste. There have recently been several attempts to create alternate extraction methods, including enzymatic extraction (Ptichkina et al. 2008), supercritical water extraction (Ueno et al. 2008), ultra-high pressure (Guo et al. 2012), microwave extraction (Fishman and Cooke 2009), high intensity (de Oliveira et al. 2015) and ultra-sound extraction (Zhang et al. 2013). Additionally, Saberian et al. (2017) made an effort to extract pectin using the ohmic heating approach. They found that this pectin has a very high emulsifying activity, which means that it can effectively create and stabilise oil-in-water emulsions. After pectin extraction, the remaining solid residue can also be used in compounded form.

3.11.3 Production of Bio-Fuel

Since 119.7 million tonnes of citrus waste are produced globally each year, there is a significant opportunity to produce bio-oil using thermochemical and biochemical processes. Alternative methods of utilising citrus trash include the biochemical synthesis of fermentable sugars from citrus peel hydrolysate and the bioconversion of methane. Citrus fruit waste can be converted into biofuels like ethanol and biogas using various polymers of soluble and insoluble carbohydrates (Mizuki et al. 1990; Wikandari et al. 2015). Citrus trash is regarded as an excellent source for the creation of biomass since it contains cellulose, hemicellulose, and lignin, which are the main energy sources in biomass (Haykiri-Acma et al. 2010; Rutkowski 2011). Biochemical processes use enzymes and microorganisms, while physico-thermal processes use crushing, heating, and pressure. With the use of a chemical catalyst and heat energy, biomass is broken down into high-energy products. While pyrolysis creates bio-oils, chemicals, and charcoal, gasification is utilised to create biofuels like gasoline and electricity (Taghizadeh-Alisaraei et al. 2017). Volpe et al. (2015) found that biochar and bio-oil produced by a slow pyrolysis process (200–650 °C) were more stable, homogeneous, and energetic. The method creates bio-char, which has a variety of uses, including decontaminating soil and water bodies that have been contaminated by metals.

3.11.4 Essential Oil Extraction

Today, many scientists are interested in the extraction of essential oils from citrus residue. Chau and Huang (2003) and Bicu and Mustata (2011) have researched the synthesis of dietary fibre, but Rivas et al. (2008) found that solid residue left over after the extraction of essential oil can also be used to produce cellulolytic fibre, which is a supporter for semi-solid fermentation. Enzyme synthesis can also be carried out using liquid residue. Essential oil's significant value in the broader medicinal sector, as a flavouring agent in the food and beverage industry, and in cleaning goods might be used to justify the economy of its extraction (Anagnostopoulou et al. 2006; Raeissi et al. 2008).

According to Bizzo et al. (2009), a maximum of 0.4 g of essential oil can be extracted from each 100 g can of pulp, or 4 kg of essential oil per tonne of pulp. Terpenes and their oxygenated derivatives, including alcohols, esters, and aldehydes (citral), are this oil's primary ingredients (Shaw 1979). Virot et al. (2008) had shown its use as a benign solvent as an alternative to dangerous petroleum solvent in the oleochemical, wax, resin, paint, and glue industries. Additionally, attempts were made to blend diesel oil and steam-distilled lemon essential oils in a 20:80 ratio (Ashok et al. 2017). The fuel properties of lemon essential oil are very similar to those of diesel, which have values of 822 kg/m^3, 3.6106, 42,700 kJ/kg At full load, the 20% blended essential oil's break thermal efficiency was comparable to that of its diesel counterpart (Ashok et al. 2017). Commercially, essential oil extraction is done using the classic method of steam distillation, but this has the drawbacks of being more expensive and taking longer to complete.

Therefore, there is a need for more efficient greener extraction methods to be developed. Recent developments based on ultrasonic extraction in this context (Vinatru et al. 1999), subcritical water extraction (Gamiz-Gracia and De Castro 2000), microwave-assistedextraction (Ferhat et al. 2007) and supercritical fluid extraction (Baysal and Starmans 1999), are currently being studied as quicker, more effective, and more affordable alternatives. According to Filly et al. (2014), the microwave technology can be used immediately, in the form of a continuous process, without any treatment (drying), and it enables continuous and quick heating. The microwave aided hydro-distillation (MAHD) approach was created by Bustamante et al. (2016) to extract essential oils from wet orange peel.

3.11.5 Preparation of Citrus Peel Packaging Films

The use of renewable resources in place of synthetic materials for food packaging and coating is in high demand on a global scale. However, although being biodegradable and having a high mechanical strength, paper and paperboard are not appropriate for packing foods with a high moisture content because of their porous nature, which renders them susceptible to moisture and some gases. Growing amounts of plastic trash have drawn attention to the need for biodegradable moisture barriers. Researchers found that zein or wax can significantly improve the kraft paper's water vapour barrier qualities (Parris et al. 1998). The mixture of wax and lipid-hydroxyl propyl methyl cellulose significantly reduces the composite film's water permeability (Sothornvit 2009). In 2017, Kasaai and Moosavi investigated the use of citrus peel and leaf hydrophobic extracts (terpene hydrocarbons and limonene) in paper and paperboard to enhance the water barrier properties without reducing mechanical strength. Kraft paper was given the treatment by being submerged in 10 mL of leaf extract (terpene hydrocarbons and limonene). The paper was then mechanically shaken dry at 25 °C after the solvent had entirely evaporated (60 rpm, 24 h). The cellulose fibres of the papers were mixed with the components of the extracts in the solution that had been partially introduced into the free volume. A thin layer of extracts from another batch formed on the paper's surface. Researchers found that kraft paper treated with peel or leaf extraction had considerably less water vapour permeability than untreated paper. Original paper's main ingredient, cellulose, is hydrophilic by nature and may absorb water. The absorption of water vapour was reduced with the use of hydrophobic substance. Citrus leaf has higher hydrophobicity than citrus peel, according to Baker and Procopiou's (2006) research. The production of biodegradable plastic has gained appeal among researchers as a result of growing worries about the nondegradable characteristics of petroleum-based polyester and their negative impact on the environment. In this regard, scientists from Cornell University in New York have created plastic produced from orange peel. Limonene oxide, a building block, was produced by oxidising orange peel oil. In order to create a novel polymer, limonene oxide from citrus peel was mixed with another building block, carbon dioxide, which is otherwise a greenhouse gas with negative environmental effects (Brand 2005). Arrieta et al. (2014) produced flexible food packaging film by Researchers assert that a newly created polymer has properties that are similar to those of polystyrene while also being biodegradable and capable of retaining carbon dioxide in the environment.

3.11.6 Encapsulating Agents

Light, heat, and oxidants can break down unsaturated bonds in the polyphenols'molecular structure. Additionally, applying phenolic extracts to food is exceedingly challenging due to their low water and liquid solubility (Spigno et al. 2013; Xueling et al. 2011). Therefore, the most common method for encapsulating phenolic compounds is spray drying, which improves their colour and makes them suitable for use as an additive (Kaderides et al. 2015). The wall material for the encapsulation must have characteristics like low viscosity even in highly concentrated solutions, efficient emulsification, efficient drying properties, film-forming capabilities, and low cost. Encapsulating agents can be different substances such gums, hydrolyzed starches, and emulsifying starches (maltodextrins, corn syrup solids). According to studies, insoluble fibre helps the digestive tract work properly whereas soluble fibre lowers blood cholesterol and is linked to the intestinal absorption of glucose (Fang and Bhandari 2010). Dietary fibre intake is advised around 30–45 g per day (Grigelmo-Miguel et al. 1999). Orange peel, pulp, and pomace, which are rich sources of fibre, make about 50% of the waste produced during the extraction of orange juice. It has a lot of promise for use in the creation of encapsulating agents. In order to create a fibre with a high moisture retention ability, citrus waste must first be washed, dried, ground, and then washed again at 90 °C for 20 min. To obtain citrus fibre with a particle size of 0.008 mm, the remaining fibre is once again dried at 60 °C (Kaderides et al. 2015). The fiber's low moisture content (8.52 g of water per 100 g of dry matter) prevents microbial growth, and the amounts of soluble and insoluble dietary fibre that were obtained were 48.9 g and 16.8 g, respectively. When encapsulating an active substance, a higher concentration of insoluble fibre might be used as the optimal material to enclose the substance inside its structure.

3.11.7 Development of Biodegradable Packages

After the high-value components from the citrus peel have been removed, the leftover residue as well as the fiber-rich pomace of the citrus can be moulded to create biodegradable packaging materials like egg trays and fruit trays, which makes it easier to recycle waste. The peel must be baked for 8–10 h at 70–80 °C. Dried materials are combined with urea-formaldehyde resin (7:3), crushed in a miller at 25,000 revolutions per minute, and then deposited on a heated extrusion mould to generate seedling plates for distributed transplanting. The manufactured distributed transplant seeding plate breaks down soon in the soil and doesn't harm the environment (Shan 2016).

3.11.8 Solid Residues as Adsorbent

Adsorption is a method that allows for the efficient removal of metals and compounds from industrial effluent. The sensitivity, capacity, and lifetime of an adsorbent have a significant impact on the technique's effectiveness (Pérez-Marín et al. 2008). Citrus pulp may have the ability to effectively eliminate these heavy metals, according to certain research. Citrus waste with a high pectin concentration (10%) contains a polysaccharide in which pectic acid is partially esterified by a methyl group. It is quickly saponified with alkalis like calcium hydroxide to produce pectic acid. Strong interactions between pectic acid and other heavy metals result in stable chelates with a five-membered ring (Dhakal et al. 2005). Citrus molasses was employed by Ajmal et al. (2000), who asserted that it removed 93% of the nickel from electroplating effluent. Pérez-Marín et al. (2008) also accomplished biosorption of lead, zinc, and cadmium with success, with a maximum sorption absorption of about 0.25 mmol/g. Biswas et al. (2008) conducted study on the removal of phosphorus from water by adsorption onto an orange waste gel loaded with zirconium and came to the conclusion that it is a promising approach of effluent treatment.

3.11.9 Development of Activated Carbon

The polymer electrolyte membrane fuel cell is regarded as the next generation of possible power sources due to its characteristics like high energy density and low emission (Avgouropoulos et al. 2016). It is observed that carbon is more advantageous for electrocatalyst support because it has a more specialised surface area than metal oxide, is more stable in acidic and basic medium, and makes it simple to recover the metal catalytic by burning off the carbon support. Numerous researchers have developed high surface area activated carbon from agricultural waste, such as groundnut shells, coconut shells, mango nuts, banana fibre, etc. In this regard, Dhelipan et al. (2017) attempted to produce activated carbon from orange peel using pyrolysis at 600 °C in conjunction with chemical activation using orthophosphoric acid (H_3PO_4) and assessed its stability as an electrolyte support. Due to its exceptional electrochemical stability and characteristics, activated carbon can be used as an electrode material in commercial supercapacitors. Bio-waste can be used as a precursor in the synthesis of activated carbon to address the rising demand as fossil fuels are nonrenewable.

3.12 Future Research Thrust

Due to the unpredictable nature of the climate, the depletion of land and water resources, the lack of high-quality planting materials, and the absence of post-harvest and marketing infrastructure, sweet lime production in India faces numerous difficulties. The difficulties of enhancing quality production and productivity are becoming increasingly apparent due to the expanding economy, growing nutritional knowledge, increased demand for fruits and beverages, and growing economic awareness of quality. These new difficulties demand a paradigm shift in manufacturing along with cutting-edge, cutting-edge technical intervention.

3.13 Conclusion

Ascorbic acid, water-soluble vitamins, folic acid, and phytochemicals are all abundant in Mosambi. Antioxidants found in abundance in it promote resistance to illnesses, especially the flu, and boost immunity. Its juice promotes digestion and detoxifies the body. Mosambi is a very hydrating fruit that also gives us important micronutrients like vitamins and minerals. It gives a lot of fullness and prevents constipation because of its high water and fibre content. Due to its high citric acid content, which minimizes hunger and cravings and speeds up metabolism, it is ideal for managing weight loss. It contains limonenes, a class of substances with potential health benefits. Mosambi is also used as an ingredient in many products to create functional foods because it is a rich source of polyphenols with high antioxidant effects. Worldwide, the excessive disposal of industrial waste from citrus processing into landfills poses a threat to both the environment and public health. Scientifically valuing these organic residues is an appealing idea that has gained traction among academics. Peel, pulp and seeds from citrus waste produced at the juice processing facility can be used as raw materials in other industries. While liquid can be used to produce enzymes, leftover solid residue can be used to extract essential oils. Hydrophobic substance from the peel can be extracted and used to make packaging materials, biodegradable polymers and food-grade kraft paper, which reduces the need for petroleum-based polyesters. Utilizing citrus residue can produce bio-oil, biogas, ethanol, and activated carbons in addition to reducing environmental pollution. Expanding the scope of product development boosts manufacturer profits overall, creates jobs, promotes sustainable socioeconomic growth, and preserves environmental stability. Although most research has not yet been widely marketed, its results are valuable for the next generation and will help the world achieve its top priority of sustainable development.

References

Ajmal M, Rao RAK, Ahmad R, Ahmad J (2000) Adsorption studies on Citrus reticulata (fruit peel of orange): removal and recovery of Ni (II) from electroplating wastewater. J Hazrd Mate 79(1–2):117–131

Ali RJ, Abass TA (2015) Study of some antioxidant compounds in some citrus. World J Pharm Res 5(1):1435–1442

Anagnostopoulou MA, Kefalas P, Papageorgiou VP, Assimopoulou AN, Boskou D (2006) Radical scavenging activity of various extracts and fractions of sweet orange peel (Citrus sinensis). Food Chem 94(1):19–25

Arias BA, Ramón-Laca L (2005) Pharmacological properties of citrus and their ancient and medieval uses in the Mediterranean region. J Ethnopharmacol 97(1):89–95

Arora RK, Sonkar RK, Kaul MK, Patel RK (2012) Rootstocks in Citrus: chronological development. In: Souvenir and abstracts national dialogue on citrus improvement, production and utilization. NRC Citrus, Nagpur, pp 130–148

Arrieta MP, López J, Hernández A, Rayón E (2014) Ternary PLA–PHB–limonene blends intended for biodegradable food packaging applications. Eur Polym J 50:255–270

Aruna T, Hemalatha G, Kumutha K, Kanchana S, Vellaikumar S (2022) Physicochemical, antioxidant and antimicrobial properties of citrus peel essential oils. J Appl Nat Sci 14(2):640–646

Ashok B, Nanthagopal K, Jathar S, Sathyanand T, Logesh S (2017) Lemon essential oil – a partial substitute for petroleum diesel fuel in compression ignition engine. Int J Renew Energy Res 7(2):467–475

Assini JM, Mulvihill EE, Sutherland BG, Telford DE, Sawyez CG, Felder SL, Huff MW (2013) Naringenin prevents cholesterol-induced systemic inflammation, metabolic dysregulation, and atherosclerosis in Ldlr mice [S]. J Lipid Res 54(3):11–724

Avgouropoulos G, Schlicker S, Schelhaas KP, Papavasiliou J, Papadimitriou KD, Theodorakopoulou E, Neophytides S (2016) Performance evaluation of a proof-of-concept 70 W internal reforming methanol fuel cell system. J Power Sources 307:875–882

Baker RA (1994) Potential dietary benefits of citrus pectin and fiber. Food Tech (Chicago) 48(11):133–139

Baker RA, Wicker L (1996) Current and potential applications of enzyme infusion in the food industry. Trend Food Sci Tech 7(9):279–284

Bampidis VA, Robinson PH (2006) Citrus by-products as ruminant feeds: a review. Anim Feed Sci Technol 128(3–4):175–217

Baysal T, Starmans DAJ (1999) Supercritical carbon dioxide extraction of carvone and limonene from caraway seed. J Supercrit Fluids 14(3):225–234

Bermejo A, Cano A (2012) Analysis of nutritional constituents in twenty citrus cultivars from the Mediterranean area at different stages of ripening. Food Nutr Sci 3(5):639–650

Bhaumik A, Nousheen H, Vennela M (2018) A potential review: phytochemical and pharmacological profile of sweet lime (MOSAMBI FRUIT). Panacea J Pharm Pharma Sci 7:01–13

Bicu I, Mustata F (2011) Cellulose extraction from orange peel using sulfite digestion reagents. Bioresour Technol 102(21):10013–10019

Biswas BK, Inoue K, Ghimire KN, Harada H, Ohto K, Kawakita H (2008) Removal and recovery of phosphorus from water by means of adsorption onto orange waste gel loaded with zirconium. Bioresour Technol 99(18):8685–8690

Bizzo HR, Hovell AMC, Rezende CM (2009) Óleos essenciais no Brasil: aspectos gerais, desenvolvimento e perspectivas. Quím nova 32:588–594

Braddock RJ (1995) By-products of citrus fruit. Food Tech (Chicago) 49(9):74–77

Braddock RJ, Cadwallader KR (1992) Citrus by-products manufacture for food use. Food Tech (Chicago) 46(2):105–110

Brand D (2005) Cornell news: making plastic from oranges. Retrieved January 30, 2005 from http://news.cornell.edu/stories/2005/01/making-plastics-oranges

Burt S (2004) Essential oils: their antibacterial properties and potential applications in foods – a review. Int J Food Microbiol 94(3):223–253

Bustamante J, van Stempvoort S, García-Gallarreta M, Houghton JA, Briers HK, Budarin VL, Clark JH (2016) Microwave assisted hydro-distillation of essential oils from wet citrus peel waste. J Clean Prod 137:598–605

Chau CF, Huang YL (2003) Comparison of the chemical composition and physicochemical properties of different fibers prepared from the peel of *Citrus sinensis L. Cv Liucheng*. J Agric Food Chem 51(9):2615–2618

Chavan P, Singh AK, Kaur G (2018) Recent progress in the utilization of industrial waste and byproducts of citrus fruits: a review. J Food Process Eng 41(8):e12895

Chen XM, Tait AR, Kitts DD (2017) Flavonoid composition of orange peel and its association with antioxidant and anti-inflammatory activities. Food Chem 218:15–21

Chinese Pharmacopoeia Commission (2015) Chinese pharmacopoeia, vol 1. China Medical Science Press, Beijing, pp 191–193

Cocke EL, Muncie FW (1976) U.S. patent no. 3,966,984. Washington, DC, U.S. Patent and Trademark Office

Codoñer-Franch P, Valls-Bellés V (2010) Citrus as functional foods. Curr Top Nutraceut Res 8(4):173–183

Crowell PL (1999) Prevention and therapy of cancer by dietary monoterpenes. J Nutr 129(3):775S–778S

Cullen AJ, Harmon DL, Nagaraja TG (1986) In vitro fermentation of sugars, grains, and by-product feeds in relation to initiation of ruminal lactate production. J Dairy Sci 69(10):2616–2621

Davis WB (1932) Deposits of oil in the juice sacs of citrus fruits. Am J Bot 19(2):101–105

de Oliveira CF, Giordani D, Gurak PD, Cladera-Oliver F, Marczak LDF (2015) Extraction of pectin from passion fruit peel using moderate electric field and conventional heating extraction methods. Innov Food Sci Emerg Technol 29:201–208

Dhakal RP, Ghimire KN, Inoue K, Yano M, Makino K (2005) Acidic polysaccharide gels for selective adsorption of lead (II) ion. Sep Purif Technol 42(3):219–225

Dhelipan M, Arunchander A, Sahu AK, Kalpana D (2017) Activated carbon from orange peels as supercapacitor electrode and catalyst support for oxygen reduction reaction in proton exchange membrane fuel cell. J Saudi Chem Soc 21(4):487–494

Dietrich DR, Swenberg JA (1991) The presence of α2u-globulin is necessary for d-limonene promotion of male rat kidney tumors. Cancer Res 51(13):3512–3521

El-Adawy TA, El-Bedawy AA, Rahma EH, Gafar AM (1999) Properties of some citrus seeds. Part 3. Evaluation as a new source of protein and oil. Food Nahrung 43(6):385–391

Elegbede JA, Elson CE, Qureshi A, Tanner MA, Gould MN (1984) Inhibition of DMBA-induced mammary cancer by the monoterpene d-limonene. Carcinogenesis 5(5):661–664

El-Sheikh SE (2005) Propagation of lime (*Citrus aurantifolia* L.) by stem cuttings technique. M.Sc. thesis, Faculty of Agriculture, University of Khartoum, Sudan

Elson CE, Maltzman TH, Boston JL, Tanner MA, Gould MN (1988) Anti-carcinogenic activity of d-limonene during the initiation and promotion/progression stages of DMBA-induced rat mammary carcinogenesis. Carcinogenesis 9(2):331–333

Fang Z, Bhandari B (2010) Encapsulation of polyphenols – a review. Trends Food Sci Tech 21(10):510–523

Ferhat MA, Meklati BY, Chemat F (2007) Comparison of different isolation methods of essential oil from citrus fruits: cold pressing, hydrodistillation and microwave 'dry' distillation. Flav Frag J 22(6):494–504

Filly A, Fernandez X, Minuti M, Visinoni F, Cravotto G, Chemat F (2014) Solvent-free microwave extraction of essential oil from aromatic herbs: from laboratory to pilot and industrial scale. Food Chem 150:193–198

Fishman ML, Cooke PH (2009) The structure of high-methoxyl sugar acid gels of citrus pectin as determined by AFM. Carbohydr Res 344(14):1792–1797

Food and Agricultural Organization Statistical (FAOSTAT) (2018). Available at: http://www.fao.org/faostat/en/#home

Gamiz-Gracia L, De Castro ML (2000) Continuous subcritical water extraction of medicinal plant essential oil: comparison with conventional techniques. Talanta 51(6):1179–1185

Gangwar RP, Singh SN (1965) Effect of type of wood, season of planting and plant growth regulators on the propagation of sweet lime (*Citrus limettioides Tanaka*) through stem cutting. Trop Agric 121:55–62

Genovese S, Fiorito S, Locatelli M, Carlucci G, Epifano F (2014) Analysis of biologically active oxyprenylated ferulic acid derivatives in citrus fruits. Plant Foods Human Nutr 69(3):255–260

Gorinstein S, Cvikrová M, Machackova I, Haruenkit R, Park YS, Jung ST, Trakhtenberg S (2004) Characterization of antioxidant compounds in Jaffa sweeties and white grapefruits. Food Chem 84(4):503–510

Grasser LA, Fadel JG, Garnett I, DePeters EJ (1995) Quantity and economic importance of nine selected by-products used in California dairy rations. J Dairy Sci 78(4):962–971

Grigelmo-Miguel N, Gorinstein S, Martín-Belloso A (1999) Characterisation of peach dietary fibre concentrate as a food ingredient. Food Chem 65(2):175–181

Guimarães R, Barros L, Barreira JC, Sousa MJ, Carvalho AM, Ferreira IC (2010) Targeting excessive free radicals with peels and juices of citrus fruits: grapefruit, lemon, lime and orange. Food Chem Toxicol 48(1):99–106

Gundgaard J, Nielsen JN, Olsen J, Sørensen J (2003) Increased intake of fruit and vegetables: estimation of impact in terms of life expectancy and healthcare costs. Public Health Nutr 6(1):25–30

Guo X, Han D, Xi H, Rao L, Liao X, Hu X, Wu J (2012) Extraction of pectin from navel orange peel assisted by ultra-high pressure, microwave or traditional heating: a comparison. Carbohydr Polym 88(2):441–448

Gupta S, Rahman MA, Sundaram S (2021) Citrus fruit as a potential source of phytochemical, antioxidant and pharmacological ingredients. J Sci Healthcare Exp:8473

Haykiri-Acma H, Yaman S, Kucukbayrak S (2010) Comparison of the thermal reactivities of isolated lignin and holocellulose during pyrolysis. Fuel Proces Tech 91(7):759–764

Ho SC, Kuo CT (2014) Hesperidin, nobiletin, and tangeretin are collectively responsible for the anti-neuroinflammatory capacity of tangerine peel (*Citri reticulatae pericarpium*). Food Chem Toxi 71:176–182

Hodgson R (1967) Horticultural varieties of citrus. In: Reuther W, Webber HJ (eds) The citrus industry, volume I: history, world distribution, botany, and varieties, rev edn. University of California, Berkeley/Los Angeles, pp 431–591

Igimi H, Hisatsugu T, Nishimura M (1976) The use of d-limonene preparation as a dissolving agent of gallstones. Am J Dig Dis 21(11):926–939

Imran M, Butt MS, Iqbal MJ, Gilani SA, Basharat S, Saeed F, Suleria HA (2016) Antioxidant potential, physico-chemical, and sensory attributes of cookies supplemented with Mosambi peel extract. Int J Fruit Sci 16(3):341–349

Inglese P, Sortino G (2019) Citrus history, taxonomy, breeding, and fruit quality citrus history, taxonomy, breeding, and fruit quality summary and keywords. Oxford University Press

Jauhari OS, Rahman SF (1959) Further investigations on rooting in cuttings of sweet lime (*Citrus limettioides*, Tanaka). Sci Cult 24:432–434

Jung UJ, Lee MK, Jeong KS, Choi MS (2004) The hypoglycemic effects of hesperidin and naringin are partly mediated by hepatic glucose-regulating enzymes in C57BL/KsJ-db/db mice. J Nutr 134(10):2499–2503

Kaderides K, Goula AM, Adamopoulos KG (2015) A process for turning pomegranate peels into a valuable food ingredient using ultrasound-assisted extraction and encapsulation. Innov Food Sci Emerg Tech 31:204–215

Kasaai MR, Moosavi A (2017) Treatment of Kraft paper with citrus wastes for food packaging applications: water and oxygen barrier properties improvement. Food Pac Shelf life 12:59–65

Kaur A, Basu T, Mittal N (2021) Evaluation of indian kangra green tea (camellia sinensis (l) o kuntze) infused with peels of local fruits (pomegranate, lemon, mosambi, orange and banana) as a natural source of antioxidants. J Adv Sci Res 12(03 Suppl 1):174–179

Ke Z, Xu X, Nie C, Zhou Z (2015) Citrus flavonoids and human cancers. J Food Nutr Res 3(5):341–351

Khan IA (2007) Citrus genetics, breeding and biotechnol. CABI

Khan MM, Iqbal M, Hanif MA, Mahmood MS, Naqvi SA, Shahid M, Jaskani MJ (2012) Antioxidant and antipathogenic activities of citrus peel oils. J Essent Oil Bear Plants 15(6):972–979

Kondo M, Goto M, Kodama A, Hirose T (2000) Fractional extraction by supercritical carbon dioxide for the deterpenation of bergamot oil. Ind Eng Chem Res 39(12):4745–4748

Kumar R, Vijay S, Khan N (2013) Comparative nutritional analysis and antioxidant activity of fruit juices of some Citrus spp. Octa J Biosci 1(1):44–53

Levigne S, Ralet MC, Thibault JF (2002) Characterisation of pectins extracted from fresh sugar beet under different conditions using an experimental design. Carbohydr Polym 49(2):145–153

Licandro G, Odio CE (2002) Citrus by-products. In: Citrus. CRC Press, pp 173–192

Malacrida CR, Kimura M, Jorge N (2012) Phytochemicals and antioxidant activity of citrus seed oils. Food Sci Tech Res 18(3):399–404

Maltzman TH, Hurt LM, Elson CE, Tanner MA, Gould MN (1989) The prevention of nitrosomethylurea-induced mammary tumors by d-limonene and orange oil. Carcinogenesis 10(4):781–783

Manthey JA, Grohmann K (2001) Phenols in citrus peel byproducts. Concentrations of hydroxycinnamates and polymethoxylated flavones in citrus peel molasses. J Agric Food Chem 49(7):3268–3273

Matlack MB (1931) The juice sac of the orange with some observations on the plastids of citrus. J Wash Acad Sci 21(17):437–440

Matsuda H, Yano M, Kubo M, Iinuma M, Oyama M, Mizuno M (1991) Pharmacological study on citrus fruits. II. Anti-allergic effect of fruit of *Citrus unshiu*. On flavonoid components. J Pharma Soc Jpn 111(3):193–198

Mesbahi G, Jamalian J, Farahnaky A (2005) A comparative study on functional properties of beet and citrus pectins in food systems. Food Hydrocoll 19(4):731–738

Middleton E Jr, Kandaswami C (1994) Potential health-promoting properties of citrus flavonoids. Food Tech (Chicago) 48(11):115–119

Miller EV, Winston JR, Schomer HA (1940) Physiological studies of plastid pigments in rinds of maturing oranges. J Agric Res 60:259–267

Mizuki E, Akao T, Saruwatari T (1990) Inhibitory effect of Citrus unshu peel on anaerobic digestion. Bio Waste 33(3):161–168

Moreira PC, REIS R, Lana AMQ (2004) Produção e composição do leite de vacas alimentadas com polpa cítrica em substituição ao milho grão. Reunião Anual Da Sociedade Brasileira De Zootecnia 41

Motial VS (1964) Fruit-set studies in sweet lime. In: Proceedings of the Indian Academy of Sciences – Section B 60, vol 6. Springer, pp 371–379

Mulvihill E, Huff M (2012) Citrus flavonoids and the prevention of atherosclerosis. Cardiovasc Haematol Disord Drug Target 12(2):84–91

Nayak SL, Sethi S, Sharma RR, Dubey AK, Bhowmik A (2020) Variation in fruit quality traits and health promoting compounds of citrus fruits grown in semi-arid region. Ind J Hortic 77(4):627–632

Nicolosi E (2007) Origin and taxonomy. Chapter III. In Khan, I.A. (ed.) Citrus genetics, breeding and biotechnology. CABI International, Wallingford, UK.

Ono E, Inoue J, Hashidume T, Shimizu M, Sato R (2011) Anti-obesity and anti-hyperglycemic effects of the dietary citrus limonoid nomilin in mice fed a high-fat diet. Biochem Biophys Res Commun 410(3):677–681

Ortuño A, Báidez A, Gómez P, Arcas MC, Porras I, García-Lidón A, Del Río JA (2006) Citrus paradisi and Citrus sinensis flavonoids: their influence in the defence mechanism against Penicillium digitatum. Food Chem 98(2):351–358

Park EJ, Pezzuto J (2012) Flavonoids in cancer prevention. Anti Cancer Agents Med Chem 12(8):836–851

Parris N, Vergano PJ, Dickey LC, Cooke PH, Craig JC (1998) Enzymatic hydrolysis of Zein– wax-coated paper. J Agric Food Chem 46(10):4056–4059

Paul A, Bhattacharyya D (2019) Analysis and study on the antioxidant of citrus fruits. Pharma Innov 8(5):397–398

Perez YY, Jimenez-Ferrer E, Alonso D, Botello-Amaro CA, Zamilpa A (2010) Citrus limetta leaves extract antagonizes the hypertensive effect of angiotensin II. J Ethnopharmacol 128(3):611–614

Pérez-Marín AB, Ballester A, González F, Blázquez ML, Muñoz JA, Sáez J, Zapata VM (2008) Study of cadmium, zinc and lead biosorption by orange wastes using the subsequent addition method. Biores Tech 99(17):8101–8106

Ptichkina NM, Markina OA, Rumyantseva GN (2008) Pectin extraction from pumpkin with the aid of microbial enzymes. Food Hydrocol 22(1):192–195

Puri M, Kaur A, Singh R, Kanwar J (2008) Immobilized enzyme technology for debittering citrus fruit juices. In: Food enzymes: application of new technologies, pp 91–103

Raeissi S, Diaz S, Espinosa S, Peters CJ, Brignole EA (2008) Ethane as an alternative solvent for supercritical extraction of orange peel oils. J Supercrit Fluid 45(3):306–313

Riccardi G, Rivellese AA, Giacco R (2008) Role of glycemic index and glycemic load in the healthy state, in prediabetes, and in diabetes. Am J Clin Nutr 87(1):269S–274S

Rivas B, Torrado A, Torre P, Converti A, Domínguez JM (2008) Submerged citric acid fermentation on orange peel autohydrolysate. J Agric Food Chem 56(7):2380–2387

Romagnolo DF, Selmin OI (2012) Flavonoids and cancer prevention: a review of the evidence. J Nutr Gerontol Gerontol 31(3):206–238

Rutkowski P (2011) Pyrolysis of cellulose, xylan and lignin with the K_2CO_3 and $ZnCl_2$ addition for bio-oil production. Fuel Proc Tech 92(3):517–522

Saberian H, Hamidi-Esfahani Z, Gavlighi HA, Barzegar M (2017) Optimization of pectin extraction from orange juice waste assisted by ohmic heating. Chem Eng Process Proc Intensif 117:154–161

Shahidi F (ed) (1997) Natural antioxidants: chemistry, health effects, and applications. AOCS Press, Champaign

Shan Y (2016) Production of biodegradable packages using citrus peel. In: Comprehensive utilization of citrus by-products, pp 91–92

Sharma G, Gupta AK, Ganjewala D, Gupta C, Prakash D (2017) Phytochemical composition, antioxidant and antibacterial potential of underutilized parts of some fruits. Int Food Res J 24(3):1167

Shaw PE (1979) Review of quantitative analyses of citrus essential oils. J Agric Food Chem 27(2):246–257

Shivankar VJ (2010) Citrus industry of India– issues and strategies. In: National seminar on citrus biodiversity for livelihood and nutritional security. NRC Citrus, Nagpur, pp 1–16

Singh JP, Dhuria HS (1960) Studies on floral biology of sweet lime (Citrus limettoides Tanaka). Indian J Hortic 17:9–20

Singh AR, Pandey NC, Pandey AK (1986) Studies on the regeneration of sweet lime (*Citrus limettioides Tanaka.*) by stem cuttings with the aid of IBA and NAA. J Hortic Sci 15(1/2):25–28

Singh E, Kumari J, Jain P, Sharma S (2014) Nutritional evaluation of dried peel of citrous species and its application in cake development. J Nutr Ecol Food Res 2(1):10–14

Sothornvit R (2009) Effect of hydroxypropyl methylcellulose and lipid on mechanical properties and water vapor permeability of coated paper. Food Res Int 42(2):307–311

Spigno G, Donsì F, Amendola D, Sessa M, Ferrari G, De Faveri DM (2013) Nanoencapsulation systems to improve solubility and antioxidant efficiency of a grape marc extract into hazelnut paste. J Food Eng 114(2):207–214

Sulieman AME, Khodari KM, Salih ZA (2013) Extraction of pectin from lemon and orange fruits peels and its utilization in jam making. Int J Food Sci Nutr Eng 3(5):81–84

Taghizadeh-Alisaraei A, Hosseini SH, Ghobadian B, Motevali A (2017) Biofuel production from citrus wastes: a feasibility study in Iran. Renew Sust Energ Rev 69:1100–1112
Taguri T, Tanaka T, Kouno I (2004) Antimicrobial activity of 10 different plant polyphenols against bacteria causing food-borne disease. Biol Pharma Bull 27(12):1965–1969
Tanaka T (1977) Fundamental discussion of Citrus classification. Stud Citrol 14:1–6
Tangirala EAS, Ramanaboina A, Rajesh P, Sujitha M, Mahaboob Basha SK (2020) Nutritive analysis of enzyme treated mosambi juice. IJCS 8(6):136–142
Teixeira JC (2001) Utilização da polpa cítrica na alimentação de bovinos leiteiros. Parte I Milkbizz Tecnol 1(3):25–28
Thakur R, Biswas PK, Singh M (2017) Comparative analysis of aqueous and ethanol extracts of citrus fruit residues. New Horiz Biotechnol:112–115
Uedo N, Tatsuta M, Iishi H, Baba M, Sakai N, Yano H, Otani T (1999) Inhibition by d-limonene of gastric carcinogenesis induced by N-methyl-N′-nitro-N-nitrosoguanidine in Wistar rats. Cancer Lett 137(2):131–136
Ueno H, Tanaka M, Hosino M, Sasaki M, Goto M (2008) Extraction of valuable compounds from the flavedo of Citrus junos using subcritical water. Sep Purif Tech 62(3):513–516
Vinatru M, Toma M, Mason TJ (1999) Ultrasonically assisted extraction of bioactive principles from plants and their constituents. Adv Sonochem 5:209–248
Virot M, Tomao V, Ginies C, Visinoni F, Chemat F (2008) Green procedure with a green solvent for fats and oils' determination: microwave-integrated Soxhlet using limonene followed by microwave Clevenger distillation. J Chromat A 1196:147–152
Volpe M, Panno D, Volpe R, Messineo A (2015) Upgrade of citrus waste as a biofuel via slow pyrolysis. J Anal Appl Pyrolysis 115:66–76
Wang L, Wang J, Fang L, Zheng Z, Zhi D, Wang S, Zhao H (2014) Anticancer activities of citrus peel polymethoxyflavones related to angiogenesis and others. BioMed Res Int 2014
Wattenberg LW (1983) Inhibition of neoplasia by minor dietary constituents. Cancer Res 43(5 Suppl):2448s–2453s
Wilkins J Jr (2002) Method for treating gastrointestinal disorder. U.S. Patent (642045)
Wing JM (2003) Citrus feedstuffs for dairy cattle. University of Florida. IFAS, p 829
Wikandari R, Nguyen H, Millati R, Niklasson C, Taherzadeh MJ (2015) Improvement of biogas production from orange peel waste by leaching of limonene. BioMed Research International 2015:494182. https://doi.org/10.1155/2015/494182
Xueling Z, Benguo L, Limin L, Xiaoai Z (2011) Microwave-assisted extraction and antioxidant activity of total phenolic compounds from pomegranate peel. J Med Plants Res 5(6):1004–1011
Yano H, Tatsuta M, Iishi H, Baba M, Sakai N, Uedo N (1999) Attenuation by dlimonene of sodium chloride-enhanced gastric carcinogenesis induced by N-methyl-N′-nitro-N-nitrosoguanidine in Wistar rats. Int J Cancer 82(5):665–668
Younis K, Islam RU, Jahan K, Yousuf B, Ray A, Yildiz F (2015) Effect of addition of mosambi (*Citrus limetta*) peel powder on textural and sensory properties of papaya jam. Cogent Food Agric 1:1023675. https://doi.org/10.1080/23311932.2015.1023675
Zhang J (2007) Flavonoids in grapefruit and commercial grapefruit juices: concentration, distribution, and potential health benefits. Proc Fla State Hortic Soc 120:288–294
Zhang L, Ye X, Ding T, Sun X, Xu Y, Liu D (2013) Ultrasound effects on the degradation kinetics, structure and rheological properties of apple pectin. Ultrason Sonochem 20(1):222–231

Chapter 4
Orange

Palvi Rawal, Isha Dudeja, Pooja, Arashdeep Singh, Sukhvinder Singh Purewal, Avneet Kaur, and Chidanandamurthy Thippeswamy Swamy

4.1 Introduction

Fruits are delicious and nutrient-dense due to their appealing appearance, unique flavour, high mineral and vitamin content, and other beneficial components. Orange, a characteristic fruit of the *Citrus* family, is grown in large quantities in Brazil, the US, China and Mexico. There are also other countries that produce a sizable number of oranges, including Spain, India, Turkey, Greece and Egypt with annual production of more than 1 million tonnes of oranges. Some countries are becoming producers as well as processing capitals (Grigelmo-Miguel and Martín-Belloso 1999). Although each cultivar of citrus fruits differs substantially in provisions of morphological and biochemical characteristics, they all produce fruit swiftly and

P. Rawal · I. Dudeja
Department of Chemistry, Punjab Agricultural University, Ludhiana, Punjab, India

Pooja
Department of Microbiology, Punjab Agricultural University, Ludhiana, Punjab, India

A. Singh (✉)
Department of Food Science and Technology, Punjab Agricultural University, Ludhiana, Punjab, India

S. Singh Purewal (✉)
University Centre for Research & Development (UCRD), Chandigarh University, Mohali, Punjab, India
e-mail: sukhvinder.e14148@cumail.in

A. Kaur
Department of Chemistry, University Institute of Sciences, Chandigarh University, Mohali, Punjab, India

C. T. Swamy
Department of Microbiology, Captain Srinivasa Murthy Central Ayurveda Research Institute, Central Council for Research in Ayurvedic Sciences, Chennai, Tamil Nadu, India

S. Singh Purewal et al. (eds.), *Recent Advances in Citrus Fruits*,
https://doi.org/10.1007/978-3-031-37534-7_4

consistently throughout a lengthy growing season (Sinclair and Jolliffe1961). In 2014, 89 million tonnes of citrus were expected to be produced, making it the world's most important fruit crop. Around 26% of the production of citrus fruits is used for industrial juicing. The most significant citrus species farmed worldwide is orange (*Citrus sinensis* (L.) Osbeck). Fruits have a sweet and pleasant taste, as well as a pleasant aroma that customers enjoy. Sucrose is the major carbohydrate in the juice, and citric acid is the dominant organic acid. It is high in antioxidants such as ascorbic acid and phenolic compounds (mostly hesperidin and narirutin). It is high in carotenoids (particularly violaxanthin, β-cryptoxanthin, and lutein). Except in a few cultivars where lycopene is the predominant pigment, the color of pigmented oranges is attributable to the presence of anthocyanins. The main amino acids discovered in the juice are proline, serine, and arginine, and the main mineral found in the juice is potassium.

According to recent literature, there are many potential uses for orange peel valorisation. Orange peel can even be used in functional meals as a dietary supplement for human or animal feed. It is also used in biochemical engineering, including the manufacturing of biofuels and biodegradable plastics by copolymerizing carbon dioxide and limonene. Orange peels have been reported to be used for extraction of pectin (Mhiri 2016). It is one of the most important fruit crops in the world and is primarily consumed fresh due to its high nutritional value and distinctive flavor. Consuming citrus juice has been found to help lower asthma and chronic heart diseases (Dugo and Di Giacomo 2002). Citrus fruit extracts also have anti-tumor, antifungal, and anti-clotting effects (Kaur and Kapoor 2001; Garg et al. 2001). Citrus fruits are good for health since, they include bioactive compounds like ferulic acid, hydrocinnamic acid, cyanidinglucoside, carotenoid, hisperidine, Vitamins C and naringin (Kelebek et al. 2007; Xu et al. 2008). Genus Citrus includes all kinds of oranges, including sweet and sour both oranges, mandarins, tangelos and tangors. Oranges are one of the most significant horticultural crops, producing more than 100 million metric tonnes of agricultural products annually worldwide. These are rich in Vitamins C, potassium, pectin, folic acid, and antioxidant phytophenolics that have well-being benefits. In reality, comprehensive research on orange and edible parts revealed that juice and extracts have significant antioxidant potential and serve as a substantial source of phenolic chemicals, particularly phenolic acids and flavonoids (Moulehi et al. 2012).

Citrus fruits of *rutaceae* family are substantial source of phenolic compounds. Phenolic compounds are widely found in plants as secondary metabolites. Broadly speaking, plant phenolics affect plant colour as well as defense against UV radiation or hostility from pathogens and parasites. They are frequently found in plant foods like fruits, vegetables, cereals, olives, legumes, cocoa, etc. as well as drinks like tea, coffee, beer, and wine, and they contribute to the whole organoleptic qualities of plant based foods (Shahidi Bonjar et al. 2004). In spite of being widely available, nutritionalists have only recently become aware of the health advantages of dietary polyphenols. Researchers and food producers have been interested in polyphenol compounds because of their antioxidant properties with abundance in the diet and possible advantages for the treatment of different oxidative stress related illnesses

(Manach et al. 2004). Plants such as large citrus bushes or small trees with evergreen leaves and prickly branches that can reach heights of 5–15 m (Ladaniya 2008) are a significant source of numerous bioactive substances, including flavanone glycosides and phenolic acids.

Considering their positive influence on boosting health, antioxidants are increasingly receiving more attention. In this regard, numerous epidemiological research has connected the ingestion of phenolics to reduced risk of a number of diseases, including cancer and cardiovascular diseases (Yang et al. 2011; Hooper et al. 2008). Moreover, antioxidants increase the shelf life of food products, preserve nutritional quality, and decrease the production of potentially dangerous oxidation products. Hence, orange being full of phenolic acids and antioxidant compounds can be incorporated in food product for achievieng beneficial health results.

4.2 Botany

Sweet orange (*Citrus sinensis* L. Osbeck) is a tiny, perennial tree that can grow as tall as 15 m in some situations. It is now commercially grown around the world in tropical and semi-tropical, making it the most normally planted fruit tree in the world. It is believed to have its origins in Southern China, (Nicolosi et al. 2000; Ehler 2011). Orange produces evergreen leaves of leathery texture that range in size from 2.5 to 9.5 cm wide and 6.5 to 15 cm long, with elliptical, oblong, and other shapes. These leaves frequently have delicate wings on their petioles. It bears fragrant, five-petalled white blooms with 20–25 yellow stamens, either arranged singly or in whorls of six, each whorl being approximately 5 cm across. The tiny, fragrant, purple or white hermaphrodite flowers provide sap for insect pollination. The globose to oval-shaped 6.5–9.5 cm in diameter fruit matures to orange or yellow.The pericarp (peel), skin, or rind, and the endocarp, sometimes known as the pulp and juice sacs, are the two separate anatomical parts of the citrus fruit (Fig. 4.1).

The skin's epidermis is comprised of epicuticular wax, and it possesses a vast number of small aromatic oil glands that produce the odour. The diversity, climate, and growth pace all influence the amount of wax. The microflora, which is mainly composed of fungus and bacteria and is more common in humid conditions, is abundant on the skin. This supports the recommendation that the fruit be thoroughly washed before eating or starting the process of extracting juice and essential oils. The peel (pericarp) is made up of the epicarp, or outer layerflavedo. Cuticle and parenchymatous cells make up the majority of its structure. Valencene, limonene, and beta/alpha sinsenal are among the terpenoid aromatic compounds produced by embedded oil glands (Goudeau et al. 2008; Sharon-Asa et al. 2003).

Flavedo has a unique yellow, green, or orange colour and present under the epidermis layer.The oliferous vesicles that are found inside the fragile, ultrafine flavedo can be collected by grating the flavedo layer. Flavedo is usually colourless, a spongy inner layer of mesophyll and thickness variations during fruit development, which

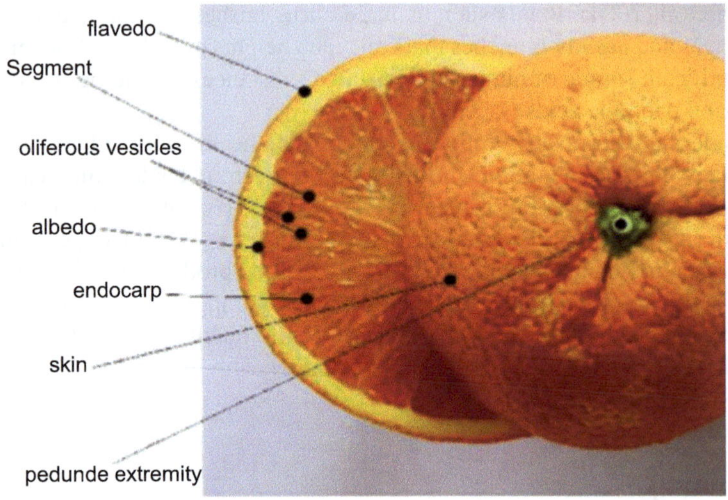

Fig. 4.1 Structure of Citrus fruit. (Goudeau et al. 2008)

have an impact on how easily the fruit peels (Webber 1989). The tissue mass known as the albedo, or mesocarp, which is located underneath the flavedo, is made up of tubular-shaped cells that have been joined together and compressed into the intercellular space. Albedo contains a number of flavonoids, which give juice its bitter flavour when added. The albedo-like tissue that makes up the majority of the fruit is spongy in texture.The Malay Archipelago, China, Southeast Asia, the Malay Archipelago, New Caledonia, and Australia all have ancient cousins of the citrus plant, yet it's probable that most citrus cultivars have an intriguing origin (Atta et al. 2012). It is supposed to have come through the interspecific hybridization of a few early citrus species, while its genetic background is uncertain (Xu et al. 2013). As a result, based on their region of origin (Mediterranean and Spanish oranges) and flavour, its numerous cultivars are grouped into four primary categories or groupings.

Although there are seedless variations, the fruit's flesh or pulp is normally juicy and delicious, separated into 10–14 segments, and comes in a variety of hues from yellow –range to red color. The thick, multiple-seeded fruit is known as a hesperidium, which is a berry when it is ripe. The accumulation of organic acids and sugars and water in fleshy juice sacs makes it difficult to extract proteins and nucleic acids, the vesicles that have juice; are located in the endocarp and carpels. From the perspective of synthetic biology, should be seen as the liquid discharged by the vacuoles and by the cytoplasm in the vesicles' interior cells.

The navel orange first arose around 1820 as a single mutation in a Brazilian convent, earning its name from the way it looks (the segmented skin mimics a human navel) (Ehsani 2007). Because they don't generate seeds and grow from graft cuttings, it is the source of all navel oranges. The popularity of these oranges increased with time, which aided in their global success. The Valencia orange was given its name by agronomist William Wolfkill in honour of Valencia, a Spanish city widely known for its orange production (Ehsani 2007). Depending on the fruit like blood oranges; named for their scarlet fleshy tissue—can be either sugary or sour. Blood oranges come in three varieties: taroco, moro, and sanguinello, each with a distinct flavour, size, and place of origin (Ehsani 2007).

The most significant percentage of the world's total citrus production, accounting for more than two thirds of the total area coverage, is sweet orange (*C. sinensis*) (FAO 2006). There are 10 species of edible citrus now known, eight of which are commercially cultivated, and five of which are quite important economically (Salunkhe and Desai 1984). Citrus is produced in excess of 104 million tonnes annually, and roughly 15 million tonnes are traded (FAO 2006). There are 1.3 million hectares of surface area in Africa used for citrus production, 44,000 of which are in South Africa and 4500 in Ethiopia. Citrus growing is dispersed across Ethiopia despite its recent introduction (Seifu 2003; Lipsky 1962; FAO 1965). The major cultivar viz. sweet orange (*Citrus sinensis*) cultivated in India are Andhra Pradesh, Maharashtra, Telangana, Mizoram,Madhya Pradesh, Punjab, Karnataka, Bihar, Assam, and Jammu & Kashmir.

4.3 Varieties

Sweet orange cultivars traditionally farmed include Mosambi (Maharashtra), Batavian (Bathayi) (Andhra Pradesh), Jaffa (Punjab)and Malta.

4.4 Cultivation

Citrus fruits are grown in India under a variety of agro-ecological environments, from the humid tropical northeast to the arid and semiarid southwest. Due to their evergreen nature, citrus trees can only be planted in real subtropical climates. Citrus trees in tropical climes, depending on the soil type and growing season, need careful regulation of water deficit stress due to their propensity for cyclical growth flushes. Between 13 and 37 °C is the best temperature range for citrus fruit growth. Temperatures below −4 °C are harmful for juvenile plants. About 25 °C can be the best soil temperature for root development.More humidity encourages the increasethe chances of many diseases. Frost is quite harmful. Summertime hot winds cause desiccation and a loss of blossoms and ripening fruits Citrus is grown in all tropical and subtropical climates, excepting these restrictions global regions that are tropical. Citrus trees thrive best in a subtropical climate and development. Mandarins from Khasi and Darjeeling are cultivated up to 2000 meters above sea level to a colder climate suited. The ideal climate has summers and winters that are clearly defined and receives between 50 and 75 cm of rain from June to September. Even at 900 m above mean sea level, sweet oranges can be grown. Extreme temperatures are required to have a higher yield. The optimal temperature is 25 °C, and both extremely low and high temperatures are inevitable.

4.5 Soil

Citrus plants can be grown in a diversity of soil types, including loamy sand and alluvial soils. North India to lateritic/acidic soils in the north-eastern hills and thick clay loam in the Deccan plateau. Light, well-draining soils are ideal for citrus orchard growth. Ideal soils are those that are deep and have a pH range of 5.5–7.5. They can, however, also grow in the pH range of 4.0–9.0. Elevated levels of calcium carbonate in the feeder root zone may have a negative impact on growth. The ideal soil types for growing sweet oranges are deep, loamy soils that drain well. If well-drained, heavy soils produce good harvests, but cultivation is challenging. The ideal pH range for soil is 6.5–7.5, and the ideal Electrical conductivity range for water is <1.0. Plants are extremely susceptible to wet soils.

4.6 Planting Material

The accessibility of high-quality planting crops is crucial for citrus farming. Citrus plants are particularly vulnerable to a variety of abiotic and biotic stresses. Therefore, choosing a perfect rootstock continues to be a problem for India's citrus industry. Currently employed rootstocks include over the past 50 years, rough lemon and Rangpur lime have undergone several changes. Consequently, National Research generated excellent choices from the usual rootstocks. State agriculture universities may grant admission to the Center for Citrus (NRCC), Nagpur, and other locations to be attained for growing top-notch plant stuff. Virus-free for bud-wood selection through shoot tip grafting, mother plants were created from the elite offspring of known pedigree. Only the NRCC, Nagpur procedure may be employed. Under shade net structures, primary nursery beds are prepared on light, rich soils or in HDPE trays. In 2–3 phases in the nursery beds, weak seedlings, off kinds, and uneven seedlings are eliminated in order to select nursery seedlings. When secondary nursery seedlings reach a height of around 30–40 cm after a year, they are prepared for planting in the main field and can be raised in polythene bags as well.

4.7 Land Preparation and Plant Density

The soil needs to be leveled and properly ploughed. In mountainous areas, planting is performed on terraces slope. High density planting is possible in these soils due to more aerial space availability than on level terrace. Since citrus plants are extremely prone to water logging and stagnation, drainage ditches that are 3–4 feet deep must be installed along the slopes surrounding the orchard during the rainy season. Sweet orange (*Citrus sinensis*) are planted at standard spacing of – 5 m × 5 m, 5.5 × 5.5 m with plant number of 400/330 per ha.

4.8 Irrigation

Citrus has to be watered throughout critical times in the first year. Fruit size is raised while fruit drop is significantly reduced. Illnesses like root rot and collar rot can emerge in places that have recently flooded. It is preferable to use high frequency light irrigation. Above 1000 ppm of salt in irrigation water is dangerous. The soil type and stage of growth determine the quantity and frequency of irrigation. Micro irrigation systems not only offer good fruit retention during crucial crop growth phases in March and April but also conserve water and nutrients, even in situations where water is not an issue.

4.9 Fertilizers and Manure

Manuring is required three times a year, in equal halves during February, June, and September.

Tables 4.1 and 4.2 provides the suggested manure and fertilizer doses.

Nitrogen must be given twice in March and October. However, a circular band of fertilizer must be given in October, 120 cm radially away from the location of tree stem, with farmyard manure, phosphorus, and potash. A basin should not have more than two or three deep placements or applications of fertiliser. In fact, in an acid lime plantation, they are more surface oriented, with 80–95% of them being identified in the top 10 cm, whereas in a sweet orange plantation, the highest activity was discovered inside a depth of 25 cm. Spray a solution (once every 3 months) at the time of fresh flesh production that contains zinc sulphate (0.5%), iron (0.25%), manganese (0.05%), boron (0.1%), magnesium (0.5%), and molybdenum (0.003%). Apply 25 g of each sulphate of zinc, manganese, and iron per tree annually.

Table 4.1 Farm Yard Manure (FYM) requirement by year (kg/plant/year)

Farmyard manure	1 year	2 years	3 years	4 years	5 years	6 years	7 years onwards
kg/plant	20	10	15	20	25	30	40

Table 4.2 Year wise requirement of various nutrients (g/plant/year)

Nutrient	1 year	2 years	3 years	4 years	5 years	6 years onwards
Nitrogen	100 g	200 g	300 g	400 g	450 g	500 g
Phosphorus	50 g	100 g	150 g	200 g	200 g	250 g
$FeSO_4$	25 g	25 g	50 g	50 g	100 g	150 g
$ZnSO_4$	25 g	25 g	50 g	50 g	100 g	150 g
$MnSO_4$	25 g	25 g	50 g	50 g	100 g	150 g
Potash	25 g	50 g	75 g	200 g	200 g	250 g

4.10 Harvesting

Sweet oranges typically reach maturity between 240 and 280 days after planting. Fruits that have reached the colour break stage are picked up every 2 or 3 days for 10–14 days. 35–45% of the overall cost of manufacturing goes on citrus fruit harvesting. Therefore, the viability and profitability of the company are significantly impacted by an increase in the efficiency of this one operation. Traditional manual harvesting requires a lot of labour and is therefore expensive. Mechanical harvesting techniques have undergone much investigation and development. A brief summary of the mechanical harvesting systems' trunk shaking, canopy shaking, air shaking, and limb shaking performances is provided. The performance of the mechanical harvester is influenced by its operating environment; hence, the outcomes of numerous studies on tree shape and orchard planning have been incorporated. The study's findings demonstrate that optimizing both harvester performance and orchard yield can be accomplished by selecting the appropriate harvesting system and orchard plan. None of the mechanical solutions under investigation could equal the hand pickers' high-quality selection capabilities, according to this research.

4.11 Yield

Sweet orange: Tends to start in the fifth year and stabilizes in the eighth year, producing 40–50 oranges per tree. After stabilization, each tree produces 500–600 fruits on average.

4.12 Post Harvesting

To give the fruits a constant yellow-orange color, it is advisable to use ethephon @ 250 ppm in combination with 1% calcium acetate as a foliar spray when the fruit is mature. Ethylene gas treatment is an alternative for de-greening and color development of mandarins with sweet oranges. A color change can be induced in a de-greening chamber in 48 h 5–10 ppm of ethylene, and 90–95% RH at 6–7 °C. For preserving particular citrus fruits for a long time, cold storage options are available. With the use of a forced air system, citrus is pre-cooled. Oranges can be packed in highly ventilated Corrugated Fibre Boxes that are 30 cm × 30 cm × 30 cm of dimensions. Sweet oranges can be kept at 7–8 °C and 85–90% RH for 4–8 weeks period.

4.13 Marketing

Due to their perishable nature, citrus fruits must be handled carefully and hygienically. Sweet oranges,lemons, and limesmay be transported to far-off locations for marketing since they stay fresh under ambient conditions. Mandarins need to be handled and transported with more care and consideration.

4.14 Nutritional Facts

Human diet must include significant micronutrients like the vitamin C and E, carotenoids, and flavonoids. These substances are available in almost all plant materials in different dietary forms (Di Majo et al. 2005). Nutritional value of food is provided by these functional food ingredients viz. antioxidants,phytochemicals or nutraceuticals. Eating phytochemicals, which are present in edible fruits and vegetables, may change a person's metabolism favorably and prevent chronic degenerative diseases (Tripoli et al. 2007). Increasing the intake of vegetables and fruits protects against degenerative diseases like cancer since, epidemiological studies have revealed an opposite link between dietary intake of flavonoids from citrus fruits and cardiovascular disorders (Di Majo et al. 2005; Hertog et al. 1993; Keys 1995).

Citruses have long been prized for their healthful nutritional and antioxidant qualities. They serve as the primary resource of significant nutrients. Scientific research has demonstrated the numerous health advantages of oranges' high vitamin and mineral content. Additionally, other biologically active, non-nutrient molecules contained in citrus fruits, like phytochemicals, antioxidants, dietary fibers—both soluble and insoluble are known to lower the threat of developing cancer and several chronic conditions, including arthritis, obesity, and coronary heart disease (Crowell 1999).

4.15 Antioxidant Properties

An orange is considered to be of great quality if it is mature and has good colour intensity dispersed evenly across the surface. These oranges must be solid, generally smooth, have the variety-typical texture and form, and be free of defects, rot, and other imperfections. When employed as antioxidants, citrus flavonoids have been demonstrated to have biological activity and positive benefits (Tripoli et al. 2007). These plant-based pigments from this group, together with anthocyanin, are what give fruits and flowers their color. Moreover, they can exercise their antioxidant activity by metal chelation and are present in dietary vegetables and fruits (Bombardelli and Morazzoni 1993). Studies demonstrate that flavonoids are effective hydroxyl radical scavengers due to their capability to suppress the free radicals and provide hydrogen atoms (Di Majo et al. 2005; Tripoli et al. 2007; Cillard and Cillard 1988; Darmon et al. 1990). Oranges contain folate, dietary fibre, and other bioactive substances like flavonoids and carotenoids that avoid cancer and other degenerative diseases in addition to being an excellent source of vitamin C. (Ejaz et al. 2006). Vitamin C-rich foods boost the body's defense system against infectious diseases and flush harmful, pro-inflammatory free radicals from blood. Phytochemicals present in sweet oranges include hesperetin and naringenin. Naringenin has a bioactive effect in terms of acting as an antioxidant, anti-inflammatory, free radical scavenger and immune system modulator.

4.16 Anti-inflammatory Properties

Citrus flavonoids have anti-inflammatory activity due to the occurrence of regulatory enzymes (protein kinase C, phospholipase, phosphodiesterase, cyclooxygenase and lipoxygenase,) that regulate the creation of the biological mediators, which are responsible for activating endothelial cells and specialized cells concerned in inflammation. It is possible to connect the inhibition of these enzymes by flavonoids to how they control inflammatory and immune reactions (Tripoli et al. 2007). In fact, phosphodiesterases and kinases required for cellular signal activation and transmission are inhibited by citrus flavonoids. They also have an impact on the activation of some immune system cells, such as B and T lymphocytes (Manthey et al. 2001). Citrus fruit flavonoids also guard against atherosclerosis by preventing atheroma development (Hertog et al. 1993). According to Tripoli et al. (2007), hesperidin produced from citrus cultures has a mild anti-inflammatory impact and serve as a beneficial precursor to produce new flavonoids (Da Silva et al. 1994).

4.17 Anti-Cancer Properties

Citrus flavonoids have the ability to inhibit cancer by selective antiproliferative actions, cytotoxicity, and apoptosis (Elangovan et al. 1994; Hirano et al. 1994). Since flavonoids can absorb ultraviolet light, they are not mutagenic and shield DNA from oxidative damage (Stapleton and Walbot 1994). Free radicals that develop close to DNA and promote mutations are neutralised by them. It has been revealed in mice with bodies exposed to X-ray radiation (Shimoi et al. 1994). Flavonoids can protect DNA by directly interacting with substances that cause tumour, such as bleomycin induced chromosomal aberrations (Heo et al. 1994). Citrus fruit flavonoids have also been shown to repress the growth of tumors and the multiplication of malignant cells in the heart and liver of syngeneic rats (Bracke et al. 1989). Citrus flavonoids'ability to perform is dependent on their ability to impede cell movement (Bracke et al. 1989). Oranges are as well high in calcium, beta-carotene, potassium, amino acids, folic acid, pectin, and fibre. They are also rich in sodium, phosphorus, iodine, salt, chlorine, manganese, zinc, and iron. One orange is thought to contain about 170 phytonutrients and 60 flavonoids with anti-tumor, blood clot inhibiting, anti-inflammatory and antioxidant qualities that promote overall health (Cha et al. 2001).

4.18 Anti-obesity Effects

Sweet oranges are low in calories, cholesterol- or saturated-fat-free, and high in pectin, a type of dietary fibre that is beneficial to obese people. The immense laxative pectin protects the mucous membrane from colon cancer by binding to injurious substances. Pectin has also known for reducing blood cholesterol levels by blocking its reabsorption in the colon (Walton et al. 1945). The alkaloid synephrine; present in orange peels, prevents the liver from producing cholesterol. Antioxidants found in oranges guard against oxidative stress, which oxidises blood LDL (low-density lipoprotein).

4.19 Antimicrobial Properties of Sweet Orange juice and Peel

Hydromethanolic extract of *Citrus sinensis* peel is known for its antibacterial against various gram positive and gram negative. The hydromethanolic extract of orange peel demonstrates antibacterial activity against *Staphylococcus aureus* (6–14 mm), *Bacillus subtilis* (6–9 mm), *Staphylococcus epidermis* (5–10 mm), *E. coli* (7–12 mm), *Shigella flexineri* (9–12 mm), and *Pseudomonas aeruginosa*(6–9 mm) (Dubey and Schares 2011).

After phytochemical research, Mostafa and Essawy (2021) discovered a range of bioactive organic components in fresh juice and methanolic peel and pulp extract. The antioxidant activity of *Citrus sinensis* L. (Sweet Orange) juice extract and crude methanolic extract of pulp and peel (MPPE) was effective against two fungal and six bacterial strains. Orange peel polyphenolic extract (OPE) is used to test for its anti-fungal properties against *Monilinia fructicola*, *Alternaria alternate* and *Botrytis cinerea*; three important post-harvest fungal diseases. Based on the dosage and the target fungus, the effect varied at lower to higher concentrations. Phenolic components in sweet orange peel like flavonoids (hesperidin, naringin, and neohesperidin) and phenolic acids (*p*-coumaric and ferulic)—were also evaluated in different studies for their anti-fungal activity. Ferulic acid and *p*-coumaric acid considerably outperformed flavonoids in terms of their capacity to inhibit growth in artificial medium. There was no indication of compound synergism, and phenolic acids might enhance OPE's inhibitory activity. Intriguingly, ferulic acid retained its effectiveness against *M. fructicola* and *A. alternata* while outperforming *p*-coumaric in the control of *B. cinerea* in peach-based medium. The potential for using ferulic acid or ferulic acid-rich based extracts as a natural substitute to other post-harvest treatments to reduce post-harvest losses and also augment fruit shelf life is suggested by these findings, which show peel orange extracts to be an excellent source of antimicrobial chemicals (Hernandez et al. 2021).

Antibacterial activity of sweet orange juice extract on diverse microorganisms using ethanol and ethyl ethanoate as solvents was also examined by researchers. Ditch method was followed to assess the sensitivity of *Escherichia coli, Staphylococcus aureus*, *Klebsiella pneumoniae,* and *Neisseria gonorrhoeae* to it. The disc method was also used to examine the effectiveness of gentamycin, streptomycin, penicillin G and ciprofloxacin, against test pathogens and positive control cultures. The Dosage effect was different on different organisms but there was a big difference between the alcohol and ethyl ethanoate extracts too. Ethyl ethanoate extract of orange peel was much more effective than the ethanol extract and the positive control (Hassan et al. 2021).

4.20 Antimicrobial Properties of Sweet Orange Essential oil and Seed Oil

According to the findings of the disc diffusion methods to determine minimum inhibitory concentration (MIC), the sweet orange essential oil shown a broad spectrum of antimicrobial activities against *Penicillium chrysogenum, Staphylococcus aureus, Escherichia coli, Bacillus subtilis,* and *Saccharomyces cerevisiae*. Their MIC ranged from 4.66 to 18.75 μL/mL, while their inhibition zones ranged from 14.57 to 23.37 mm (Tao et al. 2009).

The antibacterial activity of the orange seed oil is superior to that of the non-oil extract, according to Oikeh et al. (2020). The antifungal test findings revealed that *Aspergillus niger* and *Penicillium* sp. were not affected by the seed oil's ability to significantly slow down the expansion of *Candida albicans*. The MIC values for both extracts against bacterial and fungal strains were between 50 and 100 g/mL The minimum bactericidal and fungicidal concentration values for both extracts were in the range of 100–200 g/mL. According to the study's findings, *C. sinensis* seed oil and non-oil extracts have antioxidant, antibacterial, and antifungal qualities that can be used in numerous ways to build products.

4.21 Health Effects

Vitamin A and other flavonoid with antioxidant properties, such as zeaxanthin, lutein, beta-cryptoxanthin, alpha and beta carotenes and beta-cryptoxanthin, are also present in substantial amounts in oranges. Vitamin A is necessary for preservation of healthy skin vision and mucous membranes. The B-complex vitamins viz. pyridoxine, thiamin, folates and minerals are also abundant in it (Tables 4.3 and 4.4). These vitamins are vital for the body and it needs outer sources to replenish them. Orange fruit is a rich source of the minerals including calcium and potassium. Potassium helps control blood pressure and heart rhythm since it is an essential component of body and cell fluids. Flavonoid rich fruits aid in the body's defense against oral and lung malignancies. Potassium is a crucial part of physiological fluids and cells and helps to regulate blood pressure and heart rate. Constipation is relieved by the orange's alkaline qualities, which also stimulate the digestive juices. Orange juice consumption on a regular basis lowers kidney stone-causing calcium oxalate formation risks. Oranges contain polyphenols that protect against viral infections. Oranges keep your skin youthful and shining by protecting it from free radical damage, which keeps you looking young (Tsuda et al. 2004). Oranges can be used to make juice, concentrates, punches, orangeades, and liquors, as well as a variety of soda and cocktail beverages. Orange fruits and peels are used in several specialties, including candied peels, jams, pastries, cakes, marmalades and candies. Cooking with orange seed oil is an option, and it can also be found in polymers. As essential oil, orange peel oil is used in perfumes, along with oils made from flowers, leaves, and twigs.

4.22 Chemical Composition of Compounds in Orange Juice, Peel (Fresh Oven Dried and Microwave Dried, Seeds, Pulp)

Table 4.5 shows the collective information of the compounds present in sweet orange parts. The below table gives the information of total phenolic content (TPC), Total Flavonoid content and specific compounds present in orange.

4.23 Amino Acids Content

Many forms of orange juices, including packed orange juice manufactured from frozen concentrates and freshly squeezed oranges are available; they vary in their amino acid composition. The compositional variations observed between these juices, as assessed by their amino acid content, are presented in the form of range in Table 4.6 (Gomez-Ariza et al. 2005).

Table 4.3 Nutrient composition of sweet orange

Source	Nutrient	Amount	References
Sweet orange pulp	Energy	197 KJ	Etebu and Nwauzoma (2014), Lanza (2003) and Roussos (2016)
	Water	86.75–87.2 g	
	Carbohydrates	7.8 g	
	Sugars	9.35 g	
	Dietary fibers	1.6–2.4 g	
	Protein	0.7–0.94 g	
	Fats	0.12–0.2 g	
	Vitamin A	11–71 μg	
	Vitamin C	50 mg	
	Vitamin B_1	0.06–0.087 mg	
	Vitamin B_2	0.04–0.05 mg	
	Vitamin B_3	0.2–0.282 mg	
	Vitamin B_5	0.25 mg	
	Vitamin B_6	0.06 mg	
	Vitamin B_9	30 μg	
	Choline	8.4 mg	
	Total phenolic compounds	32.3–140.7 mg GAE/100 g	
Orange juice	Water	89.3 g	Saänchez-moreno et al. (2003), Lanza (2003), Fejzić and Ćavar (2014), AbdGhafar et al. (2010) and Irkin et al. (2015)
	Carbohydrates	8.2 g	
	Glucose	13.95–30.9 g/L	
	Fructose	15.5–33.05 g/L	
	Sucrose	32.41–59.34 g/L	
	Protein	0.5 g	
	Thiamin	0.05 mg	
	Riboflavin	0.03 mg	
	Niacin	0.4 mg	
	Vitamin A retinol equivalents	38 mg	
	Vitamin C	41.95–53.2 mg/100 ml	
	β-carotene	60.73 mg/100 ml	
	α-carotene	48.57 mg/100 ml	
	β-cryptoxanthin	314.09 mg/100 ml	
	α-cryptoxanthin	161.82 mg/100 ml	
	Zeaxanthin	345.38 mg/100 ml	
	Lutein	765.72 mg/100 ml	
	Total phenolic compounds	43.7–135 mg GAE/100 g	

(continued)

Table 4.3 (continued)

Source	Nutrient	Amount	References
Orange peel	Moisture	9.5 g/100 g	Zaker et al. (2016) and Arora and Kaur (2013)
	Protein	5.17 g/100 g	
	Fiber	74.14 g/100 g	
	Crude fat	4.41 g/100 g	
	Sugar	9.20 g/100 g	
	Total phenolic compounds	158.54–177.92 mg/100 g	

Source: USDA Nutrient Database (2014)

Table 4.4 Nutritional composition of sweet orange

Minerals (mg/100 g)	Fruit pulp	Fruit peel	Fruit juice	References
Potassium	139–2712	154	110.7–172	Roberts and Gaddum (1937), Bitters (1961), Paramasivam et al. (2000), Topuz et al. (2005), Onibon et al. (2007), De Moraes Barros et al. (2012) and Czech et al. (2020)
Sodium	0.12	0.54	0.57–0.7	
Calcium	27.9–93	41.9	5.37–9.47	
Phosphorus	23.3	25.3	12.52–32	
Magnesium	10.3–85	13.2	5–10.2	
Iron	0.37–240	0.51	0.1	
Zinc	0.17–10	0.25	0.17–0.316	
Copper	0.06	0.15	0.005–0.277	
Manganese	0.02	0.13	0.002–0.011	
Selenium	1.50	2.35	–	

4.24 Minerals Content

According to De Moraes Barros et al. (2012), eight minerals were discovered in orange juice with four major elements (K, Ca, Na, and Mg) and four trace elements (Cu, Fe, Mn, and Zn). All main minerals and trace elements had the second-highest level of Cu and K, were found in peels in higher concentrations than in pulps of oranges. The highest potassium content was found in citrus pulp and peel, followed by calcium and magnesium. Orange fruits have a low salt content (2 mg per 100 g

Table 4.5 Specific bioactive components in orange and details of extraction conditions

Source	Extraction phase	Conc.	Extraction temperature	TPC	TFC	Specific compounds	References
Orange juice	–	–	–	0.437 mg GA/mL		–	Fejzić and Ćavar (2014)
Orange peel	Ethanol	–	–	0.452 mg GA/mL		–	
Orange peel	Ethanol	80%	35 °C	1.204–2.363 g GAE/100 g	1.238–2.275 g/100 g	Hesperidin, Neohesperidin, Eriocitrin, Narirutin, Naringin, Didymin, Sinensetin, Hexamethoxyflavone, Tangeretin, Nobiletin	Mhiri et al. (2016)
Orange juice	–	–	–	135.3 mg GAE/100 mL	2.99 mg HE/100 mL	–	AbdGhafar et al. (2010)
Orange seed (immature)	Methanol	100%	–	21.64%	49.82%	Epigallocatechin, Catechin, Rutin, Naringin, Hesperidin, Quercetin, Galic acid, Syringic acid Vanillic acid, *p*-Coumaric acid, Rosmarinic acid, *trans*-2-Hydroxicinnamic acid	Moulehi et al. (2012)
Orange seed (semi-mature)				19.09%	56.95%		
Orange seed (commertial mature)				16.80%	55.97%		
Orange peel	Methanol	–	Room temp.	160.3 mg GAE per g	23.2 mg QE/g	–	Ghasemi et al. (2009)
Orange tissue				232.5 mg GAE/g	1.2 mg QE per g		
Orange pulp	Methanol	90%	55 °C	11.23 GAE/g	–	Phloroglucinol, Protocatechuic acid, *p*-Coumaric acid, Ferulic acid, Naringenin, Hesperetin, *p*-Coumaric acid, Caffeic acid, Gallic acid, Chlorogenic acid, Hesperidin Rutin	Iglesias-Carres et al. (2019)

Orange peel (fresh)	Methanol	–	–	158.54 mg/100 g	296.38 mg/100 g	Gallic acid, Resveratrol, Caffeic acid, Rutintrihydrate, Quercetin, coumaric acid, syringic acid *trans*-Ferulic, Isorhamnetin, *trans*-Cinnamic, Kaempferol, Naringenin	Musa Ozcan et al. (2020)
Orange peel (oven dried)				173.23 mg/100 g	850.54 mg/100 g		
Orange peel (microwave dried)				170.10 mg/100 g	460.44 mg/100 g		
Orange peel (infrared dried)				177.92 mg/100 g	733.04 mg/100 g		
Red blood orange pulp	Methanol	–	30, 60 °C	11.732 mg GAE/g	–	Ferulic acid, Vanillic acid, *p*-coumaric acid, Naringenin, Diosmin, Hesperidin, Rutin, Sinapic acid	Alu'datta et al. (2017)
Orange pulp	Ethanol	40%	25 °C	124.53 mg/100 g	10.57 mg/100 g	Hydroxybenzoic acids, Hydroxycinnamic acids	Stafussa et al. (2018)
Orange peel	Ethanol	50%	–	105.96 mg GAE/100 g	–	Hesperidin, ferulic acid, naringin, rutintrihydrate, *trans*-cinnamic acid, quercetin, apigenin, chlorogenic acid, hesperetin	Montero-Calderon et al. (2019)
Orange peel	Methanol	100%		36.09 mg GAE/g	4.61 (mg CE per g)	Gallic acid, Protocatechuic acid, 4-hydroxybenzoic acid, Caffeic acid, Ferulic acid, Catechin, Epigallocatechin, Vitexin, RutinLuteolin, Apigenin	Liew et al. (2018)
	Ethanol	70%		33.07 mg GAE/g	4.35 (mg CE per g)		
	Acetone	70%		38.24 mg GAE/g	5.03 (mg CE per g)		
	Water	–		12.08 mg GAE/g	1.90 (mg CE per g)		

(continued)

Table 4.5 (continued)

Source	Extraction phase	Conc.	Extraction temperature	TPC	TFC	Specific compounds	References
Orange peel	Methanol	100%	–	73.80%	23.02%	Gallic acid, Hydroxybenzoic acid, Rosmarinic acid, Catechin, Rutin, *trans*-2-hydroxycinnamic acid, acid, Syringic acid, Vanillic acid Coumaric acid, Ferulic acid, *trans*-cinnamic Naringin, Epicatechin	Karoui and Marzouk (2013)
Orange juice				71.25%	23.13%		
Orange leaves	Methanol	80%	Room temp.	12.54–44.41 mg GAE/g	0.63–3.25 mg QE/g DM	–	Lagha-Benamrouchea and Madani (2013)
Orange peel				9.61–31.62 mg GAE per g	0.56–1.29 mg QE perg DM		
Orange peel	Methanol	80%	25 °C	13.6 mg GaE/g	1.82 mg QE/g	Gallic acid, Vanillic acid, Syringic acid, Salicylic acid, Chlorogenic acid, Caffeic acid, Ellagic acid, *p*-Coumaric Quercetin, Ferulic acid, Sinapic acid, Apigenin, Kaempferol Catechin, Naringin, Rutin, Myricetin	Ghosh et al. (2019)
	Ethanol	100%		9.19 mg GaE/g	3.22 mg QE/g		
	Water	–		8.55 mg GaE/g	1.73 mg QE/g		
Orange peel	Acetone	–	Room temp.	114 mg/g	–	–	Arora and Kaur (2013)
	Methanol			158 mg/g			
	Distilled water			210 mg/g			
	Hexane			79 mg/g			
Orange pulp	Acetone			522 mg/g			
	Methanol			465 mg/g			
	Distilled water			330 mg/g			
	Hexane			201 mg/g			

Orange/peel	Methanol	80%	4 °C	11.08 mg GAE/g	5.18 mg QE/g	Chlorogenic acid, Caffeic acid, Naringin, Hesperidin	Irkin et al. (2015)
Orange juice	–	–	–	523.44 mg GAE/L	1.50 mg QE/L	Caffeic acid, Hesperidin	
Red blood orange	Methanol	100%		62.38 mg GAE/g	28.91 mg CE/g	Gallic acid, Caffic acid, Protpcatechuic acid, 3-hydroxybenzoic acid, Dihydroxylbenzoic acid	Zahoor et al. (2016)
	Methanol	80%		20.86 mg GAE/g	17.65 mg CE/g		
	Ethanol	100%		13.69 mg GAE/g	21.65 mg CE/g		
	Ethanol	80%		26.73 mg GAE/g	19.33 mg CE/g		
Orange cultivars (ValenciaValencia, Valencia Campbell, Valencia cutter, Valencia Frost, Valencia late, Valencia olinda)	–	–	–	32.3–140.7 mg GAE/100 g	1.23–425 mg CE/100 g	Naringin, Hesperidin, Ascorbic acid, Didymin	Roussos (2016)

Table 4.6 Amino acids present in sweet orange juice

Amino acids	Amount (mg/100 mL)	Reference
Alanine	8.6–22	Rockland (1961), Gomez-Ariza et al. (2005), Niu et al. (2008), Roussos (2011), Liu et al. (2012a, b) and Roussos (2016)
Arginine	54–121	
Asparagine	35–188	
Aspartic acid	7.36–114	
Glutamic acid	6.47–49	
Glutamine	15–60	
Glycine	0.93–2	
Histidine	0.18–126	
Isoleucine	0.56–7	
Leucine	0.66–8	
Lysine	3–22	
Methionine	1.90–46	
Phenylalanine	0.14–43	
Proline	64–239	
Serine	10.4–37	
Threonine	0.98–28	
Tryptophan	0.16–0.3	
Tyrosine	1.23–38	
Valine	1.08–26	

orange pulp), but they also have a high potassium content (140 mg per 100 g of orange pulp). For body's cells to maintain their electrolyte balance, the ratio of K to Na in oranges is critical.

The orange sample, according to Topuz et al. (2005), contained a variety of mineral sources, including K, Mg, Ca, and P. The Finike cultivar's juice has the highest levels of the micro components. K was a prominent element in the samples, with concentrations in the types that were examined ranging from 1011 to 1364 mg/kg. The majority of the chemical traits of the Shamouti cultivars were, on average, slightly subpar compared to those of the other types. To illustrate the concentration of minerals contained in sweet orange pulp, peel, and juice, numerous investigations were also documented in our literature and are shown in Table 4.7.

4.25 Uses

4.25.1 Food Products

Drinks, sorbets, breads, and jams all contain blood oranges (Ehsani 2007).

Table 4.7 Minerals in different parts of orange

Minerals	Fruit pulp	Fruit peel	Fruit juice	References
Potassium (mg/100 g)	139–2712	154	110.7–172	Roberts and Gaddum (1937), Bitters (1961), Paramasivam et al. (2000), Topuz et al. (2005), Onibon et al. (2007), De Moraes Barros et al. (2012) and Czech et al. (2020)
Sodium (mg/100 g)	0.12	0.54	0.57–0.7	
Calcium (mg/100 g)	27.9–93	41.9	5.37–9.47	
Phosphorus (mg/100 g)	23.3	25.3	12.52–32	
Magnesium (mg/100 g)	10.3–85	13.2	5–10.2	
Iron (mg/100 g)	0.37–240	0.51	0.1	
Zinc (mg/100 g)	0.17–10	0.25	0.17–0.316	
Copper (mg/100 g)	0.06	0.15	0.005–0.277	
Manganese (mg/100 g)	0.02	0.13	0.002–0.011	
Selenium (mg/100 g)	1.50	2.35	–	

4.25.2 *Thickening and Gelling Agent*

Residue of orange juice is also used in food products due to the material's high pectin content and probable availability in large amounts, orange juice extraction residues have the potential to be a superior source of dietary fibre (DF). Three different orange kinds' DF pulp were examined for its chemical and physical properties. The total dry matter content was relatively high, ranging between 35.4% and 36.9%. (DM). Pectins (15.7–16.3% DM), lignin (2.2–3.0% DM) and cellulose and hemicellulose (16.6–18.1% DM), were all abundant in orange DF. The product had a high water retention capacity (7.3–10.3 g water/g fibre), a significant oil absorption feature (0.9–1.3 g oil/g fibre), and a low caloric value (3519–3735 cal/g). Little amounts of protein, fat, and ash were found in orange DF concentrate (8.1–10.1%, 1.5–3.0%, and 2.6–3.1% DM, respectively) according to chemical analysis. From light orange to yellow, the DF's orange colour fluctuated. These characteristics suggested a variety of potential applications, including those for binders, texturizer, low-calorie bulk ingredients, thickening and gelling agents, as well as clouding agents in beverages (Grigelmo-Miguel and Martin-Belloso 1998, 1999).

4.25.3 Cookies

More than half of the raw materials used in the production of orange juice are converted into byproducts that are nutrient-dense and high in active chemicals. Increasing the utilisation of these byproducts could be a crucial circular economy strategy. The goal of this study was to make flour from a by-product of orange juice, define it, and then use this flour to make cookies. Chemical composition, dietary fibre, phenolic compounds, antioxidant potential, and hygroscopic characteristics of orange by-product flour (OBPF) were described. The impact of using OBPF in place of wheat flour in cookies was then assessed. Dietary fibre (73.61% dry matter (DM)), minerals (ash = 2.72% DM), and total phenolic compounds (534 mg gallic acid equivalent (GAE)/100 g of DM) were all found in very high concentrations in OBPF. In general, employing OBPF in place of wheat flour had no appreciable impact on the characteristics of cookies. Sensory evaluations revealed that cookies made with 10% OBPF scored better. As a result, OBPF displayed intriguing properties that point to potential applications in the development of fiber-rich meals like cookies. Additionally, OBPF manufacturing serves as a significant strategy for the orange juice processing sectors in the development of a circular economy in the food chain (Castro et al. 2020).

Orange peel powder can be used in baked goods like cookies and is a rich source of crude fibre, phenolic compounds, and carotene. The nutritional content of the baked cookies was examined. It was also investigated how cookies changed as they were stored. Cookies made with a blend of orange peel powder and refined wheat flour performed best when made with 2% orange peel powder and 98 percent refined wheat flour, and their chemical makeup was very constant over time. The cookies held up well during the three-month preservation period. During their three-month storage, the cookies in LDPE bags outperformed those in polypropylene (PP) in terms of quality (Belose et al. 2021).

4.25.4 Fat Substitute in Ice Creams

Crizel et al. (2013) reported that orange fibre was examined from two separate samples: F1 (peel, pulp, and seeds) and F2 (peel). The overall amount of dietary fibre in both samples was high, and the proportion of soluble to insoluble fibre was perfect. The fibres demonstrated a great ability for retaining water and oil as well as a high concentration of phenolic and carotenoids. Orange fibre was used as a fat substitute in ice cream, which resulted in a 70% reduction in fat without significantly altering the product's texture, flavour, or colour. The use of orange fibre as a fat replacement in ice cream production has proven to be a promising alternative.

4.25.5 Breads

The quantities of fibre, ash, and phytochemicals in bread are increased when orange peel flour is used. But only bread made with 3% OPF had a sensory quality that was on par with bread made with 100% wheat. In addition to giving food more value, using orange peel flour in bread production will lower environmental pollution and wheat import costs. Orange peel flour was added, which increased ash (2.3–4.3%), fibre (0.6–5.8%), and carbohydrates (59.9–62.1%), but a decreased protein (8.2–2.7%) and fat (1.7–0.8%). With higher quantities of orange peel flour in the samples, phytochemical levels also rose. Alkaloids varied from 3.6 to 4.8 mg/g, tannins from 0.9 to 1.4 mg/g, and saponins from 0.9 to 1.4 mg/g. Loaf capacity was reduced from 8.0 to 4.8 cm^3, the oven spring from 2.0 to 0.2 cm, and the specific volume decreased from 5.3 to 3.2 cm3/g. For all the studied qualities, sensory quality declined as orange peel flour content increased. However, there was no sensory quality difference between bread prepared with 100% wheat flour and that made with 3% orange peel flour (Okpala and Akpu 2014).

4.25.6 Muffins

A dietary fiber-rich orange bagasse product (DFROBP) was created, with a 41.5% total dietary fibre (TDF) content. In comparison to a control muffin, the product boosted the fibre content of the experimental muffins by 40 and 63%. Although there was no difference in the amount of immediately digested starch between the control muffin and those with DFROBP, there was an increase in the amount of slowly digestible starch after this ingredient was added. However, DFROBP addition resulted in a drop in resistant starch levels. The projected glycemic index of muffins with DFROBP added was significantly lower than expected, and the preference test revealed that the 10% replaced muffin and the control muffin was equally popular. The production of prototype baked goods with high levels of TDF and indigestible fraction was made possible by the partial replacement of wheat flour with DFROBP. These qualities may be useful in dietary plans for people with various nutritional needs (Romero-Lopez et al. 2011).

4.25.7 Candies

Orange Peel Candies are the best candy in terms of nutrient availability, flavour, and product ease. A sensory assessment test reveals that all of the goods pass muster with the panellists in terms of flavour, aroma, colour, general assessment, and appearance. Based on the results, it has been determined that the developed orange peel Candies are nutrient-rich, suitable for industrial use, and a commercially viable product. As a result, orange peels have the potential to be a significant raw resource for enterprises that manufacture food. *Citrus sinensis* (orange) peel is used to make orange peel candy, showing that it has the potential to be used in the food processing sector and lessen its negative environmental impact. The earliest method of food preservation is candy, which is a sweet confection. Sample 1 is made entirely of sugar, Sample 2 is 85% sugar to 15% honey, Sample 3 is 50% sugar to 50% honey, Sample 4 is 85% sugar to 15% molasses, and Sample 5 is 50% sugar to 50% molasses. The created orange peel candies are deemed acceptable and can be used for industrial purposes to create an economically viable product. All the prepared samples were formed to be "moderately good" (Mohanta et al. 2021).

4.25.8 Protein Feed

According to Xiao et al. (2018), orange peel can be fermented to create single cell protein feed, which can then be fermented to create ethanol and enzyme preparation. The common practice is to manufacture food from fruit peel; the majority of fruit peel pickling and fruit puree goods may be made from fruit peel, for instance, pomelo peel jam, which is a low-sugar meal with a class with nutrition and health functions. Fruit peel can be dried and used to make tea, such as dried orange peel and dried grapefruit skin. Fruit peels are now entirely utilised in industrialised nations because to the use of innovative technology for peel cleaning.

4.25.9 Jam

Jam can be made from oranges. The low gel strength of the jam can be addressed by adding pectin during processing to achieve the economically acceptable gel strength, or the deficiency can be made up for by using a combination of fruits high in pectin. The flavour and colour may be enhanced by combining the fruit with other fruits.

It is highly advised to promote jam manufacturing at the household level using regional raw materials, such as sugar beet, and rigorous guidelines must be accessible when making jam at home (Sulieman et al. 2013).

4.25.10 Colour of Orange Juice

Carotenoids, one of the major classes of natural pigments, are responsible for the colour of orange juice, although anthocyanins are primarily responsible for the colour of specific orange kinds, such as blood oranges. Food colour has an impact on consumer choices. According to certain studies, consumers' perceptions of the quality of orange juice are influenced by the colour of citrus beverages in general.

The objective evaluation of orange juice colour is highly valued in the USA, to the point where this characteristic is used to classify the product commercially based on quality. In addition to the relevance of orange juice colour to product quality, it is crucial to precisely measure this parameter because it has been shown that colour measurements can be used to quickly estimate the carotenoid concentration for quality control. Numerous methods and tools have been created over time as a result of these issues (Meléndez-Martínez et al. 2005).

Dassoff et al. (2021) reported that oranges make about 60% of the total, with juice production being their primary use. During the manufacture of orange juice, only roughly 50% of the weight of a fresh orange is converted into juice, with the remaining 50% being made up of residual (orange leaves, orange pulp, orange seeds, complete uncut orange fruits, and peel requirements). An effort has been made to deal with the tonnes of orange by-products that have resulted. Investigate potential strategies for recycling and valorizing citrus trash. However, these uses do not encompass all of the waste or fully realise the waste's potential worth. Currently, orange pomace, a by-product of the juicing process, is utilised to extract essential oils for aroma and flavour, and the remainder of the trash is used as cow feed. These by-products, which are disposed of at the owner's expense in landfills, cause global warming through carbon emissions. On the other hand, orange by-products still have a lot of beneficial nutraceutical elements, like dietary fibre and phytochemicals, which might be used to generate new products and ingredients with value-added. Making organic fertilisers and biofuels as well as extracting antioxidants, pectins, and essential oils are a few study methods in this field.Regarding using the orange pomace directly or with some basic processing, there is little information in the literature or in the food sector. This review is justified by the possibility of using orange pomace as a "clean-label," natural preservative for food product development.

References

Alu'datta MH, Rababaha T, Alhamada MN, Al-Mahasnehb MA, Ereifeja K, Al-Karakia G, Al-Duaisc M, Andraded JE, Tranchante CC, Kubowf S, Ghozlan KA (2017) Extracted free and bound phenolics profiles from Citrus fruits and their roles in biological system: content, antioxidant, anti-diabetic and anti-hypertensive properties. Food Funct 175:1–31. https://doi.org/10.1039/C7FO00212B

Arora M, Kaur P (2013) Antimicrobial & antioxidant activity of orange pulp and peel. Int J Sci Res (IJSR) 2:412–415

Atta S, Zhou CY, Zhou Y, Cao MJ, Wang XF (2012) Distribution and research advances of Citrus tristeza vírus. J Integr Agric 11(3):346–358

Belose BB, Kotecha PM, Godase SN, Chavan UD (2021) Studies on utilization of orange peel powder in the preparation of cookies. IJCS 9(1):1600–1602

Bitters WP (1961) Physical characters and chemical composition as affected by scions and rootstocks. In: Sinclair WB (ed) The orange. Its biochemistry and physiology. University of California, Division of Agricultural Sciences, pp 56–95

Bombardelli E, Morazzoni P (1993) The flavonoids: new perspectives in biological activities and therapeutics. Chim Oggi:25–28

Bracke ME, Vyncke B, Van Larebeke NA, Bruyneel EA, De Bruyne GK, De Pestel GH et al (1989) The flavonoid tangeretin inhibits invasion of MO4 mouse cells into embryonic chick heart in vitro. Clin Exp Metastasis 7:283–300

Castle WS (2010) A career perspective on citrus rootstocks, their development, and commercialization. HortScience 45(1):11–15

Castro LAD, Lizi JM, Chagas EGLD, Carvalho RAD, Vanin FM (2020) From orange juice by-product in the food industry to a functional ingredient: application in the circular economy. Foods 9(5):593

Cha JY, Cho YS, Kim I, Anno T, Rahman SM, Yanagita T (2001) Effect of hesperetin, a citrus flavonoid on the liver triacylglycerol content and phosphatidate phosphohydrolase activity in orotic acid-fed rats. Plant Foods Hum Nutr 56:349–358

Cillard J, Cillard P (1988) Composesphe' noliques et radicauxlibres. STP Pharma 4:592–596

Crowell PL (1999) Prevention and therapy of cancer by dietary monoterpenes. J Nutr 129(3):775S–778S

Czech A, Zarycka E, Yanovych D, Zasadna Z, Grzegorczyk I, Kłys S (2020) Mineral content of the pulp and Peel of various citrus fruit cultivars. Biol Trace Elem Res 193:555–563. https://doi.org/10.1007/s12011-019-01727-1

Da Silva EJA, Oliveiraand AS, Lapa AJ (1994) Pharmacological evaluation of the anti-inflammatory activity of a citrus bioflavonoid, hesperidin, and the isoflavonoids, duartin and claussequinone, in rats and mice. J Pharm Pharmacol 46:118–122

Darmon N, Ferrandiz V, Canal MT, Mitjavilla S (1990) Activiteantiradicallaire de flavonoides vis-à-vis de l'anion superoxide et Du radical hydroxyle. Liaison-Grupe Polyphenols Bull 15:158–162

Dassoff ES, Guo JX, Liu Y, Wang SC, Li YO (2021) Potential development of non-synthetic food additives from orange processing by-products – a review. Food Qual Safe 5

De Moraes Barros HR, de Castro Ferreira TAP, Genovese MI (2012) Antioxidant capacity and mineral content of pulp and peel from commercial cultivars of citrus from Brazil. Food Chem 134(4):1892–1898

Di Majo D, Giammanco M, La Guardia M, Tripoli E, Giammanco S, Finotti E (2005) Flavanones in citrus fruit: structure antioxidant activity relationships. Food Res Int 38:1161–1166

Dubey JP, Schares G (2011) Neosporosis in animals – the last five years. Vet Parasitol 180(1–2):90–108

Dugo G, Di Giacomo A (eds) (2002) Citrus: the genus citrus. CRC Press

Ehsani R (2007) In-situ measurement of the actual detachment force of oranges harvested by a canopy shaker harvesting machine, abstracts for the 2007 joint annual meeting of the Florida State Horticulture Society

Ehler S A (2011) Citrus and its benefits. J Bot 5:201–207

Ejaz S, Ejaz A, Matsuda K, Chae WL (2006) Limonoids as cancer chemopreventive agents. J Sci Food Agric 86:339–345

Elangovan V, Sekar N, Govindasamy S (1994) Chemopreventive potential of dietary bioflavonoids against 20-methylcholanthreneinduced tumorigenesis. Cancer Lett 87:107–113

Etebu E, Nwauzoma AB (2014) A review on sweet orange (*Citrus sinensis* L Osbeck): health, diseases, and management. Am J Res Commun 2(2):33–70

Fallahi E, Rodney DR (1992) Tree size, yield, fruit quality, and leaf mineral nutrient concentration offairchild'mandarin on six rootstock. J Am Soc Hortic Sci 117(1):28–31

Fallahi E, Moon JW, Rodney DR (1989) Yield and quality of 'Redblush' grapefruit on twelve rootstocks. J Am Soc Hortic Sci 114(2):187–190

Fejzić A, Ćavar S (2014) Phenolic compounds and antioxidant activity of some citruses. Bull Chem Technol Bosnia Herzegovina 42:1–4

Food and Agriculture Organization of the United Nations (2006) FAO statistics. http://faostat.fao.org/default.aspx

Food and Agricultural Organization (1965) Production year book. Rome, Italy 19, 117

Garg A, Garg S, Zaneveld LJD, Singla AK (2001) Chemistry and pharmacology of the citrus bioflavonoid hesperidin. Phytother Res 15(8):655–669

AbdGhafar MF, Nagendra Prasad K, Weng KK, Ismail A (2010) Flavonoid, hesperidine, total phenolic contents and antioxidant activities from Citrus species. Afr J Biotechnol 9(3):326–330

Ghasemi K, Ghasemi Y, Ebrahimzadeh MA (2009) Antioxidant activity, phenol and flavonoid contents of 13 citrus species peels and tissues. Pak J Pharm Sci 22:277–281

Ghosh S, Chatterjee JK, Hazra AK (2019) Comparative evaluation of in vitro antioxidant activities and high-performance liquid chromatography fingerprinting of fruit peels collected from food processing industry wastes. Pharmacogn Res 2019(11):346–351. https://doi.org/10.4103/pr.pr_13_19

Gomez-Ariza JL, Villegas-Portero MJ, Bernal-Daza V (2005) Characterization and analysis of amino acids in orange juice by HPLC-MS/MS for authenticity assessment. Anal Chim Acta 540:221–230

Goudeau D, Uratsu SL, Inoue K, daSilva FG, Leslie A, Cook D, Reagan L, Dandekar AM (2008) Tuning the orchestra: selective gene regulation and orange fruit quality. Plant Sci 174:310–320

Grigelmo-Miguel N, Martín-Belloso O (1998) Characterization of dietary fiber from orange juice extraction. Food Res Int 31:355–361

Grigelmo-Miguel N, Martín-Belloso O (1999) Comparison of dietary fibre from by-products of processing fruits and greens and from cereals. LWT Food Sci Technol 32(8):503–508

Hassan AH, Elsayed Nishawy MA, Amar MH (2021) Patterns of genetic structure and evidence of Egyptian Citrus rootstock based on informative SSR, LTR-IRAP and LTR-REMAP molecular markers. J Genet Eng Biotechnol 19(1):1–14

Heo HY, Lee SJ, Kwon CH, Kin SW, Sohn DH, Au WW (1994) Anticlastogenic effects of galangin against bleomycin induced chromosomal aberrations in mouse spleen lymphocytes. Mutat Res Fundam Mol Mech Mutagen 311:225–229

Hertog MG, Hollman PCH, Katan MB, Kromhout D (1993) Dietary antioxidant flavonoids and risk of coronary heart disease. Lancet 342:1007–1011

Hernández A, Ruiz-Moyano S, Galván AI, Merchán AV, Nevado FP, Aranda E, Serradilla MJ, de Guia Cardoba M, Martín A (2021) Anti-fungal activity of phenolic sweet orange peel extract for controlling fungi responsible for post-harvest fruit decay. Fung Biol 125(2):143–152

Hirano T, Gotoh M, Oka K (1994) Natural flavonoids and lignans are potent cytostatic agents against human leukemia HL-60 cells. Life Sci 55:1061–1069

Hooper L, Kroon PA, Rimm EB, Cohn JS, Harvey I, Le Cornu KA, Ryder JJ, Hall WL, Cassidy A (2008) Flavonoids, flavonoid-rich foods, and cardiovascular risk: a meta-analysis of randomized controlled trials. Am J Clin Nutr 88:38–50

Iglesias-Carres L, Mas-Capdevila A, Bravo FI, Aragonès G, Muguerza B, Arola-Arna A (2019) Optimization of a polyphenol extraction method for sweet orange pulp (*Citrus sinenses* L.) to identify phenolic compounds consumed from sweet oranges. PLoS One 14(1):e0211267. https://doi.org/10.1371/journal.pone.0211267

Irkin R, Dogan S, Degirmencioglu N, Diken ME, Guldas M (2015) Phenolic content, antioxidant activities and stimulatory roles of citrus fruits on some lactic acid bacteria. Arch Biol Sci 67:1313–1321

Karoui IJ, Marzouk B (2013) Characterization of bioactive compounds in Tunisian bitter orange (*Citrus aurantium* L.) peel and juice and determination of their antioxidant activities. BioMed res int 2013:1–12

Kaur C, Kapoor HC (2001) Antioxidants in Fruits and Vegetables—The Millennium's Health. Int J Food Sci Tech 36:703–725

Kelebek H, Canbas A, Cabaroglu T, Selli S (2007) Improvement of anthocyanin content in the cv. Öküzgözü wines by using pectolytic enzymes. Food Chem 105(1):334–339

Keys A (1995) Mediterranean diet and public health: personal reflections. Am J Clin Nutr 61:1321–1323

Ladaniya MS (2008) Commercial fresh citrus cultivars and producing countries. In: Citrus fruit: biology, technology and evaluation. Academic, San Diego, pp 13–65

Lagha-Benamrouchea S, Madania K (2013) Phenolic contents and antioxidant activity of orange varieties(Citrus sinensis L. and Citrus aurantium L.) cultivated in Algeria: peels and leaves. Ind Crop Prod 50(2013):723–730. https://doi.org/10.1016/j.indcrop.2013.07.048

Lanza CM (2003) CITRUS FRUITS I processed and derived products of oranges. Encycl Food Sci Nutr:1346–1354. https://doi.org/10.1016/b0-12-227055-x/00243-1

Lee HS, Coates GA (1999) Vitamin C in frozen, fresh squeezed, unpasteurized, polyethylene-bottled orange juice: a storage study. Food Chem 65(2):165–168

Liew SS, Ho WY, Yeap SK, Sharifudin SAB (2018) Phytochemical composition and in vitro antioxidant activities of Citrus sinensis peel extracts. PeerJ 6:e5331. https://doi.org/10.7717/peerj.5331

Liu K, Xu Y, Wang X (2012a) Microencapsulation of sweet orange oil terpeneless using the orifice method. J Food Eng 110(3):390–394

Liu Y, Heying E, Tanumihardjo SA (2012b) History, global distribution, and nutritional importance of citrus fruits. Compr Rev Food Sci Food Saf 11:530–545

Lipsky GA (1962) Ethiopia: its people, its society, its culture. HRAF Press, New Haven, pp. 376

Manach C, Scalbert A, Morand C, Rémésy C, Jiménez L (2004) Polyphenols: food sources and bioavailability. Am J Clin Nutr 79(5):727–747

Manthey JA, Guthrie N, Grohmann K (2001) Biological properties of citrus flavonoids pertaining to cancer and inflammation. Curr Med Chem 8:135–153

Meléndez-Martínez AJ, Vicario IM, Heredia FJ (2005) Instrumental measurement of orange juice colour: a review. J Sci Food Agric 85(6):894–901

Mhiri N, Ioannou I, Paris C, Ghoul M, MihoubiBoudhrioua N (2016) Comparison of the efficiency of different extraction methods on antioxidants of Maltease orange peel. Int J Food Nutr Sci 3(1):239–251

Mohanta V, Mukherjee I, Chottopadhyay JP (2021) Waste product utilization: preparation of candy from orange (Citrus sinensis) peel. Int J Agric Appl Sci 2(2):114–119

Montero-Calderon A, Cortes C, Zulueta A, Frigola A, Esteve MJ (2019) Green solvents and ultrasound- assisted extraction of bioactive orange (Citrus sinensis) peel compounds. Sci Rep 9:16120

Mostafa RM, Essawy HS (2021) Screening and quantification of bioactive compounds and antimicrobial activities of fresh juice, methanolic peel and pulp extract of Citrus sinensis L. (Sweet orange). Egypt Acad J Biol Sci Microbiol 13(2):1–10

Moulehi I, Bourgou S, Ourghemmi I, Tounsi MS (2012) Variety and ripening impact on phenolic composition and antioxidant activity of mandarin (Citrus reticulate Blanco) and bitter orange (Citrus aurantium L.) seeds extracts. Ind Crop Prod 39:74–80

Nicolosi E, Deng ZN, Gentile A, La Malfa S, Continella G, Tribulato E (2000) Citrus phylogeny and genetic origin of important species as investigated by molecular markers. Theor Appl Genet 100(8):1155–1166

Niu L-Y, Wu J-H, Liao X-J, Chen F, Wang Z-F, Zhao G-H, Hu X-S (2008) Physicochemical characteristics of Orange juice samples from seven cultivars. Agric Sci China 7:41–47

Oikeh EI, Oviasogie FE, Omoregie ES (2020) Quantitative phytochemical analysis and antimicrobial activities of fresh and dry ethanol extracts of Citrus sinensis (L.) Osbeck (sweet Orange) peels. Clin Phytosci 6(1):1–6

Okpala LC, Akpu MN (2014) Effect of orange peel flour on the quality characteristics of bread. Br J Appl Sci Technol 4(5):823

Onibon VO, Abulude FO, Lawal LO (2007) Nutritional and anti-nutritionals composition of some Nigerian fruits. J Food Technol 5:120–122

Ozcan MM, Ghafoor K, Al Juhaimi F, Uslu N, Babiker EE, Mohamed IA, Ahmed IA, Almusallam (2020) Influence of drying techniques on bioactive properties, phenolic compounds and fatty acid compositions of dried lemon and orange peel powders. J Food Sci Technol. https://doi.org/10.1007/s13197-020-04524-0

Paramasivam S, Alva AK, Hostler KH, Easterwood GW, Southwell JS (2000) Fruit nutrient accumulation of four orange varieties during fruit development. J Plant Nutr 23:313–327

Pupin AM, Dennis MJ, Toledo MCF (1999) HPLC analysis of carotenoids in orange juice. Food Chem 64(2):269–275

Roberts JA, Gaddum LW (1937) Composition of citrus fruit juices. Ind Eng Chem 29:574–575

Rockland LB (1961) Nitrogenous constituents. Principal juice constituents. In: Sinclair WB (ed) The orange. Its biochemistry and physiology. University of California, Division of Agricultural Sciences, pp 230–264

Romero-Lopez MR, Osorio-Diaz P, Bello-Perez LA, Tovar J, Bernardino-Nicanor A (2011) Fiber concentrate from orange (Citrus sinensis L.) bagase: characterization and application as bakery product ingredient. Int J Mol Sci 12(4):2174–2186

Roussos PA (2011) Phytochemicals and antioxidant capacity of orange (Citrus sinensis (L.) Osbeck cv. Salustiana) juice produced under organic and integrated farming system in Greece. Sci Hortic 129:253–258

Roussos PA (2016) Orange (Citrus sinensis (L.) Osbeck). Nutr Compos Fruit Cultiv: 469–496. https://doi.org/10.1016/b978-0-12-408117-8.00020-9

Salunkhe DK, Desai BB (1984) Postharvest biotechnology of fruits (vol. I). CRC Press, Florida

Saänchez-moreno C, Plaza L, De Ancos B, Cano MP (2003) Vitamin C, Provitamin A carotenoids, and other carotenoids in high-pressurized orange juice during refrigerated storage. J Agric Food Chem 2003(51):647–653

Seifu M (2003) Status of commercial fruit production in Ethiopia. Ethiopian Agricultural Research Organization, Addis Ababa

Shahidi Bonjar GH, Aghighi S, Karimi Nik A (2004) Antibacterial and antifungal survey in plants used in indigenous herbal-medicine of south east regions of Iran. J Biol Sci 4(3):405–412

Sharon-Asa L, Shalit M, Frydman A, Bar E, Holland D, Or E, Lavi U, Lewinsohn E, Eyal Y (2003) Citrus fruit flavor and aroma biosynthesis: isolation, functional characterization, and developmental regulation of Cstps1, a key gene in the production of the sesquiterpene aroma compound valencene. Plant J 36:664–674

Shimoi K, Masuda S, Furogori M, Esaki S, Kinae N (1994) Radioprotective affect of antioxidative flavonoids in c-ray irradiated mice. Carcinogenesis 15:2669–2672

Sinclair WB, Jolliffe VA (1961) Chemical changes in the juice vesicles of granulated Valencia oranges. J Food Sci 26(3):276–282

Stafussa AP, Maciel GM, Valéria Rampazzo EB, Makara CN, Junior BD, Windson Isidoro Haminiuk C (2018) Bioactive compounds of 44 traditional and exotic Brazilian fruit pulps: phenolic compounds and antioxidant activity. Int J Food Prop:1–38

Stapleton AE, Walbot V (1994) Flavonoids can protect maize DNA from the induction of ultraviolet radiation damage. Plant Physiol 105:881–889

Sulieman AME, Khodari KM, Salih ZA (2013) Extraction of pectin from lemon and orange fruits peels and its utilization in jam making. Int J Food Sci Nutr Eng 3(5):81–84

Tao N-g, Liu Y-j, Zhang M-l (2009) Chemical composition and antimicrobial activities of essential oil from the peel of bingtang sweet orange (Citrus sinensis Osbeck). Int J Food Sci Technol 44(7):1281–1285

Topuz A, Topakci M, Canakci M, Akinci I, Ozdemir F (2005) Physical and nutritional properties of four orange varieties. J Food Eng 66:519–523

Tripoli E, La Guardia M, Giammanco S, Di Majo D, Giammanco M (2007) Citrus flavonoids: molecular structure, biological activity and nutritional properties: a review. Food Chem 104:466–479

Tsuda H, Ohshima Y, Nomoto H et al (2004) Cancer prevention by natural compounds. Drug Metab Pharmacokinet 19(44):245–263

USDA Nutrient Database (2014) United States Department of Agriculture, National Nutrient Database for Standard Reference Release 26, Fruits and Fruit Juices. https://fdc.nal.usda.gov/fdc-app.html#/food-details/169097/nutrients

Walton, B. S., Bartholomew, E. T., Ramsey, R. C. (1945). Analysis of the organic acids of orange juice, Plant Physiol 20 1 3–18

Webber J (1989) The citrus industry. University of California Division of Agricultural Sciences. http://lib.ucr.edu/agnic/webber/

Xiao K, Liu H, Li Y, Yi L, Zhang X, Hu H, Yao H (2018) Correlations between hydrochar properties and chemical constitution of orange peel waste during hydrothermal carbonization. Bioresour Technol 265:432–436

Xu G, Liu D, Chen J, Ye X, Ma Y, Shi J (2008) Juice components and antioxidant capacity of citrus varieties cultivated in China. Food Chem 106(2):545–551

Xu Q, Chen LL, Ruan X, Chen D, Zhu A, Chen C, Bertrand D et al (2013) The draft genome of sweet orange (Citrus sinensis). Nat Genet 45(1):59–68

Yang H, Li Y, Wu M, Zhang ZHE, Li L, Wan S (2011) Plant community responses to nitrogen addition and increased precipitation: the importance of water availability and species traits. Glob Chang Biol 17(9):2936–2944

Zahoor S, Anwar F, Mehmood T, Sultana B, Qadir R (2016) Variation in antioxidant attributes and individual phenolics of citrus fruit peels in relation to different species and extraction solvents. J Chil Chem Soc 61:2884–2889

Zaker MA, Sawate AR, Patil BM, Sadawarte SK (2016) Studies on effect of orange peel powder incorporation on physical, nutritional and sensorial quality of cookies. Int J Eng Res Technol (IJERT) 5:78–82

Chapter 5
Pomelo

Radha Kushwaha, Vinti Singh, Prem P. Kushwaha, and Devinder Kaur

5.1 An Overview

The relationship of humans and nature led to search for different medicinal plants and their possible utilization to cure numerous diseases (Kamarudin et al. 2017). Pomelo (*Citrus maxima or Citrus grandis*) belongs to the family *Rutaceae* (Oboh et al. 2014; Vijayalakshmi and Radha 2016a; Rex et al. 2018). China is considered as native place of *Citrus grandis* (Shah et al. 2019). A variety of names are given to the fruit, which has white or pinkish flesh and a yellow to green skin. Among people from different areas, citrus fruit is delectable because of its sweetish acidic flavor (Zhang et al. 2017). Peel and flesh are two parts of the fruit (Fig. 5.1). Peel is composed of mesocarp or albedo (white, soft) and epicarp or flavedo (colored). The peel comprised maximum percentage of fruit (60.00–65.00%), followed by internal tissues (30.00–35.00%), and seeds (seedless to 10.00%) (Alnaimy et al. 2017). Delicious aroma and taste characterize the fruits. The fruit has maximum water content with small amount of carbohydrate (glucose, sucrose, and fructose), protein and fat. They have also considered as effective source of ascorbic acid and B-complexes (thiamine, niacin, pyridoxines, pantothenic acids, B_2, and B_{12}). Furthermore, citrus fruits provide an immeasurable amount of fiber, which contributes to the prevention of gastrointestinal diseases as well as the reduction of cholesterol in the body. The fruit contains several phytochemicals, like carotenoids, flavonoids, and limonoids (Liu et al. 2012).

R. Kushwaha · V. Singh · D. Kaur (✉)
Centre of Food Technology, University of Allahabad, Allahabad, Uttar Pradesh, India

P. P. Kushwaha
Department of Urology, Case Western Reserve University, Cleveland, OH, USA

S. Singh Purewal et al. (eds.), *Recent Advances in Citrus Fruits*,
https://doi.org/10.1007/978-3-031-37534-7_5

Fig. 5.1 *Citrus grandis* fruit

5.2 Botanical Description

Kingdom	:	Plantae
Order	:	Sapindales
Family	:	*Rutaceae*
Genus	:	*Citrus*
Subgenus	:	*Eucitrus*
Species	:	*Maxima or grandis*
Hindi	:	*Chakotra*
English	:	Pomelo, Pummelo, Thai grapefruit, Chinese grapefruit

The pomelo tree (C. grandis or *C. maxima*) grows two to five meters tall, has an irregular trunk and branches up to five centimeters in diameter and spines up to five centimeters long. Branchlets, limbs, and trunks of young trees are usually angular, soft, short, and densely hairy, with spines usually present. The leaves of these plants range from 5 to 20 cm long, 2 to 12 cm broad, fibrous, light green and hairy other lower side, green and glossy upwards, and they are alternate, ovate-oblong, or elliptic in shape. The petiole can be broadly wingless, or nearly wingless, up to seven centimeters wide, and contains glands that release oil when crushed. Pomelo has a *hesperidium type of fruit*, which is very fleshy, indehiscent berries that range extensively in terms of color, size, shape, and nutritional and physcio-chemical properties of juice. The fruit diameter commonly ranges from 10 to 30 cm and weighs over 1–2 kg when fully ripe. The fruits are globose to pyriform or nearly round to oblate or pear-shaped and are borne singly. The fruit of pomelo is much larger and has thick peels, and it is less juicy than other citrus fruits. A crisp carpellary membrane and juice sac surround the firm flesh. Their flesh is white or pink and they are seeded (Matheyambath 2016). A thick leathery exocarp containing many oil glands forms the rind known as the *flavedo*. A green, greenish-yellow, or pale-yellow stripe with a thickness of 1.25–2.00 cm, peeling could be difficult, clinging is possible, or it is more or less easy to remove. There are many sections within the fleshy endocarp containing stalked pulp and segregated by thin septa. Usually, the endocarp is covered in a white hard tissue called a mesocarp (known as rag or albedo).Greenish-yellow or pale-yellow juice vesicles known as pulp consists of 11–18 segments, with varying degrees of juicy and dryness but often not as acidic as traditional citrus fruit (Purseglove 1974). Depending on the variety, the flavor can be mildly sweet, slightly acidic, or quite sweet.

5.3 Chemical Constituent of Juice, Seed, and Peel

Pomelos are among the most grown and eaten citrus fruits, along with oranges, mandarins, lemons, grapefruits, and grapefruit juice (Tocmo et al. 2020). There are several proven health benefits associated with pomelo (Table 5.1) (Mäkynen et al. 2013). They ensure protection from high blood pressure, hyperglycemic response, and oxidative stress due to bioactive substances (flavonoids, carotenoids, polyphenols, lycopene, fiber, limonoids, and ascorbic acid). Food products that contain pomelo as ingredients have grown in popularity due to their health promotion properties (Reshmi et al. 2020). Pomelo can be eaten fresh or juiced (Tocmo et al. 2020), or pomelo fortified noodles can be beneficial to diabetics (Reshmi et al. 2020). However, researchers have explored alternative methods of utilization of pomelo peels into production of polysaccharides, phytochemicals, and essential oils (Tocmo et al. 2020). A huge amount of agricultural waste is generated because of juice production and fresh fruit consumption. As with other citrus fruit wastes, wet pomelo peel waste consists primarily of water, polysaccharides (hemicellulose and cellulose), soluble sugars, D-limonene, and bioactive components (like flavonoids). Pomelo (*Citrus maxima*) fruit has 87.0% moisture, 1.60% fiber, 11.5% carbohydrate, 0.6% protein 0.20% and 120 mg/100 g vitamin C (Kumar et al. 2019).

Peels were separated from fruits and dried in drier. Physicochemical characterization of peel powder of *Citrus maxima* showed that peel powder had 3.17% minerals, 0.83% acid insoluble ash, 0.67% water soluble ash, and 29.24% alcohol soluble extract along with maximum amount of flavonoids and steroids while low saponins and tannins (Sawant and Panhekar 2017). Pomelo fruit peel essential oils were found in concentrations ranging from 204 mg/100 g to 3119 mg/100 g. Pomelo peel essential oils were mainly composed of monoterpene hydrocarbons (78.19–99.54%) and sesquiterpene hydrocarbons (less than 2.09%) as well as their oxygenated derivatives (less than 9.65%) (Gonzalez-Mas et al. 2019).

It has been reported that the yields of pectin ranged from 3.11% to 39.72% and that the yields were very dependent on the material, the chemical adjuvant, and its

Table 5.1 Nutritional value of pomelo fruit

Components	Juice	Peel	References
Moisture (g/100 g)	83.3	78.0	Czech et al. (2020)
Carbohydrates (g/100 g)	5.86–9.62		Shao et al. (2017), Sawant and Panhekar (2017) and Abobatta (2019)
Dietary fiber (g/100 g)	1.00–1.90		
Fat (g/100 g)	0.04		
Protein (g/100 g)	0.76		
Minerals (g/100 g)	0.4	3.17	
Thiamine (mg/100 g)	0.034		
Riboflavin (mg/100 g)	0.027		
Niacin (mg/100 g)	0.22		
Vitamin B6 (mg/100 g)	0.036		
Vitamin C (mg/100 g)	55.25–61.00		Kumar et al. (2013)
pH	3.58		Basumatary et al. (2020)
TSS (°B)	10.00		
Titratable acidity (g/100 g)	1.57		
Fructose (g/l)	0.20–12.00		Shao et al. (2017) and Gupta et al. (2021a, b)
Glucose (g/l)	0.13–11.00		
Sucrose (g/l)	0.25–50.00		
Calcium (mg/100 g)	14.5	28.8	Sawant and Panhekar (2017) and Czech et al. (2020)
Sodium (mg/100 g)	0.10–1.0	0.68	
Potassium (mg/100 g)	104.00–216.00	127.00	
Phosphorus (mg/100 g)	17.00–18.9	21.9	
Magnesium (mg/100 g)	6.00–19.40	23.00	
Iron (mg/100 g)	0.11–0.46	0.52	
Zinc (mg/100 g)	0.08–0.10	0.12	

concentration, as well as other factors like solid: solvent ratio, extraction time, precipitant agent, and temperature. Fresh peel was rarely used as a starting material, while peel, flavedo, and albedo were commonly used (Xiao et al. 2021). As a by-product of pomelo fruits, pomelo seeds (PS) are extremely valuable. The potential of PS as an antioxidant, antibacterial, and herbicidal agent, as well as its functional components, continue to be a challenge. It is suggested from previous study by Ling et al. (2021) that naringin contributes to the antioxidant activity of seed, and the herbicidal activity could be ascribed with limonoids. Antimicrobial analysis of the water: ethanol extract of pomelo seeds and pulp showed a growth inhibitory effect on B. subtilis, S. aureus and E. coli, which makes it a natural preparation for use as an alternative preservative for food and cosmetic (Sahlan et al. 2018).

5.4 Bioactive Components

Phytochemicals present in citrus fruits have antiviral, antibacterial, antifungal, anticarcinogenic, anti-inflammatory properties, and antithrombotic, as well as lowering cholesterol levels. There are several active compounds in the plant, including ascorbic acid, phenols, flavonoids, carotenoids, sulfides, limonoids, and phytoestrogens. These compounds are antioxidative and have broad health benefits (Kuo 1996). A similar bioactive component is found in pomelo, which also provides a significant amount of protection against hyperglycemia, oxidative stress, and high blood pressure. In the development of functional foods, pomelo segments have become increasingly prominent due to their essential health-promoting properties (Reshmi et al. 2020). Fresh pomelo or juice made from pomelo may be consumed by diabetics (Tocmo et al. 2020), and pomelo fortified noodles also reported beneficial for diabetics (Reshmi et al. 2020). However, researchers have explored alternative methods to reduce postharvest loss of pomelo peels to produce essential oils, polysaccharides, and phytochemicals (Tocmo et al. 2020). A huge amount of agricultural waste is generated because of juice production and fresh fruit intake. As with other citrus fruit wastes, pomelo peel waste consists primarily of water, polysaccharides (hemicellulose and cellulose), D-limonene (lipids), soluble sugars, and bioactive components (like flavonoids).

5.4.1 Ascorbic Acid

As a citrus fruit, pomelo is a rich source of ascorbic acid. According to Haque et al. (2009) and Shao et al. (2017) pomelos have 52.3 and 54.98 mg/100 g of FW of ascorbic acid determined the presence of Vitamin C by dye titration method. Water-soluble vitamin C is helpful for the body's defense because of its antioxidant properties (Turner and Burri 2013). Basically, it's transmitted through collagen, muscle fibers, neurotransmitters, bones, and carnitine biosynthesis because these particles connect them. Consuming ascorbic acid can boost white blood cells and stimulate the immune system (Wintergerst et al. 2006). Pre-eclampsia risk can be reduced with Vitamin C during pregnancy (Chappell et al. 1999). Vitamin C supplementation seems to help reduce cold symptoms (Wintergerst et al. 2006). The results reported by Kumar and Vijay reported a comparative study of ascorbic acid content in different citrus fruits and concluded that the pomelo has lowest amount of ascorbic acid (55.25 mg/100 g) as compared to lemon, orange, and mosambi.

5.4.2 Carotenoids

Tetraterpenoids, or carotenoids, give plants their vibrant yellow, red, and orange colors. Pomelo generally falls into two categories based on their flesh color: white or yellowish flesh, and pink or red flesh. Different combinations of carotenoids produce different colors. Red fleshed pomelo has the most pigments at maturity, but white fleshed pomelo has other carotenoids, which make its pigments less intense. In addition to enhancing the immune response of the plant, carotenoids also enhance the dietary components of the plant (Alquezar et al. 2008). In the fruit, there are carotenoids viz., lycopene, α-and β-carotene, lutein, phytoene, zeaxanthin, and β-cryptoxanthin. Jiang et al. (2019) reported the carotenoid contents of fruits from different pomelo cultivars based on colors and exhibited that cultivar 'Huangjinmiyou' -red fleshed has the maximum amount of β-carotene, following others viz., 'Hongroumiyou' and 'Guanximiyou', while cultivar Hongroumiyou' showed maximum concentration of lycopene. Citrus fruits have a high carotenoid content which contributes to their sensorial and nutritional quality. Lu et al. (2012) Researchers have determined that the color of Huangjinmiyou juice is primarily due to its content of a mixture of carotenoids, including alpha-carotene, phytoene, beta-carotene, and beta-cryptoxanthin, which reach 251.77 grams per pound at full strength, which is 2.6 times greater than Hongroumiyou and 287.7 times more than Guanximiyou. Tao et al. (2010) reported that average yield of carotenoids from pomelo peel was ranged between 30 and 50μg/g. The carotenoids from pumelo peel showed efficacy against bacteria *S. aureus, E. coli,* and *B. subtilis* and a weaker but significant antifungal activity against *S. cerevisiae*, and *R. oryzae*. They had also reported the that the carotenoids extract showed no inhibition effects *on A. niger, A. flavus*, and *P. chrysogenum.* Wu et al. (2021) reported the three major types of carotenoids in mature fruits, the content of which were lycopene, β-carotene, and zeaxanthin in order from high to low from different portion of fruit juice cell, sack coat, and sponge layer. The juice cell, sack coat, and sponge layer contain lycopene (387.65μg/g), β-carotene (66.23μg/g), and zeaxanthin (98.6μg/g); contain lycopene (75.33μg/g), β-carotene (40.68μg/g), and zeaxanthin (6.80μg/g); and contain lycopene (20.74μg/g), β-carotene (26.67μg/g), and zeaxanthin (4.08μg/g), respectively. It was concluded that carotenoids were mainly accumulated in juice sacs, and lycopene was the most important pigment in juice sacs, segment wall and albedo. According to Liu et al. (2016) and Jiang et al. (2019), red-fleshed sweet pomelo (Hongrou) contains high levels of lycopene and β-carotene, with lycopene having significantly greater amounts than β-carotene. Cultivar Huangjin with orange pulp, on the other hand, produced 35 times more β-carotene than lycopene, whereas both compounds were negligible or undetectable in cv. Guanx. As reported by Liu et al. (2016), the main pigments found in Red-fleshed pummelo fruits are lycopene and beta-carotene; however, in Guanxi pummelo fruits, these pigments are too low as they cannot be detected. A study in (2006) by Xu et al. reported the carotenoids in pumelo fruit of cultivar Yuhuan, except for lycopene which was undetectable. Lutein (73.9 ng/g db) and phytoene (73.6 ng/g db) were the main flavonoid compounds, followed by zeaxanthin (21.7 ng/g db) and β-carotene (10.7 ng/g db). However, according to Yan et al. (2018) in cultivar Feicui (pale green-fleshed) only violaxanthin was detected at about 1.0μg/g db. In the flavedo, lutein, violaxanthin, α- and β-carotene were reported in both white and red fleshed pummelo fruits during development.

Its pulp and peel contain carotenoids, which give pomelos their yellow color (Liu et al. 2017a, b). Approximately 115 different carotenoids can be found in citrus fruit peel and pulp. As reported by Singh (2016), there are Aurapte, carotene, Auraptene, 5 Geranyloxy-7-methoxy-coumarin, bergamotti and roseoside, in the peel, as well as 5-methyltodannol, 6-hydroxymethylherniarin, and 5-methoxy seselin in the roots and bark of stem.

5.4.3 Total Phenolics and Flavonoids

Phytochemicals such as phenols play a critical role in preventing stress in plants (Bhattacharya et al. 2010). Khanam et al. (2014) reported a number of phenolic compounds have been isolated from its fruit, including caffeic acid, gallic acid, quercetin and trans-cinnamic acid. Fruit extract (aqueous) contain various flavonoids like naringenin, kaempferol, luteolin and quercetin (35.92%, maximum), whileapigenin and myricetin was absent phenolic acids like chlorogenic acid, 4-Hydroxy-3-methoxycinnamic, 4-Hydroxycinnamic acid, and vanillic acid were unable to detect. The major phenolic acid in pomelo was gallic acid (13.76%, maximum) followed by caffeic acid and trans-cinnamic acid. It is reported that the flavonones present in the pomelo fruit peel are mainly in the form of glycosides and aglycones (hesperetin, eriodictyol, naringenin, diosmetin, nobiletin, luteolin, sinensetin, tangeretin, and apigenin). Among the most abundant flavonones are neoeriocitrin, naringin, neohesperidine, and naringenin (Wang et al. 2008; Li et al. 2014; Xi et al. 2014; Yu et al. 2015; Lu et al. 2016). Ali et al. (2019) and Kawaii et al. (1999) reported flavonoids mainly hesperidin, neohesperidin, naringenin, naringin and rutin in pomelo fruit juice According to the Caengprasath et al. (2013), the bioactive components of pomelo fruit was 25.4 mg/g db neohesperidin, 12.04 mg/g db hesperidin, 11.90 mg/g db naringin and 9.20 mg/g db naringenin. Another study reported by Ding et al. (2013) reported high amount of the flavonoids (26.70–13.43 mg of rutin/g), and total phenolic (42.79–54.56 mg GAE/g). Whereas results reported by Chang and Azlan (2016) in different parts of fruit has different concentration of the total phenolic content like albedo (63.11 mg GAE/g FW) has high amount of phenolics as compared to flavedo (38.67 mg GAE g/1 FW) followed by segment membrane portion (30.66 mg GAE g/1 FW). While Abirami et al. (2014) reported the freshly squeezed juices from white and red pomelos showed good amount of tannins and phenols and were ranged from 333.33–523.21 mg GAE/g to 769.05–909.52 mg GAE/g, respectively. They concluded that the white pomelo has higher amounts of tannins and phenolics as compared to red fruit. Gupta et al. (2021a) reported the total phenolic contents of pomelo juice were ranged from (108.04 to 171.45μg GAE/ml) treated with different concentration of pH and reported in the reduction in their phenolic content as compared to non-treated pomelo juice. Gupta et al. (2021a) reported the total phenolic and flavonoids content of pomelo fruit at different maturity stages. The results showed that the membrane has the highest flavonoid content (1.38 mg QE/g) from other portions of the pomelo fruit growth process. However, pulp had second highest amount of flavonoid content (1233.86μg QE/g) while albedo and flavedo had minimum value. Similarly, TPC was higher in membrane followed by pulp, flavedo, albedo and seeds and minimum concentration of TPC was reported in juices. Essential oils from pomelo peels were characterized with high amount of coumarin, furanocoumarin epoxides, and oxygenated heterocyclic compounds (OHCs), mainly meranzin, oxypeucedanin, 6′,7′-epoxyauraptene and epoxybergamottin (Li et al. 2021).

The flavonoids were found to be at lower concentrations than the TPC, due to their chemical composition, and the concentration varied with the cultivar type (Mansour 2018; Samaniego et al. 2020). Depending on the genepool, the harvest time, and the parts used, pomelo's flavonoids and TPC content varies considerably (Lu et al. 2006; Ersus and Cam 2007).

5.4.4 Naringin Content

The flavanone naringenin is transformed into a glycoside by the disaccharide neohesperidose to form naringin.

Limonin and naringin are bitter compounds in pomelo fruit that affect its taste and reduce its acceptance by consumers (McIntosh and Mansell 1997). In an analysis of all parts of the fruit, Gupta et al. (2021b) found that the albedo had the highest concentration of naringin, followed by petals, peels, juice, and pulp, while the seeds had the lowest concentration. In accordance with McIntosh and Mansell (1997), bitter compounds are distributed differently in different tissues, with naringin being more abundant in the albedo and flavedo parts of the fruit. Naringin distribution is influenced by a number of processes, including transport of precursors from other plant tissues and in-situ metabolism and enzymatic activities. It has also been found that the formation of some compounds, including limonin glucoside during fruit maturation, is related to the formation of limonoid aglycones (Hasegawa et al. 1991). Sudto et al. (2009) worked on the large-scale production of naringin from albedo portion of four different cultivars of pomelo fruit and reported the naringin content ranged from 1.55% to 2.40%.

5.4.5 Essential Oils

This study found that pomelo essential oils consist of the following components: β-myrcene (00.90%), α-pinene (00.40%), β-pinene (3.71%), linalool (00.16%), Sabinine (00.93%), hexanal (00.12%), methyl heptenone (1.25%), t-Ocimine (1.19%), 1-Hexene, 3,3-dimethyl (0.67%), geranylacetate (0.82%) and geranyl formate (1.83%). Oils were analysed using GC-MS to determine their main components and they were in the following concentrations- Z-citral (13.38%), β-farnesene (0.45%), E-citral (17.75%), 4-methyl-1-hexene (15.22%), and DL-limonene

(31.83%). Furthermore, pomelo essential oil inhibited the growth of *A. flavus*. Phytochemical constituent of oils are as follows: E-Citral, 4-methyl-1-hexene, Z-Citral, limnone, Geraniol (Miller et al. 2012; Di Mola et al. 2017; Kittler et al. 2018; Medeiros et al. 2018; Anmol et al. 2021).

5.4.6 Terpenoids

Pomelo also contains terpenoids. The major terpenoids are deacetynomilin, limonin, nomilin glucoside, deoxylimonin, nomilinic acid, obacunone glucosides, and obacunone (Sawant and Panhekar 2017).

5.4.7 Steroids

There were reports of steroids in the peel, root, and fruit of this plant, including campesterol, daucosterol, β-sitosterol, and stigmasterol (Arias and Ramón-Laca 2005; Senguttuvan et al. 2014; Xu et al. 2015).

In addition to these components explained above, pomelo fruit (all the parts) had some more compounds like carbohydrate, amino acids (some essential are leucine, lysine, isoleucine, and tryptophan), acids (like citric acid, malonic, fumaric, succinic, tetra-, penta-, and hexa- decenoic acid) decyl acetate, hexanal, and α-tocopherol (Straka and Belous 2015; Athira 2017; Ani and Abel 2018; Lukitaningsih and Rumiyati 2021).

5.5 Extraction Techniques for Bioactive Compounds

Pomelo bioactive components have been extracted using a variety of techniques summarized in Table 5.2.

5.6 Health Benefits

Pomelo fruit extracts and their isolated compounds have been studied pharmacologically. This plant showed various pharmacological effects on human boy like, antioxidant, antiepileptic, antidepressant, analgesic, insecticidal, antidiabetic, anti-obesity, antimicrobial, anticancer activities, and also reduces the inflammation (Fig. 5.2). Various neurological disorders, including Alzheimer's disease, anxiety, and depression can also be treated effectively with bioactive components of this plant.

Table 5.2 Extraction techniques of bioactive components

Plant parts	Extraction methods	Bioactive components	References
Pink and white pomelo Peel	Solvent extraction	Terpinolene, α-Pinene, β-Pinene, β-Myrcene, Limonene, *trans*-β-Ocimene, Octanal, *p*-Cymene, Citronellal, Decanal, 6-Methyl-5-hepten-2-one, Hexanol, *trans*-2-Heptenal, *cis*-3-Hexen-1-ol, Nonanal, iso-Bornyl acetate, *trans*-2Hexen-1-ol, *trans*-Linalool oxide, *cis*-Linalool oxide, Octylacetate, Octanol, Linalool, Fenchol, *p*-Menthene-8-thiol, β-Caryophellene, α-Terpineol, Germacrene D, Neral, Carvyl acetate, Geranial, *trans,trans*-2,4-Decadienal, Nerol, Geraniol, *trans*-2-Dodecenal, *cis,trans*-2,6-Nonadienal, Perilla alcohol, Dodecanal, 4-Terpinenol, Carvone, Neryl acetate, α-Farnesol, *trans*-Nerolidol, Elemol, β-Sinensal, Indole, Nootkatone, Pentanethiol, *trans*-epoxyOcimene, α-Copaene, Camphor, Citronellylacetate, Carvacrol, Citronellol, Geranyl acetate, Methyl benzoate, Carveol, Benzothiazole, Perilla aldehyde	Cheong et al. (2011)
Peel	Cold-pressing, hydrodistillation, SDE, MAE, UAE, SFE	α- and β-Pinene, Sabinene, β-Myrcene, Octanal, *cis*-β-Ocimene, α-Phellandrene, Limonene, *trans*-β-Ocimene, Linalool, Octanol, *cis*-Linalool oxide, Terpinolene, Nonanal, *p*-Mentha-(*E*)-2,8(9)-dien-1-ol, Citronellal, Nonanol, 4-Terpineol, α-Terpineol, n-Decenal, *trans*-Limonene oxide, *trans*-Piperitol, *cis*-Limonene oxide, Nerol, *cis*-Carveol, Neral, Carvone, Geraniol, Geranial, Perilla aldehyde, Indole, *trans*-Carveol, δ-Elemene, β-Elemene, Dodecanal, trans-Caryophyllene, γ-Elemene, α-Humulene, Dodecenal, Germacrene D, β-Selinene, Neryl acetate, α-Copaene, Geranyl acetate, *cis*-3-Hexenyl hexanoate, Valencene, *trans*-Nerolidol, Bicyclogermacrene, α-Muurolene, α-Farnesene, δ-Cadinene, Germacrene B, Intermedeol, methyl-Hexadecanoate, Farnesol, Nootkatone	Sun et al. (2014)
Peel	Hydrodistillation, microwave assisted extraction	α- and β-Pinene, β-Phellandrene, β-Myrcene, Limonene, Undecane, β-Linalool, *cis*-ocimene,6-Isopropenyl-3-methyl-1-cyclohexen-1-ol, α-Terpineol, (-)-Carveol, Terpene-4-ol, α- and β-Citral, Perillaldehyde, δ-elemene, D, Valencene, (+)-4-Carene, neryl acetate, geranyl acetate, β-Elemene, Methyleugenol, Caryophyllene, β-Cubebene, 2-Isopropenyl-4a, 8-dimethyl-1,2,3,4,4a,5,6,7-octahydronaphthalene,Germacrene β-Patchoulene, β-Panasinsene, Selinene, (-)-α-Panasinsen, β-Neoclovene, Nootkatone, Solavetivone, Osthole	Hosni et al. (2010), Chen et al. (2016), Tuan et al. (2019) and Wu et al. (2017)

(continued)

Table 5.2 (continued)

Plant parts	Extraction methods	Bioactive components	References
Peel	Cold-pressing	*α*-and *β*-Pinene, *β*-Myrcene, *γ*-Elemene, Limonene, Ocimene, Metaraminol, Propionamide, Citral, 4-Carene, Norephedrine, Caryophyllene, Artemisia triene, Cubebene, Cathinone, *β*- Bicyclogermacrene, *β*-Farnesene, 2,6,11,15-Tetramethyl-hexadeca-2,6,8,10,14-pentaene, 7-Methoxy-6-(3-methyl-2-oxobutyl)-2H-1-benzopyran-2-one, 2-(Methylamino)-1-phenylethanol, Copaene	Lan-Phi and Vy (2015) and He et al. (2019)
Pomelo peel	Steam distillation	Mycene, Hexanal, 2-Hexanal, 3-Hexen-1-ol, *α*- and *β*-Pinene, Camphene, D-Limonene, Ocimene, *γ*-Terpinene, *cis*-Linaloloxide, Linalool, Cyclooctanone, *cis*-Limonene oxide, *trans*-Limonene oxide, 1,3,8-*p*-Menthatriene, *β*-Terpilenol, Camphor, Citronellal, 3-Methylenecyclohexene, Terpineol, (+)-Carvone, *cis*-Carveol, Citronellol, Citral, Nerol, Perillaldehyde, 4-Carene, Neryl acetateCaryophyllene, *β*-Elemene, 2,6-Nonadienal, Geranylgeraniol, Germacrene D, *α*-Farnesene, *β*-Farnesene, Isocarveol, Terpilene, Hexanoic acid (hexyl ester), 4-Decyne, Sobrerol, *β*-Bisabolene, Patchoulane, Nerolidol 2, Caryophyllene oxide, 1-(3-methylphenyl)-Ethanone, Santolina triene, Prenyl bromide, Farnesyl acetate, Perillene, Farnesol, Methyl *m*-oxybenzoate, *cis*-Carvone oxide, *p*-*α*-Dimethylstyrene, Thujanol, Nootkatone, *β*-Humulene, Cyclohexene (2-ethynyl-1,3,3-trimethyl), Dihydrocarveol, Myrcenol, *α*-Campholenal, 2-Acetylcyclopentanone, Citronellol formate, 2-Octyne, 2-Methyl-1octen-3-yne, cis-*p*-Mentha-2,8-dien-1-ol, D-Nerolidol, (-)-Spathulenol, 7-methyl-3-Octyne, *γ*-Selinene, Cyclododecyne, Longifolenaldehyde, Alloaromadendrene oxide-(1), Ledol, *p*-Mentha-3,8-diene, Cryptone, Cyclopentene, 1,5,6,7-Tetrahydro-4-Indolene, Valencen, (+)-Ledene, *cis*-*Z*-*α*-Bisabolene epoxide, Isolimonene, Octanoic acid (hexyl ester), *β*-Maaliene, *γ*-Gurjunenepoxide-(1), Verbinone, Limonene-1,2-diol, *trans*-Decalone, Santalol, Laurine, Longipinocarvone, Thunbergol, Cyclohexanepropanol, Dihydromyrcenol, Thujone, 1,3-Cycloheptadiene, 9-Dodecenol, Dihydromyrcene, Toluquinol, *cis*-2,8-Menthadiene-1-ol, *α*-Methylcinnimal, 1,4-Octadiene, Exestrol, Sabinol, *γ*-Picoline 1-oxide, 2-ethoxy-Pyridine, Geranyl acetate	Chen et al. (2018a, b, c) and Guo et al. (2018)

(continued)

Table 5.2 (continued)

Plant parts	Extraction methods	Bioactive components	References
Flavedo (essential oils)	Microwave pretreatment and hydrodistillation	*α*- and *β*-Pinene, *β*-Myrcene, *β*-Ocimene, D-Limonene, *α*Terpineol, Linalool, Benzeneacetaldehyde*cis*-Linaloloxide, furanoid, Farnesal, *trans*-*p*-Mentha-2,8-dienol, *cis*-*p*-Mentha-2,8-dien-1-ol, Terpinen-4-ol, Decanal, Nerol, Carveol, D-Carvone, Citral, Cyclohexadecane, Squalene, Geraniol, Hexyl hexanoate, Butyl caprylate, *β*-Elemene, *β*-Gurjunene, *β*-Cubebene, c-Elemene, c-Muurolene, Valencene, Germacrene D, *α*-Farnesene, Juniper camphor, 7-epi-*α*-Selinene, Nerolidol, Isoaromadendrene epoxide, *trans*-Farnesol, Nootkatone, Hexyl benzoate	Liu et al. (2017a, b)
Flower, leaf, peel and juice		Acetaldehyde, *α*- and *β*-Pinene, *α*-Thujene, Camphene, Ethyl acetate, Sabinene, *α*- and *β*- Phellandrene, *α*-Terpinene, Myrcene, Limonene, *trans*-2-Hexenal, *γ*-Terpinene, *trans*-Ocimene, *para*-Cymene, Terpinolene, 4,8-Dimethyl-1,3,7-nonatriene, *cis*-3-Hexenyl acetate, 6-Methyl-5-hepten-2-one, Hexanol, *trans*-2-Hexenol, Isoterpinolene, *cis*-3-Hexenyl propanoate, *cis*-3-Hexenol, Hexyl butanoate, *para*-Mentha-1,3,8-triene, Ethyl octanoate, *cis*-3-Hexenyl butanoate, *δ*-Elemene, *cis*- and*trans*-Linalool oxide (furanoid), Linalool, Benzaldehyde, *α*-Copaene, Perilla aldehyde, *β*-Elemene, *β*-Caryophyllene,, Terpinen-4-ol, *γ*-Elemene, *cis*-3-Hexenyl hexanoate, trans-*β*-Farnesene, *α*-Caryophyllene, *α*-Terpineol, Neral, Germacrene, *α*-Selinene, cis,trans-*α*-Farnesene, Geranial, Valencene, *α*-Muurolene, Neryl acetate, Carvone, *δ*-Cadinene, Geranyl acetate, *γ*-Cadinene, Citronellol, Geraniol, Nerol, Calamenene, *trans*- and*cis*-Carveol, Perilla alcohol, Phenol, Nerolidol, Methyl anthranilate Geranic acid, Indole, *cis*-3-Hexenyl benzoate, Nonanoic acid, Ethyl palmitate, Nootkatone Acetaldehyde, Ethyl acetate, *α*- and *β*-Pinene, *α*-Thujene, Camphene, Sabinene, *β*-Myrcene, *α*- and *β*-Phellandrene, *α*-and *γ*-Terpinene, Limonene, Decanal, trans-2-Hexenal, *trans*-Ocimene,	Goh et al. (2019)

(continued)

Table 5.2 (continued)

Plant parts	Extraction methods	Bioactive components	References
	HS-SPME	*para*-Cymene, *ortho*-Cymene, Terpinolene, Isoterpinolene, 4,8-Dimethyl-1,3,7-nonatriene, Octanal, *cis*-3-Hexenyl acetate, *cis*-2-Hexenyl acetate, 6-Methyl-5-hepten-2-one, Hexanol, *cis*-Alloocimene, *cis*-3-Hexenyl propanoate, *cis*-3- and*trans*-2-Hexenol, *para*-Mentha-1,3,8-triene, Farnesol, Indole, Ethyl octanoate, *trans*- and *cis*-furanoid, *cis*-3-and *trans*-2-Hexenyl butanoate, δ-Elemene, Hexenyl butanoate, α-Copaene, Linalool, Benzaldehyde, Octanol, β-Elemene, β-Caryophyllene, Ethyl decanoate. Hexyl hexanoate, Terpinen-4-ol, α-Terpineol, γ-Elemene, *cis*-3- and *trans*-2-Hexenyl hexanoate, Geranic acid, *trans*-β-Farnesene, α-Caryophyllene, γ-Muurolene, γ-Selinene, Neral, Germacrene, Valencene, α-Muurolene, Neryl acetate, α-Selinene, *cis,trans*-α-Farnesene, Geranial, Carvone, δ-Cadinene, Citronellol, Geranyl acetate, γ-Cadinene, Perilla aldehyde, Nerol, Calamenene, *trans*- and *cis*-Carveol, Benzyl alcohol, Nerolidol, Geraniol, Nonanoic acid, Nootkatone, Perilla alcohol, Methyl palmitate, α-Eudesmol, Methyl anthranilate, Ethyl palmitate, *trans, trans*-Farnesal	
Pomelo peels	Enzyme- and ultrasound-assisted extraction	Total phenolic content, total flavonoid content, Naringin, hesperidin	Van Hung et al. (2020)

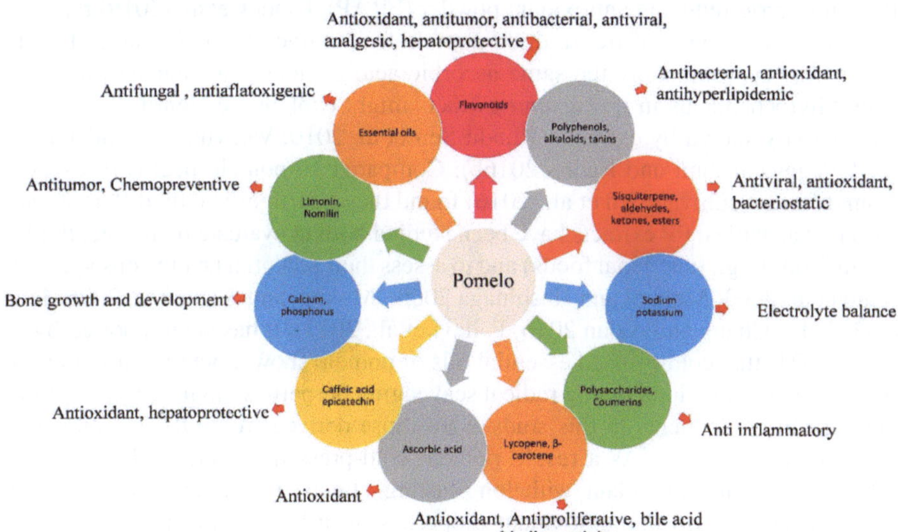

Fig. 5.2 Health benefits of pomelo

5.6.1 Antioxidant Activity (AA)

There are two types of antioxidants: direct antioxidants and indirect antioxidants. In order to maintain antioxidant activity, small molecules must be regenerated or replenished due to their redox activity and chemical modifications. In contrast, a cytoprotective protein is induced by indirect antioxidants (i.e., Phase 2 enzymes) (Dinkova-Kostova and Talalay 2008). Various secondary metabolites in pomelo can act both directly and indirectly as antioxidants (coumarins, flavonoids, carotenoids, phenolic acids, volatile compounds, and organic acids). Various methods were used to determine the antioxidant capacity like 2,2-diphenyl-1-picrylhydrazyl (DPPH), (2,2′-azino-bis(3-ethylbenzothiazoline-6-sulfonic acid) ABTS), and ferric reducing antioxidant power (FRAP).

According to Dulay and Castro (2016) reported the antioxidant activity of leaf extracts of *Citrus maxima, Citrus aurantium* and *Citrus microcarpa*, using 2,2-diphenyl-1-picrylhydrazyl (DPPH) scavenging method. It was concluded that *C. microcarpa* showed the highest DPPH activity (48.67%) followed by *Citrus maxima* (43.51%), and minimum was found in *C. aurantium.* The AA of pomelo fruit cortex and leaves extracts (in ethyl acetate) was performed by two different methods i.e., phosphomolybdenum and DPPH activity method and it was found that in DPPH scavenging activity cortex extract had minimum IC50 value, and in phosphomolybdenum method leaf extracts showed maximum IC50 value (Fidrianny et al. 2016).

The paracetamol-treated rat liver showed less lipid peroxidation than the saline-treated liver when treated with leaf extract at 400 mg/kg body weight. A paracetamol-intoxicated rat liver extract also restored catalase levels and reduced glutathione levels, indicating that extracts have in vivo antioxidant properties (Sen et al. 2011). By using ferric-reducing antioxidant powder (FRAP), Pandey et al. (2019) reported the antioxidant power of freeze-dried pomelo fruit extract was 6609 molμ. Fe^{2+}/L, which was approximately the same ascorbic acid content as in standard ascorbic acid. Phytochemicals in *C. maxima* extracts might explain the significant antioxidant activity shown by extracts (Kundu Sen et al. 2010; Vadivukarasi and Jenitha 2015; Vijayalakshmi and Radha 2016b). Compared 10 pomelo peel (PP) extracts from different cultivars, Lu et al. (2016) found they had significantly different antioxidant capacities. PP extracts have been studied both to evaluate their potential for use in foods (e.g., functional foods) and to assess their potential health consequences (Guo et al. 2003; Mokbel and Hashinaga 2006; Mokbel and Suganuma 2006; Jang et al. 2010; Chang and Azlan 2016; Pallavi et al., 2017). It has been reported by Li et al. (2021) that cold-pressed essential oils of pomelo show potent in vitro antioxidant properties and long-acting radical scavenging properties in vitro. The foremost antioxidant components in this study were those derived from OHCs (coumarins and furanocoumarins). As a result, pomelo cold-pressed essential oil may be an effective natural antioxidant with long-lasting effects. Pichaiyongvongdee et al. (2014) also assessed the antioxidant properties of different part of the fruit from different cultivars and concluded that antioxidant behavior was maximum in the seeds (79.97–85.34%), while for the other parts was -albedo (61.23–66.07%), flavedo (47.03–70.14%) and segment membranes (36.24–45.02%). Cultivar, *Thong Dee* had highest DPPH, and FRAP (antioxidant action) and TPC in the flavedo and seeds as compared to other cultivars.

5.6.2 Antimicrobial Effects

Since microbial contamination has been linked to several food poisoning outbreaks, the food industry has stepped up its efforts to improve the quality and safety of perishable foods. In addition to conventional antimicrobials, alternatives have also been explored, notably those that fall under the label of "green" (ex. Extracts from plants) (Guo et al. 2018). In order to combat these, PP waste-derived essential oils were recently evaluated as natural antimicrobial agents. It has been widely studied that pomelo has antibacterial properties.

Das et al. (2013) reported the antibacterial activity against *E. coli* and *Pseudomonas aeruginosa* from the ethanol-based extract of leaves of *Citrus maxima.* While Sahlan with other researcher in 2018 reported the antibacterial effect of ethanolic extracts of pulp and seed against *S. aureus*, *Bacillus subtilis*, and *E. coli* using disc diffusion method. Furthermore, a study performed by Singh and Navneet (2016) reported the antifungal activity of essential oil extracted from *C. sinensis and C. grandis*. Similarly, Chen et al. (2018a, b, c) investigated the antibacterial activity of essential oils of pomelo fruit against *E. coli, S. aureus, Bacillus subtilis, P. aeruginosa, Bacillus altitudinis and Bacillus licheniformis* using broth dilution method. The extract (methanolic) different part of pomelo tree showed an antifungal and aflatoxins activity against *E. coli, Staphylococcus aureus* and *Klebsiella pneumonia* (Yathiender 2017). Another study reported that naringenin and hesperidin (Gram-positive and Gram-negative) of fruit might explain its antibacterial action (Corciova et al. 2015). Similarly, Jing et al. (2014) reported that limonene present in essential oils of fruit were effective against fungus like; *A. fumigatus*, *A. flavus*, *A. terreus*, *Aspergillus niger*, *P. italicum*, *A. parasiticus*, *P. expansum*, *Penicillium chrysogenum*, *Alternaria alternata, P. digitatum*, *F. proliferatum*, and *Foxysporum.*

Ethanolic extracts of *Citrus grandis* peels were demonstrated to be effective against some kinds of mold reported a data in 2004 by Bijun. Pomelo seed extract of was found to have antifungal activity against *Aspergillus niger* and *Candida albicans* by Singh and Navneet (2016). It was found that methanolic (1:3 w/v) extract of seeds had a minimum inhibition zone of 7.66 plus 0.32 mm and MICs of 3.12–25 mg/mL against *Candida albicans*. Pomelo research provides insight into potential food applications, especially in food safety, based on its antimicrobial properties.

5.6.3 Anti-inflammatory Activity

During healing, inflammation plays an essential role in protecting the body from foreign invaders and protecting it from injury. *Citrus grandis* may be beneficial to the reduction of inflammation due to its polysaccharides (Chen et al. 2017, 2018a, b, c; Puglisi et al. 2017; Zhao et al. 2019; Fang et al. 2020). *Citrus grandis* methanolic (90%) and ethyl acetate extracts demonstrated anti-inflammatory effects using an animal model in Zhao et al. (2019). All parts of *C. maxima* have been considered to have pain-relieving and inflammation-reducing capabilities. An animal model of acute inflammation, carrageenan-induced paw edema, demonstrated the anti-inflammatory effects of pomelo peel extract. A formalin-induced licking and biting model and an acetic acid-induced writhing model were used to evaluate the pain-relieving properties of the peel methanolic extract. As compared to the standard drug diclofenac sodium at 10 mg/kg, the extract at a higher dose (500 mg/kg) showed satisfactory pain-relieving activity (73.34%) in the acetic acid-driven pain model (Liu et al. 2018). PP-derived pure compounds were also found to inhibit the production of the lysosomal enzyme elastase from activated neutrophils.

The effects of isomeranzin, epoxybergamottin, and 4-hydroxybenzaldehyde on fMLP/cytochalasin B-induced elastase release in human neutrophils were found to be potent (IC50 0.43–4.33 M) (Kuo et al. 2017). Other studies have demonstrated that 18 coumarins (mainly 6,7-furocoumarins, but also substituents of 7,8- and 7,7-furocoumarins) are anti-inflammatory (Zhao et al. 2019). Anmol et al. (2021) summarized several studies that demonstrated pomelo peels' anti-inflammatory properties are due to coumarins. The fruit's polysaccharide inhibited chronic inflammation in rabbits exposed to ammonia as well as clinical patients in a study reported by Chen et al. (2018a, b, c). Furthermore, the tail-flick method in rats, acetic acid induced writhing in mice, and the hot plate method in rats were used to compare the pain-relieving properties of different parts of the pomelo tree. A 300 mg/kg leaf extract reported the considerable amount of pain-relieving activity across all tested models (Shivananda et al. 2013). Similarly, Kundu Sen et al. (2011a, b, c) demonstrated its anti-inflammatory properties in acute rat paw edema models induced by dextran, carrageenan, and formalin.

5.6.4 Anticancer Activity

Citrus grandis natural compounds are being investigated for their ability to suppress cancer growth. Through different complex biochemical reactions, antioxidants neutralize radicals produced in the body, which are responsible for many serious and life-threatening diseases, such as cancer (Tsai et al. 2007, 2017; Ali et al. 2019). In particular, limonoids present in *Citrus grandis*, such as limonin and nomilin, are reported to exhibit anticancer properties, such as free radical scavenging and the induction of glutathione S-transferase (Yu et al. 2015; Poulose et al. 2005).

Intraperitoneal administration of *Citrus grandis* methanolic extract increased the longevity of non-viable tumor cells, decreased their count, and decreased tumor volume (Kundu Sen et al. 2011a, b, c).

Several studies have demonstrated the anticancer effects of pomelo peel (PP) extracts. There was a dose-dependent antiproliferative effect (0–400 g/mL) of supercritical CO_2 (SCO_2) extracts from whole grapefruit fruits (pulp + peel) on 4 different cancer cell lines, including human glioblastoma U373MG, gastric cancer SNU-16, cervix adenocarcinoma cells HeLa, and gastric adenocarcinoma cell line (AGS) (Gyawali et al. 2012). The cell lines showed different levels of sensitivity to the extracts, with SNU-16 showing the highest inhibition (40%) and AGS gastric adenocarcinoma showing the lowest inhibition (20%). In another study, ethanol–water extracts (0–100 mg/mL) of PP showed antiproliferative effects in four cancer cell lines including MCF-7 (human breast cancer cell line), A549 (human lung cancer cell line), HT-29 (human colon cancer cell line) and HepG2 (human hepatoma cell line). Cell lines with similar inhibition levels (90–94%) were also investigated by Li et al. (2013).

The extracts of PP contain lipid-soluble bioactive components (caryophyllene oxide, limonene, auraptene, -terpineol and hexadecenoic acid) and phenolic acids (tanteretin and nobiletin) that showed antiproliferative effects (Gyawali et al. 2012). Gyawali et al. (2012) and Li et al. (2013) did not use any positive controls (reference compounds with antiproliferative properties). In the absence of samples, cells that received culture medium containing DMSO were regarded as negative controls. Additionally, neither study used normal cells in order to establish that the extracts are only effective against cancerous cells. A number of studies have demonstrated the potency of Naringin as an anticancer agent. A 10–35 mg/kg intraperitoneal naringin treatment on Walker 256 carcinosarcoma-bearing rats (W256) showed a 75% reduction in tumor growth, and TNF-α and IL-6 levels were also reduced compared to controls (Camargo et al. 2012; Sharma et al. 2019). SK-MEL-28 human melanoma cells and B16F10 murine cancer cells were also able to respond to naringenin by proliferation and migration. A mouse colon carcinogenesis triggered by Azoxymethane (AOM) was prevented by hesperidin. Multiple AOM-induced aberrant crypt foci (ACF) and tumor incidence were significantly reduced by this hesperidin. Moreover, AOM-induced colon carcinogenesis was inhibited by lowering the proliferating cell nuclear antigen (PCNA) (Saiprasad et al. 2013; Stanisic et al. 2018). In regard to cancer prevention, limonoids, flavonoids, bioflavonoids, alkaloids, saponins, and tannins play a prominent role (Khanam et al. 2014; Kundu Sen et al. 2011a, b, c; Nair et al. 2018). Human airway epithelium-derived basal cells (BCi-NS1.1) and lung epithelial carcinoma (A549) were tested for their anticancer effects with naringenin-loaded liquid crystalline nanoparticles (LCNs). The results showed that naringenin LCNs were effective for inhibiting the migration and proliferation of cells (Wadhwa et al. 2021).

5.6.5 Anti-glycation

A variety of studies have investigated the anti-glycation properties of natural compounds. Several studies have reported that grapefruit extracts have antioxidative effects and fructose-mediated nonenzymatic glycation to bovine serum albumin (BSA) resulting in AGE (advanced glycation end products) formation in the early stages (amadori products), middle stages (protein oxidation) and last stages (cross-linking compound) (Brownlee 1995; Wu et al. 2011). Based on evidence from a study, *Citrus grandis* pulp methanolic extract decreases fructosamine levels, thus reducing production of AGEs and CMLs (Chronic myelogenous leukemia). *Citrus grandis* pulp methanolic extract inhibited protein oxidation and reduced the formation of protein carbonyls when thiol groups were lost, suggesting potential uses as anti-glycation agents and to prevent diabetic complications and aging (Caengprasath et al. 2013).

5.6.6 Antidiabetic Activity

An in vitro study was conducted to examine the inhibitory activity of the fruit juice of *C. maxima* against the enzymes α-glucosidase and α-amylase. Inhibition percentage by fruit juice against α-glucosidase and α-amylase were 70.68–72.83% and 75.55–79.75%, respectively. A streptozotocin (STZ)-induced diabetes mellitus model was used to study fruit juice's hypoglycemic properties. Experimentation reduced glucose levels compared with controls by peripheral glucose utilization or inhibition of gluconeogenic enzymes (Oyedepot and Babarinde 2013). Using glibenclamide (0.5 mg/kg, p.o.) as the reference standard, leaf extracts (200 and 400 mg/kg, b.w.) were evaluated for their antidiabetic activity in STZ-induced diabetic rats (65 mg/kg). Experimental rats had normal blood glucose levels and serum biochemical parameters compared to controls (Kundu Sen et al. 2011a, b, c). Neohesperidin improved postprandial hyperglycemic conditions via positive effects on amylase and glucose (Sinha et al. 2019). As a result of their antioxidant activity, plants may offer protection against chronic metabolic disorders (Reshmi et al. 2018). A methanolic and an ethanolic leaf extract of *C. maxima* (100 and 200 mg/kg, respectively) also exhibited similar results against the alloxan stimulated diabetes micemodel compared to the standard of glibenclamide. Experimental mice were assessed for plasma glucose level, serum lipid profile, C-reactive protein (CRP), glutamic oxaloacetic transaminase (SGOT) and glutamic pyruvic transaminase (SGPT) inhibition, and the leaf extract inhibited all these parameters compared to controls. Based on these findings, both extracts can significantly reduce glucose levels in diabetic mice and improve their lipid profiles. In addition, reports indicated that extracts from *C. maxima* leaves might improve liver function enzymes and CRP levels in diabetic mice, suggesting that this plant has hepatoprotective and cardioprotective properties (Islam et al. 2021).

5.6.7 *Anti-obesity Activity*

A study on the effects of ethanolic extract of pomelo leaves (200 and 400 mg/kg) on the prevention of obesity in rats induced by olanzapine and cafeteria diets. Diensh and Hegde (2016) evaluated and reported that serum profile, body weight and temperature, were considerably lower than those in the obese population (control). In 2013, Ding et al. (2013) administered pomelo peel extracts (ethanolic) to mice followed by chow diet till 8 weeks. As a result of the diet, the weight was reduced, fasting blood glucose stages decreased, liver lipid amount decreased, and serum insulin stages were also decreased (Ding et al. 2013). NF-B signaling is indirectly facilitated by hesperidin that helps to control inflammation and regulate lipid and glucose metabolism (Akiyama et al. 2009; Xiong et al. 2019).

5.6.8 *Hepatoprotective and Nephroprotective Activity*

The fruit of *Citrus grandis* has been used to prevent hepatitis and kidney damage. *C. maxima* extracts protected Wistar rats against carbon tetrachloride-induced liver toxicity. Alanine transaminases, alkaline phosphatases, and aspartate aminotransferases levels were significantly reduced in experimental rats, demonstrating its hepatoprotective effects (Feksa et al. 2018; Chowdhury et al. 2015). In additional research, the hepatoprotective effects of methanolic extract of pomelo leaves (200 mg/kg, body weight) were studied in rats induced by paracetamol-induced liver injury. An oral leaf extract was given for 7 days, paracetamol (2 g/kg) on the day fifth, and silymarin (100 mg/kg, body weight) at the last day of experiment. In comparison to the control, markers of liver function, hepatic antioxidant levels, total serum protein, and bilirubin, were all normal in the liver homogenate (Tabeshpour et al. 2020). The antioxidant activity of leaf extracts may be responsible for decreasing hepatocyte distortion by raising the levels of hepatic antioxidant enzymes (Abirami et al. 2015).

Citrus grandis juice reduced plasma triglyceride levels in a dose-dependent manner in a experiment conducted by Oboh et al. (2014). Due to its flavonoid content, flavonoids inhibit hepatic apolipoprotein (apo) B secretion, thus providing hepatoprotection (Lin et al. 2011).

5.7 Products and by Products

Numerous countries have documented the ethnomedical benefits of the pomelo fruit and its parts (Kundu Sen et al. 2011a, b, c; Chaudhari et al. 2016; Singh and Navneet 2016, 2017). Among the health benefits of fruits are stomach tonics, appetizers, cardiac stimulants, antipyretics, and antiemesis (Thavanapong et al. 2010; Borah et al. 2012; Sidana et al. 2013; Vijayalakshmi and Radha 2016b). The use of pulp for cosmetic purposes has been around for centuries. In addition to treating lumbago, dyspepsia, and coughs, the seeds can be used for other conditions as well. A decoction of the leaves is useful for swellings and ulcers, while the leaves are useful for epilepsy, cholera, and convulsive cough (Guo et al. 2003; Das et al. 2013).

In today's world, pulp, seeds, and peel from citrus fruits are used to make many commercially beneficial products, like as an ingredients for the product development, essential oils, pectin, enzymes, natural antioxidants, and also in packing industries in form of films.

5.8 Value Addition

Pomelo has been used to make nutritional supplements, such as noodles made from 30% fresh fruit segments and 5% dry segments. It is believed that these noodles have high acceptance people with diabetes as well as the healthy ones (Reshmi et al. 2020). Pomelo peel (50% dietary fiber) was added to a powder to create a high-fiber food. A study on rice noodles enriched with PP has been published (Wandee et al. 2014). High-rich ingredients can increase the fiber content in certain foods by incorporating PP in this manner. By stimulating the growth of good microorganism in the gastrointestinal tract (acting as prebiotics), high fiber foods can promote gut health (Slavin 2013). The polyphenols in PP are also known to play a significant role in human health, in addition to being a high fiber ingredient (Zain et al. 2014). Adding mango peel flour to soft dough biscuits and macaroni can improve the antioxidant activity of food products (Ajila et al. 2008, 2010). The antioxidant profile of food products enriched with PP has not been evaluated yet. Tea and tea additives, as well as candies and marmalade have been made using PP (Ihishara et al. 2011; Lin et al. 2012). Despite the limited scientific literature available on such food applications, grapefruit is widely used in pomelo-growing Asian countries. Comparatively to pectin and EOs, PP has yet to be fully explored for other food-related purposes.

5.8.1 Pectin

The compound pectin is widely accepted as a functional ingredient as it can emulsify, gel, stabilize, and texturize (Mesbahi et al. 2005). Besides its application in tissue engineering and drug delivery, pectin also has a variety of biomedical uses (Noreen et al. 2017). As a source of pectin, fruit peels represent about 20% of the fruit's total pectin content (Chavan et al. 2018). Pomelo has a recommendable amount of natural pectin. The pectin content of pomelo peel was 23.19% according to Methacanon et al. (2014). Roy et al. (2018) found that PP contain pectin, so they extracted it from the peels to make carrot jam.

5.8.2 Essential Oil (EO)

Essential oils are naturally occurring mixtures of aromatic oils that are volatile, complex, and obtained from plants (Figueiredo et al. 2008). Essential oils like citrus have been approved for safe public consumption for decades and are commonly established to deliver a pleasant aroma. There are many applications for essential oils throughout the world, including cosmetics, perfumery, toiletries, flavors, beverages, pharmacology, and other special hygiene items (Palazzolo et al. 2013; Sarkic and Stappen 2018; Mahato et al. 2019). There are approximately 299 recognized volatile compounds in pomelo peel (PP). In addition to volatile compounds, terpenoids make up a significant part of this group (189 volatiles, 63.2%). In addition to diterpenoids, monoterpenoids, monocyclic- and bicyclic- monoterpenoids, acyclic-, bicyclic-, and tricyclic- sesquiterpenoidsa wide variety of chemicals are released. Nonterpenoid hydrocarbons (5.0%), nonterpenoid alcohols (4.7%), esters (8.0%), and nonterpenoid aldehydes (6.0%) are other major volatile compounds in PP essential oils. Of the total volatile compounds, 11.7% (35 volatiles) were unknown. Sesquiterpenoids (17–29) and monoterpenoids (1–16) are the most prominent components of PP EOs. A variety of essential oils can be used to inhibit the growth of mold and fungi.

5.8.3 Enzyme Production

Citrus peels can be used for a variety of purposes, including the production of pectinolytic enzymes. Using pomelo peels as a substrate, *A. niger* LFP-1 was investigated by solid-state fermentation (SSF) (Darah et al. 2013). It has been demonstrated that pomelo pericarp powder is an effective substrate for *Aspergillus oryzae* JMU316 (Hussain et al. 2019). Solid state fermentation produces pectinase enzyme from pomelo t peels by *Aspergillus niger*. A number of processes such as producing sweetener precursors, preparing prunin, increasing the aroma of wine, and converting antibiotics into rhamnose require naringinase, which can be obtained from pomelo fruit (Chen et al. 2010).

5.8.4 Packaging Film Formation

Scientists have become increasingly interested in biodegradable, sustainable plastics. Researchers made plastic films using citrus peels. It is possible to make biodegradable packaging materials from citrus peels (Shinde et al. 2019). According to Das et al. (2018) chicken feather keratin blended with pomelo peel pectin to create a sustainable composite film to pack the fish fillets (fried) showed positive results by lowering the hardness, moisture loss (to maintain weight), and microbial counts on the surface.

5.8.5 *As an Adsorbent*

There is a growing concern about environmental pollution from a variety of industries, including textiles, paper, printing, and leather. Chemical dyes are discharged into water bodies during manufacturing processes such as dyeing and finishing, potentially polluting domestic, agricultural, and industrial water resources (Lellis et al. 2019; Yaseen and Scholz 2019). The PP powder was shown to be capable of absorbing azo dyes (such as RB 114 dye) with a highest (16 mg/g) adsorption power (Argun et al. 2014). Rather than physical adsorption to pores, recommended that the chemical bonds were supreme for adsorption. This mechanism is confirmed by the functional groups found on the surface of peels (such as carboxylic acids, hydroxyl groups, and polysaccharides containing OH groups). Aside from the negatively charged surfaces of PP powder, it is electrostatically attracted to positively charged dye cations at a wide range of pH (2–11) (Hameed et al. 2008; Argun et al. 2014). PP has also been proven to be an effective dye adsorbent in several other studies. PP activated carbon exhibits high dye adsorption capacity due to microporous and mesoporous structures on its surface (Hameed et al. 2008; Foo and Hameed 2011; Li et al. 2016; Nowicki et al. 2016). Chemical leaks and oil spills also pose serious threats to the environment and the water ecosystem (Kingston 2002). It is imperative to remove these pollutants from the environment to prevent damage to it. Spillage of oils and different most occurred organic pollutants could be absorbed with peels treated with acetic anhydride ($C_4H_6O_3$) and styrene (C_8H_8) as well as porous carbon aerogel of peels (Chai et al. 2015; Zhu et al. 2017). Zhu et al. (2017) also reported that carbon aerogels exhibit high sorption capacity and recyclability, making them suitable to adsorb the oil spillage.

5.9 Conclusion

Pomelo is a great source of vitamins, minerals, and fiber. For the control of chronic diseases such as diabetes, cholesterol, obesity, cardiovascular disease, and some types of cancer, bioactive and non-nutrient compounds such as saponins, flavonoids, phenols, cardiac glycosides, carotenoids, limonoids, acridone, glycosides, alkaloids, essential oils, steroids, and alkaloids are beneficial. Among the bioactive properties of components like limonene, hesperidin, naringenin, neohesperidin, and naringin are antioxidants, antidepressants, antitumors, anticancers, antimicrobials, hepatoprotective, antiobesity, relieving pain, insecticidal, as an ataractic agent, prevent from alzheimer's, regulates blood glucose level and helpful to treat ulcer. *Citrus grandis* provides health benefits to humans, and a further investigation into its various parts, such as the pulp, peels, leaves, seeds, and essential oils, may help to discover new health benefits.

References

Abirami A, Nagarani G, Siddhuraju P (2014) In vitro antioxidant, anti-diabetic, cholinesterase andtyrosinase inhibitory potential of fresh juice from *Citrus hystrix* and *C. maxima* fruits. Food Sci Hum Wellness 3:16–25

Abirami A, Nagarani G, Siddhuraju P (2015) Hepatoprotective effect of leaf extracts from *Citrus hystrix* and *C. maxima* against paracetamol induced liver injury in rats. Food Sci Human Wellness 4:35–41

Abobatta WF (2019) Nutritional benefits of citrus fruits. Am J Biomed Sci Res 3:303–306

Ajila CM, Leelavathi KUJS, Rao UP (2008) Improvement of dietary fiber content and antioxidant properties in soft dough biscuits with the incorporation of mango peel powder. J Cereal Sci 48:319–326

Ajila CM, Aalami M, Leelavathi K, Rao UP (2010) Mango peel powder: a potential source of antioxidant and dietary fiber in macaroni preparations. Innov Food Sci Emerg Technol 11:219–224

Akiyama S, Katsumata SI, Suzuki K, Ishimi Y, Wu J, Uehara M (2009) Dietary hesperidin exerts hypoglycemic and hypolipidemic effects in streptozotocin-induced marginal type 1 diabetic rats. J Clin Biochem Nutr 46:87–92

Ali M, Rumpa NEN, Paul S, Hossen M, Tanvir EM, Hossan T et al (2019) Antioxidant potential, subacute toxicity, and beneficiary effects of methanolic extract of pomelo (*Citrus grandis* L. Osbeck) in long Evan rats. J Toxicol 2019:2529569

Alnaimy A, Gad AE, Mustafa MM, Atta MAA, Basuony HAM (2017) Using of citrus by-products in farm animals feeding. Open Access J Sci 1:58–67

Alqu'ezar B, Rodrigo M, Zacar'ıas L (2008) Carotenoid biosynthesis and their regulation in citrus fruits. Tree For Sci Biotechnol 2:23–37

Ani PN, Abel HC (2018) Nutrient, phytochemical, and antinutrient composition of *Citrus maxima* fruit juice and peel extract. Food Sci Nutr 6:653–658

Anmol RJ, Marium S, Hiew FT, Han WC, Kwan LK, Wong AKY et al (2021) Phytochemical and therapeutic potential of *Citrus grandis* (L.) Osbeck: a review. J Evid-Based Integr Med 26:2515690X211043741

Argun ME, Güclü D, Karatas M (2014) Adsorption of reactive blue 114 dye by using a new adsorbent: pomelo peel. J Ind Eng Chem 20:1079–1084

Arias BA, Ramón-Laca L (2005) Pharmacological properties of citrus and their ancient and medieval uses in the Mediterranean region. J Ethnopharmacol 97:89–95

Athira U (2017) Evaluation of carbohydrate and phenol content of citrus fruits species. Int J Appl Res Nat Prod 3:160–164

Basumatary B, Nayak PK, Chandrasekar CM, Nath A, Nayak M, Kesavan RK (2020) Impact of thermo sonication and pasteurization on the physicochemical, microbiological and anti-oxidant properties of pomelo (*Citrus maxima*) juice. Int J Fruit Sci 20:S2056–S2073

Bhattacharya A, Sood P, Citovsky V (2010) The roles of plant phenolics in defense and communication during Agrobacterium and Rhizobium infection. Mol Plant Pathol 11:705–719

Borah M, Ahmed S, Das S (2012) A comparative study of the antibacterial activity of the ethanolic extracts of *Vitex negundo* L., *Fragaria vesca* L., *Terminalia arjuna* and *Citrus maxima*. Asian J Pharm Biol Res 2:183–187

Brownlee MDM (1995) Advanced protein glycosylation in diabetes and aging. Annu Rev Med 46:223–234

Caengprasath N, Ngamukote S, Mäkynen K, Adisakwattana S (2013) The protective effects of pomelo extract (*Citrus grandis* L. Osbeck) against fructose-mediated protein oxidation and glycation. EXCLI J 12:491–502

Camargo CA, Gomes-Marcondes MCC, Wutzki NC, Aoyama H (2012) Naringin inhibits tumor growth and reduces interleukin-6 and tumor necrosis factor α levels in rats with Walker 256 carcinosarcoma. Anticancer Res 32:129–133

Chai W, Liu X, Zou J, Zhang X, Li B, Yin T (2015) Pomelo peel modified with acetic anhydride and styrene as new sorbents for removal of oil pollution. Carbohydr Polym 132:245–251

Chang SQ, Azlan A (2016) Antioxidant content and activity in different parts of pomelo (*Citrus grandis* (L.) Osbeck) by-products. Acta Hortic 2016:27–34

Chappell LC, Seed PT, Briley AL, Kelly FJ, Lee R, Hunt BJ et al (1999) Effect of antioxidants on the occurrence of pre-eclampsia in women at increased risk: a randomised trial. Lancet 354:810–816

Chaudhari SY, Ruknuddin G, Prajapati P (2016) Ethno medicinal values of citrus genus: a review. Med J DY Patil Univ 9:560

Chavan P, Singh AK, Kaur G (2018) Recent progress in the utilization of industrial waste and by-products of citrus fruits: a review. J Food Process Eng 41:e12895

Chen D, Niu T, Cai H (2010) Optimizing culture medium for debittering constitutive enzyme naringinase production by *Aspergillus oryzae* JMU316. Afr J Biotechnol 9:4970–4978

Chen Q, Hu Z, Yao FYD, Liang H (2016) Study 95 of two-stage microwave extraction of essential oil and pectin from pomelo peels. LWT Food Sci Technol 66:538–545

Chen L, Lai Y, Dong L, Kang S, Chen X (2017) Polysaccharides from *Citrus grandis* L. Osbeck suppress inflammation and relieve chronic pharyngitis. Microb Pathog 113:365–371

Chen GW, Lin YH, Lin CH, Jen HC (2018a) Antibacterial activity of emulsified pomelo (*Citrus grandis* Osbeck) peel oil and water-soluble chitosan on *Staphylococcus aureus* and *Escherichia coli*. Molecules 23:840

Chen X, Lai Y, Song X, Wu J, Wang L, Zhang H et al (2018b) Polysaccharides from *Citrus grandis* associate with luteolin relieves chronic pharyngitis by anti-inflammatory via suppressing NF-κB pathway and the polarization of M1 macrophages. Int J Immunopathol Pharmacol 32:2058738418780593

Chen Y, Li T, Bai J, Nong L, Ning Z, Hu Z et al (2018c) Chemical composition and antibacterial activity of the essential oil of *Citrus maxima* (Burm.) Merr. Cv. Shatian Yu. J Biol Act Prod Nat 8:228–233

Cheong MW, Liu SQ, Yeo J, Chionh HK, Pramudya K, Curran P, Yu B (2011) Identification of aroma-active compounds in Malaysian pomelo (*Citrus grandis* (L.) Osbeck) peel by gas chromatography-olfactometry. J Essent Oil Res 23:34–42

Chowdhury MRH, Sagor MAT, Tabassum N, Potol MA, Hossain H, Alam MA (2015) Supplementation of *Citrus maxima* peel powder prevented oxidative stress, fibrosis, and hepatic damage in carbon tetrachloride (CCl4) treated rats. Evid-Based Complement Altern Med 2015:598179

Corciova A, Ciobanu C, Poiata A, Mircea C, Nicolescu A, Drobota M et al (2015) Antibacterial and antioxidant properties of hesperidin: β-cyclodextrin complexes obtained by different techniques. J Incl Phenom Macrocycl Chem 81:71–84

Czech A, Zarycka E, Yanovych D, Zasadna Z, Grzegorczyk I, Kłys S (2020) Mineral content of the pulp and peel of various citrus fruit cultivars. Biol Trace Elem Res 193:555–563

Darah I, Taufiq MMJ, Lim SH (2013) Pomelo *Citrus grandis* (L.) osbeck peel as an economical alternative substrate for fungal pectinase production. Food Sci Biotechnol 22(6):1683–1690

Das S, Borah M, Ahmed S (2013) Antibacterial activity of the ethanolic extract of leaves of *Citrus maxima* (Burm.) Merr. on Escherichia coli and Pseudomonas aeruginosa. Asian J Pharm Clin Res 6:136–139

Das P, Borah PP, Badwaik LS (2018) Transformation of chicken feather keratin and pomelo peel pectin into biodegradable composite film. J Polym Environ 26:2120–2129

Di Mola A, Massa A, De Feo V, Basile A, Pascale M, Aquino RP, De Caprariis P (2017) Effect of citral and citral related compounds on viability of pancreatic and human B-lymphoma cell lines. Med Chem Res 26:631–639

Dinesh SS, Hegde K (2016) Antiobesity activity of ethanolic extract of *Citrus maxima* leaves on cafeteria diet induced and drug induced obese rats. Res J Pharm Technol 9:907–912

Ding X, Guo L, Zhang Y, Fan S, Gu M, Lu Y et al (2013) Extracts of pomelo peels prevent high-fat diet-induced metabolic disorders in c57bl/6 mice through activating the PPARα and GLUT4 pathway. PLoS One 8:e77915

Dinkova-Kostova AT, Talalay P (2008) Direct and indirect antioxidant properties of inducers ofcytoprotective proteins. Mol Nut Food Res 52:S128–S138

Dulay RMR, Castro MEGD (2016) Antibacterial and antioxidant activities of three citrus leaves extracts. Der Pharm Lett 8:167–170

Ersus S, Cam M (2007) Determination of organic acids, total phenolic content, and antioxidant capacity of sour *Citrus aurantium* fruits. Chem Nat Compd 43:607–609

Fang J, Cao Z, Song X, Zhang X, Mai B, Wen T et al (2020) Rhoifolin alleviates inflammation of acute inflammation animal models and LPS-induced RAW264. 7 cells via IKKβ/NF-κB signaling pathway. Inflammation 43:2191–2201

Feksa DL, Coelho RP, da Costa Güllich AA, Dal Ponte ES, Piccoli JDCE, Manfredini V (2018) Extract of *Citrus maxima* (pummelo) leaves improve hepatoprotective activity in Wistar rats submitted to the induction of non-alcoholic hepatic steatosis. Biomed Pharmacother 98:338–346

Fidrianny IRDA, Sari ELVIRA, Ruslan KOMAR (2016) Phytochemical content and antioxidant activities in different organs of pomelo (*Citrus maxima* (Burm.)Merr.) using 2, 2-diphenyl-1-picrylhydrazyl and phosphomolybdenum assays. Asian J Pharm Clin Res 9:185–190

Figueiredo AC, Barroso JG, Pedro LG, Scheffer JJ (2008) Factors affecting secondary metabolite production in plants: volatile components and essential oils. Flavour Fragr J 23:213–226

Foo KY, Hameed BH (2011) Microwave assisted preparation of activated carbon from pomelo skin for the removal of anionic and cationic dyes. Chem Eng J 173:385–390

Goh RMV, Lau H, Liu SQ, Lassabliere B, Guervilly R, Sun J et al (2019) Comparative analysis of pomelo volatiles using headspace-solid phase micro-extraction and solvent assisted flavour evaporation. LWT 99:328–345

Gonzalez-Mas MC, Rambla JL, Lopez-Gresa MP, Blazquez MA, Granell A (2019) Volatile compounds in citrus essential oils: a comprehensive review. Front Plant Sci 10:1–18

Guo C, Yang J, Wei J, Li Y, Xu J, Jiang Y (2003) Antioxidant activities of peel, pulp and seed fractions of common fruits as determined by FRAP assay. Nutr Res 23:1719–1726

Guo JJ, Gao ZP, Xia JL, Ritenour MA, Li GY, Shan Y (2018) Comparative analysis of chemical composition, antimicrobial and antioxidant activity of citrus essential oils from the main cultivated varieties in China. LWT 97:825–839

Gupta AK, Yumnam M, Medhi M, Koch P, Chakraborty S, Mishra P (2021a) Isolation and characterization of naringinase enzyme and its application in debittering of Pomelo juice (*Citrus grandis*): a comparative study with macroporous resin. J Food Process Preserv 45:e15380

Gupta AK, Dhua S, Sahu PP, Abate G, Mishra P, Mastinu A (2021b) Variation in phytochemical, antioxidant and volatile composition of pomelo fruit (*Citrus grandis* (L.) osbeck) during seasonal growth and development. Plan Theory 10:1941

Gyawali R, Jeon DH, Moon J, Kim H, Song YW, Hyun HB et al (2012) Chemical composition and antiproliferative activity of supercritical extract of *Citrus grandis* (L.) Osbeck fruits from Korea. J Essent Oil Bear Plants 15:915–925

Hameed BH, Mahmoud DK, Ahmad AL (2008) Sorption of basic dye from aqueous solution by pomelo (*Citrus grandis*) peel in a batch system. Colloids Surf A Physicochem Eng Asp 316:78–84

Haque MN, Saha BK, Karim MR, Bhuiyan MNH (2009) Evaluation of nutritional and physicochemical properties of several selected fruits in Bangladesh. Bangladesh J Sci Ind Res 44:353–358

Hasegawa S, Ou P, Fong CH, Herman Z, Coggins CW Jr, Atkin DR (1991) Changes in the limonoate A-ring lactone and limonin 17-. beta.-D-glucopyranoside content of navel oranges during fruit growth and maturation. J Agric Food Chem 39:262–265

He W, Li X, Peng Y, He X, Pan S (2019) Anti-oxidant and anti-melanogenic properties of essential oil from peel of pomelo cv. Guan Xi. Molecules 24:242

Hosni K, Zahed N, Chrif R, Abid I, Medfei W, Kallel M et al (2010) Composition of peel essential oils from four selected Tunisian Citrus species: evidence for the genotypic influence. Food Chem 123:1098–1104

Hussain K, Wajid A, Babar ME, Anwar Z, Farooqi S, Siddiqa A et al (2019) Production and optimization of pectinase from pomelo by *Aspergilliusniger* through solid state fermentation. Ann Life Sci 6:21–41
Ishihara M, Toda H, Sunagane N, Ohta T (2011) Furanocoumarins contents and cytochrome P450 3A (CYP3A) inhibitory activities of various processed fruit peel products: outflow of 6′, 7′-Dihydroxybergamottin during processing treatment of peel. J Pharm Soc Jpn 131:679–684
Islam A, Tasnin MN, Bari MW, Hossain MI, Islam MA (2021) Vitro antioxidant and in vivo antidiabetic properties of *Citrus maxima* leaf extracts in alloxan-induced swiss albino diabetic mice. Asian Food Sci J 20:66–79
Jang HD, Chang KS, Chang TC, Hsu CL (2010) Antioxidant potentials of buntan pumelo (*Citrus grandis* Osbeck) and its ethanolic and acetified fermentation products. Food Chem 118:554–558
Jiang CC, Zhang YF, Lin YJ, Chen Y, Lu XK (2019) Illumina® sequencing reveals candidate genes of carotenoid metabolism in three pummelo cultivars (*Citrus maxima*) with different pulp color. Int J Mol Sci 20:2246
Jing L, Lei Z, Li L, Xie R, Xi W, Guan Y et al (2014) Antifungal activity of citrus essential oils. J Agric Food Chem 62:3011–3033
Kamarudin MNA, Sarker MMR, Kadir HA, Ming LC (2017) Ethnopharmacological uses, phytochemistry, biological activities, and therapeutic applications of *Clinacanthus nutans* (Burm. f.) Lindau: a comprehensive review. J Ethnopharmacol 206:245–266
Kawaii S, Tomono Y, Katase E, Ogawa K, Yano M (1999) Quantitation of flavonoid constituents in citrus fruits. J Agric Food Chem 47:3565–3571
Khanam Z, Ching CH, Zakaria NHBM, Sam KH, Bhat IUH (2014) Phytochemical analyses and DNA cleavage activity of *Citrus maxima* fruit. In: International conference on chemistry and environmental sciences research (ICCESR), pp 17–18
Kingston PF (2002) Long-term environmental impact of oil spills. Spill Sci Technol Bull 7:53–61
Kittler J, Krüger H, Ulrich D, Zeiger B, Schütze W, Böttcher C et al (2018) Content and composition of essential oil and content of rosmarinic acid in lemon balm and balm genotypes (*Melissa officinalis*). Genet Resour Crop Evol 65:1517–1527
Kumar R, Vijay S, Khan N (2013) Comparative nutritional analysis and antioxidant activity of fruit juices of some citrus spp. Octa J Biosci 1:44–53
Kumar D, Ladaniya MS, Gurjar M (2019) Underutilized Citrus sp. Pomelo (*Citrus grandis*) and Kachai lemon (*Citrus jambhiri*) exhale in phytochemicals and antioxidant potential. J Food Sci Technol 56:217–223
Kundu Sen S, Saha P, Bhattacharya S, Bala A, Mazumder UK, Gupta M, Haldar PK (2010) Evaluation of in vitro antioxidant activity of *Citrus limetta* and *Citrus maxima* on reactive oxygen and nitrogen species. Pharmacol Online 3:850–857
Kundu Sen S, Gupta M, Mazumder UK, Haldar PK, Saha P, Bala A (2011a) Antitumor activity of *Citrus maxima* (Burm.) Merr. leaves in Ehrlich's Ascites Carcinoma cell-treated mice. Int Sch Res 2011:138737
Kundu Sen S, Gupta M, Mazumder UK, Haldar PK, Saha P, Bhattacharya S et al (2011b) Antihyperglycemic effect and antioxidant property of *Citrus maxima* leaf in streptozotocin-induced diabetic rats. Diabetol Croat 40:1–6
Kundu Sen S, Gupta M, Mazumder UK, Haldar P, Panda S, Bhattacharya S (2011c) Exploration of anti-inflammatory potential of *Citrus limetta* Risso and *Citrus maxima* (J. Burm.) Merr. Pharmacol Online 7:702–709
Kuo SM (1996) Antiproliferative potency of structurally distinct dietary flavonoids on human colon cancer cells. Cancer Lett 110:41–48
Kuo PC, Liao YR, Hung HY, Chuang CW, Hwang TL, Huang SC, Wu TS (2017) Anti-inflammatory and neuroprotective constituents from the peels of Citrus grandis. Molecules 22:1–11
Lan-Phi NT, Vy TT (2015) Chemical composition, antioxidant and antibacterial activities of peels' essential oils of different pomelo varieties in the south of Vietnam. Int Food Res J 22:2426
Lellis B, Fávaro-Polonio CZ, Pamphile JA, Polonio JC (2019) Effects of textile dyes on health and the environment and bioremediation potential of living organisms. Biotechnol Res Innov 3:275–290

Li F, Li S, Li HB, Deng GF, Ling WH, Wu S et al (2013) Antiproliferative activity of peels, pulps and seeds of 61 fruits. J Funct Foods 5(3):1298–1309
Li PL, Liu MH, Hu JH, Su WW (2014) Systematic chemical profiling of *Citrus grandis* 'Tomentosa' by ultra-fast liquid chromatography/diode-array detector/quadrupole time-of-flight tandem mass spectrometry. J Pharm Biomed Anal 90:167–179
Li H, Sun Z, Zhang L, Tian Y, Cui G, Yan S (2016) A cost-effective porous carbon derived from pomelo peel for the removal of methyl orange from aqueous solution. Colloids Surf A Physicochem Eng Asp 489:191–199
Li G, Cheng Y, Zhang T, Li Y, Han L, Liang G (2021) Characterization of oxygenated heterocyclic compounds and in vitro antioxidant activity of pomelo essential oil. Drug Des Devel Ther 15:937
Lin SP, Chao PDL, Tsai SY, Wang MJ, Hou YC (2011) *Citrus grandis* peel increases the bioavailability of cyclosporine and tacrolimus, two important immunosuppressants, in rats. J Med Food 14:1463–1468
Lin Z, Li J, Fang S, Chen T, Chen J (2012) Analysis of volatile aroma components in pomelo peel tea by headspace-gas chromatography-mass spectrometry. Food Sci 33:256–259
Ling W, Dai T, Zhang J, Liang Y, Yin W, Zhong B, Zhang J (2021) Evaluation of pomelo seed extracts as natural antioxidant, antibacterial, herbicidal agents, and their functional components. Chem Biodivers 18:e2100679
Liu Y, Heying E, Tanumihardjo SA (2012) History, global distribution, and nutritional importance of citrus fruits. Compr Rev Food Sci Food Saf 11:530–545
Liu W, Ye Q, Jin X, Han F, Huang X, Cai S, Yang L (2016) A spontaneous bud mutant that causes lycopene and β-carotene accumulation in the juice sacs of the parental Guanxi pummelo fruits (*Citrus grandis* (L.) Osbeck). Sci Hortic 198:379–384
Liu Y, Ren C, Cao Y, Wang Y, Duan W, Xie L et al (2017a) Characterization and purification of bergamottin from *Citrus grandis* (L.) Osbeck cv. Yongjiazaoxiangyou and its antiproliferative activity and effect on glucose consumption in HepG2 cells. Molecules 22:1227
Liu Z, Zu Y, Yang L (2017b) A process to preserve valuable compounds and acquire essential oils from pomelo flavedo using a microwave irradiation treatment. Food Chem 224:172–180
Liu Y, Liu A, Ibrahim SA, Yang H, Huang W (2018) Isolation and characterization ofmicrocrystalline cellulose from pomelo peel. Int J Biol Macromol 111:717–721
Lu Y, Zhang C, Bucheli P, Wei D (2006) Citrus flavonoids in fruit and traditional Chinese medicinal food ingredients in China. Plant Foods Hum Nutr 61:55–63
Lu XK, Lin QH, Lu XM, Zhang SM, Li CS, Ye XF (2012) Comparison of carotenoid compositions and contents n different sweet pomelos. Fujianj Agric Sci 27:723–727
Lü Z, Zhang Z, Wu H, Zhou Z, Yu J (2016) Phenolic composition and antioxidant capacities of Chinese local pummelo cultivars' peel. Hortic Plant J 2:133–140
Lukitaningsih E, Rumiyati R (2021) GC-MS analysis of bioactive compounds in ethanol and ethyl acetate fraction of grapefruit (*Citrus maxima* L.) rind. Borneo J Pharm 4:29–35
Mahato N, Sharma K, Koteswararao R, Sinha M, Baral E, Cho MH (2019) Citrus essential oils: extraction, authentication and application in food preservation. Crit Rev Food Sci Nutr 59:611–625
Mäkynen K, Jitsaardkul S, Tachasamran P, Sakai N, Puranachoti S, Nirojsinlapachai N et al (2013) Cultivar variations in antioxidant and antihyperlipidemic properties of pomelo pulp (*Citrus grandis* (L.) Osbeck) in Thailand. Food Chem 139:735–743
Mansour R (2018) Determination of nutritional composition in citrus fruits (*C. aurantium*) during maturity. Nutr Food Sci. https://doi.org/10.1108/NFS-05-2018-0141
Matheyambath AC (2016) Citrus fruits. Encyclopedia of food and health. Elsevier Press, Oxford, pp 136–140
McIntosh CA, Mansell RL (1997) Three-dimensional distribution of limonin, limonoate A-ring monolactone, and naringin in the fruit tissues of three varieties of Citrus paradisi. J Agric Food Chem 45:2876–2883

Medeiros KAA, Dos Santos JR, Melo TCDS, de Souza MF, Santos LDG, de Gois AM et al (2018) Depressant effect of geraniol on the central nervous system of rats: behavior and ECoG power spectra. Biom J 41:298–305

Mesbahi G, Jamalian J, Farahnaky A (2005) A comparative study on functional properties of beet and citrus pectins in food systems. Food Hydrocoll 19:731–738

Methacanon P, Krongsin J, Gamonpilas C (2014) Pomelo (*Citrus maxima*) pectin: effects of extraction parameters and its properties. Food Hydrocoll 35:383–391

Miller JA, Thompson PA, Hakim IA, Lopez AM, Thomson CA, Chew W et al (2012) Safety and feasibility of topical application of limonene as a massage oil to the breast. J Cancer Ther 3:1–12

Mokbel MS, Hashinaga F (2006) Evaluation of the antioxidant activity of extracts from buntan (*Citrus grandis* Osbeck) fruit tissues. Food Chem 94:529–534

Mokbel MS, Suganuma T (2006) Antioxidant and antimicrobial activities of the methanol extracts from pummelo (*Citrus grandis* Osbeck) fruit albedo tissues. Eur Food Res Technol 224:39–47

Nair A, Kurup R Sr, Nair AS, Baby S (2018) Citrus peels prevent cancer. Phytomedicine 50:231–237

Noreen A, Akram J, Rasul I, Mansha A, Yaqoob N, Iqbal R et al (2017) Pectins functionalized biomaterials; a new viable approach for biomedical applications: a review. Int J Biol Macromol 101:254–272

Nowicki P, Kazmierczak-Razna J, Pietrzak R (2016) Physicochemical and adsorption properties of carbonaceous sorbents prepared by activation of tropical fruit skins with potassium carbonate. Mater Des 90:579–585

Oboh G, Bello FO, Ademosun AO (2014) Hypocholesterolemic properties of grapefruit (*Citrus paradisii*) and shaddock (*Citrus maxima*) juices and inhibition of angiotensin-1-converting enzyme activity. J Food Drug Anal 22:477–484

Oyedepot A, Babarinde SO (2013) Effects of shaddock (*Citrus maxima*) fruit juice on glucose tolerance and lipid profile in type-II diabetic rats. Chem Sci Trans 2:19–24

Palazzolo E, Laudicina VA, Germanà MA (2013) Current and potential use of citrus essential oils. Curr Org Chem 17:3042–3049

Pallavi M, Ramesh C, Krishna V, Sameera P, Swamy L (2017) Quantitative phytochemical analysis and antioxiant activities of some citrus fruits of South India. Asian J Pharm Clin Res 10:199–205

Pandey BP, Thapa R, Upreti A (2019) Total phenolic content, flavonoids content, antioxidant and antimicrobial activities of the leaves, peels, and fruits of locally available citrus plants collected from Kavre District of Nepal. Int J Pharmacogn Chin Med 3(3):1–6

Pichaiyongvongdee S, Rattanapun B, Haruenkit R (2014) Total polyphenol content and antioxidant properties in different tissues of seven pomelo (*Citrus grandis* (L.) osbeck) cultivars. Agric Nat Resour 48:989–996

Poulose SM, Harris ED, Patil BS (2005) Citrus limonoids induce apoptosis in human neuroblastoma cells and have radical scavenging activity. J Nutr 135:870–877

Puglisi I, De Patrizio A, Schena L, Jung T, Evoli M, Pane A et al (2017) Two previously unknown Phytophthora species associated with brown rot of Pomelo (*Citrus grandis*) fruits in Vietnam. PLoS One 12:e0172085

Purseglove JW (1974) Rubiaceae. In: Purseglove JW (ed) Tropical crops dicotyledons. Longman Group Limited, London, pp 451–492

Reshmi SK, Manonmani HK, Manjunatha JR, Shashirekha MN (2018) Identification of bioactive compound from *Citrus maxima* fruit against carbohydrate-hydrolysing enzymes. Curr Sci 114:2099–2105

Reshmi SK, Sudha ML, Shashirekha MN (2020) Noodles fortified with *Citrus maxima* (pomelo) fruit segments suiting the diabetic population. Bioact Carbohydr Diet Fibre 22:100213

Rex JRS, Muthukumar NMSA, Selvakumar PM (2018) Phytochemicals as a potential source for anti-microbial, anti-oxidant and wound healing – a review. MOJ Biorg Org Chem 2:61–70

Roy MC, Alam M, Saeid A, Das BC, Mia MB, Rahman MA et al (2018) Extraction and characterization of pectin from pomelo peel and its impact on nutritional properties of carrot jam during storage. J Food Process Preserv 42:e13411

Sahlan M, Damayanti V, Tristantini D, Hermansyah H, Wijanarko A, Olivia Y (2018) Antimicrobial activities of pomelo (*Citrus maxima*) seed and pulp ethanolic extract. In: AIP conference proceedings, vol 1933. AIP Publishing LLC, p 030002

Saiprasad G, Chitra P, Manikandan R, Sudhandiran G (2013) Hesperidin alleviates oxidative stress and downregulates the expressions of proliferative and inflammatory markers in azoxymethane-induced experimental colon carcinogenesis in mice. Inflamm Res 62:425–440

Samaniego I, Brito B, Viera W, Cabrera A, Llerena W, Kannangara T et al (2020) Influence of the maturity stage on the phytochemical composition and the antioxidant activity of four Andean blackberry cultivars (*Rubus glaucus* Benth) from Ecuador. Plan Theory 9:1027

Sarkic A, Stappen I (2018) Essential oils and their single compounds in cosmetics – a critical review. Cosmetics 5:11

Sawant TP, Panhekar D (2017) A brief review on recent advances of *Citrus maxima* (chakota). Int J Rec Sci Res 8:19400–19416

Sen S, Ingale SL, Kim JS, Kim KH, Kim YW, Khong C, Chae BJ (2011) Effect of supplementationof Bacillus subtilis LS 1-2 grown on citrus-juice waste and corn-soybean meal substrate ongrowth performance, nutrient retention, caecal microbiology and small intestinal morphology ofbroilers. Asian-Australas J Anim Sci 24:1120–1127

Senguttuvan J, Paulsamy S, Karthika K (2014) Phytochemical analysis and evaluation of leaf and root parts of the medicinal herb, *Hypochaeris radicata* L. for in vitro antioxidant activities. Asian Pac J Trop Biomed 4:S359–S367

Shah NNAK, Supian NAM, Hussein NA (2019) Disinfectant of pummelo (*Citrus grandis* L. Osbeck) fruit juice using gaseous ozone. J Food Sci Technol 56:262–272

Shao SY, Xu WJ, Tao J, Zhang JH, Zhou XR, Yuan G et al (2017) Glycemic index, glycemic load, and glycemic response to pomelo in patients with type 2 diabetes. Curr Med Sci 37:711–718

Sharma P, Kumar V, Guleria P (2019) Naringin: biosynthesis and pharmaceutical applications. Indian J Pharm Sci 81:988–999

Shinde M, Sonawane SK, Patil S (2019) Fruit peel utilization in food packaging. Indian Food Ind Mag 1:19–24

Shivananda A, Rao DM, Jayaveera KN (2013) Analgesic and anti-inflammatory activities of *Citrus maxima* (J. Burm) Merr in animal models. Res J Pharm, Biol Chem Sci 4:1800–1810

Sidana J, Saini V, Dahiya S, Nain P, Bala S (2013) A review on citrus – "the boon of nature". Int J Pharm Sci Rev Res 18:20–27

Singh A (2016) Evaluation of antimicrobial potential and phytochemical assessment of *Citrus maxima* Burm. seeds extracts against respiratory tract pathogens. New Y Sci J 9:4–10

Singh A, Navneet (2016) Evaluation of antimicrobial potential and phytochemical assessment of *Citrus maxima* Burm. seeds extracts against respiratory tract pathogens. New Y Sci J 9:4–10

Singh A, Navneet (2017) *Citrus maxima* (Burm.) Merr. A traditional medicine: its antimicrobial potential and pharmacological update for commercial exploitation in herbal drugs – a review. Int J ChemTech Res 10:642–651

Sinha D, Satapathy T, Jain P, Chandel JP, Sahu D, Sahu B et al (2019) In vitro antidiabetic effect of neohesperidin. J Drug Deliv Ther 9:102–109

Slavin J (2013) Fiber and prebiotics: mechanisms and health benefits. Nutrients 5:1417–1435

Stanisic D, Costa AF, Fávaro WJ, Tasic L, Seabra AB (2018) Anticancer activities of hesperidin and hesperetinin vivo and their potentiality against bladder cancer. J Nanomed Nanotechnol 9:1–6

Straka I, Belous OG (2015) Chemical composition of grapefruit fruitage (*Citrus paradise* Macf.) in terms of its nutritional value. Субтропическоедекоративноесадоводство, pp 153–160

Sudto K, Pornpakakul S, Wanichwecharungruang S (2009) An efficient method for the large scale isolation of naringin from pomelo (*Citrus grandis*) peel. Int J Food Sci Technol 44:1737–1742

Sun H, Ni H, Yang Y, Chen F, Cai H, Xiao A (2014) Sensory evaluation and gas chromatography–mass spectrometry (GC-MS) analysis of the volatile extracts of pummelo (*Citrus maxima*) peel. Flavour Frag J 29:305–312

Tabeshpour J, Hosseinzadeh H, Hashemzaei M, Karimi G (2020) A review of the hepatoprotective effects of hesperidin, a flavanon glycoside in citrus fruits, against natural and chemical toxicities. DARU J Pharm Sci 28:305–317

Tao N, Gao Y, Liu Y, Ge F (2010) Carotenoids from the peel of Shatian pummelo (*Citrus grandis* Osbeck) and its antimicrobial activity. Am-Eurasian J Agric Environ Sci 7:110–115

Thavanapong N, Wetwitayaklung P, Charoenteeraboon J (2010) Comparison of essential oils compositions of *Citrus maxima* Merr. peel obtained by cold press and vacuum stream distillation methods and of its peel and flower extract obtained by supercritical carbon dioxide extraction method and their antimicrobial activity. J Essent Oil Res 22:71–77

Tocmo R, Pena-Fronteras J, Calumba KF, Mendoza M, Johnson JJ (2020) Valorization of pomelo (*Citrus grandis* Osbeck) peel: a review of current utilization, phytochemistry, bioactivities, and mechanisms of action. Compr Rev Food Sci Food Saf 19:1969–2012

Tsai HL, Chang SK, Chang SJ (2007) Antioxidant content and free radical scavenging ability of fresh red pummelo (*Citrus grandis* (L.) Osbeck) juice and freeze-dried products. J Agric Food Chem 55:2867–2872

Tsai ML, Lin CD, Khoo KA, Wang MY, Kuan TK, Lin WC et al (2017) Composition and bioactivity of essential oil from *Citrus grandis* (L.) Osbeck 'Mato Peiyu' leaf. Molecules 22:2154

Tuan NT, Huong BTC, Danh LT (2019) One step extraction of essential oils and pectin from pomelo (*Citrus grandis*) peels. Chem Eng Process 142:107550

Turner T, Burri BJ (2013) Potential nutritional benefits of current citrus consumption. Agriculture 3:170–187

Vadivukarasi G, Jenitha XA (2015) In vitro studies on phytochemical analysis and antioxidant activity of *Citrus maxima*. Int J Res Pharmacol Pharmacother 4:245–251

Van Hung P, Yen Nhi NH, Ting LY, Lan Phi NT (2020) Chemical composition and biological activities of extracts from pomelo peel by-products under enzyme and ultrasound-assisted extractions. J Chemother 2020:1–7

Vijayalakshmi P, Radha R (2016a) In vitro anti-alzheimer and anti oxidant activity of the peels of *Citrus maxima* fruits. Res J Pharmacol Pharmacodyn 8:17

Vijayalakshmi P, Radha R (2016b) Pharmacognostical and phytochemical screening of the peels of *Citrus maxima*. Microscopy 14:16

Wadhwa R, Paudel KR, Chin LH, Hon CM, Madheswaran T, Gupta G et al (2021) Anti-inflammatory and anticancer activities of Naringenin-loaded liquid crystalline nanoparticles in vitro. J Food Biochem 45:e13572

Wandee Y, Uttapap D, Puncha-arnon S, Puttanlek C, Rungsardthong V, Wetprasit N (2014) Enrichment of rice noodles with fibre-rich fractions derived from cassava pulp and pomelo peel. Int J Food Sci Technol 49:2348–2355

Wang YC, Chuang YC, Hsu HW (2008) The flavonoid, carotenoid and pectin content in peels of citrus cultivated in Taiwan. Food Chem 106:277–284

Wintergerst ES, Maggini S, Hornig DH (2006) Immune-enhancing role of vitamin C and zinc and effect on clinical conditions. Ann Nutr Metab 50:85–94

Wu CH, Huang SM, Lin JA, Yen GC (2011) Inhibition of advanced glycation endproduct formation by foodstuffs. Food Funct 2:224–234

Wu F, Jin Y, Xu X, Yang N (2017) Electrofluidic pretreatment for enhancing essential oil extraction from citrus fruit peel waste. J Clean Prod 159:85–94

Wu Y, Huang C, Gong J, Huang J, Pan T (2021) Analysis of carotenoid components in 'Minihongyou' (*Citrus maxima* (Burm.) Merr.) fruit by HPLC. Chinese J Trop Crops 42:546

Xi W, Fang B, Zhao Q, Jiao B, Zhou Z (2014) Flavonoid composition and antioxidant activities of Chinese local pummelo (*Citrus grandis* Osbeck.) varieties. Food Chem 161:230–238

Xiao L, Ye F, Zhou Y, Zhao G (2021) Utilization of pomelo peels to manufacture value-added products: a review. Food Chem 351:129247

Xiong H, Wang J, Ran Q, Lou G, Peng C, Gan Q et al (2019) Hesperidin: a therapeutic agent for obesity. Drug Des Devel Ther 13:3855

Xu CJ, Fraser PD, Wang WJ, Bramley PM (2006) Differences in the carotenoid content of ordinary citrus and lycopene-accumulating mutants. J Agric Food Chem 54:5474–5481

Xu YR, Zhang KF, Xie QJ, Lin JX, Huan KX, Liao Y (2015) Chemical constituents from young fruits of *Citrus maxima* cv. Shatian. Zhong yaocai = Zhongyaocai = J Chinese Med Mat 38:1879–1881

Yan F, Shi M, He Z, Wu L, Xu X, He M, Chen J, Deng X, Cheng Y, Xu J (2018) Largely different carotenogenesis in two pummelo fruits with different flesh colors. PLoS One 13:e0200320

Yaseen DA, Scholz M (2019) Textile dye wastewater characteristics and constituents of synthetic effluents: a critical review. Int J Environ Sci Technol 16:1193–1226

Yathiender S (2017) A comparative study of antimicrobial activity of *Citrus maxima* and *Citrus Aurantium* plant extracts. Int J Recent Sci Res 8:18507–18509

Yu EA, Kim GS, Lee JE, Park S, Yi S, Lee SJ et al (2015) Flavonoid profiles of immature and mature fruit tissues of *Citrus grandis* Osbeck (Dangyuja) and overall contribution to the antioxidant effect. Biomed Chromatogr 29:590–594

Zain NM, Yusop SM, Ahmad I (2014) Preparation and characterization of cellulose and nanocellulose from pomelo (*Citrus grandis*) albedo. J Nutr Food Sci 5:334

Zhang M, Li L, Wu Z, Wang Y, Zang Y, Liu G (2017) Volatile composition in two pummelo cultivars (*Citrus grandis* L. Osbeck) from different cultivation regions in China. Molecules 22:716

Zhao YL, Yang XW, Wu BF, Shang JH, Liu YP, Zhi-Dai Luo XD (2019) Anti-inflammatory effect of pomelo peel and its bioactive coumarins. J Agric Food Chem 67:8810–8818

Zhu J, Liu Q, Li Z, Liu J, Zhang H, Li R et al (2017) Recovery of uranium (VI) from aqueous solutions using a modified honeycomb-like porous carbon material. Dalton Trans 46:420–429

Chapter 6
Tangerine (*Citrus reticulata*)

Cláudia Maciel, Rui Meneses, Renan Danielski, Sérgio Sousa, Norton Komora, and Paula Teixeira

6.1 Introduction

Although consumed in processed form (e.g., juices and canned fruit), the bulk of tangerineproduction is allocated toward fresh consumption. Across the United States, roughly one million tonnes of tangerines were produced in 2020/2021, with 68% of them utilized in fresh consumption and the remaining 32% designated for processing (USDA 2021a).

Nevertheless, with the rapid surge in consumption, tangerine wastages also increased, due to losses across the food chain (e.g., during production, storage, or even retail), canning, and juice processing. According to Nitayapat et al. (2015), half of the tangerine constituents are discarded as solid waste after juice extraction. In the past, the processing industries would deal with the dry solid waste residues by dumping them in factory-adjacent landfills. This procedure was particularly problematic since it would create an imbalance in carbon sources from putrefying organic matter, leading to uncontrollable methane production, and produce an imbalance in bacterial populations present in the soil while increasing the contamination of fresh groundwater (Yadav et al. 2022). Nowadays, dry residues are commonly sold for pectin extraction (Pourbafrani et al. 2010), as animal feedstuffs, or for energy production through combustion (Sridhar et al. 2021), with none of these options being extremely profitable or environmentally friendly.

C. Maciel · R. Meneses · S. Sousa (✉) · P. Teixeira
Universidade Católica Portuguesa, CBQF – Centro de Biotecnologia e Química Fina – Laboratório Associado, Escola Superior de Biotecnologia, Porto, Portugal

N. Komora
Department of Food Science and Nutrition, Faculty of Food Engineering, University of Campinas, Campinas, SP, Brazil

R. Danielski
Department of Biochemistry, Memorial University of Newfoundland, St. John's, NL, Canada

S. Singh Purewal et al. (eds.), *Recent Advances in Citrus Fruits*,
https://doi.org/10.1007/978-3-031-37534-7_6

The shed tangerine solid residue comprises components such as the peel, rag, pulp, and seeds, rich in carbohydrates, vitamins, minerals, essential oils, antioxidants, flavonoids, carotenoids, dietary fibers, and many other highly desirable bioactive compoundswith therapeutic, anti-inflammatory, antimicrobial, antiviral properties (Sagar et al. 2018; Kaur et al. 2021a).

Thus, tapping into the added value by-products of tangerine processing would reduce waste and create new high-value products while promoting circular economy models.

6.2 Economic Significance, Taxonomy, and History of Tangerines

Tangerines, *Citrus reticulata*, also known as Dancy or Beauty mandarins, are small oblate, bright orange, thin-skinned, and highly nutritious fruit desirable by the consumer owing to their sweetness and easy pealing. Due to an awareness increase in healthy eating by the average consumer, tangerines and other mandarin hybrids regained an integral part of our daily diets, becoming one of the most economically important perennial horticultural crops in the world, second only to sweet oranges (FAO 2021; USDA 2021b). For 2021/22, tangerine production expects an increase of 2.0 million tonnes from the past year, reaching a whopping 37.2 million tonnes, with both consumption and exports rising to new records (USDA 2022) (Table 6.1).

For the past decade, China has become the most prolific tangerine producer, producing approximately twenty-seven million tonnes in 2021 alone, more than three times the combined production of the European Union (EU), Turkey, Morocco, Japan, and the United States (US). However, nearly 98% of all Chinese-produced tangerines are meant for the domestic market, with China only exporting 2% of its tangerine production. Relatively to exports, Turkey is the major player, exporting

Table 6.1 Production data for 2021 of the main tangerine producers

Origin	Production (thousand tonnes)
China	27,000
European Union	3162
Turkey	1810
Morocco	1360
Japan	924
United States	758
South Africa	660
South Korea	610
Argentina	400
Australia	190

Source: (USDA, 2022)

more than half of their produced tangerines. In terms of tangerine imports, Russia, US, EU, and the United Kingdom (UK), are the major clients of imported tangerines (USDA 2022).

6.2.1 The Problematic Taxonomy of Tangerines and the Citrus Genus

The friction around the taxonomy of the Citrus genus can be attributed to a wide array of factors. Species from this genus are highly compatible among themselves. This ease in hybridization leads to the formation of exceedingly fertile natural or artificial hybrids, which become highly propagated and cultivated (e.g., tangerines). Furthermore, the selection of specific cultivars with desirable somatic mutations from hundreds or even thousands of years of human citrus cultivation hinders the identification of pure wild specimens (Bernet et al. 2010).

Additionally, although monoembryonic, mandarins and other citrus species can reproduce through nucellar embryony, a type of apomixis, where maternal clones are generated from the somatic tissue of the ovule, reproducing asexually through polyembryonic seeds (Zhang et al. 2018a). Thus, enabling the high dispersion offered by the sexual seed-based reproduction while conserving the parental genotype from the asexual reproduction. Apomixis will lead to plant populations with particular amplified phenotypes, possibly inducing taxonomists to consider these specimens separate species (Luro et al. 2017).

Some of the earliest known attempts in the western world to classify citrus fruits date back to the XVII century and are found in the classical work by Giovanni Ferrari on the citrus culture "*Hesperides sive de Malorum Aureorum Cultura et Usu*" (Ferrarius 1646), with the title as an allusion to the garden of Hesperides from Greek mythology. According to this fable, the golden fruits from this garden conferred immortality to whoever consumed them. However, in the original tale, the golden fruit was depicted as an apple (Andrews 1961). Curiously enough, this book's influence on the correlation between citrus fruits and the ancient Greek myth can be seen even today, with the botanical term hesperidium referring to fruits from citrus trees. Later in the XVIII century, Carl Linnaeus proposed our current binomial nomenclatural system and decided to organize plants based on their reproductive characteristics. Inspired by the works of de Tournefort (1700), and through the study of features in the leaves and flowers of citrus trees, in his work "*Genera plantarum*" (Linné 1737), he divided the *Citrus* genus into three different species: *Citrus medica*, encompassing both citrons and lemons; *Citrus aurantium*, which included sweet and bitter oranges along with pummelos; *Citrus trifoliata* enclosing the Poncirus. In the later work *"Species plantarum"* (Linné 1737), Linnaeus presented a radically new view on the citrus genus, only recognizing two species, *C. aurantium* and *C. medica*, which were subdivided into various varieties (e.g., *C. aurantium* var. aurantium – Sour oranges; *C. aurantium* var. sinensis – Sweet oranges). Working in collaboration with his apostle Pehr Osbeck, Linnaeus kept introducing new binomial names to the Citrus genus, many of which have since radically changed (Inglese and Sortino 2019).

Throughout the XVIII and XIX centuries, although many authors continued working and reporting on both classificational and evolutionary botany, few noteworthy radical and revolutionary Citrus classification systems were proposed. During the XX century, two pivotal taxonomic systems came to fruition the Swingle and Reece taxonomic system (Swingle and Reece 1967) and the Tanaka classification system (Tanaka 1954, 1961, 1977). Although radically different from one another and often seen as two extreme views of the Citrus genus, they are still relevant and valid systems remaining in frequent use by various authors.

Walter Swingle was one of the first botanists to propose the use of biochemical markers for taxonomy purposes. Taking into account the morphological and biochemical characteristics of various citrus, along with their history and reproduction, in 1967, Walter Swingle and Philip Reece published their proposed taxonomic system for citrus classification (Swingle and Reece 1967).

Thus, according to Swingle and Reece Citrus genus belongs to the Citreae tribe from the angiosperm subfamily Aurantioideae encompassed in the family Rutaceae in the order Sapindales (Swingle and Reece 1967; Stevens 2016). Nevertheless, even nowadays, the taxonomic condition of the Rutaceae family remains contentious, being at times outright confusing (Nicolosi et al. 2000), particularly for the tribes and subtribes of the Aurantioideae subfamily. Appelhans et al. (2021) classified nearly 90% of all genera included in the Rutaceae family and concluded that it was not yet feasible to divide the subfamily into tribes.

The Tanaka taxonomic classification system first appeared in 1954 and was proposed by the Japanese Tyôzaburô Tanaka. In its later incarnation, this system divided the Citrus genus into 157 species, of which 36 were mandarin species. However, this system overly complicated the classification of mandarins, considering numerous mandarin hybrid varieties as standalone species. For example, according to Tanaka's citrus distribution, tangerines are classed as a separate species (*Citrus tangerina*) instead of a *C. reticulata* × *C. maxima* introgression. Moreover, in this system, many other mandarin hybrids are classified as entirely distinct species from true mandarins – *Citrus reticulata*, (e.g., willowleaf mandarins – *Citrus deliciosa*; clementines – *Citrus clementina*) (Tanaka 1977). However, according to Swingle and Reece, tangerines and all the aforementioned species are classified as *Citrus reticulata* cultivars. This system's inability to recognize mandarin and other citrus admixture is one of its significant flaws.

Later proposed classification systems, such as Mabberley (2004), try to address this issue by acknowledging mandarin hybrids and other citrus hybrids. However, several hypotheses raised by Mabberley have since been proven incorrect due to new phylogenomic data (Oueslati et al. 2017; Wu et al. 2018).

During the 1970s, numerical taxonomy resulted in a better understanding of citrus domestication and the relationships between the various cultivated species of Citrus. *In the late seventies,Barret and Rhoades* (1976) *proposed the existence of only three pure ancestral citrus species (C. reticulata, C. maxima, and C. medica) from which all other known citrus species derived.*

Advances in molecular biology, with the development of Restriction Fragment Length Polymorphism (RFLP) analysis, Random Amplification of Polymorphic DNA (RAPD), and later the development of Polymerase Chain Reaction (PCR), along with the successful application of molecular markers, have shed light on the domestication and evolutionary history of various citrus fruits and their relatives (Barkley et al. 2006; Garcia-Lor et al. 2015; Curk et al. 2016; Liu et al. 2016; Yu et al. 2017). Moreover, NextGeneration Sequencing (NGS) technologies and Whole Genome Sequencing (WGS) analysis have changed the paradigm around plant evolution, in which traditional phylogenetics have slowly been replaced with new phylogenomic approaches (Egan et al. 2012; Curk et al. 2014). The capabilities provided by the new phylogenomic era allowed for the differentiation among pure ancestors and closely related hybrids, which was previously unachievable when the Tanaka and Swingle systems were proposed. Thus, nowadays, tangerines are commonly regarded as hybrids from the introgression of mandarins (*C. reticulata*) and pomelos (*C. maxima*), possessing minimal amounts of pomelo admixture (below 10 percent) (Oueslati et al. 2017; Wang et al. 2017a; Wu et al. 2018).

With the aim of reducing the entropy between the classical taxonomic systems of Swingle and Tanaka, and the new genomic data provided by new phylogenomic technologies, Ollitrault et al. (2020), proposed a new classification system in which pure ancestral species possess binomial names (e.g., *Citrus reticulata –mandarins, Citrus maxima – pomelos, Citrus medica–* citrons), while hybrids of ancestral species acquire unique species names (e.g., *Citrus aurantium which* encompasses all admixtures of *C. reticulata* × *C. maxima*), with the individual admixtures within each hybrid species being distinguished by a variety name (e.g., Tangerine – *C. aurantium* var. tangerina). However, this newly proposed system has not yet been widely adopted, with many authors still referring to tangerines as *C. reticulata*. Thus, for this chapter's purpose, tangerines will be considered *C. reticulata* × *C. maxima introgressions. Moreover, due to the minimal levels of C. maxima admixture,tangerineswill be referred to only as C. reticulata.*

Even in trivial terminology, problems arise when speaking of tangerines and mandarins, with both terms interchangeably used to describe one another, along with any easy-peel, reddish-orange mandarin-like fruit. Applying the term mandarin when referring to a tangerine is not incorrect since, in the popular meaning of the word, mandarin describes both pure *C. reticulata* specimens and the numerous varieties of *C. reticulata* × *C. maxima* admixtures. The problem comes when we use tangerine, a specific mandarin hybrid, to describe a pure mandarin or any other mandarin hybrid.

6.2.2 Historical Perspective and Phytogeography of Tangerines

The controversy around tangerines and other citrus species is not only circumscribed to taxonomy and terminology but also extends to its history and geographical origin. Although presenting itself like a seemingly simple question, due to scarce reliable paleontological information and a lackluster definition of pure citrus species, determining the birthplace of the Tangerine and other citruses have proven to be quite intricate. Only two citrus fossil specimens exist, corresponding to fossilized leaves of *Citrus meletensis* and *Citrus linczangensis* (Fischer and Butzmann 1998; Xie et al. 2015). Alongside the discovery of citrus fossil specimens, the employment of nuclear genome-based phylogeny allowed to determine when crucial tangerine precursors species emerged, with *C. reticulata* and *C. maxima* appearing in the Late Miocene, 6.9 million years (Ma) and 6.2 Ma, respectively (Wu et al. 2018).

Although poorly understood it is believed that wild mandarins firstly appeared in ancient forests across the valleys of the Nanling mountains, with two separate domestication events, leading to their expansion across the Mangshan region between Yangtze and Zhu rivers in South-China (Legge 1865; Tolkowsky 1938; Webber 1967; Wang et al. 2017a, 2018).

Throughout ancient Chinese literature, we can encounter multiple references to mandarins and other citrus fruits. In the chapter *"Tribute to Yu"* from the book *"Yu Gong"*, written 4000 years ago, mandarins appear as one of the tribute items sent to the divine Chinese emperor Yu (Medhurst 1846; Scora 1975). Moreover, in the encyclopedia "*Lu Shi Chun Qiu*", mandarins are described as a beautiful and delicious fruit produced in the Jiangsu region. The patriotic poet Qu Yuan also eulogized mandarin trees in his classical works *"Chu Ci"*, roughly translated to *"Verses of Chu"*, written around 250–300 B.C. (Zhong and Nicolosi 2020).

In 1179 A.D., Han Yen-Chih described in detail, in the three-volume collection "*Chü Lu*", viewed as one of the very first specialized books in the cultivation of citrus fruits, 27 varieties of citrus: including mandarins, tangerines, sweet oranges, bitter oranges, lemons and many more. In addition, information relevant to citrus cultivation, such as methods for fertilization, irrigation, pests, disease control, and post-treatment collection, along with illustrations, are present in this collection (Scora 1975; Inglese and Sortino 2019).

These ancient literary references, along with phylogenomic data, support the notion that the primordial demos in SouthChina, were acquainted with mandarins long before their introduction to the Middle Eastern and European civilizations.

Mandarin cultivation enabled the spread of *C. reticulata* in the SouthChina region, allowing it to come across one of its relatives *C. maxima*, which had high prevalence across Malaysia and Indochina, whit some distribution in mandarin-rich South China. These overlapping populations in the same geographical area allowed for the development of *C. reticulata* × *C. maxima* hybrids, one of which led to the modern-day tangerine.

It is believed that like other citrus, mandarins radiated through trade routes towards the West reaching the Middle East and Northern Africa. However, it is still unclear who introduced the European countries to mandarins and their hybrids, with some authors crediting British-Chinese trade for their introduction in the early eighteen-hundreds (Langgut 2017). This notion goes in accordance with Pehr Osbeck who, during the XVIII century, described mandarins in his book "*A voyage to China and the East Indies*" (Osbeck et al. 1771). However, other authors rather support the idea that mandarins arrived in the West somewhere between the seventeen and eighteen centuries as a consequence of the Portuguese-British colony trade. Thiswould possibly explainthe word mandarin, which is not a word of Chinese origin and probably evolved from Hindi, Malayan and Portuguese dialects (Ramon-Laca 2003).

Once introduced in Europe, mandarins and their hybrids quickly spread across all colonies of the European empires.

Concerning the etymology of the word tangerine, it originated in the mid-nineteen century and referred to the Moroccan mandarin hybrids imported by Major Atway. These hybrids were shipped to the mainland United States through the port of Tangiers, hence their name (Hume 1913).

6.3 Agricultural Practices in Tangerine Growth and Harvest

Citrus in general, are fond of relatively warm environments and broadly grown across tropical and subtropical regions, confined in the so-called "citrus belt" ranging from 40°N to 40°S latitude, between sea level and 600–750 m.

This current dispersion vastly contrasts the primordial citrus landscape and is a consequence of human trade and travel through the ages. Nonetheless, although the high temperature does not seem to limit tangerine development, tangerine trees are immensely susceptible to frost and chilling damage, affecting fruit quality. Thus, environments with annual temperature averages below 10–12 °C are unsuitable for tangerine growth and development.

Regarding cold temperature exposure, the topography is another variable to be considered since the topography of the region where the orchard sits will affect the overall temperatures and wind exposure. Undoubtedly cooler air will settle in lower areas while open areas are more exposed to the wind, which may damage the tangerine trees. Moreover, the tree rattling caused by stronger winds will lead to fruit scarring due to injuries caused by the abrasion of leaves, branches, and debris in the air. Usually, to prevent wind damage, windbreaks in the form of rows of trees are planted across the limits of the orchard (Norton 1988). This strategy is also known to provide some protection against freezing.

Concerning soil quality, silt soils are the most suitable for tangerine production. Nevertheless, like most citrus fruits, tangerines can grow in a wide variety of soils, ranging from coarse to loamy and even heavy soils, provided that proper water drainage and root oxygenation are available. Regarding soil pH, optimal pH growth conditions range from 5.5 to 7.0, with pH regulation performed through sulfur acidification or water irrigation. In highly alkaline soils rich in bicarbonates, water irrigation will also increase the soil's availability of micronutrients, particularly, Magnesium, Zinc, and Iron minerals (Morgan and Graham 2019). Limestone soil amendments can be employed to increase the overall pH of acidic soils.

Proper irrigation and water drainage are fundamental in obtaining good quality tangerine fruits. Traditionally, extensive fruit plantations rely on wasteful and inefficient high-volume sprinkler irrigation systems. Automated sprinkler systems can be paired with night-time irrigation to mitigate water loss due to evaporation, reducing some of the system's shortcomings (Yacoubi et al. 2010; Nagarajan and Minu 2018). Additionally, lowering the system's irrigation volume while increasing the irrigation frequency improves its efficiency and reduces water waste.

By acknowledging the increasingly high water cost and the United Nations goals 6 and 12 for sustainable development, which verse on conscious water management and sustainable production patterns (U.N. 2015) respectively, it is possible to understand how more efficient low-volume irrigation systems (e.g., drip and micro-drip irrigation) that directly deliver water to the root system (Assouline et al. 2002) have become more popular due to their efficiency, water savings, and improved nutrient transfer from the soil to the root.

Wastewater reclamation can be an alternative water source, especially in regions with frequent water shortages. Compared with well water, wastewater positively influences plant nutrition in part due to a rich constitution in macronutrients (N, P, K, S, Ca, Mg) and micronutrients (B, Cl, Cu, Co, Zn, Fe, Mn, Mo, and Ni), closer to the optimal plant growth requirements. However, due to their rich composition, particularly in micronutrients, crops irrigated with recycled water must have their nutrient supply carefully and properly balanced to avoid nutritional excess and possible toxic accumulation of micronutrients.

Along with combating water deficits, managing water surplus is also fundamental in citrus culture, especially in regions with frequent precipitation or low permeability soils. Ridges with water furrows are excellent drainage systems performing efficiently in dispersing exceeding water in orchards. Additionally, other ditch networks with water collectors and discharge pumps can be employed to reduce water availability in the soil (Dollinger et al. 2015).

Another crucial aspect of tangerine production resides in efficiently managing all the available space in an orchard with optimal tree size and spacing needing to be monitored and planned. Although tangerine trees can adapt to a wide range of spacing between trees, sunlight remains the chief concern for good tree and fruit development. Therefore, tree rows are often oriented from north to south to maximize sunlight capture.

Tree overgrowth will also directly affect sunlight exposure, leading to competition for sunlight among trees. Thus, pruning is often necessary to control the canopies of the trees and the deadwood in each tree, composed of twigs, leafless and fruitless branches. Moreover, skirting, the removal of the lower parts of the canopy, facilitates agricultural practices and harvesting, also preventing contamination with soil pathogens. Even though nifty, pruning directly affects fruit size and thus is not frequently applied in orchards.

Proper disease management and pest control are fundamental to maintaining adequate orchard sanitation and good quality fruits. Like other fruit-bearing trees, tangerine trees are susceptible to various pathogens, from insects, mites, and nematodes, to fungi, viruses, and bacteria. Combating these hardships often requires frequent pest monitoring and the usage of aerially sprayed pesticides. However, over-administration of pesticides can lead to unwanted pest resistance. Moreover, when applying pesticides, consideration must take place to ensure that by the time of harvest, the pesticide trace amounts on the fruit do not exceed the legal residue level.

Maintaining proper orchard sanitation also encompasses recurrent removal of decaying fruit and vegetation along with weed control. Weeds will directly compete with tangerine trees, depriving them of nutrients and sunlight. On top of that, weeds can also harbor numerous pests working as a means of dispersion for these pathogens.

6.3.1 Harvest

When it comes to harvesting tangerines, two options are available. Harvesting can be done manually (encompassing both traditional handpicking and mechanically assisted hand harvesting) or mechanically. As a general rule, if the fruit is destined for fresh consumption, then manual harvesting is the most desirable option since it provides less damage to the fruit when executed by experienced labor, which is particularly important in tangerines since they possess a fragile rind prone to bruises, cuts, and infections. Thus, gentle handling is essential in preserving external quality and pivotal in fruit intended for the fresh market. Moreover, traditional handpicking is more suitable for orchards sitting on uneven terrain. However, even with skillful workers, manual harvesting requires a significant amount of manpower to compete with mechanically automated approaches, which brings an added cost.

When talking about large-scale plantations, the cost/benefit relation of hand harvesting may make it less desirable as an option, with mechanical harvesting possibly becoming a more reasonable approach. There are multiple varieties of mechanical harvesters, but the most used principle relies upon the machine's ability to tremble the tree canopy enabling the ripe fruit to fall from the tree. Some of these machines possess containers or conveyor belts that catch the falling fruits, while others merely let the fruits fall into the ground to be later collected by workers (Sanders 2005). Naturally, both machines cause undesirable bruises and lacerations in the fruit and are thus reserved for fruits meant for processing.

6.4 Tangerine Phytotomy and Chemical Composition

6.4.1 Citrus Reticulata Phytotomy

The phytotomy of *C. reticulata* consists of a pericarp comprising three tissue layers, namely exocarp (flavedo), mesocarp (albedo), and endocarp (pulp, encompassing the carpellary segments and juice sacs), and seeds. Conjointly, the exocarp and mesocarp binomial composes the rind (peel) (Tadeo et al. 2008; Sadka et al. 2019).

As in other citrus fruits, the exocarp (also termed epicarp) corresponds to the outermost tissue layer (orange-colored) commonly referred to as the flavedo (derived from the Latin word *flavus*, meaning yellow). The flavedo layer represents a reservoir of carotenoids (entrapped in chromoplasts), vitamins, and essential oils entrappedin quasi-spherical volatile oil glands (Sinclair 1984; Berk 2016).

Proceeding towards the fruit's middle layer is (located) the mesocarp, also designated albedo (in Latin, the word *albus* means white) or white pith. This sponge-like white parenchymatous tissue on the inside of the rind constitutes an important source of pectin and flavanones, also comprising other compounds, namely soluble carbohydrates, amino acids and vitamins (Sinclair 1984; Berk 2016; Sadka et al. 2019). In the early stage of the *C. reticulata* development, the albedo layer fills the majority of the fruit volume, progressively thinning, concomitantly with juice sac cells growth (predominantly propelled by the vacuole expansion) as the fruit development proceeds (Katz et al. 2011). Throughout the maturation phase occurs the disintegration of albedo, remaining solely the reticula (vascular system), which is in the genesis of the species name *Citrusreticulata* (Sadka et al. 2019).

The core (inward) layer of the pericarp is the endocarp, the edible pulp. This pulp is partitioned into loculi termed segments, which are formed from carpels (1 carpel equals 1 segment) during fruit development, hence the term carpellary membranes, which is also utilized by some authors (Asencio et al. 2018). The whorl of fused carpels enclose the juice vesicles (or juice sacs), with septa running along these segments converging onto the central axis of the tangerine, in which the segments coalesce and the seeds are located. Albeit considered single structures, juice sacs are in fact composed by several cell layers, presenting specific morphologies as the fruit develops. Throughout this stage, vacuoles comprised within those cells represent most of the cell volume (>90%), discharging the content in the form of juice. Mature fruits encompass the bulk of carbohydrates and organic acids, precisely within those structures (Sadka et al. 2019). Moreover, the pulp is also a source of other nutrients which include, among others, vitamins, and minerals (Sinclair 1984; Asencio et al. 2018; Sadka et al. 2019).

The growth and development of *C. reticulata* entails three main overlapping stages, namely cell division, cellular expansion and maturation (Katz et al. 2004, 2011; Sadka et al. 2019). The first stage is initiated following fruit set triggered by phytohormones, namely gibberellin and auxin, which are highly committed to the extensive cell division (Shinozaki et al. 2020). Proceeding towards the second stage, a cessation in cell division occurs in all fruit tissues, excluding in the flavedo layer and extremities of the juice sacs. In this phase, the fruit growth ensues via an intense cell expansion, being the volume increase of juice vesicle cell mainly propelled by the vacuole enlargement, which fills nearly the whole volume of the cell. The last stage corresponds to the *C. reticulata* maturation and ripening, in which the fruit growth declines and the flesh attains its definite dimension (Katz et al. 2011). *Citrus reticulata* development encompasses profound alterations in the primary and secondary metabolisms. Concerning the former, sugars and citrate, which constitute the predominant compounds of the juice sac vesicles, are biosynthesized, achieving the desirable contents at the time-of-harvest (Bain 1958; Katz et al. 2011; Lado et al. 2018). Outwardly, the principal alteration is the transition of peel color from dark-green to the characteristic fully orange color of *C. reticulata*, which is a consequence of the biosynthesis of the secondary metabolites, carotenoids (Nawaz et al. 2020).

6.4.2 *Citrus reticulata – Chemical Profile*

Citrus reticulata is a sophisticated synthetic phytochemist uniquely endowed with an arsenal of metabolic enzymes to biosynthesize an intricate mélange of unique and chemically diverse compounds.

Hence, thishesperidium is a prolific producer of phytochemicals, biosynthesizing a valuable diversity of primary (centralcarbon metabolism) and secondary metabolites (secondary or specialized metabolism). The primary metabolites (macronutrients) encompass carbohydrates, organic acids, proteins (amino acids) and lipids (including fatty acids), while the counterparts comprise the chemical classes of flavonoids, phenolic acids, limonoids, carotenoids, essential oils, and alkaloids (Pott et al. 2019; Shorbagi et al. 2022).

The primary and secondary metabolic profiles of tangerine greatly influence the tangerines' sensory traits (particularly savour (flavor) and texture), along with nutritional and quality attributes. The chemical composition of *C. reticulata* is highly dependent on pre- and postharvest conditions, namely maturity level, storage, horticultural, and climate conditions, all of which influence the metabolic profile of the fruit (Lv et al. 2015; Sadka et al. 2019).

6.4.2.1 Citrus reticulata – Primary Metabolism

6.4.2.1.1 Carbohydrates

The carbon metabolism is closely associated to the plant growth owing to the fact that the exportation of photosynthetically biosynthesized carbohydrates from plant leaves provides the substrate for the sink tissues development and maintenance. *Citrus reticulata* fruit development is supported by carbohydrates accumulation (photosynthetically synthesized) in juice vesicles and organic acids catabolism (respiration) (Lemoine et al. 2013).

During plant development, carbon is predominantly assimilated in the form of sucrose and translocated via phloem system from the leaf mesophyll to non-photosynthetic tissues (sink tissues) (Lemoine et al. 2013). In the photosynthetic pathway, the carbon dioxide fixation occurs in the chloroplasts stroma through the Calvin cycle to generate triose phosphate (triose-P; DHAP or PGA). Subsequently, triose-P is exported to the cytosol through a triose-phosphate translocator. Once in the cytoplasm, two triose-P molecules are combined to yield one molecule of fructose 1,6-biphosphate (F 1,6 BP) in a reaction catalyzed by aldolase. This compound is subsequently hydrolyzed by fructose biphosphatase to fructose-6-phosphate, which is further isomerized to glucose-1-phosphate. Sucrose is biosynthesized from hexose phosphates in the leaves mesophyll through the combined catalysis of UTP-glucose-1-phosphate uridylyltransferase (UDP-glucose pyrophosphorylase), sucrose-phosphate synthase (SPS) and sucrose-phosphate phosphatase. Sucrose is the primary sugarimported via phloem from photosynthetic leaves to non-photosynthetic tissues (sink tissues). Sucrose may be translocated into the sink cells apoplastically by specific sucrose transporters or may be hydrolyzed into glucose and fructose by cell wall invertase (these monosaccharides may be imported to the sink cells through hexose transporters) (Ayre 2011; Lemoine et al. 2013; Ruan 2014; McClain and Sharkey 2019; Stein and Granot 2019). Once inside the juice vesicles, sucrose may be metabolized via cleavage by cytosolic invertase or sucrose synthase, or translocated into the vacuole (being stored as sucrose, converted into fructans by fructosyltransferases or hydrolyzed into hexoses by vacuolar invertase). The glycosyl transferase sucrose synthase catalyzes the conversion of sucrose into UDP-glucose and fructose (Sadka et al. 2019; Stein and Granot, 2019). In *C. reticulata* sucrose is the most abundant soluble sugar, representing more than 60% of the total content, followed by glucose and fructose, in a 2:1:1 ratio (Lado et al. 2018).

Amongst the desirable *C. reticulata* sensory traits, sweetness constitutes a pivotal taste characteristic (a key determinant of flavor acceptability). The strict balance between sweetness and sourness defines the *C. reticulata* unique taste, which is predominantly dictated by the contents and ratio of carbohydrates and organic acids in the juice vesicles (Lin et al. 2015). These compounds participate in the biosynthetic pathways of primary and secondary metabolites, namely amino acids, vitamins and volatile organic compounds, which are also pivotal determinants of *C. reticulata* quality (aroma and flavor) (Tadeo et al. 2008; Sadka et al. 2019). Throughout fruit development sugar accumulation occurs in synchrony with organic acid catabolism. Hence, the two metabolic pathways are intertwined. The photosynthetically synthesized sugar import from the leaves to the fruit is governed by the balanced interplay between biosynthesis, translocation and accumulation. Organic acids, the other prominent component of fruit taste, accumulate in the early phase of *C. reticulata* development, experiencing a content decline during the maturation stage. The regulatory mechanisms underlying organic acids catabolism remain elusive (Tadeo et al. 2008; Terol et al. 2010; Lin et al. 2015).

As abovementioned, *C. reticulata* possesses characteristic features concerning organoleptic attributes. Hence, it is of utmost importance to accurately define the timeofharvest in order to obtain high-quality fruits. The quality of *C. reticulata* is evaluated through a series of parameters (or indexes), namely, total soluble solids (TSS, °Brix), titratable acidity (TA) and the ration between both, which is termed maturity index. The TSS is a measure of the sugar content (which is usually ca. 80%) and is, therefore, intrinsically related to the sweetness, while TA conveys the acidity of the fruit which is directly associated with organic acids contents (particularly, citric acid), and is responsible for the sourness. The ratio between those parameters is indicative of the maturation stage, and is an important feature, along with color and juice content (Lado et al. 2018), to assess the overall quality of the fruits. A 7.5 minimum value of sugar/acid ratio was established in the EU concerning tangerines and other mandarins (United Nations Economic Commission for Europe (UNEC) 2018). The mentioned parameters are paramount in the commercialization of *C. reticulata* fruit regarding both consumers and industry.

6.4.2.1.2 Organic Acids

As aforementioned, citric acid is undoubtedly the most abundant organic acid in *C. reticulata* being the vacuoles of juice vesicles cells the main reservoirs. This tricarboxylic acid represents 70–90% of the total organic acids content, and the concentration is highly dependent on the equilibrium between biosynthesis, degradation, transport and vacuolar storage (Etienne et al. 2013; Sheng et al. 2017; Lado et al. 2018).

Citric acid is biosynthesized in the mitochondria (localized in the juice sacs vesicles) through the first reaction of the tricarboxylic acid (TCA) cycle or Krebs cycle, via the condensation of acetyl-CoA with oxaloacetate, catalyzed by citrate synthase. It is worthy to point out that oxaloacetate is synthesized during the glycolytic hydrolysis of sucrose through the irreversible β-carboxylation of phosphoenolpyruvate catalyzed by the cytosolic carboxy-lyase enzyme phosphoenolpyruvate carboxylase (Igamberdiev and Eprintsev 2016; Zhang and Fernie 2018; Pontiggia et al. 2019). Hence, citrate synthase and phosphoenolpyruvate carboxylase are pivotal enzymes in the citric acid biosynthetic pathway.

Organic acids are key precursors of structurally diverse primary metabolites, namely fatty acids and amino acids (Walker and Famiani 2018).

As the fruit maturation evolves, citric acid is exported from the vacuoles of juice sacs cells to the cytoplasm putatively through citrate/H^+ symporters, being catabolized through the aconitase-γ-aminobutyric acid (Aco-GABA) and/or ATP citrate lyase (ACL) pathways. Cytosolic aconitase (cyt-Aco) catalyzes the isomerization of citrate to isocitrate, which is then converted to α-ketoglutarate (2-oxoglutarate) through NADP isocitrate dehydrogenase. Afterwards, the latter compound may be metabolized to the biosynthesis of amino acids, particularly glutamine through glutamine synthase or glutamate through the catalysis of glutamate dehydrogenase and then converted into GABA by glutamate decarboxylase (Degu et al. 2011; Etienne et al. 2013; Guo et al. 2016; Sheng et al. 2017). The isomerization of citrate to isocitrate is a pivotal step in the citrate metabolism and hence a strict regulation of this conversion is of paramount importance for this tricarboxylic acid homeostasis (Degu et al. 2011). Another isoform of cyt-Aco has also been documented (and characterized), being localized in the mitochondria (myt-Aco). A partial abrogation of myt-Aco catalytic activity elicits citrate accumulation in the vacuole and cytosol. Citramalate was documented as a selective inhibitor of the enzymatic activity of the mitochondrial isoform of aconitase (Degu et al. 2011). Degu et al. (2011) advocated that citrate accumulation triggered by aconitase inhibition elicited *C. reticulata* metabolic shift, inducing amino acid biosynthesis and GABA shunt. Yun et al. (2010) performed a proteomic evaluation of *C. reticulata* cv. Egan No.1 (Ponkan fruit) and found that aconitase was upregulated throughout the post-harvest stage, being indicative of the conversion of citrate to isocitrate, hence eliciting the fleshy fruit acidity diminishment. Moreover, this effect was also attributed to an enhancement of malate dehydrogenase, which catalyzes the conversion of malate into oxaloacetate.

Lin et al. (2015) performed a transcriptomic profiling of *C. reticulata* Blanco cv. Ponkan in order to investigate the organic acids metabolism during three fruit maturation phases. The well-perceived acidic mouthfeel of this Chinese cultivar (due to the low sugar to acid ratio) negatively influences the consumers' acceptability. In an attempt to circumvent this undesirable sensory trait, the authors evaluated the impact of delaying the fruit harvest on the improvement of the taste. The study reported that during the late maturation stage, an acceleration of the Krebs cycle and glycolytic pathway occurs, suggesting a metabolic shift towards organic acids catabolism. The authors observed that the citrate degradation occurred predominantly through the GABA shunt (upregulation of aspartate aminotransferase and glutamate decarboxylase) and acetyl-CoA pathway (upregulation of ATP-citrate lyase). Concerning malic acid degradation, an upregulation of malate dehydrogenase was observed.

Albeit *C. reticulata* sourness may be mainly attributed to the citric acid content, scarce concentrations of malic acid also contribute to this sensory trait (Lado et al. 2018; Liu et al. 2021). The biosynthesis of this dicarboxylic acid occurs predominantly in the cytosol of the tangerine mesocarp cells through the carboxylation of the glycolytic intermediate, phosphoenolpyruvate, mediated by phosphoenolpyruvate carboxylase. Likewise, the catabolism of malate through decarboxylation via malate dehydrogenase enables the synthesis of phosphoenolpyruvate, which in turn triggers gluconeogenesis. As aforementioned, this phenomenon occurs predominantly during *C. reticulata* ripening (Etienne et al. 2013).

Feng et al. (2018) evaluated the metabolites profile of *C. reticulata*Blanco (Daisy) and reported a concentration of malic acid (1 g L^{-1}) 6.5-fold lower than the amount of citric acid. The authors also documented a vestigial concentration of succinic acid (10 mgL^{-1}).

6.4.2.1.3 Amino Acids

Citrus reticulata is considered a negligible protein source (Liu et al. 2012). In fact, this scarce protein content is mainly attributed to the high content of carbohydrates that comprise most of the fruits composition (Hildebrandt et al. 2015). The rind of *C. reticulata* presents a protein content of 2.3–3.1% (Pradhan and Sharma 2019), while the edible portion displays a residual value of 0.8% (USDA). Beyond the proteinaceous amino acids, the nonprotein aminoacidic residues represent most of the nitrogenous components (nearly 70%) (Liu et al. 2012). Amongst those, nonessential amino acids, namely alanine, arginine, asparagine, aspartic acid, glutamic acid, glycine, proline and serine, are prevalent in *C. reticulata* (Shorbagi et al. 2022). Wang et al. (2016) claimed that the amino acids content comprised in juice sacs was the highest in comparison to flavedo, albedo and segment membranes.

In accordance with a recent study conducted by Wallis et al. (2022), in which the amino acid profile of the leaf tissue of *C. reticulata* Blanco (W. Murcott) was evaluated, L-alanine, L-methionine, L-phenylalanine, serine, tryptophan, tyrosine and valine were found to be the most abundant amino acids (Wallis et al. 2022).

Amino acids own prominent cell functions, participating in a myriad of cell metabolic pathways, hence influencing numerous physiological processes, namely the growth and development of the fruit, the regulation of the intracellular pH and energy status, and the stress responses to abiotic and biotic stimuli (Hildebrandt et al. 2015).

The aforementioned essential amino acids provide a prominent contribution to the fruit nutritional quality. Moreover, some free amino acids present a significant impact on the fruit taste. Particularly, arginine elicits a bitter sensory perception, whilst serine, proline and alanine are significantly associated to sweetness (Keutgen and Pawelzik 2008; Pal Choudhuri et al. 2015; Zhang et al. 2022).

As formerly mentioned, amino acids metabolic pathway is intertwined with glycolysis and tricarboxylic acid cycle (Zhang et al. 2018b; Sadka et al. 2019). During the second stage of *C. reticulata* development, the citrate content declines, being metabolized into glutamate and henceforth the amino acids biosynthesis is initiated (Sadka et al. 2019).

Proline

In citrus fruits, L-proline is predominantly synthesized through the glutamate pathway, which occurs in the cytosol (Shinde et al. 2016; Boublin et al. 2022). Plants exposure to abiotic stresses, namely drought and nitrogen deprivation, triggers this route (Meena et al. 2019).

In this biosynthetic pathway, L-glutamate is initially phosphorylated to form the labile intermediate L-glutamyl 5-phosphate in a reaction mediated by the bifunctional enzyme Δ^1-pyrroline-5-carboxylate synthetase (P5CS). This enzyme comprises two catalytic domains, namely Δ^1-γ-glutamyl kinase (catalyzing the abovementioned L-glutamate phosphorylation) and glutamic-γ-semialdehyde dehydrogenase (Turchetto-Zolet et al. 2009). Subsequently, γ-glutamyl phosphate reductase of P5CS catalyzes the latter compound reduction to γ-glutamate-semialdehyde, which undergoes spontaneous cyclization be converted to 1-pyrroline-5-carboxylate. Lastly, Δ^1-pyrroline-5-carboxylate reductase catalyzes the conversion of this transient compound to L-proline. The former bifunctional protein, P5CS, is the pivotal enzyme for the L-proline synthesis, mediating the rate-limiting step in the biosynthetic route, being regulated at transcriptional and post-translation levels. This latter regulatory mechanism occurs via the feedback inhibition of the enzyme activity by L-proline (Meena et al. 2019). Alternatively, proline may be synthesized from ornithine mediated by ornithine aminotransferase (Funck et al. 2008).

Proline presents a pivotal role as a compatible osmolyte during drought (for instance), stabilizer of the intracellular cellular redox status (scavenger of reactive oxygen species), and chemical chaperone (Hildebrandt et al. 2015; Shinde et al. 2016; Boublin et al. 2022; Kishor et al. 2022). Proline catabolic pathway occurs in the mitochondria, proceeding through two oxidation reactions, catalyzed by proline dehydrogenase and 1-pyrroline-5-carboxylate dehydrogenase, to yield L-glutamate (Hildebrandt et al. 2015).

Kang et al. (2022) analysed the amino acids profile of large and small-sized *C. reticulata* cv. Shatangju after long-term cold storage. Concerning proline and leucine, the authors found that those were the only two amino acids detected in the pulp of the larger fruits. The authors reported that proline content in the pulps of the largest fruits was 10 $mgkg^{-1}$ DW while in the smallest was 55 mg kg^{-1} DW.

He et al. (2022) evaluated the amino acids content of *C. reticulata* Blanco ("Orah" mandarin) juice of distinct climatic regions and found that proline was the most abundant amino acid, with concentrations varying from 0.14 to 0.51 mg *per* mL of fresh juice. The authors documented a negative correlation between the proline concentration and the annual relative humidity. In fact, during the fruit ripening, a low relative humidity or scarce water availability triggers osmotic stress, and as a response mechanism *C. reticulata* gathers compatible solutes, namely proline (He et al. 2022). Moreover, a positive correlation between proline content and plants experiencing lower temperatures (<10 °C) was also reported. In accordance with this finding, Yun et al. (2010) observed that post-harvest cold storage (abiotic stress) of *C. reticulata* cv. Egan triggered proline biosynthesis in juice sacs through the ornithine pathway via arginine aminotransferase, which expression increased at the thirtieth day of storage, indicating proline accumulation.

Arginine

An extremely growth-constraining condition for plants is the scarcity of nutrients, particularly the essential element, nitrogen. Owing to the high nitrogen-to-carbon ratio (4:6), arginine is considered the primary organic nitrogen storage form in plants, along with asparagine and glutamine (Forde and Lea 2007; Witte 2011; Winter et al. 2015).

The biosynthetic route of arginine occurs in the chloroplast and may be partitioned in two reactions, in which glutamate is converted into ornithine, through numerous acetylated intermediates, and henceforth this latter molecule originates arginine (Slocum 2005; Winter et al. 2015). In brief, L-glutamate is first converted to *N*-acetylglutamate by *N*-acetylglutamate synthase, which is thenceforth phosphorylated via *N*-acetylglutamate kinase to yield N-acetylglutamyl-phosphate. Subsequently, this latter compound is reduced to *N*-acetylglutamate-5-semialdehyde by *N*-acetylglutamate-5-P reductase. Then, N^2-acetylornithine aminotransferase catalyses the transference of the amino moiety of L-glutamate to generate N^2-acetylornithine. At last, this compound undergoes deacetylation through *N*-acetylornithine: *N*-acetylglutamate acetyltransferase, yielding ornithine. Henceforth, ornithine undergoes a 3-step pathway, in which carbamoylated ornithine originates citrulline. This latter compound is associated to L-aspartate via argininosuccinate synthase yielding argininosuccinate, which is reversibly cleaved into fumarate and L-arginine through the catalysis of argininosuccinate lyase (Slocum 2005; Winter et al. 2015).

In order to remobilize the assimilated nitrogen, arginine must be degraded and the ammonium originated is then translocated to the sink tissues. Leaf senescence and germination activate the catalytic activity of the enzymes comprised in this catabolic pathway. In sink tissues, the guanidine moiety of arginine is mitochondrially hydrolyzed via arginase generating ornithine and urea. The hydrolysis of this latter compound, mediated by urease, culminates with the formation of ammonia and carbamate. The complete catabolism of arginine occurs via transamination of ornithine mediated by ornithine-δ-aminotransferase yieldingΔ^1-pyrroline-5-carboxylate which is subsequently oxidized to glutamate catalyzed by 1-pyrroline-5-carboxylate dehydrogenase (Witte 2011; Hildebrandt et al. 2015; Kishor et al. 2022).

Yun et al. (2010) performed a proteomic evaluation of *C. reticulata* cv. Egan (juice sacs) throughout postharvest low temperature storage. The authors found that ornithine aminotransferase (a pivotal enzyme in tangerine stress tolerance) was considerably augmented at the thirtieth day of cold storage, leading to the catabolism of arginine with putative conversion into glutamate, which presents a positive role in cold acclimation.

In citrus fruits, arginine and ornithine may be precursors of polyamines, which display a protective role to counteract cold stress owing to the antioxidant and ROS scavenging, as well as anti-senescence activity. In fact, spermidine content was found to be increased in the flavedo of a *C. reticulata* hybrid following exposure to low and high temperatures (Nehela and Killiny 2020).

Glycine

In plants, glycine may be biosynthesized through three distinct pathways, namely the photorespiratory glyoxylate route, the reverse glycine cleavage system or the threonine catabolic pathway (Joshi et al. 2006; Jander and Joshi 2010; Dellero et al. 2016; Xu et al. 2022b). Concerning this latter, the enzyme threonine aldolase catalyzes the reversible (retro-aldal) cleavage of threonine to yield glycine and acetaldehyde (Jander and Joshi 2010).

In the photorespiratory glycolate route, serine:glyoxylate aminotransferase catalyses the transamination of glyoxylate through the transfer of an amide group of serine to yield glycine and hydroxypyruvate (Dellero et al. 2016).

In the reverse glycine cleavage system, homodimeric pyridoxal 5′-phosphate-dependent glycine dehydrogenase catalyzes the conversion of N^5, N^{10}-methylene-tetrahydrofolate (5,10-CH_2-THF) into glycine (Xu et al. 2022b).

Interestingly, Degu et al. (2011) advocated that glycine biosynthesis may be triggered by aconitase inhibition in *C. reticulate*. The authors observed a prominent increase in the glycine content following tangerine treatment with the aconitase inhibitor, oxalomalate. Moreover, an improvement on the aspartate kinase enzymatic activity was reported. These observations shed light on the hypothesis that aspartate was first phosphorylated and henceforth glycine was biosynthesized viaa branch off the aspartate-derived amino acid biosynthetic pathway.

Glycine catabolic pathway occurs via the decarboxylation of the amino acid by the mitochondrial system of glycine decarboxylase, with the concomitant transfer of a methyl group to tetrahydrofolate, yielding ammonium and carbon dioxide (Hildebrandt et al. 2015).

He et al. (2022) stated that the glycine concentration of *C. reticulata* Blanco ranged between 0.032 to 0.0092 mg *per* mL of fresh juice. A positive correlation was observed between glycine concentration and the sunlight exposure, which is may be indicative of the photorespiratory glyoxylate biosynthetic route. The authors also found that a lower temperature contributed to an accumulation of L-glycine, which pointed out to a cryoprotective role, hence facilitating the fruit cold acclimation.

Asparagine

Asparagine is mainly biosynthesized through the amidation of aspartate via asparagine synthetase, in which the amide group donor is glutamine (Gaufichon et al. 2010).

In sink tissues, the complete degradation of asparagine proceeds through hydrolysis of the amide group mediated by asparaginase to yield aspartate and ammonia, eliciting the nitrogen reassimilation (Curtis et al. 2018). As aforementioned, asparagine, along with glutamine and arginine, constitutes the main nitrogen translocator and reservoir from source organs to sink tissues (Gaufichon et al. 2010).

Lin et al. (2015) performed a metabolic and transcriptomic analysis of *C. reticulate* Blanco cv. Ponkan at three different stages of the fruit maturation (early development, commercial harvest phase and delayed harvest stage). The absolute amount of asparagine was found to be 0.465, 0.001 and 0.316 mg *per* g of fresh weight, respectively. These values were highly correlated with the aspartate (asparagine precursor) content, being observed an upregulation of aspartate aminotransferase between the conventional and the delayed harvest stages. Noteworthy, all the amino acids displayed a dramatic decline from the early development to commercial harvest phase and from that phase onwards a pronounced increase was observed.

Suh et al. (2021) provided novel insights into the metabolic mechanisms underlying *C. reticulata* tolerance to Huanglongbing (a phloem-restricted bacterial disease originated by "*Candidatus* Liberibacter asiaticus"). The authors conducted a comparative metabolites evaluation of healthy and infected Huanglongbing-tolerant and -sensitive *C. reticulata,* being observed that the asparagine and glutamine contents were enhanced in the tolerant *C. reticulata*, irrespective of the healthy or infected status. The host accumulation of these metabolites in response to infection was indicative that both amino acids may predominantly constitute upstream substrates on the purine metabolic pathway, which is a core route coupled to aspartate-glutamate metabolic pathway and phytohormones biosynthesis. In this route, the downstream biosynthesized auxins and cytokinins promote the cell growth and phloem regeneration, hence conferring host tolerance to the infection (Suh et al. 2021).

Aspartate

The aspartate biosynthetic pathway intertwines the glycolytic route/TCA cycle and amino acids biosynthesis pathways, which are of utmost importance for plant nutrition, and stress responses (Galili 2011; Kirma et al. 2012). In brief, oxaloacetate undergoes a reversible transamination through the catalysis of aspartate aminotransferase (aspartate:2-oxoglutarate aminotransferase) to yield L-aspartate and 2-oxoglutarate (De La Torre et al. 2014). In this biosynthetic pathway oxaloacetate may be originated through the glycolytic pathway or via the TCA cycle. The catabolic pathway of aspartate occurs promptly through its transamination to generate oxaloacetate (Hildebrandt et al. 2015).

According to Lin et al. (2015) the absolute aspartate content of *C. reticulata* decreased from the early stage development (0.465 mg g^{-1} fw) to the delayed harvest (0.316 mg g^{-1} fw). Noteworthy, an aspartate concentration of 0.001 mg g^{-1} fw was obtained in the commercial harvest stage.

Tan et al. (2019) investigated the impact of autotetraploidization on the primary metabolism of *C. reticulata* Blanco (Ponkan mandarin) during three consecutive years. The authors found that the concentration of aspartic acid was significantly higher in the autotetraploid fruit in comparison to the parental diploid (control) variety at the 210th day post flowering. As a whole, autotetraploidization appeared to modulate *C. reticulata* metabolism, promoting an accumulation of this primary metabolite. Autotetraploidization may constitute a feasible approach to enhance plant tolerance to abiotic stressors.

Glutamate

In higher plants, glutamate displays a central role in amino acids metabolic pathway, being the α-amino moiety directly implicated in the nitrogen assimilation, and transferred to the other amino acid residues (Forde and Lea 2007; Liao et al. 2022). Moreover, glutamate is the key precursor of γ-aminobutyric acid (GABA) (Fig. 6.1), proline and arginine (Forde and Lea 2007; Majumdar et al. 2016). Glutamate may

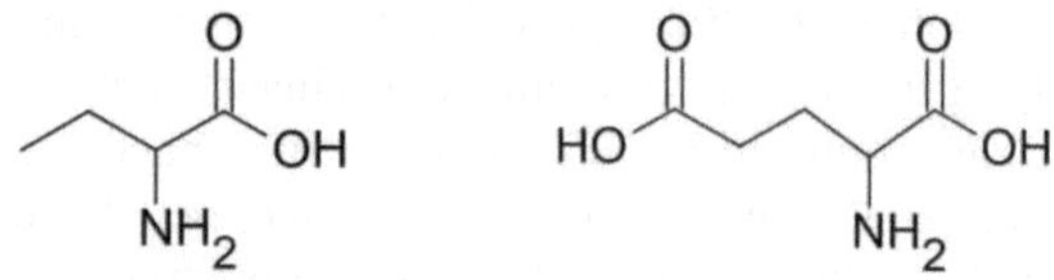

GABA L-Glutamate

Fig. 6.1 Chemical structures of GABA and L-glutamate found in *C. reticulata*

be *de novo* biosynthesized via glutamate synthase (also designated glutamine:2-oxoglutarate amidotransferase), which catalyses the transfer of the amide moiety of glutamine to 2-oxoglutarate originating two molecules of glutamate (Forde and Lea 2007).

The catabolic pathway may proceed either through oxidative deamination (aspartate aminotransferase catalyses the reversible conversion of glutamate to 2-oxoglutarate) or via irreversible decarboxylation catalysed by the cytosolic glutamate decarboxylase to generate GABA. Concerning the latter route, GABA undergoes transamination via GABA transaminase to generate succinic semialdehyde, which is then mitochondrially oxidized to succinate by succinate semialdehyde dehydrogenase. Hence, the well-documented GABA shunt circumvents two steps of the TCA cycle (Hildebrandt et al. 2015).

In order to cope with exposure to abiotic stresses, namely acidification, oxygen deprivation, heat shock, drought, citrus fruits accumulate GABA in the cytosol (Forde and Lea 2007; Chen et al. 2012; Podlešáková et al. 2019). Former studies pointed out to an activation of GABA shunt during *C. reticulata* ripening stage and postharvest storage (Chen et al. 2012; Lin et al. 2015, 2016). In fact, GABA displays a critical role in the regulation of citric acid catabolism and, consequently, in the vacuolar pH homeostasis of the juice vesicles, which influences tangerine acidity (Sadka et al. 2019; Strazzer et al. 2019). Worth noting, an upregulation of aspartate aminotransferase and glutamate decarboxylase (GAD) was documented, indicating that the GABA shunt was activated during the later phase of fruit maturation (Lin et al. 2015).

Sheng et al. (2017) also investigated the expression of GABA shunt genes in low-acid hybrids of *C. reticulata* and *Citrus grandis* throughout postharvest storage. In agreement with the former study, the authors also reported an enhancement on the GABA content concomitant with the storage time extension, being observed an upregulation of the genes encoding glutamate decarboxylase and mitochondrial GABA permease.

Serine

Serine is involved in the biosynthetic pathways of a myriad of biomolecules of paramount importance for cell development and proliferation, namely amino acids, nucleotides, phospholipids. Moreover, L-serine possesses a pivotal role in the one-carbon metabolism and in cell signalling networks (mediated via serine-kinases) (Kalhan and Hanson 2012; Ros et al. 2014).

In plants, serine may be biosynthesized through three distinct mechanisms, namely the photorespiratory glycolate route and the two non-photorespiratory pathways, the phosphorylated and glycerate routes. Notwithstanding, serine synthesis via the former pathway (glycolate) has been deemed the most significant (Benstein et al. 2013; Ros et al. 2014; Okamura and Hirai 2017). In this photorespiratory mechanism, the glycine decarboxylase multi-enzyme complex catalyzes the decarboxylation and deamination of L-glycine, originating carbon dioxide and ammonia. Henceforth, the glycine-derived methylene group is directly transferred to tetrahydrofolate yielding methylene-tetrahydrofolate. Lastly, this latter molecule is combined with a second L-glycine to form L-serine via serine hydroxymethyltransferase (Douce et al. 2001; Ros et al. 2014). Owing to its paramount role in the plant metabolism, strict regulatory mechanisms underlying serine biosynthesis are critical to warrant the amino acid homeostasis. Nonetheless, the plethora of biosynthetic routes complicates the mechanistic comprehension of this intricate regulation (Ros et al. 2014).

The complete degradation of serine occurs through transamination of the amino acid mediated by serine-glyoxylate aminotransferase to yield hydroxypyruvate, which is then reduced to 3-glycerate. Subsequently, this latter molecule undergoes phosphorylation via glycerate kinase to yield 3-phosphoglycerate, which may be reduced to glyceraldehyde 3-phosphate and utilized in carbohydrate biosynthesis or glycolysis, or may be re-introduced in the Calvin Benson cycle (Douce et al. 2001; Hildebrandt et al. 2015; Modde et al. 2017). Alternatively, serine may undergo direct deamination mediated by a dehydratase, the serine racemase localized in the cytosol to form pyruvate (Hildebrandt et al. 2015).

According to the above detailed study performed by He et al. (2022), the serine content of *C. reticulata* Blanco ranged between 0.01 and 0.025 mg *per* mL of fresh juice. The authors observed a positive correlation between the annual average sunshine hours and the serine content. The appropriate solar radiation promotes a high photosynthetic photon flow density. These results are consistent with the photorespiratory biosynthetic mechanism of serine. Interestingly, lower temperatures (<10 °C) trigger the serine biosynthesis.

Kang et al. (2022) evaluated the impact of fruit size in the physicochemical characteristics of *C. reticulata* following long-term cold storage. Serine was only absent in the largest fruits, and the highest content was observed in smallest, which was determined to be 25 mg kg^{-1} DW. The presence of lower diversity and amounts of amino acids and organic acids, both of which are responsible for flavor characteristics, in the larger specimens following long-term storage originated insipid fruits.

6.4.2.1.4 Lipids

Citrus reticulata flesh is considered to be scarcely endowed with lipids (Liu et al. 2012). Owing to the paramount role in the signal transduction, lipids are deemed to be critical for the proper plant growth and responses to abiotic stresses (Abdelrahman et al. 2021). Moreover, the tremendous importance of plants lipids, considered primary metabolites, relies on their influence in the membrane fluidity, integrity and permeability.

Citrus reticulata biosynthesizes complex blends of chemically diverse lipids, belonging to distinct classes, namely fatty acids, sterols, waxes, carotenoids, amongst others (Shorbagi et al. 2022).

In plants, fatty acids are predominantly synthesized and catabolized (β-oxidation) in chloroplasts and peroxisomes, respectively. Nonetheless, in recent years the mitochondrial contribution in the fatty acid catabolic pathway has become increasingly evident. The enzyme ATP-citrate lyase catalyzes the cytosolic conversion of this tricarboxylic acid into acetyl-CoA for *the novo* fatty acid synthesis (Daloso et al. 2015).

*Citrus reticulata*seeds and rind (with the flavedo comprising oil sacs) are the primary reservoir of fatty acids. Fatty acids profiling of *C. reticulata* unraveled that linoleic acid (C18:2) was the most abundant, followed by palmitic acid (C16:0) and α-linolenic acid (C18:3) accounting for 32.76, 19.86 and 19.77%, respectively, the of total content (Lamine et al. 2019).

The seeds of *C. reticulata* have also been observed to comprise those fatty acids, along with two others, namely, oleic (C18:1 n-9) and stearic (C18:0). In seeds oil, oleic acid was the third most abundant, while α-linolenic acid was, among the five mentioned fatty acids, the one which presented the lowest amount (Rashid et al. 2013).

Tounsi et al. (2011) examined *C. reticulata*seeds and found that linoleic, followed by oleic, palmitic, α-linolenic, stearic and palmitoleic acids were the fatty acids present in the oils (Tounsi et al. 2011). Stearic and oleic acids were also observed in the abovementioned pulp oil, in that case being the fourth and fifth most prevalent (Lamine et al. 2019).

Citrus reticulata peels were also deemed as potential sources of fatty acids, since peel oil was described to comprise similar profile to those found in pulp and seed oils, being mainly constituted by linolenic (33.71%) and palmitic acids (29.26%), also encompassing oleic, stearic, myristic and α-linolenic acids (10.8, 6.61, 5.69 and 3.49%, respectively) (Castro et al. 2018).

Lipids are also present in fruit cuticular waxes, which composition is significantly variable, depending on species, ontogeny and environmental conditions. Generally, waxes comprise mainly compounds derived from very-long-chain fatty acids, namely, fatty acids, aldehydes, alkanes, esters, primary and secondary alcohols, and ketones (Yeats and Rose 2013). Yang et al. (2022) found that *C. reticulata* cuticular wax mainly presented aldehydes (1.29 μg cm^{-3}, 45.29% of total aliphatic wax; TAW) and alkanes (1.26 μg cm^{-3}, 44.1% TAW), despite fatty acids and alcohols also being detected, although in significantly lower amounts (0.16 and 0.14 μg cm^{-3}, respectively).

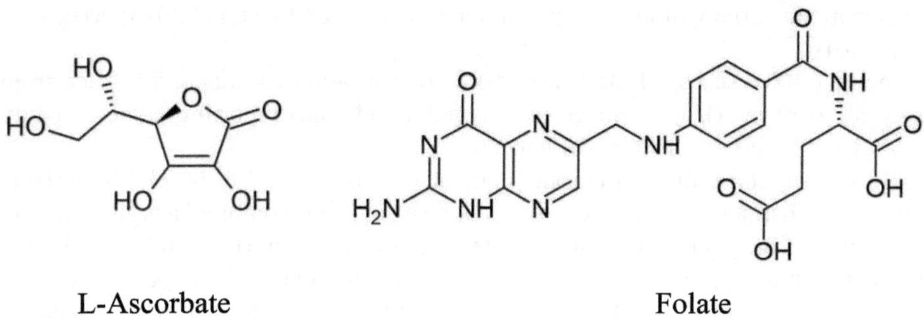

Fig. 6.2 Chemical structures of vitamins commonly found in *C. reticulata* flavedo

6.4.2.1.5 Vitamins

L-ascorbate (vitamin C)

Citrus reticulata is an important source of L-ascorbate or L-ascorbic acid (also referred as vitamin C) (Fig. 6.2). This metabolite possesses a notable antioxidant activity, being a co-factor for enzymes involved in a plethora of biochemical reactions (Asensi-Fabado and Munné-Bosch 2010; Wheeler et al. 2015; Magwaza et al. 2016).

Although elusive for many years (Wheeler et al. 1998), the biosynthetic pathway of L-ascorbic acid in plants, is still not fully deciphered (Wheeler et al. 2015). The L-D-mannose/L-galactose or Smirnoff-Wheeler route appears to be the primary pathway of the biosynthesis of L-ascorbic acid, through the conversion of GDP-D-mannose to L-ascorbic acid. It has been postulated that GDP-mannose-3,5-epimerase catalyzes the C3 and C5 epimerization of GDP-D-mannose to GDP-L-galactose, which is subsequently converted to L-galactose-1-phosphate by GDP-galactose phosphorylase. L-Galactose-1-P phosphatase catalyzes the dephosphorylation of the latter compound to generate L-galactose, which upon C-1 oxidation and cyclization is converted to L-galactone-1,4-lactone by L-galactose dehydrogenase. This 5-keto compound is finally oxidized to L-ascorbate in a reaction catalyzed by the mitochondrial FAD-linked L-galactono-1,4-lactone dehydrogenase (Laing et al. 2007; Wheeler et al. 2015; Magwaza et al. 2016; Jiang et al. 2021). In the above detailed biosynthetic pathway, the regulation of the route flow mostly relies on the GDP-L-galactose phosphorylase step (Laing et al. 2007, 2015; Bulley et al. 2012).

In citrus fruits, a plethora of factors may influence the L-ascorbate content, namely genotype, climatic conditions, preharvest (cultural) production practices, fruit maturity (time of harvest), harvesting methods and postharvest systematic practices (Lee and Kader, 2000; Magwaza et al. 2016).

Significant scientific evidences pointed out that L-ascorbate content is mostly influenced by the preharvest conditions, particularly light, temperature, nitrogen fertilization, carbohydrate content, among others (Lado et al. 2015; Magwaza et al. 2016).

Nagy (1980) surveyed the L-ascorbate concentration for mandarins juice from several countries (United States, Israel, India and Iran) reporting values ranging from 14 to 54 mg *per* 100 mL of juice.

In *C. reticulata*, the L-ascorbate content was also found to be highly variable among the distinct edible tissues. Abeysinghe et al. (2007) reported a higher vitamin C content in juice sacs (45.3 mg *per* 100 g of fresh weight (FW), followed by the segment (38.9 mg *per* 100 g FW) and segment membrane (14.2 mg *per* 100 g FW). In a study performed by (De Moraes Barros et al. 2012), the ascorbate concentrations of pulp and peel from commercial cultivars of *C. reticulata* cv. Ponkan in Brazil were compared. According to the authors, the fruit rind presented a higher vitamin content (41.1 mg *per* 100 mL) in comparison to the fleshy pulp (47.6 mg *per* mL). These values were in accordance with those previously documented by Nagy (1980), who surveyed the L-ascorbate concentration for mandarins juice from several countries (United States, Israel, India and Iran) reporting values ranging from 14 to 54 mg *per* 100 mL of juice. Noteworthy, the high variability of the reported ascorbate contents are a consequence of the aforementioned pre- and post-harvest conditions. In light of these findings, the authors postulated that the higher L-ascorbate content in the rind may be ascribed to the high requirement for antioxidant compounds with potential to efficiently scavenging reactive oxygen species (ROS) originated through the direct exposure of the fruit rind to environmental (abiotic) stresses.

Rey et al. (2020) evaluated the L-ascorbate content in *Citrus reticulata* flavedo and found that light deprivation originated a prominent decline (nearly 50%) in the vitamin amount (initial content of 2.4 g kg^{-1}).

Folate

Folate or folic acid (pteroylglutamic acid) is another example of a water-soluble vitamin occurring in citrus fruits (Brouwer et al. 1999; World Health Organization 2015). Folate, the polyglutamate form, has been argued to be the naturally occurring form of this vitamin B9 (Fig. 6.2). Similarly to L-ascorbate, folate exerts a pivotal role in the maintenance of the immune system barriers integrity and in providing the support to the proper function of several immune cells (Miles and Calder 2021; Rey et al. 2021). The vitamin is an essential cofactor for enzymes involved in the acceptance or donation of single carbon units in pivotal metabolic pathways. This coenzyme is involved in homocysteine metabolism, biosynthesis of nucleic acids (synthesizing purine and pyrimidine) and in cell growth and division (Bailey and Gregory 1999; Shang et al. 2021).

Owing to the paramount folate role in human health, disturbances in the physiological vitamin homeostasis elicit a detrimental health impact. In fact, the folate deficiency is a serious global health concern, and has long been investigated given its association with several pathologies, namely megaloblastic anaemia, inflammatory bowel disease, coronary heart disease and cancer (Rimm et al. 1998; De La Garza et al. 2007; Alam et al. 2019).

Mammalian cells are devoid of the folate biosynthesis enzymes, and hence humans must acquire this vitamin from exogenous dietary sources, particularly from plant foods (De La Garza et al. 2007; Ifergan et al. 2008).

Citrus reticulata (in its raw or juiced form) is a particularly interesting source of folate, biosynthesizing the vitamin*de novo*through an intricate metabolic pathway. The tetrahydrofolate (referred as folates) is a tripartite compound structurally composed of a pterin moiety, a *para*-aminobenzoic acid (pABA) and a γ-bounded glutamate tail (Gorelova et al. 2019; Kołton et al. 2022). The folate biosynthesis occurs in three subcellular compartments, namely cytosol (pterin biosynthesis), plastids (pABA production) and mitochondria. In this latter organelle occurs the assembly of tetrahydrofolate molecule, through the condensation of the two moieties of pterin and pABA, originating dihydropteroate, which is subsequently glutamylated via the attachment of the L-glutamate residues (De La Garza et al. 2007; Kołton et al. 2022). *Para*-aminobenzoic acid is synthesized from chorismate in a two-step reaction mediated by 4-amino-4-deoxychorismate and aminodeoxychorismate lyase. The biosynthesis of pterin initiates with the conversion of guanosine-5′-triphosphate to dihydroneopterin triphosphate mediated by GTP cyclohydrolase I. The generated molecule is then dephosphorylated in a two-step reaction through the catalysis of cytosolic Nudix hydrolase and a non-specific phosphatase. The final reaction of the pterin synthesis comprises the concomitant release of glycolaldehyde and 6-hydroxymethildihydropterin mediated by dihydroneopterin aldolase. The tetrahydrofolate biosynthesis in mitochondria initiates with the pyrophosphorylation of the latter compound catalyzed by HMDHP pyrophosphokinase, proceeding with the association with pABA, mediated by dihydropteroate synthase, originating dihydropteroate. In a subsequent reaction, this molecule is converted into dihydrofolate, mediated by dihydrofolate synthetase, which is further reduced to tetrahydrofolate by dihydrofolate reductase. The final sequential attachment of L-glutamate residues catalyzed by folylpolyglutamate synthetase leads to the conversion of the tetrahydrofolate into tetrahydrofolate polyglutamate (Gorelova et al. 2017, 2019; Kołton et al. 2022). A critical point in the regulation of folate homeostasis relies on the removal or shortening of polyglutamate tail via vacuolar gamma-glutamyl hydrolase (Gorelova et al. 2019). In plants, despite the well-described biosynthetic and metabolic pathways of folate, the underlying regulatory mechanisms and cellular compartmentalization in different plant lineages remain uncharacterized and, hence, a genetic dissection would provide valuable insights (Kołton et al. 2022).

The folate content (on average) in *C. reticulata* is quoted as 16 μg *per* 100 g of edible fruit. Noteworthy, the high variability of the vitamin concentration may be attributed to the plant physiological status, cultivar, season, post-harvest conditions, among others (Delchier et al. 2016).

Concerning the evaluation of the individual folate derivatives, 5-methyltetrahydrofolate (the fully reduced form, methylated on the nitrogen located in position 5) was found to be the most abundant vitamer in tangerine (13 ± 1 μg *per* 100 g of fresh weight), followed by the most unstable folate, tetrahydrofolate (1 ± 1 μg *per* 100 g of fresh weight) (Delchier et al. 2016). Of the total folate content in tangerine, 57% was in the form of polyglutamate (Konings et al. 2001).

6.4.2.2 Secondary Metabolism

Plants are endowed with the capability to biosynthesize a unique and complex repertoire of secondary metabolites (also termed specialized metabolites), which are low-molecular-weight molecules, biologically active, constituting a highly exploited reservoir for the synthesis of pharmacological agents (Goossens et al. 2003; Benderoth et al. 2006). This chemically diverse class of metabolites confers a selective advantage to the synthesizing organism concerning the adaptation to the surrounding environmental constraints (Craney et al. 2013). In fact, biotic and abiotic stressors trigger a plethora of physiological responses, including the biosynthesis of the aforementioned secondary metabolites (Li et al. 2016). The adaptive mechanisms, and the chemical diversity of secondary metabolites, have evolved in order to cope with the single or concurrent exposure to the abiotic and biotic stress signals/stressors (Benderoth et al. 2006; Bacete et al. 2022). This diversity may be ascribed to the plasticity of the secondary biosynthetic pathways, conferring plants the potential to generate an overwhelming array of structural and/or chemical-modified secondary metabolites (Li et al. 2016). *Citrus reticulata*is endowed with a blend of secondary metabolites, namely flavonoids (flavanones, flavones, flavonols, anthocyanins;being flavanone the most abundant), phenolic compounds or polyphenols, limonoids, volatile organic compounds, carotenoids, and alkaloids (Sanches et al. 2022). Moreover, citrus fruits are considered the most valuable dietary source of flavonoids and vitamins (Wang et al. 2016). These classes of secondary metabolites present a notable biological activity of considerable therapeutic importance to human health and well-being, with flavonoids (namely hesperidin, the most abundant flavonoid present in *C. reticulata* peel) possessing potent antioxidant, anticarcinogenic and anti-inflammatory properties, as well as neuroprotective effects (Ho and Lin 2008; Hwang et al. 2012; Ho and Kuo 2014; Zhang et al. 2014a). Furthermore, *C. reticulata* bioactive potential has been documented concerning antidiabetic, antiproliferative, antihypertensive, antiatherogenic, antiviral and antimicrobial properties, as well as skin hydrolipidic balance improvement, gut microbiota modulation and cardiomyopathy prevention (Hwang et al. 2012; Taher et al. 2020; Guo et al. 2021; Infante et al. 2022).

The tissues of *C. reticulata* (namely flavedo, albedo, segment membrane and juice vesicles) may comprise differential accumulation profiles of secondary metabolites. This was observed in the study conducted by Wang et al. (2017b), in which phenolic contents of the individual tissues of *C. reticulata* belonging to two distinct cultivars were evaluated. The analysis revealed flavedo to account for the highest content, followed by albedo, segment membranes and juice sacs, respectively. Wang et al. (2016) reported a distinctive flavonoids accumulation pattern, being observed that flavedo enclosed the highest concentration of flavonoids, followed by albedo, segment membranes and juice sacs.

A plethora of factors may impact *C. reticulata* secondary metabolites composition, in terms of profiles and abundance, being the cultivar one of those factors. In a study performed by Wang et al. (2017b), two cultivars, namely, Mantouhong and Ponkan, from a particular region in China (Zhejiang Province), were analysed in terms of phenolic contents. It was observed that, in comparison to Ponkan, Mantouhong cultivar possessed higher amounts of those compounds in flavedo and albedo, while in segment membrane there was no significant difference, and in juice sacs Ponkan phenolics concentration was higher.

6.4.2.2.1 Volatile Organic Compounds

The uniqueness of the *C. reticulata* scent is an appealing trait, co-evolving with the fruit ripeness (Melin et al. 2019). The plethora of *Citrus*aromas, namely fruity, floral, minty, lavender, lemony, citrusy, mushroom, woody, piney may be attributed to the emission of hundreds of volatile organic compounds (VOC) comprised in tangerines (Tietel et al. 2010, 2011). Owing to the inherent fruit perishability, off-flavor volatiles are biosynthesized, which may hamper the sensory acceptability (Goldenberg et al. 2016).

Volatile compounds are the main constituents of essential oils, which comprise an intricate amalgam of distinct compounds. The majority of those chemicals are comprised within the terpen class (namely, terpenoids), albeit phenylpropanoids are also abundant. Nonetheless, essential oils also encompass other compounds, namely aliphatic and aromatics (Raut and Karuppayil 2014; Dhifi et al. 2016).

The VOC profile of *C. reticulata*comprises terpenoids, alcohols, aldehydes, acids and esters, which are considered the most predominant classes (Pichersky et al. 2006; Schwab et al. 2008; Zhang et al. 2017). Among these, terpenoids represent the most prominent class in the peel of citrus fruits, particularly monoterpenes compounds (Zhang et al. 2017; González-Mas et al. 2019). Some of these compounds are considered primary metabolites, being of utmost importance for the plant growth and development (phytohormones), whilst others are secondary metabolites, namely mono-, di- and sesquiterpenes, displaying a pivotal role in surrounding environmental adaptation and chemical defense (Ward et al. 2011; Henry et al. 2015; Zhang et al. 2017). These compounds naturally occur in specialized quasi-spherical (circular cavities) oil glands in the flavedo and in oil bodies in the juice vesicles comprised in the whorl of fused segments (Schwab et al. 2008).

Terpenoids (also referred as terpenes or isoprenoids) are biosynthesized via the mevalonate pathway, hence designated owing to the formation of the intermediate mevalonic acid. Terpenoids are originated from two 5-carbon (C5) precursors, isopentenyl diphosphate (IPP) and its corresponding allylic isomer dimethylallyl pyrophosphate (DMAPP), which are biosynthesized viatwo independent canonical pathways, namely the mevalonate or mevalonic acid (MVA) (cytosolic/peroxisomal) and methylerythritol phosphate (2C-methyl-d-erythritol-4-phosphate; MEP) pathways (plastidial) (Hemmerlin et al. 2003; Sauret-Güeto et al. 2006; Kirby and Keasling 2009; Chatzivasileiou et al. 2019). The subcellular compartmentalization of MVA and MEP routes (which perform independently) comprises cytosol/peroxisomes and plastids, correspondingly (respectively) (Vranová et al. 2013). MVA pathway entails 7 sequential enzymatic reactions to synthesize IPP from acetyl-CoA. The synthetic route initiates with the condensation of two molecules of acetyl-CoA to yield acetoacetyl-CoA in a reversible reaction catalyzed by acetyl-coenzyme A (CoA) C-acetyltransferase, also designated acetoacetyl-CoA thiolase. Subsequently, acetoacetyl-CoA reacts with acetyl-CoA to generate 3-hydroxy-3-methylglutaryl-CoA (HMG-CoA) by HMG synthase. Afterwards, HMG-CoA is converted into the pathway intermediate, mevalonate (MVA), viaa reduction reaction catalyzed by 3-hydroxy-3-methylglutaryl-CoA reductase. Subsequently, the occurrence of two consecutive phosphorylation reactions converts mevalonate to MVA-5-diphosphate catalyzed by MVA kinase and phospho-MVA kinase. The final stage of the MVA biosynthetic pathway entails the decarboxylation of MVA-5-diphosphate to yield IPP via the catalytic activity of diphospho-mevalonate decarboxylase, also denominated MVA diphosphate decarboxylase. IPP may be isomerized to DMAPP via the catalysis of isopentenyl-pyrophosphate delta isomerase (McGarvey and Croteau 1995; Laule et al. 2003; Kirby and Keasling 2009; Nagegowda 2010; Hemmerlin et al. 2012; Henry et al. 2015).

As aforementioned, the non-mevalonate, MEP pathway (also termed 1- deoxy-D-xylulose 5-phosphate (DXP) pathway) comprises the synthesis of IPP and DMAPP in plastids. In this 7-reaction biosynthetic pathway, 1-deoxy-D-xylulose-5-phosphate synthase catalyzes the first reaction, the condensation of equimolar amounts of the glycolytic intermediates, pyruvate and D-glyceraldehyde 3-phosphate (GA-3P), to yield 1-deoxy-D-xylulose 5-phosphate (DPX). The latter molecule undergoes a coordinated reactional process of isomerization and reduction through the catalysis of DXP reductoisomerase to originate MEP, which in a cytidine 5′-triphosphate-dependent reaction, is subsequently phosphorylated in the hydroxyl group (C4 position) to yield 4-(cytidine 5′-diphospho)-2-C-methyl-D-erythritol (CDP-ME2P) via the catalysis of 2-C-methyl-D-erythritol 4-phosphate cytidylyltransferase. A subsequent phosphorylation of the hydroxyl group positioned on the second carbon of CDP-ME2P, through the catalysis of 2-C-methyl-D-erythritol 2,4-cyclodiphosphate synthase generates 2-C-methyl-derythritol 2,4-cyclodiphosphate (MEcPP). Subsequently, the reduction of this intermediate compound to 4-hydroxy-3-methylbut-2-enyldiphosphate (HMBPP) is catalyzed by HMBPP synthase. In the last stage of this biosynthetic route, HMBPP undergo a reduction reaction to originate IPP and DMAPP (Hemmerlin et al. 2012; Vranová et al. 2013; Henry et al. 2015).

The MVA and MEP biosynthetic pathways are coupled with the central carbon metabolism, particularly with the glycolytic route. Hence, a competition with other pivotal cellular processes for carbon sources might occur. Despite the aforementioned distinct subcellular compartmentation of MVA and MEP pathways, there is a metabolic intertwining between these two routes enabling the IPP shuttle between the plastid and cytosol across the plastidial envelope membrane (Laule et al. 2003; Kirby and Keasling 2009; Hemmerlin et al. 2012).

Isopentenyl diphosphate and DMAPP biosynthesized through the MVA route are subsequently utilized for the production of isoprenoids in the cytosol and mitochondria, while IPP and DMAPP synthesized via the MEP pathway are isoprenoids precursors in plastids (Phillips et al. 2008; Tholl 2015).

Once biosynthesized through the MVA and MEP pathways, IPP and DMAPP are subsequently utilized as key precursors (building blocks) of prenyl diphosphate (prenyl-PP) intermediates (higher C_5 homologs), through sequential condensations (via successive addition of the isoprene C_5 units) catalyzed by short-chain prenyltransferases (prenyl diphosphate synthases). In plastids, geranyl diphosphate synthase (GPPS) catalyzes the single transfer of the hemiterpene DMAPP to IPP to yield geranyl diphosphate (GPP, C_{10}), which is subsequently converted to monoterpenes (limonene, geraniol, menthol) by monoterpene synthase. In the cytosol, farnesyl diphosphate (FPP) synthase catalyzes the synthesis of FPP (C_{15}) for the subsequent production of sesquiterpenes (through sesquiterpene synthase), homoterpene, sterol (squalene synthase), brassinosteroid, and polyprenol. In plastids, geranylgeranyl diphosphate (GGPP, C_{20}) is a pivotal precursor for homoterpenes, tetraterpens (carotenoids), phytyl side-chains for chlorophyll/tocopherols/quinones, polyprenols, oligoprenols, phytohormones (abscisic acid, gibberellins, and strigolactones) (Kirby and Keasling 2009; Hemmerlin et al. 2012; Henry et al. 2015; Chatzivasileiou et al. 2019).

An in depth analysis of the volatile profile of *Citrus reticulata* unveiled a higher content of volatile compounds in peels in comparison to juice vesicles (Zhang et al. 2017). Monoterpenes were found to be the principal volatile compounds in *Citrus reticulata* peels, among which the monocyclic *d*-limonene (Fig. 6.3) was the most abundant compound (representing more than 60% of the total volatile content). Terpinolene, α-terpinene, γ- terpinene (Fig. 6.3) were also pointed out as prominent monoterpenes. Concerning sesquiterpenes, the contents of β-elemene, germacrene B, α-caryophyllene (Fig. 6.3) were higher than in other *Citrus* species. In juice vesicles, scarcer volatile compounds were identified, being the most prevalent monoterpenes (α-pinene, β-myrcene, *d*-limonene), monoterpene alcohol (β-linalool), sesquiterpenes, alkanes (decane, undecane, and dodecane) and acids (octadecanoic and n-hexadecanoic acid) (Zhang et al. 2017).

The higher volatile content in peels comparatively to juices was corroborated in the study performed by Figueira, Porto-Figueira, Pereira, and Câmara (2020), in which monoterpenes hydrocarbons were found to be the main constituents. Among the identified volatile organic metabolites, *d*-limonene, followed by ɣ-terpinene, were determined to be the two most abundant components representing, respectively, 54 and 25% of the total content. The authors reported that oxygenated terpenes were the second most prevalent volatile group, being linalool, α-terpineol and thymol the three most abundant compounds in both juices and peels, and higher contents were observed in peels, although their prevalence among those varied according to the tangerine variety. Concerning sesquiterpenes, the authors found

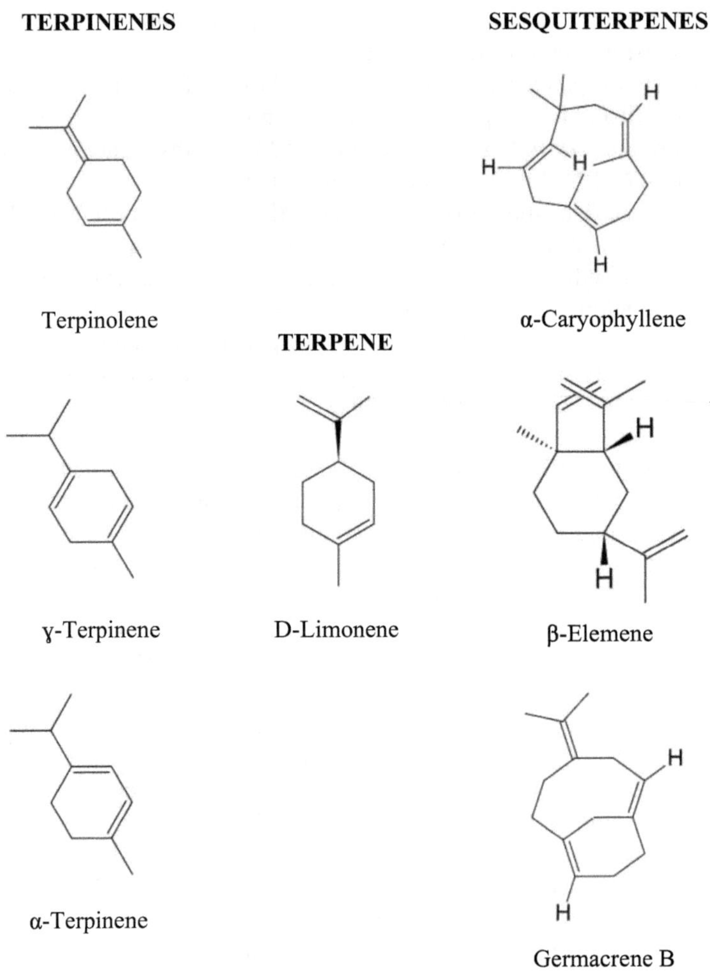

Fig. 6.3 Chemical structures of terpenoids found in peels and juice vesicles of *C. reticulata*

α-farnesene to be prevalent in peels, whilst β-caryophyllene and α-bergamotene were the most abundant in juices.

The quantitative and qualitative evaluation of the volatile compounds must be strictly performed and pondered since the chemical composition may be influenced by environmental cues, geographical distribution, cultivars, along with the fruit's maturation stage (Duan et al. 2016; Hijaz et al. 2020). Hijaz et al. (2020) assessed the impact of the time-of-harvest on the volatile profile in four tangerine cultivars. The study unveiled that monoterpenes were the main volatiles present, despite the overall tendency for their content decay (except limonene, α-pinene, o-cymene, β-myrcene, α- and γ-terpinene, α- and β-phellandrene) as fruit maturation occurred. Contrastingly, concerning off-flavor substances like acetaldehyde, ethanol, 1-hexanol, methyl acetate and ethyl acetate, an opposite correlation was observed. Recent studies provided new clues to the biochemical mechanisms underlying

C. reticulata flavor deterioration. Tietel et al. (2011) performed an analysis of the transcriptome profile of *C. reticulata* Blanco (juice sacs and membrane segments) during post-harvest storage. The authors observed an upregulation of the transcripts involved in the catabolic pathways of amino acids and fatty acids (the main precursors of the volatile compounds), and in the ethanol fermentation metabolism. Concerning this latter, the anaerobic respiration is activated, being pyruvate decarboxylated viapyruvate decarboxylase to originate acetaldehyde, which is then converted into ethanol mediated by alcohol dehydrogenase. The study documented upregulation of both enzymes transcripts, which may trigger the alcoholic fermentation, to which the generation of the "alcohol" off-flavor is attributed, indicative of this perishable fruit deterioration. The authors also proposed an activation of the amino acids catabolic pathway, in which the branched-chain amino acids, leucine and isoleucine, originated ethyl 2-methylbutanoate (isoleucine), ethyl 2-methylpropanoate and 3-methyl butanol (leucine). The accumulation of high contents of these "fruity" volatile compounds may originate an unwanted overripeness perception. Moreover, a transcription upregulation of the enzymes comprised in the fatty acid catabolic pathway, particularly acyl-CoA synthase, acyl-CoA oxidase, acetyl CoA acyltransferase and acyl-CoA dehydrogenase, was also reported. The authors postulated that the higher availability of free fatty acids entrapped in the juice sacs of 'Mor' mandarins, would provide the precursors necessary for the synthesis of fatty acid-derived volatile compounds, triggering their accumulation during fruit storage. Hence, the generated molecules would elicit an unpleasant "musty" off-flavor (Tietel et al. 2011).

6.4.2.2.2 Carotenoids

Citrus fruits are an intricate source of carotenoids, harboring the highest number of these compounds found in any fruit (Navarro et al. 2010).

Carotenoids are pigments which naturally occur in fruits, being responsible for most of the yellow to red colours (Fraser and Bramley 2004; Kultys and Kurek 2022). Hence, carotenoid contents (qualitative and quantitative profiles) of peels and pulps are paramount concerning consumer acceptance. Carotenoids, apart from conferring colour, also possess important biological activities, namely antioxidant and immunomodulatory (anti-inflammatory) activities (Kultys and Kurek 2022), prevention against ocular aging-related diseases and cancer, and improvement of cardiovascular health (Fraser and Bramley 2004).

Carotenoids are a class of lipophilic tetraterpenoids (C_{40}), which biosynthesis occurs in the plastid following the MEP pathway through the condensation of two molecules of GGPP (C_{20}) to yield phytoene (C_{40}) by phytoene synthase (Kachanovsky et al. 2012; Sun et al. 2022). Subsequently, the latter molecule undergoes a two-step reaction comprising desaturation and isomerization to generate lycopene (Van Norman et al. 2014). Afterwards, a myriad of distinct chemical forms of carotenoids are biosynthesized through cyclization, hydroxylation, epoxidation, among others (Dugo et al. 2008). In brief, phytoene may undergo two sequential desaturations catalyzed by phytoene desaturase resulting in the formation of phytofluene and ξ-carotene. Subsequently, ξ-carotene desaturase catalyzes the symmetric insertion of two double bonds in the latter molecule to yield lycopene. In this route, the lycopene cyclization constitutes a branch point, leading to the formation of either α-carotene through lycopene ε-cyclase or β-carotene by lycopene β-cyclase (Yuan et al. 2015). Henceforth, α-carotene undergoes two successive hydroxylation reactions of C-3 of β- and ε-rings through the catalysis of ε-ring hydroxylase and β-ring hydroxylase, respectively, to yield lutein (Quinlan et al. 2012). Similarly, the latter enzyme mediates the two-step hydroxylation of the carotenoid ring ofβ-carotene towards its conversion into β-cryptoxanthin and then to zeaxanthin. Lastly, the flavoprotein zeaxanthin epoxidase converts the xanthophyll zeaxanthin into antheraxanthin, which in turn is converted into violaxanthin (Jahns et al. 2009).

These terpenoids may be considered primary metabolites, displaying a pivotal role in the harvesting-light system (photosynthesis) and photo-oxidative damage protection (antioxidant effect). Moreover, carotenoids are precursors of the phytohormones abscisic acid and strigolactones, which are essential components in the regulation of cell growth and plant elongation (Van Norman et al. 2014; Yuan et al. 2015). These lipophilic tetraterpenoids are of utmost importance concerning human health, since some of these carotenoids may exert an antioxidant activity (scavenging the oxygen free radicals) (Rodrigo et al. 2013) and are precursors of vitamin A or retinoic acid (e.g., β-carotene, β-cryptoxanthin and α-carotene) (Fig. 6.4), being referred to as provitamin A carotenoids (Paine et al. 2005).

Light is a main environmental signal regulating these isoprenoid-derived metabolites production and accumulation, hence exposure to illumination trigger the carotenoids synthesis (Vranová et al. 2013).

The rind of unripe *C. reticulata* displays lutein (β,ε-xanthophyll) (Fig. 6.4) as the prevalent carotenoid. As the maturation (ripening) of the fruit evolves, a gradual

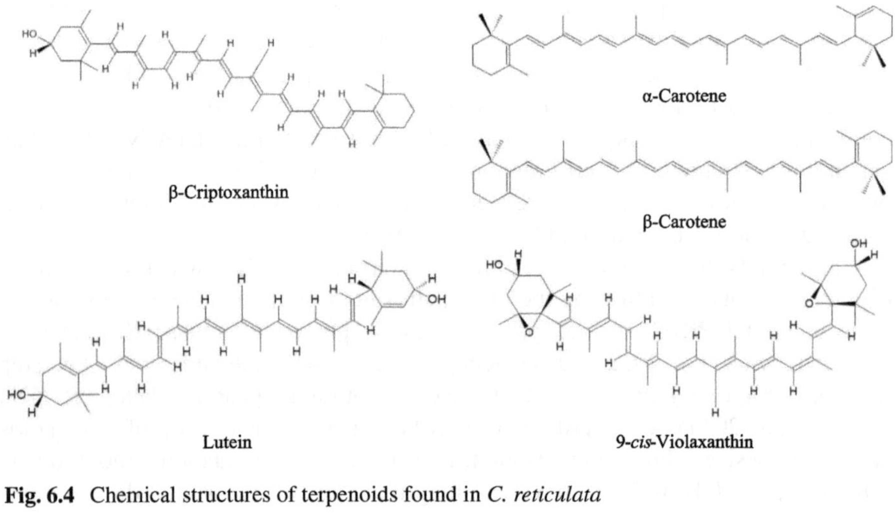

Fig. 6.4 Chemical structures of terpenoids found in *C. reticulata*

decrease in the lutein content occurs, concomitantly with the augmentation of the β,β-xanthophyll, 9-*Z*-violaxanthin, and β-cryptoxanthin (Fig. 6.4) contents (Farin et al. 1983; Alquézar et al. 2009; Rodrigo et al. 2013). The latter compound is the main carotenoid present in the endocarp pulp of the mature fruit. These shifts in the carotenoids profile (content and composition) during *C. reticulata* ripening are coordinated with the transcriptional increase of the carotenoids biosynthetic genes and the β-cyclization of lycopene appear to be a pivotal regulatory path (step) in the biosynthetic route switching the carotenoids flow during the conversion of chloroplast to chromoplast. Beyond the well-documented characteristic tetraterpenoids, the occurrence of a C_{30} apocarotenoid pigment have been documented (Curl 1965; Alquézar et al. 2009). Noteworthy, the synthesis of C_{30} apocarotenoid is a unique trait of the genus *Citrus*. These distinctive pigment seem to accumulate solely in the rind of the ripe fruit. β-citraurine (3-hydroxy-β-apo-8′-carotenal) was found to be the most abundant C_{30} apocarotenoid in tangerines' peels. In 1936, Tuzson (1936) first described this apocarotenoid, which may constitute more than 30% of the whole (total) carotenoids content in the peel of highly orange-colored mandarin (Rodrigo et al. 2013). Nonetheless, hitherto the biosynthetic pathway remains elusive.

Tocopherol (Vitamin E)

Citrus fruits peels comprise an arsenal of antioxidant compounds which entails L-ascorbate, polyphenols and carotenoids, particularly tocopherol (Zou et al. 2016).

Tocopherol is a lipophilic isoprenoid displaying vitamin E activity, well-recognized as a nature's lipophilic antioxidant (Muñoz and Munné-Bosch 2019). The tocopherol biosynthesis is initiated through the condensation of phytyl pyrophosphate (PPP) with homogentisate (HGA) in a reaction mediated by homogentisate phytyltransferase leading to the formation of 2-methyl-6-phytyl-1,4-benzoquinol (MPBQ). Subsequently, the latter compound may undergo direct cyclization to δ-tocopherol through tocopherol cyclase, or alternatively it may be initially methylated through MPBQ methyltransferase yielding 2,3-dimethyl-6-phytyl-1,4-benzoquinol (DMPBQ). This molecule may then undergo cyclization to originate γ-tocopherol by the aforementioned enzyme tocopherol cyclase and thereafter methylated resulting in the formation of α-tocopherol by tocopherol methyltransferase (Asensi-Fabado and Munné-Bosch 2010; Muñoz and Munné-Bosch 2019; Jiang et al. 2021).

The biosynthetic pathway of tocopherol is a very intricate process entailing the intersection of distinct metabolic routes, which are intertwined, providing the precursors for the vitamin synthesis. The two precursors (PPP and HGA) are synthesized in independent metabolic pathways and their availability greatly influence the final content of tocopherol synthesized and accumulated in the fruit tissues (Pellaud and Saffrané 2017; Rey et al. 2021). PPP may be biosynthesized *de novo* from GGPP (generated via the MEP pathway) by GGPP reductase. Alternatively, PPP may be originated from the recycling of phytol through the chlorophylls degradation (Dorp et al. 2015; Pellaud and Saffrané 2017). HGA is synthesized in a

Fig. 6.5 Chemical structure of terpenoids found in the flavedo of *C. reticulata*

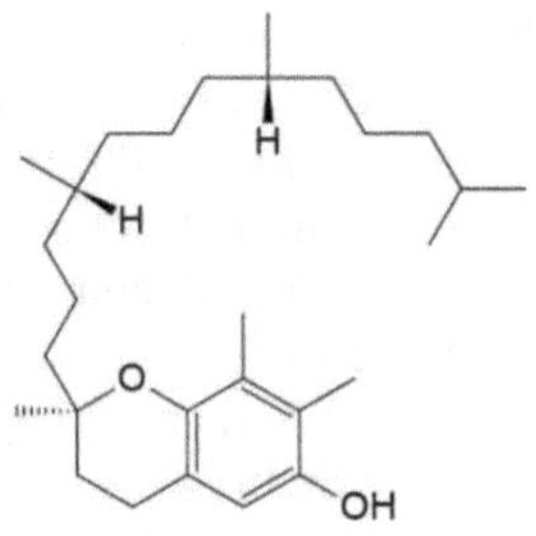

γ-Tocopherol

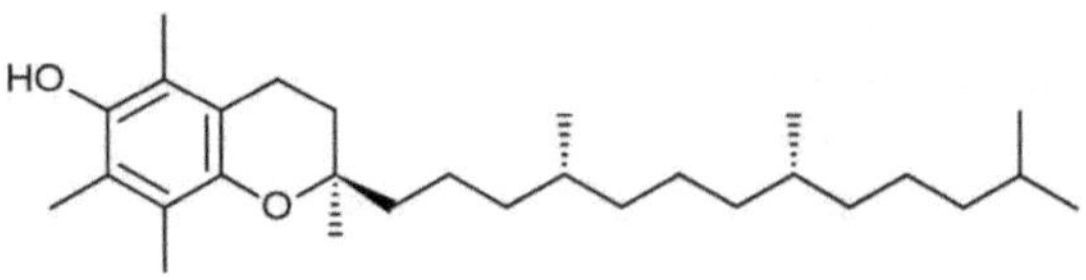

α-Tocopherol

two-reaction mechanism, through the L-tyrosine catabolism. This aromatic amino acid is produced in the shikimate pathway. Tyrosine aminotransferase catalyzes the transamination of L-tyrosine which is converted into *p*-hydroxyphenylpyruvate, which is then oxidized into HGA by 4-hydroxyphenylpyruvate dioxygenase (Muñoz and Munné-Bosch 2019; Rey et al. 2021).

In citrus fruits, and particularly in *C. reticulata*, scarce information concerning differential tissue-associated accumulation of tocopherol is available (Rey et al. 2021).

Rey et al. (2021) investigated the tocopherol biosynthesis and accumulation in the flavedo of "Nadorcott" (*C. reticulata,* Blanco). Among the naturally existing tocopherol isoforms (α, β, δ and γ), the authors found that only α- and γ-tocopherol (Fig. 6.5) were present in the flavedo of *C. reticulata* Blanco totalizing, at harvest, a content of 135 mg kg^{-1}. Of these two isoforms, α-tocopherol was the prevalent one, representing 86%, on average, of the total tocopherols.

Provitamin A

Albeit devoid of the lipophilic vitamin A, citrus fruits are considered an outstanding source of its ultimate precursors, the provitamin A carotenoids (carotenes and β-cryptoxanthin) (Fig. 6.4). Among these carotenoids, β-cryptoxanthin (an oxygenated carotenoid) is undeniably the predominant vitamin A precursor in *C. reticulata* (Liu et al. 2012). As previously detailed, β-carotene is hydroxylated into β-cryptoxanthin viaβ-ring hydroxylase (Asensi-Fabado and Munné-Bosch 2010).

These carotenoids harbor at least one retinyl group (a distinguishable feature of these compounds), which comprises a non-modified cyclic β-ionone ring and may be further utilized to synthesize vitamin A via a human carotene dioxygenase (Dela

Seña et al. 2014). Few biosynthetic mechanisms have been deciphered for the human bioconversion of dietary β-cryptoxanthin to retinol (Eroglu and Harrison 2013). The vitamin synthesis occurs primarily in the enterocytes through the initial oxidative cleavage of the centrally positioned 15–15′ double bond ofβ-cryptoxanthin by β-carotene 15,15′-oxygenase 1 (BCO1), yielding two molecules of retinal (β-apo-15-carotenal) (Seña et al. 2013). This latter is subsequently hydrolyzed to retinol by retinol dehydrogenase (Eroglu and Harrison 2013).

Interestingly, according to most dietary surveys, citrus fruits, and particularly, tangerines are considered the main dietary sources of β-cryptoxanthin (Liu et al. 2012; Burri et al. 2016). According to the United States Department of Agriculture/Agricultural Research Service National Nutrient Database For Standard Reference, raw tangerines present a β-cryptoxanthin content of 407 μg *per* 100 g of edible fruit, while the canned (juiced) fruit displays a quantity of 775 μg *per* 100 mL of beverage.

Similarly to the other carotenoids, the content of β-cryptoxanthin in *C. reticulata* is highly dependent on pre- and post-harvest conditions, namely cultivar, culture (environmental) conditions, maturity phase, post-harvest storage, and seasonal variations. Noteworthy, the concentration of this carotenoid either in citrus fruits and human blood achieves the highest value during the ripening stage (Burri et al. 2016).

β-cryptoxanthin is particularly prone to cis-isomerization and oxidation (epoxide formation) following exposure to temperature (heat treatment during food processing) and light (Li et al. 2015). Hence, the thermal and oxidative degradation of this carotenoid may alter its biological properties, namely diminishing the provitamin A activity and precluding the antioxidant capacity. Moreover, thermal processing may improve the bioavailability of β-cryptoxanthin for the digestive process (Burri et al. 2016).

In comparison with the above described provitamin A carotenoids, citrus fruits harbor vestigial contents of the naphthoquinone derivatives comprising the K-group vitamins (Liu et al. 2012).

6.4.2.2.3 Alkaloids

Alkaloids are an extremely extensive group of chemical compounds, naturally occurring in distinct organisms, namely plants, algae, fungi, and animals (Verpoorte 2005; Roy 2017). In plants, alkaloids are secondary metabolites, which are involved in the chemical defense against herbivores and pathogens (Mohan et al. 2016; Roy 2017).

An elementary definition was proposed by Pelletier (1983), which described an alkaloid as "a cyclic organic compound containing nitrogen in a negative oxidation state which is of limited distribution among living organisms" (Roberts and Wink 1998; Verpoorte 2005). The classification of alkaloids has for long been a controversial issue (non-consensual), particularly, concerning the principles which dictate the classes differentiation. Albeit predominantly biosynthesized from amino acids, namely tryptophane, lysine, phenylalanine, amongst others (Verpoorte 2005), alkaloids may also be derived from other precursors, e.g. caffeine, which is originated from a purine (Roberts and Wink 1998; Roy 2017). In this sense, some authors consider classes based on the heterocyclic ring and the biosynthetic precursor, which include indole, imidazole, quinoline, isoquinoline and pyrrolizidine alkaloids (Daly 2003; Roy 2017). Alkaloids may also be divided solely into three classes: (i) true-alkaloids (derived from amino acid and a nitrogen-containing heterocyclic ring); (ii) proto-alkaloids (the nitrogen carbon is derived from amino acid, although the latter is not comprised in the heterocyclic ring); and (iii) pseudo-alkaloids (not derived from other compounds) (Roy 2017). Nonetheless, alkaloids may also be classified according to the biological/botanical origin, namely Amaryllidaceae alkaloids and *Strychnos* alkaloids (Daly 2003; Verpoorte 2005; Berkov et al. 2020; He et al. 2020). Plant alkaloids are well recognized for their bioactive and pharmacological activities (Berkov et al. 2020; Mahmoud et al. 2021; Otimenyin 2022).

Concerning *C. reticulata*, the rind fruit was reported to possess 1.20 $mg_{alkaloids}100\ g_{DW}^{-1}$ (Okwu et al. 2007), and synephrine, the main alkaloid present in *Citrus* fruits (Shorbagi et al. 2022), was found in *C. reticulata* peels (Fu et al. 2019; Zheng et al. 2021).

Aqueous extract from dry powder of immature *C. reticulata* fruitwas found to comprise 16 mg g^{-1} of the alkaloid synephrine, which content was previously reported to be lower in mature citrus fruits than in immature (Chou et al. 2018). In a distinct part of *C. reticulata* plant, namely, the roots, in which acridone alkaloids are generally present in *Citrus* (Ye et al. 2021), two acridone–quinoline alkaloids were identified by Fomani et al. (2016). In a more recent study, Ye et al. (2021) also isolated acridone alkaloids, although, from *C. reticulata* leaves. Moreover, a novel molecule (retacarcidone) was found, which was determined to be the first pyrano[2,3-a]acridone to be identified as present in a member of the *Citrus* genus.

6.5 Bioactive Compounds

6.5.1 Phenolic Compounds

One of the major appeals of consuming fruits and vegetables is the general perception of such foods being rich sources of health-promoting substances, particularly those of antioxidant nature. Among the several classes of antioxidants found in fruits and vegetables, phenolics stand out for presenting a wealth of different compounds, as well as many biological activities (e.g., antitumor, anti-inflammatory, antiviral) other than antioxidant activity.

Phenolic compounds are naturally present in plant material as secondary metabolites. They are synthesized through two main routes – the shikimic acid pathway (predominant in higher plants) and the malonic acid pathway (significant in fungi and bacteria). In the former mechanism, carbohydrate precursors are provided by glycolysis and pentose phosphate shunt, being converted into the amino acids phenylalanine and tyrosine, from where the phenolics are derived. A key reaction involves the conversion of L-phenylalanine into *trans*-cinnamic acid through the removal of an amino group catalyzed by the enzyme phenylalanine ammonia lyase (PAL). Similarly, the enzyme tyrosine ammonia lyase (TAL) catalyzes the deamination of tyrosine, which is converted into 4-hydroxycinnamic acid. From there, a series of PAL-catalyzed reactions take place, such as the introduction of a hydroxyl group at C4 of cinnamic acid, yielding *p*- or 4-coumaric acid. The addition of a second hydroxyl group at C3 forms caffeic acid, while the *O*-methylation of this same OH group yields ferulic acid (Saltveit 2017).

By the action of 4-coumarate:CoA ligase, *p*-coumaroyl CoA is formed. This molecule is essential for the formation of several other phenolics and for the flavonoids pathway, where chalcones synthase catalyzes the combination of 4-coumaroyl CoA, yielding naringenin chalcone, later converted into naringenin by chalcone isomerase. From this point on, flavonoids are branched into subgroups, namely isoflavones, flavonols, flavones, flavanones, flavanonols, flavan-3-ols, anthocyanins, and proanthocyanidins (de Camargo et al. 2018).

Phenolic synthesis is a metabolic response to biotic (e.g., pathogen attack) and abiotic (e.g., mechanical injury, UV light) stresses, developmental stage, and cell type. These factors increase PAL synthesis, starting the reaction cascade that will result in the production of phenolic compounds. Since they result from a defense mechanism and environment adaptation, these substances are mainly concentrated on the outer layers of grains, fruit skins, as well as seeds (de Camargo et al. 2018).

The basic structure of phenolic compounds consists of a phenoxyl ring bearing one (monophenols) or more hydroxyl groups (polyphenols). As a large group comprising more than 8000 compounds, phenolics can be broken down into several subgroups, according to their nature and structural features. Phenolic acids can be either benzoic acid derivatives (hydroxybenzoic acids) or cinnamic acid derivatives (hydroxycinnamic acids). Some examples of the former group include vanillic, gallic, and 4-hydroxybenzoic acids, while ferulic, chlorogenic, caffeic, and *p*-coumaric acids are examples of hydroxycinnamic acids. Grain, legumes, and oilseeds are excellent sources of phenolic acids. In some of these raw materials, hydroxybenzoic and hydroxycinnamic acids can occur mainly in the insoluble-bound form, i.e., attached to cell wall macromolecules, such as pectin, fiber, structural protein, and cellulose (Shahidi et al. 2019).

Flavonoids, the major polyphenol group, are characterized by a C6-C3-C6 structure, where two aromatic rings are linked to a central heterocyclic ring. Based on the position of the bond between the aromatic ring and the heterocyclic ring and the functional groups' oxidation state on the heterocyclic ring, flavonoids can be further divided into their several subclasses. Their natural sources include seeds, nuts, grains, spices, wine, tea, beer, vegetables, and fruits, including tangerine. In such matrices, the flavonoids usually exist as glycoside derivatives, with a smaller fraction existing as aglycones (Shahidi et al. 2019).

Monomeric phenolic compounds can polymerize into more complex structures by either food processing or the plants themselves, forming tannins, which can be broken down into hydrolysable and condensed (proanthocyanidins). The former group consists of esters of gallic (gallotannins) and ellagic (ellagitannins) acids and can be found in grapes and wine. On the other hand, proanthocyanidins are oligomers and polymers of flavan-3-ols, particularly catechin and epicatechin. They are widely found in flowers, nuts, fruits, bark, and seeds of a variety of plants (Rauf et al. 2019). Other phenolic classes include stilbenoids, with a C6-C2-C6 basic skeleton bearing a 1,2-diphenylethylene functional group, which includes resveratrol as its main representative; hydroxytyrosol, present in olive oil and wine; lignans, cinnamic acid derivatives mainly present in sesame seeds, cereal products, tea, and coffee (Shahidi et al. 2019); and coumarins, characterized by a benzopyrone structure with multiple substitution sites (e.g., umbelliferone) (Wu et al. 2020).

The primary antioxidant mechanism displayed by phenolic compounds involves the dissociation of hydrogen atoms from their hydroxyl groups, which is transferred to free radicals, causing their stabilization and interrupting the propagation of oxidation. As a result, the phenolic becomes a phenoxyl radical, being able to dissipate its energy by the delocalization of unpaired electrons around its aromatic ring. Excessive propagation of free radicals in the body can lead to damages in the proteins, lipids, and DNA molecule, causing a myriad of diseases, which includes cardiovascular ailments, neurodegenerative diseases, cancer, among others. When exposed to a high amount of prooxidant factors, the body's endogenous antioxidant defenses are not sufficient, being necessary the supplementation with exogenous antioxidants in order to reduce the oxidative stress. Metal chelation can be performed by selected phenolics, especially those containing a galloyl moiety. In this mechanism, the phenolic chelates transition metals, making them unavailable to serve as prooxidant factors (Soobrattee et al. 2005).

Therefore, a diet rich in phenolic compounds and other antioxidant substances is a key factor in reducing theincidence of non-communicable diseases.

Besides their antioxidant function, phenolics can also exert other biological activities, such as apoptosis of cancerous cells, effect on cell differentiation, mitigation of pro-inflammatory cytokines, reduction of *N*-nitrosamine formation, antimicrobial activity, and inhibitory activity of enzymes related to metabolic syndrome diseases.

Tangerine (*Citrus reticulata*) is a rich source of phenolic compounds, as shown in Table 6.2.

According to Bentahar et al. (2020), tangerine ethanolic extracts showed a total phenolic content (TPC) of 127.33 mg of gallic acid equivalent (GAE)/g of dry weight (DW), a total flavonoid content (TFC) of 0.876 mg of quercetin equivalent (QE)/g DW, and a total tannin content (TTC) of 47.65 mg of tannic acid equivalent (TAE)/g DW.

Zhang et al. (2014) analyzed the phenolic profile of Chinese wild *C. reticulata* and found that phenolic acids are mainly present in the insoluble-bound form, both for hydroxybenzoic (soluble – 262.17–650.85 μg/g DW, insoluble-bound – 379–889.78 μg/g DW) and hydroxycinnamic acids (soluble – 381.93–56,700.01 μg/g DW, insoluble-bound – 1428.16–97,871 μg/g DW). Among hydroxybenzoic acids, the authors detected protocatechuic and *p*-hydroxybenzoic acids, while vanillic, caffeic, *p*-coumaric, ferulic, sinapic, and chlorogenic acids were some of the hydroxycinnamic acids reported. Although insoluble-bound phenolics cannot be absorbed in the small intestine, thus presenting low bioavailability, they can still present positive health effects. When released from the food matrix, these compounds reach the colon and are fermented by colonic bacteria, lowering the pH and preventing the multiplication of pathogenic microorganisms. They are also related to an increase in small chain fatty acids in this environment and reduced risk of colon cancer (Shahidi and Yeo 2016). Across 14 genotypes of the analyzed *C. reticulata*, 18 flavonoids were detected, with flavonones (e.g., hesperidin, narirutin, eriocitrin) being the predominant subclass, followed by flavones (e.g., nobiletin, tangeretin, sinensetin) (Zhang et al. 2014b). In fact, nobiletin, tangeretin, and hesperidin (Fig. 6.6) are predominant in tangerine peels, displaying antioxidant, anti-inflammatory, and anticancer activities.

Table 6.2 Phenolic content and profile of different parts of tangerine (*Citrus reticulata*) fruit

Fruit part	Total phenolic content	Phenolic acids	Flavonoids	Reference
Seeds	144.3–158.2 mgGAE/100 g	Gallic, 3,4-dihydroxybenzoic, syringic, caffeic, *p*-coumaric, *trans*-ferulic, and *trans*-cinnamic acids	(+)-Catechin, rutin trihydrate, apigenin-7-glucoside, resveratrol, quercetin, naringenin, kaempferol, and isorhamnetin	Al Juhaimi et al. (2018)
Peels	1040–5640 mgGAE/100 g	Chlorogenic, caffeic, and ferulic acids	Naringin, rutin, hesperidin, naringenin, quercetin, hesperetin, and tangeretin	Ferreira et al. (2018)
Juice	426.1–486.3 mgGAE/L	Protocatechuic, *p*-hydroxybenzoic, vanillic, caffeic, chlorogenic, *p*-coumaric, ferulic, and sinapic acids	Narirutin, hesperidin, didymin	Kelebek and Selli (2014)

GAE gallic acid equivalent

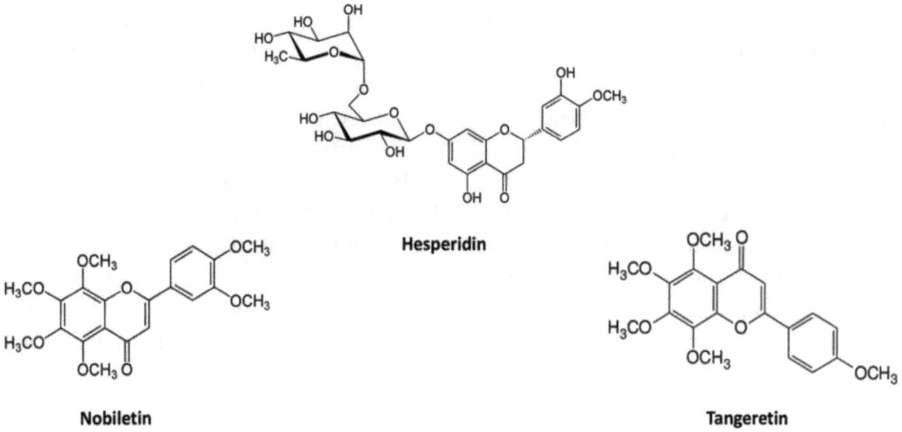

Fig. 6.6 Chemical structures of polymethoxyflavones found in tangerine peels

Besides the edible portion of tangerine, the fruit's by-products (peels, seeds, membrane, pulp leftovers) are also outstanding sources of polyphenols. The chemical profile of the peels of *C. reticulata* Blanco was studied by Ferreira, Silva and Nunes (2018), which reported that the phenolic fraction of this portion is mainly composed of the flavonoids hesperidin, naringin, tangeritin, and rutin, accounting for 86% of all phenolics detected.

Tangerineseeds have also been deemed as valuable sources of phenolics. Moulehi et al. (2012) investigated the changes in the phenolic levels of seeds according to the fruit's ripening stage. *Citrus reticulata* Blanco seeds showed a TPC of 0.68–2.11 EGA/g DW and a TFC of 1.31–2.52 mg CE/g DW, with the highest amounts detected in immature, followed by semimature seeds. However, the downward trend for phenolic content as ripening progresses was not followed by proanthocyanidins (0.12–0.37 mg CE/g DW), which did not show a significant statistical difference among ripening stages. Regardless of fruit maturity, tangerineseeds displayed a wide variety of polyphenols, including eight flavonoids, three hydroxybenzoic acids, and five hydroxycinnamic acids.

Other studies have explored the phenolic potential of tangerine's processing waste. Costanzo et al. (2022) found a higher content of total phenolics and total flavonoidsin the peels of tangerine, followed by the seeds, when compared to the fruit's pulp. Besides, proanthocyanidins were predominantly present in the seeds. A closer look into the phenolic profile revealed that this could be due to the diversity of compounds present in such fractions. A total of 28 phenolics were identified in the peels, with a predominance of delphinidin rutinoside (2109.92–2644.41 μg/L), delphinidin-3-*O*-glucoside (3255.32–4407.85 μg/L), quercetin-3-glucoside (2048.06–2654.77 μg/L), sinensetin (1769.17–4164.48 μg/L), rutin (3608.01–4055.79 μg/L), and valoneic acid dilactone (103.41–1430.20 μg/L). Most of these compounds reached their highest concentration in the semi-ripe stage. Meanwhile, 34 phenolics were detected in the immature seeds. Nevertheless, as the seeds ripened, ten of those compounds could not be detected anymore. The predominant phenolics in the seeds were valoneic acid dilactone (5947.83–13,127.81 μg/L), cyanidin-3-*O*-glucoside (174.57–345.78 μg/L), and 6-malonyldaidzin (67.11–143.32 μg/L). Contrary to what was observed for the peels, phenolic levels reached their peak in the ripe seeds, except for the ten compounds that were lost throughout the ripening process.

In a study conducted by Wang et al. (2017), polyphenol-rich extracts from the flavedo (peel's outer layer) of *C. reticulata* Blanco displayed cytotoxicity against gastric cancer cell lines SGC-7901, BGC-823, and AGS. The antitumor activity was mainly attributed to the polymethoxyflavonoids present in the extract, particularly nobiletin. Therefore, it is noteworthy that the recovery of phenolics from tangerineby-products could find multiple pharmacological and nutraceutical uses, reducing the burden of agricultural waste disposal on the environment by using a circular economy approach.

6.5.2 Limonoids

The bitter taste of *Citrus*fruits has been attributed to limonoids, a class of extensively oxygenated triterpenoids that can exist both as aglycones or glycosides. The precursor of these bioactives possesses a 4,4,8-trimethyl-17-furanylsteroid backbone, which is synthesized from the acetate-mevalonate route. The chemical structure of *Citrus* limonoids consists of a furan ring connected to a D-ring at C17, bearing functional groups at C3, C4, C7, C16, and C17. To this date, over 50 limonoid aglycones and 20 glycosides have been reported in the*Citrus*genus, with limonin, nomilin, and limonin glucoside being the most abundant compounds (Fig. 6.7). Aglycones are generally water-insoluble, occurring mainly in the peels and seeds of *Citrus*fruits, while glycosides, usually water-soluble, are found in juices and pulps. Limonoids have a great variety of bioactivities, including *in vitro*

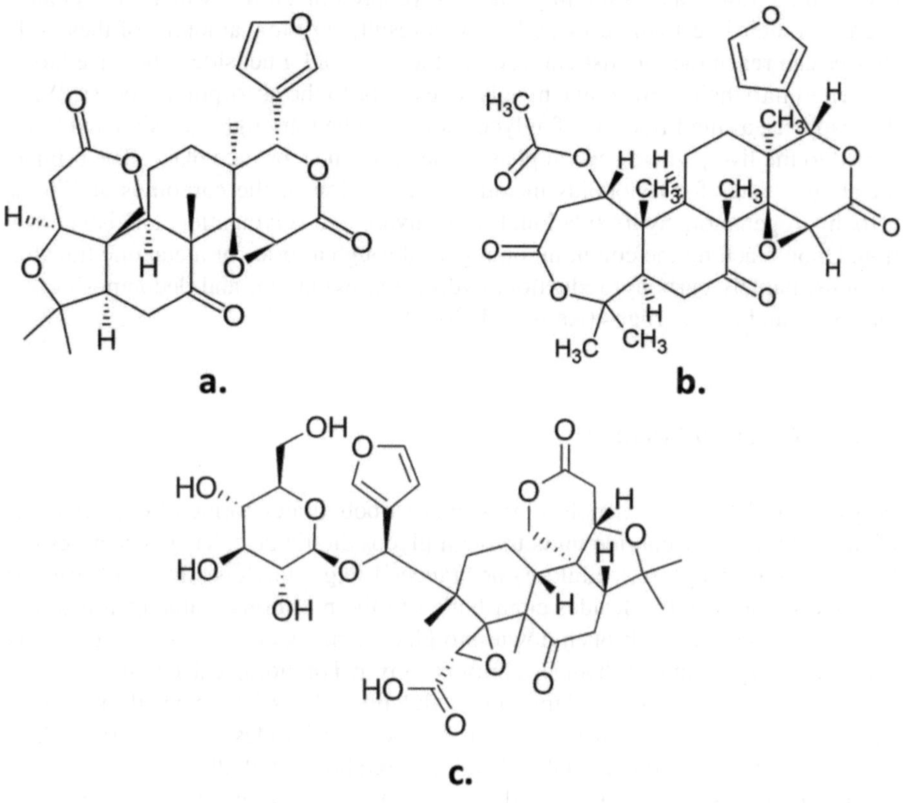

Fig. 6.7 Chemical structures of limonoids commonly-found in *Citrus*fruits, (**a**) limonin, (**b**) nomilin, (**c**) limonin glucoside

and *in vivo* antioxidant activity, anti-inflammatory, antibacterial, and antiviral effects (Shi et al. 2020).

This bioactive class can be encountered in several varieties of orange, lemon, grapefruit, lime, and tangerine (Shi et al. 2020). During fruit ripening, the enzyme limonoid glucosyltransferase (LGT) converts the insoluble limonoid aglycones into soluble glucosides, consequently decreasing bitterness in the juice and pulp (Shorbagi et al. 2022). Montoya et al. (2019) analyzed the limonoid composition of Oneco tangerineseeds and found a limonin content of 1510–2846 mgkg^{-1} and nomilin levels varying from 619 to 4651 mg kg^{-1}, according to the geographical location from where the fruits were obtained.

In another study (Vikram et al. 2007), the seeds and juice of *C. reticulata* Blanco were analyzed by high-performance liquid chromatography. The seeds showed to be rich in limonin (211.87–282.17 mg/100 g), nomilin (35.33–59.38 mg/100 g), and isolimonic acid (6.25–32.96 mg/100 g). As expected, these aglycones were not identified in the juice. Instead, limonin glucoside was prevalent (573.11 mg/100 g).

Some limonoid aglycones, particularly limonin, present poor oral bioavailability due to their low water solubility and P-glycoprotein efflux, which significantly decreases their intestinal permeability. As a result, very low amounts of these substances can reach the bloodstream. As for the limonoid glucosides, they need to be converted into their correspondent aglycones prior to the absorption process. When this happens, a small fraction of aglycones is absorbed through diffusion and transported to the liver, where several phase I metabolism steps take place. The primary metabolic routes for limonoids include the reduction of the carbonyls at C7 and C16, hydrogenation, hydroxylation, lactone hydrolysis, glycination, and isomerization. Upon reaching the colon, limonin goes through a series of metabolic transformations, namely carbonyl reduction, hydrolysis, oxidation, and decomposition of the epoxy and furan groups (Karn et al. 2021).

6.5.3 Polysaccharides

Polysaccharides are large molecular weight carbohydrates formed by the bonding of multiple monosaccharide units through glycosidic linkage. Their structures can be linear (e.g., amylose, cellulose) or branched (e.g., amylopectin), with varying degrees of complexity. Besides contributing to the nutritional value of foods, i.e., polysaccharides can be broken down into glucose molecules to provide a primary source of energy, some of them also present a myriad of biological activities, including antioxidant, antitumor, and immunomodulation effects. These health-promoting benefits have been previously reported for polysaccharides isolated from plant, edible mushrooms, and bacterial cell wall (Chakraborty et al. 2019).

Dietary fibers are polysaccharides resistant to digestion and small intestinal absorption. When consumed, they are partially or completely fermented in the large intestine. Soluble fiber is well known for slowing down sugar absorption, helping reduce blood glucose levels. They are widely found in apples, peas, beans, carrots, and *Citrus*fruits. Pectin is one of the polysaccharides that can be found in the cell wall of many fruits, such as grapefruit, lemon, apple, and tangerine. This molecule consists of a backbone of α-(1 → 4)-linked galacturonic acid or its esters and possess the ability of forming gels, being utilized in food industry as a gelling and stabilizing agent for jams, jellies, and confectionery products. *Citrus*fruits are the primary source for commercial pectin extraction. Besides its functional properties, pectin has also been related to cholesterol-lowering and anti-inflammatory effects (Chakraborty et al. 2019).

The peels of *C. reticulata* Blanco *cv. Ponkan* have been reported to yield around 37% of pectin. Colodel, Vriesmann and de Oliveira Petkowicz (2018) characterized the polysaccharides from the cell wall of this same species and found several types of pectin with different solubility characters. Water-soluble, hot-water-soluble, and chelator-soluble pectin made up the largest proportion of this polysaccharide, with 2.9, 12.4, and 7.2%, respectively. The monosaccharides composing the pectin found in tangerine peels were rhamnose, fucose, arabinose, xylose, mannose, galactose, glucose, and uronic acids. Arabinose and galactose side chains usually bind other cell wall components. The degree of methylation was also found to be high, reaching as much as 52.6%, while the acetylation degree was low in all pectin types (0.03–0.4%), being suitable for commercial application.

Non-conventional techniques have been used to extract pectin from tangerine peels. Chen et al. (2016) used microwave-assisted extraction under optimized conditions (power of 704 W, temperature of 52.2 °C, and time of 41.8 min), obtaining an extraction yield of 19.9%. The authors were also able to obtain a purified pectin polysaccharide, TPPs-2-1, composed of galacturonic acid, arabinose, galactose, rhamnose, glucose, and mannose connected mainly by α-glycosidic linkage. Tangerine peel pectin also showed *in vitro* antioxidant activity by inhibiting hydroxyl radical, 1,1-diphenyl-2-picrylhydraxyl radical, superoxide radical, and by displaying ferric-reducing antioxidant power.

6.6 Health Effects Promoted by Tangerine Bioactives

6.6.1 Antioxidant Activity

The antioxidant activity of food bioactives is one of the primary characteristics associated with such compounds and it can be directly or indirectly correlated to several other bioactivities of bioactive-rich materials. Several methods to assess the antioxidant capacity of food matrices are available, including the use of *in vitro* assays, cell line and biological models (Badarinath et al. 2010).

In vitro chemical assays can measure the ability of antioxidants to scavenge radical species. These methods usually fall into two categories, hydrogen atom transfer (HAT)-based assays and single electron transfer (SET)-based assays. The former includes oxygen radical absorbance capacity (ORAC) and chemiluminescence, while the latter includes 2,2-diphenyl-1-picrylhydrazyl (DPPH) radical scavenging and 2,2′-azinobis(3-ethylbenzothiazoline-6-sulfonic acid)) (ABTS) radical scavenging (Shahidi 2015). Some of these screening tests have been under scrutiny in the last decade for using synthetic and biologically irrelevant radical species. However, as highlighted by de Camargo et al. (2019), such methods are still able to provide valuable data on the mechanism of action of various antioxidants, especially when little or no information is available. Also, their correlation with more advanced biological models for antioxidant capacity measuring techniques is significant and should not be overlooked. Other *in vitro* assays comprise ferric reducing antioxidant power (FRAP), which measures antioxidant molecules' ability of reducing ferric ions (Fe^{3+}) to ferrous (Fe^{2+}) ions in an acidic medium, generating an intense blue-colored complex that can be measured spectrophotometrically; cupric reducing antioxidant capacity (CUPRAC), a variant of the FRAP assay that measures the reduction of cupric (Cu^{2+}) to cuprous (Cu^{+}) ions; and metal chelation capacity, measuring antioxidant's capacity of chelating ferrous ion (Badarinath et al. 2010; Shahidi 2015).

Cell cultures, such as Caco-2, HepG2, MCF-7, among others, can also be used to assess the antioxidant potential through multiple mechanisms. In this regard, the cellular antioxidant activity (CAA) assay is widely used to measure the antioxidant capacity of phenolic-rich extracts. Additionally, biological model systems are also available for this purpose. Some examples encompass LDL-cholesterol oxidation prevention and inhibition of DNA oxidation and nicking, where the capacity of antioxidants to prevent induced oxidative damage to biologically relevant molecules is quantified (Badarinath et al. 2010; Shahidi 2015).

Besides characterizing the antioxidant capacity of a food matrix, the discussed methods can also help assess the effects of processing techniques on the same materials. For instance, Shu et al. (2020) subjected tangerine peels to microwave, hot-air, and freeze-drying and compared the effect of each dehydration method on the phenolics' antioxidant capacity, measured by the ORAC assay. The results showed that microwave-dried peels exhibited higher antioxidant power than their hot-air-dried counterparts, which could be owed to phenolic degradation when the peels were exposed to a high temperature for a long time, with microwaved samples indeed displaying a higher flavonoid content. In addition, the low temperature combined with the sort time provided my microwave drying could have resulted in the release of bound phenolics, directly impacting the oxygen radical absorbance capacity of the samples. Similar results were obtained by Rafiq, Singh and Gat (2019) when evaluating the impact of tray, vacuum, and freeze-drying on the peels of *C. reticulata*, with freeze-dried peels showing the highest antioxidant capacity measured by the ABTS assay due to the preservation of polyphenols.

Animal models have been used in order to assess the physiological effects of tangerine peel phenolic-rich extracts on the oxidative stress and endogenous antioxidant defense markers. Wistar diabetic rats were treated with hydroethanolic extract of tangerine peels (100 mg/kg b.w./day for 4 weeks) and their liver oxidative stress levels were down sharply compared with the control group, presenting 39.09 and 56.82 nmol/malondialdehyde (MDA)/100 mg tissue, respectively. The same pattern was observed for the levels of antioxidant enzymes glutathione, glutathione peroxidase, glutathione S-transferase, and superoxide dismutase, which were detected in significantly higher amounts in diabetic rats treated with the extract than their non-treated counterparts. This outcome shows that exogenous antioxidants from tangerine peel also act to modulate the endogenous antioxidant defense mechanisms (Ali et al. 2020). A similar result was observed by Ke et al. (2020) in rats fed a high-fat diet, where the supplementation with tangerine peel extract restored the activity of glutathione reductase and superoxide dismutase.

In a study using the LDL-cholesterol biological model to assess the antioxidant properties of tangerine peel oil, it was observed that the oil, which was mainly composed of the monoterpene limonene, has the capacity of inhibiting LDL peroxidation, one of the primary factors contributing to the onset of atherosclerosis (Castro et al. 2020).

The antioxidant properties of tangerine and its by-products are not limited to health-related aspects but can also be used to protect food products against oxidation and extend their self-life. For instance, Bambeni et al. (2021) added tangerine-pomace extracts (450 μg/g) to beef patties as a natural antioxidant system, significantly reducing lipid oxidation, measured by thiobarbituric acid reactive substances (TBARS), at 4 °C for 9 days. The polyphenol extract also increased important sensorial parameters of the product, such asaroma intensity, beef-like aroma, and flavour. This outcome opens the possibility for tangerineby-products to become natural antioxidant systems with the ability to replace the currently used synthetic ones.

6.6.2 Anti-inflammatory Properties

Inflammation is a key biological response to infection, irritation, or injury that works through the liberation of pro-inflammatory cytokines. Although extremely important as a healing mechanism, the overproduction of some of these cytokines (e.g., interleukin (IL)-1b, tumor necrosis factor alpha (TNF-α)) may lead to allergy, arthritis, atherosclerosis, and cancer. Therefore, bioactive compounds that have the ability to inhibit pro-inflammatory cytokines can potentially help prevent such conditions (Shahidi and Yeo 2018).

Tangerine peel extract has demonstrated *in vivo*anti-inflammatory potential when administered to rats with potassium dichromate-induced hepatotoxicity. A daily dose of 400 mg kg^{-1} of tangerine peel extract significantly reduced serum TNF-α levels. This effect was owed to the high content of flavonoids, especially polymethoxyflavones, in this raw material (Bashandy et al. 2020). Besides cytokine production inhibition, other mechanisms of action have been associated with the polymethoxyflavones from tangerine, especially nobiletin, which includes mitigation of pro-inflammatory mediators, such as prostaglandin E2 and nitric oxide, suppression of T lymphocyte proliferation, and stimulation of anti-inflammatory cytokines production.

A cell study using RAW264.7 cells detected that the polymethoxyflavones nobiletin and tangeretin form tangerine peels reduced NO release generated by LPS, an inflammation-induced endotoxin (Wang et al. 2019). Nobiletin, tangeretin, and 5-demethylnobiletin supressed the production of prostaglandin E2, as well as pro-inflammatory cytokines (IL-1α, IL-1β, TNF-α, and IL-6) and the expression of cyclooxygenase-2 (COX-2). Genes of the inflammatory pathways JAK2/STAT3, NF-κB/IκBα, and TLR were also affected by tangerine's polymethoxyflavones.

Tangeretin and nobiletin, along with hesperidin have demonstrated anti-neuroinflammatory ability in another study (Ho and Kuo 2014) using the lipopolysaccharide (LPS)-activated BV2 microglia culture model. Nobiletin displayed an inhibitory effect of over 50% against NO, TNF-α, IL-6, and IL-1β at a concentration of 100 μM. Tangeretin and hesperidin alone only showed mild inhibition of pro-inflammatory cytokines. On the other hand, when acting in synergism, the level of cytokine suppression increased to 66.7–90.6% at a 2 mg/mL dose of the tangerine peel extract. Therefore, the mixture of different polymethoxyflavones has exhibited higher anti-inflammatory efficacy than its individual components, an effect that has been extensively reported for phenolic compounds (Freeman et al. 2010).

Limonene, present in the oil extracted from tangerine peels, has also been reported as a potent anti-inflammatory factor (Shorbagi et al. 2022). In a mice model, essential oil from *C. reticulata* peel, mostly composed of limonene, decreased the expression of IL-6, COX-2, and nuclear transcription factor kappa B p65 (NF-κB) (Li et al. 2022).

6.6.3 Antitumor Effects

Hesperidin, one of the major components of tangerine peels, has demonstrated a number of antitumor mechanisms. This flavonoid can promote cell cycle arrest in cancerous cells by modulating regulatory proteins, such as CDK inhibitors, cyclin-dependant kinases, and cyclins. The compound has also been shown to target cell cycle signaling pathways in cancer therapy, especially the Aberrant WNT/β-catenin pathway. Another mechanism detected is the induction of apoptosis, which can happen through the activation of caspase-9 and caspase-3, increase of reactive oxygen species, and mitigation of nuclear factor-κB. As one of the contributing factors for cancer progression, inflammation can also be targeted by hesperidin, with a considerable reduction of pro-inflammatory cytokines (Pandey and Khan 2021).

Ethanolic extracts from tangerine peel have been studied for their anticancer activity in cell line models, namely Hep G2 (liver cancer) and MCF-7 (breast cancer). A range of extract doses were tested (10–150 ug/mL) and a dose-dependent relationship was observed for both cell lines, with the inhibition of cancerous cells growth ranging from 52 to 80% (MCF-7) and 33–58% (Hep G2). This outcome was owed to the high levels of limonene, flavonoids, and other phenolic compounds found in tangerine peels (Rasool et al. 2021). Similar results were observed by Zaki and Naeem (2021), with tangerine peel powders displaying cytotoxicity against human colon carcinoma cell line HCT116.

Castro et al. (2018) detected antitumor effects of tangerine peel oil using an *in vivo* model. Lung cancer cells were implanted in nude mice, which were supplemented with tangerine oil (5.25 mg/day). The supplementation significantly suppressed tumour growth, which was detected to happen through a decrease in the levels of membrane-bound Ras protein, a key factor in cell proliferation. Another mechanism observed was the induction of apoptosis of cancerous cells by the oil, with no toxicity effect whatsoever. The reported effects were partly associated with limonene, the main monoterpene component of tangerine peel oil.

Although polymethoxyflavones encountered in tangerine have shown various bioactivities, including antitumor potential, their oral bioavailability is constrained by their poor solubility in water. Therefore, strategies have been developed to overcome this hurdle. For instance, Ting et al. (2015) synthesized an emulsion-based delivery system for tangeretin and tested its efficacy on colon cancer cell lines (HCT116 and HT29). The emulsions decreased cell viability in a dose-dependent manner. A colon tumorigenesis mice model was also used to investigate the efficacy of the emulsions, which significantly reduced tumor incidence and multiplicity by regulating tumor-related protein expression. The same was not observed for the non-emulsified tangeretin.

The anticancer mechanisms of polymethoxyflavones neohesperidin, nobiletin, tangeretin, and 5-dimethylnobiletin have been explored by Wang et al. (2020). All compounds promoted apoptosis in gastric cancer cell lines. According to the authors, the apoptotic effect is a consequence of polymethoxyflavones targeting retinoic acid receptors, associated with programmed cell death. A stimulation of gene expression for apoptosis-related proteins (caspases 3, 8, 9, and PARP1) has also been detected for this bioactive class.

6.7 Conventional and Non-conventional Extraction Techniques

As portrayed throughout the chapter, *C. reticulata* comprises a panoply of distinct compounds, such as polysaccharides, polyphenols, carotenoids, volatile compounds, lipids, and proteins, among others (Shorbagi et al. 2022; Yun and Liu 2022). Ergo, tangerine may be a source of such compounds, which are of utmost interest to the pharmaceutical, cosmetic, and food industries. Nonetheless, in order to be available for the mentioned purposes, the compounds need to be extracted, and, in this sense, distinct technologies may be applied to accomplish the highest yield and purity. Those technologies may be employed not only for tangerine but also for its wastes, which are known to be rich sources of the several distinct compounds comprised in tangerines' parts (peel, pulp, and seeds) (Shorbagi et al. 2022; Zhou et al. 2022), generating added-value products and minimizing the environmental impact of the tangerine processing industries.

A myriad of different methodologies may be applied to extract a compound from a specific matrix, varying from the most classic/conventional ones, such as mechanical or solvent extraction, to the more recently explored innovative technologies, namely microwaves, ultrasounds, and pulsed electric field, among others. Henceforth, the mechanisms underlying those extraction technologies will be briefly elucidated, and examples of the applications of the technologies in tangerine and tangerine wastes will also be presented (compiled in Table 6.2).

6.7.1 Mechanical Extraction

Mechanical extraction is one of, if not the oldest extraction techniques. As the denomination suggests, the process consists in applying a mechanical force to the matrix, leading to the expression of the compounds (mainly oils) thereof. The sample is placed between barriers which are then pressed, reducing the available space in-between them and compressing the matrix, leading to the oil being forced out (Schwartzberg 1997; Nde and Foncha 2020). Two types of equipment are utilized to perform mechanical extraction, namely, hydraulic and screw presses. Although the principle behind both is similar, the forces applied by each type of press are distinct, since, in hydraulic presses, the force is axial, while in screw presses shear forces are responsible for the expression of the oil from the matrix. Nonetheless, the end result is the same, with the oil being forced out of the raw material (e.g., seeds) and passing through a perforated barrier to be collected (Schwartzberg 1997; Chemat et al. 2015; Kemper 2020). Despite the benefits of the excellent quality of the oils obtained, and the possibility of reusing the matrix from which extraction was performed, there is the disadvantage of low yields (Kemper 2020; Nde and Foncha 2020).

Pertaining to *C. reticulata*, this technology is mainly utilized to obtain juice from the fruit pulp, as in the study of Moshonas and Shaw (1997), in which mechanical extraction was performed to evaluate volatiles content of Robinson and other mandarins juice. Differences between manual and mechanical extractions were also assessed. Results demonstrated that mechanical extraction yielded higher amounts of several lipophilic compounds than manual extraction (e.g., limonene, linalool and nonanal). Concerning volatiles contents in mandarins and orange juices, it was observed that in the latter those were generally considerably higher (with some exceptions). Nonetheless, as far as Robinson mandarin juice was concerned, regarding the main volatiles, higher limonene, similar myrcene, and lower methanol, ethanol, and valencene were observed (73, 1.4, 11.1, traces and 0.51 mg/L, in mandarin *versus* 70, 1.54, 50, 576 and 4.0 mg/L, in orange). Desai et al. (2014) also applied mechanical extraction to obtain juice from Kinnow in order to study the effect of production conditions, namely, the level of drip irrigation (water supply; 60, 80, and 100% of crop evapotranspiration) and potassium fertilizer (nutrient availability; 600, 700 or 800 g K/plant/year) on fruit quality and yield. The conditions for which juice yield was highest (100% and 800 g K/plant/year) were not the same that yielded the highest amount of fruit and total soluble solids (TSS) (80% and 700 g K/plant/year). In this sense, the authors found that 80% of crop evapotranspirationand 700 g K/plant/year were the most appropriate conditions to obtain the best amount of fruit, with the highest quality (Tables 6.3 and 6.4).

Table 6.3 Examples of extraction technologies applied to *Citrus reticulata*

Extraction Technology	Matrix	Compound	Optimum conditions	Solvent	Yield	Reference
Mechanical	Fruit	Juice	100% crop evapotranspiration	–	46.8%	Desai et al. (2014)
			800 $g_{(potassium\ fertilizer)}$ K/plant/year			
	Juice	Volatiles	–	–	Mechanical > hand	Moshonas and Shaw (1997)
	Peels	Essential oil	15,000 psi, 28–30 °C, 45 min	–	Essential Oil – 0.32%	Ahmad et al. (2016)
					Aldehyde – 42.73% of essential oil	
	Peels	Essential oil	3.68% calcium chloride	–	Essential Oil – 0.689%	Yi et al. (2018)
			2.97 h of soaking			
			0.12% sodium chloride		Limonene – 56.76% of essential oil	
Solvent	Seeds	Oil	–	Hexane	28.35%	Fernandes et al. (2002)
					d-limonene – 3.77% of volatiles	
	Peels	*d*-limonene	–	Ethyl alcohol	0.78%	Park et al. (2015)
	Seeds	Oil	–	Diethyl ether	48%	Olatunya et al. (2021)
					PUFA – 43.58% of oil	
PLE	Peels	Flavones	1500 psi, 160 °C, 20 min	Methanol	80.4%	Li et al. (2012)
			70% methanol, 25 mL/g			
	Peels	Pectin	120 °C, 5 min, 30 mL/g	Water (SWE)	Pectin – 21.95%	Wang et al. (2014)
					Galacturonic acid – 68.8%	

(continued)

Table 6.3 (continued)

Extraction Technology	Matrix	Compound	Optimum conditions	Solvent	Yield	Reference
SFE	Peels	Essential oil	14 MPa, 45 °C, 147 min	$ScCO_2$	1.34%	Xiong and Chen (2020)
	Peels	Essential oil	25 MPa, 45 °C, 120 min	$ScCO_2$	12.44%	Dong et al. (2014)
	Peels	Antioxidants	22 MPa, 80 °C, 5% Ethanol	$ScCO_2$+ Ethanol	34.76%	Franco-Arnedo et al. (2020)
	Peels	Bioactives	20 MPa, 20 °C, 20% Ethanol	$ScCO_2$+ Ethanol	17.60 g/100 g_{dw}	Romano et al. (2022)
UAE	Epicarp	Carotenoids	42 Hz, 60 °C, 60 min, 0.4 mg/mL	Sunflower oil	140.7 $mg_{\beta\text{-carotene}}$/100 g_{dw}	Ordóñez-Santos et al. (2021)
	Peels	Bioactives	31% amplitude, 41 °C, 15 min, 30 mL/g	Ethanol	36.17 $mg_{GAE}/g_{extract}$	Kaur et al. (2021b)
	Peels	Carotenoid (lutein)	32.88% amplitude, 43.14 °C, 33.71 min, 6.16 mL/g	KOH (20%) methanolic solution followed by diethyl ether	29.70 ug_{lutein}/g_{peel}	Saini et al. (2021)
	Peels	Pectin	90% amplitude, 80 °C, 15 min, 30 mL/g	Acidified water (pH 2.5)	–	Polanco-Lugo et al. (2019)
	Flowers	Volatiles	ultrasound water bath, 25 °C, 10 min, 6 mL/g	*n*-pentane:diethyl ether (1:2)	Linalool – 75.2% of volatiles	Alissandrakis et al. (2003)
MAE	Peels	Pectic polysaccharide	704 W, 52.2 °C, 41.8 min, 30 mL/g	Water	19.9%	Chen et al. (2016)
	Peels	Phenolic acids	152 W, 49 s, 16 mL/g	Methanol	Ferulic acid – 465 μg/g_{dw} *ρ*-coumaric acid – 317 μg/g_{dw}	Hayat et al. (2009)
HHP	Peels	Phenolics	300 MPa, 10 °C, 3 min	Aqueous ethanol 80% (+1% HCl)	587.28 mg_{GAE}/100 g_{fw}	Casquete et al. (2015)

PEF[a]	Fruit	Juice	3 kV/cm	Water	25–59% increase	El Kantar et al. (2018)
	Peels	Polyphenols	10 kV/cm	Water followed by ethanol 50%	22.5 mg_{GAE}/g_{dw}	
	Peels	Limonene	5 kV/cm, 40 °C, 3 h, 20 mL/g	Ethanol	33% increase	Carpentieri et al. (2022)
	Peels	Naringin	4 kV/cm, 30 pulses, 40 °C, 90 mL/g	Ethanol:water (40:60)	20% increase	Niu et al. (2021)
EAE	Peels	Phenolics	50 °C, 1.5% (w/w) Celluzyme® MX	–	115 $mg_{GAE}/100\ g_{fw}$	Li et al. (2006)
	Peels	Protein	5% (v/v) Celluclast® 1.5 L	---	0.545%	Vergara-Barberán et al. (2017)
	Pulp		5% (v/v) Palatase® 20,000 L		0.167 %	

PLE pressurized liquid extraction, *SFE* supercritical fluid extraction, *SWE* supercritical water extraction; *$ScCO_2$* supercritical carbon dioxide, *dw* dry-weight
[a]Non- *Citrus reticulata*. *UAE* ultrasound-assisted extraction, *MAE* microwave-assisted extraction, *HHP* high hydrostatic pressure, *PEF* pulsed electric field, *EAE* enzyme-assisted extraction, *GAE* gallic acid equivalents, *fw* fresh weight

Table 6.4 Tangerine products

Source	Compound	C. reticulata-based product	Function	Reference
Yerba mate (*Ilex paraguariensis* St. Hill)	Extract	Orange tangerinejuice	Antioxidant activity	Fenoglio et al. (2022)
Okra	–	Juice	Bioactive compounds	Wirivutthikorn (2019)
Plants and flours	–	Low-caloric beverages	Healthier alternatives	Contreras-López et al. (2021)
Citrus peels	–	Citrus-white tea	Odour	Wang et al. (2022)
Tangerinejuice	–	Marmalade	Alternative product	Phinney et al. (2020)
Tangerinejuice	–	Juice	Bioactive compounds and antioxidants	Noguera et al. (2021)
Yeast (*Pseudozyma* sp.)	Enzymes (cellulases)	Juice	Clarified juice	Santana et al. (2021)
Seeds of annatto (*Bixa orellana* L.)	Norbixin	Isotonic soft drink	Colored/dyed juice	Tupuna-Yerovi et al. (2020)
Source	*Compound*	*Product*	*Function*	*Reference*
Peels and bagasse	Flour	Biscuits	Increase nutritional profile	Sande et al. (2018)
Peels	Flour	Muffins	Increase nutritional profile	Khaleel et al. (2022)
Peels	Essential oils	Probiotic wort-based beverages	Probiotic delivery	Tomova et al. (2021)
Pomace	Extract	Beef patties	Nutrition	Bambeni et al. (2021)
Peels	Essential oils	Chitosan nanoparticles/fresh pork	Nutrition	Song et al. (2021)
Peels	Extract	Tilapia fillets	Nutrition	He et al. (2017)
Peels	Extract	Whey protein concentrate nanoparticles	Antioxidant activity	Hu et al. (2019)
Peels	–	Peels	Biosorbent (safranine orange)	Januário et al. (2021)
Peels	–	Peels	Biosorbent (heavy metals)	Abdić et al. (2018)
Peels	Extract	Copper nanoparticles	Anti-corrosive	Ituen et al. (2020)
Leaves	Essential oils	Preservative	Antifungal	Ali et al. (2021)

(continued)

Table 6.4 (continued)

Source	Compound	C. reticulata-based product	Function	Reference
Peels	Extract	Titanium dioxide nanocrystals	Bio-mediator of chemical reactions	Rueda et al. (2020)
Juice	–	Carbon nanodots	Ink production	Aslan and Eskalen (2021)
Peels	–	Microbial fuel cells	Bioelectricity	Rojas-Flores et al. (2020)
Peels	–	Bio-ethanol production by *Saccharomyces cerevisiae*	Bio-ethanol	Shim (2015)
Essential oils	–	Pharmaceuticals	Anxiolytic	Silveira et al. (2022)

Mechanical extraction may also be applied to extract essential oils from peels, as in the case of Ahmad et al. (2016), in which a hydraulic press (operated at 15000 psi, at 28–30 °C for 45 min) was used for that purpose. The authors compared the yield and quality traits (aldehyde content and acid number) of the essential oils obtained from six *Citrus*fruits (among which was Kinnow). Despite being second to last in terms of oil yield (0.32%), Kinnow presented the highest aldehyde content (42.73%of essential oil) and acid number (2.21), which are the main indicators of quality and flavor strength.

Yi et al. (2018) assessed the extraction of essential oils from *C. reticulata* peels by mechanical pressing, through a central composite design in which the impacts of calcium hydroxide concentration (2.7, 3.0, and 3.3%), soaking time (3, 4, and 5 h) and sodium chloride concentration (0.25, and 0.5%), in the yield were determined. The conditions that were revealed to present the highest yield (0.689%) were 3.68% calcium chloride, 2.97 h of soaking, and 0.12% sodium chloride. Thereafter, the chemical composition of the essential oil was determined, in which limonene was the most abundant compound, representing 56.76%. The extract/oil was tested for its antimicrobial activity, displaying a potent broad-spectrum antimicrobial activity, and also for antioxidant capacity (radical-scavenging activity, 1,1-Diphenyl-2-picrylhydrazyl, DPPH, and 2,2′-azino-bis(3-ethylbenzothiazoline-6-sulphonic acid), ABTS; reducing power, iron reduction), using α-tocopherol as a positive control, demonstrating moderate activities. Hence, it was concluded that the essential oil could be considered a natural alternative to synthetic additives in food products, concerning food safety and shelf-life.

6.7.2 Conventional Solvent Extraction (CSE)

Following the previously described mechanical extraction, solvent extraction (or CSE) isthe most conventionally used methodology to obtain compounds from a vast array of matrices, from polyunsaturated fatty acids in microalgae (Figueiredo et al. 2019) to *d*-limonene in the tangerine peel (Park et al. 2015). This extraction technique commonly utilizes organic solvents (*per se* or in mixtures/systems). The principle behind CSE is that molecules/compounds possess distinct polarities and, according to the polarity, have affinities towards different solvents, specifically, those with polarities similar to the compounds. The closer the polarities are between compound and solvent, the higher the affinity. This mechanism allows the extraction of the desired compound from the matrix in which it is comprised, by a phenomenon designated partitioning. As aforementioned, when in contact with two solvents with distinct polarities, the compound possesses different affinities towards them, having a higher affinity for one. Therefore, if or when the solvents are separated from each other, the target-compound will remain dissolved in the solvent for which its affinity is higher. A particular, and commonly employed, type of solvent extraction is the denominated "Soxhlet" extraction, in which an apparatus is used to extract compounds (usually lipids, with an organic solvent) from a solid matrix (Halim et al. 2012).

Citrus oils were extracted from *C. sinensis*, *C. limon* and *C. reticulata*seeds by (Fernandes et al. 2002) through "Soxhlet" extraction, using hexane as extraction solvent, to evaluate their antifungal activity against *Leucoagaricus gongylophorus*, a symbiotic fungus of the leaf-cutting ant *Atta sexdens. Citrusreticulata*seeds yielded the highest percentage (28.35%), and the oil was determined to be comprised of triglycerides and volatiles. In tangerine oil, the fatty acids present in the triglycerides were mainly palmitic, linoleic, and linolenic acids (38.6, 30.5, and 19.9%, respectively, of total fatty acids). The major volatile found was *d*-limonene, which represented 3.77% of all volatiles present in the seeds extract, which was, among the three oils, the one that displayed the most diversified volatile profile. Concerning antifungal activity, when at 200 μg/mL, the extract was able to inhibit fungal growth by 20%. Park, Ko and Kim (2015) also extracted *d*-limonene from tangerine using "Soxhlet", although the extracts were obtained from peels instead of seeds. Extraction was performed with three distinct solvents, namely, ethyl alcohol, n-hexane, and ether, which yielded 0.78, 0.049, and 0.028%, respectively, of *d*-limonene. The authors found that, in that particular case, the polarity of the solvent presented a higher impact on extraction efficiency than the boiling point thereof.

Recently, Soxhlet extraction was performed by Olatunya, Ajaja and Akintayo (2021) during 8 h at 50–60 °C, using diethyl ether as a solvent, to extract oils from the seeds of grapefruit, lime, and tangerine, which were afterwards characterized in terms of fatty acids profiles and some physicochemical properties (namely, kinematic viscosity and saponification value). In most of the parameters evaluated, tangerine oil presented intermediate values between those obtained for the two other fruits. Tangerine oil yield was 48%, while grapefruit yielded the highest (50%) and lime the lowest (40%). The main fatty acids present in all oils were found to be linoleic acid, followed by palmitic acid and oleic acid. Concerning the important polyunsaturated fatty acids, lime oil was comprised by 48.81%, and tangerine and grapefruit oils by 43.58 and 41.71%, respectively. In terms of physicochemical properties, tangerine oil was also determined to present intermediate values regarding several parameters, such as kinematic viscosity (36.50 mm^2/s), for which grapefruit presented the highest and lime the lowest (38.20 and 35.20 mm^2/s, respectively), and saponification value (195.40 mg/g_{oil}), which was highest in lime oil (212.20 mg/g_{oil}) and lowest in grapefruits' (186.80 mg/g_{oil}). The study demonstrated the potential of those oils to be used to develop novel pharmaceutical, cosmetics, and food products owing to the high fatty acid contents and physicochemical properties.

6.7.3 Pressurized Liquid Extraction (PLE)

Pressurized liquid extraction (PLE), also referred to as pressurized fluid extraction (PFE), pressurized hot-solvent extraction (PHSE) or accelerated solvent extraction (ASE), is an extraction technique in which the solvents are heated to temperatures above the boiling point, although being subjected to pressures that enable to maintain the liquids below the critical point, which results in the solvent being maintained in the liquid state throughout the extraction process (Herrero et al. 2015; Imbimbo et al. 2020). When water is utilized as the extraction solvent, despite the general principles and instrumental requirements being the same, there are other important parameters which present significant influence, and the technique can be denominated subcritical water extraction (SWE), superheated water extraction (SHWE) or pressurized hot-water extraction (PHWE) (Herrero et al. 2015; Anticona et al. 2020).

Pressurized liquid extraction conditions provide an increased mass-transfer rate, and solubility of the compounds to be extracted, as well as decreases in both solvent viscosity and surface tension. The lower solvent viscosity and surface tension will favor the easier infiltration of the solvent into the matrix, allowing it to reach more inner areas and, consequently, increasing the contact surface, which enhances the mass transfer to the solvent, resulting in an increased extraction rate. As previously mentioned, when water is utilized as the solvent there are other parameters that have to be considered, such as the dielectric constant (ε) of water. When water is kept in the liquid state while heated at high temperatures, ε, which is a measure of the polarity of the solvent, is significantly reduced and, if it is decreased to values close to the ones of organic solvents (when heated), water can be presented as an effective alternative. Even though this may not be possible for all applications, among the PLEs, SWE can be considered the "greenest" (Herrero et al. 2015).

Concerning the advantages of PLE over the conventional extraction processes, there are several, as higher selectivity, shorter extraction times, faster extraction processes and smaller amounts of organic solvents. Furthermore, there is also the possibility to automatize the process, which inherently increases reproducibility (Michalak and Chojnacka 2014; Herrero et al. 2015).

Li et al.(2012) evaluated the impact of solvent (ethanol, aqueous methanol or water), solvent:solid ratio, and extraction temperature and time, on PLE yields of flavones from *C. reticulata* peels. The process efficacy was also compared with other commonly applied extraction techniques (namely, "Soxhlet", ultrasounds and heat-reflux). The best result with PLE was obtained with 70% methanol at 25 mL/g, at a temperature of 160 °C, during 20 min at a pressure of 1500 psi, which yielded a similar amount to heat-reflux (80.4 and 80.6%, respectively), and PLE was considered to be a better choice since it was less time-consuming.

The specific PLE, SWE, was applied to the peels of *C. reticulata* by Wang, Chen and Lü (2014), in that particular case, to extract pectin. The study analyzed the extraction yields and composition of pectin obtained at different temperatures (100, 120 or 140 °C; all during 5 min with a solid to liquid ratio of 1:30), and found that the parameter did in fact influence the amount of pectin extracted, as well as that of galacturonic acid. The highest pectin and galacturonic acid yields (21.95 and 68.8%, respectively) were obtained when extraction was performed at 120 °C. Moreover, the pectins extracted at different temperatures were tested in terms of radical scavenging activities (DPPH and ABTS), as well as cytotoxicity against colon cancer cell line (HT-29). Regarding antioxidant (radical scavenging) activities, as well as anti-tumor activity (cytotoxicity), it was found that all decreased with increasing extraction temperature. The pectins were also used to prepare hydrogels, which were assessed regarding physicochemical properties. The authors reported that the hydrogel prepared with the pectin extracted using 120 °C presented the highest values concerning all the physical parameters, namely, hardness, gel strength, viscous force and stickiness.

6.7.4 Supercritical Fluid Extraction (SFE)

Supercritical fluid extraction (SFE) is, together with PLE, the most widely employed extraction technique for obtaining bioactive compounds from natural sources (Herrero et al. 2015). In SFE, the solvents are utilized at pressures and temperatures above the critical points, which confers the fluid characteristics between gases and liquids, namely, intermediate diffusivity, viscosity similar to a gas and density higher than a gas (Michalak and Chojnacka 2014; Herrero et al. 2015). Since the solvent power of a supercritical fluid is dependent on the density thereof, it can be modified through alterations in the extraction temperature and pressure, allowing it to be manipulated to extract specific compounds (Mubarak et al. 2015). The high diffusivity and low viscosity of solvents in SFE, in comparison with liquids, confer superior transport properties (mass transfer), thus enhancing extraction (Michalak and Chojnacka 2014; Ciko et al. 2018; Anticona et al. 2020). Despite the possibility for a panoply of solvents to be utilized in SFE, the most commonly used is supercritical CO_2 ($ScCO_2$ or $SC\text{-}CO_2$), due to several characteristics, namely, low critical temperature (31.1 °C) and moderate critical pressure (7.4 MPa) (Mubarak et al. 2015). Moreover, $CO_2$is nontoxic, nonflammable, nonexplosive, a noncorrosive, easily available, and cheap. When compared to alternative extraction techniques, since CO_2 is a gas at room temperature, its removal from the extract can be easily performed (Michalak and Chojnacka 2014; Herrero et al. 2015). One other advantage is that, as aforementioned, the properties of SCFs can be adjusted through pressure and temperature changes, which have a direct impact on density, allowing the technique to be very selective, which is a significant advantage when the objective is the extraction of compounds from complex matrices. This technique also provides the possibility of, during decompression, performing fractioning simply by implementing two or more decompression steps, which is useful to separate components in the extract (Herrero et al. 2015). Nonetheless, SFE also has some drawbacks, beingthe main one the low polarity of $ScCO_2$, which limits the compounds that can be extracted with this technique, which may not be able to, *per se*, extract polar compounds. However, this issue can be overcome through the addition of co-solvents (modifiers), which are used in small amounts (1–10%), during the extraction process. The co-solvents present higher polarity and change the polarity of the supercritical fluid, thus increasing the solvating power, and consequently increasing the range of compounds that can be extracted (Michalak and Chojnacka 2014; Herrero et al. 2015).

Regarding the use of SFE in tangerine matrices, the technologyhas been applied to, for example, extract essential oils from tangerine peels, as in the study of Xiong and Chen (2020), in which the impact of temperature (35–55 °C), pressure (10–30 MPa) and extraction time (0–180 min) on $ScCO_2$ extraction yields were assessed. The response surface methodology analysis indicated 45 °C, at 14 MPa during 147 min, to be the best conditions to obtain the highest yield (1.34%). The extracted oil was revealed to comprise, mainly, *n*-hexadecanoic, linoleic and oleic acids (14.62, 32.3 and 20.42%, respectively). Dong, Shao and Liang (2014) had previously applied the technology with the same purpose, also testing the effect of extraction temperature (35–55 °C), pressure (150–350 bar), and time (60–120 min). In that case, the study unveiled that the best- operating conditions were 45 °C, at 250 bar (25 MPa) during 120 min, which yielded an essential oil mainly comprising, in terms of volatiles, α-farnesene, limonene, and β-elemene (2.34, 1.55 and 1.44%, respectively).

Franco-Arnedo et al. (2020) also employed SFE, to obtain antioxidant extracts from *C. reticulata* peels, using ethanol as co-solvent. The effects of temperature, pressure, and ethanol percentage were assessed on yield, phenolic and volatiles contents, and radical scavenging activity (DPPH). The study unveiled that a similar yield to that of "Soxhlet" extraction could be achieved utilizing SFE (34.85 and 34.76%, respectively), obtaining an extract that was richer in terms of phenolics and volatile compounds, and presented higher antioxidant activity. Results also demonstrated that extraction performed with SFE was more selective, and that the extract obtained at 220 bar (22 MPa) and 80 °C, with 5% ethanol, was able to protect mayonnaise from lipid oxidation similarly to the synthetic antioxidant, butylated hydroxyanisole (BHA), used in the food industry. More recently, Romano et al.

(2022) performed a similar study, in which ethanol as a co-solvent was utilized in percentages up to 20%, which corresponded to the conditions with the highest yields (17.60 g/100 $g_{dry\text{-}weight}$). Nonetheless, in that study CO_2 was also utilized in the liquid form (also combined with ethanol as co-solvent), and better results were achieved when such methodology was applied. Despite a similar yield of 17.60 g/100 $g_{dry\text{-}weight}$, liquid CO_2-extracted oil presented higher flavonoids, naringin, and terpenes contents (including limonene), as well as superior antiradical activities (DPPH and ABTS). Supercritical extraction ($ScCO_2$ + 20% ethanol) did, however, extract more dimethylanthranilate, which was the most prevalent volatile in oils, and is a compound that may be used as a fragrance, since it is ascribed the distinctive tangerinearoma.

6.7.5 Ultrasound-Assisted Extraction (UAE)

Ultrasound-assisted extraction, or sonication, is based on the use of ultrasounds to create a cavitation phenomenon. Ultrasound frequencies above 20 kHz are utilized, and as the ultrasound waves propagate through the solvent, they create high- and low-pressure cycles in which bubbles are formed (at low-pressure) and collapse (at high-pressure). As the bubbles collapse, cavitation occurs, resulting in the formation of shear forces, consequently mechanically impacting the suspended particles in the solution. This phenomenon ruptures cell walls, thus enhancing penetration of the solvent and increasing the mass transfer, which improves the extraction efficiency (Michalak and Chojnacka 2014; Mubarak et al. 2015; Zhang et al. 2019).

This type of extraction has the advantages of reduced time of extraction, increased yields, and the reduced volume of solvent needed to perform the extraction (Mubarak et al. 2015; Polanco-Lugo et al. 2019).

Recently, Ordóñez-Santos, Esparza-Estrada and Vanegas-Mahecha (2021) utilized UAE to extract carotenoids from *C. reticulata* epicarp, using sunflower oil as solvent, with the purpose of posterior application as a natural coloring agent, to substitute tartrazine (or decrease its use), in bakery products. The authors sought to investigate the best conditions (regarding temperature, time, and solid to liquid ratio) that would yield the highest pigment content when the ultrasound frequency was 42 kHz. The response surface methodology determined 60 °C for 60 min, using 0.4 mg/mL (solid:liquid) to be the optimum parameters, which yielded 140.7 $mg_{\beta\text{-}carotene}$/100 $g_{dry\text{-}weight}$. Kaur, Panesar and Chopra (2021a) also studied the influence of several parameters (amplitude, solid:liquid ratio, temperature, and time), in that case, on the ethanolic extraction of bioactives from *C. reticulata* peels, in order to standardize the extraction process. The bioactive potential of the extracts was assessed in terms of phenolic content and antioxidant activities. The extract with the highest phenolic content (36.17 $mg_{GAE}/g_{extract}$) and antioxidant activities was obtained using 1:30 (solid:liquid), at 31% amplitude, at 41 °C for 15 min.

Saini, Panesar and Bera (2021) studied the application of UAE to extract a carotenoid (lutein) from peels of *C. reticulata*. The process was optimized through the evaluation of solvent:solid ratio, amplitude, temperature, and time, regarding their impact on lutein yield. Analysis of the results showed that lutein extraction was optimal when the process was performed with 6.16 mL/g, with 32.88% amplitude, at 43.14 °C during 33.71 min, which yielded 29.70 ug_{lutein}/g_{peel}.

Envisioning obtaining a distinct compound from the same matrix, Polanco-Lugo et al. (2019) explored the extraction of pectin from tangerine peels and found that, in comparison with conventional extraction (and commercial pectin), pectins obtained through UAE presented lower galacturonic acid percentage, as well as lower viscosity. The physicochemical characteristics of such pectins allow them to be viable options to be utilized for microencapsulation purposes. In that specific case, the extraction process was not explored for maximization of extraction yield (the parameters utilized were previously defined) since the work's aim was solely to compare the extraction techniques and not their optimization.

Ultrasound-assisted extraction was utilized by Alissandrakis et al. (2003) to extract volatiles from honey and flowers from four *Citrus* species, aiming to distinguish the honey's origin based on the volatile profile. As in the previously reported study, the aim was not to investigate the extraction process *per se*, but its practical application. The exposure to the ultrasounds was performed in a water bath with ultrasound assistance, which conferred the process swiftness and ease. The main volatile in all *Citrus* flowers was determined to be linalool (75.2% in tangerine). The authorsunveiled the viability of a simple and practical method to extract volatiles from "*Citrus* honey" and the possibility of identifying the origin of a specific honey through the analysis of the volatile profile, since the precursors of the compounds found therein were present in the flowers of the corresponding plant.

6.7.6 Microwave-Assisted Extraction (MAE)

This extraction technology is one of the innovative methodologies being extensively explored to extract compounds from different sources. The principle is that microwaves, which are electromagnetic radiation (ranging from 0.3 to 300 GHz), through dipole rotation and ionic conduction generate heat, which is uniformly distributed in the sample. The heat results from the interaction of microwaves with polar molecules (such as water), inducing vibrations that consequently raise the temperature. This temperature increase leads to evaporation of water, which in turn ruptures cell walls due to the pressure exercised therein, increasing solvent penetration and also causing compounds' leakage. These phenomena enhance mass transfer, therefore increasing extraction yields (Michalak and Chojnacka 2014; Mubarak et al. 2015; Ciko et al. 2018; Imbimbo et al. 2020).

The advantages of utilizing this technology include high yield, reduced extraction time, low solvent amount, and energy consumption (Michalak and Chojnacka 2014; Ciko et al. 2018; Anticona et al. 2020; Imbimbo et al. 2020).

Chen et al. (2016) applied MAE to tangerine peels to extract pectic polysaccharides. The study optimized the MAE conditions, namely, microwave power, temperature, and extraction time, which would yield the highest pectin amount. Response surface methodology was utilized, revealing that 704 W, at 52.2 °C for 41.8 min, were the best conditions, which were then applied to obtain a yield of 19.9%.

Similarly, Hayat et al. (2009) assessed the impact of several parameters, namely, power, temperature, liquid:solid ratio, and solvent concentration, on the efficiency of MAE in extracting phenolic acids from *C. reticulata* peels. The statistical tools utilized determined the best conditions to be 152 W for 49 s, with a liquid:solid ratio of 16, using 66% methanol. MAE performed in those conditions was able to generate an extract comprising phenolic acids, among which ferulic and ρ-coumaric acids were the prevalent (465 and 317 $\mu g/g_{dry\text{-}weight}$, respectively).

6.7.7 *High Hydrostatic Pressure (HHP)*

High hydrostatic pressure (HHP) is definitely among the emerging extraction technologies. Originally developed as an alternative to pasteurization, it has also been explored as an extraction technology to obtain compounds from a multitude of matrices. This technology is based on applying pressure to the matrix, which will enable solvent penetration and consequent mass transfer, resulting in the extraction of specific compounds. In HHP, a fluid is compressed (to pressures from 100 up to 1000 MPa), leading to the pressure being exerted instantly and uniformly throughout the fluid. Since the sample is placed within a recipient also containing the extraction solvent, as pressure is applied, there is an increase in solvent permeation into the matrix, consequence of the pressure difference between the outside and the inside of the cells, as well as of damages in their structure. In turn, these phenomena will increase mass transfer and secondary metabolites diffusion, leading to the extraction of the compounds. The main parameters that impact extraction efficiency in HHP are pressure and temperature and, as in all extraction technologies, extraction solvent and liquid to solid ratio (Anticona et al. 2020; Martín and Asuero 2021; Navarro-Baez et al. 2022).

Among the advantages of HHP extraction are the low processing time and the low temperatures utilized, which prevent the degradation of the compounds. The main constraint in utilizing HHP is the initial investment to acquire the equipment, which is considerably high (Martín and Asuero 2021).

Although, as abovementioned, this technology has been applied to extract several compounds from many distinct sources/matrices, it is not widespread concerning extraction in *Citrus*fruits. Nonetheless, Casquete et al. (2015) explored HHP to obtain phenolics from *C. reticulata* peels using ethanol as an extraction solvent. The authors assessed pressure (300 or 500 MPa) and time (3 or 10 min) as process variables, and found 300 MPa during 3 min to be the conditions that yielded the extract with the highest phenolic content (587.28 $mg_{GAE}/100\ g_{fresh\text{-}weight}$) and ABTS radical scavenging activity (658.92 $mg_{Trolox}/100\ g_{fresh\text{-}weight}$). In that study, the impact of HHP extraction when applied to peels from other *Citrus*fruits, namely, *C. limon*, *C.*

latifolia, and *C. sinensis*, was also investigated, and it was found that the extraction conditions that originated the best results were the same regarding all *Citrus*fruits (including *C. reticulata*).

6.7.8 Pulsed Electric Field (PEF)

Pulsed electric field (PEF) is, as the denomination indicates, a technology that applies an electric field (as pulses) to the sample, which causes electroporation, also termed electropermeabilization. The electric fields used in PEF are usually of high voltage, having a very short duration. The principle of the extraction process is that, upon exposure to the pulses, when the transmembrane voltage is sufficiently high, a rearrangement of the phospholipids occurs, leading to the formation of pores, which can be temporary or permanent. Such pores allow the penetration of the solvent, hence, increasing the efficiency of the extraction process and, in the cases in which the electric field is strong enough it may inclusively rupture the cells, consequently leaking intracellular compounds. The latter case, or permanent electroporation, must be attained for an effective extraction. The main parameters that have been considered to impact PEF efficiency are the field strength, pulse shape and, duration and treatment time (pulse duration × number of pulses) (Goettel et al. 2013; Geada et al. 2018; Zhang et al. 2019; Anticona et al. 2020).

Like the aforementioned concerning the use of HHP in *Citrus* matrices, PEF is also a technology that has been scarcely (to un-) explored concerning this genus. There are, however, some studies in which PEF was utilized, although mostly being conjugated with other treatments. Such an example is the study of El Kantar et al. (2018), in which PEF was applied prior to obtaining juice by compression from orange, pomelo, and limon, and extraction yield was assessed. Moreover, the impact of PEF in the extraction of polyphenols from the peels was also investigated. Results showed improved extraction yields of juices when PEF was applied at 3 kV/cm (increases ranging from 25 to 59%), as well as higher amounts of polyphenols (ca. 22.5 $mg_{GAE}/g_{dry\text{-}weight}$*versus* 12.5 $mg_{GAE}/g_{dry\text{-}weight}$ in untreated samples) in the extract obtained by performing PEF at 10 kV/cm, followed by extraction with ethanol 50%, which acted synergistically. Carpentieri et al. (2022) utilized PEF, at 5 kV/cm, as a pre-treatment in orange peels prior to solvent extraction, to obtain limonene. Results revealed the efficacy of PEF in increasing solvent permeability since higher extraction yields were obtained when solvent extraction was performed after PEF was applied as a pre-treatment (33% increase, from 3 to 4 $mg_{limonene}/100\ g_{sample}$, in ethanolic extract). Niu et al. (2021) exploited PEF *per se*to obtain naringin from pomelo peel at various electric field intensities (2–10 kV/cm) and numbers of pulses (10–50), and the impact on yield was evaluated. The authors concluded that 4 kV/cm using 30 pulses were the best PEF conditions, which were able to increase by 20% the naringin content of the extract, when compared with the extract obtained without PEF treatment.

6.7.9 Enzyme-Assisted Extraction (EAE)

The last extraction methodology addressedherein is enzyme-assisted extraction (EAE), which relies on enzymes to digest the cell walls of organisms, such as plants, thus enabling the collection of valuable compounds. Plant cell walls comprise lignocellulose and others polymers which, due to being mingled, impair the extraction process. The use of enzymes, such as pectinases, cellulases, and hemicellulases, leads to cell wall hydrolysis and degradation, which facilitates the penetration of the solvent and may even cause leakage of compounds, consequently improving extraction yields. Since enzymes have an optimum temperature at which the activity is the highest, extraction temperature is a pivotal parameter to be optimized (Michalak and Chojnacka 2014; Zhang et al. 2019; Nde and Foncha 2020). In this sense, Li, Smith and Hossain (2006) studied the impact of three food-grade enzymes (Celluzyme® MX, Celluzyme® CL, and Kleepase® AFP) in the enzyme-assisted aqueous extraction of phenolics from the peels of several *Citrus*fruits, including *C. reticulata*. The authors found that, among the various factors that influenced extraction yields was, as expected, temperature. The best results were obtained at 50 °C, which allowed the process to yield 115 $mg_{GAE}/100\ g_{fresh\text{-}weight}$. Vergara-Barberán et al. (2017) also utilized EAE, in that specific case, to extract proteins from orange and tangerine peels and pulps, although the aim of the work was not to optimize EAE parameters, but to make use of the technology to obtain the proteins, which were then profiled by capillary gel electrophoresis. Nonetheless, different enzyme combinations were evaluated to determine which combination provided the best yield. Afterwards, EAE was performed on the samples from different cultivars with the enzymes combination determined to yield the highest content and, as expected, distinct protein profiles were obtained. Hence, EAE was revealed to be an effective technology, and the authorsconcludedthat protein profiles were a viable tool to distinguish samples originating from distinct cultivars.

Throughout the current section, several distinct technologies/methodologies were addressed, which may be applied to extract a multitude of compounds from tangerine and/or its wastes, presenting a panoply of possibilities in terms of applications and activities (among which are bioactivities). Besides being utilized to obtain compounds from unprocessed tangerine, these technologies may also be applied to convert processing waste residues into added-value products, thereby minimizing the environmental impact of tangerine-processing industries. As described, some technologies are performed with concomitant, or posterior, use of solvents, and in the latter case, the application of the technologies may be considered as a pre-treatment and not as an extraction method *per se*. Nonetheless, the combination of technologies may also be (it is, in fact) a solution to obtain the highest possible amount and diversity of compounds from a single matrix. It was shown that differentmethodologies might be utilized to accomplish similar goals, such as the extraction from tangerine peels of an essential oil with antioxidant potential. In that specific case, several factors must be considered when selecting a technology to be applied, namely the yield, the antioxidant activity of the obtained extracts, and possible contaminants/interferents, among others. Moreover, when optimizing the extraction process, as reported throughout the section, physicochemical parameters, must be considered according to both the technology and the compound to be extracted. Those parameters/factors also have implications in which technologies may be utilized to extract a specific compound since the operational conditions may lead to the degradation of the target compound (e.g., some compounds are more temperature-sensitive than others, and may degrade if the temperature is raised above a specific threshold). In this sense, when seeking to explore the potential of a given matrix to extract a specific compound (or type of compound) studies have to be performed in order to define the best technology (and conditions) to be applied. Nonetheless, different technologies have the potential to process tangerine and/or its parts and wastes, to obtain compounds of utmost importance for the development of innovative products, in pharmaceutical, cosmetics, and food industries.

6.8 Tangerine-Based Products Development

In the quest for cutting-edge innovation, coping with the rapidly evolving market (trends), and seeking to address consumers' expectations, industries pursue the development of novel products and processing technologies whilst conserving fruit and fruit-based products' authenticity/identity.

Product development process entails the conceptualization/design, creation and development of an entirely novel product, or the reformulation of an already existing one (altering the composition) with the purpose of improving a particular trait (characteristic), function or property (health).

There are a fewways by which the topic of this section may be approached, namely, products in which tangerine*per se* is the matrix or a constituent thereof (for instance, tangerinemarmalade and orange-tangerine juice, respectively), and others in which its extracts may be utilized as additives or valuable sources of a specific compound (e.g. peel extract combined with copper nanoparticles).

This section will mainly address the most relevant scientific studies performed over the past 5 years to outline a perspective of the current trends. In a first stage, we sought to focus on the development of products which were based on tangerine*per se*.

Innovative food processing technologies have been exploited as promising alternatives to the conventional thermal processing. Beyond the well-documented decontamination potential, these technologies may also convey an improvement on palatability (taste, aroma and texture), edibility, and nutrients bioavailability (Carmody et al. 2011).

Noguera et al. (2021) investigated the impact of two processing methods, ultrasounds or ozone in tangerinejuice, on the antioxidantsand ascorbic acid contents, along with the effect on deteriorative enzymes activity. The authors reported that when submitted to either technology the juice's phenolics concentration increased, whilst the deteriorative enzymes were impaired. Ozone processing also improved the amount of ascorbic acid. In this sense, these technologies may, therefore, constitute viable alternatives to tangerinejuice heat processing, inclusively further improving the bioactive compounds content. Hashemi et al. (2019) explored pulsed-thermosonication to inactivate *Listeria monocytogenes*, *Shigella sonnei*, *Byssochlamys fulva* and *Saccharomyces cerevisiae* in tangerinejuice. The authors demonstrated that the association of temperature and sonication might be a feasible approach for juice decontamination, albeit further investigation must be performed in order to determine the optimum operating conditions in order to balance the microbial inactivation with the degradation of the ascorbic acid and β-carotene, which was observed and, preferably, must be absent.

Fenoglio et al. (2022) proposed the development of a technology based on the association of UV radiation and mild heat (50 °C) to inactivate potential microbial contaminants in a novel formulated orange-tangerine juice. Moreover, yerba mate (*Ilex paraguariensis* St. Hill) extract was also incorporated in the juice (in the free form or encapsulated), which provided an additional source of natural antioxidants. The treatment was effective in promoting microbial inactivation, and sensory analysis performed with consumers revealed a willingness to consume/acquire the novel product, which was higher in frequent consumers of yerba mate tea. Wirivutthikorn (2019) exploited the potential of blending tangerinejuice with an herbal drink, in which an okra-tangerine juice supplemented with lycopene was developed. The authors evaluated different percentages of tangerinejuice and the one which demonstrated the most promising results concerning sensory evaluation was comprised by 50% of each matrix (okra and tangerine). The study highlighted the potential to formulate novel juice-like beverages with combined ingredients, namely, tangerinejuice and local Thai herbs, which may ultimately be commercialized, with expected consumer acceptance. Contreras-López et al. (2021) also exploited tangerine in combination with four other fruits (guava, strawberry, pineapple, and orange) and developed low-caloric beverages utilizing assortments thereof, medicinal plants (Melissa and lemon verbena) and flowers (*Bougainvillea glabra* and *Matricaria chamomilla*), as healthier alternatives to the conventionally commercialized and consumed sweetened drinks, for overweight and obese people. Besides the low-caloric content, the beverages would also provide the beneficial effects of the bioactive compounds which the constituents comprised, namely antioxidants from tangerine. Among the four different formulations evaluated the one which comprised tangerine, bougainvillea, lemon verbena, and guava, ranked first regarding the liking of color and smell, and second concerning phenolic content and antioxidant activity. A study performed by Guo et al. (2021) evaluated the impact of the consumption of *C. reticulata*juice fermented by lactic acid bacteria (LAB), on the prevention of obesity in mice fed with a high-fat diet. Modulation of the mice gut microbiota was also assessed. The authors found positive impact of *C. reticulata*-juice in obesity-inhibition indicators (e.g. weight-gain inhibition, hepatic steatosis and fat accumulation decrease, and improved insulin sensitivity) and that, in comparison with non-fermented juice, the LAB-fermented juice presented a better overall activity against obesity. The increased inhibitory effect of the fermented juice was hypothesized to be owed to the promotion of browning of the white adipose tissue, as well as the observed increase of intestinal health-promoting probiotic bacteria. Recently, Wang et al. (2022) utilized *C. reticulata* peels for the development of an innovative citrus-white tea, which is the second worldwide most consumed non-alcoholic beverage. The manufacturing process was modified to include fresh peels after withering and previously to drying, in opposition to the conventionally exploited method in which the peels are added following the drying step. The modified process significantly influenced the volatile profile of the formulated tea, comparatively to the profile present in the citrus-white tea produced through the conventional method. Besides the characterization of the volatiles profiles, the teas were sensory-analyzed by certified professional tea-tasters, which ascribed freshness and fruity citrus-like scents to the aromas of the innovative tea, hence validating the novel product as a viable alternative to the conventionally manufactured.

Over the past decade, cutting-edge biotechnology approaches have fueled the development of novel processes to improve organoleptic features.

Recently, Santana et al. (2021) analyzed the impact of cellulases from a yeast, namely, *Pseudozyma* sp., on tangerinejuice clarification, and found that, besides being able to decrease turbidity, the enzymatic treatment also reduced the juices' viscosity. The study demonstrated that the cellulase may be a promising tool for the development of those products, regarding tangerine as well as other citrus fruits, since opacity is a major deterrent factor as far as consumers' acceptance is concerned.

The challenges posed by the global cultural diversity (indigenous food systems), climatic conditions (geographical distribution) and social structures (socioeconomic inequalities) prompted/fueled the development of more sustainable solutions to meet the highly diversified populations' necessities. This is of utmost importance in hot-environment, low-income countries, devoid of an effective cold chain to preserve perishable food products. In light of this, Phinney et al. (2020) produced tangerinemarmalade in membrane pouches through solar assisted pervaporation with the purpose of providing a product that would allow the consumption of the fruit, with the beneficial nutritional traits that it conveys, in rural Mozambique in which the harsh environmental conditions hamper the preservation/conservation of the traditionally consumed products, since the deprivation of a refrigeration system is a limitation.

In the past decade, consumers' awareness/perception concerning health and sustainability triggered a shift in the consumption pattern, which has consequently prompted investigators to explore natural sources seeking for alternatives to the commonly utilized chemical additives. With the purpose of substituting/replacing dyes for natural colorant compoundsin an isotonic tangerine soft drink, Tupuna-Yerovi et al. (2020) investigated the incorporation of microcapsules containing a carotenoid (norbixin), obtained from the seeds of annatto (*Bixa orellana* L.). The norbixin was encapsulated in order to protect this diterpenoid from degradation, which extended the shelf-life of the drink, and microcapsules were synthesized using arabic gum, hence, the goal of obtaining a more natural-like product was accomplished.

A distinct application of tangerine was explored by Sande, Moreira and Papa (2018), which produced flour from tangerine peel and bagasse, and then utilized it to replace wheat flour in a biscuit recipe. The biscuits formulated with the tangerine-based flour presented, in comparison with those manufactured solely with wheat flour, higher fiber content and antioxidant activity. Moreover, the authors reported a higher sensory acceptability of tangerine-based biscuits. Those results highlighted the potential of tangerine wastes to be converted into flour which may then be used to develop a panoply of added-value, innovative food products, accomplishing the thirteenth sustainable development goal. This approach was also explored by Khaleel et al. (2022), which developed muffins with *C. reticulata* peel flour partially substituting wheat flour. The resulting muffins possessed an improved nutritional value, in comparison with control, with higher fiber and protein contents. Those baked products could also be an interesting source of antioxidants since phenolics and flavonoids were present in the flour, which demonstrated free radical scavenging activity. Nonetheless, the incorporation of the peel flour impacted the physical properties of the muffins, negatively altering the sensorial features, except in the case of the lower percentages utilized. In this sense, the authors concluded that for obtaining a viable product only 5% of the wheat flour could be substituted by the *C. reticulata* peel flour, since the impact was imperceptible and the muffins would therefore be accepted by consumers.

Among the assortment of food products that may comprise *C. reticulata*, drinks/beverages are definitely the most common ones to which tangerine is associated. In that sense, Tomova et al. (2021) utilized tangerine and grapefruit essential oils in the formulation of probiotic wort-based beverages, and found that employing tangerine oil resulted in the best product, with the highest phenolic content, consequently presenting the highest antioxidant activity. Concerning sensory acceptability of the formulated products, between the beverages manufactured with the two oils, the one comprising tangerine oil was the preferred by the sensorial evaluation panel.

Xu et al. (2022a) reported the possibility of producing tangerine wines through fermentation utilizing alternative strains of yeasts that might improve some organoleptic traits of those wines fermented solely by the traditional *Saccharomyces cerevisiae*. The non-*Saccharomyces* strains synthesize polyphenols and volatiles, which in turn may originate wines with improved sensorial features and higher content of bioactive compounds. The authorsexploited distinct fermentation processes, namely, fermentation with pure cultures, and co-fermentation (with *S. cerevisiae* and the non-*Saccharomyces* strains), the latter using two different approaches – sequential and blended. The outcome of the study was that, through (according to) the yeast strains and the process applied in the fermentation, polyphenols and volatiles contents may be improved in tangerine wines, which may be tailored in order to fit/attain the desired features and attributes.

In a completely distinct approach, Bambeni et al. (2021) reported the incorporation of *C. reticulata* Blanco cv. clementine pomace extract in raw ground beef patties, with the purpose of providing a protective effect against the food matrix degradation throughout storage. Results demonstrated/unraveled the extracts' protective effects (of the incorporation of the extract) in terms of lipid and protein oxidation (antioxidant activity), as well as antimicrobial activity (by bacterial growth inhibition).

Moreover, amongst the several extracts evaluated, *C. reticulata* Blanco cv. clementine pomace extract was found to confer the most acceptable sensory attributes, namely, beef-like aroma and flavor, and overall aroma intensity. In a study performed by Song et al. (2021), *C. reticulata* L. essential oil (EO) was also applied to one other meat product (namely, pork), through its incorporation into chitosan nanoparticles, intending to explore the oils' antimicrobial properties. The authors documented that, when applied to fresh pork and throughout storage, the EO-entrapped nanoparticles presented improved/enhanced antimicrobial activity, in comparison with nanoparticles devoid of EO (composed of chitosan alone). Moreover, the EO-containing nanoparticles also displayed anti-biofilm formation activity, along with (as well as) mature biofilm removal (elimination) capability. With a similar purpose ofhamperingfoodstuffs deterioration, hence extending shelf-life, He et al. (2017) developed a coating from tangerine peels extract, which was utilized in tilapia fillets. The coating was applied through immersion of the fish fillets in aqueous solution containing the extract and, thereafter, samples were stored at refrigerated and freezing temperatures. The coating was capable to increase the shelf-life of frozen fillets by 35–45%, through the prevention of lipid oxidation and protein degradation, as well as suppressing water loss and alterations to the tissue structure. Within the perspective of utilizing *C. reticulata* extracts, and in an attempt to circumvent the scarce stability and bioavailability of those extracts containing flavonoids, Hu et al. (2019) studied the nanoparticles mediated delivery of flavonoids. The extract incorporation into nanoparticles composed of whey protein concentrate, as well as the stability throughout simulated gastrointestinal tract were investigated. The authors found that the nanoparticles were able to protect, to a certain extent, the flavonoids from the detrimental harsh conditions (environment) of the digestive system, evading (escaping/avoiding/impeding) degradation, and ensuring their controlled delivery.

Despite the undeniable industrial value of *C. reticulata* components and bioactive compounds towards/for the formulation of novel food products, there are other industries that may harvest the potential comprised in its components and wastes to develop new and/or alternative products.

The extensively studied peels were recently exploited by Januário et al. (2021) in an altogether distinct application from food products, as a biosorbent material for industrial wastage, namely, a cationic dye (safranin orange). It was revealed that the peels detained specific properties that promoted the fixation of such type of (micro) polluants, and that the process was, to a certain extent, reversible, which enabled the re-use of the matrix, furthering the beneficial environmental/ecological impact of the technology. Abdić et al. (2018) also utilized tangerine peels as an adsorbent to remove heavy metals (namely, Cd, Co, Cr, Cu, Mn, Ni, Pb and Zn) from contaminated water. The authors performed a chemical modification of the substrate (peels), which greatly enhanced the adsorption efficiency (ca. 40%). Despite the lower sorption ability when compared with the chemically modified version, the unmodified peels displayed, nonetheless, adsorption potential regarding all of the eight heavy metals assessed.

Focusingon the potential of tangerine wastes from an industrial standpoint, Ituen et al. (2020) utilized *C. reticulata* peels extract, in combination with copper nanoparticles, to prevent (bio- and acid-)corrosion in pipeline grade steel. The results pointed out to an increased efficiency of the extract when combined with the nanoparticles, revealing the potential of the composite to be utilized as a viable alternative for industries that use and, therefore, need to protect, such material. One other application of essential oils (in that case, from leaves) as preservatives was assessed by Ali et al. (2021) concerning fungi which cause wood deterioration. Four distinct strains were utilized, namely, three belonging to the *Aspergillus* genus and *Fusarium culmorum*, and results demonstrated/indicated antifungal activity of the essential oils on all four strains, thereby validating their potential to be used for such end.

Tangerine peels were also exploited by Rueda et al. (2020), particularly, an extract thereof, as part of a green-synthesis process to produce titanium dioxide nanocrystals. The study revealed that throughout the process by which the nanocrystals were formed, the phytochemicals present in the extract promoted condensation and controlled the crystals growth. This is another example underlining the outstanding potential of tangerine extracts for the development novel and alternative products (approaches), outside the traditionally envisioned ones.

Concerning this material-oriented product development, tangerinejuice was used by Aslan and Eskalen (2021) as a carbon source to produce carbon nanodots (carbon quantum dots). The study evaluated the potential to synthesize such dots in a one-step hydrothermal methodology which would be fast and simple, and green, as it would utilize a natural carbon source. The results indicated that the obtained carbon nanodots presented relatively good stability, therefore, revealing the viability of the process to manufacture such dots, that may be utilized to produce other materials, such as ink, which was also determined to be feasible.

Several fruits, among which are citrus fruits, have also been investigated as possible sources for alternative energy production technologies. With this intent, Rojas-Flores et al. (2020) evaluated tangerine wastes potential utilization as fuel for microbial fuel cells in order to obtain bioelectricity. Results revealed that, comparatively with lime and orange wastes, tangerines' (wastages) generated the highest voltage and power density. The study concluded that such application may be a feasible alternative as an innovative electricity generation technology. Tangerine peels were also investigated by Shim (2015) as source for bio-ethanol production by *Saccharomyces cerevisiae*, through saccharification and fermentation. The results demonstrated the feasibility of the proposed methodology, since a considerable amount of ethanol was obtained after 8 h processing.

As aforementioned, *Citrus reticulata* may also be a source of valuable compounds to be utilized by the pharmaceutical industry. In that context, Silveira, Santos Rubio and Poleti Martucci (2022) evaluated the potential of tangerine essential oil to reduce anxiety, i.e. to act as an anxiolytic, in a light-dark test using zebrafish (*Danio rerio*) as animal model. The study revealed a propensity of the zebrafish to which essential oils were administered to present reduced anxiety, without impairment of locomotive and exploratory activities. Therefore, tangerine essential oils may be a natural alternative to the currently utilized synthetic drugs which administration usually entails adverse effects.

In this section, we provided a descriptive overview of the current state of the art of *C. reticulata*-based products development. As thoroughly documented in this section, there is a myriad of products that may be developed both in terms of the hesperidium *per se* as the matrix for novel and/or improved food products, as well as by utilizing *C. reticulata* components to be integral parts in innovative products and technologies. The wealth of compounds that make up *C. reticulata* which have been formerly described, allow for such versatility, and provide a plethora of opportunities for wastes valorization, thereby increasing the possibilities of continuing to evolve society towards a greener future, in which the ecological footprint of the tangerine processing industry may be largely reduced.

References

Abdelrahman M, Nishiyama R, Tran CD, Kusano M, Nakabayashi R, Okazaki Y et al (2021) Defective cytokinin signaling reprograms lipid and flavonoid gene-to-metabolite networks to mitigate high salinity in Arabidopsis. Proc Natl Acad Sci U S A 118. https://doi.org/10.1073/pnas.2105021118

Abdić S, Memić M, Šabanović E, Sulejmanović J, Begić S (2018) Adsorptive removal of eight heavy metals from aqueous solution by unmodified and modified agricultural waste: tangerine peel. Int J Environ Sci Technol 15:2511–2518. https://doi.org/10.1007/s13762-018-1645-7

Abeysinghe DC, Li X, De Sun C, Zhang WS, Zhou CH, Chen KS (2007) Bioactive compounds and antioxidant capacities in different edible tissues of citrus fruit of four species. Food Chem 104:1338–1344. https://doi.org/10.1016/j.foodchem.2007.01.047

Ahmad MM, Salim-ur-Rehman, Qureshi TM, Nadeem M, Asghar M (2016) Variability in peel composition and quality evaluation of peel oils of citrus varieties. J Agric Res 54:747–756. https://doi.org/10.13140/RG.2.2.25607.85922

Al Juhaimi F, Özcan MM, Uslu N, Ghafoor K (2018) The effect of drying temperatures on antioxidant activity, phenolic compounds, fatty acid composition and tocopherol contents in citrus seed and oils. J Food Sci Technol 55:190–197. https://doi.org/10.1007/s13197-017-2895-y

Alam C, Aufreiter S, Georgiou CJ, Hoque MT, Finnell RH, O'Connor DL et al (2019) Upregulation of reduced folate carrier by Vitamin D enhances brain folate uptake in mice lacking folate receptor alpha. Proc Natl Acad Sci U S A 116:17531–17540. https://doi.org/10.1073/pnas.1907077116

Ali AM, Gabbar MA, Abdel-Twab SM, Fahmy EM, Ebaid H, Alhazza IM et al (2020) Antidiabetic potency, antioxidant effects, and mode of actions of citrus reticulata fruit peel hydroethanolic extract, hesperidin, and quercetin in nicotinamide/streptozotocin-induced Wistar diabetic rats. Oxidative Med Cell Longev 2020. https://doi.org/10.1155/2020/1730492

Ali HM, Elgat WAAA, El-hefny M, Salem MZM, Taha AS, Al Farraj DA et al (2021) New approach for using of Mentha longifolia L. and Citrus reticulata L. essential oils as wood-biofungicides: GC-MS, SEM and MNDO quantum chemical studies. Materials (Basel) 14:1–18. https://doi.org/10.3390/ma14061361

Alissandrakis E, Daferera D, Tarantilis PA, Polissiou M, Harizanis PC (2003) Ultrasound-assisted extraction of volatile compounds from citrus flowers and citrus honey. Food Chem 82:575–582. https://doi.org/10.1016/S0308-8146(03)00013-X

Alquézar B, Zacarías L, Rodrigo MJ (2009) Molecular and functional characterization of a novel chromoplast-specific lycopene β-cyclase from Citrus and its relation to lycopene accumulation. J Exp Bot 60:1783–1797. https://doi.org/10.1093/jxb/erp048

Andrews AC (1961) Acclimatization of citrus fruits in the Mediterranean region. Agric Hist:35–46

Anticona M, Blesa J, Frigola A, Esteve MJ (2020) High biological value compounds extraction from citruswaste with non-conventional methods. Foods 9. https://doi.org/10.3390/foods9060811

Appelhans MS, Bayly MJ, Heslewood MM, Groppo M, Verboom GA, Forster PI et al (2021) A new subfamily classification of the Citrus family (Rutaceae) based on six nuclear and plastid markers. Taxon 70:1035–1061. https://doi.org/10.1002/TAX.12543

Asencio AD, Serrano M, García-Martínez S, Pretel MT (2018) Organic acids, sugars, antioxidant activity, sensorial and other fruit characteristics of nine traditional Spanish Citrus fruits. Eur Food Res Technol 244:1497–1508. https://doi.org/10.1007/s00217-018-3064-x

Asensi-Fabado MA, Munné-Bosch S (2010) Vitamins in plants: occurrence, biosynthesis and antioxidant function. Trends Plant Sci 15:582–592. https://doi.org/10.1016/j.tplants.2010.07.003

Aslan M, Eskalen H (2021) A study of carbon nanodots (carbon quantum dots) synthesized from tangerine juice using one-step hydrothermal method. Fullerenes Nanotub Carbon Nanostruct 29:1026–1033. https://doi.org/10.1080/1536383X.2021.1926452

Assouline S, Cohen S, Meerbach D, Harodi T, Rosner M (2002) Microdrip irrigation of field crops: effect on yield, water uptake, and drainage in sweet corn. Soil Sci Soc Am J 66:228–235. https://doi.org/10.2136/SSSAJ2002.2280

Ayre BG (2011) Membrane-transport systems for sucrose in relation to whole-plant carbon partitioning. Mol Plant 4:377–394. https://doi.org/10.1093/mp/ssr014

Bacete L, Schulz J, Engelsdorf T, Bartosova Z, Vaahtera L, Yan G et al (2022) THESEUS1 modulates cell wall stiffness and abscisic acid production in Arabidopsis thaliana. Proc Natl Acad Sci U S A 119. https://doi.org/10.1073/pnas.2119258119

Badarinath AV, Rao KM, Madhu C, Chetty S, Ramkanth S, Rajan TVS et al (2010) A review on in-vitro antioxidant methods: comparisions, correlations and considerations. Int J PharmTech Res 2:1276–1285. Available at: https://www.sid.ir/en/journal/ViewPaper.aspx?ID=383951

Bailey LB, Gregory JF (1999) Recent advances in nutritional science folate metabolism and requirements. J Nutr 129:779–782

Bain JM (1958) Morphological, anatomical, and physiological changes in the developing fruit of the Valencia orange, Citrus sinensis (L) Osbeck. Aust J Bot 6:1–23

Bambeni T, Tayengwa T, Chikwanha OC, Manley M, Gouws PA, Marais J et al (2021) Biopreservative efficacy of grape (Vitis vinifera) and clementine mandarin orange (Citrus reticulata) by-product extracts in raw ground beef patties. Meat Sci 181:108609. https://doi.org/10.1016/j.meatsci.2021.108609

Barkley NA, Roose ML, Krueger RR, Federici CT (2006) Assessing genetic diversity and population structure in a citrus germplasm collection utilizing simple sequence repeat markers (SSRs). Theor Appl Genet 112:1519–1531. https://doi.org/10.1007/S00122-006-0255-9/FIGURES/2

Barret HC, Rhoades AM (1976) A numerical taxonomic study of affinity relationships in cultivated Citrus and its close relatives. Syst Bot:105–136

Bashandy SAE, Salama A, Fayed AHM, Omara EA, El-Toumy SA, Salib JY (2020) Protective effect of Mandarin (CITRUS RETICULATA) peel extract on potassium dichromate induced hepatotoxicity and nephrotoxicity in rats. Plant Arch 20:2231–2242

Benderoth M, Textor S, Windsor AJ, Mitchell-Olds T, Gershenzon J, Kroymann J (2006) Positive selection driving diversification in plant secondary metabolism. Proc Natl Acad Sci U S A 103:9118–9123. https://doi.org/10.1073/pnas.0601738103

Benstein RM, Ludewig K, Wulfert S, Wittek S, Gigolashvili T, Frerigmann H et al (2013) Arabidopsis phosphoglycerate dehydrogenase1 of the phosphoserine pathway is essential for development and required for ammonium assimilation and tryptophan biosynthesis. Plant Cell 25:5011–5029. https://doi.org/10.1105/tpc.113.118992

Bentahar A, Bouaziz A, Djidel S, Khennouf S (2020) Phenolic content and antioxidant activity of ethanolic extracts from Citrus sinensis L. and Citrus reticulata L. fruits. J Drug Deliv Ther 10:308–313. https://doi.org/10.22270/jddt.v10i5-s.4537

Berk Z (2016) Chapter 2: Morphology and chemical composition. In: Berk ZBT-CFP (ed) Citrus fruit processing. Academic, San Diego, pp 9–54. https://doi.org/10.1016/B978-0-12-803133-9.00002-3

Berkov S, Osorio E, Viladomat F, Bastida J (2020) Chemodiversity, chemotaxonomy and chemoecology of Amaryllidaceae alkaloids. Alkaloids Chem Biol:113–185

Bernet GP, Fernandez-Ribacoba J, Carbonell EA, Asins MJ (2010) Comparative genome-wide segregation analysis and map construction using a reciprocal cross design to facilitate citrus germplasm utilization. Mol Breed 25:659–673. https://doi.org/10.1007/S11032-009-9363-Y/FIGURES/4

Boublin F, Cabassa-Hourton C, Leymarie J, Leitao L (2022) Potential involvement of proline and flavonols in plant responses to ozone. Environ Res 207:112214. https://doi.org/10.1016/j.envres.2021.112214

Brouwer IA, Van Dusseldorp M, West CE, Meyboom S, Thomas CMG, Duran M et al (1999) Dietary folate from vegetables and citrus fruit decreases plasma homocysteine concentrations in humans in a dietary controlled trial. J Nutr 129:1135–1139. https://doi.org/10.1093/jn/129.6.1135

Bulley S, Wright M, Rommens C, Yan H, Rassam M, Lin-Wang K et al (2012) Enhancing ascorbate in fruits and tubers through over-expression of the l-galactose pathway gene GDP-l-galactose phosphorylase. Plant Biotechnol J 10:390–397. https://doi.org/10.1111/j.1467-7652.2011.00668.x

Burri BJ, La Frano MR, Zhu C (2016) Absorption, metabolism, and functions of β-cryptoxanthin. Nutr Rev 74:69–82. https://doi.org/10.1093/nutrit/nuv064

Carmody RN, Weintraub GS, Wrangham RW (2011) Energetic consequences of thermal and nonthermal food processing. Proc Natl Acad Sci U S A 108:19199–19203. https://doi.org/10.1073/pnas.1112128108

Carpentieri S, Režek Jambrak A, Ferrari G, Pataro G (2022) Pulsed electric field-assisted extraction of aroma and bioactive compounds from aromatic plants and food by-products. Front Nutr 8:1–12. https://doi.org/10.3389/fnut.2021.792203

Casquete R, Castro SM, Martín A, Ruiz-Moyano S, Saraiva JA, Córdoba MG et al (2015) Evaluation of the effect of high pressure on total phenolic content, antioxidant and antimicrobial activity of citrus peels. Innov Food Sci Emerg Technol 31:37–44. https://doi.org/10.1016/j.ifset.2015.07.005

Castro MA, Rodenak-Kladniew B, Massone A, Polo M, De Bravo MG, Crespo R (2018) Citrus reticulata peel oil inhibits non-small cell lung cancer cell proliferation in culture and implanted in nude mice. Food Funct 9:2290–2299. https://doi.org/10.1039/c7fo01912b

Castro MA, Llanos MA, Rodenak-Kladniew BE, Gavernet L, Galle ME, Crespo R (2020) Citrus reticulata peel oil as an antiatherogenic agent: Hypolipogenic effect in hepatic cells, lipid storage decrease in foam cells, and prevention of LDL oxidation. Nutr Metab Cardiovasc Dis 30:1590–1599. https://doi.org/10.1016/j.numecd.2020.04.033

Chakraborty, I., Sen, I. K., Mondal, S., Rout, D., Bhanja, S. K., Maity, G. N., et al. (2019). Bioactive polysaccharides from natural sources: a review on the antitumor and immunomodulating activities. Biocatal Agric Biotechnol 22, 101425. doi: https://doi.org/https://doi.org/10.1016/j.bcab.2019.101425

Chatzivasileiou AO, Ward V, Edgar SMB, Stephanopoulos G (2019) Two-step pathway for isoprenoid synthesis. Proc Natl Acad Sci U S A 116:506–511. https://doi.org/10.1073/pnas.1812935116

Chemat F, Fabiano-tixier AS, Abert M, Allaf T (2015) Solvent-free extraction of food and natural products. Trends Anal Chem 71:157–168. https://doi.org/10.1016/j.trac.2015.02.021

Chen M, Jiang Q, Yin XR, Lin Q, Chen JY, Allan AC et al (2012) Effect of hot air treatment on organic acid- and sugar-metabolism in Ponkan (Citrus reticulata) fruit. Sci Hortic (Amsterdam) 147:118–125. https://doi.org/10.1016/j.scienta.2012.09.011

Chen R, Jin C, Tong Z, Lu J, Tan L, Tian L et al (2016) Optimization extraction, characterization and antioxidant activities of pectic polysaccharide from tangerine peels. Carbohydr Polym 136:187–197. https://doi.org/10.1016/j.carbpol.2015.09.036

Chou YC, Ho CT, Pan MH (2018) Immature citrus reticulata extract promotes browning of beige adipocytes in high-fat diet-induced C57BL/6 mice. J Agric Food Chem 66:9697–9703. https://doi.org/10.1021/acs.jafc.8b02719

Ciko AM, Jokić S, Šubarić D, Jerković I (2018) Overview on the application of modern methods for the extraction of bioactive compounds from marine macroalgae. Mar Drugs 16. https://doi.org/10.3390/md16100348

Colodel C, Vriesmann LC, de Oliveira Petkowicz CL (2018) Cell wall polysaccharides from Ponkan mandarin (Citrus reticulata Blanco cv. Ponkan) peel. Carbohydr Polym 195:120–127. https://doi.org/10.1016/j.carbpol.2018.04.066

Contreras-López E, Ramírez-Godínez J, García-Martínez MM, Gutiérrez-Salomón AL, González-Olivares LG, Jaimez-Ordaz J (2021) Low-calorie beverages made from medicinal plants, flowers and fruits: characteristics and liking of a population with overweight and obesity. Appl Sci 11:1–15. https://doi.org/10.3390/app11093766

Costanzo G, Vitale E, Iesce MR, Naviglio D, Amoresano A, Fontanarosa C et al (2022) Antioxidant properties of pulp, peel and seeds of Phlegrean Mandarin (Citrus reticulata Blanco) at different stages of fruit ripening. Antioxidants (Basel, Switzerland) 11. https://doi.org/10.3390/antiox11020187

Craney A, Ahmed S, Nodwell J (2013) Towards a new science of secondary metabolism. J Antibiot (Tokyo) 66:387–400. https://doi.org/10.1038/ja.2013.25

Curk F, Ancillo G, Garcia-Lor A, Luro F, Perrier X, Jacquemoud-Collet JP et al (2014) Next generation haplotyping to decipher nuclear genomic interspecific admixture in Citrus species: analysis of chromosome 2. BMC Genet 15:1–19. https://doi.org/10.1186/S12863-014-0152-1/TABLES/7

Curk F, Ollitrault F, Garcia-Lor A, Luro F, Navarro L, Ollitrault P (2016) Phylogenetic origin of limes and lemons revealed by cytoplasmic and nuclear markers. Ann Bot 117:565–583. https://doi.org/10.1093/AOB/MCW005

Curl AL (1965) The occurrence of beta-Citraurin and of beta-Apo-8′-carotenal in the peels of california tangerines and oranges. J Food Sci 30:13–18. https://doi.org/10.1111/j.1365-2621.1965.tb00254.x

Curtis TY, Bo V, Tucker A, Halford NG (2018) Construction of a network describing asparagine metabolism in plants and its application to the identification of genes affecting asparagine metabolism in wheat under drought and nutritional stress. Food Energy Secur 7:1–16. https://doi.org/10.1002/fes3.126

Daloso DM, Müller K, Obata T, Florian A, Tohge T, Bottcher A et al (2015) Thioredoxin, a master regulator of the tricarboxylic acid cycle in plant mitochondria. Proc Natl Acad Sci U S A 112:E1392–E1400. https://doi.org/10.1073/pnas.1424840112

Daly JW (2003) Alkaloids. Nature's curse or blessing. J Nat Prod 66:901–902. https://doi.org/10.1021/np030705p

de Camargo AC, Schwember AR, Parada R, Garcia S, Maróstica MRJ, Franchin M et al (2018) Opinion on the hurdles and potential health benefits in value-added use of plant food processing by-products as sources of phenolic compounds. Int J Mol Sci 19. https://doi.org/10.3390/ijms19113498

de Camargo AC, Biasoto ACT, Schwember AR, Granato D, Rasera GB, Franchin M et al (2019) Should we ban total phenolics and antioxidant screening methods? The link between antioxidant potential and activation of NF-κB using phenolic compounds from grape by-products. Food Chem 290:229–238. https://doi.org/10.1016/j.foodchem.2019.03.145

De La Garza RID, Gregory JF, Hanson AD (2007) Folate biofortification of tomato fruit. Proc Natl Acad Sci U S A 104:4218–4222. https://doi.org/10.1073/pnas.0700409104

De La Torre F, Cañas RA, Pascual MB, Avila C, Cánovas FM (2014) Plastidic aspartate aminotransferases and the biosynthesis of essential amino acids in plants. J Exp Bot 65:5527–5534. https://doi.org/10.1093/jxb/eru240

De Moraes Barros HR, De Castro Ferreira TAP, Genovese MI (2012) Antioxidant capacity and mineral content of pulp and peel from commercial cultivars of citrus from Brazil. Food Chem 134:1892–1898. https://doi.org/10.1016/j.foodchem.2012.03.090

de Tournefort JP (1700) Institutiones rei herbariae. Editio Altera. Typ. Reg, Paris

Degu A, Hatew B, Nunes-Nesi A, Shlizerman L, Zur N, Katz E et al (2011) Inhibition of aconitase in citrus fruit callus results in a metabolic shift towards amino acid biosynthesis. Planta 234:501–513. https://doi.org/10.1007/s00425-011-1411-2

Dela Seña C, Riedl KM, Narayanasamy S, Curley RW, Schwartz SJ, Harrison EH (2014) The human enzyme that converts dietary provitamin A carotenoids to vitamin A is a dioxygenase. J Biol Chem 289:13661–13666. https://doi.org/10.1074/jbc.M114.557710

Delchier N, Herbig A-L, Rychlik M, Renard CMGC (2016) Folates in fruits and vegetables: contents, processing, and stability. Compr Rev Food Sci Food Saf 15:506–528. https://doi.org/10.1111/1541-4337.12193

Dellero Y, Jossier M, Schmitz J, Maurino VG, Hodges M (2016) Photorespiratory glycolate-glyoxylate metabolism. J Exp Bot 67:3041–3052. https://doi.org/10.1093/jxb/erw090

Desai S, Hasan M, Sonawane A, Rajurkar GB, Singh D (2014) Effect of different levels of drip irrigation and potassium fertilizer on fruit yield and quality of kinnow (citrus reticulata blanco). Int Agric Eng J 23:50–56

Dhifi W, Bellili S, Jazi S, Bahloul N, Mnif W (2016) Essential oils' chemical characterization and investigation of some biological activities: a critical review. Medicines 3:25. https://doi.org/10.3390/medicines3040025

Dollinger J, Dagès C, Bailly JS, Lagacherie P, Voltz M (2015) Managing ditches for agroecological engineering of landscape. A review. Agron Sustain Dev 35:999–1020. https://doi.org/10.1007/S13593-015-0301-6/TABLES/5

Dong ZB, Shao WY, Liang YR (2014) Isolation and characterization of essential oil extracted from tangerine peel. Asian J Chem 26:4975–4978. https://doi.org/10.14233/ajchem.2014.16277

Dorp KV, Hölzl G, Plohmann C, Eisenhut M, Abraham M, Weber APM et al (2015) Remobilization of phytol from chlorophyll degradation is essential for Tocopherol synthesis and growth of arabidopsis. Plant Cell 27:2846–2859. https://doi.org/10.1105/tpc.15.00395

Douce R, Bourguignon J, Neuburger M, Rébeillé F (2001) The glycine decarboxylase system: a fascinating complex. Trends Plant Sci 6:167–176. https://doi.org/10.1016/S1360-1385(01)01892-1

Duan L, Guo L, Dou LL, Zhou CL, Xu FG, Zheng GD et al (2016) Discrimination of Citrus reticulata Blanco and Citrus reticulata "Chachi" by gas chromatograph-mass spectrometry based metabolomics approach. Food Chem 212:123–127. https://doi.org/10.1016/j.foodchem.2016.05.141

Dugo P, Herrero M, Kumm T, Giuffrida D, Dugo G, Mondello L (2008) Comprehensive normal-phase × reversed-phase liquid chromatography coupled to photodiode array and mass spectrometry detection for the analysis of free carotenoids and carotenoid esters from mandarin. J Chromatogr A 1189:196–206. https://doi.org/10.1016/j.chroma.2007.11.116

Egan AN, Schlueter J, Spooner DM (2012) Applications of next-generation sequencing in plant biology. Am J Bot 99:175–185. https://doi.org/10.3732/ajb.1200020

El Kantar S, Boussetta N, Lebovka N, Foucart F, Rajha HN, Maroun RG et al (2018) Pulsed electric field treatment of citrus fruits: Improvement of juice and polyphenols extraction. Innov Food Sci Emerg Technol 46:153–161. https://doi.org/10.1016/j.ifset.2017.09.024

Eroglu A, Harrison EH (2013) Carotenoid metabolism in mammals, including man: formation, occurrence, and function of apocarotenoids. J Lipid Res 54:1719–1730. https://doi.org/10.1194/jlr.R039537

Etienne A, Génard M, Lobit P, Mbeguié-A-Mbéguié D, Bugaud C (2013) What controls fleshy fruit acidity? A review of malate and citrate accumulation in fruit cells. J Exp Bot 64:1451–1469. https://doi.org/10.1093/jxb/ert035

FAO (2021) Citrus fruit fresh and processed: statistical bulletin 2020

Farin D, Ikan R, Gross J (1983) The carotenoid pigments in the juice and flavedo of a mandarin hybrid (Citrus reticulata) cv michal during ripening. Phytochemistry 22:403–408. https://doi.org/10.1016/0031-9422(83)83012-X

Feng S, Niu L, Suh JH, Hung WL, Wang Y (2018) Comprehensive metabolomics analysis of mandarins (Citrus reticulata) as a tool for variety, rootstock, and grove discrimination. J Agric Food Chem 66:10317–10326. https://doi.org/10.1021/acs.jafc.8b03877

Fenoglio D, Ferrario M, Andreone A, Guerrero S (2022) Development of an orange-tangerine juice treated by assisted pilot-scale UV-C light and loaded with yerba mate: microbiological, physicochemical, and dynamic sensory studies. Food Bioprocess Technol 15:915–932. https://doi.org/10.1007/s11947-022-02775-7

Fernandes JB, David V, Facchini PH, da Silva MF d GF, Rodrigues Filho E, Vieira PC et al (2002) Extrações de óleos de sementes de citros e suas atividades sobre a formiga cortadeira Atta sexdens e seu fungo simbionte. Quim Nova 25:1091–1095. https://doi.org/10.1590/s0100-40422002000700009

Ferrarius GB (1646) Hesperides sive de malorum aureorum cultura et usu Libri Quatuor. Sumptibus Hermanni Scheus

Ferreira SS, Silva AM, Nunes FM (2018) Citrus reticulata Blanco peels as a source of antioxidant and anti-proliferative phenolic compounds. Ind Crop Prod 111:141–148. https://doi.org/10.1016/J.INDCROP.2017.10.009

Figueiredo ARP, da Costa E, Silva J, Domingues MR, Domingues P (2019) The effects of different extraction methods of lipids from Nannochloropsis oceanica on the contents of omega-3 fatty acids. Algal Res 41:101556. https://doi.org/10.1016/j.algal.2019.101556

Fischer TC, Butzmann R (1998) Citrus meletensis (Rutaceae), a new species from the Pliocene of Valdarno (Italy). Plant Syst Evol 210:51–55. https://doi.org/10.1007/BF00984727

Fomani M, Ngeufa Happi E, Nouga Bisoue A, Ndom JC, Kamdem Waffo AF, Sewald N et al (2016) Oxidative burst inhibition, cytotoxicity and antibacterial acriquinoline alkaloids from Citrus reticulate (Blanco). Bioorganic Med Chem Lett 26:306–309. https://doi.org/10.1016/j.bmcl.2015.12.028

Forde BG, Lea PJ (2007) Glutamate in plants: metabolism, regulation, and signalling. J Exp Bot 58:2339–2358. https://doi.org/10.1093/jxb/erm121

Franco-Arnedo G, Buelvas-Puello LM, Miranda-Lasprilla D, Martínez-Correa HA, Parada-Alfonso F (2020) Obtaining antioxidant extracts from tangerine (C. reticulata var. Arrayana) peels by modified supercritical CO2 and their use as protective agent against the lipid oxidation of a mayonnaise. J Supercrit Fluids 165. https://doi.org/10.1016/j.supflu.2020.104957

Fraser PD, Bramley PM (2004) The biosynthesis and nutritional uses of carotenoids. Prog Lipid Res 43:228–265. https://doi.org/10.1016/j.plipres.2003.10.002

Freeman BL, Eggett DL, Parker TL (2010) Synergistic and antagonistic interactions of phenolic compounds found in navel oranges. J Food Sci 75:C570–C576. https://doi.org/10.1111/j.1750-3841.2010.01717.x

Fu M, Zou B, An K, Yu Y, Tang D, Wu J et al (2019) Anti-asthmatic activity of alkaloid compounds from Pericarpium Citri Reticulatae (Citrus reticulata 'Chachi'). Food Funct 10:903–911. https://doi.org/10.1039/c8fo01753k

Funck D, Stadelhofer B, Koch W (2008) Ornithine-δ-aminotransferase is essential for arginine catabolism but not for proline biosynthesis. BMC Plant Biol 8:1–14. https://doi.org/10.1186/1471-2229-8-40

Galili G (2011) The aspartate-family pathway of plants: linking production of essential amino acids with energy and stress regulation. Plant Signal Behav 6:192–195. https://doi.org/10.4161/psb.6.2.14425

Garcia-Lor A, Luro F, Ollitrault P, Navarro L (2015) Genetic diversity and population structure analysis of mandarin germplasm by nuclear, chloroplastic and mitochondrial markers. Tree Genet Genomes 11:1–15. https://doi.org/10.1007/S11295-015-0951-1/TABLES/5

Gaufichon L, Reisdorf-Cren M, Rothstein SJ, Chardon F, Suzuki A (2010) Biological functions of asparagine synthetase in plants. Plant Sci 179:141–153. https://doi.org/10.1016/j.plantsci.2010.04.010

Geada P, Rodrigues R, Loureiro L, Pereira R, Fernandes B, Teixeira JA et al (2018) Electrotechnologies applied to microalgal biotechnology – applications, techniques and future trends. Renew Sust Energ Rev 94:656–668. https://doi.org/10.1016/j.rser.2018.06.059

Goettel M, Eing C, Gusbeth C, Straessner R, Frey W (2013) Pulsed electric field assisted extraction of intracellular valuables from microalgae. Algal Res 2:401–408. https://doi.org/10.1016/j.algal.2013.07.004

Goldenberg L, Yaniv Y, Choi HJ, Doron-Faigenboim A, Carmi N, Porat R (2016) Elucidating the biochemical factors governing off-flavor perception in mandarins. Postharvest Biol Technol 120:167–179. https://doi.org/10.1016/j.postharvbio.2016.06.009

González-Mas MC, Rambla JL, López-Gresa MP, Amparo Blázquez M, Granell A (2019) Volatile compounds in citrus essential oils: a comprehensive review. Front Plant Sci 10:1–18. https://doi.org/10.3389/fpls.2019.00012

Goossens A, Häkkinen ST, Laakso I, Seppänen-Laakso T, Biondi S, De Sutter V et al (2003) A functional genomics approach toward the understanding of secondary metabolism in plant cells. Proc Natl Acad Sci U S A 100:8595–8600. https://doi.org/10.1073/pnas.1032967100

Gorelova V, Ambach L, Rébeillé F, Stove C, Van Der Straeten D (2017) Folates in plants: research advances and progress in crop biofortification. Front Chem 5:1–20. https://doi.org/10.3389/fchem.2017.00021

Gorelova V, Bastien O, De Clerck O, Lespinats S, Rébeillé F, Van Der Straeten D (2019) Evolution of folate biosynthesis and metabolism across algae and land plant lineages. Sci Rep 9:1–16. https://doi.org/10.1038/s41598-019-42146-5

Guo LX, Shi CY, Liu X, Ning DY, Jing LF, Yang H et al (2016) Citrate accumulation-related gene expression and/or enzyme activity analysis combined with metabolomics provide a novel insight for an orange mutant. Sci Rep 6:1–12. https://doi.org/10.1038/srep29343

Guo X, Cao X, Fang X, Guo A, Li E (2021) Inhibitory effects of fermented Ougan (Citrus reticulata cv. Suavissima) juice on high-fat diet-induced obesity associated with white adipose tissue browning and gut microbiota modulation in mice. Food Funct 12:9300–9314. https://doi.org/10.1039/D0FO03423A

Halim R, Danquah MK, Webley PA (2012) Extraction of oil from microalgae for biodiesel production: a review. Biotechnol Adv 30:709–732. https://doi.org/10.1016/j.biotechadv.2012.01.001

Hashemi SMB, Roohi R, Mahmoudi MR, Granato D (2019) Modeling inactivation of Listeria monocytogenes, Shigella sonnei, Byssochlamys fulva and Saccharomyces cerevisiae and ascorbic acid and β-carotene degradation kinetics in tangerine juice by pulsed-thermosonication. LWT Food Sci Technol 111:612–621. https://doi.org/10.1016/j.lwt.2019.05.060

Hayat K, Hussain S, Abbas S, Farooq U, Ding B, Xia S et al (2009) Optimized microwave-assisted extraction of phenolic acids from citrus mandarin peels and evaluation of antioxidant activity in vitro. Sep Purif Technol 70:63–70. https://doi.org/10.1016/j.seppur.2009.08.012

He Q, Yang Z, Gong B, Wang J, Xiao K, Yang ST (2017) Quality evaluation focusing on tissue fractal dimension and chemical changes for frozen tilapia with treatment by tangerine peel extract. Sci Rep 7:1–8. https://doi.org/10.1038/srep42202

He W, Wang P, Chen J, Xie W (2020) Recent progress in the total synthesis of Strychnos alkaloids. Org Biomol Chem 18:1046–1056. https://doi.org/10.1039/C9OB02627D

He Y, Li W, Zhu P, Wang M, Qiu J, Sun H et al (2022) Comparison between the vegetative and fruit characteristics of "Orah" (Citrus reticulata Blanco) mandarin under different climatic conditions. Sci Hortic (Amsterdam) 300:111064. https://doi.org/10.1016/j.scienta.2022.111064

Hemmerlin A, Hoeffler JF, Meyer O, Tritsch D, Kagan IA, Grosdemange-Billiard C et al (2003) Cross-talk between the cytosolic mevalonate and the plastidial methylerythritol phosphate pathways in tobacco bright yellow-2 cells. J Biol Chem 278:26666–26676. https://doi.org/10.1074/jbc.M302526200

Hemmerlin A, Harwood JL, Bach TJ (2012) A raison d'être for two distinct pathways in the early steps of plant isoprenoid biosynthesis? Prog Lipid Res 51:95–148. https://doi.org/10.1016/j.plipres.2011.12.001

Henry LK, Gutensohn M, Thomas ST, Noel JP, Dudareva N, Croteau RB (2015) Orthologs of the archaeal isopentenyl phosphate kinase regulate terpenoid production in plants. Proc Natl Acad Sci U S A 112:10050–10055. https://doi.org/10.1073/pnas.1504798112

Herrero M, Sánchez-Camargo A d P, Cifuentes A, Ibáñez E (2015) Plants, seaweeds, microalgae and food by-products as natural sources of functional ingredients obtained using pressurized liquid extraction and supercritical fluid extraction. TrAC Trends Anal Chem 71:26–38. https://doi.org/10.1016/j.trac.2015.01.018

Hijaz F, Gmitter FG, Bai J, Baldwin E, Biotteau A, Leclair C et al (2020) Effect of fruit maturity on volatiles and sensory descriptors of four mandarin hybrids. J Food Sci 85:1548–1564. https://doi.org/10.1111/1750-3841.15116

Hildebrandt TM, Nunes Nesi A, Araújo WL, Braun HP (2015) Amino acid catabolism in plants. Mol Plant 8:1563–1579. https://doi.org/10.1016/j.molp.2015.09.005

Ho S-C, Kuo C-T (2014) Hesperidin, nobiletin, and tangeretin are collectively responsible for the anti-neuroinflammatory capacity of tangerine peel (Citri reticulatae pericarpium). Food Chem Toxicol Int J Publ Br Ind Biol Res Assoc 71:176–182. https://doi.org/10.1016/j.fct.2014.06.014

Ho SC, Lin CC (2008) Investigation of heat treating conditions for enhancing the anti-inflammatory activity of citrus fruit (Citrus reticulata) peels. J Agric Food Chem 56:7976–7982. https://doi.org/10.1021/jf801434c

Hu Y, Kou G, Chen Q, Li Y, Zhou Z (2019) Protection and delivery of mandarin (Citrus reticulata Blanco) peel extracts by encapsulation of whey protein concentrate nanoparticles. LWT 99:24–33. https://doi.org/10.1016/j.lwt.2018.09.044

Hume HH (1913) Citrus fruits and their culture. Judd Company

Hwang SL, Shih PH, Yen GC (2012) Neuroprotective effects of citrus flavonoids. J Agric Food Chem 60:877–885. https://doi.org/10.1021/jf204452y

Ifergan I, Jansen G, Assaraf YG (2008) The reduced folate carrier (RFC) is cytotoxic to cells under conditions of severe folate deprivation: RFC as a double edged sword in folate homeostasis. J Biol Chem 283:20687–20695. https://doi.org/10.1074/jbc.M802812200

Igamberdiev AU, Eprintsev AT (2016) Organic acids: the pools of fixed carbon involved in redox regulation and energy balance in higher plants. Front Plant Sci 7:1–15. https://doi.org/10.3389/fpls.2016.01042

Imbimbo P, D'Elia L, Liberti D, Olivieri G, Monti DM (2020) Towards green extraction methods from microalgae learning from the classics. Appl Microbiol Biotechnol. https://doi.org/10.1007/s00253-020-10839-x

Infante VHP, Maia Campos PMBG, Gaspar LR, Darvin ME, Schleusener J, Rangel KC et al (2022) Safety and efficacy of combined essential oils for the skin barrier properties: in vitro, ex vivo and clinical studies. Int J Cosmet Sci 44:118–130. https://doi.org/10.1111/ics.12761

Inglese P, Sortino G (2019) Citrus history, taxonomy, breeding, and fruit quality. Oxford Res Encycl Environ Sci. https://doi.org/10.1093/ACREFORE/9780199389414.013.221

Ituen E, Ekemini E, Yuanhua L, Li R, Singh A (2020) Mitigation of microbial biodeterioration and acid corrosion of pipework steel using Citrus reticulata peels extract mediated copper nanoparticles composite. Int Biodeterior Biodegrad 149:104935. https://doi.org/10.1016/j.ibiod.2020.104935

Jahns P, Latowski D, Strzalka K (2009) Mechanism and regulation of the violaxanthin cycle: the role of antenna proteins and membrane lipids. Biochim Biophys Acta Bioenerg 1787:3–14. https://doi.org/10.1016/j.bbabio.2008.09.013

Jander G, Joshi V (2010) Recent progress in deciphering the biosynthesis of aspartate-derived amino acids in plants. Mol Plant 3:54–65. https://doi.org/10.1093/mp/ssp104

Januário EFD, Vidovix TB, de Araújo LA, Bergamasco Beltran L, Bergamasco R, Vieira AMS (2021) Investigation of Citrus reticulata peels as an efficient and low-cost adsorbent for the removal of safranin orange dye. Environ Technol (United Kingdom) 0:1–15. https://doi.org/10.1080/09593330.2021.1946601

Jiang L, Strobbe S, Van Der Straeten D, Zhang C (2021) Regulation of plant vitamin metabolism: backbone of biofortification for the alleviation of hidden hunger. Mol Plant 14:40–60. https://doi.org/10.1016/j.molp.2020.11.019

Joshi V, Laubengayer KM, Schauer N, Fernie AR, Jander G (2006) Two Arabidopsis threonine aldolases are nonredundant and compete with threonine deaminase for a common substrate pool. Plant Cell 18:3565–3575. https://doi.org/10.1105/tpc.106.044958

Kachanovsky DE, Filler S, Isaacson T, Hirschberg J (2012) Epistasis in tomato color mutations involves regulation of phytoene synthase 1 expression by cis-carotenoids. Proc Natl Acad Sci U S A 109:19021–19026. https://doi.org/10.1073/pnas.1214808109

Kalhan SC, Hanson RW (2012) Resurgence of serine: an often neglected but indispensable amino acid. J Biol Chem 287:19786–19791. https://doi.org/10.1074/jbc.R112.357194

Kang C, Cao J, Sun J, Zheng G, Wang Y, Chen K et al (2022) Comparison of physiochemical characteristics of Citrus reticulata cv. Shatangju fruit with different fruit sizes after storage. Food Packag Shelf Life 31:100774. https://doi.org/10.1016/j.fpsl.2021.100774

Karn A, Zhao C, Yang F, Cui J, Gao Z, Wang M et al (2021) In-vivo biotransformation of citrus functional components and their effects on health. Crit Rev Food Sci Nutr 61:756–776. https://doi.org/10.1080/10408398.2020.1746234

Katz E, Lagunes PM, Riov J, Weiss D, Goldschmidt EE (2004) Molecular and physiological evidence suggests the existence of a system II-like pathway of ethylene production in non-climacteric Citrus fruit. Planta 219:243–252. https://doi.org/10.1007/s00425-004-1228-3

Katz E, Boo KH, Kim HY, Eigenheer RA, Phinney BS, Shulaev V et al (2011) Label-free shotgun proteomics and metabolite analysis reveal a significant metabolic shift during citrus fruit development. J Exp Bot 62:5367–5384. https://doi.org/10.1093/jxb/err197

Kaur S, Panesar PS, Chopra HK (2021a) Citrus processing by-products: an overlooked repository of bioactive compounds. https://doi.org/10.1080/10408398.2021.1943647

Kaur S, Panesar PS, Chopra HK (2021b) Standardization of ultrasound-assisted extraction of bioactive compounds from kinnow mandarin peel. Biomass Convers Biorefinery. https://doi.org/10.1007/s13399-021-01674-9

Ke Z, Zhao Y, Tan S, Chen H, Li Y, Zhou Z et al (2020) Citrus reticulata Blanco peel extract ameliorates hepatic steatosis, oxidative stress and inflammation in HF and MCD diet-induced NASH C57BL/6 J mice. J Nutr Biochem 83:108426. https://doi.org/10.1016/j.jnutbio.2020.108426

Kelebek H, Selli S (2014) Identification of phenolic compositions and the antioxidant capacity of mandarin juices and wines. J Food Sci Technol 51:1094–1101. https://doi.org/10.1007/s13197-011-0606-7

Kemper TG (2020) Oil extraction. In: Shahidi F (ed) Bailey's industrial oil and fat products, 7th edn. Wiley, pp 1–38. https://doi.org/10.1002/047167849X.bio013.pub2

Keutgen AJ, Pawelzik E (2008) Contribution of amino acids to strawberry fruit quality and their relevance as stress indicators under NaCl salinity. Food Chem 111:642–647. https://doi.org/10.1016/j.foodchem.2008.04.032

Khaleel G, Sharanagat VS, Singh L, Kumar Y, Kumar K, Kishore A et al (2022) Characterization of kinnow (Citrus reticulate) peel and its effect on the quality of muffin. J Food Process Preserv:1–14. https://doi.org/10.1111/jfpp.16716

Kirby J, Keasling JD (2009) Biosynthesis of plant isoprenoids: perspectives for microbial engineering. Annu Rev Plant Biol 60:335–355. https://doi.org/10.1146/annurev.arplant.043008.091955

Kirma M, Araújo WL, Fernie AR, Galili G (2012) The multifaceted role of aspartate-family amino acids in plant metabolism. J Exp Bot 63:4995–5001. https://doi.org/10.1093/jxb/ers119

Kishor PBK, Suravajhala P, Rathnagiri P, Sreenivasulu N (2022) Intriguing role of proline in redox potential conferring high temperature stress tolerance. 13:1–16. https://doi.org/10.3389/fpls.2022.867531

Kołton A, Długosz-Grochowska O, Wojciechowska R, Czaja M (2022) Biosynthesis regulation of folates and phenols in plants. Sci Hortic (Amsterdam) 291. https://doi.org/10.1016/j.scienta.2021.110561

Konings EJM, Roomans HHS, Dorant E, Goldbohm RA, Saris WHM, Van den Brandt PA (2001) Folate intake of the Dutch population according to newly established liquid chromatography data for foods. Am J Clin Nutr 73:765–776. https://doi.org/10.1093/ajcn/73.4.765

Kultys E, Kurek MA (2022) Green extraction of carotenoids from fruit and vegetable byproducts: a review. Molecules 27. https://doi.org/10.3390/molecules27020518

Lado J, Alós E, Rodrigo MJ, Zacarías L (2015) Light avoidance reduces ascorbic acid accumulation in the peel of Citrus fruit. Plant Sci 231:138–147. https://doi.org/10.1016/j.plantsci.2014.12.002

Lado J, Gambetta G, Zacarias L (2018) Key determinants of citrus fruit quality: metabolites and main changes during maturation. Sci Hortic (Amsterdam) 233:238–248. https://doi.org/10.1016/j.scienta.2018.01.055

Laing WA, Wright MA, Cooney J, Bulley SM (2007) The missing step of the L-galactose pathway of ascorbate biosynthesis in plants, an L-galactose guanyltransferase, increases leaf ascorbate content. Proc Natl Acad Sci U S A 104:9534–9539. https://doi.org/10.1073/pnas.0701625104

Laing WA, Martínez-Sánchez M, Wright MA, Bulley SM, Brewster D, Dare AP et al (2015) An upstream open reading frame is essential for feedback regulation of ascorbate biosynthesis in arabidopsis. Plant Cell 27:772–786. https://doi.org/10.1105/tpc.114.133777

Lamine M, Gargouri M, Rahali FZ, Mliki A (2019) Authentication of Citrus fruits through a comprehensive fatty acid profiling and health lipid indices: a nutraceutical perspectives. J Food Meas Charact 13:2211–2217. https://doi.org/10.1007/s11694-019-00141-4

Langgut D (2017) The citrus route revealed: from Southeast Asia into the Mediterranean. HortScience 52:814–822. https://doi.org/10.21273/HORTSCI11023-16

Laule O, Fürholz A, Chang HS, Zhu T, Wang X, Heifetz PB et al (2003) Crosstalk between cytosolic and plastidial pathways of isoprenoid biosynthesis in Arabidopsis thaliana. Proc Natl Acad Sci U S A 100:6866–6871. https://doi.org/10.1073/pnas.1031755100

Lee SK, Kader AA (2000) Preharvest and postharvest factors influencing vitamin C content of horticultural crops. Postharvest Biol Technol 20:207–220. https://doi.org/10.1016/S0925-5214(00)00133-2

Legge J (1865) The Chinese classics. Shoo King, or B. Hist Doc 2

Lemoine R, La Camera S, Atanassova R, Dédaldéchamp F, Allario T, Pourtau N et al (2013) Source-to-sink transport of sugar and regulation by environmental factors. Front Plant Sci 4:1–21. https://doi.org/10.3389/fpls.2013.00272

Li BB, Smith B, Hossain MM (2006) Extraction of phenolics from citrus peels: II. Enzyme-assisted extraction method. Sep Purif Technol 48:189–196. https://doi.org/10.1016/j.seppur.2005.07.019

Li W, Wang Z, Wang YP, Jiang C, Liu Q, Sun YS et al (2012) Pressurised liquid extraction combining LC-DAD-ESI/MS analysis as an alternative method to extract three major flavones in Citrus reticulata "Chachi" (Guangchenpi). Food Chem 130:1044–1049. https://doi.org/10.1016/j.foodchem.2011.07.129

Li D, Xiao Y, Zhang Z, Liu C (2015) Light-induced oxidation and isomerization of all-trans-β-cryptoxanthin in a model system. J Photochem Photobiol B Biol 142:51–58. https://doi.org/10.1016/j.jphotobiol.2014.11.003

Li D, Heiling S, Baldwin IT, Gaquerel E (2016) Illuminating a plant's tissue-specific metabolic diversity using computational metabolomics and information theory. Proc Natl Acad Sci U S A 113:E7610–E7618. https://doi.org/10.1073/pnas.1610218113

Li C, Cai Q, Wu X, Tan Z, Huang S, Wei C et al (2022) Variation in compositions and biological activities of essential oils from four citrus species: Citrus limon, Citrus sinensis, Citrus paradisi, and Citrus reticulata. Chem Biodivers 19:e202100910. https://doi.org/10.1002/cbdv.202100910

Liao HS, Chung YH, Hsieh MH (2022) Glutamate: a multifunctional amino acid in plants. Plant Sci 318:111238. https://doi.org/10.1016/j.plantsci.2022.111238

Lin Q, Wang C, Dong W, Jiang Q, Wang D, Li S et al (2015) Transcriptome and metabolome analyses of sugar and organic acid metabolism in Ponkan (Citrus reticulata) fruit during fruit maturation. Gene 554:64–74. https://doi.org/10.1016/j.gene.2014.10.025

Lin Q, Qian J, Zhao C, Wang D, Liu C, Wang Z et al (2016) Low temperature induced changes in citrate metabolism in Ponkan (Citrus reticulata Blanco cv. Ponkan) fruit during maturation. PLoS One 11:1–15. https://doi.org/10.1371/journal.pone.0156703

Linné CV (1737) Genera Plantarum

Liu Y, Heying E, Tanumihardjo SA (2012) History, global distribution, and nutritional importance of citrus fruits. Compr Rev Food Sci Food Saf 11:530–545. https://doi.org/10.1111/j.1541-4337.2012.00201.x

Liu X, Tang L, Wu H, Xi W, Yu J, Zhou Z (2016) Development of DArT markers and evaluation of phylogenetic relationship of key Citrus species. Genet Resour Crop Evol 63:1307–1318. https://doi.org/10.1007/S10722-015-0319-2/FIGURES/4

Liu, X.-C., Lin, X.-H., Liu, S.-C., Zhu, C.-Q., Grierson, D., Li, S.-J., et al. (2021). The effect of NH4+ on phosphoenolpyruvate carboxykinase gene expression, metabolic flux and citrate content of citrus juice sacs. Plant Physiol Biochem 167, 123–131. doi: https://doi.org/https://doi.org/10.1016/j.plaphy.2021.07.041

Luro F, Curk F, Froelicher Y, Ollitrault P (2017) Recent insights on Citrus diversity and phylogeny. Agrumed Archaeol Hist citrus fruit Mediterr Acclim Divers uses, 16–28

Lv X, Zhao S, Ning Z, Zeng H, Shu Y, Tao O et al (2015) Citrus fruits as a treasure trove of active natural metabolites that potentially provide benefits for human health. Chem Cent J 9:1–14. https://doi.org/10.1186/s13065-015-0145-9

Mabberley DJ (2004) Citrus (Rutaceae): a review of recent advances in etymology, systematics and medical applications. Blumea-Biodiversity Evol Biogeogr Plants 49:481–498

Magwaza LS, Mditshwa A, Tesfay SZ, Opara UL (2016) An overview of preharvest factors affecting vitamin C content of citrus fruit. Sci Hortic (Amsterdam) 216:12–21. https://doi.org/10.1016/j.scienta.2016.12.021

Mahmoud AM, Bin-Jumah M, Abukhalil MH (2021) 9 – Antiinflammatory natural products from marine algae. In: Gopi S, Amalraj A, Kunnumakkara A, S. B. T.-I, Thomas NP (eds) Inflammation and natural products. Academic, pp 175–203. https://doi.org/10.1016/B978-0-12-819218-4.00012-2

Majumdar R, Barchi B, Turlapati SA, Gagne M, Minocha R, Long S et al (2016) Glutamate, ornithine, arginine, proline, and polyamine metabolic interactions: the pathway is regulated at the post-transcriptional level. Front Plant Sci 7:1–17. https://doi.org/10.3389/fpls.2016.00078

Martín J, Asuero AG (2021) High hydrostatic pressure for recovery of anthocyanins: effects, performance, and applications. Sep Purif Rev 50:159–176. https://doi.org/10.1080/15422119.2019.1632897

McClain AM, Sharkey TD (2019) Triose phosphate utilization and beyond: from photosynthesis to end product synthesis. J Exp Bot 70:1755–1766. https://doi.org/10.1093/jxb/erz058

McGarvey DJ, Croteau R (1995) Terpenoid metabolism. Plant Cell 7:1015–1026. https://doi.org/10.2307/3870054

Medhurst WH (1846) Ancient ChinaII The Shoo King, or the Historical Classic: being the most ancient authentic record of the Annals of the Chinese Empire: illustrated by later commentators. Translated by WH Medhurst, sen. Chin. & Eng. At the Mission Press

Meena M, Divyanshu K, Kumar S, Swapnil P, Zehra A, Shukla V et al (2019) Regulation of L-proline biosynthesis, signal transduction, transport, accumulation and its vital role in plants during variable environmental conditions. Heliyon 5:e02952. https://doi.org/10.1016/j.heliyon.2019.e02952

Melin AD, Nevo O, Shirasu M, Williamson RE, Garrett EC, Endo M et al (2019) Fruit scent and observer colour vision shape food-selection strategies in wild capuchin monkeys. Nat Commun 10:1–9. https://doi.org/10.1038/s41467-019-10250-9

Michalak I, Chojnacka K (2014) Algal extracts: technology and advances. Eng Life Sci 14:581–591. https://doi.org/10.1002/elsc.201400139

Miles EA, Calder PC (2021) Effects of citrus fruit juices and their bioactive components on inflammation and immunity: a narrative review. Front Immunol 12:1–18. https://doi.org/10.3389/fimmu.2021.712608

Modde K, Timm S, Florian A, Michl K, Fernie AR, Bauwe H (2017) High serine:glyoxylate aminotransferase activity lowers leaf daytime serine levels, inducing the phosphoserine pathway in Arabidopsis. J Exp Bot 68:643–656. https://doi.org/10.1093/jxb/erw467

Mohan VR, Tresina PS, Daffodil ED (2016) Antinutritional factors in legume seeds: characteristics and determination. In: Caballero B, Finglas PM, Toldrá F (eds) Encyclopedia of food and health. Academic, pp 211–220

Montoya C, González L, Pulido S, Atehortúa L, Robledo SM (2019) Identification and quantification of limonoid aglycones content of Citrus seeds. Rev Bras Farmacogn 29:710–714. https://doi.org/10.1016/j.bjp.2019.07.006

Morgan KT, Graham JH (2019) Nutrient status and root density of huanglongbing-affected trees: consequences of irrigation water bicarbonate and soil pH mitigation with acidification. Agron 9:746. https://doi.org/10.3390/AGRONOMY9110746

Moshonas MG, Shaw PE (1997) Quantitation of volatile constituents in mandarin juices and its use for comparison with orange juices by multivariate analysis. J Agric Food Chem 45:3968–3972. https://doi.org/10.1006/fstl.1996.0207

Moulehi I, Bourgou S, Ourghemmi I, Tounsi MS (2012) Variety and ripening impact on phenolic composition and antioxidant activity of mandarin (Citrus reticulate Blanco) and bitter orange (Citrus aurantium L.) seeds extracts. Ind Crop Prod 39:74–80. https://doi.org/10.1016/j.indcrop.2012.02.013

Mubarak M, Shaija A, Suchithra TV (2015) A review on the extraction of lipid from microalgae for biodiesel production. Algal Res 7:117–123. https://doi.org/10.1016/j.algal.2014.10.008

Muñoz P, Munné-Bosch S (2019) Vitamin E in plants: biosynthesis, transport, and function. Trends Plant Sci 24:1040–1051. https://doi.org/10.1016/j.tplants.2019.08.006

Nagarajan G, Minu RI (2018) Wireless soil monitoring sensor for sprinkler irrigation automation system. Wirel Pers Commun 98:1835–1851. https://doi.org/10.1007/S11277-017-4948-Y/FIGURES/10

Nagegowda DA (2010) Plant volatile terpenoid metabolism: biosynthetic genes, transcriptional regulation and subcellular compartmentation. FEBS Lett 584:2965–2973. https://doi.org/10.1016/j.febslet.2010.05.045

Nagy S (1980) Vitamin C contents of citrus fruit and their products: a review. J Agric Food Chem 28:8–18. https://doi.org/10.1021/jf60227a026

Navarro JM, Pérez-Pérez JG, Romero P, Botía P (2010) Analysis of the changes in quality in mandarin fruit, produced by deficit irrigation treatments. Food Chem 119:1591–1596. https://doi.org/10.1016/j.foodchem.2009.09.048

Navarro-Baez JE, Martínez LM, Welti-Chanes J, Buitimea-Cantúa GV, Escobedo-Avellaneda Z (2022) High hydrostatic pressure to increase the biosynthesis and extraction of phenolic compounds in food: a review. Molecules 27. https://doi.org/10.3390/molecules27051502

Nawaz R, Abbasi NA, Khan MR, Ali I, Hasan SZU, Hayat A (2020) Color development in 'Feutrell's early' (Citrus Reticulata Blanco) affects peel composition and juice biochemical properties. Int J Fruit Sci 20:871–890. https://doi.org/10.1080/15538362.2019.1699490

Nde DB, Foncha AC (2020) Optimization methods for the extraction of vegetable oils: a review. PRO 8. https://doi.org/10.3390/pr8020209

Nehela Y, Killiny N (2020) The unknown soldier in citrus plants: polyamines-based defensive mechanisms against biotic and abiotic stresses and their relationship with other stress-associated metabolites. Plant Signal Behav 15. https://doi.org/10.1080/15592324.2020.1761080

Nicolosi E, Deng ZN, Gentile A, La Malfa S, Continella G, Tribulato E (2000) Citrus phylogeny and genetic origin of important species as investigated by molecular markers. Theor Appl Genet 100:1155–1166. https://doi.org/10.1007/S001220051419

Nitayapat N, Prakarnsombut N, Lee SJ, Boonsupthip W (2015) Bioconversion of tangerine residues by solid-state fermentation with Lentinus polychrous and drying the final products. LWT 63:773–779. https://doi.org/10.1016/J.LWT.2015.03.016

Niu D, Ren EF, Li J, Zeng XA, Li SL (2021) Effects of pulsed electric field-assisted treatment on the extraction, antioxidant activity and structure of naringin. Sep Purif Technol 265. https://doi.org/10.1016/j.seppur.2021.118480

Noguera NH, Lima DC, Filho EGA, Fonteles TV, Rodrigues S (2021) Influence of different nonthermal processing on guava, orange, and tangerine juices and the food matrix effects. Food Bioprocess Technol 14:1662–1672. https://doi.org/10.1007/s11947-021-02663-6

Norton RL (1988) 11. Windbreaks: benefits to orchard and vineyard crops. Agric Ecosyst Environ 22–23:205–213. https://doi.org/10.1016/0167-8809(88)90019-9

Okamura E, Hirai MY (2017) Novel regulatory mechanism of serine biosynthesis associated with 3-phosphoglycerate dehydrogenase in Arabidopsis thaliana. Sci Rep 7:1–7. https://doi.org/10.1038/s41598-017-03807-5

Okwu DE, Awurum AN, Okoronkwo JI (2007) Phytochemical composition and in vitro antifungal activity screening of extracts from citrus plants against Fusarium oxysporum of okra plant (Hibiscus esculentus). Pest Technol 1:145–148

Olatunya AM, Ajaja AK, Akintayo ET (2021) Comparative analysis of the lipid classes, fatty acid and phospholipid profiles of three species of citrus fruits. Sci Stud Res Chem Chem Eng Biotechnol Food Ind 22:191–199. Available at: https://www.proquest.com/scholarly-journals/comparative-analysis-lipid-classes-fatty-acid/docview/2560391149/se-2

Ollitrault P, Curk F, Krueger R (2020) Citrus taxonomy. Genus Citrus 57–81. https://doi.org/10.1016/B978-0-12-812163-4.00004-8

Ordóñez-Santos LE, Esparza-Estrada J, Vanegas-Mahecha P (2021) Ultrasound-assisted extraction of total carotenoids from mandarin epicarp and application as natural colorant in bakery products. LWT 139. https://doi.org/10.1016/j.lwt.2020.110598

Osbeck P, Torén O, Ekeberg CG (1771) A voyage to China and the East Indies. B. White

Otimenyin S (2022) Chapter 20 – Herbal biomolecules acting on central nervous system. In: Mandal SC, Nayak AK, A. K. B. T.-H. B, Dhara HA (eds) Herbal biomolecules in healthcare applications. Academic, pp 475–523. https://doi.org/10.1016/B978-0-323-85852-6.00030-5

Oueslati A, Salhi-Hannachi A, Luro F, Vignes H, Mournet P, Ollitrault P (2017) Genotyping by sequencing reveals the interspecific C. maxima/C. reticulata admixture along the genomes of modern citrus varieties of mandarins, tangors, tangelos, orangelos and grapefruits. PLoS One 12:e0185618. https://doi.org/10.1371/JOURNAL.PONE.0185618

Paine JA, Shipton CA, Chaggar S, Howells RM, Kennedy MJ, Vernon G et al (2005) Improving the nutritional value of Golden Rice through increased pro-vitamin A content. Nat Biotechnol 23:482–487. https://doi.org/10.1038/nbt1082

Pal Choudhuri S, Delay RJ, Delay ER (2015) L-amino acids elicit diverse response patterns in taste sensory cells: a role for multiple receptors. PLoS One 10:e0130088. https://doi.org/10.1371/journal.pone.0130088

Pandey P, Khan F (2021) A mechanistic review of the anticancer potential of hesperidin, a natural flavonoid from citrus fruits. Nutr Res 92:21–31. https://doi.org/10.1016/j.nutres.2021.05.011

Park SM, Ko KY, Kim IH (2015) Optimization of d-limonene extraction from tangerine peel in various solvents by using soxhlet extractor. Korean Chem Eng Res 53:717–722. https://doi.org/10.9713/kcer.2015.53.6.717

Pellaud S, Saffrané LM (2017) Metabolic origins and transport of vitamin E biosynthetic precursors. Front Plant Sci 8:1–8. https://doi.org/10.3389/fpls.2017.01959

Pelletier SW (1983) The nature and definition of an alkaloid. Alkaloids Chem Biol Perspect 1:1–31

Phillips MA, D'Auria JC, Gershenzon J, Pichersky E (2008) The Arabidopsis thaliana type I isopentenyl diphosphate isomerases are targeted to multiple subcellular compartments and have overlapping functions in isoprenoid biosynthesis. Plant Cell 20:677–696. https://doi.org/10.1105/tpc.107.053926

Phinney R, Tivana LD, Östbring K, Sjöholm I, Dhulappanavar G, Jeje I et al (2020) Concentration of citrus fruit juices in membrane pouches with solar energy. Part 2: How solar drying setup and juice pretreatment determine the microbiological quality. J Food Process Eng 43:1–21. https://doi.org/10.1111/jfpe.13377

Pichersky E, Noel JP, Dudareva N (2006) Biosynthesis of plant volatiles: nature's diversity and ingenuity. Science (80–) 311:808–811. https://doi.org/10.1126/science.1118510

Podlešáková K, Ugena L, Spíchal L, Doležal K, De Diego N (2019) Phytohormones and polyamines regulate plant stress responses by altering GABA pathway. New Biotechnol 48:53–65. https://doi.org/10.1016/j.nbt.2018.07.003

Polanco-Lugo E, Martínez-Castillo JI, Cuevas-Bernardino JC, González-Flores T, Valdez-Ojeda R, Pacheco N et al (2019) Citrus pectin obtained by ultrasound-assisted extraction: physicochemical, structural, rheological and functional properties. CYTA J Food 17:463–471. https://doi.org/10.1080/19476337.2019.1600036

Pontiggia D, Spinelli F, Fabbri C, Licursi V, Negri R, De Lorenzo G et al (2019) Changes in the microsomal proteome of tomato fruit during ripening. Sci Rep 9:1–18. https://doi.org/10.1038/s41598-019-50575-5

Pott DM, Osorio S, Vallarino JG (2019) From central to specialized metabolism: an overview of some secondary compounds derived from the primary metabolism for their role in conferring nutritional and organoleptic characteristics to fruit. Front Plant Sci 10. https://doi.org/10.3389/fpls.2019.00835

Pourbafrani M, Forgács G, Horváth IS, Niklasson C, Taherzadeh MJ (2010) Production of biofuels, limonene and pectin from citrus wastes. Bioresour Technol 101:4246–4250. https://doi.org/10.1016/J.BIORTECH.2010.01.077

Quinlan RF, Shumskaya M, Bradbury LMT, Beltrán J, Ma C, Kennelly EJ et al (2012) Synergistic interactions between carotene ring hydroxylases drive lutein formation in plant carotenoid biosynthesis. Plant Physiol 160:204–214. https://doi.org/10.1104/pp.112.198556

Rafiq S, Singh B, Gat Y (2019) Effect of different drying techniques on chemical composition, color and antioxidant properties of kinnow (Citrus reticulata) peel. J Food Sci Technol 56:2458–2466. https://doi.org/10.1007/s13197-019-03722-9

Ramon-Laca L (2003) The introduction of cultivated citrus to Europe via Northern Africa and the Iberian Peninsula. Econ Bot:502–514

Rashid U, Ibrahim M, Yasin S, Yunus R, Taufiq-Yap YH, Knothe G (2013) Biodiesel from Citrus reticulata (mandarin orange) seed oil, a potential non-food feedstock. Ind Crop Prod 45:355–359. https://doi.org/10.1016/j.indcrop.2012.12.039

Rasool S, Ahmed H, Uttra MM, Uttra AM, Khan MR, Zakir KA et al (2021) Antioxidant and anticancer effect of ethanolic extract of citrus fruits on Hep G2 and MCF-7 cell lines. J Pharm Res Int 33:84–90. https://doi.org/10.9734/jpri/2021/v33i49a33306

Rauf A, Imran M, Abu-Izneid T, Iahtisham-Ul-Haq, Patel S, Pan X et al (2019) Proanthocyanidins: a comprehensive review. Biomed Pharmacother 116:108999. https://doi.org/10.1016/j.biopha.2019.108999

Raut JS, Karuppayil SM (2014) A status review on the medicinal properties of essential oils. Ind Crop Prod 62:250–264. https://doi.org/10.1016/j.indcrop.2014.05.055

Rey F, Zacarías L, Rodrigo MJ (2020) Carotenoids, vitamin C, and antioxidant capacity in the peel of mandarin fruit in relation to the susceptibility to chilling injury during postharvest cold storage. Antioxidants 9:1–21. https://doi.org/10.3390/antiox9121296

Rey F, Zacarias L, Rodrigo MJ (2021) Regulation of tocopherol biosynthesis during fruit maturation of different citrus species. Front Plant Sci 12:1–15. https://doi.org/10.3389/fpls.2021.743993

Rimm EB, Willet WC, Hu FB, Sampson L, Colditz GA, Manson JE et al (1998) Folate and vitamin B6 from diet and supplements in relation to risk of coronary heart disease among women. JAMA 279:359–364. https://doi.org/10.1016/s1098-3597(98)80008-2

Roberts MF, Wink M (1998) In: Roberts MF, Wink M (eds) Alkaloids – biochemistry, ecology, and medicinal applications, 1st edn. Springer, New York. https://doi.org/10.1007/978-1-4757-2905-4

Rodrigo MJ, Alquézar B, Alós E, Medina V, Carmona L, Bruno M et al (2013) A novel carotenoid cleavage activity involved in the biosynthesis of Citrus fruit-specific apocarotenoid pigments. J Exp Bot 64:4461–4478. https://doi.org/10.1093/jxb/ert260

Rojas-Flores S, De La Cruz-Noriega M, Cabanillas-Chirinos L, Benites SM, Silva-Palacios F, Rosa AL et al (2020) Using lime (Citrus × aurantiifolia), orange (citrus × sinensis), and tangerine (citrus reticulata) waste as a substrate for generating bioelectricity. Environ Res Eng Manag 76:24–34. https://doi.org/10.5755/j01.erem.76.3.24785

Romano R, De Luca L, Aiello A, Rossi D, Pizzolongo F, Masi P (2022) Bioactive compounds extracted by liquid and supercritical carbon dioxide from citrus peels. Int J Food Sci Technol:1–12. https://doi.org/10.1111/ijfs.15712

Ros R, Muñoz-Bertomeu J, Krueger S (2014) Serine in plants: biosynthesis, metabolism, and functions. Trends Plant Sci 19:564–569. https://doi.org/10.1016/j.tplants.2014.06.003

Roy A (2017) A review on the alkaloids an important therapeutic compound from plants Green synthesis of nanoparticles View project Remediation of environmental contamination View project. Int J Plant Biotechnol 3:1–9. https://doi.org/10.37628/ijpb.v3i2.199

Ruan Y-L (2014) Sucrose metabolism: gateway to diverse carbon use and sugar signaling. Annu Rev Plant Biol 65:33–67. https://doi.org/10.1146/annurev-arplant-050213-040251

Rueda D, Arias V, Zhang Y, Cabot A, Agudelo AC, Cadavid D (2020) Low-cost tangerine peel waste mediated production of Titanium Dioxide Nanocrystals: synthesis and characterization. Environ Nanotechnol Monit Manag 13:100285. https://doi.org/10.1016/j.enmm.2020.100285

Sadka A, Shlizerman L, Kamara I, Blumwald E (2019) Primary metabolism in citrus fruit as affected by its unique structure. Front Plant Sci 10:1–14. https://doi.org/10.3389/fpls.2019.01167

Sagar NA, Pareek S, Sharma S, Yahia EM, Lobo MG (2018) Fruit and vegetable waste: bioactive compounds, their extraction, and possible utilization. Compr Rev Food Sci Food Saf 17:512–531. https://doi.org/10.1111/1541-4337.12330

Saini A, Panesar PS, Bera MB (2021) Valuation of Citrus reticulata (kinnow) peel for the extraction of lutein using ultrasonication technique. Biomass Convers Biorefinery 11:2157–2165. https://doi.org/10.1007/s13399-020-00605-4

Saltveit ME (2017) Synthesis and metabolism of phenolic compounds. Fruit Veg Phytochem Chem Hum Heal Second Ed 1:115–123. https://doi.org/10.1002/9781119158042.ch5

Sanches VL, Cunha TA, Viganó J, de Souza Mesquita LM, Faccioli LH, Breitkreitz MC et al (2022) Comprehensive analysis of phenolics compounds in citrus fruits peels by UPLC-PDA and UPLC-Q/TOF MS using a fused-core column. Food Chem X 14. https://doi.org/10.1016/j.fochx.2022.100262

Sande D, Moreira ACS, Papa BAJ (2018) Tangerine: converting residues into antioxidant food, rich in fiber and with sensorial acceptance. Nutr Food Sci Int J 5:1–6. https://doi.org/10.19080/nfsij.2018.05.555656

Santana ML, Bispo JAC, de Sena AR, Teshima E, de Brito AR, Costa FS et al (2021) Clarification of tangerine juice using cellulases from Pseudoyma sp. J Food Sci Technol 58:44–51. https://doi.org/10.1007/s13197-020-04511-5

Sauret-Güeto S, Botella-Pavía P, Flores-Pérez U, Martínez-García JF, San Román C, León P et al (2006) Plastid cues posttranscriptionally regulate the accumulation of key enzymes of the methylerythritol phosphate pathway in arabidopsis. Plant Physiol 141:75–84. https://doi.org/10.1104/pp.106.079855

Schwab W, Davidovich-Rikanati R, Lewinsohn E (2008) Biosynthesis of plant-derived flavor compounds. Plant J 54:712–732. https://doi.org/10.1111/j.1365-313X.2008.03446.x

Schwartzberg HG (1997) Expression of fluid from biological solids. Sep Purif Methods 26:1–213. https://doi.org/10.1080/03602549708014156

Scora RW (1975) On the history and origin of citrus. Bull Torrey Bot Club:369–375

Shahidi F (2015) 1 – Antioxidants: principles and applications. In: F. B. T.-H. of A, Shahidi FP (eds) Woodhead publishing series in food science, technology and nutrition. Woodhead Publishing, pp 1–14. https://doi.org/10.1016/B978-1-78242-089-7.00001-4

Shahidi F, Yeo J-D (2016) Insoluble-bound phenolics in food. Molecules 21. https://doi.org/10.3390/molecules21091216

Shahidi F, Yeo J (2018) Bioactivities of phenolics by focusing on suppression of chronic diseases: a review. Int J Mol Sci 19. https://doi.org/10.3390/ijms19061573

Shahidi F, Varatharajan V, Oh WY, Peng H (2019) Phenolic compounds in agri-food by-products, their bioavailability and health effects. J Food Bioact 5:57–119. https://doi.org/10.31665/jfb.2019.5178

Shang M, Yang H, Yang R, Chen T, Fu Y, Li Y et al (2021) The folate cycle enzyme MTHFD2 induces cancer immune evasion through PD-L1 up-regulation. Nat Commun 12:1–16. https://doi.org/10.1038/s41467-021-22173-5

Sheng L, Shen D, Yang W, Zhang M, Zeng Y, Xu J et al (2017) GABA pathway rate-limit citrate degradation in postharvest citrus fruit evidence from HB Pumelo (Citrus grandis) × Fairchild (Citrus reticulata) hybrid population. J Agric Food Chem 65:1669–1676. https://doi.org/10.1021/acs.jafc.6b05237

Shi Y-S, Zhang Y, Li H-T, Wu C-H, El-Seedi HR, Ye W-K et al (2020) Limonoids from Citrus: chemistry, anti-tumor potential, and other bioactivities. J Funct Foods 75:104213. https://doi.org/10.1016/j.jff.2020.104213

Shim GB (2015) A study on bio-ethanol extraction from fruit waste by simultaneous saccharification and fermentation. APEC Youth Sci J 7:154–162

Shinde S, Villamor JG, Lin W, Sharma S, Verslues PE (2016) Proline coordination with fatty acid synthesis and redox metabolism of chloroplast and mitochondria. Plant Physiol 172:1074–1088. https://doi.org/10.1104/pp.16.01097

Shinozaki Y, Beauvoit BP, Takahara M, Hao S, Ezura K, Andrieu MH et al (2020) Fruit setting rewires central metabolism via gibberellin cascades. Proc Natl Acad Sci U S A 117:23970–23981. https://doi.org/10.1073/pnas.2011859117

Shorbagi M, Fayek NM, Shao P, Farag MA (2022) Citrus reticulata Blanco (the common mandarin) fruit: an updated review of its bioactive, extraction types, food quality, therapeutic merits, and bio-waste valorization practices to maximize its economic value. Food Biosci 47:101699. https://doi.org/10.1016/j.fbio.2022.101699

Shu B, Wu G, Wang Z, Wang J, Huang F, Dong L et al (2020) The effect of microwave vacuum drying process on citrus: drying kinetics, physicochemical composition and antioxidant activity of dried citrus (Citrus reticulata Blanco) peel. J Food Meas Charact 14:2443–2452. https://doi.org/10.1007/s11694-020-00492-3

Silveira V, Santos Rubio KT, Poleti Martucci ME (2022) Anxiolytic effect of Anthemis nobilis L. (roman chamomile) and Citrus reticulata Blanco (tangerine) essential oils using the light-dark test in zebrafish (Danio rerio). J Ethnopharmacol 298:115580. https://doi.org/10.1016/j.jep.2022.115580

Sinclair WB (1984) The biochemistry and physiology of the lemon and other citrus fruits. University of California

Slocum RD (2005) Genes, enzymes and regulation of arginine biosynthesis in plants. Plant Physiol Biochem 43:729–745. https://doi.org/10.1016/j.plaphy.2005.06.007

Song X, Wang L, Liu T, Liu Y, Wu X, Liu L (2021) Mandarin (Citrus reticulata L.) essential oil incorporated into chitosan nanoparticles: Characterization, anti-biofilm properties and application in pork preservation. Int J Biol Macromol 185:620–628. https://doi.org/10.1016/j.ijbiomac.2021.06.195

Soobrattee MA, Neergheen VS, Luximon-Ramma A, Aruoma OI, Bahorun T (2005) Phenolics as potential antioxidant therapeutic agents: mechanism and actions. Mutat Res Mol Mech Mutagen 579:200–213. https://doi.org/10.1016/j.mrfmmm.2005.03.023

Sridhar A, Kapoor A, Senthil Kumar P, Ponnuchamy M, Balasubramanian S, Prabhakar S (2021) Conversion of food waste to energy: a focus on sustainability and life cycle assessment. Fuel 302:121069. https://doi.org/10.1016/J.FUEL.2021.121069

Stein O, Granot D (2019) An overview of sucrose synthases in plants. Front Plant Sci 10:1–14. https://doi.org/10.3389/fpls.2019.00095

Stevens PF (2016) Angiosperm phylogeny website. Version 13

Strazzer P, Spelt CE, Li S, Bliek M, Federici CT, Roose ML et al (2019) Hyperacidification of Citrus fruits by a vacuolar proton-pumping P-ATPase complex. Nat Commun 10. https://doi.org/10.1038/s41467-019-08516-3

Suh JH, Tang X, Zhang Y, Gmitter FG, Wang Y (2021) Metabolomic analysis provides new insight into tolerance of huanglongbing in citrus. Front Plant Sci 12:1–12. https://doi.org/10.3389/fpls.2021.710598

Sun T, Rao S, Zhou X, Li L (2022) Plant carotenoids: recent advances and future perspectives. Mol Hortic 2:1–21. https://doi.org/10.1186/s43897-022-00023-2

Swingle W, Reece P (1967) The botany of Citrus and its wild relatives of the orange subfamily (family Rutaceae, subfamily Aurantioideae)

Tadeo FR, Cercós M, Colmenero-Flores JM, Iglesias DJ, Naranjo MA, Ríos G et al (2008) Molecular physiology of development and quality of citrus. Adv Bot Res 47:147–223. https://doi.org/10.1016/S0065-2296(08)00004-9

Taher M, Shukry NASB, Susanti D, Wan Saleh WMNH, Syukri Y (2020) In: Swamy MK (ed) Citrus flavonoids in preventing cardiovascular diseases BT – plant-derived bioactives: chemistry and mode of action. Springer, Singapore, pp 495–508. https://doi.org/10.1007/978-981-15-2361-8_22

Tan FQ, Zhang M, Xie KD, Fan YJ, Song X, Wang R et al (2019) Polyploidy remodels fruit metabolism by modifying carbon source utilization and metabolic flux in Ponkan mandarin (Citrus reticulata Blanco). Plant Sci 289:110276. https://doi.org/10.1016/j.plantsci.2019.110276

Tanaka T (1954) Species problem in Citrus (Revisio aurantia cearum IX). Jpn Soc Promot Sci:1–157

Tanaka T (1961) Citologia: semi-centennial commemoration papers on Citrus Studies. Osaka Citol Support Found

Tanaka T (1977) Fundamental discussion of Citrus classification. Stud Citrol:1–6

Terol J, Soler G, Talon M, Cercos M (2010) The aconitate hydratase family from Citrus. BMC Plant Biol 10. https://doi.org/10.1186/1471-2229-10-222

Tholl D (2015) Biosynthesis and biological functions of terpenoids in plants. Adv Biochem Eng Biotechnol 148:63–106. https://doi.org/10.1007/10_2014_295

Tietel Z, Plotto A, Fallik E, Lewinsohn E, Porat R (2010) Taste and aroma of fresh and stored mandarins. J Sci Food Agric 91:14–23. https://doi.org/10.1002/jsfa.4146

Tietel Z, Feldmesser E, Lewinsohn E, Fallik E, Porat R (2011) Changes in the transcriptome of "Mor" mandarin flesh during storage: emphasis on molecular regulation of fruit flavor deterioration. J Agric Food Chem 59:3819–3827. https://doi.org/10.1021/jf104614s

Ting, Y., Chiou, Y.-S., Pan, M.-H., Ho, C.-T., and Huang, Q. (2015). In vitro and in vivo anticancer activity of tangeretin against colorectal cancer was enhanced by emulsion-based delivery system. J Funct Foods 15, 264–273. doi: https://doi.org/https://doi.org/10.1016/j.jff.2015.03.034

Tolkowsky S (1938) Hesperides. A history of theculture and use of citrus fruits

Tomova T, Petelkov I, Shopska V, Denkova-Kostova R, Kostov G, Denkova Z (2021) Production of probiotic wort-based beverages with grapefruit (Citrus paradisi L.) or tangerine (Citrus reticulata L.) zest essential oil addition. Acta Sci Pol Technol Aliment 20:237–245. https://doi.org/10.17306/J.AFS.0902

Tounsi MS, Moulehi I, Ouerghemmi I, Mejri H, Wannes WA, Hamdaoui G et al (2011) Changes in lipid composition and antioxidant capacity of bitter orange (Citrus aurantium. L) and mandarin (Citrus reticulata. Blanco) oilseeds on different stages of maturity. JAOCS J Am Oil Chem Soc 88:961–966. https://doi.org/10.1007/s11746-010-1751-2

Tupuna-Yerovi DS, Paese K, Flôres SH, Guterres SS, Rios A (2020) Addition of norbixin microcapsules obtained by spray drying in an isotonic tangerine soft drink as a natural dye. J Food Sci Technol 57:1021–1031. https://doi.org/10.1007/s13197-019-04135-4

Turchetto-Zolet AC, Margis-Pinheiro M, Margis R (2009) The evolution of pyrroline-5-carboxylate synthase in plants: a key enzyme in proline synthesis. Mol Gen Genomics 281:87–97. https://doi.org/10.1007/s00438-008-0396-4

Tuzson P (1936) Über das Polyen-Pigment der Orange (I. Mitteil). Chem Berische 69:1878

U.N. (2015) Transforming our world: the 2030 Agenda for Sustainable Development

United Nations Economic Commission for Europe (UNEC) (2018) Unece Standard FFV-04 2017 edition. Available at: http://www.unece.org/

USDA (2021a) Citrus fruits 2021 summary Available at: https://ccqc.org/wp-content/uploads/USDA-NASS-Citrus-Fruits-2021-Summary-092021.pdf

USDA (2021b) Citrus anual Available at: https://apps.fas.usda.gov/newgainapi/api/Report/DownloadReportByFileName?fileName=Citrus%20Annual_Madrid_European%20Union_E42023-0001.pdf

USDA (2022) Citrus: world markets and trade. Available at: https://public.govdelivery.com/accounts/USDAFAS/subscriber/new

Van Norman JM, Zhang J, Cazzonelli CI, Pogson BJ, Harrison PJ, Bugg TDH et al (2014) Periodic root branching in Arabidopsis requires synthesis of an uncharacterized carotenoid derivative. Proc Natl Acad Sci U S A 111. https://doi.org/10.1073/pnas.1403016111

Vergara-Barberán M, Mompó-Roselló Ó, Lerma-García MJ, Herrero-Martínez JM, Simó-Alfonso EF (2017) Enzyme-assisted extraction of proteins from Citrus fruits and prediction of their cultivar using protein profiles obtained by capillary gel electrophoresis. Food Control 72:14–19. https://doi.org/10.1016/j.foodcont.2016.07.025

Verpoorte R (2005) Alkaloids. In: Worsfold P, Townshend A, Poole C (eds) Encyclopedia of analytical science. Elsevier, pp 56–61

Vikram A, Jayaprakasha GK, Patil BS (2007) Simultaneous determination of citrus limonoid aglycones and glucosides by high performance liquid chromatography. Anal Chim Acta 590:180–186. https://doi.org/10.1016/j.aca.2007.03.029

Vranová E, Coman D, Gruissem W (2013) Network analysis of the MVA and MEP pathways for isoprenoid synthesis. Annu Rev Plant Biol 64:665–700. https://doi.org/10.1146/annurev-arplant-050312-120116

Walker RP, Famiani F (2018) Organic acids in fruits: metabolism, functions and contents. Hortic Rev (Am Soc Hortic Sci) 45:371–430. https://doi.org/10.1002/9781119431077.ch8

Wallis CM, Gorman Z, Rattner R, Hajeri S, Yokomi R (2022) Amino acid, sugar, phenolic, and terpenoid profiles are capable of distinguishing Citrus tristeza virus infection status in citrus cultivars: grapefruit, lemon, mandarin, and sweet orange. PLoS One 17:e0268255. https://doi.org/10.1371/journal.pone.0268255

Wang X, Chen Q, Lü X (2014) Pectin extracted from apple pomace and citrus peel by subcritical water. Food Hydrocoll 38:129–137. https://doi.org/10.1016/j.foodhyd.2013.12.003

Wang S, Tu H, Wan J, Chen W, Liu X, Luo J et al (2016) Spatio-temporal distribution and natural variation of metabolites in citrus fruits. Food Chem 199:8–17. https://doi.org/10.1016/j.foodchem.2015.11.113

Wang X, Xu Y, Zhang S, Cao L, Huang Y, Cheng J et al (2017a) Genomic analyses of primitive, wild and cultivated citrus provide insights into asexual reproduction. Nat Genet 495(49):765–772. https://doi.org/10.1038/ng.3839

Wang Y, Qian J, Cao J, Wang D, Liu C, Yang R et al (2017b) Antioxidant capacity, anticancer ability and flavonoids composition of 35 citrus (Citrus reticulata Blanco) varieties. Molecules 22:1–20. https://doi.org/10.3390/molecules22071114

Wang L, He F, Huang Y, He J, Yang S, Zeng J et al (2018) Genome of wild mandarin and domestication history of Mandarin. Mol Plant 11:1024–1037. https://doi.org/10.1016/J.MOLP.2018.06.001/ATTACHMENT/7095E249-6A31-43A4-9C66-33789C5C1E14/MMC4.TXT

Wang, Y., Zang, W., Ji, S., Cao, J., and Sun, C. (2019). Three polymethoxyflavones purified from Ougan (Citrus reticulata Cv. Suavissima) inhibited LPS-induced NO elevation in the neuroglia BV-2 cell line via the JAK2/STAT3 pathway. *Nutrients* 11. doi: https://doi.org/10.3390/nu11040791

Wang Y, Chen Y, Zhang H, Chen J, Cao J, Chen Q et al (2020) Polymethoxyflavones from citrus inhibited gastric cancer cell proliferation through inducing apoptosis by upregulating RARβ, both in vitro and in vivo. Food Chem Toxicol 146:111811. https://doi.org/10.1016/j.fct.2020.111811

Wang J, Shi J, Zhu Y, Ma W, Yan H, Shao C et al (2022) Insights into crucial odourants dominating the characteristic flavour of citrus-white teas prepared from citrus reticulata Blanco 'Chachiensis' and Camellia sinensis 'Fudingdabai'. Food Chem 377:132048. https://doi.org/10.1016/j.foodchem.2022.132048

Ward JL, Baker JM, Llewellyn AM, Hawkins ND, Beale MH (2011) Metabolomic analysis of Arabidopsis reveals hemiterpenoid glycosides as products of a nitrate ion-regulated, carbon flux overflow. Proc Natl Acad Sci U S A 108:10762–10767. https://doi.org/10.1073/pnas.1018875108

Webber H (1967) History and development of the citrus industry. Citrus Ind:1–37

Wheeler GL, Jones MA, Smirnoff N (1998) The biosynthetic pathway of vitamin C in higher plants. Nature 393:365–369. https://doi.org/10.1038/30728

Wheeler G, Ishikawa T, Pornsaksit V, Smirnoff N (2015) Evolution of alternative biosynthetic pathways for vitamin C following plastid acquisition in photosynthetic eukaryotes. elife 2015:1–25. https://doi.org/10.7554/eLife.06369

Winter G, Todd CD, Trovato M, Forlani G, Funck D (2015) Physiological implications of arginine metabolism in plants. Front Plant Sci 6:1–14. https://doi.org/10.3389/fpls.2015.00534

Wirivutthikorn W (2019) Optimum ratios of okra and tangerine on production of mixed juice with lycopene supplementation. Int J GEOMATE 17:8–13. https://doi.org/10.21660/2019.61.4503

Witte CP (2011) Urea metabolism in plants. Plant Sci 180:431–438. https://doi.org/10.1016/j.plantsci.2010.11.010

World Health Organization (2015) Serum and red blood cell folate concentrations for assessing folate status in populations. Vitam Miner Nutr Inf Syst:1–7

Wu GA, Terol J, Ibanez V, López-García A, Pérez-Román E, Borredá C et al (2018) Genomics of the origin and evolution of Citrus. Nature 554:311–316. https://doi.org/10.1038/nature25447

Wu Y, Xu J, Liu Y, Zeng Y, Wu G (2020) A review on anti-tumor mechanisms of coumarins. Front Oncol 10:592853. https://doi.org/10.3389/fonc.2020.592853

Xie S, Manchester SR, Liu K, Wang Y, Sun B (2015) Citrus linczangensis sp. n., a Leaf Fossil of Rutaceae from the Late Miocene of Yunnan, China. 174:1201–1207. https://doi.org/10.1086/671796

Xiong K, Chen Y (2020) Supercritical carbon dioxide extraction of essential oil from tangerine peel: experimental optimization and kinetics modelling. Chem Eng Res Des 164:412–423. https://doi.org/10.1016/j.cherd.2020.09.032

Xu A, Xiao Y, He Z, Liu J, Wang Y, Gao B et al (2022a) Use of non-Saccharomyces yeast co-fermentation with Saccharomyces cerevisiae to improve the polyphenol and volatile aroma compound contents in nanfeng tangerine wines. J Fungi (Basel, Switzerland) 8. https://doi.org/10.3390/jof8020128

Xu Y, Ren J, Wang W, Zeng AP (2022b) Improvement of glycine biosynthesis from one-carbon compounds and ammonia catalyzed by the glycine cleavage system in vitro. Eng Life Sci 22:40–53. https://doi.org/10.1002/elsc.202100047

Yacoubi S, Zayani K, Zapata N, Zairi A, Slatni A, Salvador R et al (2010) Day and night time sprinkler irrigated tomato: irrigation performance and crop yield. Biosyst Eng 107:25–35. https://doi.org/10.1016/J.BIOSYSTEMSENG.2010.06.009

Yadav V, Sarker A, Yadav A, Miftah AO, Bilal M, Iqbal HMN (2022) Integrated biorefinery approach to valorize citrus waste: a sustainable solution for resource recovery and environmental management. Chemosphere 293:133459. https://doi.org/10.1016/j.chemosphere.2021.133459

Yang H, Zou Y, Li X, Zhang M, Zhu Z (2022) QTL analysis reveals the effect of CER1-1 and CER1-3 to reduce fruit water loss by increasing cuticular wax alkanes in citrus fruit. Postharvest Biol Technol:185

Ye Y, Xu G, Li DL (2021) Acridone alkaloids and flavones from the leaves of Citrus reticulata. Nat Prod Res. https://doi.org/10.1080/14786419.2021.1876047

Yeats TH, Rose JKC (2013) Topical review on cuticle synthesis and function the formation and function of plant cuticles 1. 163:5–20. https://doi.org/10.1104/pp.113.222737

Yi F, Jin R, Sun J, Ma B, Bao X (2018) Evaluation of mechanical-pressed essential oil from Nanfeng mandarin (Citrus reticulata Blanco cv. Kinokuni) as a food preservative based on antimicrobial and antioxidant activities. LWT Food Sci Technol Food Sci Technol 95:346–353. https://doi.org/10.1016/j.lwt.2018.05.011

Yu H, Yang X, Guo F, Jiang X, Deng X, Xu Q (2017) Genetic diversity and population structure of pummelo (Citrus maxima) germplasm in China. Tree Genet Genomes 13:1–10. https://doi.org/10.1007/S11295-017-1133-0/FIGURES/6

Yuan H, Zhang J, Nageswaran D, Li L (2015) Carotenoid metabolism and regulation in horticultural crops. Hortic Res 2. https://doi.org/10.1038/hortres.2015.36

Yun D, Liu J (2022) Recent advances on the development of food packaging films based on citrus processing wastes: a review. J Agric Food Res 9:100316. https://doi.org/10.1016/j.jafr.2022.100316

Yun Z, Li W, Pan Z, Xu J, Cheng Y, Deng X (2010) Comparative proteomics analysis of differentially accumulated proteins in juice sacs of ponkan (Citrus reticulata) fruit during postharvest cold storage. Postharvest Biol Technol 56:189–201. https://doi.org/10.1016/j.postharvbio.2010.01.002

Zaki N, Naeem M (2021) Antioxidant, antimicrobial and anticancer activities of citrus peels to improve the shelf life of yoghurt drink. Egypt J Food Sci 49:249–265. https://doi.org/10.21608/ejfs.2021.58310.1092

Zhang Y, Fernie AR (2018) On the role of the tricarboxylic acid cycle in plant productivity. J Integr Plant Biol 60:1199–1216. https://doi.org/10.1111/jipb.12690

Zhang J, Wu Y, Zhao X, Luo F, Li X, Zhu H et al (2014a) Chemopreventive effect of flavonoids from Ougan (Citrus reticulata cv. Suavissima) fruit against cancer cell proliferation and migration. J Funct Foods 10:511–519. https://doi.org/10.1016/j.jff.2014.08.006

Zhang Y, Sun Y, Xi W, Shen Y, Qiao L, Zhong L et al (2014b) Phenolic compositions and antioxidant capacities of Chinese wild mandarin (Citrus reticulata Blanco) fruits. Food Chem 145:674–680. https://doi.org/10.1016/j.foodchem.2013.08.012

Zhang H, Xie Y, Liu C, Chen S, Hu S, Xie Z et al (2017) Comprehensive comparative analysis of volatile compounds in citrus fruits of different species. Food Chem 230:316–326. https://doi.org/10.1016/j.foodchem.2017.03.040

Zhang S, Liang M, Wang N, Xu Q, Deng X, Chai L (2018a) Reproduction in woody perennial Citrus: an update on nucellar embryony and self-incompatibility. Plant Reprod 31:43–57. https://doi.org/10.1007/S00497-018-0327-4/FIGURES/4

Zhang Y, Swart C, Alseekh S, Scossa F, Jiang L, Obata T et al (2018b) The extra-pathway interactome of the TCA cycle: expected and unexpected metabolic interactions. Plant Physiol 177:966–979. https://doi.org/10.1104/pp.17.01687

Zhang R, Parniakov O, Grimi N, Lebovka N, Marchal L, Vorobiev E (2019) Emerging techniques for cell disruption and extraction of valuable bio-molecules of microalgae Nannochloropsis sp. Bioprocess Biosyst Eng 42:173–186. https://doi.org/10.1007/s00449-018-2038-5

Zhang X, Su M, Du J, Zhou H, Li X, Zhang M et al (2022) Analysis of the free amino acid content and profile of 129 peach (Prunus persica (L.) Batsch) germplasms using LC-MS/MS without derivatization. J Food Compos Anal 114:104811. https://doi.org/10.1016/j.jfca.2022.104811

Zheng G, Chao Y, Liu M, Yang Y, Zhang D, Wang K et al (2021) Evaluation of dynamic changes in the bioactive components in Citri Reticulatae Pericarpium (Citrus reticulata 'Chachi') under different harvesting and drying conditions. J Sci Food Agric 101:3280–3289. https://doi.org/10.1002/jsfa.10957

Zhong G, Nicolosi E (2020) Citrus origin, diffusion, and economic importance. The Citrus Genome:5–21

Zhou X, Qin D, Xiang B, Xi J (2022) Cyclodextrin-based liquid-phase pulsed discharge extraction of flavonoids from tangerine (Citrus reticulata) pericarp: optimization, antioxidant activity and storage stability. Sep Purif Technol 278:119603. https://doi.org/10.1016/j.seppur.2021.119603

Zou Z, Xi W, Hu Y, Nie C, Zhou Z (2016) Antioxidant activity of Citrus fruits. Food Chem 196:885–896. https://doi.org/10.1016/j.foodchem.2015.09.072

Chapter 7
Etrog Citron (*Citrus medica* var. ethrog Engl)

Chelladurai Premkumar, Nartunai Govindarajan, Akansha Pandey, Prathapa Reddy Maramreddy, Kurra Sivadurga, Gokul Marimuthu, Chidanandamurthy Thippeswamy Swamy, and Sukhvinder Singh Purewal

7.1 Introduction, History and Culture

7.1.1 Introduction

Etrog citron is an evergreen small shrub or tree variety of *Citrus medica* belonging to the family Rutaceae. It has been categorized as a synonym of *Citrus medica* Engl. as per the Catalogue of Life and World Flora Online (WFO 2022; Citrus medica, Catalouge of Life 2022). IUCN has listed citron under the "Least Concern" category (Plummer 2020). There are many cultivars and hybrids of citron, varying significantly in shape and size of fruit, which is grown in specific geographical

C. Premkumar
Department of Pharmacognosy, Captain Srinivasa Murthy Central Ayurveda Research Institute, Central Council for Research in Ayurvedic Sciences, Chennai, Tamil Nadu, India

N. Govindarajan
Program Officer, MPR Section, Central Council for Research in Ayurvedic Sciences, Janakpuri, New Delhi, India

A. Pandey
Department of Botany, Captain Srinivasa Murthy Central Ayurveda Research Institute, Central Council for Research in Ayurvedic Sciences, Chennai, Tamil Nadu, India

P. R. Maramreddy · K. Sivadurga · G. Marimuthu
Department of Chemistry, Captain Srinivasa Murthy Central Ayurveda Research Institute, Central Council for Research in Ayurvedic Sciences, Chennai, Tamil Nadu, India

C. T. Swamy (✉)
Department of Microbiology, Captain Srinivasa Murthy Central Ayurveda Research Institute, Central Council for Research in Ayurvedic Sciences, Chennai, Tamil Nadu, India

S. Singh Purewal
University Centre for Research & Development (UCRD), Chandigarh University, Mohali, Punjab, India

S. Singh Purewal et al. (eds.), *Recent Advances in Citrus Fruits*,
https://doi.org/10.1007/978-3-031-37534-7_7

locations. The etrog citron grows prevalently in subtropical regions and warm countries. Etrog citron's cultivation is done commercially in Jews inhabited parts of the world. The trees are sensitive to cold and wither in cold climatic condition (Chhikara et al. 2018).

The etrog is a large shrub or small tree with stout and axillary spines throughout the stems and branches, measuring up to 4 cm in length. Leaves are simple, petiolate, and arranged alternately. The petiole is up to 5 cm long, wingless, exstipulate, and imperfectly articulated with the leaf blade. Leaf lamina is coriaceous, glabrous with elliptic-ovate to ovate-lanceolate shape, apex obtuse-rounded, margin serrate -crenate, secondary nerves 8–12 pairs, and white gland-dotted. White or pink, 5–10 sweet-scented flowers occur in axillary cymes. Mostly bisexual rarely staminate flowers also present. Calyx green and 5-lobed. Corolla 5 in number, oblong and pink or purple colored. Stamens numerous, 30–40, polyadelphous, anthers linear and yellowish. Ovary many-celled with 4–8 ovules in each cell. Fruit is a hesperidium, yellowish, ovoid-oblong or ovoid-ellipsoid, or roughly pear-shaped surface may be smooth or tuberculate, apex mamillated. The pericarp is thick, aromatic, and glandular, with microscopic pits and yellow; the mesocarp is white. The edible portion of fruit consists of multi-cellular vesicles filled with juice and is sour in taste. Seeds are non-endosperms, numerous, and smooth (Hooker 1875). In the present chapter, a comprehensive literature review, particularly regarding etrog citron's traditional uses, phytochemistry, pharmacological activities, and commercial value, has been surveyed and discussed. The cultivation and harvest practices, including etrog citron major diseases and their management, and recent research findings regarding the origin and genetic diversity of etrog citron are summarised.

7.1.2 History and Culture

A description of citron has been quoted in Theophrastus's *Historia Planatarum* chapter four (Theophrastus 1916). It is believed that etrog peel thickness and aroma made it not only expensive but also a long-distance safe trade commodity during ancient days (Liran 2013).

Etrog citron is considered native to Southeast Asia, particularly to northeast India. Archaeological digs in Mesopotamia have found seeds dating back to 4000 B.C.E. In historical records, the plant was first domesticated in Persia (Shemesh 2012). The citrus fruit was introduced into Israel only after the return of Jews from Persia or Babylon, indicating that the plant found its way to Judaism (Pierre Laszlo 2007) and this opinion was rejected by arguing that the citron arrived in Israel from Eastern Asia earlier along with the other tropical perfumes (Yehuda Felix 1997; Adess and Avraham Chaim 2004). The first evidence through palynological analysis of etrog citron cultivation in the Mediterranean basin, is found during the first and seventh centuries BC (Langgut et al. 2013). Similarly, remains of pollen, seed, and fruits have been excavated from different Roman settlements of the first and fourth centuries BC (Van der Veen 2011).

The etrog fruit has a deep-rooted connection with Judaism, and Jews consider it as the fruit of the "godly tree." The etrog fruit was mentioned in the Book of Leviticus written by Moses, and in Leviticus 23:40, it is mentioned as "And you shall take of yourselves on the first day the fruit of a goodly tree, a palm branch, the myrtle branch, and the willow of the brook; and you shall rejoice before the Lord your God seven days" (Adess and Avraham Chaim 2004; Amar Zohar 2009). The Jewish festival "Sukkot" is a 7 days agricultural harvest festival celebrated by Jews, in which four major species of plants (Etrog, palm, myrtle, and willow) are used in their celebrations (Amar Zohar 2009; Yaniv and Dudai 2014; Moster 2018). The etrog is considered the "forbidden fruit" of the garden of Eden, and it is mentioned in the Jewish literatures (Tristram 1883; Cyrus Adler and Casanowicz 1903). Etrog citron to be used for Sukkot first undergoes Rabbinical certification. The candle flame shape, flawless top-third part of the fruit, light yellow coloration of rind, and intact "pitam" (woody style) are some of the characteristics inspected and considered fit for religious purposes (Miller and Silvers 2015). Langutt have recently reviewed prehistoric texts regarding religious uses of etrog citron (Langgut 2017) (Figs. 7.1 and 7.2).

Fig. 7.1 The Four Species from Left to right (Willow, Palm branch, Myrtle and Etrog) used in "Sukkot" festival "Arbaat HaMinim - The Four Species of Sukkot - Etrog (Citron)" by Sarashira is licensed under CC BY-SA 4.0

Fig. 7.2 Image of Etrog Tree and its Habit, Leaves, Flower, Fruit and Cross section of fruit (Saville et al. 2011)

7.1.3 Taxonomic Classification of Etrog

The recent Taxonomic classification of etrog citron is given in Table 7.1.

7.2 Cultivation and Harvesting of Etrog

7.2.1 Cultivation

Etrog citron is grown basically in those areas of the world where Jewish people have settled during the course of their migration from Asia to the Mediterranean Basin and the Arabian Peninsula. Morocco, Italy, Yemen, Greece, and Israel are major etrog producing countries with annual consumer demand of 1.8 million fruits per year. The etrog citron is also under cultivation in Morocco, U.S.A., and Mexico on a small scale. During the Jewish festival season of sukkot, one fruit of etrog can sell from US$50–100 depending on the properties of the fruit (Klein et al. 2016). The demand for citron for other purposes like confectionary and beverages is approximately 7000 tons/hectare annually (Klein and Joshua 2014).

Table 7.1 Taxonomic classification of Citrus medica var. ethrog Engl (WFO 2022)

Kingdom	:	Plantae
Domain	:	Eukaryota
Kingdom	:	Plantae
Phylum	:	Spermatophyta
Subphylum	:	Angiospermae
Class	:	Dicotyledonae
Order	:	Rutales
Family	:	Rutaceae
Genus	:	*Citrus*
Botanical name	:	*Citrus medica* var. ethrog Engl

Citron is a crop of tropical and sub-tropical areas and cannot thrive well in a temperate climate. Well-drained sandy soil with heavy irrigation during fruit setting are the two basic requirements for the cultivation of etrog. Etrog citron can be propagated by seeds and vegetative cuttings. In natural conditions, etrog is propagated by seed after self-pollination. Under cultivation, the etrog is propagated by cuttings only to maintain the true breed. During grafting, rootstocks of only etrog are used for propagation as inter-specific grafts are considered misfits for religious purposes. 1.5 t/ha N, P, K (Nitrogen, Phosphorus, and Potassium) with Potassium nitrate (KNO_3) is applied during flower setting as fertilizer in citron. To manage the weeds, synthetic auxin "Picloram" is used. The application of auxin also helps in the retention of style and stigma on fruit, which is a desirable characteristic of fruit in the Jewish market (Goldschmidt and Leshem 1971). The harvesting of citron is done twice per year, one in early summer mostly for leaves, immature fruits, and flowers, and the second in late spring for mature fruits. Leaves and flowers are further used in the cosmetics industry for oil extraction, while young fruits are in the confectionery, baking, and beverages industry. Mature fruits without any blemishes, lesions, perfect color, and attached pitam are harvested packed in special citron-shaped foam rubber inserts for use in the Sukkot festival (Klein and Joshua 2014).

7.2.2 *Postharvest Management of Etrog*

The demand for a perfect etrog is very high in the market during a particular festival period, so postharvest management is detrimental in protecting the fruit from physical and physiological damage, further defining the cost. The perfect etrog is defined only by its morphological characteristics like texture and colour of skin; the cultivators usually employ those postharvest treatments which affect the external quality of fruits. To prevent water loss after harvest, the fruits are coated with natural or synthetic wax and fruits are packed in a polythene cover. The application of wax prevents fungal invasions also. The harvested fruit tends to form a callus at cut sides, which can be reduced by the application of 10 ppm copper chloride

(Hagenmaier and Baker 1994). It is suggested that hot water brushing of citron fruits at 38 °C followed by subsequent storage at 16 °C to remove pathogens from the surface without any loss in water or enhanced Carbon dioxide (CO_2) production leads to avoid the ripening of fruit (Klein et al. 2016). Etrog fruits stored below 10 °C are prone to chilling injury and produce sunken areas and lesions on fruits. A storage temperature of 17–20 °C is maintained by cultivators to avoid such chilling injuries. The second desirable quality of etrog is a fruit of a particular hue (Hue >120 °C not allowed in religious ceremonies). To maintain light yellow colour, 5pmm ethylene with 50 ppm gibberellic acid (G.A.) for 12–24 h at 20 °C is used. Ethylene induces the degreening of green fruits while G.A. induces chlorophyllase activity (Klein et al. 2013).

7.3 Disease and Its Management in Etrog Citron

Citron is a major and minor host to a number of viruses, viroids, and bacterial and fungal pathogens (CABI 2022). Table 7.2 summarises the major pathogens, diseases, symptoms, and their control affecting citron.

7.3.1 Virus and Viroid Diseases of Citron

The first case of viral disease, psoriasis in Citrus, was published in 1934 (Moreno et al. 2015) and since then, several viruses and viroid-related diseases have been reported from citron. The two most destructive viroid diseases of Citron are *Exocortis* and *Cachexi*a, causing a huge economic loss in the yield of citron worldwide. Factors like the type of rootstock, viroid, temperature, and infection status with other viroid infection to same host affect the disease epidemiology and severity (Gillings et al. 1991; Hardy 2008). The molecular basis of viroid disease pathogenesis in citron is still not very clear. The messenger RNA differential display technique has been used to demonstrate upon *Citrus viroid III* infections in etrog citron, its reported as the upregulation of genes related to plant defense, signal transduction, and RNA silencing (Tessitori et al. 2007). A microarray analysis revealed that chloroplast, cell wall, peroxidase and symporter activities are altered when etrog citron is infected with Citrus exocortis viroid (CEVd) (Rizza et al. 2012). A global transcriptome analysis of leaves of etrog citron showed that major proteins of Ribose Nucleic Acid (RNA) silencing pathway, basic defense pathway, chitinase activity, and genes of secondary metabolite pathways are upregulated upon CEVd infection (Wang et al. 2019a, b). Using site-directed mutagenesis, two nucleotides were identified in sequence of CEVd which could responsible for modulation of symptom (increased) in etrog citron (Murcia et al. 2009). It has been reported that left terminal domain of chimeric constructs of Citron viroid-V (Cvd-V) are responsible for pathogenicity (Serra et al. 2009). These studies suggest that not only host

Table 7.2 Pathogens, diseases & symptoms affecting citron

S. no	Disease	Causal organism	Symptoms	Transmission	Detection	Management	References
1	Citrus Exocortis Diseases	*Citrus Exocortis Viroid (CEVd)*	Tip browning, leaf epinasty, Stunted growth, reduced yield, Rootstock shelling	Mechanical, infected budwood, Goat horns	Biological indexing on Etrog citron, Polyacrylamide Gel Electrophoresis, Sequential-PAGE, RT-PCR, Multiplex RT-PCR, Imprinthybridization, Northern hybridization	Use of certified budwood, Eradication of infested plants, Citrus by shoot-tip grafting	Rizza et al. (2012)
2	Citrus cachexia disease.	*Hop stunt viroid (HSVd)*	Stunting, chlorosis, bark gumming, stem pitting, decline, and depressions in the wood, split bark	Mechanical, infected budwoods	Biological indexing on Etrog citron, Polyacrylamide Gel Electrophoresis, Sequential-PAGE, RT-PCR, Imprint hybridization	Viroids-free budwoods, quarantine system for new citrus varieties, shoot-tip grafting	Semancik et al. (1988)
3	Citrus dwarfing disease	*Citrus dwarfing viroid (Cvd-III)/CDVd*	Leaf epinasty, petiole ringing, necrosis, leaf drooping, midvein necrosis	Graft- and mechanical	Northern hybridization, RT-PCR, Sequential-PAGE	Budwood certification programs	Belabess et al. (2021)

(continued)

Table 7.2 (continued)

S. no	Disease	Causal organism	Symptoms	Transmission	Detection	Management	References
4	The quick decline of Citrus	*Citrus tristeza virus (CTV)*	Vein clearing, stem pitting, seedling yellows, stunting and leaf corking, quick decline	*T. citricida* Kirkaldy (brown or black citrus aphid)	Immunofluorescence, tissue staining Azure A, Light microscopic observations, Gel immunodiffusion, Enzyme-linked Immunosorbent Assay (ELISA), Electron microscopy, Immunosorbent electron microscopy (ISEM)- Decoration Technique, Molecular hybridization, Biological indexing on *C. macrophylla* and *C. medica* (Etrog)	Quarantine and budwood certification programs, elimination of infected trees, use of tristeza-tolerant rootstocks, or cross-protection with mild isolates	Iftikhar et al. (2021)
5	Citrus yellow vein clearing disease	*Yellow vein clearing (Y.V.C.)*	Vein clearing, leaf crinkling, and water-soaked veins	Whiteflies and Aphis craccivora (*Aaphis craccivora* and *A. spiraecola*)	Immunofluorescence, tissue staining Azure A, Light microscopic observations	Quarantine of infected materials., disease-free budwood, virus-free nursery trees, shoot-tip grafting, heat treatment, prevention and control of field pests	Iftikhar et al. (2021)

6	Citrus canker	*Xanthomonas citri*	Eruptive, corky, and pointed lesions, sometimes exhibiting chlorotic or water-soaked haloes, Hyperplasic spots on both leaf sides, larger corky lesions with centeredcracks on branches and fruits, Defoliation, and premature fall of fruits, Shoots, and eventual plant death, Reduction of the photosynthetic capacity	Citrus leaf miner (*Phyllocnis-tis citrella*)	Based on symptoms, Culture based isolation and identification of bacteria, molecular approaches, RFLP, hybridization, protein profiling	Balanced fertilization; Use of pathogen-free propagating material; Eradication of diseased plants, Preventive spraying with copper fungicide; Control of the citrus leaf miner (*Phyllocnis-tis citrella*); Disinfestation of shoes, equipment, and tools with quaternary ammonium solution	Ference et al. (2018)

(continued)

Table 7.2 (continued)

S. no	Disease	Causal organism	Symptoms	Transmission	Detection	Management	References
7	Huanglongbing	*candidatus liberibacter asiaticus*	Blotchy mottled leaves first single branch; Irregular yellow spots with no clear borders with green tissue; Asymmetry between the two leaf sides; Formation of "green islands" on leaves; Fruits exhibit deformed and asymmetric columella, reduced size, peel thickening, irregular ripening, seed abortion and orange-colored vessels; Leaf abscission	Insect vector (*Diaphorina* sp.)	Based on symptoms observed on plant especially on leaves, Molecular techniques, ELISA, monoclonal antibody assay	Balanced fertilization; Use of pathogen-free propagating material; Planting of windbreak trees; Eradication of symptomatic plants	da Graça and Korsten (2004)

8	Mal-secco	*plenodomus tracheiphilus*	Strands of salmon-pink to orange-red discoloration visible in stem xylem, venial chlorosis, wilt and shedding of leaves, dieback of twigs and branches	Pycnidiospores as infective propagules, rain and wind dissemination	Morphology, PCR and RT-PCR, ELISA, PAGE	Use of disease-free propagation material, Use of tolerant cultivars, Use of copper-based fungicides like zivram, Biological control by *Verticillium dahliae*	Nigro et al. (2011)
9	Phytopthoradiseases	*Phytophthora nicotianae* and *P. citrophthora.*	Gummosis, fruit brown rot, twig and leaf dieback canopy blight, Rot of seedlings, Damping-off, Fibrous root-rot, brown rot	Zoospores, water splash and occasionally by wind, within water drops, Infested nursery stocks	Fruit and leaf baiting method, Selective-media technique, ELISA	Grafting on resistant rootstocks, systemic fungicides metalaxyl and fosetyl-Al, proper irrigation practices e, Hot-water treatment of seeds	Graham and Feichtenberger (2015)

factors but viroid sequences also play an important role in pathogenicity in etrog citron.

Other viroids like *Citrus bent leaf viroid* (CBLVd), *Citrus dwarfing viroid* (CDVd), *Citrus viroid V* (Cvd-V), *Citrus viroid VI* (Cvd-VI), *Citrus viroid VII* (Cvd-VII), *Citrus barkcracking viroid* (CBCVd) have been detected during routine biological indexing studies involving etrog citron as bioamplification and bioindicator hosts. The major symptoms include localized midvein necrosis, mild petiole necrosis, epinasty symptoms, leaf bending, and curling associated with midvein necrosis, bark cracking and gummy pitting symptoms, and pronounced dwarfing. *Citrus dwarfing viroid* (CDVd) has been identified as the causal organism of epinasty and leaf necrosis in etrog citron, however co-infection with CBLVd and CTV results in characteristic "finger imprint" symptoms. This ability of CDVd is utilized in dwarfing trees of high-density citrus orchards to increase yield (Vidalakis et al. 2010). Since etrog citron is the major bioindicator host for many viroids, several viroids, viruses, and their variants have been isolated and characterized using the biological indexing process from this plant (Ito et al. 2002a, b; Chambers et al. 2017) (Table 7.2).

Citrus Tristeza virus (CTV), *Citrus leaf blotch virus* and *Citrus yellow vein clearing virus* are three major viruses affecting etrog citron and causing the quick decline, stem-pitting, and seedling yellow leading to economic losses in the citrus industry (Bové 2006; Vives et al. 2008). Several studies regarding viruses epidemiology, signal transduction and pathogenicity determinants have been published, but these are mostly based on lime which is used for biological indexing of these viruses (Moreno et al. 2008, 2015). Similar studies and research could be established in etrog citron, which will give a better understanding of host-viroid interaction and thus efficient combating strategies in the eradication of these diseases. Deep sequencing of small RNA in infected etrog citron plants allowed the identification of a novel virus provisionally named *Citrus vein enation virus* (CVEV) as the causal agent of citrus vein enation disease (Vives et al. 2013). A comprehensive list of viruses and viroids isolated from etrog plant are listed in the Table 7.3.

7.3.2 Bacterial Diseases of Citron

The two most severe bacterial diseases of Citron are *Citrus canker* and *citrus greening* (Chambers et al. 2017; Bové 2006). The bacterial diseases, causative agents, symptoms, transmission, detection and management details are tabulated in Table 7.2. The pathogen for citrus canker utilizes Type III virulent protein secretion system (Almeida and Nunney 2015), while the pathogen for *Huanglongbing* relies mainly on adhesion and biofilm formation mechanism for pathogenicity (da Graça et al. 2016). Several transporters, transcriptional regulators, and signalling molecules promoting virulence of these bacterial species in Citrus have been characterized (Kay and Bonas 2009). In an attempt to identify the potential of the host to defend against these diseases, genes like att A, CB, and FLS2 have been identified,

Table 7.3 List of viruses and viroids isolated from etrog citron

S. no.	Viroid/viruses isolated	Family	References
Viroids			
1	Hop stunt viroid like-RNA	Pospiviroidae	Sang et al. (1986)
2	Hop stunt viroid (HSV-cit)	Pospiviroidae	Sano et al. (1988)
3	CEVc-I	Pospiviroidae	Semancik et al. (1993)
4	CVd-IIIa, CVd-IIIb and CVd-IIIc	Pospiviroidae	Rakowski et al. (1994)
5	Hop stunt viroid variant	Pospiviroidae	Hsu et al. (1995)
6	CVd-Ib	Pospiviroidae	Sang et al. (1986)
7	CBLVd-225A	Pospiviroidae	Ashulin et al. (1991)
8	CVd-I-LSS	Pospiviroidae	Wei et al. (2017)
9	CBLVd variant E83AK	Pospiviroidae	Wei et al. (2017)
10	CBLVd clone A16-1-1	Pospiviroidae	Cao et al. (2009)
11	Citrus viroid O.S. (CVd-OS)	Pospiviroidae	Ito et al. (2001)
12	CEVd, CBLVd, CVd-I-LSS, CVd-OS, HSVd, CVd-III, and CVd-IV variants	Pospiviroidae	Ito et al. (2002a)
13	33 feild variants of CDVd	Pospiviroidae	Serra et al. (2009)
14	Cachexia (Ca) variants of HSVd-Ca or HSVd-Ca plus CDVd	Pospiviroidae	Eiras et al. (2013)
Viruses			
15	Citrus yellow-vein associated virus (CYVaV)	Alphaflexiviridae	Kwon et al. (2021)
16	Trifoliate citrus bud union associated virus (TCBUaV)	Alphaflexiviridae	Velázquez et al. (2017)

whose over-expression in Citrus confers resistance against citrus canker and/or *Huanglongbing* (Mendonça et al. 2017). Genome editing of susceptible genes in citrus has been emerging reported as an approach to manage citrus canker diseases (Jia et al. 2017). Similar studies aiming identification of susceptible genes, plant recognition receptors, and defense pathway regulator molecular could lead to better management of canker and *Huanlongbin*in citron plants. Other bacterial pathogens affecting Citron are Pseudomonas syringae pv. syringae causing bacterial canker or blast diseases. The main symptoms are black spots on fruits, wilting, and dieback of branches that become more severe during wet winters. The disease is usually managed by pruning of infected branches, fertilization, and Bordeaux mixture application (Xin et al. 2018).

7.3.3 Fungal Diseases of Citron

Citron is infected by most of the fungal diseases which are prevalent in other species of Citrus. The two major fungal disease-causing devastating losses are *Mal-secco* and *Phytophthora diseases*. The *Mal-secco* is a highly destructive disease of Citrus in fungus, which colonizes the vascular tissues of plants, leading to veinal chlorosis, wilt, and water stress. The presence of oxidative stress inducers and membrane transporters in fungus also facilitates the interaction with the host (Khanchouch et al. 2017). Fungal disease infection has reported toxic glycopeptides being involved in pathogenesis (Nachmias et al. 1977). The conventional disease management practices like the use of disease-free propagation material and tolerant varieties are in practice for these diseases. Pre-infection with *Citrus exocortis viroid* (CEVd) could lead to decreased severity and more tolerance towards *P. tracheiphila* infections in mal-secco disease of etrog citron (Solel et al. 1995). Further studies on molecules related to systemic-induced resistance could prove beneficial in disease management in etrog citron.

The second most destructive fungal disease of citron is soil and water-borne Phytophthora diseases, whose major symptoms include gummosis, damping off, and brown rot. In gummosis, lesions first occur on scion, slowly spreading to the cambium and inner wood, leading to cracking in bark and gum accumulation, followed by the slow death of citron. In damping off, the seedling lurches just above the ground due to fungus penetration. In brown rot, decay is accompanied by white mycelium formation on fruit, which spreads later to the canopy under favorable conditions like high moisture and temperature (Graham and Feichtenberger 2015; da Silva et al. 2021). Since Phytophthora infections are transmitted very rapidly through soil and water, and there is a strict requirement for biomonitoring of citron orchards which could lead to early disease diagnosis and its management. Biogenic copper nano-particles have been proved as potential fungicide against Phytopthora (Sawake et al. 2022). Epigenetic response of Phytopthora infection in citrus gummosis is studied and it is observed that DNA methylation strategy can be used as a defence against Phytopthora infections (da Silva et al. 2021). Similar studies targeting scion-rootstock interaction at the epigenetic level are need of the hour in etrog citron orchards (da Graça et al. 2016). Alternaria brown spot caused by *Alternaria alternata*, leaf spot, caused by *Septoria citri* Greasy spot caused by *Zasmidium citr*, Phaeoramularia fruit and leaf spot, caused by *Phaeoramularia angolens* are among emerging and endemic foliar and fruit diseases of several citrus species (Rodrigues da Silva et al. 2021), however their specific occurrence and disease distribution in etrog citron are very scanty.

Apart from these major diseases, CABI reports citron as the host of several other fungal pathogens like *Alternaria citri* causing Stalk end rot, *Aspergillus niger* causing black mold, *Phytoplasma aurantifolia* causing witches broom, *Rhizophus arrhizus* causing bark rot *Sclerotinia sclerotiorum* causing cottony soft rot diseases, *phyllosticta citricarpa* causing leaf spot and *Phomopsis theicola* causing stem and shoot canker diseases (CABI 2022; Kumaran et al. 2008). Another important pathogen is *Colletotrichum boninense* causing dieback of branches and postharvest anthracnose disease of citron, affecting fruit quality and is managed by copper-based fungicides and postharvest application of benzimidazole fungicides (Wang et al. 2019a, b).

7.3.4 *Mealybug Pest of Citrus*

Several species of mealybug have been reported to develop on *Citrus* sp. Citron is considered a major host for citrus fruit borer and scarlet mealybug. Mealybugs are usually considered minor pests of Citrus but provided suitable conditions, they can emerge as major hosts. Chemical and biological control is employed in the management of mealybug infestations. *Citripestis sagittiferella,* also called (citrus fruit borer), feeds internally on fruits causing black or brown lesions and immature dropping of fruit. *Cremastus, Rhoptromeris, Trichogrammatoidea, and Trichogrammatoidea nana* are its natural enemies which are deployed for their biological control of mealybug. *Pseudococcus calceolariae* (scarlet mealybug) feed externally on almost all the parts of the citron and causes characteristics of honeydew or sooty mould symptom. *Diadiplosis laevis* and *Midus pygmaeus* are some of its listed natural enemies of mealybug and being used as a biological control agents in Citron disease management (Franco et al. 2004; CABI 2022).

7.4 Genetic Diversity Studies of Etrog

Citron is a crop of high economic importance, and many inter and intra-species varieties are grown across the globe. However, there exists a lot of confusion among taxonomists and cultivators regarding the origin and true genetic character of Citrus. Several studies throwing light on genetic diversity and evolution and speciation of Citrus have been done (Gill et al. 2022). However, such studies, including accessions of etrog citron, are very few. Amylase polymorphism studies in etrog citron along with 16 other accessions of citrus suggested that within subgenus Citrus, only 3 basic or true species exist as, *Citrus grandis* (pummelo), *Citrus reticulata* (Mandrain) and *Citrus medica* (citron) (Scora 2010). Similarly, inter and intra-generic categorization of species and cultivars were observed in leaf and bark isozymes polymorphism studies, where etrog citron was used as one of the accessions. Etrog citron clustered under citron clade on phylogenetic analysis (Torres et al. 1978; Fang and Roose 1997; Herrero et al. 1996). Using Restriction fragment length polymorphism (RFLP), it was reported that etrog citron has low heterozygosity index suggesting its non-hybridization origin as earlier believed (Federici et al. 1998).

Etrog citron positioned as a separate cluster from other citron varieties studied, suggesting its independent origin from parent *C. medica*. Molecular markers including Random Amplified Polymorphic DNA (RAPD), Sequence Characterized Amplified Region (SCAR) and Chloroplast DNA (cpDNA) have been used to study phylogeny and origin of citrus. These studies support the previous concept of citron being as true species as citron formed a very distant cluster from its related genera. Using SCAR markers, it has also postulated that citron acted a male parent during origin of new species and hybrids like Mexican lime, Bergamot and Volkamer lemon, 'Rangpur' lime, 'Rough' lemon, and 'Palestine' lime (Nicolosi et al. 2000). Similar observation of citron forming a separate mitotype and chlorotype and cluster from other species and contributing as male parent in origin of other varieties have been reported (Froelicher et al. 2010; Curk et al. 2016). Karyotype analysis also regarded Citron as true species on the basis of homozygosity in all the chromosomes (Carvalho et al. 2005). In a similar attempt to find true citron, eight different types of citrons were studied using RAPD markers and it was observed that citrons of Mediterranen basin are phylogenetically very close. Among these etrog citron showed highest similarity index with Diamante citron from Italy (Nicolosi et al. 2005).

Seventeen citron varieties were studied using biochemical (essential oil composition) and molecular (SSR marker) methods, and low heterozygosity was observed in inter-varietal genotypes. Among the etrog genotypes, low genetic distance and low heterozygosity was observed suggesting etrog origin by self-fecundation (Luro et al. 2012). It was speculated that the polyphyletic origin of Mediterranean citron which were during course of evolution diversified due to cross-pollination among varieties and seed-propagation. A comprehensive microsatellite single-nucleotide polymorphism (SNP) marker studies on citron accessions suggested that there is a significant genetic diversity among citron varieties and citrons of Chinese and Mediterranean have different genetic component (Ramadugu et al. 2015). etrog citron formed a separate clade in the neighbour-joining tree based on nuclear SNP heterozygosity, suggesting it is one of the true citrons. Recent reviews regarding genomics studies of citrons. Suggest that more genotypes of citron including accessions from Northeast should be studied to decipher origin of Mediterranean citronie., etrog citron (Goldschmidt 2017).

Being a true species, the citron genome has been explored for parentage association of several inter and intra-varietal hybrids of Citrus. Conflicting studies regarding the parental association of etrog to *C. limonimedica* variety have been reported by different authors (Lota et al. 1999). Some studies regard etrog and *C. limonimedica* as different species, while others report *C. limonimedica* as a hybrid of lemon and etrog citron. Variability and genetic relationship were studied using RAPD fingerprint and ITS polymorphism and *C. limonimedica* was suggested as intermediate species between lemon and citron (Pessina et al. 2011). Biochemical, molecular studies involving SSR, InDel and SNP markers and cytogenetics studies have demonstrated that citron is a male contributor in the genome of all the varieties of lemons and lime (Nicolosi et al. 2000; Curk et al. 2015, 2016; Ramadugu et al. 2013).

The combination with other citrus species was evaluated on a molecular basis and it was suggested that citron as a parental ancestor of *C. aurata*, *C. excelsa*, *C. macrophylla*, and *C. aurantifolia* – 'Mexican' lime types of Tanaka's taxa, *C. limonia*, *C. karna* and *C. jambhiri* varieties of Tanaka's taxa, including popular citrus rootstocks such as 'Rangpur' lime, 'Volkamer' and 'Rough' lemons. *C. limetta* and *C. limon* – yellow lemon types – varieties of Tanaka's taxa, *C. limettioides* – 'Palestine sweet' lime types – and *C. meyeri*. Seven triploid limes, namely *C. latifolia* ('Tahiti' lime-like accessions) ('Tanepao,' 'Coppenrath,' 'Ambilobe' and 'Mohtasseb' limes, and 'Madagascar' lemon) contains genome contributed from all basic species of Citrus. *C. medica* alleles were identified in all these accessions of lime by SNP array from 454 amplicons sequencing (Curk et al. 2015, 2016).

C. medica tuberosa Risso & Poiteau is mostly cultivated in Baronia province. Earlier morphological studies found similarities in among Pompia (*C. medica*) and citron. The Pompia admixture analysis were further investigated at molecular level using multilocus marker (Carvalho et al. 2005), AFLP, ISSR, and RAPD (Ilaria Mignani et al. 2015) observing Pompia as lemon X Citron variety (Viglietti et al. 2019). ITS barcoding and cytoplasmic sequencing also revealed citron as a genetic contributor in Pompia (Liu et al. 2021; Luro et al. 2019). A study in 2019 mentioned the parentage of Pompia by nuclear and cytoplasmic SSR and InDel markers and identified Etrog citron as a male parent of Pompia (Leinonen et al. 2010). Further comprehensive molecular studies involving NGS will help in addressing the morphological and biochemical distinction of Pompia from its parents and confirmation of its inter-varietal origin.

7.4.1 Omics Studies of Etrog

The first draft genome of citron using Illumina shotgun sequencing method is available in the "Citrus Genome Database" with genome size 407 Mb and 32,579 genes (Leinonen et al. 2010). 25 germplasm has been studied under this species, among which 06 have been categorized specifically for etrog citron variety labelled as "Breeding_Research_Material". In NCBI, *C. medica* can be linked to 24 different Bio projects, of which 05 mentioned etrog Citron as a "Biosample". Most of the studies published are related to plant-viroid interaction and disease progression in Etrog citron, as this plant is a natural host for citrus viroids and is considered an indicator plant for viroid infection. One SRA data aiming analysis of interspecific admixture among secondary species and modern varieties by Genotyping by Sequencing (GBS) is available for etrog citron. Whole genome sequencing data using Illumina HiSeq 2500 platform targeting Single Nucleotide Polymorphism of different citrons linked to etrog citron are available (Wang et al. 2017; Li et al. 2020).

7.5 Traditional Uses

The etrog plant is considered to have immense medicinal properties, and since ancient times, people have extensively used this plant in their daily life to treat diseases. Various parts of the plant are used to treat ailments, and it is considered a "goodly tree" by Jews due to its immense medicinal properties (Arias and Ramón-Laca 2005). A compilation of reported traditional uses of etrog is listed in Table 7.4.

7.5.1 Seed

Few research articles mentioned that the extracts and preparations of seeds are valued for their anti-toxicity activity (Arias and Ramón-Laca 2005; Morton 1987). In the book "Treatise on healing," Maimonides (thirteenth century) mentioned that fresh or dried seeds along with poultice and colocynth root as an effective remedy against scorpion poison (Rosner 1968). The extracts of seeds are used against parasitic worm infections as a powerful vermifuge (Rosner 1968; Panara 2012). Jewish leader Rabbi Yoshef Haim from Baghdad has mentioned that the seeds are beneficial for the whole body Sarah (n.d).

7.5.2 Fruits

Fruit extracts and preparations are considered to have anti-toxic activity (Arias and Ramón-Laca 2005; Morton 1987). The whole fruit has insect repellent activity. Decoction of the fruit is considered to ward off "evil spirits" in Malaysia. People in West tropical Africa use the whole fruit as medicine and to treat rheumatism in particular. It is considered a remedy for seasickness, pulmonary troubles, intestinal ailments, and other diseases in ancient times and in the middle ages (Morton 1987). Whole fruit helps in the reduction of symptoms of memory disease (Conforti et al. 2007; Entezari et al. 2009). Jams made from the etrog fruits used during the sukkot festival are considered to ease child birth (Sofer 1969; Sperling 1982). Fruits of the etrog are considered to have special powers related to fertility and childbirth (Sperling 1982; The Saga of the Citron 2012). To have healthy child the pregnant women are advised to have etrog fruit after the sukkot festival (The Saga of the Citron 2012). The tip of fruit commonly called "pitam" by Jews is recommended to be consumed by childless women to bear child (Sperling 1982; The Saga of the Citron 2012). To ease child labour pain, the Jewish women were advised to keep the fruit tip (pitam) under the pillow (The Saga of the Citron 2012).

Table 7.4 List of used plant parts and their traditional claims

S. no	Plant part	Traditional claim	References
1	Seed	Anti-toxic activity	Arias and Ramón-Laca (2005) and Morton (1987)
		Scorpion poison treatment	Rosner (1968)
		Vermifuge	Arias and Ramón-Laca (2005) and Panara (2012)
		Whole body wellness	Sarah (n.d.)
2	Fruits	Anti-toxic activity	Arias and Ramón-Laca (2005) and Morton (1987)
		Insect repellent	Morton (1987)
		Warding off evil spirits	Morton (1987)
		Rheumatism	Morton (1987)
		Seasickness	Morton (1987)
		Pulmonary troubles	Morton (1987)
		Intestinal ailments	Morton (1987)
		Memory disease	Arias and Ramón-Laca (2005)
		Ease childbirth	Morton (1987), Sofer (1969), and Sperling (1982)
		Fertility	Sperling (1982) and The Saga of the Citron (2012)
		Healthy child at birth	The Saga of the Citron (2012)
		Labour pain easing	The Saga of the Citron (2012)
3	Peel	Stomachic	Morton (1987)
		Stimulant	Morton (1987)
		Expectorant	Morton (1987)
		Anti-dysentery	Morton (1987)
		Halitosis treatment	Morton (1987)
		Alleviate pain	Conforti et al. (2007)
		Strengthen tendons	Conforti et al. (2007)
		Ease childbirth	Sofer (1969) and Sperling (1982)
		Digestion aid	Grant (1997) and Kaufman (2006)
		Stomach strengthen	Grant (1997) and Kaufman (2006)
4	Mesocarp/Fleshy portion	Body nourishment	Grant (1997) and Kaufman (2006)
5	Fruit juice	Poison counteracting	Morton (1987)
		Improve liver function	Levey (1973), Suessmann Muntner (1955), and Fermo (1961)
		Improve heart function	Levey (1973), Suessmann Muntner (1955), and Fermo (1961)
		Eyewash	Levey (1973), Suessmann Muntner (1955), and Fermo (1961)

(continued)

Table 7.4 (continued)

S. no	Plant part	Traditional claim	References
6	Albedo	Digestion improvement	Etrog Citron (n.d.-a)
7	Essential oil	Antiseptic	Morton (1987)
		Antibiotic	Morton (1987)
		Insect repellent	Morton (1987)
8	Leaves	Vermifuge	Arias and Ramón-Laca (2005) and Panara (2012)
		Antispasmodic activity	Morton (1987)
		Memory disease	Conforti et al. (2007)
9	Flowers	Herbal medicines	Morton (1987)

7.5.3 *Peel*

In China, the candied peels are sold as stomachic, stimulant, expectorant, and tonic (Grant 1997). The peel is considered to be an effective medicine against dysentery and halitosis in India (Kaufman 2006). The peel is brewed like tea and taken along with sugar or honey to alleviate pain and is considered to strengthen tendons (Arias and Ramón-Laca 2005). Fruit peels in the raw or sugared form are considered to ease child birth (Sofer 1969; Sperling 1982). The peel is considered to aid in digestion and stomach strengthening if eaten in small quantities (Grant 1997; Kaufman 2006).

7.5.4 *Mesocarp/Fleshy Portion*

The inedible middle fleshy, white portion of the fruit is placed in vinegar and consumed later to provide nourishment to the body (Grant 1997; Kaufman 2006).

7.5.5 *Fruit Juice*

The fruit juices in the distilled form are considered to have sedative activity (Morton 1987). Saladin d'ascoli Salerno, a twelfth-century scholar, has mentioned that fruit juice has the following medicinal uses such as counteracting, improves liver and heart function, and as an ingredient in eyewash (Levey 1973; Suessmann Muntner 1955; Fermo 1961).

7.5.6 *Albedo*

The albedo presents in the epicarp of the fruit aids in digestion improvement (Fermo 1961).

7.5.7 *Essential Oil*

The essential oil extracted from the etrog citron is used for antiseptic and antibiotic purposes (Morton 1987). The aromatic oil has insect repellent activity, and the fruits are placed under clothes to repel moths (Morton 1987).

7.5.8 *Leaves*

The leaves are considered to have beneficial medicinal values such as vermifuge activity (Arias and Ramón-Laca 2005; Panara 2012) and alleviate the symptoms of memory disease (Conforti et al. 2007). Infusions of the leaves are considered to have antispasmodic activity (Morton 1987).

7.5.9 *Flowers*

Chinese consider the flowers of etrog to have high medicinal value, and they are included in various herbal remedies (Morton 1987).

7.6 Nutritional Values and Phytochemistry

The nutritional composition of etrog fruits is moisture, protein, fat, fiber (pectin), calcium, phosphorus, iron, thiamine, riboflavin, niacin, vitamin C, magnesium, potassium, copper, manganese, zinc, chromium, and other trace elemets (Etrog Citron – Specialty Produce n.d.-a; Morton 1987).

The etrog is one of the few less explored plants, and only a handful of studies have reported its phyto-constituents. Complete phytochemical screening of the plant is necessary. A compiled list of reported phytoconstituents is given in Table 7.5. Some of important phytoconstituents of etrog structures illustrated in Figs. 7.3 and 7.4.

Table 7.5 List of reported phytoconstituents present in etrog citron

S.No	Constituents	Plant part			References
		Leaves	Fruit peel	Root	
1	Limonene	✓	✓	–	Zhenia and Alexander (1991), Vekiari et al. (2002), and Meena et al. (2011)
2	Geranial	✓	✓	–	Zhenia and Alexander (1991), Vekiari et al. (2002), and Meena et al. (2011)
3	Geraniol	✓	✓	–	Zhenia and Alexander (1991), Vekiari et al. (2002), and Meena et al. (2011)
4	Neral	✓	✓	–	Zhenia and Alexander (1991), Vekiari et al. (2002), and Meena et al. (2011)
5	Nerol	✓	✓	–	Zhenia and Alexander (1991), Vekiari et al. (2002), and Meena et al. (2011)
6	n-pentanal	✓	–	–	Zhenia and Alexander (1991)
7	2-methyl-3-butene-2-ol	✓	–	–	Zhenia and Alexander (1991)
8	α-pinene	✓	✓	–	Zhenia and Alexander (1991) and Vekiari et al. (2002)
9	Methyl benzene	–	✓	–	Zhenia and Alexander (1991)
10	Camphene	✓	✓	–	Zhenia and Alexander (1991)
11	Hexanal	✓	✓	–	Zhenia and Alexander (1991)
12	β-pinene	✓	✓	–	Zhenia and Alexander (1991) and Vekiari et al. (2002)
13	1,1 –diethoxy-2-methylbutan	✓	–	–	Zhenia and Alexander (1991)
14	Sabinene	✓	✓	–	Zhenia and Alexander (1991)
15	Ethyl benzene	✓	✓	–	Zhenia and Alexander (1991)
16	1 -penten-3-01	✓	–	–	Zhenia and Alexander (1991)
17	δ-3-carene	✓	✓	–	Zhenia and Alexander (1991) and Vekiari et al. (2002)
18	Myrcene	✓	✓	–	Zhenia and Alexander (1991) and Vekiari et al. (2002)
19	α -phellandrene	✓	✓	–	Zhenia and Alexander (1991)
20	α -terpinene	✓	✓	–	Zhenia and Alexander (1991) and Vekiari et al. (2002)
21	β-phellandrene	✓	✓	–	Zhenia and Alexander (1991)
22	1,8-cineol	✓	✓	–	Zhenia and Alexander (1991)
23	Cis- β -ocimene	✓	✓	–	Zhenia and Alexander (1991) and Vekiari et al. (2002)
24	Trans- β -ocimene	✓	✓	–	Zhenia and Alexander (1991) and Vekiari et al. (2002)
25	γ -terpinene	✓	✓	–	Zhenia and Alexander (1991) and Vekiari et al. (2002)
26	Hexyl acetate	–	✓	–	Zhenia and Alexander (1991)

(continued)

Table 7.5 (continued)

S.No	Constituents	Plant part			References
		Leaves	Fruit peel	Root	
27	p-cymene	✓	✓	–	Zhenia and Alexander (1991) and Hetta et al. (2013)
28	Terpinolene	✓	✓	–	Zhenia and Alexander (1991) and Vekiari et al. (2002)
29	(Z)-3-hexenyl acetate	✓	–	–	Zhenia and Alexander (1991)
30	Octanal	–	✓	–	Zhenia and Alexander (1991)
31	6-methyl-5- hepten-2-one	–	✓	–	Zhenia and Alexander (1991)
32	5-methyl heptenone	✓	✓	–	Zhenia and Alexander (1991)
33	Hexanol	–	✓	–	Zhenia and Alexander (1991)
34	(Z)-3-hexenol	✓	✓	–	Zhenia and Alexander (1991)
35	Nonanal	✓	✓	–	Zhenia and Alexander (1991)
36	4-methyl hexanol	–	✓	–	Zhenia and Alexander (1991)
37	Ethyl octanoate	✓	–	–	Zhenia and Alexander (1991)
38	2-cyclohexenone	✓	–	–	Zhenia and Alexander (1991)
39	A limonene oxide	✓	✓	–	Zhenia and Alexander (1991)
40	Trans-sabinene hydrate	✓	✓	–	Zhenia and Alexander (1991)
41	2-ethyl furan	✓	–	–	Zhenia and Alexander (1991)
42	Citronella1	✓	✓	–	Zhenia and Alexander (1991) and Vekiari et al. (2002)
43	2,4-heptadienal	✓	–	–	Zhenia and Alexander (1991)
44	Decanal	–	✓	–	Zhenia and Alexander (1991)
45	Benzaldehyde	✓	✓	–	Zhenia and Alexander (1991)
46	Linalool	✓	✓	–	Zhenia and Alexander (1991) and Vekiari et al. (2002)
47	Cis-sabinene hydrate	✓	✓	–	Zhenia and Alexander (1991)
48	Linalyl acetate	✓	–	–	Zhenia and Alexander (1991)
49	Cis- β -bergamotene	✓	✓	–	Zhenia and Alexander (1991) and Vekiari et al. (2002)
50	Trans- β -bergamote	✓	✓	–	Zhenia and Alexander (1991) and Vekiari et al. (2002)
51	Undecanal	✓	–	–	121, 124
52	Terpinene-4-01	✓	✓	–	Zhenia and Alexander (1991) and Vekiari et al. (2002)
53	β -caryophyllene	✓	✓	–	Zhenia and Alexander (1991) and Vekiari et al. (2002)
54	Aromadendrene	–	✓	–	Zhenia and Alexander (1991)
55	Citronellyl propionate	–	✓	–	Zhenia and Alexander (1991)
56	Ethyl caprate	✓	–	–	Zhenia and Alexander (1991)
57	Nonanol	✓	–	–	Zhenia and Alexander (1991)

(continued)

Table 7.5 (continued)

S.No	Constituents	Plant part			References
		Leaves	Fruit peel	Root	
58	α -humulene	✓	✓	–	Zhenia and Alexander (1991) and Vekiari et al. (2002)
59	Neral	✓	✓	–	Zhenia and Alexander (1991) and Vekiari et al. (2002)
60	α -terpineol	✓	✓	–	Zhenia and Alexander (1991) and Vekiari et al. (2002)
61	Neryl acetate	✓	✓	–	Zhenia and Alexander (1991) and Vekiari et al. (2002)
62	β -bisabolene	✓	–	–	Zhenia and Alexander (1991)
63	Geranial	✓	✓	–	Zhenia and Alexander (1991) and Vekiari et al. (2002)
64	Bicyclogermacrene	✓	–	–	Zhenia and Alexander (1991)
65	Geranyl acetate	✓	✓	–	Zhenia and Alexander (1991)
66	Citronellol	✓	✓	–	Zhenia and Alexander (1991) and Vekiari et al. (2002)
67	β -bisabolene	–	✓	–	Zhenia and Alexander (1991)
68	γ -isogeraniol	–	✓	–	Zhenia and Alexander (1991)
69	Perillaldehyde	–	✓	–	Zhenia and Alexander (1991)
70	Ethyl phenylacetate	✓	–	–	Zhenia and Alexander (1991)
71	Anethole	✓	–	–	Zhenia and Alexander (1991)
72	Tetradecanal	–	✓	–	Zhenia and Alexander (1991)
73	Benzyl alcohol	–	✓	–	Zhenia and Alexander (1991)
74	Undecanol	✓	–	–	Zhenia and Alexander (1991)
75	Cis-carveol	✓	✓	–	Zhenia and Alexander (1991)
76	Trans-carveol	✓	–	–	Zhenia and Alexander (1991)
77	Phenethyl alcohol	✓	✓	–	Zhenia and Alexander (1991)
78	Pentadecanal	–	✓	–	Zhenia and Alexander (1991)
79	β -ionone		✓	–	Zhenia and Alexander (1991)
80	α- humulene oxide	✓	–	–	Zhenia and Alexander (1991)
81	Caryophyllene oxide	✓	–	–	Zhenia and Alexander (1991) and Vekiari et al. (2002)
82	(E)-nerolidol	✓	✓	–	Zhenia and Alexander (1991)
83	Hexadecimal	–	✓	–	Zhenia and Alexander (1991)
84	Heptadecanal	–	✓	–	Zhenia and Alexander (1991)
85	Ethyl myristate	✓		–	Zhenia and Alexander (1991)
86	β -bisabolol	✓	✓	–	Zhenia and Alexander (1991)
87	Ethyl pentadecanoate	✓	–	–	Zhenia and Alexander (1991)
88	Ethyl palmitate	✓	–	–	Zhenia and Alexander (1991)
89	Ethyl heptadecanoate	✓	–	–	Zhenia and Alexander (1991)

(continued)

Table 7.5 (continued)

S.No	Constituents	Plant part: Leaves	Plant part: Fruit peel	Plant part: Root	References
90	Neric acid	✓	–	–	Zhenia and Alexander (1991)
91	2,3-dihydrobenzofurane	✓	–	–	Zhenia and Alexander (1991)
92	Ethyl stearate	✓	–	–	Zhenia and Alexander (1991)
93	14-pentadecenoic acid	✓	–	–	Zhenia and Alexander (1991)
94	Ethyl linoleate	✓	–	–	Zhenia and Alexander (1991)
95	Phytol	✓	–	–	Zhenia and Alexander (1991)
96	Farnasene	✓	✓	–	Vekiari et al. (2002)
97	Citronellyl acetate	✓	✓	–	Vekiari et al. (2002)
98	Neryl propionate	✓	✓	–	Vekiari et al. (2002)
99	Geranyl propionate	✓	✓	–	Vekiari et al. (2002)
100	Iso-citral cis	✓	✓	–	Vekiari et al. (2002)
101	Iso-citral trans	✓	✓	–	Vekiari et al. (2002)
102	1-β-sitosterol-glucoside	✓	–	–	Hetta et al. (2013)
103	Sakuranetin	✓	–	–	Hetta et al. (2013)
104	7-O-methyl aromadendrin	✓	–	–	Hetta et al. (2013)
105	Dihydrokaem-pferide	✓	–	–	Hetta et al. (2013)
106	Hesperidin	✓	✓	–	Venturini et al. (2014) and Douglas and Walker (1983)
107	Rutin	✓	✓	–	Venturini et al. (2014) and Douglas and Walker (1983)
108	4-Desmethylsterol	–	–	✓	Douglas and Walker (1983)
109	Sitosterol	–	–	✓	Douglas and Walker (1983)
110	Stigmasterol	–	–	✓	Douglas and Walker (1983)
111	Campesterol	–	–	✓	Douglas and Walker (1983)
112	Cholesterol	–	–	✓	Douglas and Walker (1983)
113	6-methyl-hept-5-en-2-one	✓	✓	–	Luro et al. (2020)
114	Carvone	✓	✓	–	Luro et al. (2020)
115	α-copaene	✓	✓	–	Luro et al. (2020)
116	Bicyclogermacrene	✓	✓	–	Luro et al. (2020)
117	τ-cadinol	✓	✓	–	Luro et al. (2020)
118	Octanol	✓	✓	–	Luro et al. (2020)
119	Octyl acetate	✓	✓	–	Luro et al. (2020)

7.6.1 Leaves

A yield of 0.25% of essential oil was obtained from the leaves of etrog citron. The GC-MS analysis of the extracted essential oil showed five major phytoconstituents limonene (37.93%), citral (8.22%), geraniol (6.21%), neral (5.11%), and nerol (4.83%) (Zhenia and Alexander 1991).

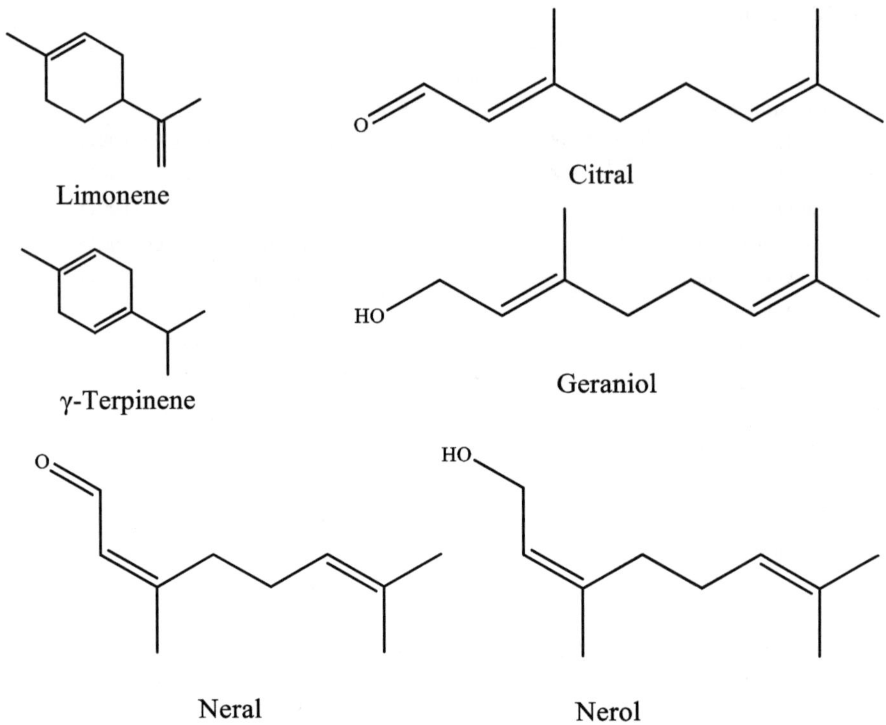

Fig. 7.3 IMajor essential oil constituents of etrog

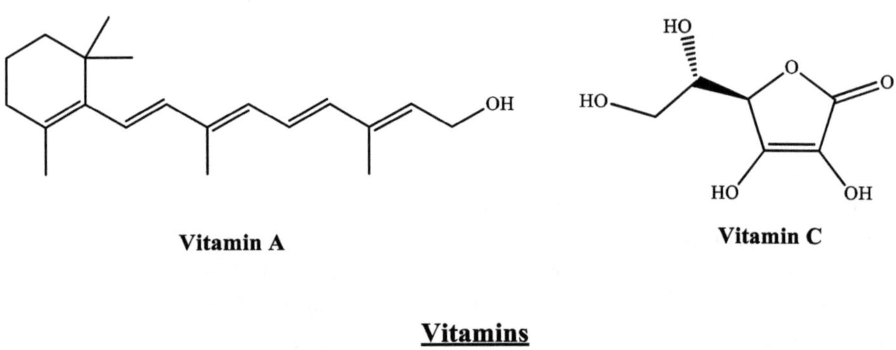

Fig. 7.4 Major vitamins present in etrog

7.6.2 Fruit Peel

A yield of 0.18% of essential oil was obtained from the fruit peel of etrog citron. The GC-MS analysis of the extracted essential oil showed five major phytoconstituents limonene (81.34%), **γ** –terpinene (4.54%), citral (4.24%), neral (2.55%), and myrcene (2.01%) (Zhenia and Alexander 1991; Vekiari et al. 2002). The other noticeable phytoconstituents like iso-limonene, phenolics, flavanones, and pectin (Zhenia and Alexander 1991; Vekiari et al. 2002; Meena et al. 2011) are present in the fruit peel. Traces of vitamin A and Vitamin E are also found in the peel (Luro et al. 2020). The majority of the fruit comprises of peel along with pith, and pectin is the major constituent present in it (Thakur et al. 1997).

7.6.3 Root

The seedlings of the plant were taken for the study, and the fibrous roots were analyzed for phytoconstituents using Gas-liquid chromatography. GC-MS analysis showed the presence of 4-desmethylsterol, sitosterol, stigmasterol and cholesterolin significant quantities (Douglas and Walker 1983).

7.6.4 Fruit Pulp

The fruit pulp comprises of fleshy pith and juicy endocarp. Vitamin-C (Shemesh 2012) and pectin (Thakur et al. 1997; M.N.I. Bhuiyan et al. 2009; Rafiq et al. 2016) are the major phytoconstituents present. Folic acid is found in minute quantities in the fruit juices of etrog citron (Sanofer 2014).

7.7 Pharmacology

Despite the fact that etrog citron has been used in various traditional medicines since ancient times, the plant is less explored for its biological activity. This lack of therapeutic exploration for bio-activity screening paves a potential research possibility for various disease treatments. Hence, this portion of the chapter deals with the reported pharmacological activities of the plant and pharmacological activities correlated with the major phytoconstituents present in the plant.

The 70% aqueous methanol extract of the leaves showed anti-hyperglycaemic activity. The extracts are orally fed in different doses (200 mg/kg & 400 mg/kg), and the drug Gliclazide is given as a treatment group. The diabetic rats were monitored for blood glucose levels, and serum lipid levels and histopathological examinations. Their blood glucose levels, serum lipids such as high-density lipoproteins (HDL), low-density lipoproteins (LDL), cholesterol, and triglycerides were significantly reduced in the treatment group. The histopathological study of Liver and kidney of the treatment group in comparison with the control group showed prevention of atrophy and necrosis of β- cells in islets of Langerhan's (Hetta et al. 2013). Preclinical and clinical evaluation of fruit juice and extracts have anticancer activity (Cirmi et al. 2017).

Ascorbic acid or commonly called Vitamin C is commonly found in citrus fruits and is reported for various pharmacological activities. Vitamin C is used in cancer treatments and cancer preventive therapy by acting as an effective antioxidant, reducing agents (Du et al. 2012; Shenoy et al. 2018), antitumor activity, cytotoxicity, antiproliferative activity, and apoptosis induction (Shenoy et al. 2018). Ascorbic acid is an essential requirement for normal physiological functions and plays an important role in the synthesis and metabolism of tyrosine, folic acid, hydroxylation of glycine, proline, lysine, carnitine, tryptophan, and catecholamine. Blood cholesterol levels are also lowered by converting them into bile acids (Chambial et al. 2013). It plays a vital role in maintaining skin health through various pharmacological activities such as promotion of collagen formation, scavenging free radicals and disposal of toxic oxidants, inhibition of melanogenesis, skin aging prevention, inhibiting U.V. radiation, preventing photo-aging, dry skin management, wrinkle management, wound healing and skin inflammation reduction (Pullar et al. 2017). Barrier integrity, wound healing, leukocyte function, and antimicrobial activity are the major immune functions that depend on Vitamin C for maintaining body homeostasis (Hemilä 2017).

Pectin, a polysaccharide polymer present in etrog fruit, has various therapeutic properties, and it is used in modern medicine. The pharmacological activities such as immune regulatory, anti-inflammatory, antibacterial, antioxidant (Minzanova et al. 2018), probiotic (Tan and Nie 2020), hypoglycemic (Brouns et al. 2012), hypocholesterolemia (Brouns et al. 2012; Zhu et al. 2017), anticancer (Bergman et al. 2010) and antidiarrheal (Behall and Reiser 1986) are reported.

Citral, a monoterpene aldehyde found in the etrog citron plant, exhibits various therapeutic activities like anti-microbial, anti-oxidant, anti-diabetic (Sharma et al. 2021), anticancer (Sharma et al. 2021; Bailly 2020), cytotoxicity (Bailly 2020), anti-inflammatory and enhancing skin health (Habib et al. 2021).

Limonoids, like D-limonene present in the fruit and leaves has beneficial effects like antioxidant, antidiabetic, anticancer, anti-inflammatory, cardioprotective, gastroprotective, hepatoprotective, immune modulatory, anti-fibrotic, anti-genotoxic (Sun 2007), gallstone dissolution, heartburn, gastroesophageal reflux treatment (Vieira et al. 2018), anti-hyperalgesic, antinociceptive, neuroprotective (Eddin et al. 2021), antiviral, antibacterial, larvicidal, insecticidal, analgesic and anti-hyperglycaemic activities (Gualdani et al. 2016). Limonene has been reported to have therapeutic activities against neurodegenerative diseases like Alzheimer, multiple sclerosis, epilepsy, anxiety, and stroke (Eddin et al. 2021).

Monoterpenoids such as geraniol, geranial, nerol, and neral are reported to have therapeutic activities such as anticancer (Silva et al. 2022), antibacterial (Mączka et al. 2020; Galappathie et al. 2017), antioxidant, antinflammatory (Mączka et al. 2020; Ye et al. 2019), antifungal (Tian et al. 2017), nematicidal (Abdel-Rahman et al. 2013), hepatoprotective (Islam et al. 2021), antimicrobial (She et al. 2020), antibiofilm (Qian et al. 2020), cardioprotective, and neuroprotective (Lei et al. 2019).

7.8 Economic Importance and Commercial Products

The etrog citron is one of the four species used in the Hebrew ritual "sukkot," and it is most prized among Jews. In Israel, the price of the etrog is around 5–100$ depending upon its quality and suitability for the ritual (Nosowitz 2018). The price of fruits grown traditionally by Jewish farmers in Israel will go up to as high as 350$ a piece (Nosowitz 2018). The Jews settled in different parts of the world prefer traditionally grown etrog citron in Israel due to the strict adherence to the Jewish customs and preferably fruits grown in Eretz Yisrael, the land of Israel (Etrog Citron n.d.-b).

7.8.1 Whole Fruits

Etrog fruits are used in making confectioneries like desserts, jams Etrog Desserts (n.d.), marmalades (Rivka 2007), pies, cakes, icing, puddings, jellies, preserves, cookies (Florence Kreisler Greenbaum 1918) and canned fruits (Rivka 2007; Florence Kreisler Greenbaum 1918). This pickled etrog is considered to have medicinal properties by Greeks (Brigand and Nahon 2016) and ancient people (Strauss 2012), which are consumed to date. Fresh fruits are used to infuse flavour into alcoholic drinks like vodka (Kitchens 2019) and other alcoholic drinks to make cocktails like etrog liqueur (Forward 2021) and etrog limoncello (Forer 2018). In the aromatic industry, the fruit is used to make perfumes and scented balls, namely Pomander balls (Forer 2018). A company named ARQUISTE Parfumeur commercially sells perfumes made from etrog fruit under the name "L'ETROG" (ARQUISTE Parfumeur n.d.). Commercially, a liqueur brand named "SUKKAH HILL SPIRITS" uses etrog fruits for flavouring their hand crafted spirits (Sukkah Hill Spirits n.d.).

7.8.2 Peels

The pith of the etrog fruit is larger in comparison with the juicy pulp, and the juice obtained from the fruits is less in comparison with other citrus fruits. The peels are candied in sugars and are sold commercially (Etrog Desserts n.d.; Rivka 2007; Florence Kreisler Greenbaum 1918). The peels are rich in essential oils, and they are used in food flavouring, food additives, cosmetics, Aromatherapy, and perfumes (Palazzolo et al. 2013).

7.8.3 *Essential Oils*

The peels, flowers, and leaves have abundant oil glands which secrete essential oils (Palazzolo et al. 2013). The essential oils are extracted traditionally by hydro-distillation, and modern methods include solvent extraction, microwave-assisted extraction, and supercritical fluid techniques (Bora et al. 2020). The extracted essential oils are used in perfumes, food flavouring, cosmetics, food preservatives, and food packaging (Neelima Mahato et al. 2017).

7.8.4 *Pectin*

Pectin is found abundant in peel and pith of citrus fruits. Pectin has wide applications in the food industry (Erich Isaac 1959), pharmaceuticalsand biomedical industries (Padma Ishwarya et al. 2022). Pectin has wide applications in food industries as an additive in food products like jams, jellies, preserves, conserves, confectionery, beverages and barbecue sauce (Padma Ishwarya et al. 2022). In the pharmaceutical industry, pectin is widely used in emulsion stabilizer, bio films (Li et al. 2021), Hemostatic, hypoglycaemic agent (Munarin et al. 2012), bulking agent (Akin-Ajani and Okunlola 2021), hydrogels, and excipients in formulations (Groult et al. 2021).

7.9 Conclusion

The etrog citron, scientifically called *Citrus medica* var. ethrog Engl is an evergreen shrub or a small tree growing in sub-tropical tropical parts of the world. In Judaism, the plant is considered one of the four species and is sought highly by Jews for their traditional uses in the religious ritual “sukkot”. The plant is mainly cultivated by Jews and strictly grown, abiding by the needs of Jewish religious practices. Species of pure breeds are considered worthy in Jewish ritual, and any hybrids are considered misfits for the religious demands. So, the vegetative propagation techniques are used for multiplication of etrog in orchards, where only the rootstocks of etrog are specifically used. The fruits with certain features are desired, and the maintenance of such physical appearance is necessary for the sukkot festival. To preserve perfect fruits, postharvest management studies have been carried out.

Etrog citron is host to a number of major and minor pathogenic organisms like viroids, viruses, bacteria, fungi and mealybug which cause huge devastation of crop under cultivation. Studies using recent technologies such as transcriptomics and microarray have been done to explore the molecular mechanism of diseases progression and defence gene expression in etrog. Multiple studies regarding prevention, control and treatment of these vulnerable diseases have been carried out but still the conventional approaches like use of certified budwood, shoot-tip grafting and eradication of infested plants are in practise for their management. Various studies have established etrog citron as a suitable bioindicator species for viruses and viroid infection and thus etrog is used not only for biological indexing of viral diseases but also in isolation of new pathogenic viroids of citrus.

Marker-based studies have been carried out to decipher the origin and phylogeny of etrog citron. Such studies have established etrog citron as one of the true or basic species among citrus genus, from which several lemon and lime varieties have evolved. Etrog has been established as male genetic contributor of Pompia and Florence lemon using not only classical molecular techniques like RAPD but also modern molecular techniques like 454 amplicon sequencing. Studies on the genetic diversity/purity of the plants have been carried out since the Jewish ritual demands the fruits to be genetically pure and there is a long debate in scientific as well as religious communities regarding origin and evolution of etrog. Earlier studies consider Mediterranean etrog to be originated from China or North-east India but recent phylogenomic studies have discarded this monophyletic theory of origin. The morphological, biochemical and genetic distinction among different citron varieties and endemic cultivars suggests a polyphyletic origin.

The etrog is considered to have high medicinal properties, and it is traditionally used by people all over the world. Etrog has a deep-rooted relationship with the ancient people belonging to different geographical regions and religions. In scriptures of Judaism, the etrog is recorded to have various therapeutic values. From China to the Middle East regions, the etrog is widely used by populations as traditional medicine. Every part of the plant is considered to treat various diseases. Even though the plant is well used in traditional medicines, the therapeutic activities of such claims are not yet properly established through pharmacological screening. The essential oils obtained from leaves and peels are analyzed by GC-MS, and their composition is reported. The essential oils obtained from the peels and leaves are used in perfumes, food flavoring, food preservatives, beverages, and other commercial products.

7.10 Future Perspectives

Regardless of the economic importance of etrog citron, not many scientific studies have been published regarding its cultivation practices, pharmacological activities, and disease management. To prevent damage of trees and maintain fruit quality, unsustainable use of insecticides and pesticides is being done by the cultivators. These types of practises certainly demand a better alternative approach in cultivation. Use of biotechnological interventions like plant tissue culture can be established for raising diseases- resistant saplings of etrog or virus-free propagules.

Several phytochemicals have been reported from etrog, but the biological activities of these phytochemicals have not been screened *in vitro* or *in vivo* biological models. Pharmacological studies targeting the bio-activity of these phytochemicals and its underlying mechanism will augment the traditional medicinal claims. The secondary metabolic pathway of essential oils biosynthesis in case of etrog have not been fully explored and future research exploring these pathways leading to metabolic engineering of essential oils holds a great promise in enhancing the commercial exploitation.

The etrog is vulnerable to many viral, fungal and bacterial diseases of Citrus, causing huge losses to cultivators. The reports regarding disease progression and pathogenicity are very limited and few hosts specific genes like defence-related or RNA silencing pathways have been identified from etrog. Functional characterization of these defence related genes and their regulatory factors like transcription factors can help in establishing a molecular tool in tackling these devastating diseases and thus their better management in etrog citron.

Since many cultivars and varieties of citron are morphologically very similar, and differences can be observed only at the genetic level, there is a requirement to create citron germplasm resources. Very little effort has been made to document the different germplasm of pure citron and related hybrids and cultivars. Such resources will strengthen future studies regarding the origin, phylogenetic relationship, and diversity of citron, particularly etrog citron, where pure germplasm is a matter of commercial importance. Many studies focussing on genetic origin of etrog citron have been done previously, but the exact origin of etrog citron and its course of evolution is still unmasked. Future phylogenomic studies including larger number of etrog germplasm accessions across different geographic regions and other varieties of citron and citrus will give a better glimpse of true origin of etrog.

Genetic studies have established etrog citron as male parent of several lemon and lime varieties, but extensive and in-depth study regarding the distinction of these varieties from their parent species in terms of biochemistry and gene expression/ genome or metabolome are still lacking. Further multi-disciplinary study including genomics, transcriptomics, proteomics and metabolomicswill pave new ways for therapeutic and commercial exploitation of these varieties. The analysis of genetic diversity will further help in conservation efforts.

References

Abdel-Rahman FH, Alaniz NM, Saleh MA (2013) Nematicidal activity of terpenoids. J Environ Sci Health B 48(1):16–22. https://doi.org/10.1080/03601234.2012.716686

Adess, Avraham Chaim (2004) The four minim: a practical illustrated guide. Feldheim, Jerusalem

Akin-Ajani OD, Okunlola A (2021) Pharmaceutical applications of pectin. In: Pectins – the new-old polysaccharides. IntechOpen, London. https://doi.org/10.5772/intechopen.100152

Almeida R, Nunney L (2015) How do plant diseases caused by Xylella fastidiosa emerge? Plant Dis 99(11):1457–1467. https://doi.org/10.1094/PDIS-02-15-0159-FE

Amar Z (2009) Four species anthology. Center of IT & IS BIU, Tel Aviv, Israel

Arias BA, Ramón-Laca L (2005) Pharmacological properties of citrus and their ancient and medieval uses in the Mediterranean region. J Ethnopharmacol 97(1):89–95. https://doi.org/10.1016/j.jep.2004.10.019

ARQUISTE Parfumeur (n.d.) L'ETROG. Retrieved August 4, 2022, from https://arquiste.com/products/letrog#:%7E:text=L'ETROG%20opens%20with%20sparkly,myrtle%2C%20date%20fruit%20and%20vetiver

Ashulin L, Lachman O, Hadas R, Bar-Joseph M (1991) Nucleotide sequence of a new viroid species, citrus bent leaf viroid (CBLVd) isolated from grapefruit in Israel. Nucleic Acids Res 19(17):4767. https://doi.org/10.1093/nar/19.17.4767

Bailly C (2020) Targets and pathways involved in the antitumor activity of citral and its stereoisomers. Eur J Pharmacol 871:172945. https://doi.org/10.1016/j.ejphar.2020.172945

Behall K, Reiser S (1986) Effects of pectin on human metabolism. Chemistry and Function of Pectins. American Chemical Society, Washington, DC, pp 248–265

Belabess Z, Radouane N, Sagouti T, Tahiri A, Lahlali R (2021) A current overview of two viroids prevailing in Citrus orchards: Citrus exocortis viroid and hop stunt viroid. Citrus Res Dev Biotechnol. https://doi.org/10.5772/INTECHOPEN.95914

Bergman M, Djaldetti M, Salman H, Bessler H (2010) Effect of citrus pectin on malignant cell proliferation. Biomed Pharmacother 64(1):44–47. https://doi.org/10.1016/j.biopha.2009.03.004

Bhuiyan MNI, Begum J, Sardar PK, Rahman MS (2009) Constituents of peel and leaf essential oils of Citrus medica L. Int J Sci Res 1(2):387–392

Bora H, Kamle M, Mahato DK, Tiwari P, Kumar P (2020) Citrus Essential Oils (CEOs) and their applications in food: an overview. Plants (Basel, Switzerland) 9(3):357. https://doi.org/10.3390/plants9030357

Bové JM (2006) Huanglongbing: a destructive, newly-emerging, century-old disease of citrus. J Plant Pathol 88(1):7–37. https://doi.org/10.4454/JPP.V88I1.828

Brigand J-P, Nahon P (2016) Gastronomy and the citron tree (Citrus medica L.). Int J Gastronomy Food Sci 3:12–16. https://doi.org/10.1016/j.ijgfs.2016.03.001

Brouns F, Theuwissen E, Adam A, Bell M, Berger A, Mensink RP (2012) Cholesterol-lowering properties of different pectin types in mildly hyper-cholesterolemic men and women. Eur J Clin Nutr 66(5):591–599. https://doi.org/10.1038/ejcn.2011.208

CABI (2022) Citrus medica (citron). CABI Compendium. Wallingford, UK: CAB International. https://www.cabidigitallibrary.org/doi/10.1079/cabicompendium.13457#tab-contributors

Cao MJ, Atta S, Liu YQ, Wang XF, Zhou CY, Mustafa A, Iftikhar Y (2009) First report of Citrus bent leaf viroid and Citrus dwarfing viroid from Citrus in Punjab. Pak Plant Dis 93(8):840. https://doi.org/10.1094/PDIS-93-8-0840C

Carvalho R, Soares Filho WS, Brasileiro-Vidal AC, Guerra M (2005) The relationships among lemons, limes and citron: a chromosomal comparison. Cytogenet Genome Res 109(1–3):276–282. https://doi.org/10.1159/000082410

Catalouge of Life: v 2022-07-12 13-08-2022. August 2022, Kew, Plants, people, possibilities, https://www.catalogueoflife.org/data/taxon/VMLC, Accessed on 13.08.2022

Chambers GA, Donovan NJ, Bodaghi S, Jelinek SM, Vidalakis G (2017) A novel citrus viroid found in Australia, tentatively named citrus viroid VII. Arch Virol 163(1):215–218. https://doi.org/10.1007/s00705-017-3591-y

Chambial S, Dwivedi S, Shukla KK, John PJ, Sharma P (2013) Vitamin C in disease prevention and cure: an overview. Indian J Clin Biochem 28(4):314–328. https://doi.org/10.1007/s12291-013-0375-3

Chhikara N, Kour R, Jaglan S, Gupta P, Gat Y, Panghal A (2018) Citrus medica: nutritional, phytochemical composition and health benefits – a review. Food Funct 9(4):1978–1992. https://doi.org/10.1039/c7fo02035j

Cirmi S, Maugeri A, Ferlazzo N, Gangemi S, Calapai G, Schumacher U, Navarra M (2017) Anticancer potential of citrus juices and their extracts: a systematic review of both preclinical and clinical studies. Front Pharmacol 8:420. https://doi.org/10.3389/fphar.2017.00420

Conforti F, Statti GA, Tundis R, Loizzo MR, Menichini F (2007) In vitro activities of Citrus medica L. cv. Diamante (Diamante citron) relevant to treatment of diabetes and Alzheimer's disease. Phytother Res PTR 21(5):427–433. https://doi.org/10.1002/ptr.2077

Curk F, Ancillo G, Ollitrault F, Perrier X, Jacquemoud-Collet J-P, Garcia-Lor A et al (2015) Nuclear species-diagnostic SNP markers mined from 454 amplicon sequencing reveal admixture genomic structure of modern citrus varieties. PLoS One 10(5):e0125628. https://doi.org/10.1371/journal.pone.0125628

Curk F, Ollitrault F, Garcia-Lor A, Luro F, Navarro L, Ollitrault P (2016) Phylogenetic origin of limes and lemons revealed by cytoplasmic and nuclear markers. Ann Bot 117(4):565–583. https://doi.org/10.1093/aob/mcw005

Cyrus Adler and Casanowicz (1903) ETROG - The Jewish Encyclopedia. Retrived August 4, 2022 from https://jewishencyclopedia.com/articles/5895-etrog

da Graça JV, Korsten L (2004) Citrus Huanglongbing: review, present status and future strategies. Dis Fruits Veg I:229–245. https://doi.org/10.1007/1-4020-2606-4_4

da Graça JV, Douhan GW, Halbert SE, Keremane ML, Lee RF, Vidalakis G, Zhao H (2016) Huanglongbing: an overview of a complex pathosystem ravaging the world's citrus. J Integr Plant Biol 58(4):373–387. https://doi.org/10.1111/jipb.12437

da Silva AR, Pinto KNS, Maserti BE, Santos-Filho HP, Gesteira A (2021) Systematic review of defense responses against Phytophthora and strategies to manage Phytophthora diseases in citrus. Funct Plant Biol FPB 48(10):963–972. https://doi.org/10.1071/FP20349

Douglas TS, Walker RR (1983) 4-Desmethylsterol composition of citrus rootstocks of different salt exclusion capacity. Physiol Plant 58(1):69–74. https://doi.org/10.1111/j.1399-3054.1983.tb04145.x

Du J, Cullen JJ, Buettner GR (2012) Ascorbic acid: chemistry, biology and the treatment of cancer. Biochim Biophys Acta 1826(2):443–457. https://doi.org/10.1016/j.bbcan.2012.06.003

Eddin LB, Jha NK, Meeran M, Kesari KK, Beiram R, Ojha S (2021) Neuroprotective potential of limonene and limonene containing natural products. Molecules (Basel, Switzerland) 26(15):4535. https://doi.org/10.3390/molecules26154535

Eiras M, Silva SR, Stuchi ES, Carvalho SA, GarcêzRM. (2013) Identification and characterization of viroids in "Navelina ISA 315" sweet orange. Trop Plant Pathol 38(1):58–62. https://doi.org/10.1590/S1982-56762013000100009

Entezari M, Majd A, Falahian F, Mehrabian S, Hashemi M, ArdeshiryLajimi A (2009) Antimutagenicity and anticancer effects of Citrus Medica fruit juice. Acta Med Iran 47(5):373–377

Erich Isaac (1959) The Citron in the Mediterranean: a study in religious influences. Econ Geogr 35(1):71–78. https://doi.org/10.2307/142080

Etrog Citron (n.d.-a) Specialty produce. Retrieved August 4, 2022, from https://specialtyproduce.com/produce/Etrog_Citron_8713.php

Etrog Citron (n.d.-b) Pearson Ranch. Retrieved August 4, 2022, from https://www.pearsonranch.com/buy-etrog-citron/

Etrog Desserts (n.d.) ETROG. Retrieved August 4, 2022, from https://www.etrog.com.tr/en/category/our-desserts-1

Fang DQ, Roose ML (1997) Identification of closely related citrus cultivars with inter-simple sequence repeat markers. TAG Theor Appl Genet 95(3):408–417. https://doi.org/10.1007/s001220050577

Federici CT, Fang DQ, Scora RW, Roose ML (1998) Phylogenetic relationships within the genus Citrus (Rutaceae) and related genera as revealed by RFLP and RAPD analysis. TAG Theor Appl Genet 96(6–7):812–822. https://doi.org/10.1007/s001220050807

Ference CM, Gochez AM, Behlau F, Wang N, Graham JH, Jones JB (2018) Recent advances in the understanding of Xanthomonas citri ssp. citri pathogenesis and citrus canker disease management. Mol Plant Pathol 19(6):1302. https://doi.org/10.1111/MPP.12638

Fermo (1961) "Compendium aromatariorum" by S. di Ascoli (c. 1430). In: Proceedings of the IV Biennial of the Marca for the history of Medicine

Florence Kreisler Greenbaum (1918) The international Jewish cook book modern "kosher cook book". Block Publishing Company, New York, Bloch publishing company

Forer R (2018) Etrog: a fruit you've probably never heard of and 3 ways to use it. Spoon University. Retrieved August 4, 2022, from https://spoonuniversity.com/lifestyle/3-ways-to-use-the-etrog

Forward T (2021) When life gives you etrogs, make etrog liqueur. The Forward. Retrieved August 4, 2022, from https://forward.com/food/476017/when-life-gives-you-etrogs-make-etrog-liqueur/

Franco JC, Suma P, da Silva EB, Blumberg D, Mendel Z (2004) Management strategies of mealybug pests of citrus in mediterranean countries. Phytoparasitica 32(5):507–522. https://doi.org/10.1007/bf02980445

Froelicher Y, Mouhaya W, Bassene J-B, Costantino G, Kamiri M, Luro F et al (2010) New universal mitochondrial PCR markers reveal new information on maternal citrus phylogeny. Tree Genet Genomes 7(1):49–61. https://doi.org/10.1007/s11295-010-0314-x

Galappathie S, Edwards DJ, Elliott AG, Cooper MA, Palombo EA, Butler MS, Mahon PJ (2017) Antibacterial Nerol Cinnamates from the Australian plant Eremophila longifolia. J Nat Prod 80(4):1178–1181. https://doi.org/10.1021/acs.jnatprod.6b00888

Gill K, Kumar P, Kumar A, Kapoor B, Sharma R, Joshi AK (2022) Comprehensive mechanistic insights into the citrus genetics, breeding challenges, biotechnological implications, and omics-based interventions. Tree Genet Genomes 18(2). https://doi.org/10.1007/s11295-022-01544-z

Gillings M, Broadbent P, Gollnow B (1991) Viroids in Australian citrus: relationship to exocortis, cachexia and citrus dwarfing. Funct Plant Biol 18(5):559–570. https://doi.org/10.1071/PP9910559

Goldschmidt EE (2017) New insights in Citron (Citrus medica L.) genomics and fruit development. HortScience 52(6):823–826. https://doi.org/10.21273/hortsci11142-16

Goldschmidt EE, Leshem B (1971) Style abscission in the citron (citrus medica l.) and other citrus species: morphology, physiology, and chemical control with picloram. Am J Bot 58(1):14–23. https://doi.org/10.1002/j.1537-2197.1971.tb09940.x

Graham J, Feichtenberger E (2015) Citrus phytophthora diseases: management challenges and successes. J Citrus Pathol 2(1). https://doi.org/10.5070/c421027203

Grant M (1997) Dieting for an emperor. A translation of Books 1 and 4 of Oribasius' medical compilations with an introduction and commentary. Stud Anc Med 15:1–388

Groult S, Buwalda S, Budtova T (2021) Pectin hydrogels, aerogels, cryogels and xerogels: influence of drying on structural and release properties. Eur Polym J 149:110386. https://doi.org/10.1016/j.eurpolymj.2021.110386

Gualdani R, Cavalluzzi MM, Lentini G, Habtemariam S (2016) The chemistry and pharmacology of citrus limonoids. Molecules (Basel, Switzerland) 21(11):1530. https://doi.org/10.3390/molecules21111530

Habib S, Gupta P, Bhat SS, Gupta J (2021) In silico, in-vitro and in vivo screening of biological activities of citral. Int J Vitam Nutr Res Internationale Zeitschrift fur Vitamin- und Ernahrungsforschung Journal international de vitaminologie et de nutrition 91(3–4):251–260. https://doi.org/10.1024/0300-9831/a000625

Hagenmaier RD, Baker RA (1994) Wax microemulsions and emulsions as citrus coatings. J Agric Food Chem 42(4):899–902. https://doi.org/10.1021/JF00040A012/ASSET/JF00040A012.FP.PNG_V03

Hardy S (2008) Citrus exocortis. Prime facts. Retrieved August 4, 2022, from https://www.dpi.nsw.gov.au/__data/assets/pdf_file/0009/228852/Citrus-exocortis.pdf

Hemilä H (2017) Vitamin C and infections. Nutrients 9(4):339. https://doi.org/10.3390/nu9040339

Herrero R, Asíns MJ, Carbonell EA, Navarro L (1996) Genetic diversity in the orange subfamily Aurantioideae. I. Intraspecies and intragenus genetic variability. Theor Appl Genet 92(5):599–609. https://doi.org/10.1007/bf00224564

Hetta M, El-Alfy TS, Yassin N, Abdel-Rahman R, Kadry EM (2013) Phytochemical and antihyperglycemic studies on Citrus medica L. leaves (Etrog) growing in Egypt. Int J Pharmacogn Phytochem Res 5:271–277

Hooker JD (1875) The flora of British India, vol I. L. Reeve & Co. L.T.D./The Oast House Publications, London, p 514

Hsu YH, Chen W, Owens RA (1995) Nucleotide sequence of a hop stunt viroid variant isolated from citrus growing in Taiwan. Virus Genes 9(2):193–195. https://doi.org/10.1007/BF01702664

Iftikhar Y, Majeed MZ, Vadamalai G, Sajid A (2021) Indexing virus and virus-like diseases of citrus. Citrus Res Dev Biotechnol. https://doi.org/10.5772/INTECHOPEN.95897

Ilaria M, Maurizio M, Renato M, Nadia L, Alberto S, Elisabetta N, Daniele B (2015) Characterization by molecular markers of "POMPIA", a natural citrus hybrid cultivated in SARDINIA. Acta Hortic 1065:165–172. https://doi.org/10.17660/actahortic.2015.1065.16

Islam MT, Quispe C, Islam MA, Ali ES, Saha S, Asha UH, Mondal M, Razis A, Sunusi U, Kamal RM, Kumar M, Sharifi-Rad J (2021) Effects of nerol on paracetamol-induced liver damage in Wistar albino rats. Biomed Pharmacother 140:111732. https://doi.org/10.1016/j.biopha.2021.111732

Ito T, Ieki H, Ozaki K, Ito T (2001) Characterization of a new citrus viroid species tentatively termed Citrus viroid OS. Arch Virol 146(5):975–982. https://doi.org/10.1007/s007050170129

Ito T, Ieki H, Ozaki K, Iwanami T, Nakahara K, Hataya T et al (2002a) Multiple citrus viroids in citrus from Japan and their ability to produce exocortis-like symptoms in citron. Phytopathology 92(5):542–547. https://doi.org/10.1094/phyto.2002.92.5.542

Ito T, Ieki H, Ozaki K (2002b) Simultaneous detection of six citrus viroids and Apple stem grooving virus from citrus plants by multiplex reverse transcription polymerase chain reaction. J Virol Methods 106(2):235–239. https://doi.org/10.1016/s0166-0934(02)00147-7

Jia H, Zhang Y, Orbović V, Xu J, White FF, Jones JB, Wang N (2017) Genome editing of the disease susceptibility gene CsLOB1 in citrus confers resistance to citrus canker. Plant Biotechnol J 15(7):817–823. https://doi.org/10.1111/pbi.12677

Kaufman CK (2006) Cooking in ancient civilizations. Greenwood Publishing Group, California, USA, p 176

Kay S, Bonas U (2009) How Xanthomonas type III effectors manipulate the host plant. Curr Opin Microbiol 12(1):37–43. https://doi.org/10.1016/j.mib.2008.12.006

Khanchouch K, Pane A, Chriki A, Cacciola SO (2017) Major and emerging fungal diseases of citrus in the Mediterranean region. Citrus Pathol. https://doi.org/10.5772/66943

Kitchens JGT (2019) Etrog Vodka. Jamie Geller. Retrieved August 4, 2022, from https://jamiegeller.com/recipes/etrog-vodka/

Klein, Joshua (2014) Citron cultivation, production and uses in the Mediterranean region. In: Medicinal and aromatic plants of the Middle-East, vol 2. Springer, Dordrecht, p 199. https://doi.org/10.1007/978-94-017-9276-9_10

Klein JD, Hebbe Y, Shapovalov A, Shklyar G, Korol L, Cohen S (2013) Changes in peel color of citron fruits from different genetic origins in response to postharvest copper and gibberellic acid treatments. Acta Hortic 1012:385–390. https://doi.org/10.17660/ActaHortic.2013.1012.48

Klein JD, Raz Shalev Y, Cohen S, Fallik E (2016) Postharvest handling of "etrog" citron (Citrus medica, L.) fruit. Isr J Plant Sci 63(1):64–75. https://doi.org/10.1080/07929978.2016.1159409

Kumaran RS, Muthumary J, Hur BK (2008) Taxol from Phyllostictacitricarpa, a leaf spot fungus of the angiosperm Citrus medica. J Biosci Bioeng 106(1):103–106. https://doi.org/10.1263/jbb.106.103

Kwon SJ, Bodaghi S, Dang T, Gadhave KR, Ho T, Osman F, Al Rwahnih M, Tzanetakis IE, Simon AE, Vidalakis G (2021) Complete nucleotide sequence, genome organization, and comparative genomic analyses of Citrus Yellow-Vein Associated Virus (CYVaV). Front Microbiol 12:683130. https://doi.org/10.3389/fmicb.2021.683130

Langgut D (2017) The Citrus route revealed: from Southeast Asia into the Mediterranean. HortSciencehorts 52(6):814–822. Retrieved August 10, 2022, from https://journals.ashs.org/hortsci/view/journals/hortsci/52/6/article-p814.xml

Langgut D, Gadot Y, Porat N, Lipschits O (2013) Fossil pollen reveals the secrets of royal Persian garden in Ramat Rahel (Jerusalem). Palynology 37:115–129

Lei Y, Fu P, Jun X, Cheng P (2019) Pharmacological properties of geraniol – a review. Planta Med 85(1):48–55. https://doi.org/10.1055/a-0750-6907

Leinonen R, Sugawara H, Shumway M (2010) The sequence read archive. Nucleic Acids Res 39(Database):D19–D21. https://doi.org/10.1093/nar/gkq1019

Levey M (1973) Early Arabic pharmacology: an introduction based on ancient and medieval sources. Brill, Leiden

Li Q, Qi J, Qin X, Dou W, Lei T, Hu A, Jia R, Jiang G, Zou X, Long Q, Xu L, Peng A, Yao L, Chen S, He Y (2020) CitGVD: a comprehensive database of citrus genomic variations. Hortic Res 7:12. https://doi.org/10.1038/s41438-019-0234-3

Li DQ, Li J, Dong HL, Li X, Zhang JQ, Ramaswamy S, Xu F (2021) Pectin in biomedical and drug delivery applications: a review. Int J Biol Macromol 185:49–65. https://doi.org/10.1016/j.ijbiomac.2021.06.088

Liran N (2013) The Etrog in the Jewish culture: interdisciplinary study of the ritual object throughout the ages. University Haifa, Haifa, PhD disseration in Hebrew, with English Abstract

Liu M, Wang K, Chen B, Cai Y, Li C, Yang W, Wei M, Zheng G (2021) Intraspecific DNA barcoding and variation analysis for CitriReticulataePericarpium of Citrus reticulata "Chachi". Evid Based Complement Altern Med eCAM 2021:2609935. https://doi.org/10.1155/2021/2609935

Lota M-L, de Rocca Serra D, Tomi F, Bessiere J-M, Casanova J (1999) Chemical composition of peel and leaf essential oils of Citrus medica L. and C. limonimedica Lush. Flavour Fragr J 14(3):161–166. https://doi.org/10.1002/(sici)1099-1026(199905/06)14:3<161::aid-ffj801>3.0.co;2-8

Luro F, Venturini N, Costantino G, Paolini J, Ollitrault P, Costa J (2012) Genetic and chemical diversity of citron (Citrus medica L.) based on nuclear and cytoplasmic markers and leaf essential oil composition. Phytochemistry 77:186–196. https://doi.org/10.1016/j.phytochem.2011.12.013

Luro F, Viglietti G, Marchi E, Costantino G, Scarpia MG, Tomi F et al (2019) Genetic, morphological and chemical investigations reveal the genetic origin of Pompia (C. medica tuberosa Risso& Poiteau) – an old endemic Sardinian citrus fruit. Phytochemistry 112083. https://doi.org/10.1016/j.phytochem.2019.1126

Luro F, Garcia Neves C, Costantino G, da Silva Gesteira A, Paoli M, Ollitrault P et al (2020) Effect of environmental conditions on the yield of peel and composition of essential oils from citrus cultivated in Bahia (Brazil) and Corsica (France). Agronomy 10(9):1256. https://doi.org/10.3390/agronomy10091256

Mączka W, Wińska K, Grabarczyk M (2020) One hundred faces of geraniol. Molecules (Basel, Switzerland) 25(14):3303. https://doi.org/10.3390/molecules25143303

Meena A, Kandale A, Meda M (2011) A review on citron-pharmacognosy, phytochemistry and medicinal uses. J Pharm 2:14

Mendonça LBP, Zambolim L, Badel JL (2017) Bacterial citrus diseases: major threats and recent progress. J Bacteriol Mycol Open Access 5(4):340–350. https://doi.org/10.15406/jbmoa.2017.05.00143

Miller NCF, Silvers S (2015) The prospects of commercial Etrog production in South Florida, a culturally significant Citron. Proc Fla State Hortic Soc 128:20–24

Minzanova ST, Mironov VF, Arkhipova DM, Khabibullina AV, Mironova LG, Zakirova YM, Milyukov VA (2018) Biological activity and pharmacological application of pectic polysaccharides: a review. Polymers 10(12):1407. https://doi.org/10.3390/polym10121407

Moreno P, Ambrós S, Albiach-Martí MR, Guerri J, Peña L (2008) Citrus tristeza virus: a pathogen that changed the course of the citrus industry. Mol Plant Pathol 9(2):251–268. https://doi.org/10.1111/j.1364-3703.2007.00455.x

Moreno P, Guerri J, García ML (2015) The psorosis disease of citrus: a pale light at the end of the tunnel. J Citrus Pathol 2(1):1–18. https://doi.org/10.5070/c421028860

Morton J (1987) Citron. In: Fruits of warm climates. Julia F. Morton, Miami, pp 179–182. http://www.hort.purdue.edu/newcrop/Morton,1987/citron.html#Food Uses

Moster DZ (2018) Introduction. In: Etrog. Palgrave Pivot, Cham. https://doi.org/10.1007/978-3-319-73736-2_1

Munarin F, Tanzi MC, Petrini P (2012) Advances in biomedical applications of pectin gels. Int J Biol Macromol 51(4):681–689. https://doi.org/10.1016/j.ijbiomac.2012.07.002

Murcia N, Bernad L, Serra P, Bani Hashemian SM, Duran-Vila N (2009) Molecular and biological characterization of natural variants of Citrus dwarfing viroid. Arch Virol 154(8):1329–1334. https://doi.org/10.1007/s00705-009-0430-9

Nachmias A, Barash I, Solel Z, Strobel GA (1977) Purification and characterization of a phytotoxin produced by Phomatracheiphila, the causal agent of mal secco disease of citrus. Physiol Plant Pathol 10(2):147–157. https://doi.org/10.1016/0048-4059(77)90018-2

Neelima Mahato KS, Rakoti Koteswararao MS, RajBaral E, Cho MH (2017) Citrus essential oils: extraction, authentication and application in food preservation. Crit Rev Food Sci Nutr 59:611–625. https://doi.org/10.1080/10408398.2017.1384716

Nicolosi E, Deng ZN, Gentile A, La Malfa S, Continella G, Tribulato E (2000) Citrus phylogeny and genetic origin of important species as investigated by molecular markers. TAG Theor Appl Genet 100(8):1155–1166. https://doi.org/10.1007/s001220051419

Nicolosi E, La Malfa S, Mohamed E-O, Negbi M, Goldschmidt E (2005) The search for the Authentic Citron (Citrus medica L.): historic and genetic analysis. HortScience Publ Am Soc Hortic Sci 40:1963–1968. https://doi.org/10.21273/HORTSCI.40.7.1963

Nigro F, Ippolito A, Salerno MG (2011) Mal secco disease of Citrus: a journey through a century of research. J Plant Pathol 93(3):523–560

Nosowitz D (2018) The mostly inedible fruit that can sell for hundreds of dollars in Brooklyn. Modern Farmer. Retrieved August 4, 2022, from https://modernfarmer.com/2014/10/mostly-inedible-fruit-can-sell-hundreds-dollars-brooklyn/

Padma Ishwarya S, Sandhya R, Nisha P (2022) Advances and prospects in the food applications of pectin hydrogels. Crit Rev Food Sci Nutr 62(16):4393–4417. https://doi.org/10.1080/10408398.2021.1875394

Palazzolo E, Laudicina VA, Germanà MA (2013) Current and potential use of citrus essential oils. Curr Org Chem 17:3042–3049. https://doi.org/10.2174/13852728113179990122

Panara K (2012) A review on phytochemical and pharmacological properties of Citrus medica Linn. Int J Pharm Biol Archive 3(6):1292–1297

Pessina D, Gentili R, Barcaccia G, Nicolè S, Rossi G, Barbesti S, Sgorbati S (2011) DNA content, morphometric and molecular marker analyses of Citrus limonimedica, C limon and C medica for the determination of their variability and genetic relationships within the genus Citrus. Sci Hortic 129(4):663–673. https://doi.org/10.1016/j.scienta.2011.05.012

Pierre Laszlo (2007) Citrus: a history. University of Chicago Press, Chicago, pp 12–13

Plummer J (2020) The IUCN red list of threatened species. Retrieved August 4, 2022, from https://www.iucnredlist.org/species/62041221/62041228

Pullar JM, Carr AC, Vissers M (2017) The roles of vitamin C in skin health. Nutrients 9(8):866. https://doi.org/10.3390/nu9080866

Qian W, Liu M, Fu Y, Wang T, Zhang J, Yang M, Sun Z, Li X, Li Y (2020) Antimicrobial and antibiofilm activities of citral against carbapenem-resistant Enterobacter cloacae. Foodborne Pathog Dis 17(7):459–465. https://doi.org/10.1089/fpd.2019.2751

Rafiq S, Kaul R, Sofi SA, Bashir N, Nazir F, Ahmad Nayik G (2016) Citrus peel as a source of functional ingredient: a review. J Saudi Soc Agric Sci 17:351–358. https://doi.org/10.1016/j.jssas.2016.07.006

Rakowski AG, Szychowski JA, Avena ZS, Semancik JS (1994) Nucleotide sequence and structural features of the group III citrus viroids. J Gen Virol 75(Pt 12):3581–3584. https://doi.org/10.1099/0022-1317-75-12-3581

Ramadugu C, Pfeil BE, Keremane ML, Lee RF, Maureira-Butler IJ, Roose ML (2013) A six nuclear gene phylogeny of citrus (Rutaceae) taking into account hybridization and lineage sorting. PLoS One 8(7):e68410. https://doi.org/10.1371/journal.pone.0068410

Ramadugu C, Keremane ML, Hu X, Karp D, Federici CT, Kahn T et al (2015) Genetic analysis of citron (Citrus medica L.) using simple sequence repeats and single nucleotide polymorphisms. Sci Hortic 195:124–137. https://doi.org/10.1016/j.scienta.2015.09.004
Rivka (2007) Etrog marmalade, Not Derby Pie. Retrieved August 4, 2022, from http://notderbypie.com/etrog-marmalade/
Rizza S, Conesa A, Juarez J, Catara A, Navarro L, Duran-Vila N, Ancillo G (2012) Microarray analysis of Etrog citron (Citrus medica L.) reveals changes in chloroplast, cell wall, peroxidase and symporter activities in response to viroid infection. Mol Plant Pathol 13(8):852–864. https://doi.org/10.1111/j.1364-3703.2012.00794.x
Rodrigues da Silva A, da Costa Silva D, Dos Santos Pinto KN, Santos Filho HP, Coelho Filho MA, Dos Santos Soares Filho W, Ferreira CF, da Silva Gesteira A (2021) Epigenetic responses to Phytophthora citrophthora gummosis in citrus. Plant Sci Int J Exp Plant Biol 313:111082. https://doi.org/10.1016/j.plantsci.2021.111082
Rosner F (1968) Moses Maimonides' treatise on poisons. JAMA 205(13):914–916
Sang T, Hataya TJI, Sasaki A, Shikata E (1986) Etrog Citron is latently infected with hop stunt viroid-like RNA. Proc Jpn Acad 62B(8):325
Sano T, Hataya T, Shikata E (1988) Complete nucleotide sequence of a viroid isolated from Etrog citron, a new member of hop stunt viroid group. Nucleic Acids Res 16(1):347. https://doi.org/10.1093/nar/16.1.347
Sanofer A (2014) Role of citrus fruits in health. J Pharm Sci Res 6:121–123
Sarah Oren (n.d.) Citron/Etrog (Citrus Medica) NeotKedumim. Retrieved August 4, 2022, from https://www.neot-kedumim.org.il/index-plant-of-the-month/citron-etrog-citrus-medica/
Saville AC, Krings A, Kahn T, Trice MD, Redford AJ (2011) Citrus ID, Edition 2. USDA APHIS PPQ CPHST, Fort Collins and North Carolina State University. [04.08.2022]
Sawake MM, Moharil MP, Ingle YV, Jadhav PV, Ingle AP, Khelurkar VC, Paithankar DH, Bathe GA, Gade AK (2022) Management of Phytophthora parasitica causing gummosis in citrus using biogenic copper oxide nanoparticles. J Appl Microbiol 132(4):3142–3154. https://doi.org/10.1111/jam.15472
Scora RW (2010) Torrey botanical society on the history and origin of citrus. Society 102(6):369–375
Semancik JS, Roistacher CN, Rivera-Bustamante R, Duran-Vila N (1988) Citrus cachexia viroid, a new viroid of citrus: relationship to viroids of the exocortis disease complex. J Gen Virol 69:3059–3068
Semancik JS, Szychowski JA, Rakowski AG, Symons RH (1993) Isolates of citrus exocortis viroid recovered by host and tissue selection. J Gen Virol 74(Pt 11):2427–2436. https://doi.org/10.1099/0022-1317-74-11-2427
Serra P, Bani Hashemian SM, Pensabene-Bellavia G, Gago S, Duran-Vila N (2009) An artificial chimeric derivative of citrus viroid V involves the terminal left domain in pathogenicity. Mol Plant Pathol 10(4):515–522. https://doi.org/10.1111/j.1364-3703.2009.00553.x
Sharma S, Habib S, Sahu D, Gupta J (2021) Chemical properties and therapeutic potential of citral, a monoterpene isolated from lemongrass. Med Chem (Shariqah, United Arab Emirates) 17(1):2–12. https://doi.org/10.2174/1573406416666191227111106
She QH, Li WS, Jiang YY, Wu YC, Zhou YH, Zhang L (2020) Chemical composition, antimicrobial activity and antioxidant activity of Litsea cubeba essential oils in different months. Nat Prod Res 34(22):3285–3288. https://doi.org/10.1080/14786419.2018.1557177
Shemesh AO (2012) The fingered citron and the dibdib citron for the ritual of the four species in medieval and modern literature. Torah U-Madda J 16:173–185. http://www.jstor.org/stable/23596061
Shenoy N, Creagan E, Witzig T, Levine M (2018) Ascorbic acid in cancer treatment: let the Phoenix Fly. Cancer Cell 34(5):700–706. https://doi.org/10.1016/j.ccell.2018.07.014
Silva G, Marques J, Linhares E, Bonora CM, Costa ÉT, Saraiva MF (2022) Review of anticancer activity of monoterpenoids: geraniol, nerol, geranial and neral. Chem Biol Interact 362:109994. https://doi.org/10.1016/j.cbi.2022.109994
Sofer YC (1969) Kaf HaChaim M Sofer pub, Jerusalem. Section 764:60. (Hebrew)

Solel Z, Mogilner N, Gafny R, Bar-Joseph M (1995) Induced tolerance to mal secco disease in Etrog citron and Rangpur lime by infection with the citrus exocortis viroid. Plant Dis 79(1):60–62. https://doi.org/10.1094/pd-79-0060

Sperling AY (1982) TaameiHaMinhagim (sources of [Jewish] customs). Eshkol Publishing, Jerusalem, p 67. (Hebrew)

Strauss M (2012) Etrog drop candy cookies with Etrog salt. Melinda Strauss. Retrieved August 4, 2022, from https://melindastrauss.com/2012/10/12/etrog-drop-candy-cookies-with-etrog-salt/

Suessmann, M. (1955) G. Sarton, S. di Ascoli (c. 1430). Compendium aromatariorum. The book of the pharmacists. First edition in Hebrew by Vol 10, No 1, pp 133–138

Sukkah Hill Spirits (n.d.) Retrieved August 4, 2022, from https://sukkahhill.com/products/

Sun J (2007) D-limonene: safety and clinical applications. Altern Med Rev J Clin Ther 12(3):259–264

Tan H, Nie S (2020) Deciphering diet-gut microbiota-host interplay: investigations of pectin. Trends Food Sci Technol 106:171–181. https://doi.org/10.1016/j.tifs.2020.10.010

Tessitori M, Maria G, Capasso C, Catara G, Rizza S, De Luca V, Catara A, Capasso A, Carginale V (2007) Differential display analysis of gene expression in Etrog citron leaves infected by Citrus viroid III. Biochim Biophys Acta 1769(4):228–235. https://doi.org/10.1016/j.bbaexp.2007.03.004

Thakur BR, Singh RK, Handa AK (1997) Chemistry and uses of pectin – a review. Crit Rev Food Sci Nutr 37(1):47–73. https://doi.org/10.1080/10408399709527767

The Saga of the Citron (2012) Reform Judaism. Retrieved August 4, 2022, from https://reformjudaism.org/jewish-holidays/sukkot/saga-citron

Theophrastus (1916) Enquiry into plants. English translation by A.F. Hort, Loeb Classical Library, Cambridge

Tian J, Lu Z, Wang Y, Zhang M, Wang X, Tang X, Peng X, Zeng H (2017) Nerol triggers mitochondrial dysfunction and disruption via elevation of Ca2+ and ROS in Candida albicans. Int J Biochem Cell Biol 85:114–122. https://doi.org/10.1016/j.biocel.2017.02.006

Torres AM, Soost RK, Diedenhofen U (1978) Leaf isozymes as genetic markers in citrus. Am J Bot 65(8):869–881. https://doi.org/10.2307/2442183

Tristram HB (Henry Baker) (1883) The natural history of the Bible: being a review of the physical geography, geology, and meteorology of the Holy Land: with a description of every animal and plant mentioned in Holy Scripture, London: Society for Promoting Christian Knowledge. E. & J.B. Young, New York, p 347 et seq

van der Veen M (2011) Consumption, trade and innovation: exploring the botanical remains from the Roman and Islamic ports at Quseir al-Qadim, Egypt. Africa Magna Verlag, Frankfurt

Vekiari SA, Protopapadakis EE, Papadopoulou P (2002) Gas chromatography-mass spectroscopy analysis of aromatic compounds of leaves and peel from healthy and viroid-infected citron plants. In: International organization of citrus virologists conference proceedings (1957–2010), vol 15(15). https://doi.org/10.5070/C50872d83b

Velázquez K, Hervàs J, Guerri J, Vives MC (2017) Molecular characterization of new viral agents causing old citrus diseases by deep sequencing of small R.N.A.s. In: V Simposio Internacional de Fruticultura Tropical y Subtropical. Retrieved August 4, 2022, from https://redivia.gva.es/handle/20.500.11939/5825

Venturini N, Barboni T, Curk F, Costa J, Paolini J (2014) Volatile and flavonoid composition of the peel of Citrus medica L. var. Corsican fruit for quality assessment of its liqueur. Food Technol Biotechnol 52(4):403–410. https://doi.org/10.17113/ftb.52.04.14.3717

Vidalakis G, Pagliaccia D, Bash JA, Semancik JS (2010) Effects of mixtures of citrus viroids as transmissible small nuclear R.N.A. on tree dwarfing and commercial scion performance on Carrizo citrange rootstock. Ann Appl Biol 157(3):415–423. https://doi.org/10.1111/J.1744-7348.2010.00430.X

Vieira AJ, Beserra FP, Souza MC, Totti BM, Rozza AL (2018) Limonene: aroma of innovation in health and disease. Chem Biol Interact 283:97–106. https://doi.org/10.1016/j.cbi.2018.02.007

Viglietti G, Galla G, Porceddu A, Barcaccia G, Curk F, Luro F, Scarpa G (2019) Karyological analysis and DNA barcoding of Pompia Citron: a first step toward the identification of its relatives. Plan Theory 8(4):83. https://doi.org/10.3390/plants8040083

Vives MC, Martín S, Ambrós S, Renovell Á, Navarro L, Pina JA et al (2008) Development of a full-genome cDNA clone of Citrus leaf blotch virusand infection of citrus plants. Mol Plant Pathol 9(6):787–797. https://doi.org/10.1111/j.1364-3703.2008.00501.x

Vives MC, Velázquez K, Pina JA, Moreno P, Guerri J, Navarro L (2013) Identification of a new Enamovirus associated with citrus vein enation disease by deep sequencing of small RNAs. Phytopathology 103(10):1077–1086. https://doi.org/10.1094/phyto-03-13-0068-r

Wang X, Xu Y, Zhang S, Cao L, Huang Y, Cheng J, Wu G, Tian S, Chen C, Liu Y, Yu H, Yang X, Lan H, Wang N, Wang L, Xu J, Jiang X, Xie Z, Tan M, Larkin RM et al (2017) Genomic analyses of primitive, wild and cultivated citrus provide insights into asexual reproduction. Nat Genet 49(5):765–772. https://doi.org/10.1038/ng.3839

Wang T, Xiang S, Ren DD, Zhu PK, Xu L (2019a) First report of Colletotrichum gloeosporioides causing postharvest fruit rot on Citrus aurantifolia in China. Plant Disease, PDIS–02–19–0413. https://doi.org/10.1094/pdis-02-19-0413-pdn

Wang Y, Wu J, Qiu Y, Atta S, Zhou C, Cao M (2019b) Global transcriptomic analysis reveals insights into the response of 'Etrog' Citron (Citrus medica L.) to citrus Exocortis viroid infection. Viruses 11(5):453. https://doi.org/10.3390/v11050453

Wei Y, Yasir K, Ling L, Ganesan K (2017) Citrus Bent Leaf Viroid; 3(December), pp 31–40

WFO (2022) Citrus medica var. ethrog Engl. Retrived August 4, 2022 from http://www.worldflora-online.org/taxon/wfo-0000608147

Xin XF, Kvitko B, He SY (2018) Pseudomonas syringae: what it takes to be a pathogen. Nat Rev Microbiol 16(5):316–328. https://doi.org/10.1038/nrmicro.2018.17

Yaniv Z, Dudai N (eds) (2014) Medicinal and aromatic plants of the Middle-East, Medicinal and aromatic plants of the world 2. Springer, Dordrecht/Heidelberg/New York/London. https://doi.org/10.1007/978-94-017-9276-9

Ye CJ, Li SA, Zhang Y, Lee WH (2019) Geraniol targets KV1.3 ion channel and exhibits anti-inflammatory activity in vitro and in vivo. Fitoterapia 139:104394. https://doi.org/10.1016/j.fitote.2019.104394

Yehuda Felix (1997) "Fruit of the citrus tree: the Etrog" (Heb.). Beit Mikra 42:288–292

Zhenia F, Alexander F (1991) The essential oils of Etrog (Citrus medica L. Var. ethrog Engl). J Essent Oil Res 3(5):377–379. https://doi.org/10.1080/10412905.1991.9697965

Zhu RG, Sun YD, Hou YT, Fan JG, Chen G, Li TP (2017) Pectin penta-oligogalacturonide reduces cholesterol accumulation by promoting bile acid biosynthesis and excretion in high-cholesterol-fed mice. Chem Biol Interact 272:153–159. https://doi.org/10.1016/j.cbi.2017.05.018

Chapter 8
Jabara (*Citrus jabara*)

Chidanandamurthy Thippeswamy Swamy ,
Chakra Siddappa Prashantkumar, and Sukhvinder Singh Purewal

8.1 Introduction

The Japanese archipelago is a citrus island paradise, and the nation is blessed with one of the most large, exotic, and diversified citrus plantations on earth. Almost in every direction one can cite, citrus trees, filling the air with their gently scented blooms and enhancing the scenery with their brightly coloured spherical fruits and glossy evergreen leaves. They can be found in the backyards of suburban homes, in the courtyard gardens of urban households, and on tiny family farms built on historic stone terraces bordering mountainsides and buried deep into protected mountain valleys (Tom 2017). A rare indigenous sour citrus fruit produced in Kitayama village, Wakayama, Japan, is commonly called Jabara, and its approved scientific name is *C. jabara* hort. ex Y.Tanaka. Due to the fact that jabara is still solely grown in this region, its output is constrained. In 1971, yuzu's (*Citrus junos* Siebold ex Tanaka), closely related species *C. jabara* was certified as a new species and given a botanical name by Tanaka, & registered in 1979 (Tanaka 1948; Miyake et al. 1990). *C. jabara*, which is a member of the Rutaceae family, was first produced by natural crossbreeding of citrus grown locally in Kitayamamura, Japan and its nearby areas (Jabara Laboratory 2022). *C. jabara* is a type of Japanese citrus fruit, similar to the yuzu fruit, made from a cross of the yuzu and pomelo-hybridized mandarin (DB Pedia 2022).

C. T. Swamy (✉)
Department of Microbiology, Captain Srinivasa Murthy Central Ayurveda Research Institute, Central Council for Research in Ayurvedic Sciences, Chennai, Tamil Nadu, India

C. S. Prashantkumar
Department of Studies in Microbiology, Davangere University, Davangere, Karnataka, India

S. Singh Purewal
University Centre for Research & Development (UCRD), Chandigarh University, Mohali, Punjab, India

S. Singh Purewal et al. (eds.), *Recent Advances in Citrus Fruits*,
https://doi.org/10.1007/978-3-031-37534-7_8

Fig. 8.1 Fruit of *Citrus jabara* and cross-section. (Source: https://herbalistics.com.au/product/citrus-jabara-plant/)

Jabara is a rare citrus fruit related to yuzu, and it has a unique flavour profile that is similar to yuzu. It is juicier than yuzu and has a robust flavour that gives a very good fragrance and taste to meats and other foods. However, the Jabara fruit weighs more than the yuzu fruit, and the seeds are less. The Jabara fruit juice yield, soluble sugar content, pectin content, oil content, and major volatile components were recorded to be more than that of yuzu fruit. Jabara extract is used in cosmetics and has many benefits for the skin (Miyake et al. 1990; DB Pedia 2022). The image of the fruit and cross section showed in Fig. 8.1.

C. jabara is a cold-tolerant plant and yields during December to February season. Common fertilizers used for other citrus plants are also used for Jabara cultivation, and the use of any special fertilizers is not reported (Herbalistics Pty Ltd. 2022). Japan has a unique citrus family called "kizu" that includes fragrant and exotic yuzu, as well as sudachi, kabosu, yuko, jabara, and yuzu kichi. These citrus fruits have a mild tartness that is perfect for the soft Japanese palate and light Japanese cuisine. Their fragrances and flavours range from slightly floral to woody, peppery, and spicy. Jabara has a rich, peppery, sour acidity that goes well with chunkier, late-fall and winter soups, stews, and other dishes. It matures around month of November and December. Additionally, it pairs nicely with the meats found in the savoury mountain cuisine of its native Wakayama prefecture, including as duck, deer, and boar meat. Yuzu juice, konbu seaweed, and fermented dashi stock are the main ingredients of the traditional Japanese condiment known as ponzu. This ponzu can also be made from the Jabara (Tom 2017). The presence of flavonoid narirutin in Jabara may be used as medicine for pollen allergies. Jabara has been used to eat with matsutake mushrooms, so it is called an "indispensable companion" to consume this particular mushroom. The ponzu is a pon vinegar made with the extracts of Jabara used as an alternative to sudachi and other oranges.

The Jabara fruit is not only a valuable part; the peel of the Jabara fruit is also used after the proper process. The peel is a bitter and astringent taste; therefore, its process operation is complicated. Some innovative patented methods are available to reduce or remove bitterness and astringent without affecting the highly functional compounds and making them ready to eat. The aroma of the citrus peel is similar to cedar and cypress, having the property of healing and is also accepted by health-conscious people. Marmalade is a fruit preservative made from the finely chopped peels of the Jabara fruit and is rich in pectin and other natural functional ingredients (Japan External Trade Organization 2022).

The scientific classification or taxonomy of the Jabara fruit as per the NCBI taxonomy browser has Taxonomy ID of 1263158, Division eudicots and the currently used name is *C. jabara* hort. ex Y.Tanaka (Schoch et al. 2020; Anonymous 2022). However, GBIF record *Citrus x jabara* Tanaka with Taxon Id of 44661346 with the same taxonomic details (Döring M and Global Biodiversity Information Facility 2022).

Kingdom: Plantae
Phylum: Streptophyta
Class: Magnoliopsida
Order: Sapindales
Family: Rutaceae
Genus: Citrus
Species: *Jabara*

Several aroma compounds, such as myrcene, limonene, pinene etc., are reported from *C. jabara* using Gas Chromatography-Mass spectrometry (GC-MS), Gas Chromatography-Olfactometry (Omori et al. 2011). Further confirmation, amounts of hesperidin, naringin, neohesperidin, and naringenin present in *C. jabara* fruit peels were reported by HPLC analysis (Azuma et al. 2020). The *C. jabara* possess compounds exhibiting several biological activities.

Major bioactive compounds are found in the fruit peel of a *C. jabara* reported anti-allergic and anti-degranulation activity (Azuma et al. 2020), reduction in rhinorrhea (dose-dependent)/allergy-like symptoms (Azuma et al. 2021), and suppression of MUC5AC mucin production in human lung epithelial cells (Iwashita et al. 2017). In addition, peel powder cream is effective against atopic dermatitis (Inaba et al. 2020).

The present chapter focus on the *C. jabara* in general detail, chemical compounds, biological activities of these compounds, and a special focus on flavonoids, characterization and activities. The range of products available is also focused.

8.2 Cultivation

In Japan, Wakayama is generally referred to as "The Fruit Kingdom." In this prefecture's wide varieties of juicy, sweet mandarins, oranges, and tangelos, as well as yuzu (lemon-like), sudachi (lime-looking), jabara (mandarin-like), hassaku (grapefruit-like), bitter-orange daida and exquisitely delicious sanbokan are found (Tom 2020).

The *C. jabara* grows naturally in Kitayama village, Wakayama prefecture of Japan. Jabara plants are also grown in the same condition as other citrus plants. Like other citrus, *C. jabara* prefers a location with full sun to part shade, sufficient moisture, and periodic fertiliser. They are drought, cold-tolerant, and easy to maintain; pruning practices are advisable until the plant reaches to desired form and height (Herbalistics Pty Ltd. 2022). Around the month of May, Jabara flowers blossoms, produce fruit and are harvested between the end of November and December (Jabara Laboratory 2022).

Although the origin of the jabara is uncertain, it is assumed that it is a hybrid of other citrus species that were widely dispersed in the Wakayama area, including Yuzu, Kunenbo (*C. nobilis* Lour.), and Kishu-mikan or Kishu (*C. kinokuni* Hort. ex Tanaka) (Yuichirō Tanaka 1980). A molecular study inferred that Kunenbo-A is a seed parent of *C. jabara* and yuzu is a pollen parent. The genome composition of jabara is consistent with parent varieties, and the observed fluctuation of genome composition with offspring derived from same parents. The pummelo genome fluctuated in four siblings of kunenbo-A × yuzu (henka mikan, jabara, kabosu and mochiyu) henka mikan and jabara showed 3.7% and 20.5% fluctuation, respectively (Shimizu et al. 2016). In the Cleaved Amplified Polymorphic Sequences (CAPS) analysis of citrus varieties, in this researchers, 26 CAPS markers that were previously used to identify table citrus cultivars were examined to determine thirty-three acid citrus cultivars and their lines. It showed that despite having various phenotypic and local origins, all accessions of the citrus trees Yuzu (*Citrus junos* hort. ex Tanaka), Yuko (*C. yuko* hort. ex Tanaka), Jabara (*C. jabara* hort. ex Tanaka), and Sudachi (*C. sudachi* hort. ex Shirai), shared the same genetic makeup. In acid citrus cultivars, five CAPS markers detected novel alleles that were not found in table citrus cultivars. A minimal marker set consisting of 6 CAPS markers can distinguish all 35 citrus cultivars according to the minimal marker program's analysis of 25 CAPS markers genotypes, with the exception of Tf0001/Msp I (Niimi et al. 2021).

A recent US patent claimed that some bioactive priming peptides and priming compositions are useful with agricultural formulations. These peptides or compositions are perhaps used externally, like the surface of a plant or internal use such as the membrane of a plant cell or endogenously to the interior of a plant or a plant cell. Applying these formulations to plants on the surface or rhizosphere region may increase plant growth, yield, longevity, health, and productivity. Further, the vigour of the applied plant or part of a plant increases and protects the whole plant or part from diseases. These formulations also improve the innate immunity, quality, and quantity of juice obtained from the plant or plant part. These peptide formulations are applied as a prophylactic agent to agronomically essential citrus varieties comprised of hybrid and non-hybrid plants (including *C. jabara*) to prevent the onset of infection or a choice of treatment for infected citrus plants (Brian and Michelle 2020).

8.3 Phytochemicals of *Citrus jabara*

8.3.1 Estimation of Jabara Compounds

8.3.1.1 Isolation of the Volatile Compounds

In a study, aromatic compounds reported from the Flavedo layers of Jabara fruits around 420 g flavedo layers obtained from ~2 kg of fruits were soaked in 1600 ml of dichloromethane for 2 h at ambient temperature. Further, the filtrate was concentrated using the Vigreux column at 45 °C, and volatile oil was isolated from the extract using the solvent-assisted flavour evaporation (SAFE) method at 30 °C. The isolated volatile oil was further concentrated to 15 ml by using anhydrous sodium sulphate and Vigreux column at 45 °C. Furthermore, the isolate was concentrated using a nitrogen gas evaporator to produce peel extract. The obtained extracts were analysed with Gas chromatography-mass spectrometry (GC-MS), Gas chromatography-Olfactometry (GC-O) etc. (Omori et al. 2011).

8.3.1.2 Gas Chromatography-Mass Spectrometry Analysis of Jabara Compounds

Gas chromatography-mass spectrometry (GC-MS) analysis using Agilent 6890 GC equipped with an HP 5973 mass spectrometer for identifying volatile oil constituents of *C. jabara*. DB-Wax column with dimensions of 60 m × 0.25 mm i.d. was used to isolate the volatile constituents. The column oven conditions were maintained at 40 °C for 2 min and gradually increased 2 °C/min to achieve 200 °C, and the sample injector was maintained at 250 °C. The flow rate of carrier gas (helium) was 28 cm/s. The mass spectrometer was set to a scan range of f m/z 35–350 with and 70 eV ionization voltage and ion source temperature of 230 °C. The peel extract was injected in the split mode of ratio 10:1, and the injection volume was 0.2 μl. A total of 49 compounds were identified, and the name of the compounds are tabulated in Table 8.1. The major volatile oil components are Myrcene, Limonene, γ-Terpinene, β- Phellandrenec, α-Pinene and β-Pinene (Omori et al. 2011).

8.3.1.3 Gas Chromatography–Flame Ionization Detection Analysis of Jabara Compounds

Gas Chromatography–Flame Ionization Detection (GC-FID) was used to quantify the significant constituents in the Jabara peel extract. 100 mg of peel extract was dissolved in 1 ml of dichloromethane, and ethoxybenzene (20 mg) was used as an internal standard. The major constituents of peel extract, such as myrcene, limonene, and γ-terpinene, were identified and quantified by GC-MS. The standard curve was plotted using different standard concentrations of myrcene, limonene and γ-terpinene spiked with ethoxybenzene as internal standard.

Table 8.1 Various phytoconstituents of *Citrus jabara*

S. no	Phytoconstituents	Part	References
1	Myrcene	Fruit peel, Juice	Miyake et al. (1990) and Omori et al. (2011)
2	Limonene	Fruit peel, Juice	Miyake et al. (1990) and Omori et al. (2011)
3	γ-Terpinene	Fruit peel, Juice	Miyake et al. (1990) and Omori et al. (2011)
4	β-Phellandrenec	Fruit peel, Juice	Omori et al. (2011)
5	α-Pinene	Fruit peel, Juice	Miyake et al. (1990) and Omori et al. (2011)
6	β-Pinene	Fruit peel, Juice	Miyake et al. (1990) and Omori et al. (2011)
7	Terpinolene	Fruit peel, Juice	Miyake et al. (1990) and Omori et al. (2011)
8	α-Thujene	Fruit peel	Omori et al. (2011)
9	α-Phellandrene	Fruit peel	Omori et al. (2011)
10	α-Terpinene	Fruit peel	Omori et al. (2011)
11	Sabinene	Fruit peel	Omori et al. (2011)
12	1,8-Cineole	Fruit peel	Omori et al. (2011)
13	(Z)-β-Ocimene	Fruit peel	Omori et al. (2011)
14	(E)-β-Ocimene	Fruit peel	Omori et al. (2011)
15	p-Cymene	Fruit peel, Juice	Miyake et al. (1990) and Omori et al. (2011)
16	(Z)-3-Hexenyl acetate	Fruit peel	Omori et al. (2011)
17	Hexanol	Fruit peel	Omori et al. (2011)
18	(Z)-3-Hexenol	Fruit peel	Omori et al. (2011)
19	Myrcene-6,7-epoxide	Fruit peel	Omori et al. (2011)
20	(E)-2-Hexenol	Fruit peel	Omori et al. (2011)
21	Linalool oxide cis-furanoid	Fruit peel	Omori et al. (2011)
22	*Cis*-Limonene-1,2-epoxide	Fruit Peel	Omori et al. (2011)
23	*Trans*-Limonene-1,2-epoxide	Fruit peel	Omori et al. (2011)
24	δ-Elemene	Fruit peel	Omori et al. (2011)
25	Trans-Sabinene hydrate	Fruit peel	Omori et al. (2011)
26	Linalool	Fruit peel, Juice	Miyake et al. (1990) and Omori et al. (2011)
27	Cis-Sabinene hydrate	Fruit peel	Omori et al. (2011)
28	*trans-p*-Menth-2-en-1-ol[c]	Fruit peel	Omori et al. (2011)
29	Terpinen-4-ol	Fruit peel	Omori et al. (2011)
30	Neral	Fruit peel	Omori et al. (2011)
31	2-Methyl-6-methylene-2,7-octadien-4-ol	Fruit peel	Omori et al. (2011)
32	p-Mentha-1,8-dien-4-ol	Fruit peel	Omori et al. (2011)
33	Germacrene D	Fruit peel	Omori et al. (2011)

(continued)

Table 8.1 (continued)

S. no	Phytoconstituents	Part	References
34	α-Terpineol	Fruit peel, Juice	Miyake et al. (1990) and Omori et al. (2011)
35	Bicyclogermacrene[c]	Fruit peel	Omori et al. (2011)
36	Neryl acetate	Fruit peel	Omori et al. (2011)
37	Geranial	Fruit peel	Omori et al. (2011)
38	(E,E)-3,6-α-Farnesene	Fruit peel	Omori et al. (2011)
39	Trans-Piperitol	Fruit peel	Omori et al. (2011)
40	Geranyl acetate	Fruit peel	Omori et al. (2011)
41	Perillaldehyde	Fruit peel	Omori et al. (2011)
42	Nerol	Fruit peel	Omori et al. (2011)
43	Germacrene B[c]	Fruit peel	Omori et al. (2011)
44	Trans-Carveol	Fruit peel	Omori et al. (2011)
45	Geraniol	Fruit peel	Omori et al. (2011)
46	p-Mentha-1,8-dien-9-yl acetate	Fruit peel	Omori et al. (2011)
47	Perillyl alcohol	Fruit peel	Omori et al. (2011)
48	2,6-Dimethyl-6-acetoxy-2,7-octadienoic acid	Fruit peel	Omori et al. (2011)
49	4′-Hydroxy-3,5,6,7,8,3′-hexamethoxyflavone	Fruit peel	Rie et al. (2012)
50	3′, 4′, 5, 6, 7, 8-heptamethoxyflavone	Fruit peel	Rie et al. (2012)
51	Natsudaidain (3-hydroxy-3′, 4′, 5, 6, 7, 8-hexamethoxyflavone)	Fruit peel	Rie et al. (2012) and Azuma et al. (2020)
52	5-hydroxy-3, 3′, 4′, 6, 7, 8-hexamethoxyflavone	Fruit peel	Rie et al. (2012) and Azuma et al. (2020)
53	3, 3′, 4′, 5, 6, 7-hexamethoxyflavone	Fruit peel	Rie et al. (2012)
54	8-hydroxy-3, 4′, 5, 6, 7-pentamethoxyflavone	Fruit peel	Rie et al. (2012)
55	4′, 5-dihydroxy-3, 3′, 6, 7, 8-pentamethoxyflavone	Fruit peel	Rie et al. (2012)
56	Retusin (5-hydroxy-3, 3′, 4′, 7-tetramethoxyflavone)	Fruit peel	Rie et al. (2012)
57	Nobiletin (3′, 4′, 5, 6, 7, 8-hexamethoxyflavone)	Fruit peel	Rie et al. (2012)
58	Narirutin (Naringenin-7-O-rutinoside)	Fruit peel	Rie et al. (2012)
59	3′, 4′, 5, 6, 7, 8-hexamethoxyflavanone	Fruit peel	Rie et al. (2012)
60	3′, 4′, 5, 6, 7-pentamethoxyflavanone	Fruit peel	Rie et al. (2012)
61	4′, 5, 6, 7-tetramethoxyflavanone	Fruit peel	Rie et al. (2012)
62	2′-hydroxy-3, 3′, 4, 4′, 5′, 6′-hexamethoxychalcone	Fruit peel	Rie et al. (2012)
63	2′-hydroxy-3, 4, 4′, 5′, 6′-pentamethoxychalcone	Fruit peel	Rie et al. (2012)
64	2′-hydroxy-4, 4′, 5′, 6′-tetramethoxychalcone	Fruit peel	Rie et al. (2012)
65	5,6,7,8,3′,4′-hexamethoxyflavanone	Fruit peel	Rie et al. (2012)
66	5,6,7,4′-tetramethoxyflavanone	Fruit peel	Rie et al. (2012)

(continued)

Table 8.1 (continued)

S. no	Phytoconstituents	Part	References
67	5,7,4-trihydroxy-8,3′-dimethoxy- flavanone	Fruit peel	Rie et al. (2012)
68	Scopoletin	Fruit peel	Rie et al. (2012)
69	Sucrose	Juice	Miyake et al. (1990)
70	Glucose	Juice	Miyake et al. (1990)
71	Fructose	Juice	Miyake et al. (1990)
72	Citric acid	Juice	Miyake et al. (1990)
73	Malic acid	Juice	Miyake et al. (1990)
74	Pectin	Juice	Miyake et al. (1990)
75	Ascorbic acid	Juice	Miyake et al. (1990)
76	Carotenoid	Juice	Miyake et al. (1990)
77	Flavonoid	Juice	Miyake et al. (1990)
78	β Farnesene	Juice	Miyake et al. (1990)

The GC equipped with FID was used to quantify and the column and oven conditions were set as same for GC-MS analysis. The sample injector was maintained at 250 °C, and the flow rate of carrier gas (helium) was 32 cm/s. The aliquot peel extract was injected in the split mode of a ratio of 30:1, and the injection volume was 1 μl (Omori et al. 2011).

8.3.1.4 Gas Chromatograph–Olfactometry for Analysis of Aroma Compounds

Aroma Extract Dilution Analysis (AEDA) is a technique to assess potent odorant compounds by Serial dilution of the sample and evaluated with GC-O. The highest dilution (lower concentrations) at which the odourant compound is detected is called Flavour Dilution (FD).

The GC equipped with TCD and DB-Wax column with dimensions of 30 m × 0.32 mm (i.d.,) and DB-1 column 30 m × 0.53 mm (i.d.,) is used for the analysis of odourant compounds. The column oven temperature is maintained at 60 °C for 2 min, and a temperature of 220 °C was achieved by an increase of temperature to 5 °C/min. The sample injector and detector temperature were maintained at 250 °C. The helium carrier gas flow was maintained at 31 cm/s, and 1 μl of the peel extract was injected by splitless mode for analysis. The Peel extracts showed 22 odour-active regions with the FD factor ranging from 8 to 16,384. Many odourant compounds with unique smells were identified by comparison with GC-MS spectra. Myrcene was the most abundant compound and showed the highest FD factor reported. Myrcene has a metallic and resinous odour responsible for the characteristic odour of Jabara peel extract (Omori et al. 2011).

8.3.2 Fruit Juice Constituents

Jabara fruit juice yields around 27.5–30.1% in rotary press extractors and 32.2–41.7% in In-line extractors. The soluble sugars of jabara fruit juice were 3.15–3.21%. The significant sugar constituents are sucrose, glucose and fructose. Citric acid, ascorbic acid, malic acid, carotenoids, flavonoids, recoverable oils and pectin are found in the juice. It also contains nitrogen, ash, and minerals such as potassium, sodium, calcium, magnesium and iron. Volatile components such as myrcene, limonene, γ-terpinene, α-pinene, β-pinene, terpinolene, Linalool and other phytochemicals reported from the fruit extract (Miyake et al. 1990).

8.3.3 Volatile Oils from the Peel

420 g of peeled flavedo layers were soaked in dichloromethane for 2 h at room temperature. The filtered extract was concentrated, and volatile components were isolated using a solvent-assisted flavour evaporation technique. The isolated volatile components were further concentrated using anhydrous sodium sulphate and then a gentle stream of nitrogen. The yielded jabara peel extract was subjected to GC-MS, GC-FID to identify and quantify volatile compounds. The major volatile oil components are Myrcene, Limonene, γ-Terpinene, β-phellandrene, α-Pineneand β-Pinene (Omori et al. 2011).

8.3.4 Polymethoxy Flavonoids from Jabara Peel

Unripe peels (1.143 Kg) were successively extracted with hexane and 99% ethanol. The yield of ethanolic extract was 192.72 g and subjected to chromatographic separations through reverse phase column material made of adsorbent resin (Diaion HP-20). The column was eluted by highly polar to non-polar solvents such as water (H_20), 50% methanolic solution (50% MeOH-H_20), Methanol (MeOH), ethanol, ethyl acetate (AcOEt) and Chloroform ($CHCl_3$). The MeOH extracts were selected and eluted by silica gel column and preparative TLC. The eluents were subjected to High-resolution electrospray ionisation mass spectrometry (HRESI-MS) analysis. The results showed 17 known compounds (11-Flavones, 5-Flavanones and 1-Coumarin) and 1 new compound (colourless oil). The newly isolated compound showed strong absorptions at 1635 cm^{-1} in IR and 253, 268, and 344 nm in UV wavelength by indicating the presence of a flavone skeleton. The new molecule was subjected to Nuclear Magnetic Resonance Spectroscopy (^1H NMR) to elucidate the molecular structure, and the results indicated the structure for 4'Hydroxy-3,5,6,7,8, 3'hexamethoxyflavone. This newly isolated compound is the first reported natural compound extracted from *C. jabara* (Rie et al. 2012).

8.4 Methods of Analysis of Flavonoids from *Citrus jabara*

The *C. jabara* fruits harvested from Kitayama village, Wakayama prefecture. The residue from after juicing the *C. jabara* fruits was used as fruit peels. These peels were dried and then chopped before being sealed in an aluminium puck for later use (Azuma et al. 2020). HPLC analysis was employed to confirm the chemical compounds in Jabara. An ultrasonic extraction process was employed in mixing Jabara powder with 50% of methanol for 20 min, and later, extracts were filtered in a syringe filter of 0.45 μm size and stored in a glass vial. 5.0 μL of extract subjected to HPLC separation using a COSMOSIL 5C18-AR-II column (φ 4.6 × 250 mm). The operational conditions employed were 40 °C temperature, 20% acetonitrile containing 0.8% acetic acid as a mobile phase, 1.0 mL/min set as flow rate and UV wavelength, 280 nm. There are four flavanones, eight flavones and three chalcones reported. A standard curve was plotted for each flavanone (hesperidin, neohesperidin, naringin, and naringenin), respectively, using analytical standards. The Limit of Detection (LOD) and Limit of Quantification (LOQ) of each flavanone is hesperidin, 40 ng/mL & 120 ng/mL; neohesperidin, 92 ng/mL & 280 ng/mL; naringin, 65 ng/mL & 200 ng/mL and naringenin, 10 ng/ml & 30 ng/mL. The amount of flavanones was below the detection limit (Azuma et al. 2020).

In another study, from Kitayamamura Jabararamura Center, Jabara juice was purchased (Wakayama, Japan; lot number: 2018.12.16 C; production year: 2018) to analyse the contents by HPLC method. For HPLC sample preparation, the jabara juice was centrifuged at 16,000 g for 15 min at 4 °C. The supernatant was passed through a 0.22 μm membrane filter to remove any insoluble materials before HPLC analysis, the filtrate was diluted to 10%, and a 20 μL sample was injected for analysis. Further, narirutin, naringin and naringenin HPLC grade analytical standards were used as comparative agents to identify compounds in jabara juice extracts.

Shimadzu HPLC model Prominence-i LC-2030C pump with UV detector and C18 (octadecylsilyl silica-ODS) column (Cosmosil, 5C18-MSII, 4.6 × 250 mm, 5 μm; NacalaiTesque) was used for analysis with following chromatographic conditions. The total method runtime was 33.5 min consisting of mobile phase acetonitrile (A) and water (B) with gradient flow (0–14 min: 21% A; 14–21 min: 21–65% A, linear; 21–26 min: 65–70% A, linear; 26–27 min: 70–21% A, linear; 27–33.5 min: 21%). The other chromatographic conditions were column oven temperature, 25 °C; flow rate, 1 ml/min; UV wavelength, 280 nm and injection volume, 20 μL. Standard curves for narirutin, naringin and naringenin were plotted with quantification ranges of 2–100 μg/mL, 4–200 μg/mL, and 1–50 μg/mL and analysed the constituents in Jabara juice (Morita et al. 2020).

8.4.1 Toxicity of Flavonoids

Cytotoxicity of Jabara flavonoids was analysed on RAW 264.7 macrophage cell line using cell counting kit-8. In this assay, ~1.0 × 10^5 RAW 264.7 cells were inoculated on a 96-well plate containing Dulbecco's Modified Eagle Medium (DMEM) with 10% Fetal Bovine Serum (FBS) and antibiotics. Various flavonoids are inoculated to wells and incubated at 37 °C for 24 h. After the incubation, cells were washed with 200 μL of DMEM supplemented with 10% FBS and 10 μL of WST-8 reagent. Later, incubated for 2 h, and cell viability was measured at 450 nm.

Flavonoids like narirutin and naringenin showed cell viability without toxicity up to 500 μM and 400 μM, respectively. However, at 500 μM concentration, naringenin only exhibited cell viability at 85.8 ± 6.0%. There was mild toxicity reported from three flavones and one flavanone. Further, one chalcone showed substantial toxicity against RAW 264.7 cell line reported (Azuma et al. 2020).

8.5 Uses of *Citrus jabara* in Disease Management

8.5.1 Atopic Diseases

Atopic dermatitis (AD) is characterized by abnormally rough and dry skin with dry scales and small cracks (Xerosis), intense itchy skin (Pruritus), and Eczematous lesions leading to chronic inflammatory skin disease, symptoms of AD relapse and chronically fluctuate. The incidence of AD is increasing worldwide, and it is usually developing in infants or early childhood, and it may relapse in childhood and in some cases, the recurrent relapses lead the AD into adulthood (Arima et al. 2018). AD leads to other allergic diseases like asthma and allergic rhinitis; the patient suffering from AD during his childhood is said to be at risk of 50% and 75%, respectively (Spergel and Paller 2003; van der Hulst et al. 2007; Thomsen 2015a). AD development is commonly observed in individuals with a personal or familial history of eczema, susceptible to overproduction of immunoglobulin E (IgE). Studies even predicted genetic predisposition and some environmental factors play a role in causing AD (Thomsen 2015b; Lavery et al. 2016). The pathophysiology of AD is complex as it has multiple etiologies, making it a multifactorial disease as there is no particular hierarchy among the etiologies, which makes AD with diverse symptoms, however generally in chronic AD, primarily type 2 inflammation is prevalent due to immune disorders and skin barrier dysfunction (Langan et al. 2020; Katoh et al. 2020).

According to the report by the International Study of Asthma and Allergies in Childhood (ISAAC) phase three, the prevalence of AD varies from country to country, and even within a country. In this study, the researchers investigated a diverse cohort, comprising two age groups: children aged 6 to 7 years (385,853 participants from 60 countries) and adolescents aged 13 to 14 (663,256 participants from approximately 96 countries). The findings revealed a broad range of prevalence rates, ranging from 0.9% in India to 22.5% in Ecuador for the children's age group. Conversely, the adolescent group exhibited prevalence rates ranging from 0.2% in China to 24.6% in Columbia (Odhiambo et al. 2009). A Japanese study showed that in 2005 age groups of 6–10, 12–13, and 14–15 years had a prevalence rate of 19.6%, 13.6%, and 10.9%, respectively (Furue et al. 2011; Takizawa et al. 2018).

Due to horny cell layer dysfunction, the stratum corneum becomes leaky, and allergen sensitization is increased, resulting in inflammation. This tends to happen due to an abnormal decrease in ceramide content in the stratum corneum, and a mutation in the horny cell filaggrin protein gene (FLG) leads to inflammation (Melnik et al. 1988; Cabanillas and Novak 2016). Because allergens such as proteins, dust mites, and many other allergens enter through a disrupted skin barrier, exacerbating type 2 immune reactions, these allergens indirectly activate Th2 cells, which are linked to allergic reactions, by producing interleukin (IL) IL-33 and IL-25, and this immune response also activates allergen-specific IgE, causing inflammation. Itching is triggered due to chemicals such as cytokines and chemokines, IL-31, IL-4, and thymic stromal lymphopoietin (TSLP). Hypersensitivity results from skin lesions that cause pruritus and act on nerves, triggering scratching (Tominaga and Takamori 2014; Katoh et al. 2020).

Genetic linkage studies have provided sufficient solid evidence regarding the genetic background of AD; also, the mutation of the FLG gene accounts for 40% of cases in patients with AD. Other than the FLG gene, many different genes have been reported based on genome-wide linkage analysis from Japanese samples, are CARD4, CARD15, CARD11, SCYA11, CCDC80, GM-CSF, TLR2, TLR9, IL4, IL13, IL18, GLB1, RANTES, TGFb1 many more other genes have been reported (Palmer et al. 2006; Hirota et al. 2012; Katoh et al. 2020). In addition to genetics, environmental factors and epigenetics also significantly contribute to the aetiology of AD. There is sufficient evidence to conclude that environmental influences such as temperature, cold and dry air, microbial exposure, improved agricultural cleanliness, broad-spectrum antibiotics, urban life, a western diet, air pollution, and cigarette smoke can affect epigenetic changes (Flohr and Mann 2014; Sabounchi et al. 2015; Perkin et al. 2016; Takizawa et al. 2018).

Yet another cause of AD is the dysbiosis of the normal microbiome of the skin in cases of AD; this is prevalent, typically when an infant is born with normal delivery harbours with the microbiota of mother vaginal canal microbiota predominating with, *Prevotella, or Sneathia and Lactobacillus* spp., on the other hand, the baby born by cesarean section harbours the skin microflora dominated by *Staphylococcus, Propionibacterium, and Corynebacterium* spp. (Gevers et al. 2012; Iwatsuki et al. 2018). The abundance and diversity among each and every individual is also based on the cutaneous conditions of the individuals. The microbiota change can be seen in sebaceous and moist regions of the body colonized by the *Propionibacterium, Staphylococcus* and *Corynebacterium* species (Costello et al. 2009). 90% of patients suffering from AD harbour *S. aureus* in the lesional skin. As discussed, there are gene mutations of skin barrier functions, including filaggrin (FLG) and with the serine protease inhibitor Kazal-type 5 (SPINK-5), which encodes the protease inhibitor lymphoepithelial Kazal-type-related inhibitor (LEKTI). After colonization,

S. aureus secrets different types of virulence factors such as α-, β-, δ-, and γ-toxins which are hemolysins and a group of enzymes like collagenase, hyaluronidase, protease, nuclease etc. these virulence factors further deteriorate the skin barrier function and illicit inflammatory response by activation of various immunes cells and cytokines.

Diagnosis of AD is carried out based on two widely accepted diagnostic criteria by the Japanese Dermatological Association (JDA) and Diagnostic Criteria by Hanifin and Rajka. AD is a multifactorial disease, and completely curing AD is not possible. Controlling the aggravating factors associated with pathogenesis and also the reduction of irritability, inflammation, and other factors is the treatment procedure, which starts by assessing the patient's grade of symptoms. Drug treatment is used wherein topical anti-inflammatory drugs such as topical corticosteroids (TCS) are chosen. These TCS are graded and ranked into 5 different groups based on the effect shown, with Group 1 being the strongest and Group 5 being the weak. The group is chosen based on the severity of the AD eruption in the patients based on lesion characteristics. TCS can be used in the form of an ointment, cream, lotion, and tape preparations. Over the period of 4 weeks, TCS is applied; if there is no significant effect on eruptions, then it is recommended to change the treatment method (Katoh et al. 2020). TCS indirectly affects inflammation-inducing transcription factors such as AP-1, NF-κB, and C/EBP, which glucocorticoid receptors activation by TCS inhibits. Local side effects of the usage of TCS are bacterial, fungal and viral skin infections along with acne, hypertrichosis, dryness, skin atrophy and striae distensae (Cain and Cidlowski 2017).

Topical tacrolimus (TTS) is another topical inflammatory drug used to treat AD. The molecular weight of TTS is around 822 Da dosage is recommended in Japan for adults and children at 0.1% and 0.03%, respectively, with an upper limit of applying TTS for adults being 5 g, and in children, it varies on the body weight, and age factor. 1 g is recommended for 2–5 years aged with an average body weight of less than 20 kg. Children aged 6–12 years with an average weight of 20–50 kg can use the ointment 2–4 g. The mechanism of action of TTS is different from TCS. It suppresses the mast cell degranulation, reduces the expression of receptors like IL-31, IL-33 and ST2, and eosinophil activation. There is suppression of expression of genes of Cytokines such as IL-4, IL-5, IL-12, and IFN- γ responsible for inflammation. Along with this, T-cells activation is also suppressed by binding with FK506 protein in T-cells, thus inhibiting the dephosphorylation of the NF-AT transcription factor. TTS over usage cause side effects burning and irritation symptoms at the topical part, and viral, bacterial skin infections are apparent (Arellano et al. 2007; Miyano and Tsunemi 2021).

Topical delgocitinib inflammatory drug with a molecular weight of 310 da smaller than TCS and TTS 0.5% of ointment with an upper limit of 5 g is used to treat AD. The mechanism of action includes decreased filaggrin expression, inhibiting the expression of cytokines that act on sensory nerves, leading to pruritis. It also inhibits the JAK subtypes, a transcription factor of various cytokine-expressing genes. Side effects include contact dermatitis, acne, and herpes virus infection, which can be used by pregnant and lactating mothers (Katoh et al. 2020; Miyano and Tsunemi 2021).

In the severe case of AD, treatment with topical anti-inflammatory drugs alone is not sufficient in controlling the AD, systemic therapy can be used considering the patient profile and the effect which may be caused due to systemic therapy on the patients, and the topical treatment is also followed along with the systemic therapy and the focus of the treatment resolve the exacerbating factors to reduce the dependence on the drugs. The medication which is used for systemic treatment is Cyclosporin which acts as an inhibitor of calcineurin, and its use is widely practised in Europe, The USA, Japan and other countries. Cyclosporin is only approved in severe AD cases, with more than 30% of marked eruptions with inflammation on the body surface. The dosage recommended is 3 mg/kg/day, and the course of therapy should be completed within 8–12 weeks patients with nephropathy, hypertension, and other infections must be subjected to screening before starting the systemic therapy (Yu et al. 2018; Katoh et al. 2020; Miyano and Tsunemi 2021).

Even though oral corticosteroids are used, they are not recommended as a treatment for long-term usage of oral corticosteroids leads to adverse systemic reactions. One such drug mostly used for a short period of time is Dupilumab. Many randomized controlled phase trials have been conducted on this drug. It showed promising results in widespread skin infections, systemic infections, herpesvirus infection and Kaposi's varicelliform eruption was significantly reduced compared to the placebo group. Even improved dermatitis eruption and pruritus were reported, and this drug can be used in severe AD as a multidrug treatment schedule (de Bruin-Weller et al. 2018; Katoh et al. 2020).

If the above topical, anti-inflammatory, anti-histamines and moisturizer treatment option fails to contain the severity of AD and its remission, then Ultraviolet therapy is advised for such patients. UV light of different bands like Narrow-band UV B (NB-UVB) with wavelength 311 nm, medium-wave UV (290–320 nm) and long-wave UV (320–380 nm) is used. Among them, NB-UVB is widely used in clinical practice, apoptotic effect on T-cell infiltrating dermis leads to a reduction in inflammatory cytokines production, leading to an immunosuppressive impact on patients, UV treatment on pregnant women's and children below 10 years age group is not yet studied and recommended (Patrizi et al. 2015; Fujita et al. 2018; Katoh et al. 2020).

The anti-inflammatory activity of seven compounds isolated from *C. jabara* peels (4-Flavones, 2- Flavanones and 1-Chalacones) were evaluated by cell-based assay with murine macrophage-like cell line (RAW 264.7) (RCB0535). The anti-inflammatory activity was evaluated by measuring levels of various inflammatory factors (nitric oxide (NO), IL-6, and TNF-α), protein expression of iNOs by western blot analysis and enzyme inhibitory assay (soluble epoxide hydrolase (sEH), hyaluronidase and chymase). RAW 264.7 cells were cultured with different concentrations of isolated flavonoid compounds and inflammatory responses were induced by Lipopolysaccharide (LPS) (Azuma et al. 2020).

Nitrous Oxide (NO) Production and iNOS Protein Expression

The level of NO production was measured by Griess reagent and 1 μg/mL LPS was given as inflammatory producing agent. The iNOS protein expression was compared with β-actin protein expression using western blot analysis. The flavonoids were given in 50, 100, and 200 μM concentrations and dose-dependently inhibition of NO productions and iNOS protein expression were observed (Azuma et al. 2020).

IL-6, and TNF-α Production Inhibition The RAW 264.7 cells were cultivated with 1 μg/mL LPS as inflammatory inducing agent. The seven isolated flavonoids of varying concentrations (50, 100, and 200 μM) are tested for its IL-6, and TNF-α production inhibition and the IL-6, and TNF-α were measured using mouse ELISA kits. The isolated flavonoids showed dose dependent inhibitory action. The inhibitory action for IL-6 secretion among the isolated compounds, narirutin and FLV3 showed minimum activity but naringenin, FLV1 and FNN2 showed highest activity. For TNF-α inhibitory activity, only FNN2 showed highest inhibitory activity (Azuma et al. 2020).

Enzyme Inhibition Assay Soluble epoxide hydrolase (sEH), hyaluronidase and chymase related to inflammation reaction. RAW 264.7 cells were cultivated with 1 μg/mL LPS as inflammatory inducing agent and the enzymes inhibitory activity were evaluated for the isolated compounds. Dose dependent inhibitory activity was observed and naringenin showed significant enzyme inhibitory activity while narirutin showed least activity (Azuma et al. 2020).

In Japan, Jabara fruits has been considered as anti-allergic functional foods. Narirutin, a flavonoid isolated from *C. jabara* fruit peels showed anti-inflammatory activity. Narirutin is metabolized into aglycone naringenin and showed the following inhibitory effects such as nitric oxide synthesis, nitric oxide synthase induction, Interleukin-6 synthesis and inducible soluble epoxide hydrolase activities. In a study conducted in Japan on the fruit peel powder cream produced by *C. jabara* was evaluated on 20 different patients suffering from AD. Patients treated with 5% CJ peel powder cream for weeks showed a significant decrease in AD severity. Around 11 patients have experienced an improvement in skin lesions, and 16 out of 20 recommended it as safe to use and valuable cream for treating AD (Inaba et al. 2020). CJ fruit is anti-allergic in nature. One of the studies evaluated the flavonoids present in the CJ fruit. It was found that it is abundant in narirutin flavonoid, and aglycone naringenin of narirutin showed epoxide hydrolase activity, Interleukin-6 synthesis, nitric oxide synthase induction inhibitory effects, which proves that ingestion of CJ induces anti-inflammatory effects in patients suffering from AD and other diseases (Azuma et al. 2020). A randomized double-blind parallel-group comparative study conducted by Azuma et al. (2021) reported that a daily intake of CJ peel powder for 4 weeks would alleviate the allergy-like symptoms in patients compared to placebo group patients who reported a dose-dependent reduction in symptoms of rhinorrhea.

The juice of CJ is bitter, so Uchida et al. (2020) fermented it and made it more palatable. The fermented *C. jabara* juice evaluation revealed that it contains 5-hydroxymethylfurfural (5-HMF), and it increases as the fermentation time increases. It was noted that it is the active component in the juice responsible for regulating mast cell degranulation. Apart from cytokine reduction, it also reduced the expression of phospho-PLCγ1 and phospho-ERK1/2 because of FcεRI stimulation by 5-HMF.

Japanese Cedar Pollinosis

A tree called *Cryptomeria japonica* (Japan cedar) in Japan is the root cause of a disease named Japanese cedar pollinosis (JCP), which is caused by the inhalation of the pollen produced by Japan cedar tree causing an allergic condition, and it is a major reported health problem in Japan, due to its high commonness in the Japan, poor impulsive recover rate and severe symptoms make it as health problem with grave concern and economic burden on the patient (Okuda 2003).

Mucus is produced by the goblet or submucosal glands of epithelia; usually, a small amount is helpful to keep the airway moist, but in the case of allergic asthma or COPD, there is an exacerbation in the secretion leading to breathing difficulties. Around twenty different genes are responsible for the production. Still, it is noted that in patients, there is overexpression of MUC5AC, another factor that causes excessive mucin production is the lipopolysaccharide layer of microorganisms that enter the airways, which causes inflammation and activates proinflammatory cytokines, which indirectly causes the overexpression of the MUC5AC gene and excessive mucus secretion. *C. jabara* fruits contains a range of flavonoids, including narirutin, a particularly abundant flavonoid. *C. jabara* extracts have been used to test on NCI-H292 human airway epithelial cells, and it was observed that there was a significant decrease in the activity of the ERK pathway in NCI-H292 human airway epithelial cells, inhibition of ERK leads to the deactivation of the transcription factors NF-κB or Sp1 that induce the transcription of the MUC5AC gene, repressing the overproduction of MUC5AC induced by LPS in cells and in medium, which might be a promising natural way of treating allergic asthma and COPD (Iwashita et al. 2017).

Apart from the anti-inflammatory properties of *C. jabara* extracts, there is a report in conference proceedings by (Shiho and Yoshiharu 2016) that *C. jabara* water extracts showed a strong lipolysis proving to be an anti-obesity candidate. They have isolated three different compounds from *C. jabara* water extracts 3,5,6,7,8,3′,4′-Heptametoxyflavone, 3-hydroxy-5,6,7,8,3′,4′-hexametoxyflavone and 3,5,6,7,3′,4′-hexametoxyflavone and each one of them with concentration of 200 μM were tested against in vitro cell line 3T3-L1 adipocytes and the lipid accumulation reduction was assessed using Oil red O staining showed significant percentage of reduction that is 50%, 50%, and 10% respectively when compared with untreated adipocytes, to conform the antiadipocytic effect and which compound is showing highest antiadipocytic effect, glycerol-3-phosphate dehydrogenase (GPDH) activity was assessed and 3,5,6,7,8,3′,4′-Heptametoxyflavone compound showed intense lipolysis activity among other two compounds. The development of Jabara and the uses of jabara compounds pictorially presented in Fig. 8.2.

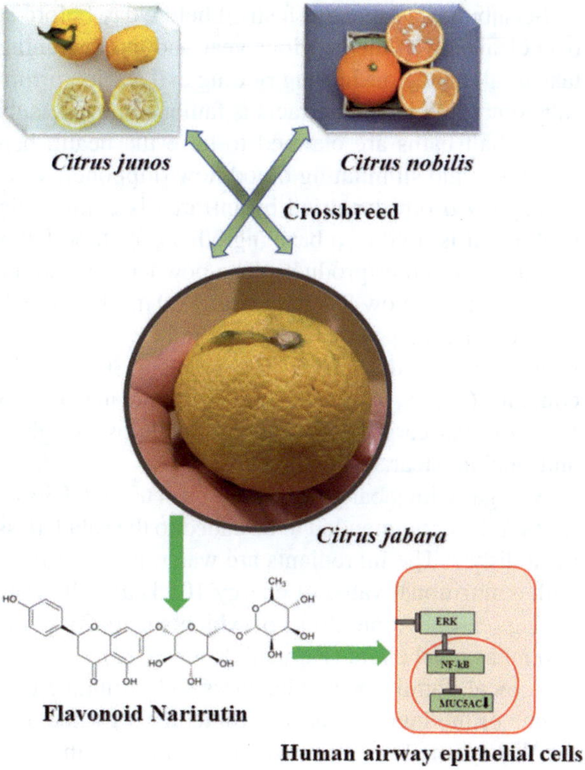

Fig. 8.2 Pictorial representation of natural hybrid of Jabara and uses of jabara compounds

8.6 Products of *Citrus jabara*

C. jabara fruit has a unique taste and has been consumed by Japanese people since ancient times, and Jabara-flavoured products are marketed due to their popularity. Sugar-free mint candies (Japan Candy Store Kawaii Group 2022a), chewy candies (Japan Candy Store Kawaii Group 2022b) and gummies (Meccha Japan 2022) made from these fruits are popular among Japan population. Marmalades made from finely minced fruit peels and fruit extracts are marketed. The product claims to treat bitterness and astringency (Narurich Corporation 2022).

Liqueurs made from soaking Jabara fruits in sake for flavour enhancement are marketed all over Japan (Yoshimura Hideo Shoten Co. 2022). A unique product, fruit liquor containing 7% alcohol content, uses Shimokitayama village-grown *C. jabara* in the marketed alcohol preparation (Umeshuthai store 2022).

C. jabara concentrates in pill form as supplements are sold in the market containing 270 as a pack (Goods Of Japan 2022). Jabara powder with Shin-Sanmi – Sansho (Japanese Pepper) and Red Chili is used as one seasoning powder for various dishes. This product was made in southern Japan, where the Jabara is native and supplied to Michelin-star restaurants (Michelin-Kaiseki Supplies 2022).

Setsubun is a Japanese festival held a day before spring, and people believe it to be a cleansing of the previous year and the beginning of a new year. The Japanese take a salt bath, considering ridding evil spirits (prtimes.com Japan 2022). The bath salt containing jabara extracts is famous in the setsubun festival and sold widely in japan. Salt baths are claimed to have the health benefits of miniaturization, skin softening, and stimulating blood flow (nipponeffect.com 2022).

C. jabara powder dried by infrared is commercially sold in aluminium pouch packing. It is used as a base ingredient for food flavouring, confectionaries, health products and other products. The powdered jabara claims to contain narirutin and water content below 8% with below 60 mesh size (Chokyu Souyaku Co. 2022).

The pericarp powder is commercially sold as a seasonal allergy relief supplement in a pack dosage of 60 tablets for 30 days (2 tablets per day). Each tablet contains *C. jabara* powder 150 mg and narirutin 6.25 mg. The ingredients are *C. jabara* Pericarp Powder, maltitol/cellulose, brightener, silicon oxide nanopowder and calcium stearate (Koplina 2022).

Vinegar with jabara flavour has been used in Kitayama local cuisine since ancient times. It is recommended to be added to the salad, desserts, fresh beverages, fish and meat dishes. The ingredients are water, honey, 15% jabara juice and apple vinegar with a nutritional value of energy 105 kcal (446 kJ); fat <0.5 g, of which saturates <0.1 g; carbohydrate 26 g, of which sugars 21 g; protein <0.5 g; salt <0.01 g per 100 ml as label claim (Kumano Kodo 2022).

A local, family-owned business sells a unique tea blend containing jabara peels as one of the ingredients. The ingredients for the tea blend are locally produced in the Wakayama Prefecture with cultivation methods free of pesticides and chemical fertilizers. It is claimed to have mellow sweetness and refreshing aroma, which creates a calming sensation (Yunomi Newsletter 2022).

A Japanese company has a variety of skincare products like moisturisers, skin softeners, de-stressor, skin soothers, antioxidants and depigmentation. The label claims that these products contain 2% jabara extract as one of the main active ingredients. The company claims that extract is rich in antioxidants, Vitamin C and polyphenols, which helps to reduce pigmentation (beautypie.com 2022).

The *C. jabara* has many patents filed in Japan, US and WIPO for their broad therapeutic uses. Japan patent application (JP2011287558A) published July 18, 1995; has applied for their antiallergic composition to treat type I allergic diseases, including allergic rhinitis, hay fever, bronchial asthma and atopic dermatitis (Konaga Yanagisawa et al. 2012).

The WIPO patent application dated 06 Aug 2009 (WIPO – WO 209/096988 A1) on *C. jabara* extracts has mentioned that it has skin smoothening and anti-histamine activity which can be used to treat skin allergy, sensitive skin, skin inflammation and skin irritation (Goyarts et al. 2009).

8.7 Conclusion

C. jabara is one of the citrus fruits usually grown locally in Wakayama prefecture, Kitayamamura, Japan and its nearby areas. Jabara assumed that it is a hybrid of other citrus species which are present in the Wakayama area, including Yuzu, Kunenbo (*C. nobilis* Lour.), and Kishu-mikan or Kishu (*C. kinokuni* Hort. ex Tanaka). Some molecular studies predicted that Kunenbo-A is a seed parent and yuzu is a male parent. Jabara grows naturally in Wakayama prefecture, and there is no need for any special kind of fertilizers other than those used to cultivate other citrus species. Jabara flowering time is around the month of May, and the development of fruit and their harvesting takes between the end of November and December month starting. *C. jabara*juice mainly contains up to 50% of its total soluble sugars like glucose and fructose, along with sugars, pectin oil, and citric acids. Further, forty-nine different compounds are reported from fruit and peel of Jabara, among which myrcene, limonene, and terpinene are found majorly. These compounds have various biological activities, such as their use against atopic dermatitis, anti-inflammatory and allergic asthma and COPD problems. In Japan, Jabara used for various allergy problems as a natural preventive substance. From Jabara various products are also prepared such as candies, drinks toffies, powder in confectionaries, and vinegar.

References

Anonymous (2022) Citrus jabara Taxonomy ID. NCBI taxonomy database

Arellano FM, Wentworth CE, Arana A et al (2007) Risk of lymphoma following exposure to calcineurin inhibitors and topical steroids in patients with atopic dermatitis. J Invest Dermatol 127:808–816. https://doi.org/10.1038/sj.jid.5700622

Arima K, Gupta S, Gadkari A et al (2018) Burden of atopic dermatitis in Japanese adults: analysis of data from the 2013 National Health and Wellness Survey. J Dermatol 45:390–396. https://doi.org/10.1111/1346-8138.14218

Azuma S, Murakami Y, Azuma E et al (2020) Anti-inflammatory effects of flavonoids in Citrus jabara fruit peels. Trends Immunother 4:5. https://doi.org/10.24294/ti.v4.i1.844

Azuma S, Murakami Y, Taniguchi M et al (2021) Daily intake of Citrus jabara fruit peel powder (Japanese Patent No. 5,323,127) improves allergy-like symptoms: a randomized double-blind parallel-group comparative study. Trends Immunother 5(21). https://doi.org/10.24294/ti.v5.i2.1390

beautypie.com (2022) Beauty Pie jabara. https://www.beautypie.com/us/shop/search?term=jabara

Brian T, Michelle L (2020) Compositions for treating citrus disease and promoting yield increase in row crops. U.S. Patent Application 16/749,576, filed August 13, 2020.

Cabanillas B, Novak N (2016) Atopic dermatitis and filaggrin. Curr Opin Immunol 42:1–8. https://doi.org/10.1016/j.coi.2016.05.002

Cain DW, Cidlowski JA (2017) Immune regulation by glucocorticoids. Nat Rev Immunol 17:233–247. https://doi.org/10.1038/nri.2017.1

Chokyu Souyaku Co L (2022) Citrus jabara powder. https://chokyusouyaku.com/en/products/jabarapowder/

Costello EK, Lauber CL, Hamady M et al (2009) Bacterial community variation in human body habitats across space and time. Science 326:1694–1697. https://doi.org/10.1126/science.1177486
DB Pedia (2022) About: Jabara (citrus). https://dbpedia.org/page/Jabara_(citrus)
de Bruin-Weller M, Thaçi D, Smith CH et al (2018) Dupilumab with concomitant topical corticosteroid treatment in adults with atopic dermatitis with an inadequate response or intolerance to ciclosporin A or when this treatment is medically inadvisable: a placebo-controlled, randomized phase III clinical trial (LIBERTY AD CAFÉ). Br J Dermatol 178:1083–1101. https://doi.org/10.1111/bjd.16156
Döring M, Global Biodiversity Information Facility (2022) Citrus ×jabara Tanaka. https://www.gbif.org/species/165581432
Flohr C, Mann J (2014) New insights into the epidemiology of childhood atopic dermatitis. Allergy 69:3–16. https://doi.org/10.1111/all.12270
Fujita H, Terui T, Hayama K et al (2018) Japanese guidelines for the management and treatment of generalized pustular psoriasis: the new pathogenesis and treatment of GPP. J Dermatol 45:1235–1270. https://doi.org/10.1111/1346-8138.14523
Furue M, Chiba T, Takeuchi S (2011) Current status of atopic dermatitis in Japan. Asia Pac Allergy 1:64–72. https://doi.org/10.5415/apallergy.2011.1.2.64
Gevers D, Knight R, Petrosino JF et al (2012) The human microbiome project: a community resource for the healthy human microbiome. PLoS Biol 10:e1001377. https://doi.org/10.1371/journal.pbio.1001377
Goods of Japan (2022) JABARA supplements 270 seeds: miracle citrus Jabara concentrate blue melon ingredients 1 PC: 1 month supply. https://goodsofjapan.com/products/jabara-supplements-270-seeds-miracle-citrus-jabara-concentrate-blue-melon-ingredients-1-pc-1-month-supply-41443
Goyarts E, Mammone T, Ranieri-Chamberlain A, et al (2009) Topical composition containing Citrus Jabara extract. U.S. Patent Application 12/024,710, filed August 6, 2009
Herbalistics Pty Ltd (2022) Citrus jabara – Jabara (plant) QLD ONLY! https://herbalistics.com.au/product/citrus-jabara-plant/
Hirota T, Takahashi A, Kubo M et al (2012) Genome-wide association study identifies eight new susceptibility loci for atopic dermatitis in the Japanese population. Nat Genet 44:1222–1226. https://doi.org/10.1038/ng.2438
Inaba Y, Furukawa F, Azuma S et al (2020) Evaluation of the safety and usefulness of citrus jabara fruit peel powder cream for patients with atopic dermatitis. Trends Immunother 4:42. https://doi.org/10.24294/ti.v4.i1.1230
Iwashita J, Iguchi N, Takashima A et al (2017) Citrus jabara extracts suppress MUC5AC mucin production in human lung epithelial cells. Adv Biol Chem 07:139–150. https://doi.org/10.4236/abc.2017.73009
Iwatsuki K, Yamasaki O, Morizane S (2018) Microbiome, dysbiosis, and atopic dermatitis. In: Evolution of atopic dermatitis in the 21st century. Springer, Singapore, pp 141–155
Jabara Laboratory (2022) Jabara Laboratory. https://www.jabaralaboratory.com/Jabaralaboratory/English.html
Japan Candy Store Kawaii Group (2022a) Mintia Asahi Mintia Mints – Jabara Citrus. https://www.japancandystore.com/asahi-mintia-mints-jabara-citrus
Japan Candy Store Kawaii Group (2022b) Puccho chewy candy – Jabara. https://www.japancandystore.com/puccho-chewy-candy-jabara
Japan External Trade Organization (2022) Citrus Jabara Marmalade. JETRO, Los Angeles, https://foodanddrinkjapan.com/products/citrus-jabara-marmalade/
Katoh N, Ohya Y, Ikeda M et al (2020) Japanese guidelines for atopic dermatitis 2020. Allergol Int 69:356–369. https://doi.org/10.1016/j.alit.2020.02.006
Konaga Y, Ryo S, Akinori K, Hideki I (2012) JP5909795B2 - Antiallergic Composition. Japan Patents, JAPAN PATENT OFFICE, 27 Apr. 2016, patents.google.com/patent/JP2013136528A/en. Accessed 11 July 2023

Koplina (2022) Jabara Tablet Citrus Jabara, 60 tablets, 30 days, Seasonal allergy relief supplement. https://www.lazada.com.my/products/jabara-tablet-citrus-jabara-60-tablets-30-days-seasonal-allergy-relief-supplement-i2123616836-s8654395966.html?spm=a2o4k.tm80167375.4138544390.1.35794516Q6TozQ.35794516Q6TozQ

Kumano Kodo (2022) Jabara citrus vinegar. https://www.nishikidori.com/en/condiment/525-jabara-citrus-vinegar-3701184001612.html

Langan SM, Irvine AD, Weidinger S (2020) Atopic dermatitis. Lancet 396:345–360. https://doi.org/10.1016/S0140-6736(20)31286-1

Lavery MJ, Stull C, Kinney MO, Yosipovitch G (2016) Nocturnal pruritus: the battle for a peaceful night's sleep. Int J Mol Sci 17:425. https://doi.org/10.3390/ijms17030425

Meccha Japan (2022) Gummies citrus Jabara Honpo. https://meccha-japan.com/en/gummies/64478-gummies-citrus-jabara-honpo.html

Melnik B, Hollmann J, Plewig G (1988) Decreased stratum corneum cerarnides in atopic individuals—a pathobiochemical factor in xerosis? Br J Dermatol 119:547–548. https://doi.org/10.1111/j.1365-2133.1988.tb03262.x

Michelin-Kaiseki Supplies (2022) Shin-Sanmi: Sansho (Japanese Pepper) + Red Chili + Citrus Jabara Powder seasoning Grill BBQ teriyaki Kaiseki fried chicken yakitori beef pork unagi don yaki soba udon ramen noodle oyakodon egg. https://www.amazon.com/%E4%B8%89%E5%91%B3-Shin-Sanmi-Japanese-seasoning-teriyaki/dp/B07PYNFYRZ

Miyake M, Inaba N, Maeda H, Ifuku Y (1990) Quality characteristics of Jabara (Citrus jabara Hort. ex Tanaka) and Yuzu (Citrus junos Sieb. ex Tanaka) fruit juice. Nippon Shokuhin Kogyo Gakkaishi 37:346–354. https://doi.org/10.3136/nskkk1962.37.5_346

Miyano K, Tsunemi Y (2021) Current treatments for atopic dermatitis in Japan. J Dermatol 48:140–151. https://doi.org/10.1111/1346-8138.15730

Morita T, Akiyoshi T, Sato R et al (2020) Citrus fruit-derived flavanone glycoside narirutin is a novel potent inhibitor of organic anion-transporting polypeptides. J Agric Food Chem 68:14182–14191. https://doi.org/10.1021/acs.jafc.0c06132

Narurich Corporation (2022) Citrus Jabara marmalade. https://foodanddrinkjapan.com/products/citrus-jabara-marmalade/

Niimi E, Fujii H, Ohta S et al (2021) Development of acid citrus cultivar identification by CAPS markers. Hortic Res (Japan) 20:17–27. https://doi.org/10.2503/hrj.20.17

nipponeffect.com (2022) Jabara Fruit Bath Salt – 25g Sachet. https://www.nipponeffect.com.au/jabara-fruit-bath-salt-25g-sachet

Odhiambo JA, Williams HC, Clayton TO et al (2009) Global variations in prevalence of eczema symptoms in children from ISAAC Phase Three. J Allergy Clin Immunol 124:1251–8.e23. https://doi.org/10.1016/j.jaci.2009.10.009

Okuda M (2003) Epidemiology of Japanese cedar pollinosis throughout Japan. Ann Allergy Asthma Immunol 91:288–296. https://doi.org/10.1016/S1081-1206(10)63532-6

Omori H, Nakahara K, Umano K (2011) Characterization of aroma compounds in the peel extract of Jabara (Citrus jabara Hort. ex Tanaka). Flavour Fragr J 26:396–402. https://doi.org/10.1002/ffj.2066

Palmer CNA, Irvine AD, Terron-Kwiatkowski A et al (2006) Common loss-of-function variants of the epidermal barrier protein filaggrin are a major predisposing factor for atopic dermatitis. Nat Genet 38:441–446. https://doi.org/10.1038/ng1767

Patrizi A, Raone B, Ravaioli GM (2015) Management of atopic dermatitis: safety and efficacy of phototherapy. Clin Cosmet Investig Dermatol 8:511–520. https://doi.org/10.2147/CCID.S87987

Perkin MR, Craven J, Logan K et al (2016) Association between domestic water hardness, chlorine, and atopic dermatitis risk in early life: a population-based cross-sectional study. J Allergy Clin Immunol 138:509–516. https://doi.org/10.1016/j.jaci.2016.03.031

prtimes.com Japan (2022) Get rid of evil spirits with Jabara Bath for Setsubun! Rare Fruit Spa leading Smooth Skin. https://prtimes.com/en/art/1517444374/

Rie M, Hiromi Y, Motoharu J et al (2012) Polymethoxyflavonoids from the peel of Citrus jabara. Shoyakugaku Zasshi 66:23–25

Sabounchi S, Bollyky J, Nadeau K (2015) Review of environmental impact on the epigenetic regulation of atopic diseases. Curr Allergy Asthma Rep 15:33. https://doi.org/10.1007/s11882-015-0533-1

Schoch CL, Ciufo S, Domrachev M et al (2020) NCBI taxonomy: a comprehensive update on curation, resources and tools. Database (Oxford) 2020:baaa062. https://doi.org/10.1093/database/baaa062

Shiho O, Yoshiharu O (2016) Antiobesity compounds from Citrus jabara. In: Proceedings of the II international conference on food chemistry & technology. J Food Chem Nanotechnol, Las Vegas, USA, pp S29–S30

Shimizu T, Kitajima A, Nonaka K et al (2016) Hybrid origins of citrus varieties inferred from DNA marker analysis of nuclear and organelle genomes. PLoS One 11:e0166969. https://doi.org/10.1371/journal.pone.0166969

Spergel JM, Paller AS (2003) Atopic dermatitis and the atopic march. J Allergy Clin Immunol 112:S118–S127. https://doi.org/10.1016/j.jaci.2003.09.033

Takizawa T, Morikawa A, Arakawa H (2018) Epidemiology of atopic dermatitis in Japan. In: Evolution of atopic dermatitis in the 21st century. Springer, Singapore, pp 13–24

Tanaka Y (1948) An iconograph of Japanese citrus fruits: a monographic study of species and varieties of citrus fruits grown in Japan. Yôkendô Company Limited, Tokyo

Thomsen SF (2015a) Epidemiology and natural history of atopic diseases. Eur Clin Respir J 2. https://doi.org/10.3402/ecrj.v2.24642

Thomsen SF (2015b) The contribution of twin studies to the understanding of the aetiology of asthma and atopic diseases. Eur Clin Respir J 2:27803. https://doi.org/10.3402/ecrj.v2.27803

Tom S (2017) Citrus 柑橘系. https://www.oishisojapan.com/home/2017/11/22/citrus-japanese-fruit-kankitsukei

Tom S (2020) Sanbokan: Japan's rare, sour citrus fruit

Tominaga M, Takamori K (2014) Itch and nerve fibers with special reference to atopic dermatitis: therapeutic implications. J Dermatol 41:205–212. https://doi.org/10.1111/1346-8138.12317

Uchida R, Kato M, Hattori Y, Kikuchi H, Watanabe E, Kobayashi K, Nishida K. (2020) Identification of 5-Hydroxymethylfurfural (5-HMF) as an Active Component Citrus Jabara That Suppresses FcεRI-Mediated Mast Cell Activation. Int J Mol Sci 21:7, 2472. https://doi.org/10.3390/ijms21072472

Umeshuthai store (2022) Shimokitayama Mura Jabara Orange/yuzu Liqueur. https://www.umeshuthai.com/umeshu/shimokitayama-mura-jabara-orange-yuzu-liqueur-720-ml/

van der Hulst AE, Klip H, Brand PLP (2007) Risk of developing asthma in young children with atopic eczema: a systematic review. J Allergy Clin Immunol 120:565–569. https://doi.org/10.1016/j.jaci.2007.05.042

Yoshimura Hideo Shoten Co. Ltd (2022) Liqueurs. https://nihonsyu-nihonjyou.co.jp/en/sake/liqueur/

Yu SH, Drucker AM, Lebwohl M, Silverberg JI (2018) A systematic review of the safety and efficacy of systemic corticosteroids in atopic dermatitis. J Am Acad Dermatol 78:733–740.e11. https://doi.org/10.1016/j.jaad.2017.09.074

Yuichirō T (1980) In an iconograph of Japanese citrus fruits: a monographic study of species and varieties of citrus fruits grown in Japan. Yokendo Co. Ltd, Tokyo

Yunomi Newsletter (2022) Wakayama Tamaryokucha green tea with Yuzu & Jabara Citrus. https://yunomi.life/products/dokodemosora-chamomile-sencha

Chapter 9
Kumquat

Secil Karahuseyin and Merve Nenni

9.1 Kumquat

Citrus, which is among the fruits with high nutritional value, grows in tropical and subtropical regions of the world. The genus *Citrus*, belonging to the Rutaceae family, includes more than 162 species, and kumquat (*Citrus japonica)* is one of the most important. Its name comes from Chinese, consisting of two words; the first is "Kum" meaning golden, and the second is "Quat" meaning good fortune (Safana et al. 2022).

9.2 History, Origin and Distribution

The kumquat tree is an extremely little tropical fruit tree. It comes from China, where it has been farmed since the twelfth century. Given that its fruit resembles a very small orange, its name, "kam kwat," is Cantonese for "golden orange." The fruit has a sweet and tart flavor and is eaten whole, including the peel. It is typically made into a liqueur, marmalade, or sweet treat (Pawełczyk et al. 2021). The fruit was widely popular in China throughout the Tang (618–907) and Song (960–1279) dynasties and was first referenced in Chinese literature in AD 1178. Since 1712,

S. Karahuseyin
Department of Pharmacognosy, Faculty of Pharmacy, Cukurova University, Adana, Türkiye
e-mail: skarahuseyin@cu.edu.tr

M. Nenni (✉)
Department of Analytical Chemistry, Faculty of Pharmacy, Cukurova University, Adana, Türkiye

S. Singh Purewal et al. (eds.), *Recent Advances in Citrus Fruits*,
https://doi.org/10.1007/978-3-031-37534-7_9

when they were identified as cultivable plants in Asia and Japan, kumquat cultivations have expanded to encompass Australia, Brazil, Argentina, California, Florida, South Africa, and the Mediterranean region, albeit as a minor crop (Palma and D'Aquino 2018).

The genus *Fortunella* Swingle belongs to the Rutaceae Juss. and Subfamily Aurantioideae Engl., and its species can grow to be shrubs or small trees. *Fortunella*'s taxonomy and systematics have long been a mystery, with numerous writers agreeing that there are six recognized species within the genus, all of which are found in Southeast Asia. Robert Fortune, a Scottish explorer and botanist, brought the genus to Europe in 1846. Following its continued cultivation, in the 1880s, it gained widespread recognition in Europe and was given his name. Swingle first classified the genus *Fortunella* in 1915, at which time he identified four species based on physical characteristics of the leaves, fruits, and seeds. *F. crassifolia* Swingle, *F. hindsii* (Champ. ex Benth.) Swingle, *F. japonica* (Thunb.) Swingle, and *F. margarita* (Lour.) Swingle were the four species that were described. However, later research produced a variety of viewpoints, with other authors releasing their insights into the novel species to increase the number. *Fortunella swinglei* Tanaka was printed in Malaysia in 1928 (Wang et al. 2022).

Fortunella japonica Swingle,"Round" or "Marumi" kumquat, and *Fortunella margarita* Swingle, "Oval" or "Nagami" kumquat, are the two most significant cultivars. The "Meiwa" kumquat (*Citrus crassiflora* Swingle), which is regarded as a natural hybrid between "Nagami" and "Marumi" kumquats, is another widely grown species, particularly in China, Australia and the United States. According to legend, the term "kumquat"derives from "chin kan" the original Chinese name, which has the meaning "gold orange". The "Marumi" kumquat's Japanese name is "kin kit" or "kin kan". The trade name for kumquats in Brazil is "kunquat", "kumquat"or "laranja de ouro dos orientais" whereas the name for them in Southeast Asia is "kin kuit." The most typical name in the Mediterranean region and the United States is kumquat. The seedless kumquats, pomelos, and other minor citrus fruits contributed roughly 10% of the world's total citrus production (FAO 2006; Ladaniya 2008). According to estimates, Korea produced 3589 tons of kumquats in 2001. However, trustworthy statistics sources for the global production of kumquats or for specific nations are not readily available (Palma and D'Aquino 2018).

9.3 Characteristics of Kumquat

The smallest citrus tree species are kumquats. The plant is a tiny, evergreen tree that is 2.5–4.5 m tall, with a small crown and spine-free limbs. The leaves are lance-shaped, alternating, and resemble miniature mandarin leaves. With a delicate scent, the white flowers are typically solitary and bisexual (Morton 1987).

From a taxonomic standpoint, all kumquats were formerly grouped under the *Citrus* genus, but as they adapted to life in Europe, a new genus called *Fortunella* was established specifically for them. The primary interspecies variations within this genus are seen in the physical traits of the fruits, particularly in their shape, intensity of color, and size. These factors have led to the separation of plants into two major categories: those with fruits in both a more rounded and an oval shape. The basic European species *Fortunella crassifolia,* also known as Meiwa kumquat, and *Fortunella japonica* Swingle (*Citrus japonica)*, also known as Morgani kumquat or Marumi kumquat, are included in the first category. The species *Fortunella obovata* (*Citrus obovata*), also known as Fukushu kumquat or Jiangsu kumquat, *Fortunella margarita* Swingle (*Citrus margarita*), also known as Nagami kumquat, and *Fortunella polyandra,* popularly known as Malayan kumquat, are included in the group of ovoid fruit trees. *Fortunella hindsii,* also referred to as the Hong Kong kumquat in trade, is a typical ornamental kumquat with small fruits. Although there is currently a widespread belief that the systematics of this plant should once again place it in the genus *Citrus*, practically all of the recent scientific literature on the subject still refers to it by the genus name *Fortunella*. Regardless of their genus or name, all kumquat species are members of the well-known Rutaceae family, the subfamily Aurantioideae, the tribe Citreae, and the subtribe Citrinae. Instead of being classified as cultivars, the numerous kumquats are identified as botanical species (Chen et al. 2017; Pawełczyk et al. 2021; Yasuda et al. 2016).

Botanically speaking, the kumquat is divided into two species: *Eufortunella* and *Protocitrus*. *Fortunella hindsii* Swingle (primitive Hong Kong kumquats) are part of the *Protocitrus* species, but the *Eufortunella* genus also includes Meiwa kumquats (*F. crassifolia* Swingle), round kumquats (*F. japonica* Swingle), oval kumquats (*F. margarita* Swingle), Malayan kumquat (*F. obovata* Swingle). The kumquat is indigenous to China, particularly Taiwan, and has also been grown in Southeast Asia, the Philippines, Japan, and India (Li et al. 2022).

The typical height of kumquat trees is 2.5–4.5 m, and they have a low-set crown with dense branches, glossy, dark-green leaves, and thorny plant developing at the base of the leaves. Like the majority of *Citrus* fruits, kumquat tree blooms are petite, white, also have five petals, and a potent scent. During the vegetative period, kumquat trees need very high temperatures for the greatest growth and fructification. When the temperature is between 25 and 32 °C, they grow and bloom most vigorously; below 18 °C, they stop blooming. Kumquat trees are less resilient to long-lasting droughts as well as abundant water in the form of torrential rain or flooding. They can, however, resist brief frosts of up to 10 °C during the winter. Even though kumquats are now grown in practically all of South and Central Asia's nations as well as regions of America with similar climatic conditions, only four of the six prominent types are actually employed in food and cosmetic products. When cultivated in a container system (pots), kumquats are more sheltered from the elements and produce more fruit. Kumquat can produce numerous so-called mixed hybrid variants since individual *Citrus* plant species are so similar to one another (Jayaprakasha et al. 2013). The kumquat fruit is about the size of a large olive or a Hungarian plum, however it can occasionally have a little more spherical form. A single fruit typically weighs around 12 g and the cross-sectional fruit's circumference is roughly 3 cm. Kumquats were identified as the smallest known *Citrus*fruit through an examination of its physical traits (Pawełczyk et al. 2021).

Kumquat trees are capable of surviving severe winters with temperatures as low as 21 °C and sweltering summers (25–38 °C). They can therefore be grown in locations where it is too cold for other *Citrus* fruits to thrive. Due to the prolonged wintertime state of dormancy, which is only broken by the appearance of new shoots or blossoms after many weeks of warm weather, plants are able to withstand temperatures below zero degrees for an extended period of time. Despite this, warmer climates are where kumquat trees thrive and yield the highest-quality fruits (Morton 1987).

It is uncommon to raise kumquats from seeds.Typically, they are grafted onto trifoliate oranges (*Poncirus trifoliata*), which has been shown to be the finest rootstock not just for field crops but also for potted crops. Although they have limited uses, bitter orange and grapefruit make good rootstocks. Kumquats can be planted in orchards at great densities. Fruit that can be picked by hand from the ground is made easier to harvest by reducing the canopy's growth (Palma and D'Aquino 2018).

The kumquat fruit (hesperidium) is a modified berry that develops from a single ovary. Even after the fruit is ripe, the calyx is still securely attached to the peduncle and does not often abscise naturally because of this (Ladaniya 2008).

Kumquats typically weigh 12 g, a diameter of around 3 cm, and a slightly oblate or obovate form. In the northern hemisphere, kumquats reach maturity from late November until the beginning of February. The kumquat's peel changes from green to orange as fruit ages and eventually takes on a golden yellow reddish-orange hue at its peak of maturity. The huge oily glands are seen at the same time that the peel becomes shiny and smooth. The finest organoleptic and nutritional qualities of the kumquat can only be attained on the tree, thus it must be harvested fully ripe. Fruit harvest, however, can be postponed for a few weeks once it has reached the most optimal maturation stage because prolonged retention on the plant does not result in substantial changes to the fruit's physical, microbiological, or nutritional makeup (Kassim et al. 2016; Ladaniya 2008).

The thin peel of the kumquat, which has a more delicate and sweet flavor than the fruit's pulp, is the essentially sole *Citrus* fruit that is consumed with the fruit (Abobatta 2018). Its core is separated into segments that are 3–7 in number and often contain one or two bitter, essentially inedible seeds. This plant is also frequently referred to as aeither a golden orange or a miniature orange because of the color and look of its fruit. One little tree can produce up to several thousand fruits, demonstrating the tremendous plant fertility. They are only used after the third year of growing, when the fruit has developed its best flavor and contains the most desirable elements (Pawełczyk et al. 2021).

Different characteristics set kumquats apart from other *Citrus* fruits. The flavedo, which is made up of the pericarp's outermost tissue, is pigmented, sweet, fleshy, tightly-clinging, thick, and edible and has oil glands. Below it, the albedo, which is made up of parenchymatous colorless cells, is seldom present. The endocarp of kumquats is sparse and split radially into 3–6 parts, each of which is separated by fine membranes and contains a few small, acidic to subacid, juicy seeds, in contrast to most other citrus fruits where the endocarp is the main and edible portion. Since the rind of kumquats is sweet and fragrant and has these qualities, they can be eaten whole, leaving a flavor that is both pleasant and enduring (Ladaniya 2008; Palma and D'Aquino 2018; Tadeo et al. 2008).

9.4 Products, Both Fresh and Processed

The kumquat fruit should be eaten whole, including the peel. This makes it possible to employ them in a variety of ways. It is said that gently squeezing the fruit in the hands enhances the aroma and the flavor of the fruit by allowing the essential oils (EOs) to escape from the oil vessels. Since neither the inside of the fruit nor the flavor of the fruit pulp are altered by this method, the flavor of the freshly prepared fruit is generally a little different from that of processed goods made from crushed fruits. The whole fruit must be chewed for a longer period of time and kept in the mouth to get the most nuanced flavor effect. Because of this, the initially perceived taste, which is a little bit bitter, gradually transforms into an increasingly palatable sweetness (Pawełczyk et al. 2021).

Kumquat fruits are occasionally candied to maintain their original form and look during lengthier durations of shipping and storage. Additionally, they are eaten whole in this fashion. Kumquat fruit can be used to decorate food products both fresh and candied. In addition to giving food products their own flavor, fruits also allow for the creation of processed preserves with an extremely unique, highly orange color. When flavor and aesthetics are united, it is possible to enjoy the use of processed preserves such as chopped fruit artificially finely or other extracts with a unique sweet aroma as well as entire fruits or their pieces as ornamental ingredients of prepared dishes (Pawełczyk et al. 2021).

It is possible to consume kumquat fruits raw or to prepare them in a variety of ways. The raw fruit is typically taken whole in order to completely experience their organoleptic qualities. It is advised, however, to roll the fruit between your hands before eating in order to release the rind's essential oils. By doing so, it will be simpler to distinguish between the sweet flavor of the peel's essential oils and the acidic flavor of the pulp's acidity. The distinctive berry flavor also lingers in the mouth for a considerable amount of time. Fresh fruit can be used to make a variety of dishes, such as salads, cakes, and sweets, as well as drinks, such "Kumquat mojitos," a twist on the classic mojito in which kumquats are used in place of the typical mojito's lime (Small 2011).

Fruit salads, marmalades, jams, jellies, juices or syrups, desserts and sweets are some of the most significant food items made from kumquat fruits (Liu et al. 2018; Pieniążek 2003). The fruits are frequently used as garnishes for milkshakes, mixed drinks, and cold beverages. Kumquat fruit mousse makes a fantastic addition to or ingredient in ice cream. For main courses made with delicate meats and fish, fruits that have been mechanically prepared and cooked can be utilized as a decorative appetizer (seasoning) (Rodrigues et al. 2018). Kumquat fruit is occasionally used in the brewing and preparation of fruit teas, but not very regularly (Jayaprakasha et al. 2013). The kumquat fruit is safe to eat while pregnant or nursing in all of these forms (Pawełczyk et al. 2021).

The confectionary business enjoys popularity with kumquats, where they are used to make marmalades or crystalline sugar-glazed candies, pectin, chocolate, or edible coating. Kumquats can also be cooked and added as a garnish to a range of fish and meat dishes, and when cooked in syrup, they go particularly well with ice cream as a dessert. Kumquats can flavor liqueurs when macerated in vodka or another clear alcohol. If they are canned, they can be kept for a long time and sold in nations outside of the production areas (Small 2011).

Along with being eaten as fresh fruit, kumquats are frequently used to produce kumquat fruit tea by adding boiling water to a glass of kumquats (Yasuda et al. 2016). Due to its unique flavor and aroma, the resulting beverage is especially cherished by the Taiwanese people in the winter. Chinese traditional folk medicine has also utilized kumquat fruit to treat respiratory tract inflammation, colds, and coughs, as well as to avoid blood vessel rupture and increase vascular permeability. Kumquat fruit is frequently transformed into a variety of items in the food production industries, including sweets, jams, syrups, fruit in spirits, and liqueur garnishes (Barreca et al. 2011; Rodrigues et al. 2018). However, an enormous amount of peels, pulp, and seed-containing byproducts are produced as a result of industrializing the kumquat procedure. Environmental pollution and a harm to aquatic life are caused by improper waste management techniques, such as the disposal of kumquat fruit waste or byproducts into natural water sources. Due to expensive disposal costs and environmental protection laws, kumquat fruit waste disposal is therefore restricted. This in turn spurred creative recycling of kumquat waste and byproducts into useful resources like biogas. Due to its high nutritional content, kumquat peel waste has historically been fed to ruminants in Italy. The high expense of transportation, the bitter taste of the byproducts that make them less palatable for animals, and the ignorance of animal keepers on the nutritional potential of kumquat byproducts, however, prevent those processing enterprises from being profitable over the long term (Pawełczyk et al. 2021).

Finally, kumquats are employed as a component to provide various drinks with low and moderate alcohol content their flavor, scent, and look. The most popular among them are the kumquat liqueurs (Summo et al. 2016), which are available in orange and colorless varieties. They can be made in a number of ways, but typically by macerating whole or crushed fruits with varying concentrations of spirits (Rodrigues et al. 2018) before sweetening the resulting macerate. Terpenes and flavonoids (Choi et al. 2000) are mostly to blame for the very sweet flavor and perfume of extracts used to make liqueurs, as well as for giving them a distinct green-herbal scent (Pha et al. 1996). But among kumquat liqueurs, the one made only from fruit peels (Summo et al. 2016) is thought to be superior (Pawełczyk et al. 2021).

The mono- and disaccharides, which make up about 9% of the fresh fruit's weight, also affect the flavor of kumquat fruits. The accompanying dietary fibre, which is contained in one of the largest amounts among *Citrus* fruits and amounts to roughly 6.5%, has an impact on how well their nutrients are absorbed. In contrast, the pungency and distinctiveness of the fruit flavor are brought about by the sesquiterpene α-muurolene. Only 0.06% of its substance is typically reached. The kumquat fruit initially seems astringent and mildly spicy, but over time, it gets sweeter due to the combination of several chemicals that are rich in significant polyphenolic and fatty compounds (Pawełczyk et al. 2021).

The various groups of compounds included in kumquat fruit are crucial in producing its characteristic, pleasant flavor, even if the organic acids in it give it its notably sour character, the most significant of which are: Malic acid, citric acid, and oxalic acid are all present in fresh juice in amounts of 2.8 g, 0.35 mg, and 0.02 mg, respectively. Ascorbic acid, or vitamin C, is present in quantities of roughly 44 mg per 100 g of fresh fruit. Dietary fiber and vitamin C are two of the fruits' top nutritional components. They also stand out for having low levels of lipids and proteins, despite the latter having a very specific profile with, no matter what kind of fruit it is, more than 50% polyunsaturated fatty acids with a predominance of α-linolenic acid. However, they only provide a small amount of carbohydrates compared to other fruits, which have significantly higher concentrations of carbohydrates (Pawełczyk et al. 2021).

According to numerous reports, kumquat fruit is a rich source of various vitamins and phytochemicals, including polyphenols (including flavonoids), which have medicinal capabilities like antioxidant, anticancer, and antibacterial activity. Thus, it is crucial to discuss the biochemical makeup of the different kumquat fruit components in order to select the best processing method for turning waste kumquats into beneficial byproducts that are packed with nutrients and phytochemicals (Li et al. 2022).

9.5 Benefits of Kumquat Fruit for Health

A review of the literature on kumquat fruit has been done. The facts and data collected demonstrate that this plant has significant chemical, biological, or nutritional potential. Kumquats and their constituents' antioxidant, antibacterial, anticancer, and anti-inflammatory properties have been researched (Lou and Ho 2017), but little attention has been paid to identifying the fruit's bioactive chemicals. Phenolic chemicals, particularly flavonoids, are the chemical components of the kumquat that have been most thoroughly explained in the literature (Pawełczyk et al. 2021). In order to gather, present, and summarize numerous pieces of information, particularly chemical composition analysis on *Citrus* species, this book was decided to conduct a thorough review.

Several bioactive chemicals, including those listed in the previous section, are abundant in kumquat fruit. These bioactive chemicals constitute the basis for numerous medicinal qualities.

9.6 The Significance of the Kumquat's Chemical Composition

Kumquats are high in nutrients and phytochemicals when compared to other *Citrus* fruits. To learn more about the kumquat's nutritional and nutraceutical composition, several studies have been carried out. The outcomes described in the literature vary widely, not only in terms of nutritional composition but also in terms of secondary metabolite quantity and quality. This is due to a variety of factors including different cultural management, genetic diversity, climate conditions, production areas, and nonstandardized extraction and analysis methods (Palma and D'Aquino 2018).

What stands out in the rather extensive body of literature on the chemical structure of the kumquat fruit and the associated nutritional qualities is the huge variety of data that is cited. This is caused by a number of factors, the most significant of which are by far the location and method of the climate and cultivation, quantity of rain, insolation, and period during which the daily average temperature sharply drops. Other factors include genetics, the techniques used to obtain plant raw materials, and the analytical techniques used in laboratories (Pawełczyk et al. 2021).

Fruits rich in sugar, fiber, and microelements like kumquats are nourishing. According to the USDA's Agricultural Research Service, a 100 g serving of fresh kumquats contains the following nutrients: 80.85 g of water, 1.88 g of protein, 0.86 g of total lipid, 0.52 g of ash, 15.9 g of carbohydrates, 6.5 g of total dietary fiber, 9.36 g of total sugars, 0.62 mg of calcium, 0.86 mg of iron, 20 mg of magnesium, 19 mg of phosphorus, 186 mg of potassium, 10 mg of sodium, 0.17 mg of zinc, 0.135 mg of manganese, and 0.095 mg of copper together contribute 71 kcal (296 kJ) of energy (Palma and D'Aquino 2018).

Sucrose, which has a concentration of 5.66 g/100 g of fresh weight, and fructose and glucose, which are found in comparable amounts (2.5 g/100 g of fresh weight), are the sugars that are mostly concentrated in the peel (Palma and D'Aquino 2018).

The primary organic acid is citric acid, which has a concentration of roughly 2.8 g/100 mL juice. Ossalic and malic acids are present, although in small amounts, at 0.35 and 0.02 mg/100 mL juice, respectively (Palma and D'Aquino 2018).

Dietary fiber, carbs, lipids, proteins, carotenoids, phytosterols, flavonoids and other phenolic compounds, organic acids, vitamins, and mineral components are the most crucial nutrients found in all portions of the kumquat fruit (Rodrigues et al. 2018). Compared to the other fruit portions, the peel has a substantially higher concentration of antioxidant phenolic chemicals. The flavor, taste, fragrance, additionally to dietary and cosmetic uses for specific fruit sections or - as is more often the case for reasons of practicality - the fruit as a whole, are also influenced by a variety of oil vessels and glands where special EOsare maintained (Pawełczyk et al. 2021).

9.7 Determination of Kumquat Composition

Citrus is a frequently preferred fruit. Although kumquat is included in this *Citrus* group, it is a fruit that is not consumed frequently, but has just begun to be consumed. Kumquat is in rising consumer interest due to its exotic taste and potential to be healthy to consumers. Traditionally, it is consumed as a whole fruit, that is, with its peel and pulp, and kumquat fruit has a unique taste. For such reasons, the physical and chemical properties of the kumquat shell and pulp were analyzed in studies with kumquats (Souza et al. 2020). There are few studies on the determination of the chemical contents of kumquat fruit. One of these studies belongs to Souza et al. (2020) in Brazil, and the skin and pulp of kumquat fruit were used in these studies. The United States Department of Agriculture (USDA) ("USDA FoodData Central ")has conducted studies on *Fortunella* sp., and a comparison of these studies is included in a new source (Li et al. 2022). Comparative data of these studies are given in Table 9.1. Total energy value, carbohydrates, moisture, sugars, total including NLEA, and total fiber values were found to be higher than the data in USDA, while lipids, protein, and total ash values were found to be lower. However, differences were also observed in minerals (calcium, sodium, potassium, phosphorus, magnesium, sulfur, iron, zinc, copper, chrome, manganese, cadmium, nickel, aluminum, and lead), vitamins (vitamin A, C, and E) and other compounds (flavanones, flavones, and total phenolics). The moisture content of kumquat is lower than that of orange (*Citrus sinensis*), while the ash, fibers and lipids are higher than oranges("Oranges, raw, all commercial varieties").

The kumquat Ca, K, Mg, and Pb are the chemical elements that are expressed the most. When compared to *C. sinensis*, the ascorbic acid concentration in kumquats was lower ("Oranges, raw, all commercial varieties"). *Citrus* fruits'vitamin C content varies based on a number of variables, including species, ripening time, and harvest time. Even analysis methods can cause variability in vitamin C values. For these reasons, it is natural to observe changes in vitamin C values. Souza et al. found

Table 9.1 Chemical compositions of kumquat fruits

Chemical compositions	*Fortunella* sp. ("USDA FoodData Central")	*F. margarita*; peel and pulp (Souza et al. 2020)
Total energy value	+	+
Carbohydrates	+	+
Insoluble fiber	–	+
Lipids	+	+
Moisture	+	+
Protein	+	+
Sugars, total including NLEA	+	–
Soluble fiber	–	+
Total ash	+	+
Total fiber	+	+

α-tocopherol (major component), β-tocotrienol, α-tocotrienol, γ-tocopherol, respectively, from high to low concentrations of vitamin E compounds in kumquats. Kumquats also contain vitamin E, which primarily consists of α-tocopherol. Although there are not many sources about the vitamin E value in citrus fruits, there are few studies on the fact that they contain α- and β-tocopherol from vitamin E compounds. The total vitamin E value was found to be many times higher than that of oranges (de Oliveira et al. 2010). Apigenin from flavonoids and eriodictyol from flavanones were also detected in similar kumquat studies (Lou and Ho 2017). Total phenolic substances concentration in kumquat was lower than that of *F. crassifolia* (Chen et al. 2017). Typically, a fruit's various portions contain diverse phytochemical substances. The guardian who keeps the fruit safe is the peel. As a result, the amounts of the various substances in the fruit and peel are varied. In comparison to the pulp, the peel often contains larger concentrations of phenolic and other bioactive substances (Manzoor et al. 2012). Variations in chemical composition may occur depending on a number of factors, including planting site, sowing methods, subspecies, ripening period, length of harvest, storage method, and storage period. It is therefore recommended to conduct further research into the biological variation in kumquat fruits from diverse subtypes grown in various farmed places. In light of this information, finding greater fruits that produce more completed items with strong therapeutic effects will be made simpler. Kumquat peel and pulp contain fruit, carotenoids, ascorbic acid, flavonoids (apigenin and eriodictyol), and significant amounts of total phenolic components, making them a potent source of antioxidants. It has a lot of vitamins and fiber. As a result, the kumquat has begun to establish itself in the food and nutrition industries as a new additional option of production and food economy. By fact, kumquat indicates strong consuming capacity, but still more study is required (Souza et al. 2020).

9.7.1 Flavonoid Compounds

Kumquats are a significant source of secondary metabolites with health benefits, such as flavonoids, vitamins, carotenoids, and terpenoids, and they also have increased antioxidant activity (Palma and D'Aquino 2018).

The polyphenolic chemicals, which are present in kumquat fruits in substantial levels and include an abundant set of flavonoids along with phloretin, which resembles them strongly, and have a substantial impact on the fruit's nutritional, health-related, taste-related, and cosmetic qualities of the fruits. The most significant flavonoid compounds found in fruits of kumquat are kaempferol, luteolin, rutin, margaritene, fortunellin, rhoifolin, narirutin, hesperidin, poncirin, didymin, naringin, phloretin, phlorizin, phloretin-3′,5′-di-C-glucoside (Pawełczyk et al. 2021).

The significant antioxidant activity of kumquat fruit is due to all these compounds. As a result, the quantity and nature of these compounds are essential for the actual use of this raw material, as recognized by various researchers who, in their study, several other factors that can affect the concentration of these molecules were taken into consideration. These factors include, among others: the plant's regional background, species, utilized anatomical part (entire fruit, fruit without peel, fruit pulp, fruit peel only, or fruit juice), fruit processing intensity (fresh, seedless, dried), fruit ripeness scale, the harvesting season, extraction solvent (ethanol, diluted methanol or ethanol, water-glycerol mixtures, water), and method for identifying phenolic compounds (Pawełczyk et al. 2021).

Gallic acid equivalents were used to express the overall number of polyphenolic compounds in each of this research, making it much simpler to compare the data and group them for comparison. The range of numerical numbers used to indicate the concentration of these chemicals was quite wide, but this was entirely justified by the multiplicity of variables at play (Lou et al. 2016; Pinheiro-Sant'Ana et al. 2019; Sadek et al. 2009). The amount of phenolic chemicals found in the plant raw material ranged from 2.9 to 106.2 mg/g, however it most frequently oscillated between 10 and 30 mg/g. Based on the information gathered, in particular the work of Lou et al. (Lou et al. 2016), it can be said that fruit peels contain phenolic composition at levels that are significantly higher than those found in fruit pulp. Unripe fruits also contain phenolic compounds at levels that are nearly twice as high as ripe fruits that are ready for harvest. If the harvest date is put off long enough, the polyphenol content begins to rise once more. These chemicals were found at lower concentrations in fruit juice when measured and reported as absolute values. When the type of solvent is taken into consideration, it has been found that water, maybe with the addition of glycerol, is the most efficient for this purpose. Methanol as well asits aqueous solutions were found to be the least effective, with ethanol coming in third in terms of the volume of isolated polyphenols, followed by its mixture with water (Lou et al. 2016; Pinheiro-Sant'Ana et al. 2019; Sadek et al. 2009). The estimated antioxidant activity of the raw material is provided by a trustworthy, decent understanding of the quantity of phenolic compounds in the material. However, the results are very difficult to compare and contrast due to the wide range of techniques used to determine this quality and the correspondingly broad range of ways to describe it. As flavonoid compounds make up a portion of all phenolic compounds, calculating the total of their content and attempting to identify them allow for an approximation of their projected biological activity's directions. This method is another way to assess the phenolic content in a plant (Pawełczyk et al. 2021).

Similar broad correlations to those of phenolic compounds can be seen for flavonoids. Preserves made from immature fruit typically contain twice as much of them as preserves made from mature fruits (Kawaii et al. 1999; Lou et al. 2016; Ramful et al. 2011). An analysis of three different species of kumquat produced in Japan reveals that the variation known as *F. japonica* is the most abundant in flavonoid compounds. Flavonoids are primarily present in fruit peel. Despite the aforementioned reasons, narirutin is the flavonoid that is found in the greatest amount overall. Rhoifolin, kaempferol, and luteolin occupy the next three positions in the list. Even though they are present in much smaller concentrations than the ones found in essential oils, other chemicals of this kind can be crucial to the antioxidant qualities of these fruits. Phloretin, its monoglycoside phlorizine, and 3′,5′-di-C-glucoside, which are substances with strong flavonoid similarities and fall within the category of dihydrochalcones, are regarded as taxonomically distinctive of the genus *Fortunella*. Depending on the type of plant and the section of the plant being used, these chemicals can be found in quantities ranging from 1.5 to 60.2 mg/g, with leaves accounting for a substantial portion of their presence (Lou et al. 2015). Fruits are dried at 110–130 °C yield extracts with the maximum flavonoid content; use of a higher temperature results in a substantial reduction in the concentration of active ingredients and tannish color of the raw material (Ramful et al. 2011).

The total phenolic and flavonoid content of kumquat is higher in peel extracts than in pulp or overripe kumquats. Immature kumquat has a total phenolic content that is two times higher than ripe fruit, ranging from 2346–3000 mg gallic acid equivalent/100 g dry extract. In contrast to Italian-grown kumquats, which had total phenol levels that ranged between 290 and 253 mg gallic acid equivalent/100 fresh weight, Greek and Egyptian kumquats had noticeably lower levels of total phenols, between 80 and 40 and between 100 and 40 mg gallic acid equivalent/g of peel dry fraction, respectively (Palma and D'Aquino 2018).

Kumquats contain the following flavonoids: kaempferol, luteolin, poncirin, narirutin, rhoifolin, apigenin, hesperidin, isorhoifolin, didymin, eriocitrin, neoponcirin, and quercetin. 3′,5′-di-C-β-glucopyranosylphloretin, a dihydrochalcone derivative, was also discovered in the genus *Fortunella* (peel, 6.5–15.2 mg/g dry weight; juice, 1.5–10.5 mg/g). The two flavonoids Schirra et al. (2008) found to be most prevalent were rhoifolin and narirutin (37 and 107 mg/100 g of fresh weight, respectively). Similar to this, Lou and Ho (2017) identified narirutin (2082–1348 mg/100 mg dried sample) as a significant flavonoid, which upon further examination revealed to be a 3′,5′-di-C-β-glucopyranosylphloretin. Fortunellin levels are also rather high (234–97 mg/100 mg in dried sample) although other flavonoid typically only appear in minute quantities (Cho et al. 2005; Lou and Ho 2017; Schirra Mario et al. 2008).

9.7.2 Carotenoids

The carotenoids found in mature kumquat fruits are important for cosmetics as well as nutrition and health. Although the majority of sources state that they only make up a small portion, their distinct color is crucial to the instrumental and organoleptic properties of the goods made from kumquat fruits. Except for the genres known as Hong Kong kumquat (Zhang et al. 2009) and the Brazilian variant (Pinheiro-Sant'Ana et al. 2019), these *Citrus* are completely devoid of β-carotene, which is arguably the most prevalent carotenoid pigment in the plant kingdom. However, the kumquat fruit's remaining carotenoids can give foods, drinks, and cosmetics made from them the golden orange color. According to literature data, the most important carotenoids present in kumquat fruits are violaxanthin, β-citraurin, β-cryptoxanthin, lutein, neochrome, mutatoxanthin and auroxanthin. Violaxanthin, citraurine, and cryptoxanthin are the most prevalent carotenoid components. The latter two, which take the form of fatty acid esters, are discovered to be the most prevalent in the

Brazilian type. The two chemicals with the highest quantities in this group, lutein and β-cryptoxanthin, are those that are reported the most frequently (Pawełczyk et al. 2021).

Carotenoids are more abundant in the peel, whereas the pulp contains only a trace amount. Schirra et al. (2008) discovered 1.27 mg of total carotenoids per 100 g of edible portion. Agócs et al. (2007) discovered that the peel contains 16.9% violeoxanthin, 16.6% β-citraurin, 11.4% β-cryptoxanthin and 9.8% violaxanthin as main components and 5.7% cryptochromes, 5.5% lutein, 3.7% luteoxanthin, 2.8% β-cryptoxanthin, 2.4% β-citraurins, 1.7% violaxanthin 1.5% neochrome, 0.9% ξ-carotene and 0.2% α-cryptoxanthin as minor components (Agócs et al. 2007; Palma and D'Aquino 2018; Schirra Mario et al. 2008).

9.7.3 Phytosterols and Related Compounds

Phytosterols and their related chemicals constitute a significant group of substances, despite their minor presence. The sterol fraction contained sitosterol, stigmasterol, and campesterol at concentrations of 7.04, 1.33, and 1.02 g/g of fruit dry matter, respectively. These compounds prevent cholesterol from being absorbed and raise its level. Additionally, this percentage contains lupenone and amyrin, each present in amounts of 10.45 and 8.43 g/g of dry fruit mass. Another very significant truth is that the kumquat fruit doesn't contain any cholesterol at all, not even in trace amounts, as numerous authors of thorough research have noted (Pawełczyk et al. 2021).

Dodecanol-1 (correspondingly 20.8% and 12.9%) and linoleic acid (correspondingly 16.3% and 13.1%) had been discovered in the hexane extract of the fruit peel and pulp (Koyasako and Bernhard 1983). Additionally, the essential oils from kumquats have limonoids, which can be found in concentrations of up to 0.095% and are modest but significant. The main ones are limonin, which has a concentration of 1.44 g/g, and nomilin, which has a concentration of 0.16 g/g. However, they are chiefly accountable for their insecticidal function, which gives the oils and fruit their initially slightly bitter flavor (Pawełczyk et al. 2021).

9.7.4 Vitamins

Vitamins areone of the mostsignificant secondary metabolites in kumquat fruits. Vitamin C (43–20 mg), total folate (17 g), vitamin B3 (0.04 mg), vitamin B5 (0.037 mg), vitamin B1 (0.03 mg), vitamin B6 (0.03 mg), vitamin A (290 UI), vitamin E as α-tocopherol (0.15 mg), and total vitamin E (1.19 mg) make up the vitamins in a 100 g serving of fresh kumquat (Palma and D'Aquino 2018).

9.7.5 Essential Oils (Volatile Oils)

The essential oils (EOs) that kumquat and other citrus fruits possess, in addition to the aforementioned components, affect their flavor. The essential oil from kumquats is a clear, translucent liquid with a faint golden hue. The ingredients in it have an impact on almost all of the fruit's functional qualities, including flavor, aroma, and health-promoting qualities as well as any possible cosmetic uses (Choi 2005; Koyasako and Bernhard 1983; Quijano and Pino 2009a, b).

It consistently held the highest level oflimonene, which has far more of than all the other components combined, despite the fact that numerous research organizations have discovered and recognized a huge number of separate components in the EOs. Since it does not contribute to the flavor, fragrance, or scent of the essential oil, this monoterpene does not significantly contribute to its composition. This is so that this characteristic effect can only be produced by the combination of the ingredients, which are present in minute amounts and determine these features (Schirra et al. 2011). The most significant constituents of kumquat EOsare α-terpinene, geraniol, sabinene, γ-terpinene, citronellal, 1-p-menthene-8-thiol, terpinolene, terpinene-4-ol, 1-p-menthene-9-thiol, myrcene, p-mentha-1-en-9-yl-acetate, β-pinene, neral, citronellyl acetate, geranial, camphene, α-muurolene, γ-muurolene, bicyclogermacrene, β-elemene, δ-elemene, nootkatone and hinesol. These elements are in charge of giving the oil its smell and other useful qualities (Pawełczyk et al. 2021).

Although kumquats only contain a little quantity of citronellyl acetate, it is the essential oil that gives it its distinct scent in the case of kumquats (Choi 2005). The following compounds are also important for kumquat fruit fragrance: sabinene, α-pinene, myrcene, camphene, β-elemenen, bicyclogermacrene, terpinen-4-ol, and p-menth-1-en-9-yl acetate. All of these compounds have a high relative fragrance activity (RFA) value of over 2.5 and a high dilution factor (FD) value of over 5, which point to the essential oils'beneficial fragrance qualities as a blend of various elements (Pawełczyk et al. 2021).

Kumquat essential oils have a strong, expressive scent that is a little evocative of many spice scents used in cooking. The most significant chemicals that are thought to make such an impact are as follows: γ-terpinene, citral, nootkatone, geraniol, citronellal, terpinolene, and α-terpinene. The elements of kumquat EOsalso affect their antioxidant activity (Choi et al. 2000). Despite being present in extremely minute amounts in the essential oils, the latter compounds significantly affect their anti-free radical activity.

Although its general profile is rather stable (Schirra et al. 2011), the composition of EOsis not consistent and changes depending on a variety of conditions. These factors can be the kind and species of fruits from which the EOswere extracted, their regional background, the anatomical component of the fruit that served as the oil's source material, the fruit's level of ripeness, or the method used to extract the oil. When the latter element was examined, it was discovered that steam distillation followed by 15 min of 90 °C water heating of the raw material was the most efficient way to extract essential oils from kumquat fruits (Peng et al. 2013). There have been attempts to extract the essential oils using carbon dioxide under supercritical conditions, however the procedure had no discernible impact on the product's composition. The thin kumquat fruit peel is the most effective plant component for extracting food oils (Pawełczyk et al. 2021).

Additionally, depending on the manner of processing, the literature's data on the volatile compounds of goods made from candied kumquats varies (Hu et al. 2019).

The essential oil content of kumquat tree leaves was investigated (Quijano and Pino 2009a; Satyal et al. 2012; Sicari and Poiana 2017). The findings of these investigations suggest that the average amount of EOsin the leaves is slightly lower and that their composition is slightly different. This oil does not have many practical uses because of the composition's impact on the important functional organoleptic qualities.

In the literature on the essential oil composition of kumquat leaves reveals that the small changes in the composition of substances that are present in low concentrations have a big impact on the essential oil's aroma and fragrance as well as its potential for use in both food and cosmetic products.

9.7.5.1 Chromatographic Analysis of Essential Oils

The field of analytical chemistry gives both qualitative and quantitative information about chemical properties of a substance. In 1900, chromatography was invented. Tsvet, a botanist, researched plant pigments called chlorophylls. Tsvet created a method for extracting specific chlorophylls from a plant extract. His approach was known as "chromatography" (from the Greek khroma). In a little more than a 1900s, chromatography has developed into a number of different subtypes, each of which enables us to separate molecules for a wide range of pharmacological, environmental, chemical, and biological purposes.

Separating, identifying, and quantifying an essential oil's constituent parts are typically necessary for analysis. The analysis method of choice for essential oil components is capillary gas chromatography due to their volatility and polarity. Even though essential oils are typically complicated blends of elements with comparable physico - chemical properties. Typically, a comprehensive essential oil separation can be achieved by mixing two stationary phases with distinct polarities (Rubiolo et al. 2010).

According to international definitions, an essential oil is a substance made through hydrodistillation via an optimal mechanical technique without warming a plant or a portion of it, whether it would be steam distillation, dry distillation, or another method(" European Pharmacopoeia" 2008). A combination of volatile chemicals in an element of vegetable origin which may be analyzed due to their capacity to vaporize naturally and/or under the right circumstances or methods is generally referred to as a "volatile components."

Furthermore, EOs and headspace contents have frequently been contrasted and just not identified in a number of papers. The amounts of a sampleacquired with the two procedures could not be matched, despite the fact that the compositions of an EOsand the corresponding headspace sampled by various methods are occasionally identical. Given that they were collected using completely distinct methods, this had a significant impact on both the qualitative and quantitative compositions that were produced. That point becomes evident when comparing the descriptions of EOsand headspace that were previously mentioned. One such method is headspace sampling, which takes samples of the fluid or gaseous phase that is (or is not) in balance with a liquid or solid media in analysis to define its content (Rubiolo et al. 2010).

Prepare a Sample Main methods used in sample preparation: vacuum distillation, hydrodistillation, steam distillation, solvent extraction offline combined with distillation, simultaneous distillation with hydrodistillation, extraction, supercritical fluid extraction, and microwave assisted extraction, static, dynamic, and high concentration capacity headspace sampling. The volatile percentage can be sampled using a variety of traditional methods. It needs to be emphasized once more. Using steam distillation, cold expression for citrus, and hydrodistillation, essential oil is produced (Rubiolo et al. 2010).

Kumquat essential oils have reportedly been shown to have strong pharmacologic and wellbeing properties. Numerous studies on the chemical components of EOsfrom kumquat fruit that contribute to its biological properties have been carried out during the past few years. Therefore, it is essential to examine the biological makeup of EOsfrom various kumquat segments.

When compounds approach the detector, they are disassembled into smaller pieces that are then individually sorted and counted based on their mass. The entire process happens so quickly, multiple times every second. Very huge number of volatile compounds can be found using GC/MS. Several fragments of a molecule will result based on its conformation. Additionally, it's crucial to keep in mind that each component that enters the detector represents thousands of identical molecules doing so simultaneously. For a particular molecular structure, some fragments will statistically appear more frequently than others because they are simpler to synthesize.

The ultimate result is a mass spectrum that summarizes the ratios of all mass pieces detected by the detector at a particular instant in time. Next, this spectrum can be evaluated by comparing to standard chemical libraries. The same molecule will produce highly similar mass spectrum when handled by equivalent MS equipment, that may be thought like a fingerprint. MS is incredibly effective at performing identification checks. For one molecule, the fingerprint is frequently highly particular. Datasets of reference substances had already been created. Additionally, a lot of laboratories gradually develop its own datasets using standards or by comparing them to the scientific literature, which also includes a lot of spectral references (Adams 2007; Bicchi et al. 2008; St-Gelais).

Nouri and et al., in their study, investigated the content and antioxidant activity of *C. japonica* peel and seed. This essential oil obtained by hydrodistillation of *C. japonica* peel and seed is yellow in color. The findings obtained from the analysis of the EOs of *C. japonica*peels and seeds by GC/MS are given in Table 9.2. While limonene, germacrene, and β-myrcene have been identified as the primary

Table 9.2 Chemical compounds and groups identified by chromatographic separation and analysis of essential oils obtained from different parts of kumquat species

Plants	Part	Extraction	Analysis	Volatile oil componuds	References
Citrus japonica Thunb.	Peel	Clevenger apparatus (Hydrodistillation) and Soxhlet extraction (with hexane 70 °C 2 h)	GC/MS	**Monoterpenoid**	Nouri and Shafaghatlonbar (2016)
				β-Myrcene, D-Limonene, β-Phellandrene, α-Pinene, (+)-Carvone, Geranyl acetate, (+)-Carvone	
				Sesquiterpenoids	
				Germacrene D, δ-Cadinene, Germacrene B, γ-Cadinene, γ-Elemene, γ-Muurolene	
				Fatty acid	
				Myristicin, Linolenic acid	
				Hydrocarbons	
				Nonene-4-methyl-5, Decene-2-(Z), Decene-5-(Z), 1,6-Octadien-3-ol,3,7,dimethyl, Dodecene-1, Hexadecene-3(Z)	
				Others	
				Acetic acid, octyl ester, Heptadecyl 2,2-dichloroacetate	
	Kernel	Clevenger apparatus (Hydrodistillation) and Soxhlet extraction (with hexane 70 °C 2 h)	GC/MS	**Monoterpenoid**	Nouri and Shafaghatlonbar (2016)
				β-Myrcene, D-Limonene, β-Phellandrene, α-Pinene, (+)-Carvone, Geranyl acetate, (+)-Carvone	
				Sesquiterpenoids	
				Germacrene D, δ-Cadinene, Germacrene B, γ-Cadinene, γ-Elemene, γ-Muurolene	
				Fatty acid	
				Myristicin, Linolenic acid	
				Hydrocarbons	
				Nonene-4-methyl-5, Decene-2-(Z), Decene-5-(Z), 1,6-Octadien-3-ol,3,7,dimethyl, Dodecene-1, Tetradecene-7-(Z), Hexadecene-3(Z)	
				Others	
				Acetic acid, octyl ester, Heptadecyl 2,2-dichloroacetate	

(continued)

Table 9.2 (continued)

Plants	Part	Extraction	Analysis	Volatile oil componuds	References
Fortunella crassifolia Swingle	Flavedo	Headspace solid phase microextraction	GC/MS	**Monoterpenoid**	Liu et al. (2019)
				α-Terpinene, γ-Terpinene, α-Pinene, Sabinene, Myrcene, Limonene, β-Ocimene, Terpinolene, 3-Carene, Linalool, 4-Terpineol, α-Terpineol, cis-Carveol, d-Citronellol, Geraniol, Perillo, β-Citral, Citral, Perillylaldehyde, 1-Perillaldehyde, Carvomenthenal, Nerol acetate, Geranyl acetate, trans-Dihydrocarvone, Dihydrocarvone, Piperitone, Carvone, Camphor, Myrcene acetylated	
				Sesquiterpenoids	
				Cosmene, 3,4-Dimethyl-2,4,6-octatriene, Alloocimene, Neoalloocimene, 1,3,8-Menthatriene, Isolimonene, α-Cubebene, Ylangene, α-Copaene, cis-3-Tetradecene, 2-Isopropyl-5-methyl-9-methylene[4.4.0], Dec-1-ene, Germacrene B, β-Cubebene, α-Guaiene, β-Patchoulene, Isoledene, Humulene, Bicyclosesquiphellandrene, γ-Muurolene, α-Amorphene, Germacrene D, α-Selinene, Valencene, β-Maaliene, α-Muurolene, δ-Cadinene, α-Farnesene, Cada-1,4-diene, α-Gurjunene, β-Cadinene, Selina-3,7(11)-diene, Elixene, β-Panasinsene, α-Elemene,β-Elemene,γ-Elemene, Alloaromadendrene, β-Gurjunene, β-Guaiene, Epizonarene, Elemol, Nerolidol,Guaiol,Agarospirol, β-Eudesmol, Bulnesol	
				Hydrocarbons	
				cis-3-Hexenol, Cyclohexanol, 1-Octanol, trans-2-Hexenal, Decanal, Undecanal, Dodecanal, Octyl acetate, Nonyl acetate, Decyl acetate, Methoxy-phenyl oxime, 4-Acetyl-1-methylcyclohexene, Tridecane, Tetradecane l	
	Peel	Clevenger apparatus (Hydrodistillation)	GC/MS	**Monoterpenoid**	Wang et al. (2012)
				Limonene, γ-Terpinene, α-Pinene, Myrcene, Camphene, p-Cymene, p-Mentha-2,8-dien-1-ol, p-Mentha-1,5-dien-8-ol, Piperitenone, Cis-Myrtanyl acetate, trans-Myrtanyl acetate, Perillyl acetat, Linalool, Carveol, Perilla aldehyde	
				Sesquiterpenoids	
				Carvone, α-Copaene, β-Elemene, α-Selinene, Bicyclogermacrene, Spathulenol	
				Hydrocarbons	
				2,7-Dimethyl-1,6-octadione, 3,4-Dimethyl styrene, Octyl acetate, Isopropyl cinnamate	

	Whole	Clevenger apparatus (Hydrodistillation)	GC/MS	**Monoterpenoid**	Chen et al. (2017)
				Limonene, α-pinene, β-myrcene, α-terpineol, Geranyl acetate	
				Sesquiterpenoids	
				Germacrene-D	
Fortunella japonica Swingle	Fruits	Clevenger apparatus (Hydrodistillation)	GC/MS	**Monoterpenes**	Schirra Mario et al. (2008)
				β-Pinene, Myrcene, α-Terpinene, p-Cymene, Limonene, γ-Terpinene, α-Phellandrene, p-Cymenene, α-Thujene, Sabinene, trans-Ocimene, Terpinolene, cis-Limonene oxide, cis-p-menth-2,8-dien-1-ol, p-mentha-1,5-dien-1-ol,p-cymen-8-ol, Linalol, Terpinen-4-ol, α-Terpineol, trans-Carveol, cis-Carveol, Carvone, β-Caryophyllene	
				Hydrocarbons	
				Octyl acetate	
				Sesquiterpenes	
				Germacrene B, cis-M uurola-4 (14,5-diene), B icyclogermacrene, γ-G ujunene, α-C opaene, 7-epi-α-selinene, Galaxolide	
	Peel	Clevenger apparatus (Hydrodistillation)	GC/MS	**Monoterpenes**	Al-Saman et al. (2019)
				α-Pinene, Sabinene	
				Sesquiterpenes	
				Limonene, d-Germacrene, Bicyclogermacrene	
Fortunella margarita Swingle	Flavedo	Hydrodistillation and Supercritical carbon dioxide (SC-CO_2) extractions	GC/MS	**Monoterpenes**	Sicari and Poiana (2017)
				α-Pinene, Sabinene, β-Pinene, Myrcene, p-Cymene, Limonene, γ-Terpinene, Terpinolene, Linalool, Terpinen-4-ol, α-Terpineol, Carveol, Geraniol, Geranial, Geranyl acetate, Linalyl acetate	
				Sesquiterpenes	
				β-Cariofillene, α-Humulene, Germacrene-D, β-Bisabolene, Nerol, Neryl acetate	
				Hydrocarbons	
				Octyl acetate	

(continued)

Table 9.2 (continued)

Plants	Part	Extraction	Analysis	Volatile oil componuds	References
Fortunella margarita Swingle	Fruits	Steam distillation	GC/MS	**Monoterpenes**	Liu et al. (2012a, b)
				α-Pinene, β-Myrcene, D-Limonene, Terpinolene, Citral,Trans -(+) -carveol,β-Carveol,Linalool, α-terpinol, Terpenol, β-terpinol, Carvol, Geraniol, β-Ocimenum,γ-terpinene, Perilla alcohol, Geranyl acetate, Perilla aldehyde	
				Sesquiterpenoids	
				γ-Elemene, β- Elemene, α-Farnesene, δ-cadinen, Elemol, α-cadinol, caryophyllene, α-caryophyllene, Aristolene, Nerol, Cis-Neral	
				Hydrocarbons	
				Octanol, Pentylcyclopropane, Octanal	
				Fatty acid	
				Palmitic acid	
	Fruits	Steam distillation	GC/MS	**Monoterpenes**	Ibrahim et al. (2015)
				Linalool, α-Terpineol, Carveol, Carvone, Limonene, Citronellal, Bisabolene trans-gamma	
				Sesquiterpenes	
				δ-Cadinene, α-Cadinene, γ-Muurolene, Germacren B, Galaxolide, Cedrol	
				Others	
				Diethylphthalate	
	Leaves	Steam distillation	GC/MS	**Monoterpenes**	Liu et al. (2012a, b)
				α-pinene, Linalool, Terpenol, α-Terpinol, Geraniol, Cis-terpineol, Geranyl acetate, Citronellal, Citronellol, Thymol methyl ether, Citral, β-canjophyllene, Carene	
				Sesquiterpenoids	
				β-Elemene, γ-Elemene, δ-cadinene, Neryl acetate, Trans-β-Farnesene, Aristolene, γ-Gurjunene, α-canjophyllene, Cadinol, α-cadinol	
				Hydrocarbons	
				1-methyl−4-(1-isopropyl)-2-cyclohexenone-1-ol,4-isopropyl−3-methylphenol,	
				Others	
				Pohytol, N-methyl anthranilic acid methyl ester	

Fortunella margarita Swingle	Leaves	Steam distillation	GC/MS	**Monoterpenes**	Ibrahim et al. (2015)
				Dihydrocarveol, Bisabolene	
				Sesquiterpenes	
				δ-Elemene, δ-Cadinene, γ-Cadinene, α-Gurjunene, β-Gurjunene, α-Muurolene, β-Eudesmol, γ-Eudesmol, Germacren D-4-ol, γ-Eudesmol, α-Humulene, Cadinol, β-Cedrene, α-Cubebene, β-Cubebene, α-Guaiene, β-Bourbonene,	
	Peel	Clevenger apparatus (Hydrodistillation)	GC/MS	**Monoterpenes**	Fitsiou et al. (2016)
				α-Pinene, Sabinene, β-Pinene, Myrcene, α-Phellandrene, limonene, cis-Ocimene, trans-Ocimene, γ-Terpinene, Terpinolene, Linalol, trans-p-Menthene-2.3-dien-1-ol, cis-p-Menthene-2.8-diene-1-ol, Terp-1-ene-4-ol, α-Terpineol, trans-Carveol, cis-Carveol, Perilla aldehyde, geranyl acetate, Carvone, p-Menth-1-en-9-yl acetate, δ-β-Bisabolene	
				Sesquiterpenes	
				δ-Germacrene, Valencene, Cadinene, Neryl acetate, β-Elemene, δ-Elemene, Humulene, Bicyclogermacrene, Caryophyllene, α-Copaene	
				Hydrocarbons	
				Octyl acetate, Decanal	

components in peel oil, both of these compounds were also found in varying amounts in the EOs of the seed, along with β-phellandrene. In the end, it was discovered that the EOs extracted from kumquat peel contained six monoterpenoids (limonene, β-myrcene, β-phellandrene, (+)-carvone, and α-pinene), five sesquiterpenoids (germacrene D, δ-cadinene, germacrene B, and γ-cadinene), two oxygenated monoterpenes (geranyl acetate, (+)-carvone), one oxygenated aromatic compound (myristicin), and oil constituents. Five monoterpene hydrocarbons limonene, β-myrcene, β-phellandrene, (+)- carvone, α-pinene), six sesquiterpenoids (germacrene D, δ-cadinene, germacrene B, γ -cadinene, γ-elemene, γ-muurolene), and two oxygenated monoterpenes (geranyl acetate and (+)- carvone) were the main components of the kumquat kernel essential oil. Limonene has been discovered as the primary oil ingredient in the EOsextracted from the peel and kernel of *C. japonica*. These oils were distinguished by significant concentrations of sesquiterpenes (germacrene D) and monoterpene hydrocarbons (limonene and -myrcene). Also, there are many studies on terpenes having antioxidant activity. Unsaturated fatty acids are another typical constituent of volatile oil. In the food, cosmetics products, fragrance, and drug companies, unsaturated fats are utilized as flavoring and aroma. This study demonstrated that the volatile oil of *C. japonica*peel and seed included major portion of the unsaturated fats alpha-linolenic and linoleic from unsaturated fats. Unsaturated fats decrease blood cholesterol and protect against cardiovascular disease. It can be said that the whole fruit of *C. japonica*, which is often eaten as a food in Iran, is a readily available naturally occurring source of antioxidants and may have beneficial benefits on health (Nouri and Shafaghatlonbar 2016).

Wang et al. used their hands to remove the peels from kumquat fruits from the *Fortunella crassifolia* Swingle species. Fresh peelings were blendedbefore being hydro-distilled for 4 h with a Clevenger-style device. The EOs was extracted using water, dried using anhydrous sodium sulfate, and then kept in a freezer at 4 °C for analysis. The obtained essential oil performed by GC-MS analysis. Table 9.2 lists the twenty-five components that have been determined. Like in other citrus fruits, terpenoid hydrocarbons were the significant ingredients. The content of limonene (approximately 75%), which is the major component in the EOs obtained from the peel of *F. crassifolia*, is high. This content is followed camphene, myrcene, α-pinene, α-selinene, β-elemene, and 3,4-dimethyl styrene with low amounts of p-cymene, α-copaene γ-terpinene, bicyclgermacrene, and 2,7-dimethyl-1,6-dioctadione (Wang et al. 2012). In the studies of Moufida et al. in blood orange, one of the citrus species, the limonene content was found to be approximately 63% (Moufida and Marzouk 2003). In research by Choi, a GC-MS analysis revealed that limonene was present in approximately 93% of the samples. This study examines the aroma elements in the essential oil that was extracted using the cold pressing technique from the peel of *F. japonica*. Limonene served as the report's main component (Choi 2005). The odor profile of the bark oil is fruity, sweet. Top notes are woody and base notes are oily. The aroma threshold levels of the volatile substances that are eluted from the gas chromatography column are the basis of gas chromatography–olfactometry (GC-O) procedures. The peel oil of *F. japonica* Swingle contained 82 distinct components. 27 hydrocarbons, 8 aldehydes, 25 alcohols, 4 ketones, 10 esters,

5 oxides and epoxides, and 3 acids were all present in the essential oil. There were found to be several monoterpenes in peel essential oil. The main substance is limonene, even though other frequent molecules include myrcene and ethyl acetate (Choi 2003; Minh Tu et al. 2002). Although limonene made up a significant fraction of the peel oil generally, as observed in many other *Citrus* fruits, its distinctive scents were found to have little effect on the flavour effect. According to this, the tiny ester molecule citronellil acetate may be what gives kumquat peel oil its distinctive scent. However, when the works of Wang (Wang et al. 2012) and Choi (Choi 2005) were evaluated, it was discovered that the amount of limonene in the peel oil of *F. crassifolia* was significantly lower than that of *F. japonica* peel oil. The amount of α-pinene in the EO was similar to that reported for *F. japonica* peel oil. But it was discovered that the levels of myrcene, camphene, α-selinene, and alcohols (carveol, linalool, p-mentha-1,5-diene-8-ol and spatulenol, p-mentha-2,8-dien-1-ol) were higher than those previously reported for *F. japonica* peel oil. Similar to the peel oil from *F. japonica*, there were little ketones, esters, and aldehydes present (Choi 2005; Wang et al. 2012).

Cells in the fruit surface that create essential oils are found there. These oils vaporize in the presence of wind and warmth. Consequently, they were known as volatile oils (Chen et al. 2016). *Fortunella crassifolia* Swingle fruit essential oil extraction by steam distillation was 0.71% (w/w) in a study by Chen et al. The volatile compounds of volatile oils were identified using GC-MS. There were 48 substances found. These substances were divided into groups based on their chemical structures: 11 monoterpenes, 13 sesquiterpenes, 11 aldehydes, 16 alcohols, 6 esters, and a ketone. The largest concentration of these chemicals was found in limonene, which was accompanied by concentrations of -myrcene, -pinene, germacrene-D, -terpineol, and geranyl acetate. The flavour and aroma of essential oils are not derived from their terpene components. These terpene molecules, which made up about 98% of the EO in the *F. crassifolia* Swingle fruits, were discovered to be the primary constituents of the oleoresin in the studied samples. Due to the fact that terpenes are both light and temperature sensitive. Terpenes can degrade when they are stored. Limonene undergoes oxidative destruction as a consequence of thermal stress. Due to the fact that terpenes are both light and temperature sensitive. Terpenes can degrade when they are stored. Limonene undergoes oxidative destruction as a consequence of thermal stress. Due to this circumstance, oxidized monoterpenes or monoterpene hydrocarbons are formed.

The oxidative breakdown of limonene results in the formation of the terpineol and carvone molecules (Wang et al. 2012). The molecules terpineol and carvone are both widely recognized for their role in the deterioration of the flavour and taste of *Fortunella* juices.

However, it has been determined that linalool plays a significant role in the *Citrus/Fortunella* fruits' distinctive flowery scent (Rouseff et al. 2009). The aroma of linalool is flowery. Linalool works in concert with some other substances to enhance the general flowery fragrance. The base odor cannot exist without limonene. Aliphatic aldehydes are among the citrus's fortunella bioactive substances, and limonene is one of them. Additionally, limonene emits a sweet waxy scent and a scent resembling *Fortunella* peel (Liu et al. 2012a, b). The distilled process includes a crucial component called deterpenation. Deterpenation improves flavour stability and resistance to oxidation. Additionally, it affects the amount of volatile chemicals. As the condensation increases, the terpene compounds decrease and hence an increase in the oxygenated compound is caused (Torres-Alvarez et al. 2017). While this study is supported, several physiological responses are independent of a particular chemical. Different substances have combined effects that are more powerful than the impacts of any one ingredient by itself. In this study, the essential oils, flavonoids, and phytosterol compositions of the two fruits (calamondin and kumquat) were examined. The current study's findings not only reveal the nutritious content of kumquat and calamondin when they are eaten as fresh fruits, in addition they serve as a foundation for evaluating their numerous by-products (Chen et al. 2017).

Liu et al. (2019) conducted a study on the chemical contents of *Fortunella crassifolia* Swingle fruits at five different stages of development. Depending on how many days had passed since flowering, samples of fruits were taken from the top layer of the canopy at five different phases of development ((Stage 1 (S1): 30 days after flowering, Stage 2 (S2): 60 days after flowering, Stage 3 (S3): 90 days after flowering, Stage 4 (S4): 120 days after flowering, and Stage 5 (S5): 150 days after flowering). Each fruit's flavedo portion of the peel was removed. Followed by headspace solid phase micro-extraction gas chromatography mass spectrometry analysis of the fruits (HS-SPME-GC/MS). The samples were put in a tube for HS-SPME analysis. After that, samples were turned into a powder using a blender after being frozen in liquid nitrogen. Between the stages of fruit growth, various differences in volatile compounds were seen. However, abundance of a number of significant compounds was seen at a high level at fivephases. In five stages, limonene predominated, followed by myrcene and germacrene-D. This outcome is consistent with findings that have already been reported (Sicari and Poiana 2017). Additionally, some of the more plentiful components were linalool, γ-elemene, δ-cadinene, α-amorphene, geranyl acetate, 2-isopropyl-5-methyl-9-methylene, β-elemene, dec-1-ene, and α-cubebene were also among the more abundant components. The pattern of essential oil buildup and fruit growth were not in lockstep. According to the findings of this study, fruit growth and surface area gain occurred more quickly throughout fruit development rather than during the maturationstage. Also, they argued that the two key elements governing volatile oils production in nature were the distribution of carbohydrates and the biological role of volatile components. There are numerous competing needs for carbs during the developmental phase. As a result, photosynthesis primary role should continue to be the production of fruit peel. When a fruit is ripening, its growth progressively stops, and the production of chlorophyll finally stops altogether (Iglesias et al. 2007). At every stage of fruit growth, limonene was consistently present in the highest concentration in the essential oils, with a steady ratio of roughly 70%. In the final phases of fruit growth, the amount of chemicals linked to disease resistance and rich aroma decreased (Liu et al. 2019).

In their study, Sicari et al. (Sicari and Poiana 2017). compared the principal components of the EO obtained from the kumquat (*Fortunella margarita* Swingle) peel using solvent extraction, hydrodistillation, and supercritical carbon dioxide extraction. The hydrodistillate of kumquats was a colorless oil. The perfume of kumquat hydrodistillate essential oil is similar to that of the fruit. The extract of pentane was a reddish oil. The aroma of pentane extract essential oil is particularly potent. Red oils made up the supercritical extracts. White substance precipitates from the supercritical extraction of EO. The substances discovered by GC/MS analysis. Six terpene alcohols, one terpene ketone, one terpene aldehyde, four terpene esters, eight monoterpene hydrocarbons, and four sesquiterpene hydrocarbons were all quantified, for a total of 24 different substances. According to previous published papers, the main ingredient, limonene, has a high concentration in the essential oil made from kumquat peels (around 96%), followed by myrcene and germacrene-D (Chen et al. 2013). In comparison to *Fortunella japonica* peel oil (Choi 2005) and *Fortunella crassifolia* peel oil (Wang et al. 2012), *Fortunella margarita* peel oil (in this study) had a higher limonene content. According to Schirra et al. (2008), monoterpene compounds were the most prevalent in the oils extracted using the four distinct procedures because of the high limonene concentration. Limonene inhibited the development and spread of human gastric cancer transplanted in mice via having antiangiogenic and proapoptotic effects (Lu et al. 2004).

In comparison to essential oils derived through steam distillation and solvent extraction, those obtained through supercritical carbon dioxide extraction exhibit higher concentrations of linalyl acetate, neryl acetate, and geranyl acetate. This condition is unquestionably brought on by these chemicals' solubility in water. A comparable pattern for the sesquiterpene content may be seen. Consequently, the GC/MS analysis of kumquat (*F. margarita*) essential oil revealed that monoterpenes are the class of component found in the oil in the greatest number, with a value of roughly 98% in all oils studied. The monoterpene component with the highest concentration is limonene. The variations in the extraction techniques utilized are not particularly evident in these statistics. The essential oil extracted by supercritical carbon dioxide has a higher concentration of esters and sesquiterpenes, in comparison. The taste and scent of the extract may significantly increase as a result of the increasing ester concentration (Sicari and Poiana 2017).

Fitsiou and et all. (Fitsiou et al. 2016) shared the findings of another research. For extracting the fruit peel from *F. margarita* fruits, they used hydro-distillation. They used GC/MS analysis. They found 45 different chemicals. The major component (93.8%) was limonene. Other significant substances included myrcene (2.7%) and δ-germacrene (1.34%) (Fitsiou et al. 2016).

Again, using the steam distillation process, Liu and et al. (2012a, b) extracted EOsfrom *F. margarita* leaves and fruits. The EOsfrom *F. margarita* leaves were also used to identify 27 chemical according to earlier research by compounds using the GC/MS method. 10 alcohol, 10 terpenes, 2 aldehydes, 3 esters, 1 ether, and 1 phenol are among the constituents. Linalool had the largest relative level in our experiment, at around 27%, followed by citronellal and elemi, at about 5% and 9%, respectively. Linalool is a key basic element of spice in the perfume business and is utilized extensively in both every day and culinary perfume formulas. It also has certain therapeutic benefits, including antibacterial, antiviral, and sedative properties. Citronellal has inhibitory effects on fungi, *Staphylococcus aureus* Rosenbach, and *Salmonella typhi* in addition to acting as an insect repellent and perfume. Elemi is a brand-new non-cytotoxic anticancer medication that is now being used in clinical settings, particularly for the treatment of cerebroma, liver cancer, and lung cancer. Citral also effectively kills and drives away insects, preserves, and kills and depresses fungus. The pharmacological activity of *F. margarita* leaves depends on the aforementioned chemical components that were previously discussed.

A total of 34 chemical constituents were found in the essential oils found in the fruit peels of *F. margarita*. 4 terpenes, 12 alcohol, 4 aldehydes, geranyl acetate, carvol, pentylcyclopropane, and palmitic acid are among the constituents. Wherein terpenes made up 89% of the total and were the predominant constituents. D-limonene, p-myrcene, a-pinene, p-caryophyllene, and linalool were among those present in them, with respective contents of around 73%, 7%, 2%, 2%, and 1% being higher than 1% and others being below 1%. D-limonene has a number of therapeutic properties, including cholagogue, gallstone dissolving, energy regulating, appetite stimulating, pain relieving, and tumor preventing actions. Myrcene, which has anacidic flavor of pine, is frequently used to make tropical fruit juices like papaya, mango, carrot, bergamot, orange, and orange juice.

In this study, when the essential oil components from leaves and fruits were compared, 13 chemicals were found to be the same in terms of content, but their amounts were different. These compounds: α-pinene, linalool, geranyl acetate, p-elemene, geraniol, terpene, 3-caryophyllene, α -caryophyllene, 6-cadinene, α -terpinol, aristolene, α -cadinol and citral.

In comparison to *F. margarita* leaves, the relative number of other components was lower in the fruit peels. D-limonene concentrations in essential oils from *F. margarita* fruit peels were greater than those found in citrus fruit peels, which is consistent with previous findings. D-limonene was not discovered in the essential oils extracted from *F. margarita* leaves, though. Linalools relative concentration in *F. margarita*leaf essential oils was 26.55%, compared to just 1.62% in fruit peels. The chemical composition and relative concentration of fruit peels and leaves of *F. margarita*also differed significantly (Liu et al. 2012a, b).

According to Nabaweya et all (Ibrahim et al. 2015). study, hydrodistillation was used to create the essential oils from the fresh *F. margarita* Lour. Swingle (Family: Rutaceae) leaves and fruits. Twenty substances, or 86.96% of the total leaves oil, were found in the oils of both organs after their oils underwent GC/MS analysis. The principal chemicals were eudesmol, murolene, and gurjunene. There were discovered to be fourteen fruit oil-derived chemicals in the oil of the fruits. Terpineol, t-carveol, limonene, muurolene, and cadinene are some of these ingredients that are important ones. Terpineol has been obtained from a number of naturally occurring monoterpene alcohols, including petitgrain oil, cajuput oil, and pine oil. Alpha-, beta-, and gamma-terpineol are its three isomers, with the final two changing only in where the double bond is located. α-terpineol is often the main component of terpineol, which is typically a combination of these isomers. It is frequently used as a component in tastes, cosmetics, and fragrances because of its lovely lilac-like aroma (Yao et al. 2005). There were 27 chemicals found in the leaves of *F. margarita*, according to earlier research by (Liu et al. 2012a, b), while 34 compounds were found in fruit peels. Both parts differed in their primary structural elements. Using headspace solid phase micro-extraction in combined with GC/MS, the volatile contents of 3 types of kumquats (*F. crassifolia, F. japonica, F. margarita*) were identified. The 3 kumquat types had 28 different ingredients. The amount of D-limonene, which is the major component, is *F. margarita, F. japonica, F. crassifolia*, from largest to smallest. Also present in all sample essential were α-pinene, β-elemene, D-limonene, δ-cadinene, cis-α-cadinene, iso-caryophyllene, and acetic acid esters. Additionally, each type of kumquat has its own parts. i.e., *F. crassifolia* Swingle includes 4-carene; *F. margarita* Swingle includes D-germacrene, cedrene and linalool; *F. japonica* Swingle includes ocimene, isopropyl palmitate, 2,6,10,14-tetramethyl-heptadecanoic, 2,6-ditert-butyl-4-butyl phenol, cubaene, and limonene (Ibrahim et al. 2015; Zhonghai et al. 2009).

In this work (Al-Saman et al. 2019), hydro distillation was used to get EOs from the peels of kumquat fruit. To identify the volatile compounds of the essential oil, GC/MS analysis was performed. Six substances were recognized and categorized as monoterpenes and sesquiterpenes based on their chemical configurations. Four monoterpenes (limonene, α-myrcene, α-pinene and sabinene) and two sesquiterpenes (D-germacrene and bicyclogermacrene) were present in the essential oil (Al-Saman et al. 2019).

In another study (Güney et al. 2015), one of the significant tropical fruit species which already lately been invented to Turkey is the kumquat. It fits the Turkish Mediterranean region beautifully. Quite several researches have concentrated on the kumquats chemical components; the majority of early study on this fruit focused on adaptability and pomological features. Farmers in Turkey began their farms after realizing the nutritional benefits of kumquats. Accordingly, the lipids, fatty acid, and volatile components of the fruits of five different kumquat species *F. crassifolia* Swingle, *F. hindsii* (Champ. ex Benth.) Swingle, *F. margarita* (Lour.) Swingle, *F. obovata* Hort. ex Tanaka, and limequat [*Citrusaurantifolia* × *F. japonica* (Thumb.)] were compared in this study. Gas chromatography with flame ionization detection (GC/FID) and headspace gas chromatography mass spectrometry (HS-GC/MS) techniques were used to identify volatile patterns. By using a headspace approach, the volatile components of kumquat fruits were extracted. As a result, 39 substances were discovered, and terpenes emerged as the dominant chemical class across all kumquat genotypes (Table 9.3). Additionally found were esters, aldehydes, alcohols, and ketones. The results of the current study showed that the kumquat fruit genotypes varied according to their fatty acid, lipid, and volatile ingredient compositions, and that the fruit had a significant amount of wellbeing chemo-preventive effects (Güney et al. 2015).

Table 9.3 Major compounds and groups of essential oils obtained from kumquat species cultivated in Turkey

	Compounds	*Fortunella crassifolia*	*Fortunella hindsii*	*Fortunella japonica* (Thumb.)	*Fortunella margarita*	*Fortunella obovata*
Hydrocarbons	Methylbutenol	+	+	−	+	+
	n-decanal	−	−	−	−	+
	n-Decyl acetate	−	+	−	−	+
	n-Hexanal	+	+	−	+	+
	n-nonanal	+	−	−	−	+
	n-Octanal	−	−	−	−	+
	n-Pentanal	−	+	−	−	+
	Octyl acetate	−	−	+	−	−
	Octyl alcohol	−	−	−	−	−
	Pent-1-en-3-one	−	−	−	+	−
	Trans-2-Pentenol	−	+	−	−	−
	1-Pentene-3-ol	−	+	−	−	−
Monoterpenes	Camphene	−	+	+	+	+
	Carvol	+	+	−	−	+
	Cymenene	+	+	+	+	−
	D-limonene	+	+	+	+	+
	Linalyl acetate	−	−	+	−	+
	Linalyl alcohol	+	−	−	−	+
	Perillyl acetate	−	−	−	−	+
	Pinene	+	+	+	+	+
	Trans-Carveol	−	+	−	+	−
	Terpinolene	+	+	+	+	+
	α-Phellanderene	−	−	+	+	+
	α-Terpinene	+	+	−	+	+
	α-trans-Bergamotene	+	−	+	−	−
	β-Bisabolene	+	−	+	−	−
	β-Myrcene	+	+	+	+	+
	β-Phellandrene	+	+	+	+	+
	γ-Terpinene	+	+	+	+	+
	(−)-β-Pinene	+	−	+	+	+
	4-Terpineol	+	+	−	+	+
Sesquiterpenoids	Aromandendrene	−	−	+	+	−
	Caryophyllene	−	−	+	−	−
	D-GermacreneN	+	−	+	−	+
	Neryl acetate	−	−	+	−	+
	α-Cubebene	−	−	+	−	−
	β-Elemene	+	−	+	+	−
	δ-Cadinene	+	+	+	+	+

9.8 Biological Activities of Kumquat

The scavenging of free radicals and the regulation of oxidation processes, both of which are linked to cellular aging and the emergence of most diseases, are activities that naturally take place in the human body and are aided by food antioxidants. Over the last few decades, new information has been available regarding the function of fruits and vegetables' radical scavenging activity in the variety of chronic diseases'treatment and prevention. Studies have found that kumquat extracts have important antioxidant and antibacterial effects. In addition to their capacity to influence numerous enzyme systems' productiveness in a positive way, phenols and flavonoids in particular demonstrate a variety of biological consequences, such as antioxidant activity. The primary ingredients in kumquat extracts that contribute to the free radical scavenging activity are flavonoids. This is supported by the observation that during the ripening process, antioxidant activity falls at the same rate as flavonoid concentrations. Antioxidant capabilities are also present in essential oils, but the relative contributions of the various components vary greatly. For instance, the radical scavenging activity of limonene and myrcene was low, whereas that of α-terpinene, citral citronellal, geraniol, nootkatone, and terpinolene was higher. Despite making up a small portion of the essential oils from kumquats, the terpenes γ-terpinene and terpinolene are thought to be significant contributors to the fruit's overall capacity to scavenge free radicals (Palma and D'Aquino 2018).

Despite recent scientific evidence supporting the kumquat's anti-aging properties and the healthful effects of its many bioactive compounds, traditional medicines in areas where kumquats are commonly grown, like China, have long recognized the fruit's role in promoting health and lowering the risk of disease. These properties have been linked to the prevention of cancers, infectious diseases, cardiovascular disorders, as well as antiallergic, antiinflammatory, antibacterial, and vasodilatory properties. Kumquat extract also appears to normalize metabolic issues brought on by obesity, indicating intriguing potential usage as a nutritional supplement. (Palma and D'Aquino 2018).

The presence of active substances like terpenes, carotenoids, flavonoids, and many others is what gives kumquat fruit its biological activity. The primary chemical constituents of essential oils, monoterpenes, exhibit a wide range of pharmacological properties, including antioxidant, antibacterial, antifungal, anticancer, antiaggregating, antiarrhythmic, local anesthetic, anti-inflammatory, antinociceptive, antispasmodic, and antihistaminic activities. (Koziol et al. 2014) Carotenoids are a group of naturally occurring, lipid-soluble pigments. Epidemiological studies have shown a link between eating a diet high in carotenoids and a decreased risk of numerous diseases, including certain malignancies and eye conditions. Carotenoids are vital elements of the antioxidant network, and this is supported by compelling research. The most prevalent carotenoids thought to act as photoprotectants are lutein and zeaxanthin (Krinsky and Johnson 2005).

Biological activity exhibited by flavonoids range greatly. From a therapeutic perspective, their antioxidant capabilities and ability to block the many enzymes are most crucial. Inhibiting cyclooxygenase and lipoxygenase, as well as scavenging superoxide anions, appear to be crucial from a pharmacological perspective. Flavonoids work as both antithrombotic and anti-inflammatory substances (Robak and Gryglewski 1996).

Due to the widespread cultivation of this plant, medical uses for kumquats have been developed based on traditional Chinese medicine and are now being used in other nations as well, providing the necessary circumstances for acquiring high-quality raw materials. Even though this tree's leaves also contain essential oils, only its fruits are employed for certain applications (Pawełczyk et al. 2021).

Only EOs, mostly their active terpene compounds, water extracts and water-alcohol extracts, and other natural substances are employed as ingredients in pharmaceutical and parapharmaceutical products for technological and functional reasons. Owing to their large quantity of difficult to extract fiber, whole fruits are utilized as a solution that regulates and promotes digestion, enhances digestive tract function, and eliminates gas and constipation. The beneficial qualities of the fruit that aid in weight loss are determined by the same nutritious component, which has a low sugar level. Kumquat fruit extract has a positive impact on maintaining a healthy blood cholesterol concentration and, in particular, recovers and keeps the ratio of HDL to LDL in check by encouraging the production of beneficial HDL cholesterol. It can be applied to help treat metabolic illnesses (Kawaii et al. 1999; Tan et al. 2014).

Although the EOsfrom the kumquat fruit contain lipolytic capabilities, their tempo of this range of activity is comparable to that of other *Citrus* fruits (Choi 2006). Due to its unique composition and the ratios between the numerous active ingredients, such as its low sugar and sodium content, minimal fat content, and complete lack of cholesterol, it has the ability to lower the risk of developing diabetes. Kumquat fruits are suggested as a beneficial nutritional element in both forms of diabetes for the aforementioned reasons (Allam et al. 2015; Zeng et al. 2016).

The body regulates the ratio of potassium to sodium ions because potassium ions are far more abundant than sodium ions, which lowers blood pressure. In addition, foods with high iron contents help the body make more red blood cells, which has a preventative effect on anemia.

The kumquat fruit is a good source of vitamins A and C. The former impacts vision quality along with β-carotene (found in just two varieties) (Jayaprakasha et al. 2013), while the latter works in tandem with flavonoids and polysaccharide percentage (Zeng et al. 2015).Vitamin C has a specific function in the body, acting as an antioxidant to stop the onset of many diseases, including those caused by free radicals (Sadek et al. 2009) and as a stimulant of the immune system to stop many types of inflammation.

The kumquat fruit peel contains carotenoid compounds that function as immune system activators by increasing the synthesis of γ-interferon (IFN- γ). Both in vivo and in vitro experiments show this impact. The chemical working in this way that proved to be most effective was β-cryptoxanthin (Nagahama et al. 2015; Terao et al. 2019).

Several writers have reported that the EOsextracted from kumquat fruits and, a lesser frequency, leaves exhibit average or moderate antibacterial activity. Its antifungal action has been seen against *Candida fungus* as well as *Aspergillus niger*and *Saccharomyces cerevisiae* (Bunrathep et al. 2015; Fitsiou et al. 2016; Quijano and Pino 2009b) *Pseudomonas aeruginosa*and*Escherichia coli* are unaffected, although it prevents the growth of both gram negative and positive bacteria (Bunrathep et al. 2015; Wang et al. 2012). Due to the presence of phloretin and, to a lesser extent, its glycosidic derivatives (such as phlorizin and its isomer), it prevents the growth of *Listeria monocytogenes, Salmonella typhimurium, and Staphylococcus aureus* (Terao et al. 2019). Whole fresh fruit EOsexhibit notable antiviral action, particularly against the H5N1 influenza virus, perhaps as a result of their high α-terpineol concentration (Sicari and Poiana 2017). It has been determined that the essential oils taken from the fruit are far more potent than those extracted from the leaves in terms of their antibacterial activity (Barreca et al. 2014; Bunrathep et al. 2015; Quijano and Pino 2009b).

The limonene concentration and cell line used to make the determination, as well as other factors, all play a significant role in the antiproliferative activity of kumquat EOs, which can be characterized as moderate. Its antioxidant and anti-free radical activity is, however, more potent than that of other essential oils derived from herbaceous plants that have a comparable effect. It inhibits DPPH in 34.5% when used at a dosage of 43 mg/mL (Fitsiou et al. 2016). The entire spectrum of chemical compounds found in essential oils work as both an expectorant and a remedy for a variety of ailments, including sore throats and upper respiratory tract infections (Choi et al. 2000; Jayaprakasha et al. 2013).

Kumquat fruits are essential for aiding weight loss in obese individuals, as a cardioprotective agent, and as a digestive aid, especially in the course of metabolic illnesses, according to findings derived from "traditional medicine"(Rodrigues et al. 2018). They are also employed as an antidepressant and a detoxifying agent in Asian nations where alcohol consumption is prominent. (Pawełczyk et al. 2021; Zhang et al. 2009).

9.8.1 Cosmetic Potential

Kumquats are utilized in cosmetics, just like many other *Citrus* fruits. The effect of the pulp and the peel on kumquat fruit, however, can hardly be distinguished from that of the other *Citrus* fruits, unlike with the other *Citrus* fruits. Whole fruits are utilized to make pulp for cosmetic treatments, and essential oils acquired via the use of various techniques are also used.

Regarding the use of these fruits in the production of cosmetics, several of the vitamins and minerals also play a significant role (Pawełczyk et al. 2021).

Most frequently, water and glycerol are combined as the solvent to create an extract for cosmetic applications. These extracts are typically put to cosmetics at a concentration of 5–10% that are created as moisturizing tonics, creams, or lotions.

Even at higher concentrations, it can be used to deodorize (Choi et al. 2000) and odorize. The International Fragrance Association (IFRA) recommends that the concentration of EOsderived from whole fruits in products intended for use on the skin not exceed 5%.

Flavonoids are multi-functional ingredients that are mostly employed in conventional cosmetics for their calming and antioxidant properties. The primary benefits of kumquat fruits for skin care are their antioxidant activity, which prevents individual skin cells from aging negatively. The presence of dihydrochalcone phloretin and flavonoids in high concentrations is primarily responsible for this action. The actions of the enzymes are improved by these compounds (Rodrigues et al. 2018).

The complete kumquat fruits, but especially the essential oils that are produced from them, are a great source of carotenoids, and the Brazilian and Hong Kong types also have vitamin A and β-carotene. The materials used as raw are employed as ingredients in hair care treatments because they include these compounds, which work by lowering sebum production and improving hair moisturization (Jayaprakasha et al. 2013).

As antioxidants, the flavonoids and carotenoidsfound in the parts of kumquat fruitare important components of nutricosmetics and other products that enhance the natural beauty of the skin, nails, and hair.

References

Abobatta W (2018) Overview of kumquat tree-management practice and varieties. Open Acc J Agric Res 1:1–4

Adams RP (2007) Identification of essential oil components by gas chromatography/mass spectrometry, vol 456. Allured Publishing Corporation Carol Stream

Agócs A, Nagy V, Szabó Z, Márk L, Ohmacht R, Deli J (2007) Comparative study on the carotenoid composition of the peel and the pulp of different citrus species. Innov Food Sci Emerg Technol 8(3):390–394

Allam M, Khedr A, El-Beltagy A (2015) Kumquat as a potent natural material to improve lipid profile of hypercholestrolemic rats. Biolife 3(1):171–181

Al-Saman MA, Abdella A, Mazrou KE, Tayel AA, Irmak S (2019) Antimicrobial and antioxidant activities of different extracts of the peel of kumquat (Citrus japonica Thunb). J Food Measur Charact 13(4):3221–3229

Barreca D, Bellocco E, Caristi C, Leuzzi U, Gattuso G (2011) Kumquat (Fortunella japonica Swingle) juice: flavonoid distribution and antioxidant properties. Food Res Int 44(7):2190–2197

Barreca D, Bellocco E, Laganà G, Ginestra G, Bisignano C (2014) Biochemical and antimicrobial activity of phloretin and its glycosilated derivatives present in apple and kumquat. Food Chem 160:292–297

Bicchi C, Liberto E, Matteodo M, Sgorbini B, Mondello L, Zellner BA et al (2008) Quantitative analysis of essential oils: a complex task. Flavour Fragr J 23(6):382–391

Bunrathep S, Soodvilai S, Settharaksa S, Palanuvej C (2015) Chemical constituents and biological activities of Fortunella japonica essential oils. Bullet Health Sci Technol 13:43–49

Chen H-C, Peng L-W, Sheu M-J, Lin L-Y, Chiang H-M, Wu C-T et al (2013) Effects of hot water treatment on the essential oils of calamondin. J Food Drug Anal 21(4):363–368

Chen Q, Hu Z, Yao FY-D, Liang H (2016) Study of two-stage microwave extraction of essential oil and pectin from pomelo peels. LWT-Food Sci Technol 66:538–545

Chen M-H, Yang K-M, Huang T-C, Wu M-L (2017) Traditional small-size citrus from Taiwan: essential oils, bioactive compounds and antioxidant capacity. Medicine 4(2):28
Cho J-Y, Kawazoe K, Moon J-H, Park K-H, Murakami K, Takaishi Y (2005) Chemical constituents from tile fruit peels of Fortunella japonica. Food Sci Biotechnol 14(5):599–603
Choi H-S (2003) Character impact odorants of citrus Hallabong [(C. unshiu Marcov× C. sinensis Osbeck)× C. reticulata Blanco] cold-pressed peel oil. J Agric Food Chem 51(9):2687–2692
Choi H-S (2005) Characteristic odor components of kumquat (Fortunella japonica Swingle) peel oil. J Agric Food Chem 53(5):1642–1647
Choi H-S (2006) Lipolytic effects of citrus peel oils and their components. J Agric Food Chem 54(9):3254–3258
Choi H-S, Song HS, Ukeda H, Sawamura M (2000) Radical-scavenging activities of citrus essential oils and their components: detection using 1, 1-diphenyl-2-picrylhydrazyl. J Agric Food Chem 48(9):4156–4161
de Oliveira RP, Schroder EC, EDS S, Scivittaro WB, de Castro LAS, da Rocha PSG (2010) Laranjeiras sem acidez. Pelotas: Embrapa Clima Temperado Pelotas
European Pharmacopoeia (2008) EDQM, 6th edn. (Council of Europe), Strasbourg
FAO (2006) Fresh and processed citrus fruit annual statistics. Commodities and Trade Division Food and Agriculture Organization of the United Nations, Rome
Fitsiou E, Mitropoulou G, Spyridopoulou K, Tiptiri-Kourpeti A, Vamvakias M, Bardouki H et al (2016) Phytochemical profile and evaluation of the biological activities of essential oils derived from the Greek aromatic plant species Ocimum basilicum, Mentha spicata, pimpinella anisum and Fortunella margarita. Molecules 21(8):1069
Güney M, Oz AT, Kafkas E (2015) Comparison of lipids, fatty acids and volatile compounds of various kumquat species using HS/GC/MS/FID techniques. J Sci Food Agric 95(6):1268–1273
Hu X, Wang R, Guo J, Ge K, Li G, Fu F et al (2019) Changes in the volatile components of candied kumquats in different processing methodologies with headspace–gas chromatography–ion mobility spectrometry. Molecules 24(17):3053
Ibrahim NA, El-Hawary SS, Mohammed MM, Farid MA, AbdelWahed NA, Ali MA, El-Abd EA (2015) Chemical composition, antiviral against avian influenza (H5N1) virus and antimicrobial activities of the essential oils of the leaves and fruits of Fortunella margarita, lour. Swingle, growing in Egypt. J Appl Pharm Sci 5(1):006–012
Iglesias DJ, Cercós M, Colmenero-Flores JM, Naranjo MA, Ríos G, Carrera E et al (2007) Physiology of citrus fruiting. Braz J Plant Physiol 19:333–362
Jayaprakasha G, Kumar CK, Chaudhary P, Murthy KC, Patil BS (2013) Identification of volatiles from Kumquats and their biological activities. In: Tropical and subtropical fruits: flavors, color, and health benefits. ACS Publications, pp 63–92
Kassim A, Workneh TS, Laing M, Basdew I (2016) The effects of different pre-packaging treatments on the quality of kumquat fruit. CyTA-J Food 14(4):639–648
Kawaii S, Tomono Y, Katase E, Ogawa K, Yano M (1999) Quantitation of flavonoid constituents in citrus fruits. J Agric Food Chem 47(9):3565–3571
Koyasako A, Bernhard R (1983) Volatile constituents of the essential oil of kumquat. J Food Sci 48(6):1807–1812
Koziol A, Stryjewska A, Librowski T, Salat K, Gawel M, Moniczewski A, Lochynski S (2014) An overview of the pharmacological properties and potential applications of natural monoterpenes. Mini Rev Med Chem 14(14):1156–1168
Krinsky NI, Johnson EJ (2005) Carotenoid actions and their relation to health and disease. Mol Aspects Med 26(6):459–516
Ladaniya MS (2008) Commercial fresh citrus cultivars and producing countries. Citrus fruit: biology, technology and evaluation. Academic Press, San Diego, pp 13–65
Li X, Meenu M, Xu B (2022) Recent development in bioactive compounds and health benefits of kumquat fruits. Food Reviews International, pp 1–21
Liu C, Cheng Y, Zhang H, Deng X, Chen F, Xu J (2012a) Volatile constituents of wild citrus Mangshanyegan (Citrus nobilis Lauriro) peel oil. J Agric Food Chem 60(10):2617–2628

Liu S, Liu H, Zhang L, Huang Y (2012b) Comparative analysis on chemical components in essential oils from the leaves and fruit peels of Fortunella margarita. Med Plant 3(1):43–46
Liu Y, Liu Y, Liu Y, Liu H, Shang Y (2018) Evaluating effects of ellagic acid on the quality of kumquat fruits during storage. Sci Hortic 227:244–254
Liu X, Liu B, Jiang D, Zhu S, Shen W, Yu X et al (2019) The accumulation and composition of essential oil in kumquat peel. Sci Hortic 252:121–129
Lou S-N, Ho C-T (2017) Phenolic compounds and biological activities of small-size citrus: kumquat and calamondin. J Food Drug Anal 25(1):162–175
Lou S-N, Lai Y-C, Huang J-D, Ho C-T, Ferng L-HA, Chang Y-C (2015) Drying effect on flavonoid composition and antioxidant activity of immature kumquat. Food Chem 171:356–363
Lou S-N, Lai Y-C, Hsu Y-S, Ho C-T (2016) Phenolic content, antioxidant activity and effective compounds of kumquat extracted by different solvents. Food Chem 197:1–6
Lu X-G, Zhan L-B, Feng B-A, Qu M-Y, Yu L-H, Xie J-H (2004) Inhibition of growth and metastasis of human gastric cancer implanted in nude mice by d-limonene. World J Gastroenterol: WJG 10(14):2140
Manzoor M, Anwar F, Saari N, Ashraf M (2012) Variations of antioxidant characteristics and mineral contents in pulp and peel of different apple (Malus domestica Borkh.) cultivars from Pakistan. Molecules 17(1):390–407
Minh Tu NT, Onishi Y, Choi H-S, Kondo Y, Bassore SM, Ukeda H, Sawamura M (2002) Characteristic odor components of Citrus sphaerocarpa Tanaka (Kabosu) cold-pressed peel oil. J Agric Food Chem 50(10):2908–2913
Morton J (1987) Kumquat. In: Fruits of warm climates. Creative resource systems, Miami, pp 182–185
Moufida S, Marzouk B (2003) Biochemical characterization of blood orange, sweet orange, lemon, bergamot and bitter orange. Phytochemistry 62(8):1283–1289
Nagahama K, Eto N, Shimojo T, Kondoh T, Nakahara K, Sakakibara Y et al (2015) Effect of kumquat (Fortunella crassifolia) pericarp on natural killer cell activity in vitro and in vivo. Biosci Biotechnol Biochem 79(8):1327–1336
Nouri A, Shafaghatlonbar A (2016) Chemical constituents and antioxidant activity of essential oil and organic extract from the peel and kernel parts of Citrus japonica Thunb.(kumquat) from Iran. Nat Prod Res 30(9):1093–1097
Oranges, Raw, All Commercial Varieties (n.d.). Retrieved from https://fdc.nal.usda.gov/fdc-app.html#/food-details/169097/nutrients
Palma A, D'Aquino S (2018) Kumquat—Fortunella japonica. In: Exotic fruits. Elsevier, pp 271–278
Pawełczyk A, Żwawiak J, Zaprutko L (2021) Kumquat fruits as an important source of food ingredients and utility compounds. Food Reviews International, pp 1–21
Peng L-W, Sheu M-J, Lin L-Y, Wu C-T, Chiang H-M, Lin W-H et al (2013) Effect of heat treatments on the essential oils of kumquat (Fortunella margarita Swingle). Food Chem 136(2):532–537
Pha NM, Lô VN, Dũng NX, Leclercq PA (1996) Constituents of the fruit peel oil of Citrus japonica L. from Vietnam. J Essent Oil Res 8(4):415–416
Pieniążek JPS (2003) Kumkwat. In: Szweykowscy AJ (ed) Botanical dictionairy. PW Wiedza Powszechna, Warszawa, pp 434–435
Pinheiro-Sant'Ana HM, Anunciação PC, Souza CSe, de Paula Filho GX, Salvo A, Dugo G, Giuffrida D (2019) Quali-quantitative profile of native carotenoids in kumquat from Brazil by HPLC-DAD-APCI/MS. Foods 8(5):166
Quijano CE, Pino JA (2009a) Volatile compounds of kumquat (Fortunella margarita (Lour.) Swingle) leaf oil. J Essent Oil Res 21(3):194–196
Quijano CE, Pino JA (2009b) Volatile compounds of round kumquat (Fortunella japonica Swingle) peel oil from Colombia. J Essent Oil Res 21(6):483–485
Ramful D, Tarnus E, Aruoma OI, Bourdon E, Bahorun T (2011) Polyphenol composition, vitamin C content and antioxidant capacity of Mauritian citrus fruit pulps. Food Res Int 44(7):2088–2099
Robak J, Gryglewski R (1996) Bioactivity of flavonoids. Pol J Pharmacol 48(6):555–564

Rodrigues S, de Oliveira SE, de Brito ES (2018) Exotic fruits reference guide. Academic
Rouseff RL, Ruiz Perez-Cacho P, Jabalpurwala F (2009) Historical review of citrus flavor research during the past 100 years. J Agric Food Chem 57(18):8115–8124
Rubiolo P, Sgorbini B, Liberto E, Cordero C, Bicchi C (2010) Essential oils and volatiles: sample preparation and analysis. A review. Flavour Fragr J 25(5):282–290
Sadek ES, Makris DP, Kefalas P (2009) Polyphenolic composition and antioxidant characteristics of kumquat (Fortunella margarita) peel fractions. Plant Foods Hum Nutr 64(4):297–302
Safana HS, Ibrahim MA-HI, Abd A-KM (2022) Effect of NAA and chitosan in rooting branches resulting from stem nodes plantation of Kumquat (Citrus japonica) in vitro. J Kerbala Agric Sci 9(1):7–18
Satyal P, Paudel P, Limbu K, Setzer WN (2012) Leaf essential oil composition of Citrus japonica from Nepal. J Essen Oil Bear Plant 15(3):357–359
Schirra M, Palma A, D'Aquino S, Angioni A, Minello EV, Melis M, Cabras P (2008) Influence of postharvest hot water treatment on nutritional and functional properties of kumquat (Fortunella japonica Lour. Swingle Cv Ovale) fruit. J Agric Food Chem 56(2):455–460
Schirra M, Angioni A, Cabras P, Daquino S, Palma A (2011) Effects of postharvest hot water and hot air treatments on storage decay and quality traits of kumquat. (Fortunella japonica Lour. Swingle, cv Ovale) fruit
Sicari V, Poiana M (2017) Comparison of the volatile component of the essential oil of kumquat (Fortunella margarita swingle) extracted by supercritical carbon dioxide, hydrodistillation and conventional solvent extraction. J Essen Oil Bear Plant 20(1):87–94
Small E (2011) Top 100 exotic food plants. Crc Press
Souza CSe, Anunciação PC, Lucia CMD, Dôres RGRd, Milagres RCRdM, Sant'Ana HMP (2020) Kumquat (Fortunella margarita): a good alternative for the ingestion of nutrients and bioactive compounds. Paper Presented at the Proceedings
St-Gelais A (n.d.) The highs and lows of GC-MS in essential oil analysis. Retrieved from https://tisserandinstitute.org/highs-lows-gc-ms-essential-oil-analysis/
Summo C, Trani A, Faccia M, Caponio F, Gambacorta G (2016) Volatiles and acceptability of liqueurs from kumquat and grapefruit. Ital J Food Sci 28(2):258–270
Tadeo FR, Cercos M, Colmenero-Flores JM, Iglesias DJ, Naranjo MA, Rios G et al (2008) Molecular physiology of development and quality of citrus. Adv Bot Res 47:147–223
Tan S, Li M, Ding X, Fan S, Guo L, Gu M et al (2014) Effects of Fortunella margarita fruit extract on metabolic disorders in high-fat diet-induced obese C57BL/6 mice. PloS One 9(4):e93510
Terao R, Murata A, Sugamoto K, Watanabe T, Nagahama K, Nakahara K et al (2019) Immunostimulatory effect of kumquat (Fortunella crassifolia) and its constituents, β-cryptoxanthin and R-limonene. Food Funct 10(1):38–48
Torres-Alvarez C, Núñez González A, Rodríguez J, Castillo S, Leos-Rivas C, Báez-González JG (2017) Chemical composition, antimicrobial, and antioxidant activities of orange essential oil and its concentrated oils. CyTA-J Food 15(1):129–135
USDA FoodData Central (n.d.). Retrieved from https://fdc.nal.usda.gov/fdc-app.html#/food-details/168154/nutrients
Wang Y-W, Zeng W-C, Xu P-Y, Lan Y-J, Zhu R-X, Zhong K et al (2012) Chemical composition and antimicrobial activity of the essential oil of kumquat (Fortunella crassifolia Swingle) peel. Int J Mol Sci 13(3):3382–3393
Wang T, Chen L-L, Shu H-J, You F, Liang X-L, Li J et al (2022) Fortunella venosa (Champ. ex Benth.) CC Huang and F. hindsii (Champ. ex Benth.) Swingle as independent species: evidence from morphology and molecular systematics and taxonomic revision of Fortunella (Rutaceae). Front Plant Sci 13
Yao S-S, Guo W-F, Lu Y, Jiang Y-X (2005) Flavor characteristics of lapsang souchong and smoked lapsang souchong, a special Chinese black tea with pine smoking process. J Agric Food Chem 53(22):8688–8693
Yasuda K, Yahata M, Kunitake H (2016) Phylogeny and classification of kumquats (Fortunella spp.) inferred from CMA karyotype composition. Hortic J 85(2):115–121

Zeng H, Zhang Y, Zhao Y, Tian Y, Miao S, Zheng B (2015) Extraction optimization, structure and antioxidant activities of Fortunella margarita Swingle polysaccharides. Int J Biol Macromol 74:232–242

Zeng H, Miao S, Zhang Y, Lin S, Jian Y, Tian Y, Zheng B (2016) Isolation, preliminary structural characterization and hypolipidemic effect of polysaccharide fractions from Fortunella margarita (Lour.) Swingle. Food Hydrocoll 52:126–136

Zhang J, Tao N, Xu Q, Zhou W, Cao H, Xu J, Deng X (2009) Functional characterization of citrus PSY gene in Hongkong kumquat (Fortunella hindsii Swingle). Plant Cell Rep 28(11):1737–1746

Zhonghai L, Jie B, Jilie L, Haiyan Z (2009) Analysis of volatile components of three kumquat species by HS-SPME/GC-MS. J Chinese Cereals Oils Assoc 9

Chapter 10
Kinnow Mandarin (*Citrus reticulate* L.)

Isha Dudeja, Arashdeep Singh, Ramandeep Kaur Mankoo, Avneet Kaur, and Sukhvinder Singh Purewal

10.1 Introduction

Kinnow is a citrus fruit that belongs to the "Mandarin" family and is widely grown in India and Pakistan. The fruit was first created in 1935 at the University of California Citrus Experiment Station and was first introduced in India in the early 1940s (Mahawar et al. 2020).

Kinnow is a cross between the citrus cultivars "King" (*Citrus nobilis*) and "Willow Leaf" mandarin (*Citrus deliciosa*) (Rattanpal et al. 2017). In India, the orange family, comprising mandarin and kinnow, produced around 4.75 million tonnes from an area of 0.43 million hectares (Mahawar et al. 2019). After its discovery, it was released for commercialization in 1935. However, the developed hybrid did not make a breakthrough in the United States, but when introduced in other citrus-producing countries like India, Pakistan and Bangladesh, a revolution in the citrus industry resulted owing to its stunning golden-orange colour, great yield and economic returns. Citrus cultivation is a successful enterprise that contributes

I. Dudeja · R. K. Mankoo
Department of Chemistry, Punjab Agricultural University, Ludhiana, Punjab, India

A. Singh (✉)
Department of Food Science and Technology, Punjab Agricultural University, Ludhiana, Punjab, India

A. Kaur
Department of Chemistry, Chandigarh University, Mohali, Punjab, India

S. Singh Purewal (✉)
University Centre for Research & Development (UCRD), Chandigarh University, Mohali, Punjab, India
e-mail: sukhvinder.e14148@cumail.in

S. Singh Purewal et al. (eds.), *Recent Advances in Citrus Fruits*,
https://doi.org/10.1007/978-3-031-37534-7_10

considerably to the economies of countries such as the United States, Mexico, Greece, China, Spain, India, and Iran, among others. India is fifth in the world in terms of citrus fruits production, with important fruits including oranges, limes, grapefruits, mandarins, lemons, and tangerines. Citrus fruits occupy 10% of the total production land among fruit crops in India, and their output ranks third behind mango and banana. Citrus fruits mainly Kinnow are mostly planted for juice extraction and processing. The citrus fruit processing business generates a lot of trash in the form of seeds, skins, and fruit peel and pomace.Citrus peels and seeds may include useful chemicals such as polyphenols, flavonoids, antioxidants, limonoids, carotenoids, and tocopherols. Citrus processing companies discard those Kinnow mandarin wastes, which contain a wealth of bioactive chemicals with antioxidant potential. Such organic matter has a significant potential for transformation into nutritionally valueable commodities using a variety of extraction processes. The extraction of bioactive phytochemicals from byproducts, which may then be isolated and used as nutritional supplements in food production systems.

10.2 Cultivation

Citrus being the third most consumed fruit covers 1.078 million ha and accounts for 12.5% of total fruit cultivation. Kinnow accounts for 4.6% of the citrus area and 3.6% of total fruit production. H.B. Frost was the first to develop Kinnow(*Citrus nobilis* × *Citrus deliciosa*) (Rattanpal et al. 2017) and for cultivating Kinnow specific temperature, soil and nutrient conditions are requisite. The temperature necessary for cultivation ranges in 10–35 °C. Kinnow can withstand temperatures as high as 40 °C in summer and as low as 0 °C in winter. An ideal site for Kinnow cultivation requires clay loam soils with a pH ranging between 6.0 and 7.5. It is usually propagated via budding onto specific rootstock around August–September and February–March. Irrigation is done using the drip method and the frequency depends on the climatic factors (Kumar et al. 2016).

10.3 Taxonomical Classification

Kinnow is taxonomically classified with the scientific name *Citrus nobilis* x *Citrus deliciosa*, having common name Kinnow mandarin, and belongs to the domain of Eukaryota, Kingdom of Plantae, Phylum of Spermatophyta, Subphylum of Angiospermae and Class of Dicotyledonae.

10.4 Physiological Traits of Kinnow

Kinnow fruit is oblate, flattened, and bright orange-yellow in color. The two main sections of the fruit are the pericarp (peel) and endocarp, while the pericarp branches into the epicarp (flavedo) and mesocarp (albedo). This mid-season cultivar has a few seeds adhering to 9–10 portions, a silky peel that becomes deep orange as it ripens, and tasty juice (Ladaniya 2008). When the total soluble solids (TSS)/acid ratio reaches 12:1–14:1 and the exterior color turns to orange, the fruit is considered fully mature (Table 10.1). The best time for Kinnow picking ranges from November to February (Anonymous 2018).

10.5 Harvesting, Postharvest Losses, Waxing, and Packaging

Kinnow is traditionally picked utilizing clippers, then dropped to the ground and gathered in boxes or bags, resulting in severe post-harvest loss. Fruit that has been damaged during harvesting is unsuitable for eating or marketing. Picking (19.6%), packaging (3.5%), carrying (2.2%), loading and transportation (7.1%) accounts for 32.4% of Kinnow post harvest wastage (Ahmad et al. 2015). Total harvest and post-harvest losses in Kinnow, according to Singh et al. (2016), range from 25% to 30%. Waxing is an important unit process that is commonly utilised in industrial catchment areas. As a result, Kinnow seems to have an extremely short shelf life of 8–10 days in moderate environments, that can be prolonged to 20 days with appropriate planning. Fruits are typically manually packaged in corrugated fiberboard (CFB) racks or boxes with a capacity of around 10 kg per box.

10.6 Varieties – Local, PAU Kinnow, Daisy, W. Murcott

Local The fruits of this variety mature in December and January. They have characteristic shapes and possess small furrowed necks at the base. It has excellent cadmium yellow color and contains 3–7 seed per fruit making it more convenient for

Table 10.1 Characteristic of Kinnow at maturity

Characteristic	Quality indices of Kinnow
Colour	Golden yellow (peel), deep yellowish orange (pulp)
TSS/acid ratio	12:1–14:1
Size	5.0–9.7 cm
Shape	Moderate to oblate; both base and apex flattened or slightly depressed
Appearance	Very smooth, glossy and sometime faintly pitted
Firmness	Firm, not soft and easily peel able

Fig. 10.1 Varieties of Kinnow (Source: Rattanpal et al. 2017)

out of hand eating. The acidity of juice is slightly high as compared to others. This is mainly cultivated in small regions in Hoshiarpur, Ropar and Gurdaspur districts of Punjab.

PAU Kinnow It is reported to be produced via mutation breeding. The fruits appear as globose to oblate in shape. The color is usually golden orange after attaining maturity. The Kinnow of this variety is said to have fewer seeds (0–9 to fruit) as compared to other varieties. The TSS and acidity value range between 10–11° Brix (B) and 0.7–0.9%.

Daisy It matures in first to third week of November. The skin of fruit appears to be glossy reddish golden. It has high sugar to acid ratio and possesses good flavour. The fruits are medium to large in size. The daisy consists of a moderate number of seeds (10–14/fruit).

W. Murcott The fruits are small to medium size with flat shape and easily peelable skin. The fruits are rich in flavour with perfect balance of sugars and acid. They contain less seed when self-pollinated as compared to cross pollinated. It is mid-season variety.

The different varieties of Kinnow are shown in Fig. 10.1.

10.7 Parts of Kinnow- Peel, Pulp and Pomace, Juice, Seed and Leaves

The peel, pomace, seed and juice of Kinnow is filled with wide range of nutrients having role in functional properties. The following description of each part of *Kinnow* clearly justifies their role:

Peel Fruit juice extraction is done on the same scale as fresh fruit consumption. The peels and pith are byproducts of juice extraction or fresh consumption. The Kinnow peel waste several bioactive chemicals (flavanones, polyphenolic compounds, and carotenoids) and essential oils that have applications in a variety of industries. Based to certain findings, the Kinnow peels, which account for 30–40%

of the total fruit, have more polyphenols and bioactive chemicals than the remainder of the fruit (Lim et al. 2007; Godara et al. 2020). As a result of the significant features of *Kinnow* peel, researchers have decided to investigate its potential as a fundamental element for various industries (Babbar et al. 2011; Rafiq et al. 2018). Albedo (inner layer) and flavedo (outer layer) are the two tissues that make up the *Kinnow* peel. Polyphenolic compounds present in the peels possess anti-inflammatory and anti-cancer effects (Tripoli et al. 2007; Xu et al. 2019). These bioactive components have a high nutritive value and are useful in industries such as pharmaceuticals, food processing, and biofuel production. The useful compounds can be extracted using various techniques. Albedo is primarily made up of different nutrients like sugars, proteins, minerals and fibre (Marin et al. 2007).

Furthermore, the flavedo generates essential oils that are used in the flavour and fragrance industries (Bejar et al. 2012). They are abundant in carotenoids, terpenes, and linalool (Mondello et al. 2005), and numerous studies have discovered anti-oxidant and antibacterial capabilities (Tepe et al. 2006; Jayprakasha et al. 2008; Viuda-Martos et al. 2008). Because the peel portion contains relatively more phenolics and has a higher potential for useful substances than the pulp has led researchers to investigate further. Approx 32–33% peel portion was observed in Kinnow. Kinnow peel comprises of 22.45% total solids, 12.50 °B TSS, 1.38% acidity, 41.57 mg/100 g vitamin C, 6.23% total sugars, 5.99% reducing sugars, 0.67% ash, 13.65 mg/100 g carotenoids, 7.43 mg/100 g β-carotene, 1.85% pectin, 0.420 mg/g naringin, 4.69 mg/g limonene and 0.77% fat content (Aggarwal and Sandhu 2003; Prem et al. 1994). The yield of ascorbic acid, pectin, naringin and limonin in Kinnow peel was reported to be 47.52 mg/100 g, 18.56%, 358μg/g and 60.75μg/g, respectively (Sidhu et al. 2016). Fresh Kinnow peel had 40.7 mg/100 g ascorbic acid and 374 g naringin, according to Maan et al. (2013). In Kinnow peel, Aggarwal and Michael (2016) observed 459 ppm limonin and 40.7 mg/100 g ascorbic acid. The total gallic acid, *p*-hydroxybenzoic acid, vanilic acid, *p*-Coumaric acid, Ferulic acid and the sum of other phenolic acid content in Kinnow peel was 252, 290, 664, 823, 1556 and 3585 μg/g, respectively.

Pulp and Pomace The pomace, or byproduct of processing Kinnow fruits to generate juices and various products, is utilized to produce pectin and syrup. The Kinnow pomace includes a variety of physiologically bioactive constituents such as antioxidant properties, polyphenolic compounds, and flavonoids, although their quantity is lesser than that found in the peels (Yaqoob et al. 2020). As a result, Kinnow pomace can be utilised as a primary component in the value addition of a variety of culinary items (Hayat et al. 2010). Natural antioxidants are becoming more popular as an alternative to artificial antioxidants in foods and pharmaceuticals. Considering chemical constituents have adverse health effects, the user's safety issue promotes the concept of natural antioxidants (Dilas et al. 2009). Consuming meals high in natural antioxidants helps to avoid illness induced by oxidative stress (Dilas et al. 2009). Oxidative damage is created by an unbalance among pro-oxidant and anti-oxidant molecules, that can be generated by an over-abundance of pro-oxidant substances, a deficiency of antioxidants, or even both.

Kinnow pomace contains the most polyphenolic compounds in a confined state. The proportion of polyphenol content has indeed been found to be affected by geographical indication, cultivars, harvest time, warehousing, and processing parameters (Mallavadhani et al. 2006). After juice processing, the pomace has a significant concentration of glycosylated flavones, flavones, flavonoids, and polyphenols.The juice extraction residue (pomace) is abundant in glycosylated flavones, flavones, flavonoid, and polyphenols. According to Hayat et al., the Kinnow pomace possesses 18.4% and 30% in 2, 2-diphenyl-1-picrylhydrazyl (DPPH) radical scavenging activity and hydroxyl radical scavenging activities, correspondingly (2010). Research have been conducted to investigate the link between antioxidant properties and polyphenols in fruits and vegetables, and a linear association has been established (Minatel et al. 2017). The bulk of the polyphenol chemicals in Kinnow pulp are flavonoids and their derivatives such as flavones, flavonols, flavanones, flavonols, and anthocyanidins. According to the studies of Singh et al. (2016), the Kinnow pulp had 354.9 mg GAE/100 DW total phenolic contents and 261.3 mg QE/100 DW of total flavonoid content. In terms of DPPH and 2,2′-azino-bis (3-eth ylbenzothiazoline-6-sulfonic acid (ABTS), the antioxidant activity (mM trolox equivalent or TE/g) was found to be 2.9–3 and 3.3–3.5, respectively. Caffeic acid (16.8 mg/100 g), ferulic acid (12.3 mg/100 g), sinapic acid (14.4 mg/100 g), and kaempferol acid (14.2 mg/100 g) are among the polyphenolic found in Kinnow pulp.

Juice The Kinnow fruit is famous throughout India but is particularly well-known in Punjab and Rajasthan due to its unusual flavour and the presence of significant phytochemical components that are linked to a variety of health advantages. The Kinnow fruit juice content typically depends on a number of attributes, including cultivar, environment, cultural practices, etc. The juice is typically extracted by hand operating the machine on a smaller scale, which involves dividing the fruits in two and reaming the juice from the halves onto an appropriate rosette. Typically conical in shape, the rosette comprises ribbed or grooved sides. The oil cells in the peel are not damaged, and neither are the seeds, hence this method of juice extraction is the best. Contrarily, mechanical juice extractors are also available for the extraction of juice, in which the peel of the *Kinnow* may occasionally be manually removed before being placed into the screw-type juice extractor. Mechanical juice extraction method may produce a yield of about 60% whereas the manual method can produce about 40% of the juice. Due to changes in taste preferences, dietary customs, and consumer behaviour in recent years, there has been a surge in demand for processed Kinnow goods, primarily juice, as a result of increased Kinnow production in India (Bhardwaj and Pandey 2011). According to Bhardwaj and Mukherjee's (2011) studies, Kinnow juice was observed to possess 11.50°B TSS, 0.76% acidity, 21.15 mg/100 ml ascorbic acid content, 7.50% total sugars, 0.22 mg/ml limonin and 0.08 non-enzymatic browning. The total phenolic content, DPPH and ascorbic acid activity of Kinnow juice was reported to be 91.8 mg/100 ml, 59.19% and 52–53 mg/100 ml, respectively (Al Juhaimi et al. 2018). The physicochemical parameters of Kinnow juice are enlisted in Table 10.2. In developing the antioxidant potential of Kinnow juice, ascorbic acid also plays a crucial function in addition to phenolic components (Sun et al. 2002; Ghafoor, Al-Juhaimi and

Table 10.2 Physicochemical parameters of juice

Physiochemical parameters	Quantity/100 gm
Vitamin C	31.0 (mg/100 ml juice)
Ca	40.0 (mg/ 100 ml)
Fe	0.4 (mg/ 100 ml)
P	18.0 (mg/ 100 ml)
TSS	11.5 (%)
Acidity	0.9 (%)
TSS/acid ratio	12.0–14.0:1

Choi 2011). Kinnow juice is packed with vitamins such as ascorbic acid (31.66 mg/100 g), B1 (0.12 mg/100 g), B2 (0.01 mg/100 g), and niacin (0.43 mg/100 g). Natural Kinnow juice drink is much more nutritional, calorie-dense, and medicinal than manufactured drinks. Kinnow juice contains high levels of vitamin C, carotenoid, and polyphenols, as well as modest amounts of vitamin A, calcium, phosphorus, and iron, all of which assist considerably to the juice's total antioxidant potential (Peterson et al. 2006; Dhaka et al. 2016).Kinnow juice contains vitamin C, which serves as an antioxidant in the epidermis by intercepting and absorbing free radicals created by UV rays and reducing their oxidative stress (Okwu 2008). Because Kinnow juice cannot be marketed in their truest form due to quick bittering, attempts have been made to utilise Kinnow as a basic ingredient in the production of mixed juices as well as other products with additional value. Processing and storage are also required to reduce post - harvest loss owing to the highly perishable crop and the abundance of output throughout seasonal peaks (Mahawar et al. 2020).

Seeds Throughout the processing of Kinnow fruit for the production of various food items, a substantial quantity of waste in the form of Kinnow seed was discovered. Each Kinnow fruit has 20–25 seeds, and all these seeds can be utilised to produce limonin. These Kinnow seeds wreak havoc on the ecology and garbage handling (Matthaus and Ozcan 2012). Kinnow seeds are commonly accessible, and they possess biologically active compounds which can be employed as antioxidants. Several scholars studied the composition of Kinnow seeds in order to properly comprehend how to utilize them. Anwar et al. (2008) determined that the oil, protein, fibre, and ash content of Kinnow seed were 31.15%, 9.56%, 6.50%, and 5.60%, respectively. Babbar et al. (2011) determined that the antioxidant capacity of Kinnow seeds was 20.50 mg trolox equivalent or TE/g DW and total phenols were 3.68 mg GAE/g DW. Liu et al. (2012) employed an alkaline environment to extract limonin in their investigation, and the ideal pH is close approximately 11, the ideal temperature was 70 °C, the alkaline solution/seeds ratio was 20:1, and the ultrasonic power requirements used were 800 W for 30 min. The yield of limonin from citrus seeds obtained with 98% purity was 7.5 mg/g. According to Al Juhaimi et al. (2018), Kinnow seeds are a substantial source of essential oils such as palmitic, stearic, oleic, linoleic, arachidic, linolenic, behenic, and arachidonic in proportions of 15.77, 2.62, 23.53, 45.92, 0.36, 4.92, 0.29, and 0.10%. Essential oils are utilised in the culinary, pharmaceutical, and beauty sectors. Polyphenols, carotenoids, and

Table 10.3 Elemental composition of Kinnow seeds

Elements	mg/kg DW
Microelements	
Ca	7619
Mg	1186
K	10,334
P	3119
S	1132
Microelements	
B	12.91
Cr	0.396
Cu	9.30
Fe	40.50
Mn	6.08
M	0.318
Ni	0.51
Zn	14.67

tocopherols are also abundant in these seeds.The main phenolic chemicals identified in Kinnow seed were kaempferol, isorhamnetin, catechin, and 1, 2-dihydroxybenzene (Al Juhaimin et al. 2018). Kinnow seeds'antioxidant activity and total phenolic content were reported to be 20.50 mg TE/g-DW and 3.68 mg GAE/g-DW, respectively. Furthermore, macroelements and microelements were observed to be present, as shown in Table 10.3. Furthermore, the estimated sugar content was 5.30, 5.83, 4.86, 5.59, 5.29, and 3.59 g/kg for glucose, fructose, raffinose, staxioz, saccharose, and galactose, respectively. In comparison to other fruit parts, the largest concentrations of limonoids, an important class of compounds, have also been detected (Yaqoob et al. 2020).

Leaves Owing of their possible central nervous network involvement, natural compounds can be employed as a reservoir of biomolecules to generate new analgesics that operate versus discomfort and include characteristics such as pungency, tingling, and needles. Although the leaves of *C. reticulata* Blanco (Rutaceae) are often swallowed to flavour the palate and produce this sort of impact on lips and tongues, the natural compound's composition may function as the foundation for various types of analgesics. Sharma and Tyagi (2019) tested numerous solutions extracted from different Kinnow preparations for phytochemical activity. The content of alkaloids, cardiac glycosides, flavonoids, steroids, saponins, and tannins was investigated in Kinnow leaf, peel, and pulp extracts.

10.8 Chemical Components – Pectin, Polyphenols and Flavonoids, Carotenoids, Essential Oil, Seed Oil

The chemical composition of Kinnow fruit is summarized in Table 10.4. Kinnow is rich in bioactives and fibres detailed information is presented below:

Table 10.4 Chemical composition

Part	Properties	Content	Reference
Whole fruit 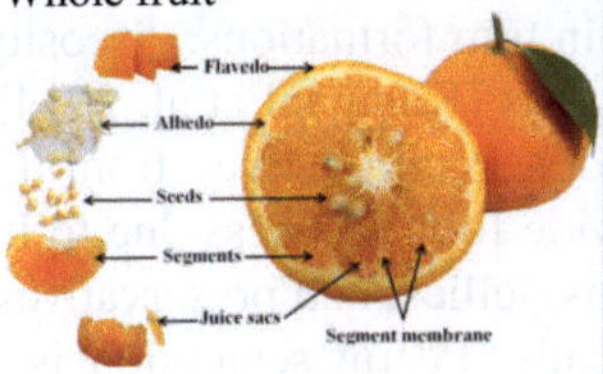	Weight	177.62 g	Mahawar et al. (2019)
	Volume	218.40 ml	
	Peel	29%	
	Pomace	25%	
	Juice	38%	
	No. of Seed	19.40	
	Firmness	5.86 kgf	
Peel	Thickness of peel	4.01 mm	Mahawar et al. (2019)
	Weight of flavedo	51.30 g	
	Weight of Pomace	44.20 g	
	Pectin	16.1%	Sharma et al. (2013)
	Total solids	22.5%	
	TSS	12.50°B	
	Acidity	1.38%	
	Ascorbic acid	41.57 mg/100gDW	
	Total sugars	6.23%	
	Reducing sugars	5.99%	
	Ash	0.67%	
	Carotenoids	13.65 mg/100 g DW	
	β- carotene	7.43 mg/100 g DW	
	Naringin	0.420 mg/g	
	Limonene	4.69 mg/g	
	Fat	0.77%	
	Total phenolic content	28.30 mg GAE/g DW	Saini et al. (2019)
	Total flavonoid content	4.40 mg QE/g DW	
	Antioxidant capacity	81.5%	Rafiq et al. (2021)
Pulp	Total phenolic content	345.9mgGAE/100gDW	Singh et al. (2016)
	Total flavonoid content	261.3mgQE/100gDW	
	Antioxidant capacity	50%	Mathur et al. (2011)
Seeds	Oil	31.15%	Anwar et al. (2008)
	Protein	9.56%	
	Fibre	6.50%	
	Ash	5.60%	
	Antioxidant capacity	20.50 mg TE/gDW	Babbaret al. (2011)
	Total phenolic content	3.68 mg GAE/gDW	
Juice	TSS	11.50°B	Mahawaret al. (2019)
	Acidity	0.90%	
	Turbidity	227.60 NTU	
	Vitamin C	23.50 mg/100 ml juice	

Pectin Pectin is a structural heteropolysaccharide that is typically derived from citrus fruit peels and is used primarily in the food industry as a gelling agent and stabilizer. It is made up of α-(1–4)-D-galacturonic acid units joined in a linear fashion. It finds its application as a biodegradable polymer in film formations, viscosity modifiers, coating agents, emulsifiers and chelating agents (Kanmani et al. 2014). Pectin appears as a white or brown powder with no odor. Pectin derived from the Kinnow peel has a wide variety of uses in the worldwide food business due to its usage in the creation of various food items such as jams, jellies, and preservatives. Pectin is typically removed chemically or enzymatically. Pectin separation is a multi-step procedure in which the material mixture is acidified with a powerful acid, the combination is maintained in a shaker, and pectin in the precipitate is produced when ethanol (96%) is added to the filtrate obtained in the shaker. Pectin was also recovered from Kinnow peel under various experimental circumstances such as temperature, time and pH using HNO3.

The yield of pectin obtained varied with these conditions and the highest yield observed was 16.1% in Kinnow peel at a temperature 60 °C, pH 1.75 and extraction time of 70 min. Pectin obtained from pomace (6.2%) was less at the same experimental conditions (Sharma et al. 2013). Ghoshal and Negi (2020) extracted pectin from Kinnow peel by ethanol precipitation method. The properties of pectin were assessed as a function of temperature, time and pH. pH was reported to be the determining factor for pectin yield and properties. The optimum pH observed was 5 and the yield obtained was 6.13%. In another study, Khule et al. (2012) reported the extraction of pectin peel and its utilization as a binding agent in paracetamol tablets. 18.21% pectin was extracted from Kinnow peel using ethanol treatment at pH 2 after 120 min of extraction. *In vivo* dissolution studies showed that 81.88% drug release was observed in pectin-bound paracetamol, indicating the potential use of Kinnow peel pectin as a binder. Pectin was isolated from pre-harvest fallen Kinnow by Bhatlu et al. (2016). The moisture, methoxyl content, ash content, equivalent weight, anhydrounic acid, and degree of esterification of the recovered pectin were all determined. The resultant pectin recovery yield of 57.5%, moisture 9.6%, methoxyl content 5.9%, degree of esterification 53.3%, and anhydrouronic acid concentration were 70%.

Polyphenols and Flavonoids Polyphenols are natural antioxidants by nature and are present in fruits, vegetables and some of their residues such as Kinnow peel. These are very beneficial for human health as it chelates with pro-oxidant ions present in the human body and acts as free radical scavengers (Wojdylo et al. 2007; Osman et al. 2009). Polyphenols are usually extracted by conventional methods such as solvent extraction method involving maceration with methanol, ethanol, etc. as solvents. However, as technology advances, many novel techniques of phenolic extraction, such as microwave-Assisted extraction (MAE), ultrasound-assisted extraction (UAE), supercritical fluid extraction, and accelerated solvent extraction, are being investigated for improved yield. Flavonoids, that are polyphenolic compounds, have a phenyl benzopyrone structure that consists of two benzene rings (C6) joined by a linear three-carbon chain (C3) with a carbonyl group at the C position. While flavonoids are commonly considered non-nutritive agents, they have piqued the attention of many researchers due to their possible significance in the prevention of significant chronic illnesses. Citrus flavonoids comprise polymethoxylated flavones (PMFs) such as nobiletin and tangeretin, which are comparatively two common ones, as well as a family of glycosides known as hesperidin and naringin. Citrus fruit peels contain the greatest quantities of PMFs when compared to other palatable segments of the fruit. According to Liu et al. (2012), for the recovery of flavones from *C. reticulate* using accelerated solvents extraction, methanol appears to be the most efficient extraction solvent when compared to ethanol and water.

Sharma et al. (2016) studied the pattern of polyphenol constituents with antioxidants activity in Kinnow mandarin juicy sacs, both granular and non-granulated. Based on their observations, granulated Kinnow has a low concentration of polyphenol components (4.3 mg GAE/100 ml), ascorbic acid (16.2 mg/100 ml), and antioxidant activity (1.78 mol Trolox/g). Nongranulated Kinnow extracts, on the contrary hand, indicated significantly greater levels of polyphenol components (11.3 mg GAE/100 ml), ascorbic acid (28.6 mg/100 ml), and antioxidant activity (9.51 mol Trolox/g). MAE of polyphenols from Kinnow peels was performed in research (Hayat et al. 2009), and the results were matched to UAE and rotational extraction. The best extraction parameters were discovered to be 152 W microwave power, 49 s extraction duration, a liquid -to -solid ratio of 16, and a level of 66% methanol (solvent). Gallic acid, *p*-hydroxybenzoic acid, vanillic acid, *p*-coumaric acid, and ferulic acid were amongst the polyphenols acids recovered. These phenolics were examined for antioxidant efficacy and matched to various traditional procedures such as ultrasound and rotary extraction, with MAE being shown to be the most efficient approach due to its low time utilization and high efficiency. Polyphenols were extracted from Kinnow peel using maceration and UAE methods using different solvents and results showed that maceration was least competent than UAE procedure in extraction of polyphenols. In extracts prepared using maceration methanol (80%) extract had a higher content of polyphenols than ethanol, while ethyl acetate extract (80%) had the lowest polyphenol content. Safdar et al. (2017a, b) showed that when polyphenols were extracted from kinnow peel that 80% ethanol produced hesperidin and ferulic acid at concentrations of 92.9 and 65.2 g/g dry sample, respectively. Tumbas et al. (2010) identified hesperidin (80.9 mg/g extract) and narirutin (15.3 mg/g extract) in mandarin peel extract, which displayed high DPPH and hydroxyl free radical scavenging capabilities of 0.179 mg/ml and 0.415 mg/ml, respectively. The total polyphenolic content in Kinnow peel and pulp recorded was 148 and 76 μg/g GAE, respectively. The superoxide anion radical scavenging capability is maximum for ethanol extract of Kinnow peel (87%) and pulp (50%) among extracts prepared in water and chloroform (Mathur et al. 2011). Hayat et al. (2011) discovered phenolic components were found in concentrations ranging from 32.9 to 63.2 mg GAE/g in different kinnow cultivers among which ponkan variety showed maximum concentration while Kinnow had lesser phenolic and flavonoids compounds. The total flavonoid content observed in mandarin peel was 4.20 mg QE/g of extract powder. The mandarin peel extract possessed 84.61% nitric oxide scavenging capacity at 200μg/mL and 73.9% at 800μg/mL DPPH scavenging capacity (Olyad et al. 2020). Safdar et al. (2017a, b) in their study revealed presence of eleven polyphenolic compounds from Kinnow peel extract and the most abundant were ferulic acid and hisperdin. Rafiq et al. (2021) in their study reported that total phenols and flavonoids present in the extract were 24.51 mg GAE/g and 9.12 mg QE/g, respectively while the freeze-dried peel extract has DPPH radical scavenging capacity of 81.5%. Microwave heating increased the yield of flavanone glycosides from Kinnow peels having 4975.4 g/g DW of naringin, 820.7 g/g DW of naringenin, and 53.0 g/g DW of hesperidin, whereas non-heated Kinnow peel powder was 3915.8 g/g DW of naringin, 655.3 g/g DW of naringenin, and 42.5 g/g DW of hesperidin (Hayat et al. 2010). Saini et al. (2019) estimated polyphenolic content, flavonoids content and antioxidant activity from citrus peels. The extraction was done using maceration and UAE; it was observed that the higher yield from Kinnow peel was obtained *via* UAE method (5.85%) rather than maceration (5.20%) furthermore the phenolic and flavonoid contents was maximum in UAE sample (28.30 mg GAE/DW and 4.40 mg CE/g DW, respectively) and lowest when extraction was done using maceration (23 mg GAE/g DW, 3.75 mg CE/g DW, respectively).

Naringin Naringin is a flavonoid found in Kinnow peel, especially in young fruits; it, together with other compounds such as limonene, is chiefly accountable for the bitterness of Kinnow peel. It is widely used in the pharmaceutical and food sectors. Naringin has antioxidant and anticancer characteristics, which make it useful in the pharmaceutical sector. It is said to lower levels of cholesterol, defend from carcinogenic compounds, and fight against poisons during chemo (Gorinstein et al. 2006; Jiang et al. 2006). Researchers employed solvent extraction (Bhatlu et al. 2017; Bhatlu et al. 2016), supercritical carbon dioxide extraction (Yu et al. 2007), precipitation technique (Tripodo et al. 2007), and adsorption techniques to retrieve naringin from peels (Jiang et al. 2006). Because the naringin concentration in immature fruits is higher, pre-harvest fallen fruits may be used in the recovery method, which is commercially favourable to producers. Researchers looked at removing bitterness from Kinnow peel and pomace by separating naringin and limonene (Singla et al. 2021). The acetone solventogenesis approach yielded the maximum yield of naringin and limonene from Kinnow peel and pomace. This technique employs a by-product and offers a long-term remedy (Singla et al. 2019). In Kinnow, pre-harvest dropping is an issue, and the fruit that drops off might be utilised to recover pectin and naringin (Bhatlu et al. 2016). The unripe Kinnow pulp was boiled, and the resultant liquid was passed across Indion PA 800, a native raisin, whereupon naringin is captured. After desorbing the immobilized naringin with ethanol, the mixture was screened and naringin was obtained by boiling the residue. The naringin retrieved was 52% pure, with an eventual purity of 91–93%. In a research on the chemical characterisation of naringin retrieved from Kinnow peel waste, Puri et al. (2011) discovered that rhamnose (an essential component for the food industry) can be generated from naringin using α-l-rhamnosidase. These studies demonstrate the usefulness of naringin in a variety of industries, as well as a method for using Kinnow pulp waste.

Carotenoids Kinnow mandarin accumulates carotenoid in a similar but more complicated pattern than orange fruit. Mandarin peel and pulp contain more than twenty different carotenoids and xanthophylls at maturity, which is higher than oranges. Clementine mandarin peel contains four times less total carotenoids than Satsuma peel. Mandarins, like orange fruits, accumulated carotenoids primarily in the peel (up to 94% of total carotenoids in Dancy mandarin), with β,β-xanthophylls being the main constituents (68–90% of total carotenoids). The violaxanthin concentration varied between cultivars and tissues, with the *cis*-isomer outnumbering the *trans*-isomer.

The chloroplastic carotenoids were the most prevalent in the green fruit peel, while the concentration of β,β-xanthophylls rose following the colour break in both the peel and the pulp. The increased accumulation of xanthophyll β-cryptoxanthin and apocarotenoids in mandarin fruits distinguishes it from oranges. β-Cryptoxanthin accumulates in the peel and pulp of mandarin fruit, contributing to their strong colouring. The C-30 β-apocarotenoids β-citraurin and β-citraurinene are especially plentiful in the peel, which gives mandarin fruits their unique orange-reddish colour. The relative proportions of red β-citraurin and orange β-cryptoxanthin are regarded to be the key contributors to the colour in Dancy mandarin, despite the fact that violaxanthin accounted for more than half of total carotenoids (Gross 1987). Another example of rich colouring is Michal mandarin juice, which appears to be due to high levels of β-cryptoxanthin and zeaxanthin in the pulp (Farin et al. 1983). It is worth noting that the differences in carotenoid composition between flavedo and pulp are more pronounced in mandarin fruits than in orange fruits. Carotenoids isolated from Kinnow peels utilising a cellulolytic enzyme and improved carotenoids stability (Nadeem et al. 2018). The combination of CMCase and pectinase (250 IU/100 g peel) produced the maximum carotenoid production (8.60 mg/100 g peel). Furthermore, the recovered carotenoid pigments were highly stable in the darkness at 30 °C compared to the open, and because resilience diminishes with rising temperature, freeze-drying the pigments culminated in better durability. The lutein was recovered utilizing the UAE technique with methanol as the extraction solvent, with the optimal parameters being 43.14 °C, 6.16 mL/g solvent/solid ratio, and 33.71 min time duration (Saini et al. 2021). These circumstances resulted in the highest lutein output (29.70 g/g).

Essential Oils Several research findings have advocated various ways for obtaining essential oils from Kinnow peel. Hydro distillation, centrifugation, hydrodistillation of fresh peel, hydrodistillation of manually dried peel powder, and hydrodistillation of sun dried peel powder are all methods of extracting oil from Kinnow peel. Javed et al. (2014) extracted essential oils from fresh Kinnow peel using the hydro-distillation method, yielding 0.33% and taking 210 min. Kamal et al. (2011) employed the hydro-distillation procedure to extract essential oils in a comparable manner. Their research revealed that the essential oil content ranged from 0.30 to 0.50 g/100 g in fresh peels, 0.24–1.07 g/100 g in ambient peels and 0.20–0.40 g/100 g in oven-dried peels.

The mechanically dried peel powder obtained by hydro-distillation contains the highest oil concentration and comprises higher essential oils than fresher peel, most probably due to oil pore bursting through the powdering process. It also meant that a physical process such as centrifugation created a high redness value in a short period of time.

Terpenes were the compounds present in the oil of *C. reticulate* leaves that give kinnow and other leaves of citrus plants their pungent flavor and it was discovered to be a methyl-N- methyl anthranilate compounds as confirmed by 1D and 2D NMR spectroscopy, which is responsible for this bitterness. This chemical resembles another type of molecule with antinociceptive effects. These compounds with this action might be utilised to discover new analgesics for pain relief (Correa et al. 2016). The essential oil obtained from hydro-distillation of *C. reticulate* Blanco var. Kinnow (seedless and seeded) leaves revealed that oil contains 62.00% of β- phellandrene, 6.53% of β-pinene, 2.81% of β-myrcene and limonene both and 0.51% of caryophyllene as the prime components. In another study, its was reported that the essential oil obtained from low-seeded kinnow contains 37.35% β-phellandrene, 2.79% α-pinene, 3.26% β-pinene, 4.16% β-myrcene, 5.77% limonene and 1.41% caryophyllene as the main component when quantified by GC-MS. These were the five main components that accounted for about 100% of the total oil composition. Furthermore in their study, the researchers showed that microwave-assisted hydrodistillation (MAHD) had higher extraction efficiency of 6.8%, followed by microwave extraction (ME) having an extraction efficiency of 5.5% and hydrodistillation (HD) with the least extraction efficiency of 3.6%, when all the techniques were compared. While there were some quantitative differences, all three techniques appeared to yield essential oils with a composition that was qualitatively similar. Mandarin peel essential oil has five distinct components, the majority of which were monoterpene hydrocarbons. While computing the data by GC-MS their results revealed oil compositions had 0.54%, 0.375% of α-pinene,0.414%, 0.284% of β-pinene, 1.405%, 1.461% of β-myrcene, and 97.64%, 97.88% of limonene as the major four compounds when extraction was performed by for HD and ME, respectively. Whereas, when oil was extracted using a combination of techniques, the oil contains 0.518% of α-pinene, 0.317% of β-pinene,1.104% of β-myrcene, 97.94% of limonene, and 0.122% of sabinene as major five components found in mandarin peel essential oil. The findings revealed that the combination approach dealing with MAHD was a feasible alternative solution to HD for shortening the time since it seemed to generate essential oils with a greater yield and comparable chemical characteristics and value to those generated using conventional techniques. According to reports of Sultana et al. (2012) essential oil recovered from Indian mandarin by hydrodistillation method showed limonene as the prime component (87.45%).

Caryophyllene oxide, caryophylla-3 (Correa et al. 2016), 7(14)-dien-6-ol, α-pinene and 2,6-dimethyl-1,5,7-octatrien-3-ol are the major compounds of essential oil obtained via hydrodistilled from Nigerian *C. reticulata* Blanco and *Citrus aurantifolia* Swingle leaves (Lawal et al. 2014), whereas the chemical composition of *C. aurantifolia* (Ikotun) essential oil comprises of limonene and geranial. While the primary ingredients detected in the C. *reticulata* oil sample from Ikotun were 38.1% of citronellal, 25.9% of (Z)-β-ocimene, 14.5% of linalool, and 12.2% of limonene (12.2%), while the oils from Ijanikin included 22.7% of pinocarvone, 20.0% of *trans*-pinocarveol acetate, and 12.8% of β -thujone. The essential oil from *C. reticulata* Blanco cv. Kinokuni peels by mechanical pressing method under optimized conditions yield 0.689% that closely resembles the predicted value with limonene as the prime constituent having 56.76% configuration. GC-MS was also used to identify 64 components, accounting for 96.34% of the total oil. These essential oil at a concentration of 7.5–60 mg/mL as minimum inhibitory concentration exhibits broad-spectrum antimicrobial effects against different microorganisms with a zone of inhibition values ranging from 9.24 to 16.35 mm. *C. reticulata* (Kinnow), *Citrus sinensis* (Mussammi), and *Citrus x sinensis (Red Blood Orange)*, yield 0.86, 1.70, and 1.07%, respectively of essential oils, when extracted using Hydro-distillation extraction method and furthermore the chemical composition of these extracted essential oils revealed that they comprise of 46.30–54.57% of limonene, 10.02–24.00% of geraniol, and 10.05–14.00% of citraniol as major components of essential oils from three different citrus cultivars peels as stated above along with great antimicrobial effect with higher zone of inhibition against tested bacterial strains (Qadir et al. 2018). The steam distillation process extracted 0.34% of the essential oil from the huge Kinnow fruits' green peel. GC identified 24 compounds among the different oil components, which were made up of 15.33% of 6-methyl-5-heptene-2-one, 13.8% of carvone, 10.04% of *cis*-carveol, and 4.55% of thujanol, while the other twenty identified chemicals, accounting for 35.84% of the total oil, were only found in trace levels. The percentage of limonene in the Kinnow orange green peel oil was relatively low (2.76%), compared to the bulk of citrus oils (35–85%) (Mahmud et al. 2005).

Seed Oil Kinnow seeds are recognised as a possible oil resource due to unique fatty acid profile and substantial tocopherol content (Juhaimi et al. 2016). Moreover, since it is high in protein, minerals, and fibre, it might be utilised to make high-value items in addition to culinary purposes. Among the fatty acids, pammitic, stearic, oleic, *cis*-vaccenic, linoleic, linolenic, arachidic, elcosenoic, and behenic acid comprise 98.6% of its total fatty acid content. The seed oil exhibited an iodine value of 104.80 g of I/100 g, 1.4650.927 mg/ml of density, saponification value of 186 mg of KOH/g of oil, an unsaponified matter of 0.48%, acid value of 1.30 mg KOH/g of oil, color red color value of 2.50 and yellow value of 20.00. The oil had excellent oxidative stability of 2.64 at 232 nm and 0.81 at 270 nm as measured by specific extinctions along with 3.15 of *p*-anisidine value and 2.40 mequiv/kg of oil of peroxide value.

10.9 Antimicrobial Effect

Mathur et al. (2011) investigated the antibacterial properties of different peel and pulp extracts of Kinnow, orange, and shaddock against *A. niger*, *C. albicans*, *S. aureus*, and *S. pyogens*. The most effective antibacterial agent against *S. pyogens* was determined to be an aqueous extract of Kinnow pulp (zone of inhibition, ZOI = 25 mm), which was superior to aqueous extracts of pulp from orange and shaddock. In comparison to chloroform extracts of orange and shaddock, extract from Kinnow pulp was shown to be the utmost effective antibacterial agent against *S. pyogens* (ZOI = 17 mm), whereas Kinnow peel chloroform extract had the highest antimicrobial bacterial activity against *S. pyogens* (ZOI = 18 mm). The minimum inhibitory concentration (MIC) and inhibitory effect of the various solvent extracts of the leaves, pulp, and peel of *Citrus nobilis* and *Citrus sinensis* were evaluated against bacteria and fungi species. The MIC was found to be 18 g mL^{-1} for *B. cereus*, 25 g mL^{-1} for *P. aeruginosa*, and 30 g mL^{-1} for *C. albicans* in the ethanolic extract of Kinnow peel. The MIC was found to be 18 g mL^{-1} for *S. epidermidis* and 40 g mL^{-1} for P. vulgaris, in ethanolic extract of leaves; 20 g mL^{-1} and 25 g mL^{-1} for *S. typhimurium* and *T. viride*, respectively, in ethanolic extract of pulp; and 25 g mL^{-1} for *E. coli*, respectively, in benzene extract of Kinnow (Sharma and Tyagi 2019).

10.10 Debittering

Kinnow peels and seeds contain the main bitterness-producing substances (Naringin and Limonin). Development of limonin is the cause of delayed bitterness. In acidic circumstances, an enzyme (limonoate-D-ring lactone hydrolase, found predominantly in seeds) catalyses the conversion of limonoate A-ring lactone, a non-bitter precursor, to bitter limonin, resulting in prolonged bitter taste of juice after 3–4 h of extracting. Fruit elements such as the peel and seeds include varying quantities of bitter chemicals. Premi et al. (1994) reported that the peel has highest level of narginin order with concentration of 0.422 mg/g, followed by juice having narginin concentration of 0.230 mg/g while seed exhibits the lowest levels of narginin having a concentration of 0.134 mg/g, while seeds showed highest levels of lignin with a concentration of 9.50 mg/g followed by peel with lignin concentration of 4.69 mg/g and lowest in with juice lignin concentration of juice 0.218 mg/g. Ilame and Singh (2018) identified the hollow fibre membrane with a molecular weight of 30 kDa as the best choice for increasing the storage period of ultra-filtered Kinnow juice beyond 60 days without any usage of extra regulators. This showed that by physically removing seeds and peel (albedo + flavedo) prior to juice extraction, bitter taste levels might be reduced owing to less tissue disturbance. Around the globe, genuine attempts have really been undertaken to reduce or eradicate the bitter taste that is now found in Kinnow juice. A variety of compounds, such as naringin dihydrochalcone and neohesiperidin dihydrochalcone, have indeed been identified to hide bitterness of juice through their sweetening index. Debittering could be performed by absorbing bitter compounds on vinyl-dodecylbenzene resins. Upon running bitter taste juice of Kinnow through sheets of polymeric adsorbent resins, Singh et al. (2009) noticed enhanced sensory qualities during storage. Kaur (2017) investigated the enzymatic debittering and fragrance improvement of Kinnow juice using the enzymes limonin dehydrogenase, naringinase, and b-glucosidase. The results showed an increment in glucose levels by up to 4.38 lg/ml, acidity by 30.13%, and total sugar content by 42.97 lg/ml, whereas there is a significant drop in the bitterness components such as limonin by 87.34% and naringin by 58.41%. Singla et al. (2021) investigated the debittering of pulp residue from *C. reticulata* (Kinnow). Naringin, a substance that causes bitterness, was converted by an enzymatic process into naringenin, a substance that does not cause bitterness, to track the optimization of debittering.

The naringinase enzyme was found to be successful in reducing the bitterness of Kinnow pulp residue in which naringin content was decreased by 66.19% along with an increase in naringenin content a non-bitter compound by 52.38%, when enzymes were used at concentration of 1 U/mg, and incubated for4 h at 50 °C temperature range, at pH of 4.5. A Naringinase immobilisation test on various matrices such as alginate, k-carrageenan and polyacrylamide at varied concentration was done (Puri et al. 1996) to ensure its applicability in Kinnow juice debittering. The results showed that the optimal matrix was 2% sodium alginate. The expansion of pH optima gave desired versatility for debittering Kinnow juice of varied pH, and temperature gradients showed enhanced thermo-stability, which may be advantageous if debittering costs decreased. After immobilisation, 30 U of naringinase hydrolyzed 82% of the naringin in 3 h. Alginate enabled simple homeostasis with no impediments to naringin inflow or naringenin/prunin outflow, in addition to good physical durability, indicating the possibility of its industrial use. Debittering was decreased by 60% once kinetic models were introduced to Kinnow juice and optimised with pure naringin. The immobilisation of naringinase on glutaraldehyde-coated hen egg white (1 g HEW beads, 10 U of naringinase, 37C, pH 4.0, and 48 h) was accomplished by optimising a technique including 1% glutaraldehyde cross-linking. Immobilisation was 140% efficient under optimum circumstances (5 U = g HEW, pH 3.0, 60C, and 5 h), but soluble naringinase gave 91% efficiency for conventional naringin hydrolysis. It delivered a 68% efficient debittering when used to debitter Kinnow mandarin juice (Puri et al. 2001).

10.11 Applications

10.11.1 Food Applications

The valorization of Kinnow in food products such as jams, fruit bars, bakery products is well pronounced and it is summarized as below (Fig. 10.2):

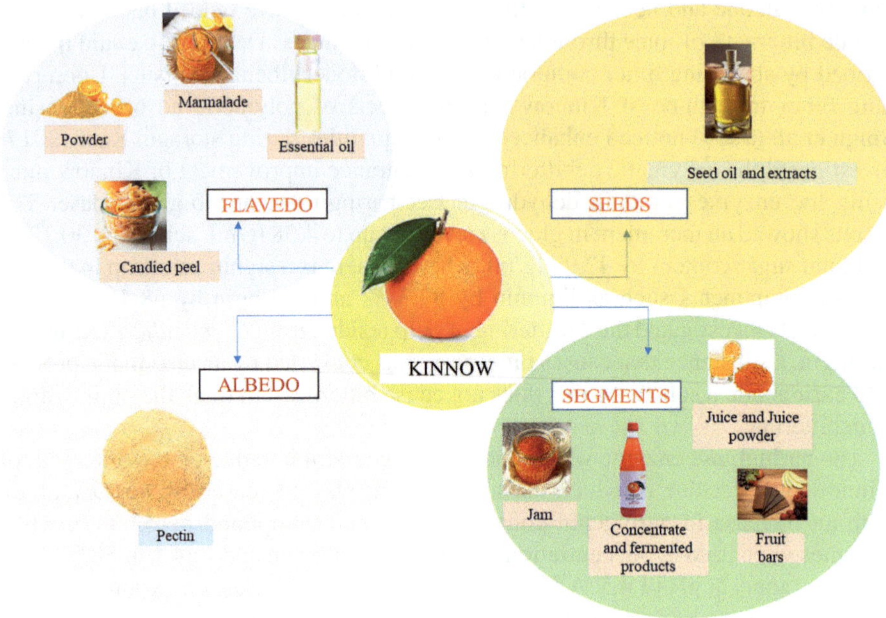

Fig. 10.2 Food applications of Kinnow

Bakery Products Kinnow residues (peel and pomace) and their phytochemicals were used to make functional cookies in place of wheat flour. The flour was replaced with dry peel and pomace powder at 5–20% levels, respectively, while phytochemicals were added at 1–4% levels. Cookies containing 10% peel, 5% pomace, and 4% phytochemicals received the highest sensory evaluations. The addition of Kinnow peel and pomace to cookies increased the crude fibre, ash, and L* value while lowering the spread ratio. Because of their strong DPPH and FRAP scavenging activity values, supplemented cookies performed better. Cookies containing peel phytochemicals had the greatest levels of phenolics 12.02 0.11 mg GAE/g, 0.47 0.06 mg QE/g flavonoids, and 0.76 0.03 mg/100 g carotenoids,. The produced cookies were more oxidatively stable than the cookies containing chemical antioxidants such as Butylated hydroxyanisole (Yaqoob et al. 2021). Similarly, Yaqoob et al. (2019) prepared muffins incorporating Kinnow peel and pomace by replacing wheat flour at concentrations *viz* 5, 10, 15, and 20%. In comparison to BHA substituted muffins (control), the supplemented muffins showed high levels of crude fibre i.e., 3.57% (peel) and 2.82% (pomace). Phytochemicals incorporating muffins were also prepared and at 4% phytochemical concentration higher phenolics content (12.34 mg/g of GAE), flavonoids (0.50 mg/g of QE), carotenoids (0.82 mg/100 g), and antioxidant activities i.e., DPPH (59.71%) and FRAP (23.46%).

Rafiq et al. (2022) studied the techno-functional, sensorial and nutritional characteristics of freeze-dried Kinnow peel powder incorporated soup sticks and it was observed that as the substitution level of the peel powder was increased in wheat

Table 10.5 Food applications

Part	Application	Reference
Peel and pomace	Cookies and muffins Improved crude fibre and ash Decreased spread ratio High DPPH scavenging and FRAP power High phenolic, flavonoid and carotenoid content Better oxidative stability	Yaqoob et al. (2019, 2021)
Peel (blanched and unbalanced)	Ice cream Better appearance, taste and overall acceptabilty. High contents of naringin, total soluble solids (TSS) and ascorbic acid	Mann et al. (2013)
Peel	Soup sticks Increased breaking strength, ash content, yellowness, visco-elasticity, baking and storage stability Decreased moisture content, crude protein, crude fat, pasting parameters, physical parameters, L* (lightness), and a* (redness) values.	Rafiq et al. (2022)
Peel	Candy Less bitterness development Appreciable content of bioactive components	Sidhu et al. (2016) and Sogi and Singh (2001)
Segments	Jam Less bitternes High overall acceptability	Sogi and Singh (2001)
Juice	Fruit bars High overall acceptability Good economic viability	Kaur (2017)

flour from 0 to 9% caused notable increase in properties such as breaking strength, ash content, yellowness, visco-elasticity, baking and storage stability. Furthermore, there was a significant decrease in moisture, protein, fat, pasting characteristics, physical properties, L* (lightness), and a* (redness) values. Furthermore, the digesting time for the prepared sticks was longer than that of the reference, and a study of the look and composition demonstrated a decrease in pore diameter. Raising the substitution amount boosted the strength of scent and colour but decreased taste, according to the sensory attributes. The application of Kinnow in food products is summarized in Table 10.5.

Candy According to Sidhu et al. (2016), osmotically dried Kinnow peel candy has ascorbic acid content of 30 mg/100 g, naringin content of 170 g/g, and limonin content of 38 g/g while peel powder had 40 mg/100 g of ascorbic acid 165 g/g of naringin, and 38 30 g/g of limonin content. According to Aggarwal and Michel (2016), sugar and fructose were utilised to turn the whole Kinnow with peel into candy. Kinnow candy has an ascorbic acid content of 11.2–11.4 mg/100 g and a limonin value of 410–412 ppm. Candied kinnow segments were placed in 30 B syrup and slowly cooked to 80 B in a study conducted by Sogi and Singh (2001). After being drained of the syrup, the sections were dry (at 55 °C for 5 h) and sprinkled with icing sugar. The stated quality criteria for the generated product were TSS (83.50 °B), acidity (0.65%), ascorbic acid (18.34 mg/100 ml of juice), total sugars (63.45%), reducing sugars (2.82%), and overall acceptability (7.73). There was no bitterness accumulation in candied portion during the storage time. Alam et al. (2019) utilised the osmotic drying process to make sweets from entire Kinnow fruit. The optimal configuration of more loss of moisture and lesser osmotic absorption was found to be the osmotic process temperature of 65 °C, sugar solution concentration of 65–75 °B, solution to fruit ratio of 5:1, and immersion duration of 270 min.

Beverages and Ice Cream Market-available blended juice are made from the concentrates of two or more different varieties. Kinnow juice is combined with citrus juice that is less acidic and bitter. The idea of blending juice has become popular recently, and there are numerous studies available in the same direction. A study conducted by Bhardawaj and Mukherjee (2011) showed the production of Kinnow: amnla: ginger (92:5:3) blended juice which possessed 12°B, 0.80% acidity, 45.30 mg/100 g ascorbic acid and 7.44% total sugars. The bacterial, yeast and mould population recorded was 7.69×10^3, 3.49×10^3 and 2.99×10^3 cfu/ml of juice, respectively. Mann et al. (2013) investigated the creation of ice cream using frozen Kinnow peel (unblanched and blanched) at concentrations of 1, 3, and 5%. Sensory and chemical evaluations revealed that ice creams with Kinnow peel supplements had a higher look, flavour, and overall acceptability. The levels of naringin, TSS, and ascorbic acid in made ice cream were high and increased as the amount of peel added increased. According to sensory assessment, the optimal concentrations of frozen Kinnow peel in ice cream were 3% for unblanched and 5% for blanched-5%.

Jam Sogi and Singh (2001) created jam from unpeeled and lye peeled Kinnow chunks and the prepared jams was concentrated to 70°B and stored at room temperature (16–20 °C) for 105 days in airtight glass jars. According to the sensory data, bitterness did not occur in jam created using lye-peeled segmentation until 105 days, although bitterness did show up in jam prepared from unpeeled segments after 30 days and increased with time. The jam has TSS of 70°B, acidity of 1.34%, ascorbic acid content of 26.33 mg/100 ml, total sugars content of 61.05% and sensory scores of 8.33 for flavour, 8.13 for colour, and 8.33 for overall acceptability.

Fruit Bars Kaur (2017) created a composite fruit bars with residual Kinnow, grape, and guava juice in various ratios, such as 100% Kinnow, 50% Kinnow and grapes, 50% Kinnow and guava, and 50% Kinnow and grapes. The juice waste was concentrated in an open pan technique at 80 °C till 40 °B using 20% sugar and 0.2% citric acid. The resultant mixture was spread in a thin layer (4–5 mm) on an aluminum tray and dried for 12–18 h at 50 °Cto a moisture content of 17–18%. Fruits bars prepared from Kinnow, grape, and guava were demonstrated to be good for storage for up to six months (14–32 °C). The average cost of the manufactured product was determined to be Rs 60/kg, confirming its economic feasibility from the perspective of both makers and consumers. The developed bar moisture content of 17.63%, total solids content of 82.37 °B, TSS of 69.97 °B, acidity of 1.30%, 64.95% of total sugars, 11.12% of crude fibre, 275.20 mg/100 g ascorbic acid, 920 mg/100 g of total phenolics, 4.60 mg/100 g of total anthocyanins.

Powdered Juice and Concentrates Due to its many benefits in terms of ease of handling, transportation, and storage, the production of fruit juice powder has grown significantly in recent years. A useful auxiliary for the ice cream, bread, and confectionery sectors is juice powder. The most popular technique for drying liquid foodstuffs is spray drying. Juyal et al. (2015) investigated the spray drying technique's capacity to create Kinnow juice powders and found recovery factor of 36.45% at a 60:40 (Kinnow juice: maltodextrin) ratio at 146.5 °C inlet temperature, and 26 °C feed temperature. The powder recoveries were 37.66% in a 40:10:50 mixture of maltodextrin, sucrose, and juice. The final optimized powder has 4.60–5.74% moisture, 86.07–87.24% of dispersibility, L* value of 85.01–85.89, a* value of 1.61–2.06, b* value of 12.48–13.88, colour change value of 52.31–53.18.

Concentrating liquid foods lower the costs associated with handling, storage, and transportation activities. The procedure also makes sure that fruit juice is used and accessible all year long. The concentration of Kinnow juice was carried out in a rotary vacuum evaporator at temperature of 50–60 °C under vacuum pressure of 27–28 inch Hg. The 72 °B juice concentrate was packed in a glass container with 700 ppm SO2 and kept at −18 °C without experiencing any noticeable changes in its physicochemical properties, nor did it exhibit any signs of fermentation or an off-smelling odour (Thakur et al. 2000). Khamrui and Pal (2002) used reverse osmosis to concentrate Kinnow juice while optimising temperature (40 °C) and operation pressure (40 bar). The resultant concentrated Kinnow juice has TSS of 23 °B, pH of 3.20, acidity of 2.49%, total carbohydrates content of 17.25%, reducing sugars content of 5.85%, sucrose content of 6.79%, viscosity of 6.21 cP, and sugar-to-acid ratio 9.23. The following percentages of acid components, total soluble solids, total carbohydrates, reducing sugars, and sucrose were recovered during the concentration process: 89.68%, 96.02%, 95.50%, 94.32%, and 98.22%, respectively. Reverse osmosis (RO) and vacuum thermal evaporation were used to concentrate Kinnow juice (Thakur et al. 2004). They also examined the effects of centrifugation (CF) and ultrafiltration (UF) (VTE). They indicated that juice may be clarified with UF and condensed with RO up to 26°B before being clarified with CF. Juice could also be clarified with UF up to 24–26°B. The corresponding quality parameters for VTE treatment were 43°B of TSS, 2.33% acidity, 56.48 mg/100 ml of ascorbic acid content, 19.0% reducing sugar, of total sugar and 1.79 mg/100 ml of carotenoid content. The statistical difference in the sensory properties of juice concentrate obtained using various methods was insignificant.

10.11.2 Green Catalyst

The powdered kinnow peel is said to be a cost-effective, environmentally benign, and long-lasting catalyst for the synthesis of Schiff base. N-Benzalideneaniline and its descendants were synthesised from benzaldehyde and aniline variants using powder of Kinnow peel acting as a catalyst. In tidy circumstances, the reaction yield was 78–85%(Verma et al. 2022).

10.11.3 Bioethanol Production

Ethanol was produced from Kinnow garbage and banana peels in 4:6 ratio at 30 °C temperature, with 6% of inoculum using concomitant saccharification and fermentation with cellulase and co-culture of Saccharomyces cerevisiae G and Pachysolentannophilus MTCC 1077 for incubation time of 48 h, and agitation for the first 24 h were shown to be optimum for ethanol production employing two byproducts. Following enzymatic saccharification, the pretreated steam exploded biomass containing 63 gL^{-1} reducing sugars were fermented under optimised reaction situations both with hexose and pentose fermenting yeast strains, yielding 26.84 gL^{-1} of ethanol, a yield of 0.426 g/g, and fermentation efficiency 83.52%, respectively (Sharma et al. 2007).

In a step of producing ethanol from Kinnow peel via simultaneous saccharification of crude filtrate extract produced by *Aspergillus oryzae* and fermentation with *Pichia kudriavzevii* revealed that pre-hydrolysis of Kinnow peel with crude filtrate extract at three cellulase filter paper units/g dry substrate (FPU/g-ds) at 50 °C resulted in glucose (24.870.75 g/l), fructose (21.980.53 g/l), sucrose (10.860.34 g/l), and galacturonic acid (6.560.29 g/l) and galacturonic acid (6.560. After 12 h, ethanol production declined, culminating in ethanol levels of 33.87 g l-1 and efficiency of 2.82 g l-1 h-1, respectively.

10.11.4 Other Applications – Cellulase Production, Reducing and Capping Agent, Nanocellulase Production, Animal Feed

Cellulase Production An research was conducted to see whether *Trichoderma reesei* Rut C-30 could manufacture cellulases from Kinnow pulp. Amongst some of the different permutations assessed, dehydrated Kinnow pulp blended with wheat bran in a 4:1 ratio generated the highest filter paper cellulase (FPase) activity of 13.4 IU/gds, but the better endo-1, 4-glucanase (CMCase) activity was revealed once Kinnow pulp was accompanied with wheat bran in a 3:2 proportion utilising Mandel Weber (MW) mode. The 3:2 ratio of Kinnow pulp to wheat bran in MW media resulted in the maximum amounts of β-glucosidase activity (18 IU/gds). In the case of pretreatment lignocellulosic substrate, an FPase:-glucosidase ratio of roughly 1:1 was deemed to be the most optimal for reaching the ideal saccharification effectiveness, which was achieved when water was employed as a vehicle with wheat bran to Kinnow pulp ratio of 3:2. The findings of this study explored the feasibility of employing Kinnow pulp for cellulase production and demonstrated that a substrate with little economic relevance and that causes environmental contamination due to inappropriate management may be utilised to make cellulases (Oberoi et al. 2011).

Reducing and Capping Agent for Nanoparticle Production Naz et al. (2017) demonstrated an effective approach for manufacturing nanoparticles (AgNPs) using Kinnow peel extracts as a reducing and capping agent in their study. AgNPs were synthesised utilising dilute and concentrated peel extract. For nano-particles (NPs) produced from diluted and concentrated extracts, ultra-violet-visible spectroscopic technique revealed different absorption maxima at 425 and 400 nm, respectively. The X-ray diffraction investigation of nano-fabricated silver revealed a pure face-centered cubic crystal structure with dimensions determined using the Scherrer formula of 27.4 and 18.1 nm. The synthesised NPs had a constant spherical form, according to SEM analysis. Antioxidant, bioactive components, and antimicrobial investigations indicate that the these AgNPs could be employed well in biological purposes. The AgNPs that were created have excellent anti - oxidative, phytochemical, and antibacterial activities. An ecologically benign approach was used to generate silver nano-particles by using Kinnow mandarian aqueous peel extract as the reducing and capping agents (AgNPs). A simple process was utilised to manufacture persistent silver nano-particles at ambient temperature using organic wastes from fruit peel. The shampoo containing synthesised AgNPs was developed, and its physical characteristics such as foaming ability, surface tension, pH, solid content, dirt dispersion test, and anti-microbial action against *Escherichia coli* and *Candida albicans*, two harmful microorganisms, were evaluated (Bala et al. 2017).

Nanocellulose Production Cellulose was extracted by immersing Kinnow peels in toluene: ethanol (2:1) solvent system followed by bleaching with H_2O_2 and alkaline treatment with 6% NaOH. The cellulose was converted into nanocellulose by acid hydrolysis with a mixture of HCl and H_2SO_4 and thee formed nanocellulose was characterized using XRD and SEM. The crystalline behaviour of nanocellulose obtained from Kinnow peel extract was evaluated using X-ray diffraction method and it was revealed that a single diffraction peak at 2Θ value of 22.34° confirmed the crystallinity of nanocellulose. The SEM micrographs revealed the dense arrangement of cellulose in the form of nanorods and needle-like structures depending upon acid mixture concentrations.

Animal Feed Kour et al. (2016) investigated the effect of Kinnow mandarin waste (KMW) added in the diet on feed consumption and nutrient absorption in goats. It was revealed that 40% Kinnow trash incorporation expanded the Ca: P ratio because of its high calcium content, demonstrating that preserving the ratio required careful thought when administering. Kinnow waste incorporation in the diet showed no detrimental impacts on the goats' general health, according to blood biochemical indicators.

10.12 Development of Natural Coatings for Kinnow

Edible coatings are less harmful to the ecosystem and people's health than petrochemical waxes. Baswal et al. (2020), performed research to examine the influence of different edible coatings on the fruit freshness and cold rooms lifetime of

'Kinnow'mandarin. On Kinnow fruit, uncoated (control) fruits were contrasted to treatment options with 1–2.0 g/L carboxymethylcellulose (CMC), 0.5–1.5 g/L chitosan, and 5–15 g/L beeswax. Both coated and untreated fruit were kept in refrigerated stores for a maximum of 75 days (5–7 °C and 90–95% RH). Fruit samples were collected following 30, 45, 60, and 75 days of cold storage and evaluated for several quality attributes on the peel surface employing SEM. The results demonstrate variance among the polymer coatings in terms of fruit's physical quality parameters, compositions, bioactive compounds and sensory attributes characteristics. All coating techniques increased fruit's shine, however, when the cold storing duration was prolonged, the chitosan and carnauba wax fruit exhibited rind breaches. In refrigerated temperature, the CMC (2.0 g L^{-1}) coating excelled overall in terms of maintaining fruit quality metrics such as SSC total soluble solids, acidity, ascorbic acid, and olfactory qualities, in addition to postponing the activity of cell wall-degrading enzymes such as pectin methylesterase and cellulase. The impacts of CMC and guar gum-based coatings incorporating silver nano-particles on the post-harvest preservation sustainability of Kinnow mandarin was studied for 120 days (85–90% relative humidity) at 4 °C and 10 °C (Shah et al. 2015). Physiological and microbial characteristics were assessed every 15 days during storage. Total solids, sugars, and loss of weight all rose generally, but in coated fruits held at 4 °C, the rise was less evident. Wrapped fruits with 4 °C storage exhibited considerably greater amounts of vitamin C, phenolic content, and antioxidant properties. In contrast to coated Kinnow stored at 4 °C, titrable acidity reduced dramatically. Positive control held at 10 °C showed significant levels of fruit withering but no chilling harm. Aerobic plate thermophilic microbes, fungi, and molds were discovered in all storage settings, although proliferation was low in coated fruits at 4 °C. Kinnow fruit may be maintained in excellent condition over 4 months at 4 °C and 2 months at 10 °C after coating. A research looked at the use of cellulose coatings with Calcium and Magnesium ions for Kinnow. The results revealed that by utilising 5% cellulose covering and 1% Ca and Mg salts, optimum conservation was accomplished, and Kinnow's quality and durability were superior to the control. The films deposited samples had the lowest physiology loss of weight, juice %, and chilling damage, as well as higher vitamin C, acidity, and lower total solids. In addition, fruit firmness was also maintained inhibiting ethylene production (Randhawa et al. 2018).

10.13 Conclusion

Kinnow production grows year after year, and it is distributed around the world due to its nutritious benefits. Because the fruit is high in bioactive components, it may be used in a variety of ways to maximise its value. New extraction procedures are being investigated in order to increase the yield and productivity of bioactive components that may be used in food items. However, debittering research is still in its early stages, and researchers are looking for new ways to remove bitterness. The current research trends using Kinnow or its byproducts are mostly focused on the extraction of fibres and bioactives found in it and their use in diverse industries.

References

Aggarwal P, Michael M (2016) Effect of replacing sucrose with fructose on the physico-chemical sensory characteristics of Kinnow candy. Czech J Food Sci 32:158–163

Aggarwal P, Sandhu KS (2003) Effect of harvesting time on physico-chemical properties of juice and non-juice components of 'Kinnow'. J Food Sci Technol (Mysore) 40:666–668

Ahmad UI, Ying L, Mushtaq K, Bashir MK (2015) An econometric estimation of post-harvest losses of Kinnow in Pakistan. Int J Econ Commer Manag 3(5):773–783

Alam M, Kaur M, Ramya HG (2019) Mass transfer kinetics for osmotic dehydration of Kinnow fruit in sugar solution. Proc Natl Acad Sci India Sect B Biol Sci 89:361–370

Al Juhaimi F, Özcan MM, Uslu N, Ghafoor K (2018) The effect of drying temperatures on antioxidant activity, phenolic compounds, fatty acid composition and tocopherol contents in citrus seed and oils. J Food Sci Tech 55:190–197

Anwar F, Naseer R, Bhanger MI, Ashraf S, Talpur FN, Aladedunye FA (2008) Physico-chemical characteristics of citrus seeds and seed oils from Pakistan. J Am Oil Chem Soc 85:321–330

Anonymous (2018) Value chain strategy of kinnow in Fazilka district of Punjab. Report by NHRDF, Ministry of Agriculture and Farmers Welfare; Department of Agriculture, Cooperation & Farmers Welfare

Babbar N, Oberoi HS, Uppal DS, Patil RT (2011) Total phenolic content and antioxidant capacity of extracts obtained from six important fruit residues. Food Res Int 44:391–396

Bala R, Madaan R, Arora S (2017) Green synthesis and characterization of silver nanoparticles using Kinnow mandarian peels extract and its application in shampoo formulation. Res J Pharm Technol 10:2461–2466

Baswal AK, Dhaliwal HS, Singh Z, Mahajan BVC, Kalia A, Gill KS (2020) Influence of carboxy methylcellulose, chitosan and beeswax coatings on cold storage life and quality of Kinnow mandarin fruit. Sci Hortic 260:108887

Bejar AK, Mihoubi NB, Kechaou N (2012) Moisture sorption isotherms–experimental and mathematical investigations of orange (*Citrus sinensis*) peel and leaves. Food Chem 132:1728–1735

Bhardwaj RL, Mukherjee S (2011) Effects of fruit juice blending ratios on Kinnow juice preservation at ambient storage condition. Afr J Food Sci 5:281–286

Bhardwaj RL, Pandey S (2011) Juice blends—a way of utilization of under-utilized fruits, vegetables, and spices: a review. Crit Rev Food Sci Nutr 51:563–570

Bhatlu M, Laxmi D, Katiyar P, Singh SV, Verma AK (2016) Pre-harvest dropped *Kinnow* (*Citrus reticulata* Blanco) waste management through the extraction of naringin and pectin from their peels using indigenous resin. J Inst Eng India Ser A 97:285–290. https://doi.org/10.1007/s40030-016-0173-z

Bhatlu M, Singh SV, Verma AK (2017) Recovery of naringin from *Kinnow* (*Citrus reticulata* Blanco) peels by adsorption–desorption technique using an indigenous resin. Sādhanā 42:85–94

Correa E, Quiñones W, Echeverri F (2016) Methyl-N-methylanthranilate, a pungent compound from *Citrus reticulata* Blanco leaves. Pharm Biol 54(4):569–571

Dhaka A, Sharma M, Singh SK (2016) Use of additives to reduce browning, microbial load and quality loss of Kinnow juice under ambient storage. Indian J Sci Technol 9:1–11

Dilas S, Čanadanović-Brunet J, Ćetković G (2009) By-products of fruits processing as a source of phytochemicals. CICEQ 15:191–202

Farin D, Ikan R, Gross J (1983) The carotenoid pigments in the juice and flavedo of a mandarin hybrid (*Citrus reticulata*) cv Michal during ripening. Phytochemistry 22:403–408

Ghafoor KASHIF, Al-Juhaimi F, Choi YH (2011) Effects of grape (*Vitis labrusca* B.) peel and seed extracts on phenolics, antioxidants and anthocyanins in grape juice. Pak J Bot 43:1581–1586

Ghoshal G, Negi P (2020) Isolation of pectin from Kinnow peels and its characterization. Food Bioprod Process 124:342–353

Godara A, Kumar NV, Sharma A, Hudda J, Bakshi M (2020) Beneficial ingredients from Kinnow peel-extraction and uses: a review. Int J Curr Microbiol App Sci 9:2401–2411

Gorinstein S, Huang D, Leontowicz H, Leontowicz M, Yamamoto K, Soliva-Fortuny R, Martín Belloso O, Matínez Ayala AL, Trakhtenberg S (2006) Determination of naringin and hesperidin in citrus fruit by high-performance liquid chromatography. The antioxidant potential of citrus fruit. Acta Chromatogr 17:108–124

Gross J (1987) Pigments in fruits. Academic, London, p 303

Hayat K, Hussain S, Abbas S, Farooq U, Ding B, Xia S, Xia W (2009) Optimized microwave-assisted extraction of phenolic acids from citrus mandarin peels and evaluation of antioxidant activity in vitro. Sep Purif Technol 70:63–70

Hayat K, Zhang X, Chen H, Xia S, Jia C, Zhong F (2010) Liberation and separation of phenolic compounds from citrus mandarin peels by microwave heating and its effect on antioxidant activity. Sep Purif Technol 73:371–376

Hayat K, Abbas S, Jia C, Xia S, Zhang X (2011) Comparative study on phenolic compounds and antioxidant activity of Feutrell's early and Kinnow peel extracts. J Food Biochem 35:454–471

Javed S, Javaid A, Nawaz S, Saeed MK, Mahmood Z, Siddiqui SZ, Ahmad R (2014) Phytochemistry, GC-MS analysis, antioxidant and antimicrobial potential of essential oil from five citrus species. J Agric Sci 6:201

Jayaprakasha GK, Girennavar B, Patil BS (2008) Radical scavenging activities of Rio red grapefruits and sour orange fruit extracts in different in vitro model systems. Bioresour Technol 99:4484–4494

Jiang X, Zhou J, Zhou C (2006) Study on adsorption and separation of naringin with macroporous resin. Front Chem China 1:77–81

Juhaimi FA, Matthäus B, Özcan MM, Ghafoor K (2016) The physico-chemical properties of some citrus seeds and seed oils. Z Naturforsch C 71:79–85

Juyal D, Singh A, Kumar M (2015) Production of powder from Kinnow juice using spray drying technology. Agric Res J 52(1):37–41

Kamal GM, Anwar F, Hussain AI, Sarri N, Ashraf MY (2011) Yield and chemical composition of citrus essential oils as affected by drying pretreatment of peels. Int Food Res J 18

Kanmani P, Dhivya E, Aravind J, Kumaresan K (2014) Extraction and analysis of pectin from citrus peels: augmenting the yield from *Citrus Limon* using statistical experimental design. Iranian (Iranica) J Energy Environ 5

Kaur M (2017) Enzymatic debittering and aroma enhancement of Kinnow juice. M.Sc Thesis submitted to Punjab Agricultural University, Ludhiana

Khamrui K, Pal D (2002) Application of reverse osmosis for concentration of Kinnow mandarin juice. J Food Sci Technol-Mysore 39:310–312

Khule NR, Mahale NB, Shelar DS, Rokade MM, Chaudhari SR (2012) Extraction of pectin from citrus fruit peel and use as natural binder in paracetamol tablet. Pharm Lett 4:558–564

Kumar S, Awasthi OP, Singh A (2016) Kinnow mandarin enriching citrus industry. https://krishi.icar.gov.in/

Lawal OA, Ogunwande IA, Owolabi MS, Giwa-Ajeniya AO, Kasali AA, Abudu FA, Sanni AA, Opoku AR (2014) Comparative analysis of essential oils of *Citrus aurantifolia* Swingle and *Citrus reticulata* Blanco, from two different localities of Lagos state, Nigeria. Am J Essent Oils Nat Prod 2(2):8–12

Ladaniya MS (2008). Introduction. Citrus Fruit: Biology, Technology and Evaluation. Academic Press, San Diego, pp 1–2

Lim YY, Lim TT, Tee JJ (2007) Antioxidant properties of several tropical fruits: A comparative study. Food Chem 103:1003–1008

Liu C, Liu J, Rong Y, Liang N, Rong L (2012) Aqueous extraction of limonin from C*itrus reticulate* Blanco. Czech J Food Sci 30:364–368

Mahawar MK, Bibwe B, Jalgaonkar K, Ghodki B M (2019) Mass modeling of kinnow mandarin based on some physical attributes. J Food Process Eng 42(5):e13079

Mahawar MK, Jalgaonkar K, Bibwe B, Bhushan B, Meena VS, Sonkar RK (2020) Post-harvest processing and valorization of Kinnow mandarin (C*itrus reticulate* L.): A review. J Food Sci Technol 57:799–815

Mahmud S, Waheed A, Nazir T, Khanum R (2005) Essential oil composition of green peel of the inter-varietal mandarin hybrid, Kinnow orange. Pak J Sci Indu Res 48:329–333.

Mallavadhani UV, Narasimhan K, Sudhakar AVS, Mahapatra A, Li W, van Breemen RB (2006) Three new pentacyclic triterpenes and some flavonoids from the fruits of an Indian Ayurvedic plant Dendrophthoe falcata and their estrogen receptor binding activity. Chem Pharm Bull 54:740–744

Mann S, Minhas KS, Aggarwal P (2013) Development of phytochemical rich ice cream incorporating Kinnow peel. Glob J Sci Front Res 13:1–3

Marín FR, Soler-Rivas C, Benavente-García O, Castillo J, Pérez-Alvarez JA (2007) By-products from different citrus processes as a source of customized functional fibres. Food Chem 100:736–741

Mathur A, Verma SK, Purohit R, Gupta V, Dua VK, Prasad GBKS, Mathur D, Singh SK, Singh S (2011) Evaluation of in vitro antimicrobial and antioxidant activities of peel and pulp of some citrus fruits. J Biotechnol Biotherapeutic 1:1–17

Matthaus B, Özcan MM (2012) Chemical evaluation of citrus seeds, an agro-industrial waste, as a new potential source of vegetable oils. Grasas Aceites 63:313–320

Minatel IO, Borges CV, Ferreira MI, Gomez HAG, Chen CYO, Lima GPP (2017) Phenolic compounds: functional properties, impact of processing and bioavailability. Phenolic Compd Biol Act 8:1–24

Mondello L, Casilli A, Tranchida PQ, Dugo P, Dugo G (2005) Comprehensive two-dimensional GC for the analysis of citrus essential oils. Flavour Fragr J 20:136–140

Nadeem MT, Saima H, Afzaal M, Yasmeen A, Ahmad A, Shahid MZ, Aamir M (2018) Cellulolytic treatment: A competent approach to improve extraction and storage stability of carotenoids from Kinnow (*Citrus reticulate*) peel. Asian J Chem 30:603–606

Naz S, Shams F, Tabassum S, Ul-Haq I, Ashraf M, Zia M (2017) Kinnow peel extract as a reducing and capping agent for the fabrication of silver NPs and their biological applications. IET Nanobiotechnol 11:1040–1045

Oberoi HS, Vadlani PV, Nanjundaswamy A, Bansal S, Singh S, Kaur S, Babbar N (2011) Enhanced ethanol production from Kinnow mandarin (*Citrus reticulata*) waste via a statistically optimized simultaneous saccharification and fermentation process. Bioresour Technol 102:1593–1601

Olyad ERBA, Atomsa D, Chimdessa M, Gonfa T (2020) Determination of flavonoid contents and evaluation of in vitro antioxidant activities of the extract of selected citrus fruit peel. Int J Sec Metabol 7(1):8–18

Okwu DE (2008) Citrus fruits: A rich source of phytochemicals and their roles in human health. Int J Chem Sci 6:451–471

Osman H, Rahim AA, Isa NM, Bakhir NM (2009) Antioxidant activity and phenolic content of *Paederiafoetida* and *Syzygiumaqueum*. Molecules 14:970–978

Peterson JJ, Beecher GR, Bhagwat SA, Dwyer JT, Gebhardt SE, Haytowitz DB, Holden JM (2006) Flavanones in grapefruit, lemons, and limes: A compilation and review of the data from the analytical literature. J Food Compos Anal 19:S74–S80

Premi BR, Lal BB, Joshi VK (1994) Distribution pattern of bittering principles in Kinnow fruit. J Food Sci Technol-Mysore 31:140–141

Puri M, Kaur A, Schwarz WH, Singh S, Kennedy JF (2011) Molecular characterization and enzymatic hydrolysis of naringin extracted from Kinnow peel waste. Int J Biol Macromol 48:58–62

Qadir R, Farooq Anwar TM, Shahid M, Zahoor S (2018) Variations in chemical composition, antimicrobial and haemolytic activities of peel essential oils from three local citrus cultivars. *Pure Appl Biol (PAB)* 7:282–291

Rafiq S, Kaul R, Sofi SA, Bashir N, Nazir F, Nayik GA (2018) Citrus peel as a source of functional ingredient: A review. J Saudi Soc Agric Sci 17:351–358

Rafiq S, Sofi SA, Kaul R, Dar BN (2022) Effect of freeze-dried Kinnow peel powder incorporation on nutritional, quality characteristics, baking, sensorial properties, and storage stability of traditional wheat-based soup sticks. J Food Process Preserv 46:e16652

Rattanpal HS, Singh G, Singh S, Arora A (2017) Citrus cultivation in Punjab. Punjab Agricultural University, Ludhiana, India

Safdar MN, Kausar T, Jabbar S, Mumtaz A, Ahad K, Saddozai AA (2017a) Extraction and quantification of polyphenols from Kinnow (*Citrus reticulate* L.) peel using ultrasound and maceration techniques. J Food Drug Anal 25:488–500
Safdar MN, Kausar T, Nadeem M, Jabbar S, Ahmed S, Taj T, Mehmood W (2017b) Cardioprotective effect of mango and Kinnow peel extracts on doxorubicin-induced cardiotoxicity in albino rats: cardioprotective effect of mango and Kinnow peel extracts on doxorubicin. Proc Natl Acad Sci India Sect B Biol Sci 54:219–235
Saini A, Panesar PS, Bera MB (2019) Comparative study on the extraction and quantification of polyphenols from citrus peels using maceration and ultrasonic technique. Curr Res Nutr Food Sci 7:678–685
Saini A, Panesar PS, Bera MB (2021) Valuation of *Citrus reticulata* (Kinnow) peel for the extraction of lutein using ultrasonication technique. Biomass Conver Bior 11:2157–2165
Shah SWA, Jahangir M, Qaisar M, Khan SA, Mahmood T, Saeed M, Farid A, Liaquat M (2015) Storage stability of Kinnow fruit (*Citrus reticulata*) as affected by CMC and guar gum-based silver nanoparticle coatings. Molecules 20:22645–22661
Sharma P (2016) Development of oil extraction method from Kinnow peel. M. Tech (Doctoral dissertation, Thesis, Department of Processing and Food Engineering, Punjab Agricultural University, Ludhiana
Sharma AMIT, Tyagi SAPNA (2019) Antimicrobial screening and phytochemical analysis of citrus cultivars growing in vivo & in vitro. J Phytologic Res 32:35–42
Sharma N, Kalra KL, Oberoi HS, Bansal S (2007) Optimization of fermentation parameters for production of ethanol from Kinnow waste and banana peels by simultaneous saccharification and fermentation. Indian J Microbiol 47:310–316
Sharma H, Bhatia S, Alam MS (2013) Studies on pectin extraction from Kinnow peel and pomace. J Res, Punjab Agr Univ 50:128–130
Sidhu N, Arora M, Alam MS (2016) Biochemical, microbial stability and sensory evaluation of osmotically dehydrated Kinnow peel candy and peel powder. Int J Sci Res 5:1428–1436
Singh SV, Jain RK, Gupta AK (2009) Changes in quality of debittered kinnow juice during storage. J Food Sci Tech 46(6):598–600
Singh JP, Kaur A, Shevkani K, Singh N (2016) Composition, bioactive compounds and antioxidant activity of common Indian fruits and vegetables. J Food Sci Technol 53:4056–4066
Singla G, Krishania M, Sandhu PP, Sangwan RS, Panesar PS (2019) Value addition of Kinnow juice processing industry by-products using green solvents. J Pharm Phytochem SP1 19:22–62
Singla G, Singh U, Sangwan RS, Panesar PS, Krishania M (2021) Comparative study of various processes used for removal of bitterness from Kinnow pomace and Kinnow pulp residue. Food Chem 335:127643
Sogi DS, Singh S (2001) Studies on bitterness development in Kinnow juice, ready-to-serve beverage, squash, jam and candy. J Food Sci Technol (Mysore) 38:433–438
Sultana HS, Ali M, Panda BP (2012) Influence of volatile constituents of fruit peels of *Citrus reticulata* Blanco on clinically isolated pathogenic microorganisms under in–vitro. Asian Pac J Trop Biomed 2:S1299–S1302
Sun JYF, Chu XW, Liu RH (2002) Antioxidant and antiproliferative activities of common fruits. J Agric Food Chem 50:7449–7454
Tepe B, Akpulat HA, Sokmen M, Daferera D, Yumrutas O, Aydin E, Polissiou M, Sokmen A (2006) Screening of the antioxidative and antimicrobial properties of the essential oils of *Pimpinella anisetum* and *Pimpinella flabellifolia* from Turkey. Food Chem 97:719–724
Thakur NK, Kaushal BB, Joshi VK (2000) Effect of different conditions of storage on physicochemical and microbiological qualities of debittered Kinnow juice concentrate. J Food Sci Technol (Mysore) 37:415–418
Thakur V, Manjunath SS, Das Gupta DK (2004) Comparative evaluation of clarification and concentration technique for Kinnow fruit juice. J Food Sci Technol-Mysore 41:382–385
Tripodo MM, Lanuzza F, Mondello F (2007) Utilization of a citrus industry waste: bergamot peels. In Forum Ware Int 2:20–26

Tripoli E, La Guardia M, Giammanco S, Di Majo D, Giammanco M (2007) Citrus flavonoids: molecular structure, biological activity and nutritional properties: A review. Food Chem 104:466–479

Tumbas VT, Ćetković GS, Đilas SM, Čanadanović-Brunet JM, Vulić JJ, Knez Ž, Škerget M (2010) Antioxidant activity of mandarin (*Citrus reticulata*) peel. Acta Periodica Technol 41:195–203

Verma R, Lamba NP, Dandia A, Srivastava A, Modi K, Chauhan MS, Prasad J (2022) Synthesis of N-Benzylideneaniline by Schiff base reaction using Kinnow peel powder as green catalyst and comparative study of derivatives through ANOVA techniques. Sci Rep 12:1–12

Viuda-Martos M, Ruiz-Navajas Y, Fernández-López J, Pérez-Álvarez J (2008) Antifungal activity of lemon (*Citrus lemon L.*), mandarin (*Citrus reticulata* L.), grapefruit (*Citrus paradisi* L.) and orange (*Citrus sinensis* L.) essential oils. Food Control 19:1130–1138

Wojdyło A, Oszmiański J, Czemerys R (2007) Antioxidant activity and phenolic compounds in 32 selected herbs. Food Chem 105:940–949

Xu M, Ran L, Chen N, Fan X, Ren D, Yi L (2019) Polarity-dependent extraction of flavonoids from citrus peel waste using a tailor-made deep eutectic solvent. Food Chem 297:124970

Yaqoob M, Aggarwal P, Ahluwalia P (2019) Enhanced functional properties and shelf stability of muffins by fortification of Kinnow derived phytochemicals and residues. Think Ind J 22:1422–1445

Yaqoob M, Kaur M, Aggarwal P, Ahluwalia P (2020) Kinnow- in antioxidants in fruits: properties and health benefits. Springer, Singapore, pp 417–433

Yaqoob M, Aggarwal P, Rasool N, Baba WN, Ahluwalia P, Abdelrahman R (2021) Enhanced functional properties and shelf stability of cookies by fortification of Kinnow derived phytochemicals and residues. J Food Measur Charact 15:2369–2376

Yi Z, Yu Y, Liang Y, Zeng B (2008) In vitro antioxidant and antimicrobial activities of the extract of Pericarpium *Citri Reticulatae* of a new citrus cultivar and its main flavonoids. LWT-Food Sci Technol 41:597–603

Yu J, Dandekar DV, Toledo RT, Singh RK, Patil BS (2007) Supercritical fluid extraction of limonoids and naringin from grapefruit (*Citrus paradisi* Macf.) seeds. Food Chem 105:1026–1031

Chapter 11
Post Harvest Handling of Citrus Fruits

G. Rajiv, K. Elango, M. Kalaimani, R. P. Jegadeeshkanth, M. Kannan, and P. Arunkumar

11.1 Introduction

A product is harvested when it is separated from the mother plant at the right maturity stage, using the right technology, as quickly as possible, and with the least amount of loss or damage to the commodity. This procedure is done at a relatively cheap cost. Hand harvesting is the optimum method when quality, not price, not speed, and not time, is the main factor. Fruits meant for the fresh market are often hand-harvested. When a crop is harvested, the process of deterioration begins; the longer it is held (stored) until usage, the worse the crop will be. Exceptions can be granted for fruits that reach their best quality after a ripening time. Produce is

G. Rajiv
Department of Horticulture, SRM College of Agricultural Sciences,
Chengalpattu, Tamil Nadu, India

K. Elango (✉)
Department of Agricultural Entomology, Kumaraguru Institute of Agriculture,
Tamil Nadu Agricultural University, Erode, Tamil Nadu, India

M. Kalaimani
Department of Horticulture, MIT College of Agriculture and Technology,
Musiri, Tamil Nadu, India

R. P. Jegadeeshkanth
Department of Horticulture, Kumaraguru Institute of Agriculture,
Tamil Nadu Agricultural University, Erode, Tamil Nadu, India

M. Kannan
Department of Nano Science and Technology, Tamil Nadu Agricultural University,
Coimbatore, India

P. Arunkumar
Department of Agricultural Entomology, Centre for Plant Protection Studies,
Tamil Nadu Agricultural University, Coimbatore, Tamil Nadu, India

S. Singh Purewal et al. (eds.), *Recent Advances in Citrus Fruits*,
https://doi.org/10.1007/978-3-031-37534-7_11

extremely perishable, so there will always be some loss. Postharvest loss levels vary depending on the crop, handling process, storage, and environmental circumstances. Some products cultivated in remote locations could never even make it to the market, while others might only suffer a very little loss if they are consumed nearby. The most commonly used projections for postharvest losses of the most widely consumed fruits and vegetables range from 25% to 30% (Ramaswamy 2015).

The actual pipeline of food from harvest to consumer may be more precise or complex than that depicted. Still, it represents a loss during handling operations like preprocessing, transportation, storage, processing, packaging, and marketing. Estimating the magnitude of food losses has been a contention for decades. However, there is very little reliable information on perishable produce postharvest losses (Chakraverty and Singh 2014). Post harvest operations in citrus fruits are depicted in Fig. 11.1.

Their wide-scale cultivation evidence of the value of citrus in the global economy under tropical and sub-tropical climate conditions, where oranges (*Citrus sinensis* L.), mandarin orange (*Citrus reticulata* L.), lime and lemons (*Citrus aurantifolia and Citrus limon* L.) exist the more cultured citrus groups. Citrus has a significant socioeconomic and cultural impact on society as a whole. The fruit's numerous nutritional and medicinal properties make it widely regarded as being necessary. Citrus is mainly valued as a fruit that can be consumed raw, juiced, or added to dishes and beverages. Citrus comprises many species of economic commercially cultivated groups including lemon, limes, sweet oranges, mandarins and grape fruit. Citrus group fruits are having more amounts of Vitamins C, A, B and phosphorus. Citrus rind (pericarp) is divided into two parts: the exocarp (flavado) is the exterior, pigmented region with oil glands, while the albedo is the interior, white section. The cuticle, a thin continuous polymer composed of waxes, is the flavedo's outermost constituent. Before applying postharvest wax, it serves as the main barrier between the fruit and its surroundings. In addition, it also contributes significantly to gas exchange and keeps tissues high water content, both of which are essential for regular metabolism (Maria et al. 2017).

Citrus fruits are popular among consumers for their flavor and health benefits, as they are high in bioactive substances like vitamin C and phenolic compounds like hydroxycinnamic acids and flavonoids. Flavanones and anthocyanins are the most important flavonoids in citrus and are mostly found in pigmented varieties. Anthocyanins provide important sensory input and play an important role in their

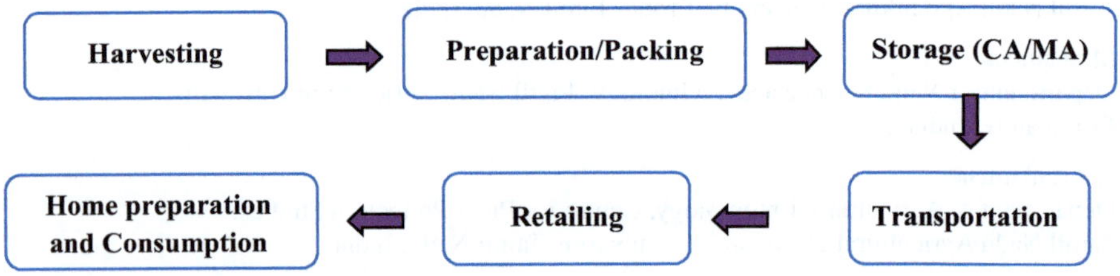

Fig. 11.1 Post harvest operations in fruits

pharmacological and antioxidant properties. In addition to their salutistic properties, fresh fruit consumption is associated with a reduced risk of cardiovascular disease, stroke, and cancer (Ramaswamy 2015). Citrus fruits have the unique characteristics of being highly seasonal, perishable, and bulky. As a result, citrus fruits require special attention during the harvest and post-harvest periods. As the product moves from citrus growing to the final consumer reach, citrus fruit post-harvest losses are estimated to be between 25% and 35%.

Citrus fruits does not respond to climate, so they are different from fruits that do (like mango, sapota, and bananas) in as they mature, they don't produce more ethylene or take in more oxygen. Compared to other tropical fruits, they have a relatively prolonged shelf life, but if handled and stored properly, they would be suitable for marketing. Fruits' perishability is inversely correlated with their rate of respiration and the amount of heat they emit, which impacts postharvest technology decisions like estimating the need for refrigeration and ventilation. Fruits degrade consequently due to the loss of nutritional value, textural softening, and wilting and shriveling in appearance. Cultural practices, including harvesting technique, timing, and pre- and postharvest elements, influence citrus fruit's postharvest decay (Beghin 2014). The transpiration rate is influenced by rinsing injuries, maturity level, and environmental factors. Surface coatings are employed to regulate the procedure and the storage environment (low temperatures, high relative humidity, and control of air circulation). Diseases, physiological problems, fruit senescence, and physical damage are the primary causes of postharvest losses (Kassim et al. 2020). Citrus postharvest decay causes considerable waste and quality degradation, and because it renders fresh fruit unsuitable for eating, it causes enormous economic losses. In developed and developing nations, losses can amount to 30–50% of total production. Various factors, including high storage temperatures, increased respiration and transpiration, as well as harvesting and postharvest handling stress causes physiological disorders and fruit senescence. Because they hasten water loss, increase respiration and ethylene production, and promote the growth of pathogens, rinsing injuries and impact bruising affect fruit deterioration (Youssef et al. 2014). Currently, citrus is picked by hand to minimize fruit damage, lower the chance of early decay, and improve postharvest quality (Sanders 2005). Because it requires more flexibility and the ability to select specific fruits, mechanical harvesting has not been widely used in the citrus industry. The desired selection criteria for specific citrus fruits are now used in more automated systems that have been developed. Yellow to orange fruit is ready for picking (Beghin 2014).

Fruits must be fully ripe before being picked. A sour fruit may lose acidity after being picked, especially if allowed to dry out, but it will still have a bland flavor. Immature fruits have poor quality and erratic ripening when they are harvested. On the other hand, delaying harvesting increases their susceptibility to decay, leading to poor quality and a low market value. For the fruits to have a good flavor, they should be allowed to mature on the tree. Citrus fruit maturity can be assessed using a variety of indices. The two most important characteristics used by small orchards to assess maturity are fruit color and size. Fruit size and color are the two most important factors to consider when determining maturity. However, this technique is only effective with certain citrus species (Kassim et al. 2020).

11.2 Harvest of Citrus Fruits

Citrus fruits are harvested and handled carefully to maintain their quality. Injuries and bruising provide entry points for microorganisms, which lead to rotting. The idea that citrus fruit can withstand rough handling is a common misconception. Compared to many other fruits, citrus is more resilient to bruising. In addition, Citrus bruises may not show up for a few days, by which time the fruit is typically transported. Fruit harvesting avoids inappropriate handling and pulling from the branches because the skin at the stem end can rupture. More importantly, stems left on the fruit can damage other fruits, decreasing packing efficiency and resulting in spoilage and fruit loss. Fruits born on high branches are challenging to harvest. Such fruits are harvested using a long pole with a curved blade. Inappropriate fruit handling accelerates physiological deterioration, which is reflected in increased microbiological activity, accelerated ripening, and accelerated decay. This can have other market-related impacts, such as lower farm income and a negative importer perception of exporting countries. The most harmful post-harvest fungal pathogens for citrus have been identified by research as *Penicillium digitatum* and *Penicillium italicum* (Ashebre 2015).

11.3 Physical Quality and Chemical Quality Parameters

11.3.1 Skin Colour

The color perception of citrus fruit influences a customer's willingness to purchase. Color can be measured subjectively or objectively, just like firmness. The eye determines subjective color measurement (Ladaniya 2008). The colour scale system divides samples into deep green, light green, yellowish-green, greenish-yellow, yellowish-orange, orange, and deep orange. Depending on the citrus cultivar, this scale may differ (Singh and Reddy 2006).

11.3.2 Weight Loss

Loss of weight, which frequently results in a loss of firmness, is a significant contributor to citrus fruit deterioration (Chien et al. 2007). The pulp and peel of citrus fruits contain a lot of moisture. After harvest, moisture is rapidly lost through transpiration and respiration, which encourages fruit decay (Ghanema et al. 2012).

11.3.3 Firmness

Fruit firmness is a mechanical characteristic indicated by resistance to piercing and deformation. Fruit firmness is frequently used as a standard to assess the effects of storage and shelf life (Singh and Reddy 2006). Citrus fruit firmness is primarily controlled by cell turgidity, which is linked to moisture content.

11.3.4 Total Titratable Acid

The organoleptic properties of citrus fruit are influenced mainly by organic acids. Citric acid accounts for approximately 80–95% of citrus fruit's total titratable acids (TTA). The TTA of citrus fruit typically decreases as it ripens, depending on the cultivar (Ladaniya 2008).

11.3.5 Total Soluble Solids

Citrus fruit, which makes up about 10–20% of the fresh weight, comprises total soluble solids (TSS). Carbohydrates make up roughly 70–80% of the TSS. Organic acids, proteins, lipids, and minerals are additional minor components of TSS. TSS is computed using a refractometer and the fruit juice's refractive index (Iglesias et al. 2007; Olmo et al. 2000).

11.3.6 Maturity Index

To choose the ideal time for harvest, commercial maturity indices are crucial. Nevertheless, this depends on the citrus species and varieties as well as external elements like growing regions and destinations (Lado et al. 2014). The ratio of TSS to TTA can be used to calculate the maturity index (Iglesias et al. 2007).

11.4 Challenges for the Citrus Handler

- The causes of peel damage are unclear.
- Damage caused by low temperature, high temperature, methyl bromide fumigation, and other factors is frequently similar.
- It is challenging to determine how physical injury interacts with other postharvest treatments.

Postharvest physiological problems are brought on by the climate, transportation, treatment, and storage conditions that fruits are subjected to.Citrus fruit loss when maintained over several months or exported over large distances are mostly caused by the low temperatures required for cold storage. Chilling injury (CI) can occur when citrus fruits are stored at low, non-freezing temperatures (0–10 °C). It depends on the crop's location, cultivar, season, and maturity stage. Temperature and exposure duration have an impact on CI severity. Recent studies show that oxidative stress and CI are related (Lafuente and Zacarias 2006). The symptoms of these physiological disorders included a variety of signs and symptoms. Still, the ones that were most frequently present were areas of the rind that collapsed and darkened to form pits (unrelated to the oil glands) and areas of brown color. The mold development on the fruit surface may have increased due to CI. The shriveling and collapse of the button tissue, brought on by prolonged storage at low temperatures, are physiological symptoms of aging.

11.5 Storage Temperature Requirement to Citrus

1. Varies depending on the variety and citrus species (Fig. 11.3)
2. Degrees from roughly 0 to 15 °C

11.6 Citrus fruit Handling and Preventions

Citrus fruits transported from the orchard to the packing house are exposed to field heat and ambient temperatures, raising the fruit's temperature and promoting the growth of microorganisms, which hasten the decay process (Brosnan and Sun 2001). Citrus fruit was given pre-packaging treatments like hot water, surface coatings, ultraviolet irradiation, chlorinated water, biocontrol agents, carbonate, and bicarbonate salts to enhance the quality of the fruit after harvest (Porat et al. 2000; Njombolwana et al. 2013). Because heat shock proteins have evolved, heat treatments have been used to help fruit become more resistant to cold damage and pathogens. The fruit's shelf life is improved by surface coatings or wax coatings, which enhance its aesthetic appearance and reduce moisture loss (Johnston and Banks 1998). Due to its germicidal properties and capacity to increase the fruit's tolerance to deteriorate, ultraviolet irradiation prevents corrosion in citrus fruit. Citrus treated with carbonate and bicarbonate salts have a defense mechanism that can delay postharvest deterioration (Youssef et al. 2014). Similar to this, the citrus fruit industry uses chlorine (hypochlorite) extensively as a disinfectant to increase the shelf life of citrus fruits.

11.7 Packinghouse Practices and Treatments Reduce Decay By

- Eliminating the microorganism on the fruit surface
- Impeding the development of invisible disorders
- Prevention of pathogens that invade wounds infection
- Fruit surface protection through wounding against a subsequent infection
- Preventing the sporulation and spread of disease from infected to healthy fruit

Mechanical injury has the following adverse effects; therefore, care should be taken when harvesting fruit in the field:

- Risen decay
- A more significant loss of water
- Peel breakdown in ensuing handling

The most common type of injury was made by the clippers many were injured by stem punctures, while others showed scratches from thorns. Other common injuries were from gravel and twigs in the bottom of boxes and cuts by the finger nails of the pickers.

11.8 Post Harvest Manageable Strategic of Citrus Fruits

1. Fruits wash with Sanitation solutions
2. Conventional and alternative treatments
3. Physical treatments
4. General recommendation for fruit safe
5. Wax Coatings
6. Cold storages

11.9 Fruits Wash with Sanitation Solutions

The packing line and surroundings are thoroughly sanitized to benefit the fruits to reduce the spore inoculum density. Sanitizers used for fruit texture sterilization try to control the disease by lowering the initial high level of microbes in the products. The primary products used in packing houses right now are chlorine solutions. The classic cleaning approaches involve using sodium hypochlorite solutions (100–150 ppm) before rinsing with potable water (Rouissi et al. 2009). This method has some limitations because it requires constant adjustments to the amount of chlorine available and monitoring the pH of the solution to ensure proper treatment. In addition, when reacting with organic materials, compounds that release chlorine can create hazardous breakdown byproducts (halogenated) (Ladaniya 2010a, b). water disinfection methods used in citrus fruit packinghouses, including peroxyacetic acid (PAA), ozone, and electrolyzed water (EW). Hydrogen peroxide and acetic acid are combined to create peroxyacetic acid, a potent oxidizer effective against various microorganisms (Lanza 2006, 2007). Ozone is a potent oxidizing agent that occurs naturally and can kill various inoculums, including bacteria, viruses, and fungal spores. Citrus process water's microbial pathogen population has decreased due to the use of electrolytic water, which is it was created by putting a liquid salt solution into an electrolytic cell (Fallanaj et al. 2011).

11.10 Conventional and Alternative Treatments

The current methods for controlling post-harvest decay rely on synthetic chemical fungicides, which are readily available, inexpensive, and have a wide range of beneficial effects against pre-existing and newly emerging infections. Fungicide addition is the method of application that is most frequently used, such as imazalil (IMZ) to commercial fluid wax sprinkled on the fruit before the fruit storehouse. Also, when applying IMZ in a wax mixture, some issues can arise, such as nozzle obstruction, which reduces treatment effectiveness and uniformity of dispersion, and loss of fungicide and wax on brushes, which causes problems with waste treatment and spore accumulation in the brushes to the point where routine cleaning is required (Sapers 2011). The chemical makeup of wax is crucial because it includes some alkaline soluble compounds that may result in an ideal partition of IMZ in the wax and, as a result, minimal aqueous phase penetration through the peel. In light of all these issues, fungicide concentrations must be raised. When preventing decay, fruit dipping is more productive than wax dipping because pathogens can grow in tiny divots in the fruit's skin and water solution can more easily penetrate than wax's greater density. Further, as the amount of fungicide in a water emulsion slowly reduces and the number of microbes increases (Renzo et al. 2006), there is a chance that there will be issues with treatment uniformity, fruit cross-contamination, and fungicide stability in the actuality of sanitizing mechanisms. Only three fungicides—IMZ, thiabendazole (TBZ), and sodium-orthophenylphenate (SOPP)—have received government approval for postharvest applications (Smilanick et al. 2006). These treatments leave a residue on fruit and the environment even though they are strictly regulated and thoroughly tested for side effects. To control postharvest diseases in fresh produce, there is an increasing interest in creating and implementing eco-friendly, nontoxic methods that are effective and independent of standard fungicide applications.

Recent registrations for more minimal-risk fungicides that pose a low risk to human and environmental health, including Trifloxystrobin, Azoxystrobin, Fludioxonil, Cyprodinil and Pyrimethanil, prevent citrus postharvest decay (Schirra et al. 2005). It has been difficult to find both widely accepted and economically viable alternatives. Alternatives to fungicide treatments include physical treatments with generally recognized as safe (GRAS) substances, biological control agents, and naturally occurring antimicrobial substances. Their inhibitory impact on degradation control is primarily due to pathogen inhibition (Wisniewski et al. 2016). Some of these techniques, though, have demonstrated the capacity to strengthen citrus fruit tissues' defense mechanisms, resulting in the development of disease resistance and significantly contributing to the prevention of *Penicillium* decay. Nevertheless, it needs to be included in a comprehensive management strategy that can increase fruit shelf life and reduce postharvest decay.

11.11 Physical Treatments

Physical remedies are regarded as an intriguing and helpful alternative to chemicals because they have no effect on the outcomes of treatment and have the potential to create resistance to subsequent infections. While technological challenges can be resolved and understood to enable commercial applications, some antagonistic effects that could affect fruit quality could occur when not properly applied. The most common form of physical treatment used in postharvest applications is heating, They are applied through curing, hot dry air, and vapour heat (Fallik 2004).

A hot water dip involves completely submerging fruit for 2–3 min in water heated above 40 °C. Numerous researchers found that a hot water dip (HWD) on organic citrus fruits decreased rot growth without adversely affecting fruit traits related to appearance and quality. This technique in storage rooms, where it's crucial to process significant volumes of product quickly, could be improved by the length of the fruit immersion period. When it comes to stopping the germination of *P. digitatum* spores, treatments at 56 °C for 20 s are superior to those at 52 °C for more prolonged exposure times (Strano et al. 2014). Tap water is sprayed through a nozzle during SHWB (short hot water brushing). The fruits are wetted as they pass over brushes on the sorting and grading line, then flushed with pressurized, superheated water at a temperature (60 °C) for a brief period (20–60 s), after which they are dried with forced air (Fallik 2004). In Europe, superheated water application to apples (*Malus domestica*) and citrus fruit green mold has been reported to be consistently reduced by applying SHWB at 60 °C for 20 s (Sapers 2011 and Lanza et al. 2006).

In recent years, promising technologies have gained interest in managing postharvest fruit diseases. The usage of microwave heating, hypobaric and hyperbaric pressure and physical mutagens such as ultraviolet-light therapy (UV-C irradiation) has also shown the potential to induce resistance in the fruits (Thompson 2011; Romanazzi et al. 2016).

11.12 Wax Coatings

The anionic microemulsions used to make synthetic coatings contain petroleum, shellac, carnauba, beeswax, polyethylene, and other resins and waxes. Improvement of fruit gloss, reduction of weight loss, shrinkage, and chilling damages, as well as reduction of transpiration and respiration, are the main objectives of wax layering. Alternative, edible coatings and films have been created as environmentally friendly technology. Food is coated with edible wax or films, creating a semi-permeable barrier to water vapor, oxygen and carbon dioxide between the food and the environment, preventing physical damage, chemical deterioration, and microbiological damage, which extends product shelf life (Dhall 2013). In the global citrus trade's high economic value, much research is being done on creating novel edible antifungal coatings for citrus fruit. More research has been done on biopolymers and cellulose derivatives that are generally considered safe (GRAS) (Velásquez et al. 2014; El-Mohamedy et al. 2015).

11.13 Cold Storage

Temperature and relative humidity play a crucial role in regulating the rate of citrus fruit deterioration during the postharvest handling chain (Fig. 11.2). According to research by several authors, the primary method for extending shelf life during postharvest storage is to maintain the temperature level. The ideal conditions for storing citrus fruit include a high relative humidity (RH) of 90–95% (Tables 11.1 and 11.2). RH regulates shrinkage by preventing host tissues from losing moisture. Different citrus species and varieties require different temperatures for storage. Mandarins, lemons, limes and sweet oranges are among the citrus groups more sensitive to low temperatures, so store the fruit above 9 °C to prevent the resurgence of chilling damages. The majority of citrus varieties can

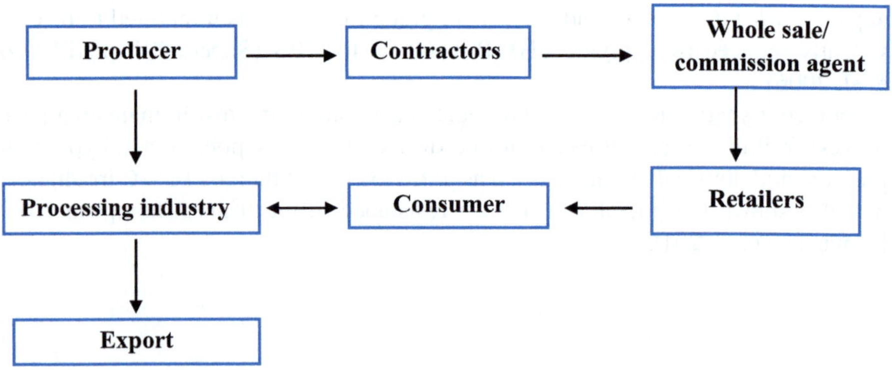

Fig. 11.2 Supply chain of citrus groups

Table 11.1 Recommended temperature ranges, relative humidity (RH) values and approximate storage lives for citrus fruit species

Citrus species and variety	Temperature range (°C)	RH %	Approximate storage life
Grapefruit	10–15	85–90	6–8 weeks
Lemon	10–13	85–90	1–6 months
Limes	9–10	85–90	6–8 weeks
Mandarin hybrids (fortune, Nova)	8–9	85–90	4–6 weeks
Mandarin, tangelo	5–6	90–95	2–4 weeks
Clementine, Satsuma	4–5	90–95	2–4 weeks
Kumquat	4–5	90–95	2–4 weeks
Pigmented orange	6–8	90–95	3–8 weeks
Blond orange	3–9	85–90	3–8 weeks

Table 11.2 Controlled atmosphere (CA) requirements for some citrus fruits

Species	Temperature range (°C)	Relative humidity (%)	CA	
			% O_2	% CO_2
Lemon	10–15	90–95	5–10	0–10
Lime	10–15	90–95	5–10	0–10
Orange	5–10	90–95	5–10	0–5

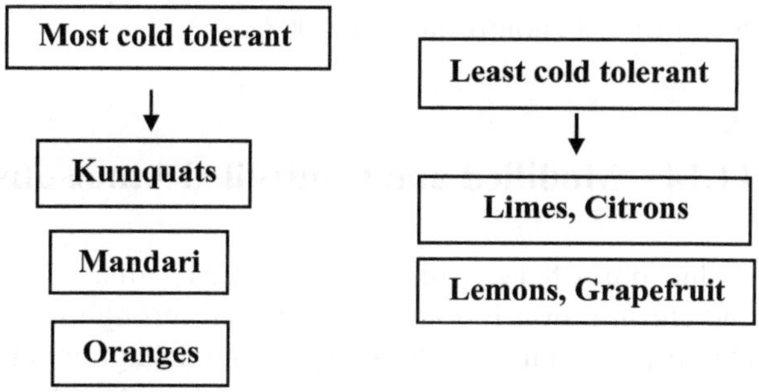

Fig. 11.3 Temperature requirements for different citrus fruits

withstand low temperatures during long storage (Lafuente et al. 2005). Typical chilling injuries include the emergence of a brown depression resembling a pit in the flavedo (in grapefruits and mandarins) and the development of a superficial burn in some oranges. Rapid cooling is necessary to reduce the temperature suitable for storage, cold treatment, and shipment to market, significantly reducing weight loss and deterioration. Various quick cooling systems are available with the heat removal system, including forced air cooling, hydro cooling (chilled water), vacuum cooling, and room cooling (chilled air). The most effective way to cool citrus fruits is with an air-cooling system, particularly when forcing air through the fruits as they are stacked on a pallet (Thompson et al. 2008). The "room-cooling" system blows cold air directly into cold storage rooms. Fruits packed in cartons, sacks, or bins are exposed to heat exchangers that transfer heat to the cool air at a speed of between 1 and 2 m/s. The room cooling system is still the most popular, primarily for economic reasons. This system, however, results in excessive water loss and inefficient cooling (due to the prolonged drying of citrus skin). The distance between box stacks within the storage room is critical for minimizing cooling time. Cold air can circulate between pallets or boxes at a maximum distance of 10–15 cm (4–6 inches). Furthermore, using vented boxes improves cooling efficiency over using unvented boxes. These factors indicate that a conventional cold storage room only suits already cooled products. The cooling mechanism is slow and inconvenient for the fast removal of "fruit heat." A pre-cooling plant with fast cooling power is needed to quickly remove the "field heat" from fruit after harvesting. The forced ventilation system was created to improve citrus storage rooms' temperature and cooling uniformity. When using forced air cooling, heat is removed from the product by chilled air from an evaporator of a mechanical refrigeration system passing through the product mass through a fan commonly operating under pressure. After pre-cooling, keeping the area around the fruits humid is crucial. Plastic films play a particularly important role among the techniques used for citrus fruit preservation and commercial distribution. Such films can lead to carbon dioxide and oxygen decrease accumulation, reducing respiratory activity and reserve substance degradation. Further, the fruits' transpiration promotes the development of high moisture levels, slows aging processes and weight loss, restricts ethylene production, and lengthens the "shelf life." (Thompson et al. 2008).

11.14 Modified and Controlled Atmosphere Storage

Reduced postharvest disorders in citrus fruits, such as stem-end rind breakdown and chilling injuries, may be achieved through modified atmosphere packaging (MAP), which uses different types of packaging materials, such as polyethylene (PE) and polyvinyl chloride (PVC) films. MAP replaces the air concentration with the O2 and CO2 levels required to reduce pathological rot and physiological disorders in the fruit inside a sealed box (Techavises and Hikida 2010). MAP can be applied commercially to prolong the shelf life of citrus fruits by generating low O_2 and medium to low CO_2 partial pressures, which can slow respiration and prevent fungal growth, and maintaining high relative humidity, which prevents water loss and weight loss. This keeps the fruit fresh and maintains its original fruit quality throughout storage (Dhillon et al. 2016). Selecting films with a low moisture transmission rate and selective permeability to O_2 and CO_2 is essential to control pathological decay without causing physiological harm or injuries (Aquino et al. 2001). CA storage is generally used as an additional technology to refrigerated storage to extend the shelf life of fruits. Still, a more profound understanding is required for its application in producing citrus fruits.

Numerous attempts have been made to use CA to maintain the quality indices connected to flavor, odor and nutritive profile. Most citrus fruit postharvest pathogens must be controlled using CA storage and high CO_2 concentrations, but fruit cannot tolerate this, leading to the growth of rind disorders or off-flavors associated with the accumulation and threshold levels of anaerobic respiration products (Ladaniya 2010a, b).

11.15 Fruit Quality Evaluation- Non-destructive Methods

Fresh citrus fruit must be of high internal and external quality from harvest to consumption. The assortment of fruit qualities that affect consumer acceptance can be referred to as fruit attributes or characteristics. To ensure that they meet minimum requirements for maturity (internal quality) and grade (external quality), fresh citrus fruits are continually checked for quality control. Various laws specify minimum grade requirements, and minimum maturity requirements are unique to each type of citrus. Quality inspectors evaluate the product's quality and determine its utility and marketability based on these standards. Quality assessment and control are also necessary when deciding on a fruit price. Both subjective and objective methods can be used to evaluate quality attributes. In contrast to subjective methods, which rely on human senses, accurate methods are precise and use instruments. Further, the most effective manual or automated systems for classifying citrus fruit quality are founded on some tried-and-true conventional techniques and methods for judging fruit quality. These procedures could be more laborious, sample-destructive, and time-consuming, restricting their usefulness in online and in-line quality checking. The need for novel tools in the industry has also grown: evaluating and monitoring citrus fruit quality can be done fast and affordably (Wang et al. 2016).

11.16 Grading of Citrus

- The fruit is examined and unripe, immature, undersized, damaged, or decayed fruits are discarded.
- Citrus fruits are divided into small, medium, and large grades depending on size for local markets (Table 11.3).
- The following list of recommended grades applies to the export of Nagpur mandarins.

Table 11.3 Grade size

Grade size (mm)	No. of fruits/10 Kg packing
50–55	115
50–60	98
60–65	84
65–70	76
70–75	64

11.17 Packaging

Citrus fruits are packaged for shipment to local markets in sacks, bags, bamboo baskets, and wooden boxes. Fruits of the citrus family are packaged in telescoping cardboard boxes for urban markets. When transporting fruits, corrugated trays are a good option for packaging. Because they can be reused, using such trays is economical.

For the export of Nagpur mandarin, two-piece, telescoping corrugated fiber board boxes with a three- or five-ply construction are typically used. The box size may vary depending on the requirements of the importing country. A box with dimensions of 49.5 × 29.5 × 17.5 cm and a capacity of 10 kg is usually recommended. The containers must have 5% of their surface area punched as ventilation holes. To prevent the movement of the fruits inside the box, three-ply wax-treated dividers with ventilation holes are used.

11.18 Innovative Packaging

When citrus fruits are packaged properly, they may be more resistant to mechanical damage during storage and transportation, have a longer shelf life, and have a higher market value. Depending on the storage mode, citrus fruits are packaged in corrugated fiberboard (CFB) boxes and net or film bags (Berk 2016). Package design in the citrus industry because it significantly decreases the forced-convective precooling time and improves citrus fruit quality by providing the fastest and most uniform cooling without chilling-induced damage. Packaging must combine mechanical strength for stacking containers and protecting fruit with the best ventilation.

An emerging technology called antimicrobial active food packaging extends the shelf life of food products by releasing encapsulated antimicrobial compounds under controlled conditions (Khaneghah et al. 2018). Citrus fruits are frequently packaged using corrugated cardboard, an environmentally friendly material that can be used for both boxes and trays.

11.19 Conclusions

Citrus fruits are widely grown worldwide due to their beneficial health attributes, particularly their bioactive components, vitamin C, and phenolic compounds. Citrus fruits are frequently packaged using corrugated cardboard, an environmentally friendly material that can be used for both boxes and trays. Therefore, encapsulated in cyclodextrin is added to produce antimicrobial active cardboard packaging and extend the shelf life of fruits and vegetables. Maintaining a high temperature and relative humidity when storing citrus fruits in the cold is important. Citrus fruit postharvest life is severely affected by fungi, especially if it is stored for a long period or is shipped far. There have been several applications that aim to reduce the risk of fungal inspections to both the environment and human health. In the postharvest industry, identifying and using novel methods to assess citrus fruit quality is a fundamental focus. Based on the current understanding of citrus postharvest and extending fruit shelf life, the following issues may arise in the future: reducing mechanical damage during handling and packing operations; researchers to find non-chemical fruit treatments.

References

Ashebre KM (2015) Pre-harvest and post harvest factors affecting citrus fruit and post-harvest treatments. Int J Agric Biol 5:19–29

Beghin S (2014) Personal communication. Premier Fruit Exports (Pty) Ltd, Durban

Berk Z (2016) Citrus fruit processing. Academic, London

Brosnan T, Sun DW (2001) Precooling techniques and applications for horticultural products-a review. Int J Refrig 24:154–170

Chakraverty A, Singh RP (2014) Postharvest technology and food process engineering. CRC Press, Boca Raton

Chien PJ, Sheu F, Lin HR (2007) Coating citrus (*Murcott tangor*) fruit with low molecular weight chitosan increases postharvest quality and shelf life. Food Chem 100:1160–1164

D'Aquino S, Angioni M, Schirru S, Agabbio M (2001) Quality and physiological changes of film packaged 'Malvasio' mandarins during long term storage. LWT-Food Sci Technol 34:206–214

Dhall RK (2013) Advances in edible coatings for fresh fruits and vegetables: a review. Crit Rev Food Sci Nutr 53:435–450

Dhillon WS, Mahajan BVC, Chahal TS, Kumar M, Sidhu MK, Singh SP (2016) Effect of different packaging films on shelf-life and quality of daisy mandarin under ambient conditions. Indian J Hortic 73:262–266

Di Renzo GC, Altieri G, Lanza G, Di Martino AE, Strano MC (2006) Continuous monitoring of ozone and Imazalil used in citrus post harvest. Italus Hortus 13:48–53

El-Mohamedy RS, El-Gamal NG, Bakeer AR (2015) Application of chitosan and essential oils as alternatives fungicides to control green and blue moulds of citrus fruits. Int J Curr Microbiol Appl Sci 4:629–643

Fallanaj F, Sanzani SM, Zavanella C, IPPOLITO A. (2011) Activity of electrolysis in reducing microbial population and controlling postharvest diseases of citrus fruits. J Plant Pathol 93:24–31

Fallik E (2004) Prestorage hot water treatments (immersion, rinsing and brushing). Postharvest Biol Technol 32:125–134

Ghanem N, Mihoubi D, Kechaou N, Mihoubi NB (2012) Microwave dehydration of three citrus peel cultivars: effect on water and oil retention capacities, color, shrinkage and total phenols content. Ind Crop Prod 40:167–177

Iglesias DJ, Cercós M, Colmenero-Flores JM, Naranjo MA, Ríos G, Carrera E, Ruiz-Rivero O, Lliso I, Morillon R, Tadeo FR, Talon M (2007) Physiology of citrus fruiting. Braz J Plant Physiol 19:333–362

Johnston JW, Banks NH (1998) Selection of a surface coating and optimisation of its concentration for use on 'Hass' avocado (*Persea americana* Mill.) fruit. New Zeal J Crop Hort 26:143–151

Kassim A, Workneh TS, Laing MD (2020) A review of the postharvest characteristics and pre-packaging treatments of citrus fruit. AIMS Agri Food 5:337–364

Khaneghah AM, Hashemi SM, Limbo S (2018) Antimicrobial agents and packaging systems in antimicrobial active food packaging: an overview of approaches and interactions. Food Bioprod Process 111:1–9

Ladaniya MS (2008) Citrus fruit: biology, technology and evaluation. Elsevier, London

Ladaniya M (2010a) Fruit morphology, anatomy, and physiology. In: Citrus fruit: biology, technology and evaluation. Academic/Elsevier, London, pp 103–124

Ladaniya MS (2010b) Citrus fruit: biology, technology and evaluation. Elsevier, Amsterdam, pp 311–318

Lado J, Rodrigo MJ, Zacarías L (2014) Maturity indicators and citrus fruit quality. Stewart Postharvest Review 10:1–6

Lafuente MT, Zacarias L (2006) Postharvest physiological disorders in citrus fruit. Stewart Postharvest Review 2:1–9

Lafuente MT, Zacarias L, Sala JM, Sanchez-Ballesta MT, Gosalbes MJ, Marcos JF, Gonzalez-Candelas L, Lluch Y, Granell A (2005) Understanding the basis of chilling injury in citrus fruit. Acta Hortic 682:831–842

Lanza G, Di Martino Aleppo E, Strano MC, Calandra M, Aloisi V. (2006) Evaluation of new treatments to control postharvest decay of citrus. In: Proceedings of the 12th Congress of the Mediterranean Phytopathological Union. pp. 172–174

Lanza G, di Martino AE, Strano MC, Aloisi V, Privitera D (2007) Effectiveness of peracetic acid in integrated control strategies of *Penicillium* decay in tarocco orange fruit. In: Proceedings of the COST 924 international congress on novel approaches for the control of postharvest diseases and disorders. Bologna, Italy, pp 60–65

Maria CS, Altieri G, Admane N, Genovese F, Di Renzo GC (2017) Advance in citrus postharvest management: diseases, cold storage and quality evaluation, IntechOpen publisher

Njombolwana NS, Erasmus A, Van Zyl JG, du Plooy W, Cronje PJ, Fourie PH (2013) Effects of citrus wax coating and brush type on imazalil residue loading, green mould control and fruit quality retention of sweet oranges. Postharvest Biol Tec 86:362–371

Olmo M, Nadas A, García JM (2000) Nondestructive methods to evaluate maturity level of oranges. J Food Sci 65:365–369

Porat R, Daus A, Weiss B, Cohen L, Fallik E, Droby S (2000) Reduction of postharvest decay in organic citrus fruit by a short hot water brushing treatment. Postharvest Biol Tec 18:151–157

Ramaswamy S (2015) Post-harvest Technologies of Fruits & vegetables. DEStech Publications, Inc, Lancaster

Romanazzi G, Smilanick JL, Feliziani E, Droby S (2016) Integrated management of postharvest gray mold on fruit crops. Postharvest Biol Technol 113:69–76

Rouissi W, Sanzani SM, Ligorio A, Khamis Y, Yaseen T, Cherif M, D'Onghia AM, Ippolito A (2009) Application of salts and natural substances to reduce incidence of *Penicillium* rot on Maltaise and Valencia late oranges in Tunisia. In: Proceedings of the 10th Arab congress of plant protection. Arab Journal of Plant Protection, Beirut, p E-71

Sanders KF (2005) Orange harvesting systems review. Biosyst Eng 90:115–125

Sapers GM (2011) Efficacy of washing and sanitizing methods for disinfection of fresh fruit and vegetable products. Food Tech Biotech 39:301–311

Schirra M, D'Aquino S, Palma A, Marceddu S, Angioni A, Cabras P, Scherm B, Migheli Q (2005) Residue level, persistence, and storage performance of citrus fruit treated with fludioxonil. J Agric Food Chem 53:6718–6724

Singh KK, Reddy BS (2006) Post-harvest physico-mechanical properties of orange peel and fruit. J Food Eng 73:112–120

Smilanick JL, Brown GE, Eckert JW (2006) The biology and control of postharvest diseases. In: Wardowski WF, Miller WM, Hall DJ, Grierson W (eds) Fresh citrus fruits, 2nd edn. Florida Science Source, Inc, Longboat Key, pp 339–396

Strano MC, Calandra M, Aloisi V, Rapisarda P, Strano T, Ruberto G (2014) Hot water dipping treatment on Tarocco orange fruit and their effects on peel essential oil. Postharvest Biol Technol 94:26–34

Techavises N, Hikida Y (2010) Prevention of citrus rind injury by utilizing modified atmosphere packaging with perforation. Acta Hortic 857:381–388

Thompson AK (2011) Fruit and vegetable storage: hypobaric, hyperbaric and controlled atmosphere. In: Springer briefs in food, health, and nutrition. Springer, p 126

Thompson JF, Mitchell FG, Rumsey TR, Kasmire RF, Crisosto CH (2008) Commercial cooling of fruits, vegetables, and flowers, Revised edn. UCANR Publications, p 61

Velásquez MA, Passaro CP, Lara-Guzmán OJ, Álvarez R, Londono J (2014) Effect of an edible, fungistatic coating on the quality of the 'Valencia' orange during storage and marketing. Acta Hortic 1016:163–169

Wang NN, Sun DW, Yang YC, Pu H, Zhu Z (2016) Recent advances in the application of hyperspectral imaging for evaluating fruit quality. Food Anal Meth 9:178–191

Wisniewski M, Droby S, Norelli J, Liu J, Schena L (2016) Alternative management technologies for postharvest disease control: the journey from simplicity to complexity. Postharvest Biol Technol 122:3–10

Youssef K, Sanzani SM, Ligorio A, Ippolito A, Terry LA (2014) Sodium carbonate and bicarbonate treatments induce resistance to postharvest green mould on citrus fruit. Postharvest Biol Tec 87:61–69 [illegible] J., Brown GE, [illegible] (20[illegible]) The biology and control of postharvest diseases. In: Wardowski WF, Miller WM, Hall DJ, Grierson W (eds) Fresh citrus fruits, 2nd edn. Florida Science Source, Inc., Longboat Key, pp 339-396

[illegible] M, [illegible] M, [illegible] V, [illegible] P, [illegible] (20[illegible]) [illegible] treatment on Tarocco orange fruit and their effect on [illegible]. Postharvest Biol Technol 94:[illegible]

[illegible] N, [illegible] V (20[illegible]) [illegible] technology by utilizing modified atmosphere packaging with perforation. [illegible]

Thompson AK (2010) Fruit and vegetable storage: hypobaric, hypoxia and controlled atmosphere. [illegible]

[illegible] (2005) [illegible] fruits, vegetables and [illegible]

[illegible] (20[illegible]) [illegible] storage [illegible]

[illegible]

[illegible]

Chapter 12
Citrus Based Food Products and Their Shelf Life

Monika Mahajan and Rohit Sadana

12.1 Introduction

Citrus fruits are most abundantly cultivated in subtropical and tropical regions globally. However, the authentic origin of citrus genus is reported to be southern slopes of the Himalayas in north eastern India and also along with Northern Myanmar. Even some species of citrus are reported to be originated in the Yunnan province in China (Gmitter and Hu 1990; Ziegler and Wolfe 1961; Liu et al. 2012a, b). The citrus fruits are of different sizes and shapes (round, oblong etc.). Citrus genus comprises of following species as sweet orange (*Citrus sinensis*) known to have 61.1% share, tangerine (*Citrus reticulata*) with19.9%, lime and limon (*Citrus aurantifolia and Citrus limon*) as 1% and grapefruit (*Citrusparadise*) has 5% share in global citrus production. There are other minor types of Citrus that includes sour orange (*Citrus*quarantium), Citrus (*Citrus* medica), Shaddocks (*Citrus* grandis) etc. and they constitute the remaining 2.0% share (Mamma and Christakopoulos 2014). In general citrus fruit mainly comprises of two portions i.e., peel and flesh along with seeds. The peel comprises of 30–35% of total fruit. Its skin ranges from rough to robust and bright skin colour from green to yellow. This peel which is also known as epicarp, covers the fruits to save it from external damages. The glands present on the peel are source of various essential oils that contributes to its fragrance. There is a thick white colored spongy mesocarp in between in the internal tissues and epicarp. This mesocarp and epicarp are collectively known as pericarp which is the peel of Citrus fruit (Alnaimy et al. 2017). The internal tissues comprise of 60–65% of total fruit. Basically, the internal tissues are the pulp divided into different

---

M. Mahajan (✉)
Punjab Agricultural University (PAU), Regional Research Station, Bathinda, Punjab, India

R. Sadana
Sant Longowal Institute of Engineering and Technology (SLIET), Sangrur, Punjab, India

S. Singh Purewal et al. (eds.), *Recent Advances in Citrus Fruits*,
https://doi.org/10.1007/978-3-031-37534-7_12

segments. These are also known as juice sacs. These sacs may be without or with seeds as per the varieties. They are also covered with a thick radical layer known as endocarp. These tissues are rich in ascorbic acid, soluble sugars, fibers, potassium salts, pectin and other organic acids. The citrine flavor present in the fruit is because of these chemicals. On the other hand, the seeds comprise 0–10% of total fruit. As per an estimate, out of total citrus fruit production globally 33% is used for the purpose of juice production along with essential oils (Castro 2014). Citrus fruits are rich source of sucrose, fructose and glucose. There is immense quantity of dietary fiber present in fresh citrus fruits. This is helpful in prevention of gastrointestinal disease and also help in lowering cholesterol levels. The aroma of citrus fruit is very distinct and taste is delicious accompanied by low fat and low protein content. In addition to this, citrus fruits are having high nutritive value as they are rich in Vitamin C and Vitamin B (Niacin, Folate, Thiamines, Pantothenic acids, Riboflavin, Pyridoxines etc.). Certain, phytochemicals, such as flavonoids, limonoids and carotenoids are also present in citrus fruits (Ziegler and Wolfe 1961; Mamma and Christakopoulos 2014). These phytochemicals are reported to have anti-carcinogenic, antibacterial, antiviral, antifungal or antiinflammatory agents (Abobatta 2019). Citrus fruits are used for its juice and also as fresh goods globally. The juice can be converted into wine through fermentation. The candies, jam, jellies, cakes, squash, pies, marmalades, etc. are other processed food products made from citrus. Besides that, the large waste generated in the form of peel, seeds, pomace from citrus processing industries has immense application in food sector. The peel is a significant source of antioxidant components. (Manthey and Grohmann 2001). Further, the Polymethoxy flavones (PMFs) which is a flavonoid and is also conveyed to be existing in citrus fruits. It is further reported that it contains having antioxidant properties well as anticancer properties. (Wang et al. 2014; Chen et al. 2017). During the processing of citrus fruit, certain by products such as pectin, various enzymes, ethanol, antioxidants and essential oils are produced. Essential oils of citrus fruits can be represented as an alternate to the chemicals for industrial usage (Viuda-Mart et al. 2008). Pectin which is a byproduct of citrus peel is having usage in manufacturing of jams, jellies as a gelling agent and further as an emulsifier, thickner and stabilizer in various dairy products. It is useful in cosmetic and pharmaceutical industry because of its jellifying properties (Pagan et al. 2001). The shelf life of the citrus juice, concentrates and its processed food products is of major concern. As improper storage temperature, packaging can lead to various spoilage that can be enzymatic, non-enzymatic, microbial, chemical and physical. This spoilage deteorates the taste, aroma, color and bioactive components of citrus products. The present book chapter compiles the phytochemical and nutritional significance of citrus genus fruits, the processing of its various parts into value-added products and as a rich source of functional compounds those are used food sector. Along with this, the type of packaging and storage temperature through which the shelf life of citrus products can be extended is also considered.

12.2 Citrus Genus Fruits

The family to which shelf life of citrus products belongs is Rutaceae. The genus citrus contains trees, shrubs as well as herbs in its species. Citrus is the globally cultivated as well as traded fruit variety as a garden plant. There are various species of Citrus fruit such as Pomelo (*Citrus maxima* (Burm.) Merr.), Mandarin (*Citrus reticulata* Blanco), Lemon (*Citrus limon* (L.) Osbeck), Orange (*Citrus sinensis* (L.) Osbeck), Grapefruit (*Citrus paradisi* Macfad.), lime (*Citrus aurantiifolia* (Christm.) Swingle) and Citron (*Citrus medica* L.) (Wang et al. 2019a) (Fig. 12.1).

All these species have been explained as follows:

1. ***Citrus limon***: It is a small thorny evergreen plant and usually cultivated in subtropical regions. It contains an acidic juice that is a source of potassium, vitamin C, flavonoids and other essential oils (Garcia et al. 2017; Chekani et al. 2021; Ehiobu et al. 2021). Further to this, phytochemical components like terpenoid, saponin, alkaloid, steroid, flavonoid and cardiac glycosides, etc. are also found in its peel and leaf parts. These chemicals have antiviral, antioxidant, anticancer, antimicrobial, antifungal properties (Chekani et al. 2021; Ehiobu et al. 2021).
2. ***Citrus grandis* (L.) Osbeck and *Citrus limetta* Risso:** These are cultivated in parts of Central and Southeast Asia. Further these are reported to contain vitamin C along with pectin, folic acid and potassium (Gupta et al. 2021).
3. ***Citrus aurantium* L.**: It is commonly called bitter orange or sour orange also. It has commercial value and is mainly used for medicinal purposes or for nutritional supplement. However, its usage as fruit is not common because of its bitter and sour taste. The reason behind its sour taste is mainly the presence of neohesperidin and naringin. It is used as a lemon juice alternative in vegetable salads and as an appetizer in Turkey. Further to this, its by-product which is its bark, is very commonly used as an animal feed component. Its fruits are used in jams and

Fig. 12.1 Various citrus fruit species cultivated globally

flavoring activities as they are of antioxidant nature and causes various cardio protective, anticancer, antiobesity and antidiabetic effects (Hosseini et al. 2016; Ersus and Cam 2007; Wang et al. 2019b).

4. ***Citrus sinensis***: It is commonly known as blood orange and this variety is commonly grown in Italy. It contains anthocyanins, Vitamin C, hydroxycinnamic acid and flavanones. In comparison to normal oranges, the quantity of these chemicals is much higher in case of blood orange (Rapisarda et al. 2001). As per reports, the main anthocyanins commonly found in blood oranges are cyanidin 3-(600 malonyl)-glucoside and cyanidin 3-glucoside. Its higher antioxidant ratio in comparison to an ordinary orange is directly linked to the level of present anthocyanins. (Rapisarda et al. 2009)
5. ***Citrus reticulate:*** Its common name is Mandarin orange. The fruit is consumed solo or along with salad. It is orange colored, oblate small. The taste is a little sweeter than the common orange and flesh is soft. The peel has little white mesocarp, further it is loose and thin so, it is easy to peel (Morton 1987).
6. ***Citrus paradise:*** It is commonly known as grapefruit and is the most consumed species of *Citrus* genus. The taste of grapefruit is little bit bitter, so its crossing with orange fruit, to reduce its bitterness was desirable, but as a result of this crossing, its ascorbic acid ratio is decreased however the fructose ratio was enhanced. Further to this, it contains high amount of flavonoids, naringenin and naringin (Sicari et al. 2018).
7. ***Citrus clementina* Hort. ex Tan.**: It is also known as Clementine as this species of Citrus genus is grown in the southern Italy. It contains bioactive compounds like phenolic compounds and Vitamin C (Strano et al. 2021).

12.3 Phytochemical and Nutritional Profile of Citrus Fruits

I. Nutritional Profile

Citrus fruits are the rich source of various natural chemicals; those are very valuable. Certain Vitamins such as Vitamin C, Vitamin B and other elements such as phosphorus, potassium etc. are also present in citrus (Nagy et al. 1980; Ting et al. 1986; Waleed 2019). *The chemical composition of citrus genus has been discussed in* Table 12.1.

II. Phytochemicals in Citrus Fruits

Phytochemicals are the compounds having high antioxidant activities and helps in scavenging of free radicals generated in the body. Phytochemicals are having antiviral, anti-carcinogenic, antifungal, antibacterial, anti-inflammatory or antithrombotic properties (Fattouch et al. 2007). Further they are helpful in controlling the cholesterol levels (Samman et al. 1996). The chemicals such as carotenoids, phenols, sulphides and phytoestrogen possess anti-oxidative potential and hence are regarded as health promoter because of their wide applicability in preventing risk of

Table 12.1 Citrus fruits chemical composition (quantity per 100 gm of its edible portion)

Particulars	*C. paradisi*	*C. sinensis*	*C. reticulate*	*C. limon*	*C. aurantium*
Protein (g)	1	0.8	0.9	1	0.7
Carbohydrates (g)	10	9.3	10.6	11.1	10.9
Fibre (g)	–	0.5	–	1.7	0.3
Fat (g)	0.1	0.3	0.3	0.9	0.2
Carotene, μg	–	–	350	–	1104
Iron (mg)	0.2	0.7	0.1	2.3	0.3
Phosphorous (mg)	30	30	20	10	20
Calcium (mg)	30	40	50	70	26
Vitamin C (mg)	–	50	68	39 (in juice)	30

Data reported as per Waleed (2019); Nagy et al. (1980); Ting et al. (1986)

Table 12.2 Carotenoids in citrus fruits

Carotenoids (mg/100 g)	*C. paradisi*	*C. sinensis*	*C. reticulate*	*C. limon*
Cis β-carotene	–	–	11	–
Auroxanthin	–	0.23	–	–
Cryptoxanthin	3.3	–	20	–
α-Carotene	1	19–20	20	–
Lutein	9.5	27	20–50	–
β-Cryptoxanthin	150	–	–	–

Data reported as per Matsumoto et al. (2007)

chronic diseases in the human body (Kuo 1996). Citrus fruits are enriched in phytochemicals (Marti et al. 2009). The various phytochemicals in citrus fruits are discussed as follows:

(i) **Flavonoids:** Flavonoids are having certain physiological properties those are helpful in controlling anti-microbial and antiviral activity. They are present in citrus plants in good quantity. The glycoside is the main form of their presence. Further it contains certain glycosylic derivatives i.e., flavonoid glycosides that have various health benefits (Loizzo et al. 2012). This is helpful in repressing the infections and also controlling chronic diseases in human body. Polioviruses, parainfluenza and herpes virus are reported to be suppressed by Quercetin and Hesperidin (Kandaswami 1991). Further it is reported that naringin metabolites are very effective against the negative and positive bacteria because of natural antimicrobials activity present in them (Calomme et al. 1996). The highest flavonoid content is reported in the mandarin fruit. Common oranges and grapefruit are also having some group of flavonoids but lemon fruit lacks it.

(ii) **Carotenoids:** Basically, carotenoid pigments are helpful in protection from various human diseases. They are majorly found in grapefruits, mandarin and sweet oranges. Further, zeaxanthin and lutein are some of the carotenoids found in grapefruits (Matsumoto et al. 2007). They are important for eyes and for boosting the immune system of the body. Table 12.2 depicts the comparison of carotenoids available in citrus fruits. It is evident that highest quantities are present in mandarins followed by grapefruit and other citrus fruits. However, lemon is not having any carotenoids contents in them.

(iii) **Essential oil:** Various aromatic compounds are present in in Citrus fruits peel. There are about 400 compounds of volatile and nonvolatile nature available in citrus fruits.Since ancient times, the essential oils are being used for aromatic as well as medicinal purposes. These days the usages of these compounds is notably increasing in pharmaceutical as well as cosmetic industries due to their antimicrobial and antispasmodic activities (Jung et al. 2004).

III. Health Benefits of Citrus Fruits

Citrus fruits are very helpful in maintaining wide range of various biological activities of human body. Approximately 200 bioactive compounds, those are present in lime juice are reported to regulate human body. Also, Majority of bioactive compounds are found in lemon and sweet orange which are reported to be helpful in controlling 60–70% of liver diseases (Sidana et al. 2013).

IV. The Major Health Benefits of Citrus Fruits

1. **Anti-carcinogenic properties:** Flavonoids present in citrus fruits are having anticarcinogenic as well as and antitumor properties (Li et al. 2006).
2. **Hyperglycemia:** Hyperglycemia is a condition in which the level of glucose rises to undesirable levels. Citrus flavonoids helps in controlling its progression through binding with starch. Thus, they help in enhancing the hepatic glycolysis and thus results in lower down the hepatic gluconeogenesis process (Shen et al. 2012).
3. **Cardiovascular properties:** The antiaggregation and antiadhesive action of citrus flavonoids against red cell clumping are helpful in controlling cardiovascular diseases (Robbins 1974).
4. **Anti-microbial activity:** Antiviral and antifungal properties of flavonoids in citrus are responsible for their physiological action in the body (Calomme et al. 1996).
5. **Antiallergic, Anti-inflammatory and Analgesic activity:** The flavonoids reported in Citrus fruits e.g. diosmin, hesperidin, quercetin etc. are reported for their dose-dependent anti-inflammatory activity. This occurs because of their influence on arachidonic acid metabolism that helps in histamine release (Galati et al. 1994).
6. **Lipids control:** The usefulness of citrus juice in controlling higher cholesterol levels is reported in various studies. The presence of higher quantity of insoluble and soluble fiber in citrus fruit juice, is the main reason behind this. (Kurowska et al. 2000).
7. **Anti-depressant and Anti-anxiety:** A flavoinoid named Apigenin present in citrus fruits is known to have antidepressant activity (Matsuda et al. 1991).

12.4 Citrus Fruit Derivatives and Its Utilization in Food Sector

As per the report of 2019, global citrus production is reported as 91,879 thousand tons. Out of which, the orange comprises of 46,062 thousand tons. The share of tangerine is 31,568 thousand tons. The lemon worldwide production is 7550 thousand tons and that of grapefruit is 6699 thousand tons. As per consumption data of citrus genus is concerned, it is 70,177 thousand tons during 2019, in which the share of oranges is 28,324 thousand tons and tangerine comprises 30,117 thousand tons in consumption. The consumption of lemon in whole world was reported to be 5664 thousand tons whereas grapefruit is reported to be 6072 thousand tons. The total export of citrus fruits during 2019 was ten million tons. The highest share in exports is of orange fruits i.e. 45%. The share of other citrus fruits is also significant, as mandarin (27%), lemon (19%) and grapefruit (9%) are also exported worldwide (Ordu 2021).

Citrus fruits are being consumed in fresh form as well as processed citrus products. As per an estimate, 30% of citrus consumption is consumed through processing. This proportion is a little higher in caser of oranges. The major processed citrus fruit product is in the form of orange juice.More than 40% of the oranges of worldwide production are used in processing of different commercialized products such as jams, marmalades, juice dehydrated products and as a flavouring agent in beverages (Fig. 12.2). There are three main derivatives of Citrus fruits namely Citrus Juice, Citrus pulp and citrus puree. These derivatives and their usages in food sector are explained as follows:

12.4.1 Citrus Juice

Because of higher Vitamin C content of Citrus fruits, its juice is recommended for body dehydration. There are two types of juices i.e., 100% natural fruit juices and concentrated forms of juices. In general, the natural juices are prepared from inner parts of fruits and the when these juices are evaporated after extraction, they become concentrated and lead to the formation of concentrated juices.

12.4.1.1 Processing

The technology being used globally for citrus juice processing is almost similar. However, there may be differences in equipment or processing systems as per their regional availability. Before the extraction of juice, washing of citrus fruits and their selection is done. Juice extraction is done at small scale through mechanical processes. The juice is extracted from the juice sacs in the endocarp of citrus fruit (Vishal and Shreran 2009). In Line system of juice extraction is used for citrus

Fig. 12.2 Food products made from Citrus

fruits. In this system, fruits of uniform size are selected and then 3–8 fruits per cycle are processed. There are 3–8 lower as well as upper cups in the system. The upper cups are made movable and further are go downhill towards its adjacent lower cups. There are small knives of circular shape mounted on lower cups those are already fixed. The knives in the cups are of finger shape and on up down movement they intersect with each other. Further the cups when got fitted in each other the peel is removed from the fruit. The fruit pulp and remaining parts are collected in a collector for final operations. The juice extraction is also carried out with help of equipment such as "Brown extractor". In this extractor, the fruit is cut into two halves and then they are located into special cups designed for the purpose. The extraction is carried out with help of reamers i.e. special bulbs (Ikechukwu and Okonkwo 2014). These reamers revolve during the squeezing process. As per requirement, the distance between the reamers can be adjusted. The extraction pressure has impact on the composition of the juice but not on the quantity and yield. After this the extracted juice passes through the vibrating filter that removes the cells, seeds and other coarse materials. Thus, extracted juice is collected. The extracted juice is de-aerated that helps to control the oxidation of ascorbic acid present in the juice. The juice is usually pasteurized through tubular heat exchangers. This pasteurization helps to protect the juice from the microbial growth. Hence after this, the juice is preserved. For juice concentrate, evaporators are used with adequate temperature settings to perverse the aroma of juice (Domenico et al. 2010).

12.4.1.2 Food Applications

Citrus fruit juice has various food applications as:

1. Citrus fruit Juice: There are two forms of Citrus fruit juice i.e. 100% natural fruit juice and fruit juice concentrates. The fruit juices available in the market may contain added sugars as they contain small quality of real juice along with fruit juice concentrates which are prepared through evaporation. Citrus juice is very popular among the consumers that account for 2/3 of the global market. But undesirable bitterness is the major hindrance in its global acceptance. The bitterness is either due to the flavonoids and their derivatives or it might be because of the conversion of non-bitter precursors into bitter compounds. So, researchers are focussing on debittering the citrus juices in order to increase their consumer acceptance (Kore and Chakraborty 2015; Purewal and Sandhu 2021).

   Now a days, a novel product of citrus juice containing probiotics like *Lactobacillus fermentium, L. casei, L. plantarum, and L. pentosus* are coming in the market, which is received a huge attention. They have the properties of both citrus as well as citrus strains like antipathogenic, chronic disease prevention, immune regulation etc. (Yu et al. 2015; Yuasa et al. 2021).
2. Jellies: The jellies are also prepared from citrus juice. Rubio-Arraez et al. (2016a, b) have replaced the sucrose in the orange jelly with other sweetners or their mixtures like oligofructose, isomaltulose, tagatose having low glycemic index. The jellies prepared with the different formulations were screened for TSS, pH, moisture, color, antioxidant capacity and mechanical properties. The results showed the reduced antioxidant activity during storage in all the formulations. The addition of low GI sweeteners produced the jellies with low a, b and crome values i.e. the product was lighter in color than the sucrose added product. The jelly prepared with either oligofructose or tagatose or with isomaltulose and tagatose mixture, have mechanical properties similar to that of the control and the later combination has got the maximum score in sensory analysis.
3. Ice lollies: These are ice cubes formed from citrus juices. They are cool and tasty.
4. Sweet Pastries: 100% citrus fruit juice forms are having their usage in manufacturing of cakes, biscuits etc. Orange burfi is also an example of barfi made from milk and orange juices in Nagpur (India).
5. Blended juices are mainly manufactured by adding concentrates of two or more than two varieties. The blending gives the juice which is nutritionally and organoleptically superior. For example: In grapefruit juice about 15–35% blending is possible. Wagner and Shaw (1978) showed that blending of grapefruit juice with orange and tangerine juice is less acidic and bitter. Grapefruit juice is also blended with other non-fruit juices. Inyang and Abah (1997) found that the orange juice blended with the steamed cashew apple juice is stable and have good sensory acceptance. Selinium enriched probiotic Streptococcus thermophilus was added to the fermented juice blended with orange, carrot, apple and ber juice. A significant increase in of about 13-fold Se was observed in fermented beverage with unique flavour characteristics, aroma and nutrition (Xu et al. 2019).

6. Fortified juice: Fortification of commercial orange juice with encapsulated orange pulp extract is being carried out to increase the antioxidant potential of the final product. The fortification has increased total phenolic content from 10–22% and improved the antioxidant potential of the final product. Islam et al. (2014) has evaluated the quality of mixed juice prepared by fortifying the orange juice with pineapple juice in different ratios during storage. The product was pasteurized for 5 min at 97 °C and stored in PET bottles for 35 days at 4 °C. The juice that was fortified in 1:1 ratio was found to have maximum acidity and ascorbic acid as 0.785% and 37.1 mg/100 ml respectively. The decrease in TSS, ascorbic acid and acidity was observed in the blended juice with storage and the aerobic plate count was found to be in range. The juice was stable for 3 weeks without any preservative and for 5 weeks after adding sodium benzoate at the conc. of 0.6 mg/L. The overall acceptability score was found to be more than 8, which suggests commercial scope of the mixed orange juice blended with pineapple juice.

   Furthermore, Biancuzzo et al. 2010 have added Vitamin D in calcium fortified orange juice for minimizing the deficiency of vitamin D in school children and adults. They have compared the juice fortified with 1000 IU Vitamin D2 to that of vitamin D supplements and found that the blood serum vitamin D was more in adults who have taken this fortified orange juice rather than oral supplements.
7. Fermented juices: Citrus juices can be used for preparing fermented beverages. Escudero-López et al. (2013) prepared a product i.e. fermented orange juice through the process of alcoholic fermentation. The fermented juices have health beneficial properties as they have high flavonoid and carotenoid content. Nagpal et al. (2012) and Yuasa et al. (2021) have prepared citrus juice beverages through process of fermentation by making the use of probiotic bacteria *Lactobacillus plantarum* and *L. pentosus*. The viable cell count was found to be similar to the probiotic foods. As per chemical composition is concerned, the malic acid was found to decrease in fermented juice whereas lactic acid increased but no difference was observed in sensory evaluation before and after the fermentation.

   Citrus wine is the another product made from citrus. It adds value addition to citrus and promotes citrus industry. Citrus wines hold the original flavour and functional component of citrus like flavonoids, polyphenols, carotenoids, pectins etc. and has low alcohol content (Selli 2007). These nutrients not only give unique flavour but also provide the nourishment and relaxing properties. From last so many years, citrus wine industry is gaining a lot of attraction. The unique flavour of wine is because of the substances which are produced during fermentation by microbes. However the problem of bitterness in wines is also producing a problem during storage. Moreover, the blending of fermented wine with other fruits and vegetables such as apple, pineapple, ber, carrot etc. is also practiced in recent years. This blending produces the quality wines with multiple health benefits (Archibong et al. 2015).
8. Ready to serve beverage: Fruit intake in the form of beverages is liked by people of all age groups. Ready to serve beverages use the blending of two or more

juices with the purpose of making the final product better in terms of nutrition and organoleptic properties. Kausar et al. 2020 has prepared ready to serve beverage containing aloe vera gel (0, 5, 10, 15, 20%) in orange juice with desirable quality, taste, physiochemical and microbial properties. It has been found that the addition of aloe vera gel to orange juice, increase the antioxidant activity, reducing power and phenolic content. The total plate count was also reduced after blending and RTS was acceptable even after 3 months of storage. The effect of storage of 3 months was also studied in resulted RTS beverage.

12.4.2 Citrus Pulp

Citrus fruit pulp is separated from seed and the peel portion of the fruit through mechanical or manual methodology. In general terms, that part of a citrus fruit which is neither fermented nor concentrated or diluted, is known as citrus pulp. When a mixture of peel and pulp is available, it is considered very good for health as it provides health related benefits of both ingredients.

12.4.2.1 Processing

Citrus pulp is extracted from any fruit of citrus family i.e. limes, oranges, lemons, grapefruits etc. The main product of citrus processing is citrus juice and the co-product left after juice extraction is citrus pulp. In fact, the solid residue, left over after the extraction of juice from the fruit, is the citrus pulp. Pulp contains seeds, internal tissues along with peel. It is approximately 3/4th in weight to the total fruit weight. It is having very limited shelf life. It is high energy food and can be used for animal feed purpose. For making use of pulp, it is dried and then transported as per requirement. Sometimes pellets of pulp are also made. During drying of pulp, the moisture content of the pulp is reduced and further lime is also added to offset the acidity.

12.4.2.2 Food Applications

Pulp is considered as most versatile food derivative that can further be used for manufacturing various fruit products, those are as follows:

1. Jam and Marmalade: Jam is a processed product made from fresh fruits by adding sugar, thickners etc. This aims to reduce the post-harvest losses by extending the preservation life in the form of processed product. The main characteristic of citrus fruit jam is its unique flavor, bright color, taste and nutrition. Today orange and grapefruit jams are used in the breakfast, confectionary, bakery and dairy products. The sugar content of the jam has to be 50–70% in order to obtain the

better texture. But the high sugar content in jam is a matter of health significance. Therefore today, technologists are coming up with the idea of replacing sugar in jams with natural sweeteners like tagatose, isomaltulose (Rubio-Arraez et al. 2016a, b). It has been observed that during the conventional method of jam preparation, there is a loss of beneficial nutrients due to heat that can be prevented by using novel ways of other processing methods like microwave and osmotic dehydration (Igual et al. 2010; Igual et al. 2013).
2. Ice Cream: These days, orange pulps are being used in manufacturing of ice creams. They are source of nutrients as well as vitamins. Thus ice cream manufactured from pulp by addition of milk are very healthier as well as nutritive.
3. Frozen Citrus Pulp: This product is used in bakery, confectionary etc. This is used as raw material in many fruit products.
4. Smoothies: It is kind of fruit shakes made from citrus pulp and is in demand these days

There are other products which are made from citrus pulp are salad toppings, sweet pastries etc.

12.4.3 Citrus Puree

Citrus puree (grapefruit, lemon, orange purees, etc.) is obtained by smashing the inner portion of the fruit. It is reported to be very versatile. It can be manufactured from different fruits. It is reported to be rich in Vitamin C, minerals, antioxidants etc. Fiber is also present in Puree that is helpful in combating constipation related problems.

12.4.3.1 Processing

In general, for preparation of puree from any citrus fruit, whether its lemon or orange, the procedure is almost similar. Firstly, the whole fruit is either crushed or sliced and pureed. Puree is basically a semi processed product useful in production of various bakery items. The main method to preserve this is through freezing. The puree produced from fruit is around 50–60% of its weight.

12.4.3.2 Food Applications

Citrus puree is used in manufacturing of various food items or as food ingredients. These can be explained as

1. Citrus Puree: Citrus puree especially orange puree is very popular in children globally. These purees are natural and free from color preservatives.

2. Desserts: Citrus puree is used in various deserts because of its unique flavor and taste.
3. Savoury dishes. Mixture of different flavours of citrus puree are being used by chefs to make different combinations to allure the food lovers.

12.5 Citrus Fruit Waste/By-products

The fruits of citrus genus are usually marketed as either fresh fruits or processed juice. This includes *C. sinensis* (sweet orange), *C. reticulata* (tangerine), *C. aurantifolia* (lemon and lime) and *C. paradisi* (grapefruit) while some are minor Citrus species as shaddock (*C. grandis*), citron (*C. medica*), sour orange (*C. quarantium*) etc. Citrus fruits are mainly fragmented into flesh and peel. The peel is further divided into Flavedo and albedo. The outermost layer of the fruit is flavedo which is also known as exocarp, consists of oil glands, cellulosic material, and various pigments. Whereas, alvedo is the inner portion of the citrus fruit, that is next to flavedo. It is enriched with pectin and is removed before consuming the fruit. The endocarp is alienated into juice sacs segments or vesicles.

12.5.1 Citrus Peel

Large amount of byproducts and residues are generated by peels during the processing of citrus fruits. About 40 million tons of citrus waste is produced world-wide every year and for 50% of the original mass of citrus fruits contributes towards the fruit residues (Ferguson et al. 1990). The main solid waste or residues produced during the citrus processing are peels, membranes and seeds. Liquid waste includes cannery effluents, fruit washing waste waters, peeling, table waste water during sectioning, and water during flushing the floor. Distillery effluents involve citrus molasses effluents, pectin and citric along with citrus oil plants. The other semisolid waste consists of residual membranes of endocarp, some extent of exocarp and albedo. Other wastes which are produced during citrus processing are pressed pulps generated by juice manufacturing industries, expired, different sizes and damaged fruits.

Earlier these wastes disposal was a big issue as it was discarded on land and over the time the increasing putrefying bacteria on it increase the risk to the local water resources and also led to methane production which was uncontrollable. The direct discharge of wastes into lake water cause pollution and damage to the aquatic life. Therefore, citrus waste management is a big challenge. In recent years, researchers are focussing on developing the innovative methods to exploit the citrus waste into useful and value-added products (Tripodo et al. 2004; De Gregorio et al. 2002; Lo Curto et al. 1992). The solid and semisolid waste produced during the processing of citrus is rich in sugars, organic acids, amino acids, oils, minerals, fibers, flavonoids and vitamins (Braddock et al. 1999). The different fractions of the waste has these compounds in the varied amounts. Also, the methods employed for the juice extraction affect the proportion of these components (Marin et al. 2002). The best use of citrus waste is to get fiber and other food ingredients like pectin and mucilages. In addition to this, many commercially important phytochemicals are also extracted from the citrus waste.

12.5.2 *Citrus Seeds*

The seed content in grapefruit and oranges is found to be 7% and 2% respectively. Finishers can easily separate seeds from the extracted juice. Seeds of citrus are very important source of oils and proteins. The oil is mainly extracted through pressing it in expeller or through solvent extraction method (Belshaw 1978; El-Adawy et al. 1999). The citrus seed crude oil is used in making soaps because it is deeply bitter in taste. But, it is purified into edible oil which is quite healthy. The defatted seed oil contains about 43% protein. The flour prepared from orange seeds was found to have 28.5% carbohydrates, 3.1% crude protein, 5.5% fiber and also found to be enriched in calcium and potassium (Akpata and Akubor (1999). Studies have shown that 41.9% and 33.4% oil was extracted from citrus seeds grown in Iran and Kerman respectively. The main unsaturated fatty acid which is found in citrus seed oil is linoleic acid (33–36%). Thus, the results showed that citrus seed oil can be consumed by human (Reazai et al. 2014).

12.5.3 *Citrus Pomace*

During the production of citrus juice, the solids that is separated from juice is known as pomace. During mechanical juice production, the adjustment of machinery parameters decides the yield of juice and dryness of citrus pomace. Citrus pomace mainly comprises of various essential components such as pectin, hemi cellulose, other sugars and cellulose (Zannini et al. 2021). The annual production of citrus pomace is estimated to be one million ton (approx.). Citrus pomace is also enriched in polyphenols that can be used as nutraceuticals in beverages as well as in various functional foods. The extraction conditions decide the quality of various oils produced from citrus pomace (Caballero et al. 2021).

12.6 Utilization and Valorisation of Citrus By-Products

The effective extraction methods that are being employed today for extracting valuable by-products from citrus waste include reflux, stirring, shaking, ultrasonic extraction (Jeong et al. 2004; Xu et al. 2008). Today, researchers are developing more efficient extraction methods for industries that are energy saving, safer, compact and sustainable. These include subcritical water extraction, supercritical fluid extraction, ultrasound extraction, controlled pressure drop method. Other extraction techniques like microwave assisted extraction, accelerated solvent extraction and ultrasonic extraction are also being used today and are emerged as the effective methods that also deals with various environmental problems. Amongst them, microwave assisted extraction is the best method for extraction as it uses microwave energy to carry out the extraction that ensures ease, high extraction rate, rapidity, less energy consumption and better production yield with less inputs (Spigno et al. 2009). Another novel eco-friendly green technology is Microwave hydro diffusion and gravity. It is highly advantageous over the available conventional methods because of its economic viability and shorter extraction time. It is mainly used in polyphenols extraction from the onion (Bousbia et al. 2009). It works at atmospheric pressure under the principle of microwave and gravity without the involvement of solvents (Chemat et al. 2006). It involves the heating of fresh plant material at a certain temperature without adding any solvent. The direct interaction of biological water in the fresh plant material with microwaves led to the release of the phytochemicals. These compounds come out in crude juice or hot water and then settle down in a spiral condenser under the effect of gravity.

12.6.1 Source of Bioactive Compounds

The waste from citrus fruits is rich in bioactive compounds like phenolics, flavonoids etc. These compounds have high antioxidant i.e. free radical scavenging activity, anti-inflammatory, anticarcinogenic and antilipid peroxidation properties (Ghasemzadeh and Ghasemzadeh 2011; Kumar and Goel 2019). Mainly two groups of flavonoids are reported in citrus fruits. First are the glycosylated flavones such as luteolin, apigenin and diosmin glucosides and the other is polymethoxylated flavones. The amount of these flavonoids varies among the citrus species. The glycosylated flavanones are convertible into dihydrochalcones and further they are the strong natural sweeteners. Hesperidin, naringin, eriocitrin and narirutin are the major flavonoids reported in citrus species. It has been observed that the peel and solid waste of lemon contains only hesperidin and eriocitrin whereas the liquid residue has naringin and eriocitrin as predominant flavonoids (Zhao et al. 2017). In the same way, the seed waste of citrus is also enriched in flavonoids and phenolics, but the amount is more in peel compared to seeds. The amount of these flavonoids varies in peel and seeds.The lemon peel is reported to be rich in naringenin, neohesperidin and neoeriocitrin, whereas seeds have principal amount of hesperidin and eriocitrin. The major phenolic acids reported in citrus peel of are ferulic, caffeic, p-coumaric and sinapic acids. An oprimized microwave assisted extraction method is utilised for phenolic compounds extraction from citrus peels (Hayat et al. 2009).

12.6.2 Application in Food Industry

Pectin

Pectin is one of the valuable products of citrus processing industries. It is used as gellifying agents in jams and stabilizer in many food products. About 40–50 thousand tones production of pectin occurs worldwide and 85% of its production come from citrus waste and only 15% is contributed by apple pomace. Mainly fresh or dry peels of lemon, grapefruit and oranges are used for pectin production. About 9–18% and 1–3% pectin is obtained from dried and fresh citrus peels. It is further reported that the lemon albedo dried peel has highest pectin content about 30–40%.The process of dehydration of citrus peel for pectin extraction is quite different from dehydration done for animal feed where peels are limed. For extraction of pectins from peel, firstly leached fresh or dehydrated peels portion are dipped in acidified boiling water. During this the pectin get solubilized and thus the viscosity of the extract is increased. The extract is then filtered and the process is repeated again. Ultimately the pressed cake obtained after filtration is dried, grinded and sell as insoluble fibers. The extract is further concentrated by evaporation and the pectin is precipitated by adding alcohol to 60%. The pectin obtained as fibrous mass is dried and grinded to fine powder.

Citrus Fiber

It is a dietary supplement and is in a great demand today because of its beneficial physiological effects. Consumers want to take a diet rich in fibers for a healthy well-being. So, industries are including more and more fibers in their formulations. Fibers are important in food products because of its functional properties and citrus fiber is best known for its functional action (Lundberg et al. 2014). Two types of fibers are there. One is soluble fibers which include pectin, alginates, inulin, gums etc. and the other is insoluble fibers that include cellulose, chitin, lignin etc. Several food products are available in the market that include citrus peel fiber as their ingredient where most of the fiber is contributed by albedo (Marín et al. 2007). The study conducted by Larrauri et al. (1997) has shown that the fiber content in citrus peel decrease with the maturation of fruit. They observed uronic acids, galactose, arabinose, glucose, and xylose as the main soluble fibers present in grapefruit peel. The residues which are left after pectin extraction are good source of insoluble fibers that have excellent fat and water absorption capacity. It has its main utilization in meat and bakery products (Vergara 2013). Citrus fiber is also used as fat replacer in many food products like ice-creams. Orange fiber can replace about 70% fat in ice-creams. The another use of citrus fiber is its utilization in emulsion foods and salad dressings as an emulsifier or stabilizer (Chatsisvili et al. 2012). Studies have reported that the products rich in citrus fiber have a probiotic effect produces healthy bacteria by fermentation during digestion. It is also reported to reduce the blood glucose and cholesterol level.

Essential Oils

Citrus peel as well as juice are rich source of essential oils and they can be extracted from both of them. The monoterpene d-limonene, a colorless liquid is the main component of various essential oils those are extracted from citrus peel and Juice. About 90% of orange oil mass is represented by essential oils. These essential oils are reported to be extracted simultaneously along with juice and mainly marketed as "cold pressed oils". A study conducted by Tongnuanchan and Benjakul (2014) has emphasize the role of citrus essential oils as food preservative having good antioxidant activity. The cols pressed oil extracted from lemon peel is quite expensive compared to that extracted from orange peel. The essential oils are concentrated by vacuum distillation where more volatile components gets separated from terpenes. The solvent partition method using aqueous ethanol is used for limonene from terpene less oil. These essential oils are degraded easily during storage, so they are packed in small metal containers with no headspace or purged with nitrogen gas. Exposure to sunlight also causes the flavor deterioration in essential oils, so they are often encapsulated to prevent oxidation (Liu et al. 2012a, b; Tackenberg et al. 2014). Steam distillation is used to recover essential oil from peels and frit (Chemat 2010; Kashiwagi and Sawamura 2010). This oil is used for the production of food grade limonene and in toiletry. Food grade Limonene has an orange like odor and is used in pharmacy, foods and beverages. Removal of d limonene from the essential oil increase its stability as d limonene has a property to get oxidise and racemize to other compounds that have terpentine like odor. Such oils with no d limonene are called terpeneless oils (Schmidt 2010) and are known to be more expensive than originals. D limonene has other applications like they are used as cosmetic solvent and cleanser.

12.6.3 Animal Feed

In citrus industries, the citrus pulp waste is used for animal feed after alkaline or enzymatic treatment. It has a good digestibity because of the ruminant's ability to sollublize the fibers present in the pulp by fermentation process (Tripodo et al. 2004). It has a high energy content but lower in nitrogen content which is supplemented by urea, liquid ammonia and ammonium salts. Citrus peels addition to the silage ingredients like legumes and grass help to enhance its performance (Bampidis and Robinson 2006). Citrus waste is also used as an organic fertilizer in agriculture industry. Lim & Matu 2015 reported the production of citrus waste fertilizer using fermentation technology. But the main concern with these fertilizers were their low pH which increase the acidity of the soil. Therefore, in orange peels, the pH, moisture and C/N ratio was adjusted to 6.3; 60%; 24:1 respectively before it was used as a fertilizer. Another study has used citrus peel powder with high nitrogen along with alkaline peel powder with high phosphorous and potassium content of 4.2 mg/g and 2.1 mg/g respectively. The acidic peel powder reduces the soil salinity whereas alkaline citric powder minimizes acidity content of soil.

12.7 Shelf Life of Citrus-Based Products: Storage and Packaging

Shelf life is the length of time (post harvesting) until the safety and quality of the food product is reduced under a quantified level. Further, spoilage is the process that reduce the safety and quality level of any food item or product. The spoilage may be physical, chemical, enzymatic or microbial. In case of Citrus fruit products, the main spoilage is of chemical nature that may occur because of color changes, non-enzymatic browning, deterioration of aroma and taste and loss of ascorbic acid etc. Further texture-related spoilage is also reported in products those are in solid form e.g. softening of food products.The shelf life of citrus based products in terms of storage and packaging has been discussed as:

12.7.1 Single Strength Juices

Citrus is seasonal fruit and principle of not-from concentrate (NFC) production for juice is followed. There is demand for the product full year around. So, storage of NFC juice for long term is the demand of the industry (WFLO 2008). As for storage a enormous quantity of citrus juices, large storage facilities are required. During earlier times, for storage of juices, technique used was freezing the juice and this freezed juice was stored as large frozen slabs till they are required for processing purpose. This method is difficult to perform and costly. So to overcome these difficulties, large sized refrigerators were used for juice storage to enhance its shelf life. Further, the juice was pasteurized, then deaerated. After this the juice is chilled to 0–1 °C before pumping into presterilized refrigerated tanks.in general the capacity of the storage tanks is approx. 3500 cub. meters (one million gallons) or more. These tanks are made up of carbon steel that is lined with epoxy resin. The group of tanks is known as tank farms. In these tanks the temperature slightly above the freezing point is maintained and oxygen availability is competently omitted. This system guarantees a shelf life of one year.

Many studies have shown the processing and storage effect on the citrus juice shelf life (Perez-Cacho and Rouseff 2008). The quality of hot filled bottled and aseptic bottled orange juice was compared by Mannheim and Havkin 1981 and has been found that aseptic juices had better quality but degraded after storage. Another research conducted by Sandler et al. (1992) has compared the microbial, enzymatic and biochemical composition of orange juice store at 4 °C with no pasteurization, with little pasteurization and with full pasteurization. They found that the permeability of oxygen did not affect the un pasteurizated juice. The studies reported that the low or fully pasteurized juice had little microbial growth, had better retention of ascorbic acid and less cloud loss even after 21 days of storage. There was not much difference observed in low or fully pasteurizated juice. UV treatment of orange juice and grapefruit juice was studied by Uysal Pala and KırcaToklucu (2013); La Cava and Sgroppo (2015) followed by refrigeration. This treatment increases the storage stability of juices. Other studies have shown that the packaging films containing silver and titanium compounds also enhance the shelf life of citrus juice (Peter et al. 2015). Further the impact of packaging practices on the shelf life of single strength orange juice was also studied. Multilayer and monolayer PET bottles are also used for storing orange juice. It has been found that monolayer PET bottles showed less retention of ascorbic acid, but use of other protective agents like addition of oxygen scavengers, purging with liquid nitrogen, can extend the shelf life of citrus juice in monolayer PET bottles to the level that is found with multilayer or glass bottles. Packaging usually consists of four layers. First layer is of polythene present internally for sealability. Second layer is the layer of aluminium foil for light and gas impermeability. Third layer consists of paper for printability and mechanical strength and the last layer is of polythene for external protection. The storage behaviour of reconstituted orange juice processed by either conventional thermal processing or by high hydrostatic pressure stored in polypropylene bottles or laminated pouches was studied at 0–15 °C. The juice given hydrostatic pressure treatment retain more ascorbic acid compared to conventional heat treatment. The retention was more i.e. 65% at 0 °C compared to 11% at 15 °C. In pouches it was 57% and 24% at 0 °C and 15 °C respectively.

Addition of thiols (glutathione, N acetyl L cysteine and L cysteine) in pasteurized orange juice reduce the formation of p-vinyl guaiacol which degrades the taste of orange juice. Addition of thiols also prevent the ascorbic acid degradation during storage. Non enzymatic browning of orange and grapefruit juice is quite common during the storage as it produces the off flavour. They investigated browning in juices stored in Tetra brick cartons. They found release of carbonyl compounds due to the degradation of ascorbic acid is responsible for browning not Millard reaction. Nagy et al. (1990) has studied browning in canned/bottled grapefruit juices stored for 18 weeks at 10–50 °C. They observed that reducing power of tin at acidic medium prevents the browning in canned juices that were stored at 10, 20 °C.

Colorimetry method using CIELAB system to detect the change in color and degradation of carotenoids which is mainly affected by non-enzymatic browning. Kenedy et al. (1992) evaluated the rate of vitamin C degradation in blood orange juice stored in Tetrabrik cartons. Amount of dissolved oxygen affects the aerobic and anaerobic deteoration of ascorbic acid. Finally L-ascorbic acid aerobic degradation occurs till the dissolved oxygen reaches the equilibrium. Soares and Hotchkiss (1999) reported that the aerobic degradation is inversely proportional to oxygen permeability. Therefore, juices are deaerated before they are aseptically stored in tanks. The loss of aroma due to the adsorption during packaging is a matter of concern. About 19% and 10% compounds of aroma are lost during polyethylene and polypropylene.

12.7.2 Citrus Concentrates

Aseptic packaging is commonly used to store the concentrated citrus juices. Burdurlu et al. (2006) investigated the storage effect (28–45 °C) on the stability of vitamin C in lemon, orange, grapefruit, tangerine juice concentrates. It was found that the retention of vitamin C after 2 months of storage at 28 was 55–84% whereas 24–27% and 15–20% retention in vitamin C was found at 37 and 45 °C respectively. Ascorbic acid loss was correlated with Hydroxymethyl furfural accumulation in all juice concentrates at different storage temperatures (Kanner et al. 1981). It was found that furfural formation was quite rapid in orange juice (12°Brix) compared to its concentrate (58°Brix) stored for 3 months at 17 °C. Other finding was the non-enzymatic browning was found to be less in concentrates stored below 12 °C. Further the flavour change in concentrates (58°Brix) was quite low even after 12 and 10 months of storage at 5 °C and 12 °C respectively. So, it has been concluded that aseptic citrus juice concentrates can be stored at 5 °C for 1 year without the need of freezing. In industries, the juice concentrates are stored in refrigerated tanks with huge capacity. During transport, these concentrates are filled aseptically in steel drums with polymer lining, refrigerated tanks and in Scholle bag pouches. These pouches come in different sizes, are multi-layered, presterilized and have aseptic filling nozzles. They have standalone version as well as come in bag-in-the box version. For retail selling, these concentrates are store at sub-freezing temperature packed in plastic cups or small cans.

12.7.3 Citrus by Products

As it is already discussed in previous section, that essential oils undergo oxidation, racemization, polymerization that degrades its pharmaceutical and sensory quality. The main factors responsible for its quality loss is oxygen, light and some catalysts. But it has been found through studies that the effect of these factors decreases when low temperature is used for oil is storage (Turek and Stintzing 2013). Citrus essential oils are generally stored in tin and aluminium containers or in dark colored glass bottles and stored at low temperature even for a year. Studies have shown that the fresh orange aroma can be restored with any significant change by storing at 0 °F with shelf life of approx. 3 months (Guadagni et al. 1970). So low temperature is important for retaining the essence and aroma of oranges. Other products like pectin and citrus fibre are usually maintained as dry powders which are shelf stable. Only dry and cool places are recommended for their storage.

12.7.4 Other Miscellaneous Products

The segments of grapefruit when canned are shelf stable even for one year when stored at low temperature. High temperature storage can cause the rapid browning of the segments and can degrade its quality. Many packaging is tried for grapefruit segments like mono and multilayer plastic trays, tinplate cans etc. But the best packaging for grape fruit segments are plain, non-lacquered cans (Miltz et al. 1995). For candied fruit and peels, hermetic containers or polymer pouches impermeable to water are suggested. Because at high humidity stickiness, caking and attack by mold can occur. Jams and jellies are hot filled in presterilized/sanitized glass or plastic jars in order to prevent the sugar crystallization and severe browning.

12.8 Future Directions

Citrus fruits are being used in various manners such as citrus juice, puree, other waste by products globally. In spite of the probable usage of citrus fruits and its waste products, still there is much scope through valorization for production of various essential oils and cattle feed pellets etc. These innovative ideas need structural changes in the existing phenomenon as well as adoption of useful implementations. Cost effective, eco-friendly and sustainable procedures for extraction of biochemical such as Vitamins, essential oils etc. is the need of hour. Further, there is an urgent requirement for improvement in process optimization and production efficiency of enzymes and other chemicals of citrus fruit ingredients. Improvement in conversion efficiency, cost reduction, along with prohibition of intermediate may be the futuristic options available. For sustainable societal development, process efficiency need to be improved. The implementation of valorization of citrus waste in future biorefineries can be a great invention. Usage of citrus waste as biomanure can be a future source of fertilisation, advanced research in this regard must be carried out. Collaborations in between different industrial partners as well as Govt Agencies along with advanced research, awareness, legislation can be helpful in building new era of sustainable society and more specifically a bio-based developed economy.

12.9 Conclusion

Citrus genus fruits are one of the most commonly consumed commodities globally. The genus includes various species that are enriched in Vitamin, Minerals, fibers and antioxidants, important for human wellbeing. Every part of citrus fruit can be processed into value added products. The flesh can be juiced and the pulp which is left can be used for the production of jams, jellies, marmalades, cakes, pies, smoothies etc. Even the waste part of the fruit in the form of peel, seeds and pomace, can also be processed into valuable products like pectin, essential oils, fiber etc. Furthermore, proper storage at low temperature and deaerated or aseptic packaging is very important to maintain the quality of citrus based products for longer duration of time.

References

Abobatta FW (2019) Nutritional benefits of citrus fruits. Amer J Biomed Sc Res 3:303–306
Akpata MI, Akubor PI (1999) Chemical composition and selected functional. Properties of sweet orange (Citrus sinensis) seed flour. Plant Foods Hum Nutr 54:353–362
Alnaimy A, Gad AE, Mustafa MM, Atta MAA, Basuony HAM (2017) Using of citrus by-products in farm animals feeding. Open Access J Sci 1:58–67
Archibong EJ, Ezemba CC, Chukwujama IC, Archibong UE (2015) Production of wine from mixed fruits: pineapple (Ananas comosus) and Orange (Citrus sinensis) using yeast isolated from palm wine. World J Pharm Pharm Sci 4:126–136
Bampidis VA, Robinson PH (2006) Citrus by-products as ruminant feeds: a review. Anim Feed Sci Technol 128:175–217
Belshaw R (1978) Citrus flour: a new fiber nutrient source. Food Prod Dev 12:36–38
Biancuzzo RM, Young A, Bibuld D, Cai MH, Winter MR, Klein EK, Ameri A, Reitz R, Salameh W, Chen TC, Holick MF (2010) Fortification of orange juice with vitamin D2 or vitamin D3 is as effective as an oral supplement in maintaining vitamin D status in adults. Amer J Clin Nutr 91:1621–1626
Bousbia N, Vian MA, Ferhat MA, Meklati BY, Chemat F (2009) A new process for extraction of essential oil from citrus peels: microwave hydrodiffusion and gravity. J Food Eng 90:409–413
Braddock RJ, Braddock RJ, Weiss E (1999) Handbook of citrus by-products and processing technology. Wiley, New York
Burdurlu HS, Koca N, Karadeniz F (2006) Degradation of vitamin C in citrus juice concentrates during storage. J Food Eng 74:211–216
Caballero S, Li YO, McClements DJ, Davidov-Pardo G (2021) Encapsulation and delivery of bioactive citrus pomace polyphenols: a review. Crit Rev Food Sci Nutr 13:1–17
Calomme M, Pieters L, Vlietinck A, Berghe DV (1996) Inhibition of bacterial mutagenesis by Citrus flavonoids. Planta Med 62:222–226
Castro MDLD (2014) Towards a comprehensive exploitation of agrofood residues: olive tree—olive oil as example. Compt Rendus Chim 17:252–260
Chatsisvili NT, Amvrosiadis I, Kiosseoglou V (2012) Physicochemical properties of a dressing-type o/w emulsion as influenced by orange pulp fiber incorporation. LWT Food Sci Technol 46:335–340
Chekani R, Akrami R, Ghiasvand Z, Chitsaz H, Jorjani S (2021) Effect of dietary dehydrated lemon peel (*Citrus limon*) supplementation on growth, hemato-immunolological and antioxidant status of rainbow trout (*Oncorhynchus mykiss*) under exposure to crowding stress. Aquaculture 539:736597
Chemat F (2010) Steam and hydrodistillation. In: Flavor, Fragrance. In: Sawamura M (ed) Citrus essential oils. Wiley, New York
Chemat F, Lucchesi ME, Smadja J, Favretto L, Colnaghi G, Visinoni F (2006) Microwave accelerated steam distillation of essential oil from lavender: a rapid, clean and environmentally friendly approach. Anal Chim Acta 555:157–160
Chen XM, Tait AR, Kitts DD (2017) Flavonoid composition of orange peel and its association with antioxidant and anti-inflammatory activities. Food Chem 218:15–21. https://doi.org/10.1016/j.foodchem.2016.09.016
De Gregorio A, Mandalari G, Arena N, Nucita F, Tripodo MM, Lo Curto RB (2002) SCP and crude pectinase production by slurry-state fermentation of lemon pulps. Bioresour Technol 83:89–94
Domenico C, Luigi S, Alfonso G (2010) Enzymes in citrus juice processing In Enzymes in Fruit and Vegetable Processing: Chemistry and Engineering Applications (Alev Bayindirdli) eds. Pp 197-214 CRC Press
Ehiobu JM, Idamokoro ME, Afolayan AJ (2021) Phytochemical content and antioxidant potential of leaf extracts of *Citrus Limon* (L) Osbeck collected in the Eastern Cape Province. South Africa South Afric J Bot 141:480–486

El-Adawy TA, Rahma EH, El-Bedawy AA, Gafar AM (1999) Properties of some citrus seeds. Part 3. Evaluation as a new source of protein and oil. Food Nahrung 43:385–391
Ersus S, Cam M (2007) Determination of organic acids, total phenolic content, and antioxidant capacity of sour *Citrus aurantium* fruits. Chem Nat Compd 43:607–609
Escudero-López B, Cerillo I, Herrero-Martin G, Hornero-Mendez D, Gil-Izquierdo A, Medina S, Ferreres F, Berná G, Martin F, Fernández-Pachón M-S (2013) Fermented orange juice: source of higher carotenoid and flavanone content. J Agric Food Chem 61:8773–8782
Fattouch S, Caboni P, Coroneo V, Tuberoso CIG, Angioni A, Dessi S, Marzouki, Cabras P (2007) Antimicrobial activity of Tunisian Quince. (Cydonia oblonga miller) Pulp and peel polyphenolic extracts. J Agric Food Chem 55:963–969
Ferguson U (1990) Citrus fruits processing. Horticultural Sciences Department, Florida, pp 117–118
Galati EM, Monforte MT, Kirjavainen S, Forestieri AM, Trovato A (1994) Biological effects of hesperidin, a Citrus flavonoid (note I): Antiinflammatory and analgesic activity. Farmaco 40:709–712
Garcia Beltran JM, Espinosa C, Guardiola FA, Esteban MA (2017) Dietary dehydrated lemon peel improves the immune but not the antioxidant status of gilthead seabream (*Sparus aurata*). Fish Shellfish Immunol 64:426–436
Ghasemzadeh A, Ghasemzadeh N (2011) Flavonoids and phenolic acids: role and biochemical activity in plants and human. J Med Plant Res 5:6697–6703
Gmitter FG, Hu X (1990) The possible role of Yunnan, China, in the origin of contemporary citrus species (Rutaceae). Econ Bot 44:267–277
Guadagni OG, Bomben JL, Mannheim CH (1970) Effect of temperature on stability of orange aroma solution. J Food Sci 35:279–281
Gupta AK, Mishra P, Senapati M, Sahu PP (2021) A novel electrochemical device for naringin quantification and removal from bitter variety of *Citrus* fruits. J Food Eng 306:110637
Hayat K, Hussain S, Abbas S, Farooq U, Ding B, Xia S (2009) Optimized microwave-assisted extraction of phenolic acids from citrus mandarin peels and evaluation of antioxidant activity in vitro. Separ Purif Technol 70:63–70
Hosseini SS, Khodaiyan F, Yarmand MS (2016) Optimization of microwave assisted extraction of pectin from sour orange peel and its physicochemical properties. Carbohydr Polym 140:59–65
Igual M, Contreras C, Martínez-Navarrete N (2010) Non-conventional techniques to obtain grapefruit jam. Innov Food Sci Emerg Technol 11:335–341
Inyang UE, Abah UJ (1997) Chemical composition and organoleptic evaluation of juice from steamed cashew apple blended with orange juice. Plant Foods Human Nutri 50:295–300
Igual M, García-Martínez E, Camacho MD, Martínez, Navarrete N (2013) Physicochemical and sensorial properties of grapefruit jams as affected by processing. Food Bioprocess Technol 6:177–185
Ikechukwu GA, Okonkwo CA (2014) "Improved mechanized fruit juice extracting technology for sustainable economic development in Nigeria". Proceedings of the world congress on engineering and computer science. Volume II WCECS San Francisco, USA
Islam MA, Ahmad I, Ahmed S, Sarker A (2014) Biochemical composition and shelf life study of mixed fruit juice from orange & pineapple. J Environ Sci Nat Res 7:227–232
Jeong SM, Kim SY, Kim DR, Jo SC, Nam KC, Ahn DU (2004) Effect of heat treatment on the antioxidant activity of extracts from citrus peels. J Agric Food Chem 52:3389–3393
Jung UJ, Lee MK, Jeong KS, Chol MS (2004) The hypoglycemic effects of hesperidin and naringin are partly mediated by hepaticglucose regulating enzymes in C57 BL/KsJ-db/db mice. J Nutr 134:2499–2503
Kandaswami C, Perkins E, Soloniuk DS, Drzewiecki G, Middleton E (1991) Antiproliferative effects of citrus flavonoids on a human squamous cell carcinoma in *vitro*. Cancer Lett 56:147–152
Kanner J, Harel S, Kanner J, Fishbein Y, Shalom P (1981) Furfural accumulation in stored orange juice concentrates. J Agric Food Chem 29:948–949

Kausar T, Shamim F, Iftikahr F (2020) Preparation and quality evaluation of ready to serve beverage (RTS) from orange juice and Aloe vera gel during storage. Pure App Biol 9:219–228

Kashiwagi T, Sawamura M (2010) Eco-conscious system of oil extraction from citrus peel wastes. In: Sawamura M (ed) Citrus essential oils flavor and fragrance. Wiley, New York

Kennedy JF, Rivera ZS, Lloyd LL, Jumel K (1992) L-ascorbic acid stability in aseptically processed orange juice in TetraBrik cartons and the effect of oxygen. Food Chem 45:327–331

Kore VT, Chakraborty I (2015) Efficacy of various techniques on biochemical characteristics and bitterness of pummelo juice. J Food Sci Technol 52:6073–6077

Kumar N, Goel N (2019) Phenolic acids: natural versatile molecules with promising therapeutic applications. Biotechnol Reports 24:e00370

Kuo S (1996) Ant proliferative potency of structurally distinct dietary flavonoids on human colon cancer cells. Cancer Lett 110:41–48

Kurowska EM, Spence JD, Jordan J, Wetmore S, Freeman DJ (2000) HDL cholesterol raising effect of dietary orange juice in subjects with hypercholesterolemia. Am J Clin Nutr 72:1095–1100

La Cava ELM, Sgroppo SC (2015) Evolution during refrigerated storage of bioactive compounds and quality characteristics of grapefruit [Citrus paradisi (Macf)] juice treated with UV-C light. LWT Food Sci Technol 63:1325–1333

Larrauri JA, Rupérez P, Borroto B, Saura-Calixto F (1997) Seasonal changes in the composition and properties of a high dietary fibre powder from grapefruit peel. J Sci Food Agric 74:308–312

Li S, Lo CY, Ho CT (2006) Hydroxylated polymethoxy flavones and methyl flavonoids in sweet orange Citrus sinensis flavedo. J Agric Food Chem 54:4176–4185

Lim SF and Matu SU (2015) Utilization of agro-wastes to produce biofertilizer. Int J Energy Environ Eng 6:31–35

Liu K, Xu Y, Wang X (2012a) Microencapsulation of sweet orange oil terpeneless using the orifice method. J Food Eng 110:390–394

Liu Y, Heying E, Tanumihardjo SA (2012b) History, global distribution, and nutritional importance of citrus fruits. Compr Rev Food Sci Food Saf 11:530–545

Lo Curto RB, Tripodo MM, Leuzzi U, Giuffre D, Vaccarino C (1992) Flavonoids recovery and SCP production from orange peel. BioresourTechnol 42:83–87

Loizzo MR, Tundis R, Bonesi M, Menichini F, De Luca D (2012) Evaluation of Citrus aurantifolia peel and leaves extracts for their chemical composition, antioxidant and anti-cholinesterase activities. J Sci Food Agric 92:2960–2967

Lundberg B, Pan X, White A, Chau H, Hotchkiss A (2014) Rheology and composition of citrus fiber. J Food Eng 125:97–104

Mamma D, Christakopoulos P (2014) Biotransformation of citrus by-products into value added products. Waste Biomass Valorization 5:529–549

Mannheim CH, Havkin M (1981) Shelf-life of aseptically bottled orange juice. J Food Proc Pres 5:1–6

Manthey JA, Grohmann K (2001) Phenols in citrus peel byproducts. concentrations of hydroxycinnamates and polymethoxylated flavones in citrus peel molasses. J Agric Food Chem 49:3268–3273. https://doi.org/10.1021/jf010011r

Marın FR, Martınez M, Uribesalgo T, Castillo S, Frutos MJ (2002) Changes in nutraceutical composition of lemon juices according to different industrial extraction systems. Food Chem 78:319–324

Marín FR, Soler-Rivas C, Benavente-García O, Castillo J, Pérez-Alvarez JA (2007) By-products from different citrus processes as a source of customized functional fibres Food Chem 100:736–741

Martí N, Mena P, Cánovas JA, Micol V, Saura D (2009) Vitamin C and the role of citrus juices as functional food. Nat Prod Commun 4(5):677–700

Matsuda H, Yano M, Kubo M, Iinuma M, Oyama M et al (1991) Pharmacological study on citrus fruits II anti-allergic effects of fruits of Citrus unshiu Markovicj (2) on flavonoids components. Yakugaku Zasshi 111:193–198

Matsumoto H, Ikoma Y, Kato M, Kuniga T, Nakajima N, Yoshida T (2007) Quantification of carotenoids in citrus fruit by LC-MS and comparison of patterns of seasonal changes for carotenoids among citrus varieties. J Agric Food Chem 55:2356–2368

Miltz J, Raz I, Passy N (1995) Shelf-life of grapefruit segments in multilayer plastic trays. LWT Food Sci Technol 28:442–447

Morton JF (1987) Mandarin orange; in: fruits of warm climates. New Crop Resource Online Program Center for New Crops and Plant Products, Purdue University, pp 142–145

Nagpal R, Kumar A, Kumar M (2012) Fortification and fermentation of fruit juices with probiotic lactobacilli. Ann Microbiol 62:1573–1578

Nagy S (1980) Vitamin C contents of citrus fruit and their products: a review. J Agric Food Chem 28:8–18

Nagy S, Lee H, Rouseff RL, Lint JCC (1990) Nonenzymic browning of commercially canned and bottled grapefruit juice. J Agric Food Chem 38:343–346

Ordu TB (2021) Narenciye report https://www.ordutborgtr/wp-content/uploads/2021/01/Narenciye-Raporupdf

Pagan J, Ibarz A, Llorca M, Pagán A, Barbosa-Cánovas GV (2001) Extraction and characterization of pectin from stored peach pomace. Food Res Int 34:605–612

Peter A, Mihali-Cozmuta L, Mihali-Cozmuta A, Nicula C, Indrea E, Barbu-Tudoran L (2015) Testing the preservation activity of *Ag-TiO2-Fe* and *TiO2* composites included in the polyethylene during orange juice storage. J Food Proc Eng 37:596–608

Perez-Cacho PR, Rouseff R (2008) Processing and storage effects on orange juice aroma: a review. J Agric Food Chem 56(21):9785–9796

Purewal SS, Sandhu KS (2021) Debittering of citrus juice by different processing methods: a novel approach for food industry and agro-industrial sector. Sci Hortic 276:109750

Rapisarda P, Bellomo SE, Intrigliolo F (2001) Anthocyanins in blood oranges: composition and biological activity. In: Pandalai SG (ed) Recent research developments in agricultural and food chemistry, vol 5. Research Signpost, Trivandrum, pp 217–230

Rapisarda P, Fabroni SP, Russo S, Mock HP (2009) Juice of new citrus hybrids (*Citrus clementine* Hort ex Tan *C sinensis* L Osbeck) as a source of natural antioxidants. Food Chem 117:212–218

Reazai M, Mohammadpourfard I, Nazmara S, Jahanbakhsh M, Shiri L (2014) Physicochemical characteristics of citrus seed oils from Kerman, Iran. J Lipids 2014:1–3

Robbins RC (1974) Action of flavonoids on blood cells: trimodal action of flavonoids elucidates their inconsistent physiologic effects. Intern J Vitamin Nutr Res 44:203–216

Rubio-Arraez S, Capella JV, Castello ML, Ortola MD (2016a) Physicochemical characteristics of citrus jelly with non cariogenic and functional sweeteners. J Food Sci Technol 53:3642–3650

Rubio-Arraez S, Capella JV, Ortola MD, Castello ML (2016b) Effect of replacing sucrose with tagatose and isomaltulose in Mandarin orange marmalade on rheology, colour, antioxidant activity, and sensory properties. Acta Aliment 45:406–415

Samman S, Wall PML, Cook NC (1996) Flavonoids and coronary heart disease: dietary perspectives. In: Manthey JA, Buslig BS (eds) Flavonoids in the living system. Plenum Press, New York, pp 469–481

Sandler GD, Wicker PL (1992) Microbial, Enzymatic, and Chemical Changes During Storage of Fresh and Processed Orange Juice. J Food Sci 57(5):1187–1197

Schmidt E (2010) Production of essential oils. In: Baser KHC, Buchbauer G (eds) Handbook of essential oils. CRC Press, Boca Raton

Selli S (2007) Volatile constituents of orange wine obtained from moro oranges (Citrus sinensis [L.] osbeck). J Food Qual 30:330–341

Shen W, Xu Y, Lu YH (2012) Inhibitory effects of citrus flavonoids on starch digestion and antihyperglycemic effects in HepG2 cells. J Agric Food Chem 60:9609–9619

Sicari V, Pellicanò TM, Giuffrè AM, Zappia C, Capocasale M, Poiana M (2018) Physical chemical properties and antioxidant capacities of grapefruit juice (*Citrus paradisi*) extracted from two different varieties. Inter Food Res J 25:978–984

Sidana J, Saini V, Dahiya S, Nain P, Bala S (2013) A review on citrus – "the boon of nature". Int J Pharm Sci Rev Res 18:20–27

Soares NFF, Hotchkiss JH (1999) Comparative effects of de-aeration and package permeability on ascorbic acid loss in refrigerated orange juice. Packag Technol Sci 12:111–118

Spigno G, Faveri DM (2009) Microwave-assisted extraction of tea phenols: a phenomenological study. J Food Eng 93:210–217

Strano MC, Timpanaron N, Allegra M, Foti P, Pangallo S, Romeo FV (2021) Effect of ozonated water combined with sodium bicarbonate on microbial load and shelf life of cold stored clementine (citrus clementina Hort ex Tan). Sci Hortic 276:109775

Tackenberg MW, Marmann A, Thommes M, Schuchmann HP, Kleinebudde P (2014) Orange terpenes, carvacrol and a-tocopherol encapsulated in maltodextrin and sucrose matrices via batch mixing. J Food Eng 135:44–52

Ting SV, Roussef RL (1986) Citrus fruits and their products: analysis and technology. Marcel Dekker, New York

Tongnuanchan P, Benjakul S (2014) Essential oils: extraction, bioactivities, and their uses for food preservation. J Food Sci 79:R1231–R1249

Tripodo MM, Lanuzza F, Micali G, Coppolino R, Nucita F (2004) Citrus waste recovery: a new environmentally friendly procedure to obtain animal feed. Bioresour Technol 91:111–115

Turek C, Stintzing FC (2013) Stability of essential oils: a review. Comp Rev Food Sci Food Saf 12:1541–1546

Uysal Pala C, KırcaToklucu A (2013) Microbial, physicochemical and sensory properties of UV-C processed orange juice and its microbial stability during refrigerated storage. LWT Food Sci Technol 50:426–431

Vergara TC (2013) Insoluble dietary citrus fibers. Food Market Technol 2013:14–18

Vishal SB, Shreran SK (2009) Standardization of extraction methods and preservation techniques of hill lemon juice. J Sci Indus Res 68:608–610

Viuda-Martos M, Ruiz-Navajas Y, Fernández-López J, Pérez-Álvarez J (2008) Antifungal activity of lemon (Citrus lemon L), mandarin (Citrus reticulate L), grapefruit (Citrus paradisi L) and orange (Citrus sinensis L) essential oils. Food Control 19:1130–1138

Wagner CJ, Shaw PE (1978) Sensory evaluations on tangerine-grapefruit juice blends. J Food Sci 43:267–269

Waleed FA (2019) Nutritional benefit of citrus fruits. Amer J Biomed Sci and Res 3(4):303–306

Wang L, Wang J, Fang L, Zheng Z, Zhi D, Wang S, Li S, Ho CT, Zhao H (2014) Anticancer activities of citrus peel polymethoxyflavones related to angiogenesis and others. BioMed Res Inter 2014:1–10

Wang Y, Ji S, Zang N, Cao J, Li J, Sun C (2019a) Identification of phenolic compounds from a unique *Citrus* species, finger lime (*Citrus australasica*) and their inhibition of LPS-induced NO-releasing in BV-2 cell line. Food Chem Toxicol 129:54–63

Wang R, Hassan W, Ahmad FD, Jabeen Q, Ahmed H, Iqbal O (2019b) *Citrus aurantium* ameliorates cisplatin-induced nephrotoxicity. BioMed Res Inter 3960908. https://doi.org/10.1155/2019/3960908

WFLO (2008) Commodity storage manual—fruit juices, Citrus juices. World Food Logistics Org, Farmville

Xu G, Ye X, Liu D, Ma Y, Chen J (2008) Composition and distribution of phenolic acids in Ponkan (Citrus poonensis Hort. ex Tanaka) and Huyou (citrus paradise Macf. Changshanhuyou) during maturity. J Food Comp Anal 21:382–389

Xu X, Bao Y, Wu B, Lao F, Hu X, Wu J (2019) Chemical analysis and flavor properties of blended orange, carrot, apple and Chinese jujube juice fermented by selenium-enriched probiotics. Food Chem 289:250–258

Yu Y, Xiao G, Xu Y, Wu J, Fu M, Wen J (2015) Slight fermentation with *Lactobacillus fermentium* improves the taste (sugar: acid ratio) of citrus (Citrus reticulata cv. Chachiensis). Juice J Food Sci 80:M2543–M2547

Yuasa M, Shimada A, Matsuzaki A, Eguchi A, Tominaga M (2021) Chemical composition and sensory properties of fermented citrus juice using probiotic lactic acid bacteria. Food Biosci 39:100810

Zannini D, Dal Poggetto G, Malinconico M, Santagata G, Immirzi B (2021) Citrus pomace biomass as a source of pectin and lignocellulose fibers: From waste to upgraded biocomposites for mulching applications. Polymers 13:1280

Zhao Z, He S, Hu Y, Yang Y, Jiao B, Fang Q (2017) Fruit flavonoid variation between and within four cultivated citrus species evaluated by UPLC-PDA system. Sci Hortic 224:93–101

Ziegler LW, Wolfe HS (1961) Citrus growing in Florida. University Press of Florida, Gainesville

Chapter 13
Citrus Waste: A Treasure of Promised Phytochemicals and Its Nutritional-Nutraceutical Value in Health Promotion and Industrial Applications

Chidanandamurthy Thippeswamy Swamy, Kurra Sivadurga, M. Prathapa Reddy, Gokul Marimuthu, Chakra Siddappa Prashantkumar, Chelladurai Premkumar, and Sukhvinder Singh Purewal

13.1 Introduction

The improvement in life quality over the past few decades has resulted in a constant rise in the global population, resulting in excessive resource consumption and, as a result, a sizeable creation of garbage. Anything that is underutilised or not fully utilised is referred to as "waste". Therefore, excessive waste products may be caused by ineffective waste management. Due to the high expenses associated with this procedure and the fact that garbage is either dumped in landfills or burned to create energy, it may also harm the environment and the health of living things. Therefore, a recycling/reuse approach is required to decrease generated waste volume and

C. T. Swamy (✉)
Department of Microbiology, Captain Srinivasa Murthy Central Ayurveda Research Institute, Central Council for Research in Ayurvedic Sciences, Arignar Anna Government Hospital Campus, Arumbakkam, Chennai, Tamil Nadu, India

K. Sivadurga · M. P. Reddy · G. Marimuthu
Department of Chemistry, Captain Srinivasa Murthy Central Ayurveda Research Institute, Central Council for Research in Ayurvedic Sciences, Arignar Anna Government Hospital Campus, Arumbakkam, Chennai, Tamil Nadu, India

C. S. Prashantkumar
Department of Studies in Microbiology, Davangere University, Davangere, Karnataka, India

C. Premkumar
Department of Pharmacognosy, Captain Srinivasa Murthy Central Ayurveda Research Institute, Central Council for Research in Ayurvedic Sciences, Arignar Anna Government Hospital Campus, Arumbakkam, Chennai, Tamil Nadu, India

S. Singh Purewal
University Centre for Research & Development (UCRD), Chandigarh University, Gharuan, Mohali, Punjab, India

S. Singh Purewal et al. (eds.), *Recent Advances in Citrus Fruits*,
https://doi.org/10.1007/978-3-031-37534-7_13

associated socio-economic costs. This might reduce the overuse of raw materials and enable the trash's potential economic and biological worth to be used.

Fruit and vegetable-rich diets are highly recommended for good health due to various phytoconstituents, vitamins, minerals, electrolytes, fibres, and antioxidants (Slavin and Lloyd 2012). Citrus culture or citriculture has been observed for more than 4000 years, and it has been assumed that it is originated from South East Asia. Various kinds of citrus fruits and their varieties are reported worldwide (El-Otmani et al. 2011). *Citrus, Clymania, Microcitrus, Poncirus,Fortunella, and Eremocitrus* are the six genera in the Rutaceae family. Among six genera, only three genera, such as *Citrus, Poncirus,* and *Fortunella,* produce fruits of commercial importance (Ramana et al. 1981).

Citrus is a genus exhibiting characteristic features of evergreen aromatic shrubs and small trees, mostly branched, distributed throughout the tropical and temperate regions. Among the citrus fruits, mandarin, sour oranges, and sweet oranges are the most important as fresh fruits, and they contribute to roughly 80% of the world's citrus fruit production (Anonymous 1992). According to a recent Food and Agriculture Organization (FAO) report, around 143755.6 thousand tonnes or ~143 million tonnes of citrus fruits are produced worldwide. Further, Oranges, Tangerines, Lemons and Limes, and Grapefruits were reported to produce 76292.6, 37429.3, 20529.5, and 9504.1 thousand tonnes, respectively in the year 2019. The zone-wise production of citrus fruits is illustrated in Fig. 13.1 (FAO 2021). The demand for citrus fruits is increasing massively due to the rising global population (Sharma et al. 2017). The different major citrus fruit's world region-wise production is illustrated in Fig. 13.2 (FAO 2021).

Citrus fruits are a rich source of phytochemicals that play a vital role in health promotion. Fruit processing produces significant volumes of waste that are either used as fodder for animals or dumped, which increases the environmental burden

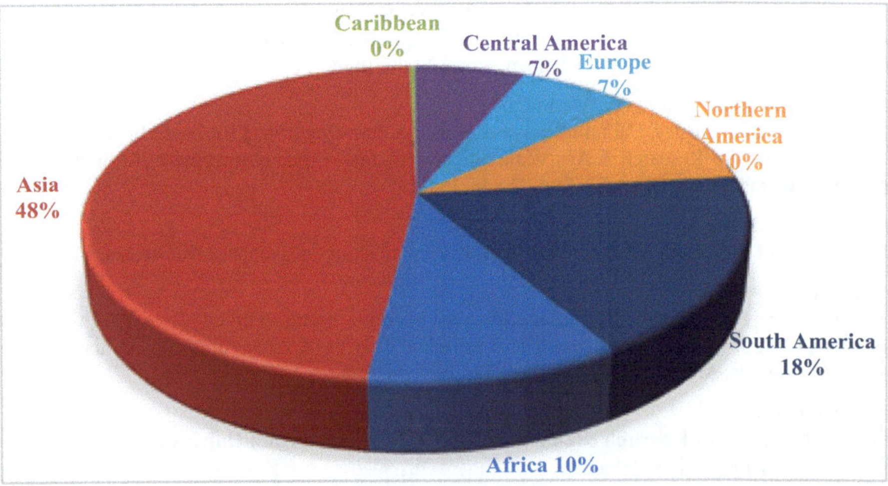

Fig. 13.1 Total production of citrus fruits worldwide

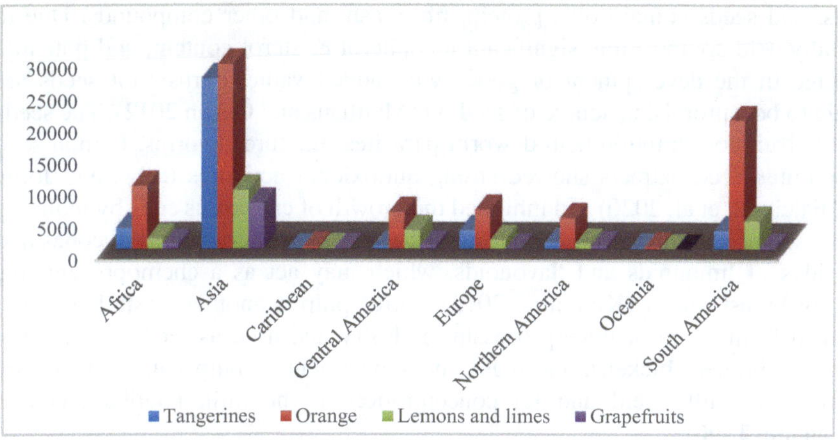

Fig. 13.2 Major citrus fruit types area-wise production

(Suri et al. 2022). Various food and non-food industries use citrus waste, such as peel, seeds, pome, pulp, etc., due to its valuable bioactive compounds. Citrus waste contains naturally occurring bioactive substances that can be used as food additives, encapsulating materials, nanoparticle synthesis, prebiotics, pectin sources, essential oils extraction, polyphenols, carotenes, or dietary fibre. Additionally, it can be utilised as a natural component in packaging, medications, cosmetics, and synthetic fuels. Whatever the case, citrus waste is regarded as an ecological danger due massive pile-up of untreated waste, producing an unpleasant odour by colonising microorganisms, including pathogens. Recently, some tactics to lessen its adverse effects have been developed in light of this concern (Suri et al. 2022).

The citrus fruits extraction or processing industry utilises the fruits to develop value-added products like juice. Every year, the citrus industries process citrus fruits and produce enormous amounts of trash. Among these, citrus peel alone was found to be half of the wet fruit mass. With an annual yield of 110–124 million tonnes, citrus is the world's most widely produced fruit crop. Food processing companies often discard 45–55% of the whole fruit after processing as waste (Mahato et al. 2021). The major citrus wastes such as peel, seeds, and pulp are produced in large amounts by the food processing industries as byproducts. Citrus peel consists of pectin, essential oils, flavonoids, carotenoids, vitamin C, and phenolic acids. Polyphenols and other bioactive compounds of citrus waste are well-known antioxidant substances and helpful for developing natural antioxidant formulations. Phenolic compounds of citrus waste are used in various fields like preparation of balanced diet products, cosmetic production, and pharmaceuticals manufacturing (M'hiri et al. 2017).

Citrus peel is also rich in pectin content; therefore, it is a potential alternative source for the lignocellulose compounds in biofuel production (Jeong et al. 2021; Saini et al. 2022). Another waste product in the bioprocessing of the citrus fruit industry is seeds. Due to specific compounds, seeds are bitter and effective against insects, and seeds contain oils, protein, fibre, ash, and other compounds. Due to their fatty acid composition, significant tocopherol & sterol content, and potential for usage in the development of goods with added value, citrus fruit seeds are thought to be a promising source of seed oil (Matthaus and Özcan 2012). The seeds of citrus fruit consumption help deworm parasites like threadworms. Lemon seed and sprouted seed extracts showed strong antioxidant activities (Silva and Jorge 2019; Falcinelli et al. 2020) and inhibited the growth of cancerous cells by inducing apoptosis. These findings imply that MeOH: water extract contains aglycones and glucosides of limonoids and flavonoids, which may act as a chemopreventative agent for breast cancer (Kim et al. 2012). Citrus pulp is another waste byproduct generated in the food or juice processing industry and used as feed for ruminant organisms, broiler chickens, and pigs. The supplement of pulp with regular feed improves the milk yield and fat concentration in the milk (Belibasakis and Tsirgogianni 1996).

The pulp of citrus fruits is a rich source of macro and micronutrients. Macronutrients such as potassium, sodium, phosphorus, calcium, and magnesium; micronutrients like iron, zinc, copper, manganese, and selenium are more affluent than fruit (Czech et al. 2020). Citrus waste (Peel) is used in the production of biogas, bioethanol, and biodiesel. The raw material pretreatment is done with physical, chemical, and biological methods. Hydrolysis processes such as chemical and enzymatic hydrolysis were commonly used to convert complex polysaccharides to simple sugars, and fermentation of simple sugar will give bio-ethanol (Mahato et al. 2021). A citrus fruit diet supplement in the feed of lactating Ewes at 30% reduces the milk fat and total solids only but not the coagulation property of milk (Jaramillo et al. 2009). *Citrus aurantium L.* flowers are a novel source of milk coagulants used in the cheese industry instead of calf rennet. Extract of citrus flowers enhances the ripening time and flavour of cheese (Nasiri et al. 2020)The citrus fruit processing industry produced a lot of waste and was used as manure. Organic manure used in the study increases the crop yield by up to 400%, as reported by Mediterranean soil (Tuttobene et al. 2009).

In this chapter, we discussed the application of significant citrus waste. Herein, we covered a brief discussion about phytochemical compounds present in citrus waste, followed by extraction methods used to obtain phytochemical compounds from citrus waste. Furthermore, this chapter covers nutritional and nutraceutical aspects, health benefits, asunder applications, and the importance of citrus waste.

13.2 Phytochemicals of Citrus Waste

Phytocompounds are naturally present in citrus juices, grouped into diverse chemicals, and play a role in the human body's physiological functions and metabolic changes. Plants rely on phytochemicals to protect them from environmental stress by producing proactive molecules. A broad range of phytochemicals and their distribution are reported in asunder parts of citrus. A few vital phytochemicals found in citrus waste are shown in Figs. 13.3, 13.4 and 13.5. Citrus waste comprises of diverse phytochemicals such as phenolic compounds, essential oils, pigments, carotenoids, limonoids, etc. Thus, this section briefly discusses various phytochemicals present in citrus waste.

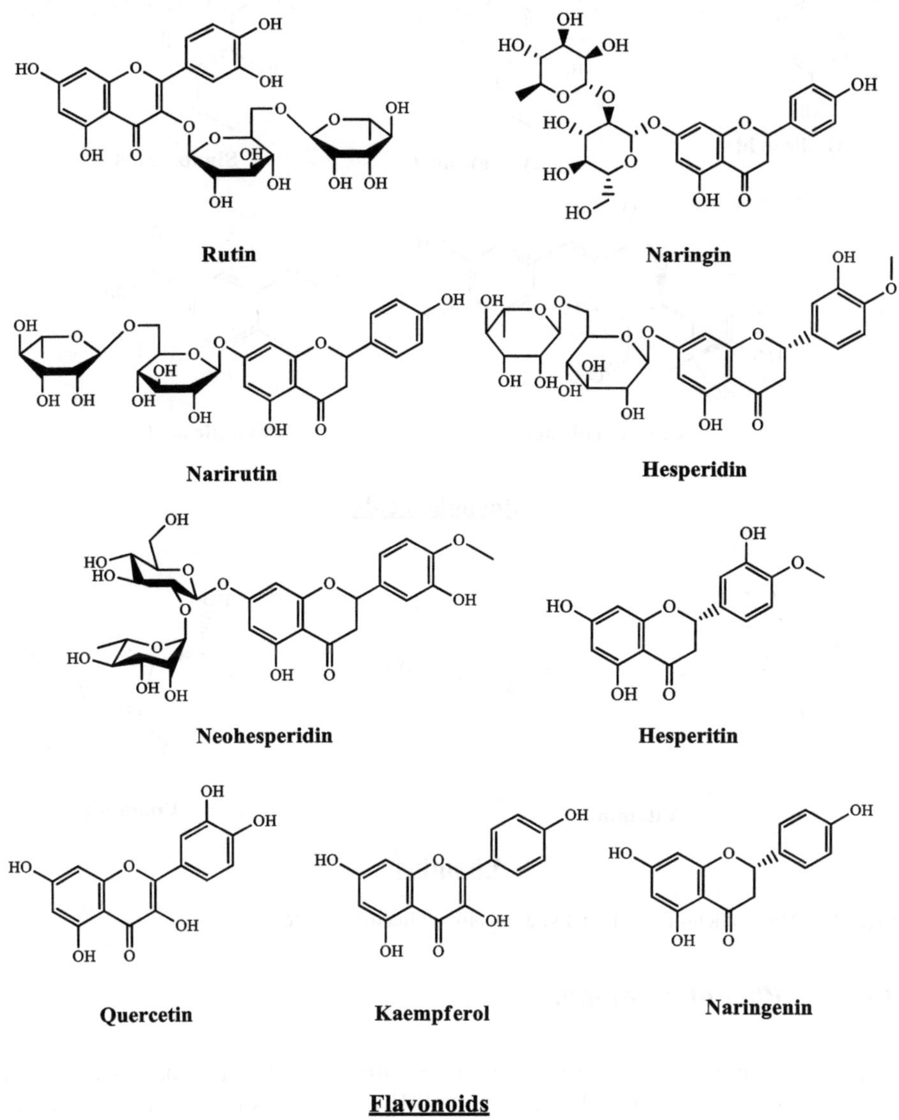

Fig. 13.3 Major flavonoids found in citrus waste

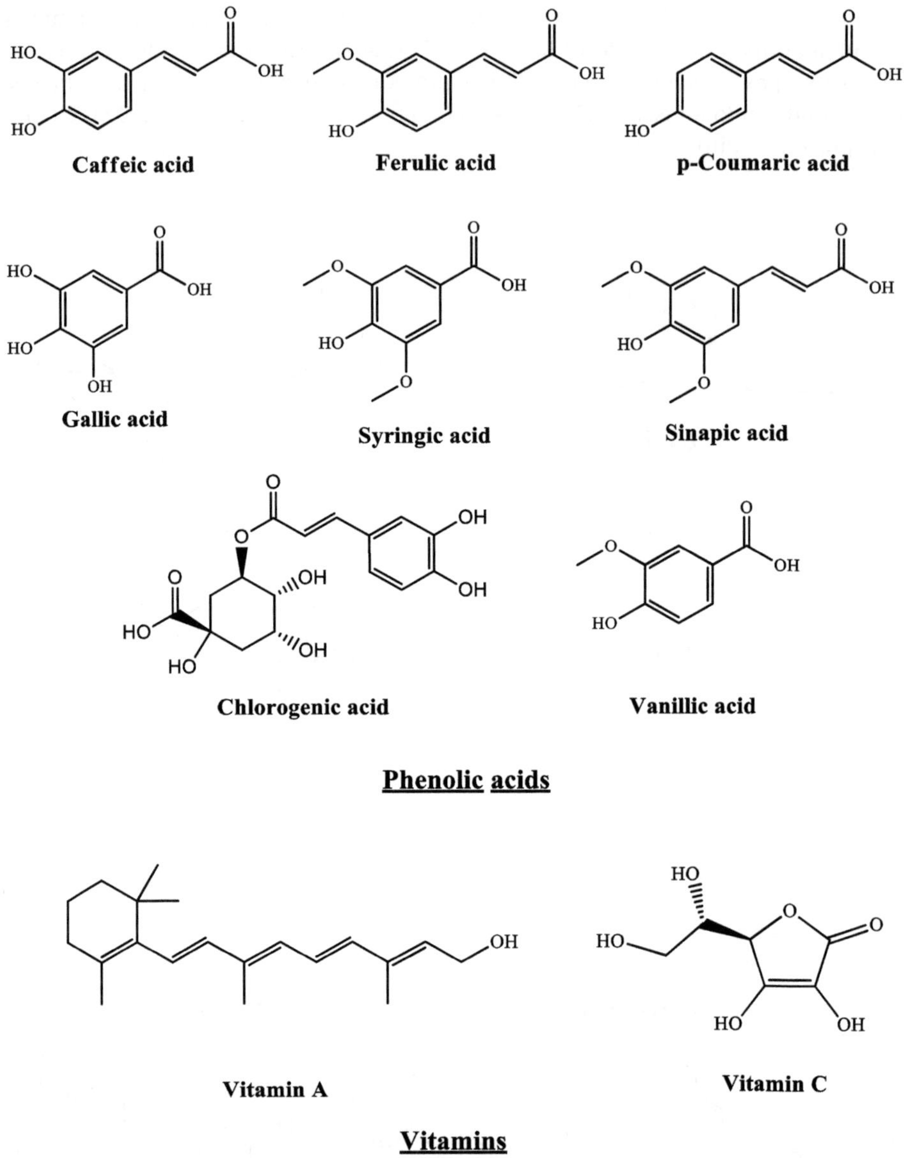

Fig. 13.4 Major phenolic acids and vitamins found in citrus waste

13.2.1 Phenolic Compounds

Phenolic compounds are well reported in the citrus peel than in other citrus waste, such as pulp and seed (Sir Elkhatim et al. 2018). The citrus peel contains phenolic acids such as p-coumaric, sinapic, ferulic, and caffeic acids. Flavanones and

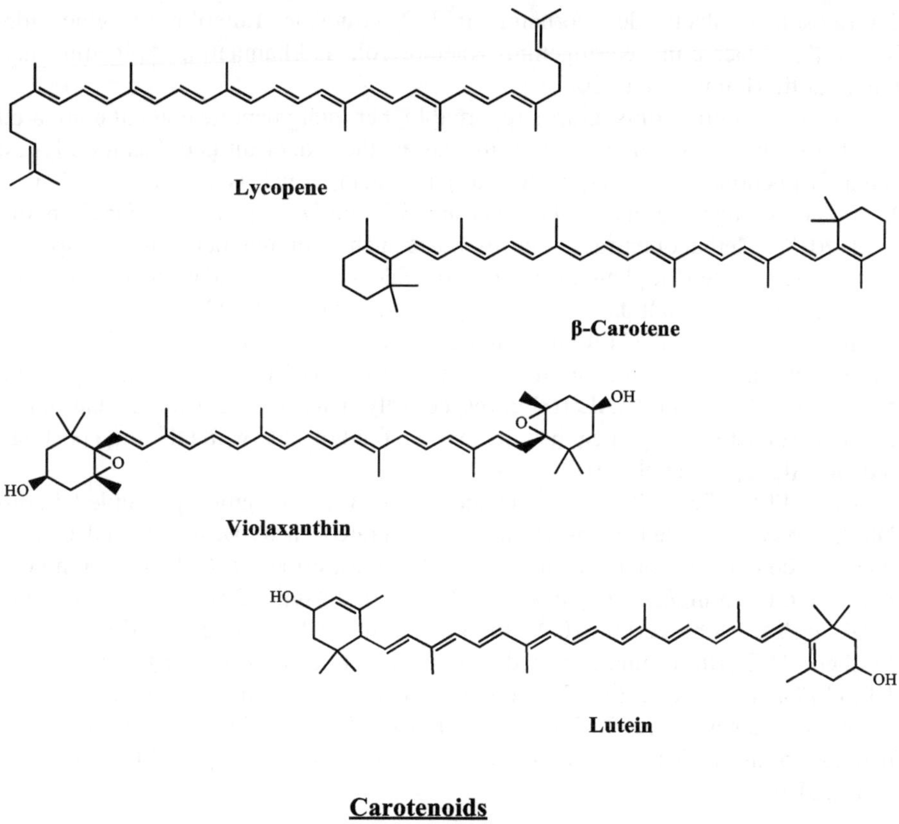

Fig. 13.5 Major carotenoids found in citrus waste

polymethoxylated flavones are another class of phenolic compounds consisting of naringin & hesperidin and nobiletin & tangeretin (Singh et al. 2020). Phenolic compounds (PCs) are chemicals found in all parts of plants, including fruits and vegetables, and are made through the shikimic acid and phenylpropanoid pathways (de la Rosa et al. 2019). These are a type of organic compounds that contains an aromatic hydrocarbon ring with one or more hydroxyl groups attached to it. They can be found in various natural products and are commonly used as dietary supplements or food additives. The most common phenolic compounds include flavonoids, phenolic acids, tannins, stilbenes, and lignans (Ayad and Akkal 2019). Hydroxy benzoate and hydroxy cinnamate derivatives of phenolic acids & flavonoids such as flavonols, flavanones, flavones, isoflavones, flavanols, and anthocyanins are found in citrus waste majorly.

The phenolic compounds are found in different varieties and concentrations in citrus waste. The individual phenolic compounds in citrus waste comprise of Gallic acid, Chlorogenic acid, Syringic acid, p-Coumaric acid, Sinapic acid, Ferulic acid, Ellagic acid, Catechin hydrate, Rutin, Amentofavone, Myrcitrin, Kaempferol

3-rutinoside-7-galactoside, Isorhamnetin-3-O-rutinoside, Luteolin-7-O-glucoside, Myricetin, Quercetin, Naringenin, Kaempferol, Isorhamnetin, Apigenin, and Cirsimaritin (Lamine et al. 2021).

Among the citrus fruits, grapes reported higher total phenolic content compared to other citrus fruits. Among the citrus wastes, the grapefruit peel has the highest reported phenolic content equivalent to gallic acid, which is 77.3 mg/g, whereas lemon and orange peel have 49.8 mg/g and 35.6 mg/g, respectively (Sir Elkhatim et al. 2018). Methanol extract of citrus waste such as orange peel, citrus pulp, and grape marc reported the phenolic content equivalent to tannic acid is 1583.0 ± 154.35, 565.6 ± 106.80, and 4480.5 ± 886.58 mg/100 g, respectively (Castrica et al. 2019). In another study, the TPC (Total Phenolic Content) in control samples of fresh peel extracts of lemon, lime, mandarin, and orange peel was found to be 222.76, 362.98, 530.05, and 284.19 mg GAE/100 g, respectively. This study concludes that out of the four fruits, the orange peel had the least amount of TPC while the lime peel had the most (Casquete et al. 2015).

The UPLC-PDA (Ultra-Performance Liquid Chromatography coupled Photo Diode Array) mediated quantification of macerated citrus waste showed diverse phenolic compounds such as catechin (0.20 mg/g), eriocitrin (5.12 mg/g), p-coumaric acid (0.63 mg/g), sinapic acid (0.71 mg/g), diosmin (2.00 mg/g), hesperidin (2.01 mg/g), neohesperidin (0.27 mg/g), naringin (0.25 mg/g) in (dry weight). Further, UAE (Ultrasound-Assisted Extraction) of same citrus waste analysed in UPLC-PDA record catechin (2.92 mg/g), eriocitrin (20.71 mg/g), p-coumaric acid (2.86 mg/g), sinapic acid (3.67 mg/g), diosmin (18.59 mg/g), hesperidin (7.30 mg/g), neohesperidin (1.63 mg/g), naringin (0.44 mg/g) in dry weight (Medina-Torres et al. 2019).

13.2.1.1 Phenolic Acids

Phenolic acids, such as hydroxycinnamic and hydroxybenzoic acid derivatives, are present in citrus peels. Seven phenolic acids were reported from the citrus waste (huyou peel); these included benzoic acids (p-hydroxybenzoic acid, vanillic acid), cinnamic acids (caffeic acid, p-coumaric acid, ferulic acid, sinapic acid), and chlorogenic acid (a representative of phenolic esters) (Xu et al. 2007).

Chlorogenic acid, p-coumaric acid, and caffeic acid are reported from citrus waste by employing the ultrasound-assisted extraction (UAE) method. In the UAE approach, decreasing the sample particle size increases the recovery of the phenolic acids from citrus waste (Papoutsis et al. 2018). Benzoic acid, Protocatechuic acid, p-coumaric acid, Ferulic acid and Gallic acid are reported in Kumquat peel. Benzoic acid, Ferulic acid, Protocatechuic acid, p-coumaric acid, Chlorogenic acid, Caffeic acid, and Gallic acid were reported in sweet orange. In addition, Benzoic acid, Protocatechuic acid, Chlorogenic acid, & p-coumaric acid were reported in lemon. From Mandarin peel, Benzoic acid, Protocatechuic acid, Chlorogenic acid, Ferulic acid, & p-coumaric acid were reported (Chen et al. 2021).

13.2.1.2 Flavanoids

A class of phenolic compounds known as flavonoids is extensively distributed. It has various biological effects, including inhibiting vital enzymes involved in mitochondrial respiration, defence against coronary heart disease, and anti-inflammatory, anti-cancer, and antibacterial actions (Im et al. 2014). The colourimetric-based estimation procedures were employed to determine the total flavonoid content in the citrus fruits and wastes. The total flavonoid content is rich in the peel of grapefruit (80.8 mg/g) and orange (83.3 mg/g) compared to lemon (59.9 mg/g) (Sir Elkhatim et al. 2018). The main flavonoids in citrus waste reported are hesperidin, narirutin, and rutin. The hesperidin content in citrus waste reported 305.7 mg/100 g as one of the significant flavonoid contents in the citrus peel (Im et al. 2014). The most common flavonol compounds in citrus waste or peels of lemon and sweet orange are rutin, quercetin, and kaempferol, which are in the form of glycosidic in nature (Chen et al. 2021). A flavonol compound, rutin reported from citrus waste and its hydrolysed content (Im et al. 2014). Around 3.3–4.7 mg/g of rutin flavonol was reported from the citrus peel (Gómez-Mejía et al. 2019). There are six flavanones: naringenin, hesperidin, naringin, neohesperidin, narirutin, and hesperetin, and one flavonol (rutin) were reported from the hydrolysed citrus peel extract (Im et al. 2014). Hesperidin is a prominent flavanone found in orange peel, mandarin peel, and kumquat peel, along with hesperidin, naringin, narirutin, and neohesperidin (Chen et al. 2021).

The flavanone glycosides such as narirutin, naringin, hesperidin, and neohesperdin are reported from citrus peel, and the thermal treatment of peel showed the reduced content of these compounds (Xu et al. 2007). The ultrasound-assisted extraction (UAE) method was employed to recover the hesperidin, where the size of peel particles was 2.00 and 1.40 mm, yielding 6.44 and 6.27 mg/g of hesperidin, respectively (Papoutsis et al. 2018). The highest hesperidin content reported from citrus peel is 280–673 mg/g (Gómez-Mejía et al. 2019).

Diosmetin-6, 8-di-C-glucoside (Lucenin-2,4′-methyl ether), Apigenin-6,8-di-C-glucoside (vicenin-2), Eriocitrin, Chrysoeriol-6,8-di-C-glucoside (stellarin-2), Vitexin 2″-xyloside, Diosmetin-7-neohesperidoside, Rhoifolin-4-glucoside, Quercetin-3-O-neohesperidoside, and others are reported from citrus waste by using UPLC-PDA instruments (Sanches et al. 2022). Citrus and mandarin peels reported flavone glycosides or aglycones such as diosmin, rhamnosylvitexin, and accetin. Polymethoxylated flavones such as nobiletin and tangeritin were found in the mandarin peel (Chen et al. 2021).

13.2.2 *Essential Oils (EO)*

Citrus waste is a significant source of essential oils commonly used in the food industry as flavouring agents. EO's are found mostly in oil sacs or glands of fruit peel. EOs are extracted by steam distillation, solvent extraction with hexane or CO_2, and cold pressing (Ángel Siles López et al. 2010). Pomelo (*Citrus maxima*) peel wastes are rich in essential oils and yield 0.58% of EOs obtained through hydro-distillation (Teigiserova et al. 2021). Further, EO reported from various citrus peel wastes such as orange peel, tangerine peel, sweet lemon, and sour lemon contained 1–15% non-volatile and 85–99% volatile components (Patsalou et al. 2020). Five major classes of compounds are present in citrus essential oils, namely Monoterpene hydrocarbons, Oxygenated Monoterpene hydrocarbons, Sesquiterpene hydrocarbons, Oxygenated Sesquiterpene hydrocarbons, and other hydrocarbon constituents (Badalamenti et al. 2022).

13.2.3 Pigments

Natural colouring substances called pigments or biochromes are abundant in citrus wastes, with wide industrial applications such as food additives, colouring agents, and preservatives. The pigments are extracted using conventional (cold press, maceration, solvent extraction, etc.), nonconventional methods (ultrasound-assisted extraction, supercritical fluid extraction), and other methods (Boukroufa et al. 2017). Citrus peels are rich in phytopigments like chlorophyll, carotenoids (Costanzo et al. 2020), Anthocyanins, Hydroxycinnamates (Scordino et al. 2005), and betalains (Sharma et al. 2021). Anthocyanin pigments present in citrus waste include Malvidin, Delphinidin, Cyanidin, Petunidin, Petunidin, andpelargonidin (Fabroni et al. 2016; Sharma et al. 2021). A pigment anthocyanin content reported from citrus waste such as orange peel, citrus pulp, and grape marc are as follows 5.2, 3.4, and 29.6 mg/100 g respectively (Castrica et al. 2019). Narirutin, neoeriocitrin, hesperidin, poncirin, and naringin are the few glycosylated flavones considered as pigments present in citrus plants (Lu et al. 2021). Carotenoids like α and β-carotene, antheraxanthin, lutein, violaxanthin, and β-cryptoxanthin from the citrus peel wastes are the few pigments isolated belonging to this class (Lu et al. 2021).

13.2.4 Vitamins

Citrus waste contains a considerable amount of water-soluble and fat-soluble vitamins. Vitamin A & E (fat-soluble vitamins), Vitamin B complexes & Vitamin C (water-soluble vitamins) (Khan et al. 2021) are found in citrus wastes. Precursors of Vitamin K1 are reported to be found in citrus wastes (Costanzo et al. 2020). Further, in the Vitamin B complex, Vitamin B1 (thiamin), Vitamin B2 (riboflavin), Vitamin B3 (niacin), Vitamin B6 (pyridoxine), and Vitamin B9 (folic acid) are reported to be present in citrus waste (González-Molina et al. 2010). The 110 mg/100 g of vitamin C was reported from the citrus peel (Sir Elkhatim et al. 2018). Vitamin B12 can be produced by the mediation of *Propionibacterium freudenreichii* and

Propionibacterium shermanii via anaerobic digestion of citrus waste (Pérez-Mendoza and García-Hernández 1983).

13.3 Extraction Techniques

Extraction separates medicinally active organic material using specific solvents and established processes. This process aims to isolate the dissolvable vital phytocomponents from the insoluble cellular debris. In the crude extracts, these procedures produce plant metabolites such as terpenoids, alkaloids, phenolics, flavonoids, and glycosides.

13.3.1 Conventional Extraction Techniques

Traditional techniques have been practised for traditional medicine preparations for centuries. With industrialisation, extraction gained solid ground in the isolation and production of phytocomponents. Through research, various extraction alterations and innovations have been introduced to make the process more effective and sustainable. This section of the chapter will focus on applying conventional extraction techniques used in research and industry for separating and identifying phytochemical components from citrus waste. The main conventional techniques include maceration, cold press, soxhlet, and hydro-distillation are widely used to extract the bioactive compounds in citrus waste.

13.3.1.1 Maceration, Percolation, Infusion, and Decoction

Maceration is a technique that has been used for a long time to prepare tonics. It is composed of several steps. First, plant samples are ground into tiny particles. Second, a suitable amount of solvent is poured into a closed vessel. The liquid is discarded in the third step of this process, and a larger yield is achieved by pressing the solid residue (Sagar et al. 2018). Percolation is also based upon a similar principle involving the percolator equipment. The procedure is usually performed at a moderate rate for 2–3 h before evaporation to obtain concentrated extracts.

Moreover, decoction and infusion also share the same principle as maceration. For decoction, boiling of the sample is carried out for a period of time in a specific volume of water, whereas, for infusion, the soaking period is short. Additionally, the decoction is suitable only for thermostable substances and hard plant materials like barks and roots.

These are the most straightforward methods. On the other hand, organic waste becomes challenging because of the increasing volume of solvents, demanding appropriate waste administration. Using different temperatures and solvents may improve the extraction process and reduce time.

13.3.1.2 Soxhlet Extraction

The soxhlet extraction method is one of the old conventional types of extraction procedures in this large quantity of solvents required to extract the desired compound and time-consuming process. However, it is still used to compare with the modern mode of extraction techniques. Commonly a known quantity of samples is mixed with solvent and run for hours to extract the compounds. In a study, the 2 g of dried citrus peel was mixed with hexane solution in a ratio of 1:25 (matrix: solvent) at solvent boiling temperature for 04 h (Lopresto et al. 2014; Sagar et al. 2018).

Although this approach used a tiny amount of solvent, it had several drawbacks, including exposure to flammable and hazardous organic solvents and poisonous gas emissions during extraction. Additionally, this approach requires high purity of solvent, which adds to the high cost. Furthermore, a dry and fine coarse solid is the perfect sample to be used for Soxhlet extraction, along with several factors such as solvent-sample ratio, agitation speed, and temperature.

13.3.1.3 Cold Press

The cold press method employs to extract the bioactive components from citrus waste. Here, the best way for isolating heat-labile compounds from substrates is not to heat the substances at higher temperatures. The cold-pressed extract of lemon seeds at the laboratory-scale cold-press machine. The seed moisture content was first adjusted (12%) before cold pressing, and cold pressing was applied at 30 rpm rotation speed, 10 mm exit die at 40 °C temperature of oil exit (Guneser and Yilmaz 2017).

13.3.1.4 Hydro-Distillation

The oldest and easiest method for obtaining essential oils from plants is hydro-distillation. Before dehydrating the plant sample, the traditional method of hydro-distillation is used to extract significant oils and different bioactive components from plant sources. There are three different methods of hydro-distillation: direct steam distillation, water distillation, and water and steam distillation (Sagar et al. 2018). Packing the plant sample in a still compartment is the first stage of hydro-distillation. After that, enough water is added and cooked. The substitute is steam, which is also an option. Steam and hot water are efficient in removing bioactive substances from plant cells. Indirect water cooling is used to condense the vaporised mixture of water and oil. The condensed mixture is sent to a separator from the condenser. Here, the water and bioactive compounds spontaneously separated from the oil. Heat-induced breakdown, hydrolysis, and hydro diffusion are the three significant physicochemical processes involved in hydro-distillation. Heat-labile chemicals may be lost or damaged due to the high extraction temperature, which is bad for them (Sagar et al. 2018).

13.3.2 Nonconventional Extraction Methods

Several process modifications have been reported to improve the performance of simple extraction techniques. The application of such nonconventional methods for the extraction of phytocompounds from citrus waste is discussed in this segment.

13.3.2.1 Ultrasound-Assisted Extraction (UAE)

UAE is widely used to extract phenolic compounds or polar compounds from citrus waste. In the ultrasound-assisted extraction process, place citrus waste in a beaker with ethanol at a 1 gm: 10 ml ratio and immerse it in ice to not exceed 40 °C during the ultrasonic process after different watts of ultrasonic power at the different times employed based on nature of the sample (Montero-Calderon et al. 2019). Another study used a temperature-controlled ultrasound bath at 37 kHz with an ultrasonic power of 50 W to extract bioactive components from citrus waste (Orange peel). Remove 01 g of the solid residue of supercritical CO_2 extraction was collected and mixed with ethanol in different ratios to obtain phenolic compounds (Jokić et al. 2020). This practice is easy and cost-efficient. Ultrasound energy more significant than 20 kHz, on the other hand, may affect active phytochemicals by forming free radicals.

13.3.2.2 Microwave-Assisted Extraction

Microwaves are electromagnetic radiation with a frequency range of 300 MHz to 300 GHz. The heating principle that microwaves use is based on their immediate effects on polar materials. When microwave energy hits these materials, it's transformed into heat through dipole rotation and ionic conduction mechanisms (Sagar et al. 2018). Since the transmission of microwave energy is only conducted via dielectric absorption, the nonpolar solvents conduct poor heating of the samples. This method required less time, a low quantity of solvent, and significant reproducibility and recovery of analytes were observed. Nevertheless, it favours polar molecules, requires a solvent with a high dielectric constant, and is usually beneficial for extracting phenolic compounds.

13.3.2.3 Sequential Ultrasound-Microwave Assisted Acid Extraction (UMAE)

In the UMAE method, 10 grams of sample (dried pomelo powder) was weighed and mixed with 290 mL of distilled water. The pH of the mixture was then adjusted by using citric acid before being placed in an ultra-sonic bath at 40 kHz. The solution was left to sit for varying periods of time before being transferred to a microwave oven. The extraction process is always performed three times to ensure accuracy (Liew et al. 2016).

13.3.2.4 Supercritical Fluid Extraction

Supercritical CO_2 extraction is important for obtaining nonpolar essential oil compounds from citrus waste. For each experiment, citrus waste, such as mandarin peel 100 g, was freeze-dried and used for further experimentation. The peel is milled in a laboratory mill for making a powder and placed into a stainless-steel extractor vessel with a bar, an outer diameter of 100 mm, and a height of 500 mm. The extraction may carry out under different pressure conditions (100 and 300 bar) at a constant temperature of 40 °C and a CO_2 mass flow rate of 2 kg/h controlled through a flowmeter. The extraction process was carried out for 90 min to collect the total extracts (Šafranko et al. 2021).

13.3.2.5 High Pressure – High-Temperature Extraction (HPTE)

The HPTE Instrument reactor (100 mL) is employed to extract bioactive compounds in a study. The reactor has equipped with a mechanical stirrer, high-temperature fabric heating mantle, and sturdy aluminum shell. Hexane was used as the solvent, lemon peels as the matrix, and extractions were performed at different extraction times (30, 105, and 180 min), temperatures (100 °C, 150 °C, and 200 °C), and matrix/solvent ratios (1:4, 1:10, and 1:15 w/w). The study was conducted under a nitrogen atmosphere to avoid limonene oxidation (Lopresto et al. 2014).

13.4 Nutritional and Nutraceutical Aspects of Citrus Waste

The development and survival of living things are significantly influenced by nutrition. Because of the high antioxidant, anti-mutagenic, and anti-cancer activities, nontoxic phytochemicals are crucial for human health. Citrus extracts, for instance, have proved to minimise the chances of diseases including heart disease, cancer, and diabetes. With the help of antioxidants' actions on the body, they stop tissue damage, inflammatory process, and oxidative stress. Citrus fruit byproducts such as squeezed secondary juices, seeds, pulp, leaves, and peel, contain proteins, organic acids, polyphenols, lipids, sugars (namely glucose, sucrose, and fructose), monoterpenes (linalool and limonene), dietary fibres, vitamins, and carotenoids (Mohanty et al. 2015). The kind of cultivar, the technique of cultivation, the time of harvest, and the level of maturity of the fruit can all affect the molecular composition of the byproducts. After the juice is extracted, citrus fruit peel accounts for over half of the fresh fruit mass (Sharma et al. 2017). It is notably rich in dietary fibres, polyphenols, aromatic compounds, pectin, and natural pigments (Rafiq et al. 2018).

Plants produce a wide variety of flavonoids to protect themselves from pathogens or UV radiation (Russo et al. 2021). Phenolic acids can be found in small quantities and are reported to exhibit free radical scavenging action. p-Coumaric acid, sinapic, ferulic, and caffeic acids are the various examples of hydroxycinnamic acids & vanillic, gallic, and syringic acids are examples of hydroxybenzoic acids (Kim and Kim 2016; Kumar and Goel 2019). Juice extraction separates the seeds, which are a valuable source of phenolic substances, proteins, oil, limonoids, and especially the flavonoids eriocitrin and hesperidin (Rosa et al. 2019).

Citrus fruit essential oils can also be found in much lower amounts in the leaves, seeds, and oil sacs of the peels and cuticles. Essential oils are composed of oxygenated derivatives (like esters, aldehydes, alcohols, and ketones), monoterpenes, and sesquiterpenes (whose structure has two to three isoprene units). β-ocimene, sabinene, and β-pinene are the constituents of essential oils extracted from citrus leaves, while the primary component of essential oils produced from citrus byproducts is limonene (Chi et al. 2020). For years, citrus fruit essential oils have been employed as flavourings in food making and medicinal & cosmetic items and recently undergone a new evaluation for their health-promoting qualities (Dosoky and Setzer 2018; Bruni et al. 2019). Exhausted citrus fruit peels possess dietary fibres and pectin, which can also be found in pulp and juice (Dimopoulou et al. 2019). Pectin, gum, and a small amount of cellulose are soluble forms of dietary fibres, whereas hemicellulose, cellulose, and lignin are insoluble forms. Dietaryfibres are a group of non-starch polysaccharides having at least ten carbohydrate molecules; they are challenging to digest and problematic for the gut to absorb (Dimopoulou et al. 2019). Pectin is made up of D-galacturonic acid groups connected by α-1,4glycosidic bonds and is also known as a complex polysaccharide that is partly esterified with methanol or acetic acid. Since it is a naturally gelating substance, it is employed to thicken, emulsify, texturise, and stabilise confectionery, jams, jellies, and biodegradable products. It often appears in complex or insoluble forms, and its colour is between white and light brown.

Carotenoids and flavonoids found in peels are abundant in secondary citrus fruit juices. Carotenoids, the biosynthetic pigments widely present in many vegetables and fruits, are categorised into two groups: Xanthophylls and Carotenes. Lutein and Violaxanthin are examples of oxygenated carotenoids generally referred to as "Xanthophylls", whereas Lycopene and β-Carotene are examples of hydrocarbon carotenoids generally referred to as "Carotenes" (Saini and Keum 2018; Sharma et al. 2021). They are precursors to vitamin A, which improves vision, strengthens the immune system, and encourages the growth of epithelial tissues (Widjaja-Adhi et al. 2018).

Waste products from the processing of citrus waste, such as peels, and seeds, can be effectively employed as a source of phytochemicals that are essential to nutraceuticals. The development and production of functional and nutraceutical products is a developing trend in the food market today. Due to rising customer interest in "healthy" foods, this new category of food items has garnered much demand in the food sector. Therefore, the mutual interest of both the food and pharmaceutical sectors is to obtain novel and naturally derived bioactive compounds that could be employed as nutraceuticals, medicines, and functional food additives or ingredients (Kumar et al. 2017). Citrus waste may be utilised to develop functional foods and nutraceuticals since it includes bioactive components that can be extracted.

13.5 Application of Citrus Waste in Human Health

Numerous studies have demonstrated that consuming foods like fruits and vegetables, which are high in flavonoids and low in fatty acids, can help to lower the prevalence of metabolic illnesses in people (Espín et al. 2007). Citrus fruits are the world's most produced fruit sector;waste is the predominant byproduct of the industries that process citrus (Raimondo et al. 2018). These leftover citrus fruit components, which are typically thrown away as waste, could serve as resources for nutraceuticals. These wastes have the potential to provide essential low-cost nutritional supplements due to their wide availability and low price. Using such bioactive-rich citrus leftovers can offer a productive, affordable, and environmentally sustainable basis for implementing new nutraceuticals or advancing existing ones. Thus, we discussed citrus waste's biological activities and potential advantages in this segment.

13.5.1 Antimicrobial Activity

Hesperetin and naringenin exhibited the most substantial antibacterial effects, as per researchers' study on the antimicrobial effects of flavonoids on *Helicobacter pylori* strain (Moon et al. 2013). Another study indicated that bergamot extract's antibacterial effects are more potent against Gram-negative bacteria (Zhang et al. 2013). A lime leaf extract has shown an *in vitro* antibacterial action against *Bacillus, Streptococcus faecalis*, *Escherichia coli*, *Salmonella* spp., and *Staphylococcus aureus* (Oboh and Abulu 1997),while the development of antimicrobial drugs is aided by the discovery that oil from the orange seed and non-oil extracts exhibit promising antifungal and antibacterial characteristics in addition to antioxidant properties (Oikeh et al. 2020). Other studies have validated that naringin and its analogues' showed antimicrobial efficacy on Gram-positive bacteria (Değirmenci and Erkurt 2020). In China, citrus fruits were discovered to be a powerful antioxidant, tyrosinase inhibitor, and antibacterial agent (Guo et al. 2020).

Notably, the recycled *C. limetta* pulp has antibacterial action toward Gram-negative and Gram-positive bacteria, making it a valuable resource for the economy (Thakur et al. 2019). Lemon leaf essential oils, which are high in citronellal, limonene, and citronellol, have antibacterial action against pathogenic bacteria and possess insecticidal potential also (Asker et al. 2020; Wu et al. 2021). The majority of bacterial pathogens were shown to be susceptible to silver nanoparticles produced from *C. lemon* peel waste (Alkhulaifi et al. 2020). Additionally, it was observed that yellow lemon peel worked well against *Klebsiella pneumonia* (Saleem and Saeed 2020). This indicates that citrus waste is beneficial in developing a successful therapy for some multidrug-resistant microorganisms. More studies and clinical trials are still advised, though.

13.5.2 Neuroprotective Activity

Neurological problems are becoming an increasingly significant and widespread issue. Ischemic brain damage, Huntington's disease, Parkinson's, and Alzheimer's diseases are the most prevalent neurological problems, and it has been shown that the pathogenesis of these disorders is caused by neuroinflammation and oxidative stress (Gaur et al. 2011). In experimental models using both cell-based and cell-free models, orange juice extract was investigated for its antioxidant activity and found to be rich in a flavonoid (Barreca et al. 2014; Ferlazzo et al. 2015). Moreover, the orange essential oil was shown to have promising anti-anxiety characteristics (Mannucci et al. 2018), while orange juice extract exhibited anti-epileptic properties in *in-vivo* studies (Citraro et al. 2016). It is reported that hesperidin has the ability to prevent the formation of Reactive Oxygen Species, increase intracellular free calcium levels, and open mitochondrial permeability transition pores may all be employed in combination to prevent mitochondrial malfunction (Gaur et al. 2011).

The important factor in the antioxidant activities of extract made from *C. sinensis* peel is the phenolic and flavonoid components and other phytochemicals (Liew et al. 2018). A flavonoid, quercetagetin, was already found to be a radical scavenger with the ability to lower ROS levels and protect against H_2O_2-induced DNA damage in Vero cells. This flavonoid was extracted from the methanolic extract of the powdered peel of "Satsuma Mandarin" (Yang et al. 2011). Natal, Perario, Valencia, and Hamlin orange species residual seeds are rich in essential oils, and this byproduct exhibits free radical scavenging activity due to the high availability of carotenoids, tocopherols, phytosterols and phenolic compounds (Jorge et al. 2016). *Citrus aurantium's* major essential oils are limonene (97.83%) and mirsen (1.43%), which are found in around a tenth of these quantities and can fight against central nervous system depression (Guillon and Champ 2000). Bergamot juice extract in one study suppressed both lipopolysaccharide-induced inflammatory responses by controlling the NF-κB system through the AMPK/SIRT1 axis and β-amyloid-induced inflammatory responses by controlling MAPK/AP-1 pathway in THP-1 monocytes (Risitano et al. 2014; Currò et al. 2016; Maugeri et al. 2019). Citrus waste can aid in the treatment of neurological disorders; however, more *in-vitro*, *in-vivo,* and specific research is highly recommended.

13.5.3 Anti-cancer Activity

Uncontrolled cell division and proliferation are the main causes of the large group of disorders known as cancer. The aberrant energy metabolism of these abnormal cells requires a source, and glucose is thought to be their preferred metabolic substrate. Oxidative stress and its effects, such as DNA alterations and the emergence of cancer, can be avoided by consuming foods high in antioxidants that contain various types of antioxidant phytochemicals in addition to phenolic compounds. Hypoglycemic, antimicrobial, anti-cancer, antioxidant, anti-inflammatory, and hypolipidemicfunctions of pomelo peel extracts have recently been mentioned in a summary of the health advantages and therapeutic potential of this fruit (Tocmo et al. 2020). Waste from *C. maxima*, including the carpels, flesh, essential oil, and peels, can treat metabolic diseases, including obesity and hyperlipidemia. Indeed, research involving the AMPK-SREBP-PPARS pathway in Wistar rats indicated its anti-lipogenic capabilities. Additionally, important phytonutrients from pomelo, including γ-terpinene, β-sitosterol,limonene,hesperidin, and p-synephrine decreased the crucial enzyme activityin HepG2 cells, which are essential in the *de novo* formation of cholesterol and triacylglycerides (Lin et al. 2020).

Citrus oils have been shown to exhibit an anti-proliferative impact in murine B16F10 melanoma cells as well as an inhibition action on the activity of the enzyme tyrosinase in mushrooms (Rosa et al. 2019). Waste from *C. limon* has previously been thought of as a potential resource for reuse. The reduction of plasma and hepatic cholesterol in hybrid F1B hamsters by lemon peel waste stream has been proven in this sector (Terpstra et al. 2002). A study showed citrus flavonoids' ability to suppress oral carcinogenesis and their anti-cancer effects in hamsters. Further, naringin and naringenin were found to be active against oral carcinogenesis, whereas hesperetin, neohesperidin, tangeretin, and nobiletin were not found effective in the same study (Aranganathan et al. 2008). Previous studies have shown hesperetin's ability to prevent several cancer types, including breast and colon cancer (Mulder and Ouwerkerk-Mahadevan 1997). Cancers connected to inflammation, namely colon cancer and skin cancer, have been treated utilising citral as a chemopreventive medicine for cancer (Beecher 2009). Additionally, it has been confirmed that dietary fibre and vitamin C have a positive impact on cancer prevention and therapy (Diplock 1993; Platz et al. 1997; Fuchs et al. 1999; Moon et al. 2013).

In research, flavonoids in *Citrus aurantium* L. extract caused human leukaemia cells to undergo apoptosis by inhibiting protein kinase B (Han et al. 2012). In a different investigation, the same fruit extract triggered apoptosis and cell cycle arrest in non-small-cell lung cancer cells (A549) (Han et al. 2012). Two flavonoids, namely hesperetin and naringenin, were attributed to the anti-cancer activities of this fermented extract, according to a more thorough *in-silico* investigation (Kim et al. 2017). Regarding seed debris, the extract from *C. aurantiifolia* Swingle seeds has cytotoxic effects on the L5178Y human lymphoma cell line (Castillo-Herrera et al. 2015). Though clinical trials are going on to evaluate the use of citrus waste for treating cancer, still more research is required for prominent knowledge.

13.5.4 Cardioprotective Effect

A major cause of mortality worldwide is cardiovascular disease. Hesperidin's antioxidant, lipid-lowering, antihypertensive, anti-inflammatory, and insulin-sensitivity-improving activities have been proven to show significant cardiovascular therapeutic potential. Numerous epidemiological research studies have revealed a clear correlation between dietary fibre intake with a lower risk of numerous diseases, particularly cardiovascular diseases (Ellegård and Andersson 2007). By enhancing both bile acids and cholesterol defecation, soluble dietary fibre might lower the absorption quantity of cholesterol (Akiyama et al. 2010). Hesperidin can be used to prevent and cure cardiovascular disease because, in diabetic rats, it can raise both diastolic and systolic blood pressure and can modulate vasomotor function. Hesperidin's hypotensive effects are connected with increased NO bioavailability, endothelial malfunction correction, and NADPH oxidase expression (Yamamoto et al. 2008).

Furthermore, researchers observed that hesperidin protected diabetic rats from experimentally induced myocardial infarction (Kakadiya et al. 2010). Additionally, the orange essential oil was considered to have significant anti-anxiety characteristics (Mannucci et al. 2018) whereas; orange juices shown anti-epileptic (Citraro et al. 2016) and anti-inflammatory (Fusco et al. 2017; Cirmi et al. 2021) characteristics in *in-vivo* models. In a similar manner, a lemon and red-orange extract derived from pulp manufacturing waste showed promising antiallergic efficacy, primarily by lowering basophil activation, degranulation and also reducing the release of inflammation-causing mediators triggered by allergic stimuli (Caruso et al. 2021). In a mice model of dextran sulphate sodium-induced colitis, various waste products of the commercial isolation of orange juice exhibited anti-inflammatory effects (Pacheco et al. 2018). In HepG2 cells, hesperidin and β-sitosterol suppress crucial enzymes involved in the de novo production of triacylglycerides and cholesterol (Lin et al. 2020). Even though there have been many studies on using citrus waste to treat cardiac disorders, more direct investigations are still needed for a clear understanding.

13.5.5 Miscellaneous Biological Activities

Dietary fibre-rich byproducts of the lemon industry have an important role in reducing the growth of variety of digestive diseases, including constipation, haemorrhoids, hypercholesterolemia, and colorectal cancer (Lipkin et al. 1999; Ferguson et al. 2004). Citrate's inherent capacity to suppress urine crystallisation makes it possible to employ citrus fruit acid, citric acid to treat calcium (Pfaltzgraff et al. 2013). The flavonoids melitidin and brutieridin have been proven to be helpful in the treatment of hyperlipidaemia, albedo is a byproduct produced by bergamot industries that could be recycled (Nauman and Johnson 2019). Using mice infected with *Plasmodium berghei* as well as LPS-stimulated macrophage cells, a lime peel extract is said to have the ability to combat malaria even more and this ability attributed to the both anti-inflammatory and antioxidant activities of its flavonoid contents (Mohanty et al. 2015). Food fibres, both soluble and insoluble, have a number of positive benefits on human health. Both soluble and insoluble citrus fruit fibres are good for health and offer advantages including toxin clearance, improved nutrient absorption in the gastrointestinal tract, reduced energy absorption, and support healthy liver and bile duct function (Fouad Abobatta 2019). In one study, inflammation in mice's ear produced by TPA (12-O-Tetradecanoylphorbol-13-acetate) was lowered when lime essential oils were applied topically (Maurya et al. 2018). Leaf fraction of a hybrid species of *C. reticulate* and *C. sinensis* tends to possess gastroprotective as well as anti-ulcerogenic actions against stomach ulcers in rats caused by alcohol because of its anxiolytic, anti-apoptotic, anti-inflammatory and anti-oxidant actions (Hamdan et al. 2020).

According to the research reviewed, citrus waste is an effective antibacterial, anti-cancer, anti-inflammatory, antidiabetic, cardioprotective, and neurological agent. However, much clinical trial and *in-vivo* research are still needed to gain a deeper understanding. Researchers should also prioritise the creation of unique methods for investigating various uses for the compounds obtained from citrus waste.

13.6 Citrus Waste Used in Various Ways

According to the FAO, a total of 1,43,755.6 thousand tons of citrus fruits are produced. In India alone, 13,314.3 thousand tons of citrus fruits are produced (FAO 2021). After using citrus fruits, the waste produced worldwide is around 117.9 million tons. In India, eight million tons of citrus waste is produced, which, when dumped without proper treatment, leads to environmental pollution. Besides that, if citrus waste is appropriately valorised, it can be utilised in different ways to reduce pollution and boost the economy. The residue of citrus fruits such as pulp, peel, and seeds can be used for production of biogas, bioethanol, cosmetics, essential oils, coloring agents, and medical product (Chavan et al. 2018). This section of the chapter contains various other applications of citrus waste.

13.6.1 Biofuel Production

Due to the increase in population and energy demand, the use of fossil fuels causes the emission of a high level of carbon which leads to drastic changes in climate. Therefore, there is a severe need to replace fossil fuels with biofuels which are much more eco-friendly with fewer emissions. Worldwide citrus waste can be refined to obtain biofuels like bioethanol, biodiesel, and biogas. A massive amount of simple sugars and complex polysaccharides with less lignin in citrus waste peel (CPW) makes it for biofuel production using the anaerobic digestion and fermentation technologies (Ángel Siles López et al. 2010; Mahato et al. 2021).

13.6.1.1 Bioethanol Production

Bioethanol is a promising alternative to petrol, or it can be used by mixing with petrol, which reduces the carbon emission due to its high-octane number. It can be easily produced by fermenting sugar biomass using citrus peel as a fermenting substrate (İçöz et al. 2009). However, D-limonene in the peel, about 0.8–1.6%, acts as an inhibitor of yeast fermentation. Thus, pretreatment is required to reduce the D-limonene content to 0.21–0.01% (John et al. 2017). Pre-treated CPW using a steam explosion process (CPW was subjected to steam in a continuous tube reactor for 2–3 min at 150–160 °C) leads to a reduction of D-Limonene up to 90% (Wilkins et al. 2007).

Similarly, another study found a pretreatment technique called biomass-popping pretreatment. The popping reactor was modified, and the reaction time was increased to 10 min, with a pressure of 15 kgf/cm at 150 °C. This study reduced the D-limonene concentrations to 0.01%, and using *Saccharomyces cerevisiae, fermentation* was accomplished, yielding an ethanol concentration of about 46.2 g/L from raw peel concentration of 39.8 g/L (Choi et al. 2013). Further, another study reports that in comparison *Kluyveromyces marxianus*, and *Saccharomyces cerevisiae, Pichia kudriavzevii* showed the highest ethanol production (Patsalou et al. 2019).

13.6.1.2 Biodiesel Production

Biodiesel is the long chain fatty acid of monoalkyl ester, which can be produced by transesterification of methanol catalysed by acid, alkali, or enzymatic treatment with vegetable oil, seed oil, animal oil, etc. Recently oils made from citrus byproducts, primarily seeds, are often trans-esterified with alcohol to create biodiesel, a nontoxic biofuel. This eco-friendly biorefinery product has gained significant attention. Due to its high biodegradability, it can be used directly in an existing engine or blended with Diesel and other fuels. When biodiesel is produced, the bottom layer of the reactor contains crude glycerol with a purity of 55%, which can be used for making soap and other organic products (Taghizadeh-Alisaraei et al. 2017; Kumar et al. 2019). The following study reported biodiesel production from *Citrus sinensis,* in which the seeds were dried, dehulled, and the oil was extracted using an oil expeller. The extracted oil was mixed with methanol in various ratios. They found that methanol to oil ratio (6:1) is optimal with 1.0 wt.% of potassium hydroxide to give the maximum yield at a temperature of 60 °C. They compared the Fatty acid methyl esters (FAME) of produced biodiesel with the European Committee for Standardization (EN 14214), which satisfied the fluidity nature of fuel (Dhanasekaran et al. 2016). Biodiesel production from the Mandarin (*Citrus reticulata*) was reported, which has around 24–34 seeds per fruit with rich oil content and non-edible seeds, which were dried. Oil was extracted and mixed with 98% n-Hexane solution. The biodiesel produced 96.82% conversion productivity from oil to bio-diesel, and the biodiesel showed 84.48% energy efficiency from total biomass. They matched the physicochemical parameter concerning American Society for Testing and Materials standards (ASTM D6751) and European Committee for Standardization (EN 14103), respectively (Azad 2017).

In one study, a new biodiesel production method was reported to overcome the problems faced during the batch, continuous process, high-cost purification steps, and complex equipment. In their proposed approach, high energy consumption was tamed using a new nanohybrid electrocatalytic biodiesel production. Here, they used the trimetallic nanoalloy of Platinum, Iridium, and Ruthenium (PtIrRu) on the lemon seed oil, with methanol to oil ratio of 10:1 with 0.3 wt% of NaCl at 20 °C with the conversion value of 98.2% and yield of 80% is recorded and compared with EN14214 standards (Sarno and Ponticorvo 2020).

13.6.1.3 Biogas Production

Biomass, when digested with an anaerobic fermentation process, leads to the formation of a combination of gases like Methane (CH_4), Hydrogen (H_2), Hydrogen sulfide (H_2S), carbon dioxide (CO_2), and oxygen (O_2). Biogas is the anaerobic or thermochemical breakdown of biomass that produces biogas, an energy-rich gas. Methane and carbon dioxide makes up most of the biogas. Another great advantage of producing biogas is the digested sludge (rich in nitrogen) which can be treated and used directly onto the soil to increase its nitrogen content. This is also used to make the soil fertile. In this regard, the CPW is the best suitable material for producing biogas. Alternatively, pollution from citrus peel waste is reduced with the usage of biogas which is clean energy leading to the reduction of carbon emission (Mahato et al. 2021). An evaluation of the co-digestion of sludge with citrus waste (CW) by adding biochar positively affected methane production. Adding biochar can benefit us by reducing the costs of pretreatments to remove microbial inhibitors. On the other hand, biochar acts as an immobilising agent and provides a greater surface area for digestion. In its anaerobic digestion of 20–30 days, there was a 60% improvement in methane yields of standards above 500 L CH_4 kg VS^{-1} at an OLR of 1.49 kg VS m^{-3} d^{-1} (Martínez et al. 2018). Another study used orange peels (OPs) and extracted limonene first from Ops and then subjected to Anaerobic digestion at a pilot scale for 1 week at 37 °C, which produced a yield of 365 L_{CH4}/kg_{TVS} methane (TVS- volatile solids) (Battista et al. 2020).

Further, one study also focused on bypassing the pretreatment and removal of limonene. It employed a strategy where the polyvinylidene difluoride (PVDF) membrane in a biogas reactor was used limonene is hydrophobic. A hydrophilic PVDF membrane repels it; thus, Archaea producing more sensitive methane were encapsulated in the PVDF membrane. The other digesting bacteria were left to digest the CW and produce soluble compounds that can pass through the PVDF membrane. Encapsulated archaea utilise it and produce biogas. The membrane bioreactor yielded about 73% of the theoretical methane production (Wikandari et al. 2014).

13.6.2 Biofertilizer Production

Farmers use chemically manufactured fertilisers, which have disastrous impacts on the environment. As mentioned, the total waste generated by the consumption of citrus fruit can also be used as fertiliser with proper treatment because it is rich in nutrients. Similarly, the sludge produced by the anaerobic digestion of CPW for biogas production is also rich in nitrogen. It can be used as manure to increase soil fertility after treatment (Durán-Lara et al. 2020).

Muhammad Aon et al. (2015) evaluated the efficacy of citrus peel biochar with green waste biochar. They found the ratios C: N, C: P, and C: S was greater than green waste biochar. Both biochars tested the improvement of maise crop growth in calcareous soil. Indeed, maise growth, physiology, N and P uptake, and recovery are all improved by CPB more effectively than green waste biochar. A laboratory scale study was conducted by (Stella Mary et al. 2016) on three different biochars of wastes from leaf (*Brassica oleracea*), pod (*Pisum sativum*), and peel (*Citrus sinensis*) wastes. Biochar was produced at temperatures ranging from 100 °C to 600 °C for 1 h. They found an increase in levels of mineral content, water retention capacity, total surface anions, and organic carbon in all three biochars. However, citrus peel had less amendment than the other two biochars, but this approach can be used and eliminate the citrus waste with the additional benefit of soil fertility.

Another study was conducted using olive waste and citrus pulps different mixed proportions of manure, maise silage, and milk serum were evaluated. Firstly, they were subjected to biogas production. Later the digestate, which can be used as a biofertiliser, was checked for phytotoxicity, chemical, and biological composition. They suggested that according to the digestate composition, the nutrient value and microbial population depend, so one must keep all these parameters in check and correlate with the plant or the optimal crop growth enhancing digestates used (Panuccio et al. 2016).

13.6.3 Citrus Waste in the Food and Dairy Industry

The waste produced after using citrus fruits serves as a cheap source of nutritional bioactive compounds with disease alleviating traits as they have wound healing, antimicrobial properties, vitamins, and dietary fibers and are rich in antioxidants. This makes citrus waste suitable for valorising these compounds and use in food and dairy supplements (Ademosun 2022). Different food products have been introduced into the market due to their potential other beneficial properties. Flavouring agents made from citrus essential oils can be directly used in food and diluted to drink. Similarly, granules of dried citrus peel are used for culinary purposes. Chocolates containing orange peel or orange oils are also present in markets (Wedamulla et al. 2022).

A study focused on isolating pectin from red and white grapefruit peels which were treated with 95% alcohol, hot distilled water, and acid. This study reports producing 25% of high ester pectin with more than 7% methoxy content, and the obtained pectin can be used in gelling, similar to commercial pectin (Mohamed 2016). A comparative study on producing pectin from *Citrus maxima* peel from different extraction methods were assessed. They conclude that UME is the best method of extracting HM pectin, and SWE is better for extracting LM pectin, which is good grade pectin (Liew et al. 2019).

The diversified plant fibers can play asunder role in production of prebiotics and probiotics. As discussed in this chapter, CPW is a rich source of dietary fibers, and the pectic oligosaccharides derived from citrus fruit waste can also be used as prebiotics. In research reports, POS made from low-methoxy apple pectin, high-methoxy citrus pectin, and orange peel promotes the growth of lactobacilli and bifidobacteria while inhibiting the growth of pathogenic genera (Mamma and Christakopoulos 2014).

Antioxidants fortified panzer, a fat-rich dairy product, antioxidants from three different fruit peels of lemon, orange, and pomegranate, enhance the shelf life of paneer by preventing the peroxide activity and preventing rancidity. Thus, using bioactive compounds from waste helps protect and increases the shelf life of fat and oil-rich foods (Singh and Immanuel 2014). Nasser (2022) conducted similar research who used lemon peel to enhance the yogurt drink with immune booster vitamin C and other compounds were analysed. The discharge of wastewater from industry leads the pollution. The biological means of wastewater treatment are eco-friendly and cheaper. The bio-enzymes produced from citrus waste can be used to treat dairy wastewater. Before reusing it in farmlands, they found that 6% crude bio-enzyme was sufficient and made the wastewater with permissible parameters (Srimathi et al. 2020).

13.7 Conclusion and Future Outlook

Processing citrus waste helps to save the environment from pollution risks, lower the cost of manufactured products, and utilise less synthetic chemicals. Additionally, it will benefit the developing economies of tropical and subtropical nations, which are the emerging markets for nutraceuticals, as well as the stressed global market following the recession. The pharmaceutical, nutraceutical, food, health beverage, and various other sector say all benefit from the important chemicals found in citrus byproducts, which are regarded as a cost-effective and renewable resource. Plant-derived value-added chemicals are used as natural additives and flavouring agents in all of these systems. Additionally, these plant-derived bioactive chemicals are effectively included into both therapeutic and pharmaceutical formulations with the intention of offering humans a high degree of protection.

However, there are several difficulties and problems that must be resolved or handled in the next research on citrus. Future citrus resource exploitation should be focused on the premise of precise phytochemical and nutritional substance profiling due to the vast changes in chemical composition between citrus varieties, fruit portions, maturation, and production regions. More research on the conservation of these bioactive substances is crucial for increasing the economic value of citrus since citrus firms are now faced with the conundrum that its phytochemicals and nutrients are sensitive when consumed and absorbed by humans. Additionally, clinical research on citrus now focuses mostly on the structure-activity connections of certain chemicals isolated from citrus, but overall, citrus' positive impact on human health deserves more attention. To address the market's increasing demand for high-quality there is a citrus goods need to create unique farming practises and processing methods that preserve the fruit's taste, phytochemicals, and nutritional qualities to the fullest extent possible. The foundation for a more effective circular economic system to create wealth from waste will be laid by industry and research working together because one man's trash is another man's treasure.

References

Ademosun AO (2022) Citrus peels odyssey: From the waste bin to the lab bench to the dining table. Appl Food Res 2:100083. https://doi.org/10.1016/j.afres.2022.100083

Akiyama S, Katsumata S-I, Suzuki K et al (2010) Dietary hesperidin exerts hypoglycemic and hypolipidemic effects in streptozotocin-induced marginal type 1 diabetic rats. J Clin Biochem Nutr 46:87–92. https://doi.org/10.3164/jcbn.09-82

Alkhulaifi MM, Alshehri JH, Alwehaibi MA et al (2020) Green synthesis of silver nanoparticles using Citrus limon peels and evaluation of their antibacterial and cytotoxic properties. Saudi J Biol Sci 27:3434–3441. https://doi.org/10.1016/j.sjbs.2020.09.031

Ángel Siles López J, Li Q, Thompson IP (2010) Biorefinery of waste orange peel. Crit Rev Biotechnol 30:63–69. https://doi.org/10.3109/07388550903425201

Anonymous (1992) The Wealth of India. Publications & Information Directorate, CSIR, New Delhi

Aranganathan S, Selvam JP, Nalini N (2008) Effect of hesperetin, a citrus flavonoid, on bacterial enzymes and carcinogen-induced aberrant crypt foci in colon cancer rats: a dose-dependent study. J Pharm Pharmacol 60:1385–1392. https://doi.org/10.1211/jpp/60.10.0015

Asker M, El-gengaihi SE, Hassan EM et al (2020) Phytochemical constituents and antibacterial activity of Citrus lemon leaves. Bull Natl Res Centre 44:194. https://doi.org/10.1186/s42269-020-00446-1

Ayad R, Akkal S (2019) Phytochemistry and biological activities of algerian Centaurea and related genera. In: Studies in natural products chemistry. Elsevier, pp 357–414

Azad A (2017) Biodiesel from Mandarin Seed Oil: A Surprising Source of Alternative Fuel. Energies (Basel) 10:1689. https://doi.org/10.3390/en10111689

Badalamenti N, Bruno M, Schicchi R et al (2022) Reuse of food waste: the chemical composition and health properties of pomelo (Citrus maxima) cultivar essential oils. Molecules 27. https://doi.org/10.3390/molecules27103273

Barreca D, Bellocco E, Leuzzi U, Gattuso G (2014) First evidence of C- and O-glycosyl flavone in blood orange (Citrus sinensis (L.) Osbeck) juice and their influence on antioxidant properties. Food Chem 149:244–252. https://doi.org/10.1016/j.foodchem.2013.10.096

Battista F, Remelli G, Zanzoni S, Bolzonella D (2020) Valorization of residual orange peels: limonene recovery, volatile fatty acids, and biogas production. ACS Sustain Chem Eng 8:6834–6843. https://doi.org/10.1021/acssuschemeng.0c01735

Beecher GR (2009) Phytonutrients' role in metabolism: effects on resistance to degenerative processes. Nutr Rev 57:3–6. https://doi.org/10.1111/j.1753-4887.1999.tb01800.x

Belibasakis NG, Tsirgogianni D (1996) Effects of dried citrus pulp on milk yield, milk composition and blood components of dairy cows. Anim Feed Sci Technol 60:87–92. https://doi.org/10.1016/0377-8401(95)00927-2

Boukroufa M, Boutekedjiret C, Chemat F (2017) Development of a green procedure of citrus fruits waste processing to recover carotenoids. Resour Eff Technol 3:252–262. https://doi.org/10.1016/j.reffit.2017.08.007

Bruni R, Barreca D, Protti M et al (2019) Botanical sources, chemistry, analysis, and biological activity of furanocoumarins of pharmaceutical interest. Molecules 24:2163. https://doi.org/10.3390/molecules24112163

Caruso M, Fabroni S, Emma R et al (2021) A new standardized phytoextract from red orange and lemon wastes (red orange and lemon extract) reduces basophil degranulation and activation. Nat Prod Res 35:5354–5359. https://doi.org/10.1080/14786419.2020.1761355

Casquete R, Castro SM, Martín A et al (2015) Evaluation of the effect of high pressure on total phenolic content, antioxidant and antimicrobial activity of citrus peels. Innovative Food Sci Emerg Technol 31:37–44. https://doi.org/10.1016/j.ifset.2015.07.005

Castillo-Herrera GA, Farías-Álvarez LJ, García-Fajardo JA et al (2015) Bioactive extracts of Citrus aurantifolia swingle seeds obtained by supercritical CO2 and organic solvents comparing its cytotoxic activity against L5178Y leukemia lymphoblasts. J Supercrit Fluids 101:81–86. https://doi.org/10.1016/j.supflu.2015.02.026

Castrica M, Rebucci R, Giromini C et al (2019) Total phenolic content and antioxidant capacity of agri-food waste and by-products. Ital J Anim Sci 18:336–341. https://doi.org/10.1080/1828051X.2018.1529544

Chavan P, Singh AK, Kaur G (2018) Recent progress in the utilization of industrial waste and by-products of citrus fruits: a review. J Food Process Eng 41:e12895. https://doi.org/10.1111/jfpe.12895

Chen Y, Pan H, Hao S et al (2021) Evaluation of phenolic composition and antioxidant properties of different varieties of Chinese citrus. Food Chem 364:130413. https://doi.org/10.1016/j.foodchem.2021.130413

Chi PTL, van Hung P, le Thanh H, Phi NTL (2020) Valorization of citrus leaves: chemical composition, antioxidant and antibacterial activities of essential oils. Waste Biomass Valoriz 11:4849–4857. https://doi.org/10.1007/s12649-019-00815-6

Choi IS, Kim J-H, Wi SG et al (2013) Bioethanol production from mandarin (Citrus unshiu) peel waste using popping pretreatment. Appl Energy 102:204–210. https://doi.org/10.1016/j.apenergy.2012.03.066

Cirmi S, Randazzo B, Russo C et al (2021) Anti-inflammatory effect of a flavonoid-rich extract of orange juice in adult zebrafish subjected to *Vibrio anguillarum* -induced enteritis. Nat Prod Res 35:5350–5353. https://doi.org/10.1080/14786419.2020.1758096

Citraro R, Navarra M, Leo A et al (2016) The anticonvulsant activity of a flavonoid-rich extract from orange juice involves both NMDA and GABA-benzodiazepine receptor complexes. Molecules 21:1261. https://doi.org/10.3390/molecules21091261

Costanzo G, Iesce MR, Naviglio D et al (2020) Comparative Studies on Different Citrus Cultivars: A Revaluation of Waste Mandarin Components. Antioxidants (Basel) 9. https://doi.org/10.3390/antiox9060517

Currò M, Risitano R, Ferlazzo N et al (2016) Citrus bergamia juice extract attenuates β-amyloid-induced pro-inflammatory activation of THP-1 cells through MAPK and AP-1 pathways. Sci Rep 6:20809. https://doi.org/10.1038/srep20809

Czech A, Zarycka E, Yanovych D et al (2020) Mineral content of the pulp and peel of various citrus fruit cultivars. Biol Trace Elem Res 193:555–563. https://doi.org/10.1007/s12011-019-01727-1

de la Rosa LA, Moreno-Escamilla JO, Rodrigo-García J, Alvarez-Parrilla E (2019) Phenolic compounds. In: Postharvest physiology and biochemistry of fruits and vegetables. Elsevier, pp 253–271

Değirmenci H, Erkurt H (2020) Relationship between volatile components, antimicrobial and antioxidant properties of the essential oil, hydrosol and extracts of Citrus aurantium L. flowers. J Infect Public Health 13:58–67. https://doi.org/10.1016/j.jiph.2019.06.017

Dhanasekaran K, Musthafa MM, Dharmendirakumar M (2016) Processing and characterization of biodiesel from sweet orange (*Citrus sinensis*) seed oil. Energy Sources Part A Recover Util Environ Eff 38:2582–2589. https://doi.org/10.1080/15567036.2015.1075089

Dimopoulou M, Alba K, Campbell G, Kontogiorgos V (2019) Pectin recovery and characterization from lemon juice waste streams. J Sci Food Agric 99:6191–6198. https://doi.org/10.1002/jsfa.9891

Diplock AT (1993) Antioxidant Nutrients and Disease. Nutr Health 9:37–42. https://doi.org/10.1177/026010609300900105

Dosoky NS, Setzer WN (2018) Biological Activities and Safety of Citrus spp. Essential oils. Int J Mol Sci 19. https://doi.org/10.3390/ijms19071966

Durán-Lara EF, Valderrama A, Marican A (2020) Natural organic compounds for application in organic farming. Agriculture 10:41. https://doi.org/10.3390/agriculture10020041

Ellegård L, Andersson H (2007) Oat bran rapidly increases bile acid excretion and bile acid synthesis: an ileostomy study. Eur J Clin Nutr 61:938–945. https://doi.org/10.1038/sj.ejcn.1602607

El-Otmani M, Ait-Oubahou A, Zacarías L (2011) Citrus spp .: orange, mandarin, tangerine, clementine, grapefruit, pomelo, lemon and lime. In: Postharvest biology and technology of tropical and subtropical fruits. Elsevier, pp 437–516

Espín JC, García-Conesa MT, Tomás-Barberán FA (2007) Nutraceuticals: facts and fiction. Phytochemistry 68:2986–3008. https://doi.org/10.1016/j.phytochem.2007.09.014

Fabroni S, Ballistreri G, Amenta M, Rapisarda P (2016) Anthocyanins in different *Citrus* species: an UHPLC-PDA-ESI/MS ^{n}-assisted qualitative and quantitative investigation. J Sci Food Agric 96:4797–4808. https://doi.org/10.1002/jsfa.7916

Falcinelli B, Famiani F, Paoletti A et al (2020) Phenolic compounds and antioxidant activity of sprouts from seeds of citrus species. Agriculture 10:33. https://doi.org/10.3390/agriculture10020033

FAO (2021) Citrus Fruit Statistical Compendium 2020

Ferguson LR, Philpott M, Karunasinghe N (2004) Dietary cancer and prevention using antimutagens. Toxicology 198:147–159. https://doi.org/10.1016/j.tox.2004.01.035

Ferlazzo N, Visalli G, Smeriglio A et al (2015) Flavonoid fraction of orange and bergamot juices protect human lung epithelial cells from hydrogen peroxide-induced oxidative stress. Evid Based Complement Alternat Med 2015:1–14. https://doi.org/10.1155/2015/957031

Fouad Abobatta W (2019) Nutritional benefits of citrus fruits. Am J Biomed Sci Res 3:303–306. https://doi.org/10.34297/AJBSR.2019.03.000681

Fuchs C, Giovannucci E, Colditz G (1999) Dietary fiber and the risk of colorectal cancer. South Med J 92:733–734. https://doi.org/10.1097/00007611-199907000-00025

Fusco R, Cirmi S, Gugliandolo E et al (2017) A flavonoid-rich extract of orange juice reduced oxidative stress in an experimental model of inflammatory bowel disease. J Funct Foods 30:168–178. https://doi.org/10.1016/j.jff.2016.12.038

Gaur V, Aggarwal A, Kumar A (2011) Possible nitric oxide mechanism in the protective effect of hesperidin against ischemic reperfusion cerebral injury in rats. Indian J Exp Biol 49:609–618

Gómez-Mejía E, Rosales-Conrado N, León-González ME, Madrid Y (2019) Citrus peels waste as a source of value-added compounds: Extraction and quantification of bioactive polyphenols. Food Chem 295:289–299. https://doi.org/10.1016/j.foodchem.2019.05.136

González-Molina E, Domínguez-Perles R, Moreno DA, García-Viguera C (2010) Natural bioactive compounds of Citrus limon for food and health. J Pharm Biomed Anal 51:327–345. https://doi.org/10.1016/j.jpba.2009.07.027

Guillon F, Champ M (2000) Structural and physical properties of dietary fibres, and consequences of processing on human physiology. Food Res Int 33:233–245. https://doi.org/10.1016/S0963-9969(00)00038-7

Guneser BA, Yilmaz E (2017) Bioactives, aromatics and sensory properties of cold-pressed and hexane-extracted lemon (Citrus Limon L.) seed oils. J Am Oil Chem Soc 94:723–731. https://doi.org/10.1007/s11746-017-2977-z

Guo C, Shan Y, Yang Z et al (2020) Chemical composition, antioxidant, antibacterial, and tyrosinase inhibition activity of extracts from Newhall navel orange (*Citrus sinensis* Osbeck cv. Newhall) peel. J Sci Food Agric 100:2664–2674. https://doi.org/10.1002/jsfa.10297

Hamdan DI, El-Shiekh RA, El-Sayed MA et al (2020) Phytochemical characterization and anti-inflammatory potential of Egyptian Murcott mandarin cultivar waste (stem, leaves and peel). Food Funct 11:8214–8236. https://doi.org/10.1039/D0FO01796E

Han MH, Lee WS, Lu JN et al (2012) Citrus aurantium L. exhibits apoptotic effects on U937 human leukemia cells partly through inhibition of Akt. Int J Oncol 40:2090–2096. https://doi.org/10.3892/ijo.2012.1350

İçöz E, Mehmet Tuğrul K, Saral A, İçöz E (2009) Research on ethanol production and use from sugar beet in Turkey. Biomass Bioenergy 33:1–7. https://doi.org/10.1016/j.biombioe.2008.05.005

Im SJ, Kim J-H, Kim MY (2014) Evaluation of bioactive components and antioxidant and anti-cancer properties of citrus wastes generated during bioethanol production. Nat Prod Commun 9:1934578X1400900. https://doi.org/10.1177/1934578X1400900413

Jaramillo DP, García T, Buffa M et al (2009) Effect of the inclusion of whole citrus in the ration of lactating ewes on the properties of milk and cheese characteristics during ripening. J Dairy Sci 92:469–476. https://doi.org/10.3168/jds.2008-1293

Jeong D, Park H, Jang B-K et al (2021) Recent advances in the biological valorization of citrus peel waste into fuels and chemicals. Bioresour Technol 323:124603. https://doi.org/10.1016/j.biortech.2020.124603

John I, Muthukumar K, Arunagiri A (2017) A review on the potential of citrus waste for D-Limonene, pectin, and bioethanol production. Int J Green Energy 14:599–612. https://doi.org/10.1080/15435075.2017.1307753

Jokić S, Molnar M, Cikoš A-M et al (2020) Separation of selected bioactive compounds from orange peel using the sequence of supercritical CO_2 extraction and ultrasound solvent extraction: optimization of limonene and hesperidin content. Sep Sci Technol 55:2799–2811. https://doi.org/10.1080/01496395.2019.1647245

Jorge N, da Silva AC, Aranha CPM (2016) Antioxidant activity of oils extracted from orange (Citrus sinensis) seeds. An Acad Bras Cienc 88:951–958. https://doi.org/10.1590/0001-3765201620140562

Kakadiya J, Mulani H, Shah N (2010) Protective effect of hesperidin on cardiovascular complication in experimentally induced myocardial infarction in diabetes in rats. J Basic Clin Pharm 1:85–91

Khan UM, Sameen A, Aadil RM et al (2021) Citrus genus and its waste utilization: a review on health-promoting activities and industrial application. Evid Based Complement Alternat Med 2021:1–17. https://doi.org/10.1155/2021/2488804

Kim J, Kim M (2016) The potential use of citrus juice waste as sources of natural phenolic antioxidants. J Appl Pharm Sci 6:202–205. https://doi.org/10.7324/JAPS.2016.601230

Kim J, Jayaprakasha GK, Uckoo RM, Patil BS (2012) Evaluation of chemopreventive and cytotoxic effect of lemon seed extracts on human breast cancer (MCF-7) cells. Food Chem Toxicol 50:423–430. https://doi.org/10.1016/j.fct.2011.10.057

Kim SS, Park KJ, An HJ, Choi YH (2017) Phytochemical, antioxidant, and antibacterial activities of fermented Citrus unshiu byproduct. Food Sci Biotechnol 26:461–466. https://doi.org/10.1007/s10068-017-0063-9

Kumar N, Goel N (2019) Phenolic acids: natural versatile molecules with promising therapeutic applications. Biotechnol Rep 24:e00370. https://doi.org/10.1016/j.btre.2019.e00370

Kumar K, Yadav AN, Kumar V et al (2017) Food waste: a potential bioresource for extraction of nutraceuticals and bioactive compounds. Bioresour Bioprocess 4:18. https://doi.org/10.1186/s40643-017-0148-6

Kumar S, Kaur G, Joshi A, Joshi HC (2019) Production of biodiesel from citrus maxima (Chakotara) seed oil, a potential of non-food feedstock and its blends with n butanol-diesel and purification, utilization of glycerol obtained as by-product from biodiesel. Energy Sources Part A: Rec Util Environ Eff 41:1508–1517. https://doi.org/10.1080/15567036.2018.1549125

Lamine M, Gargouri M, Rahali FZ, Mliki A (2021) Recovering and Characterizing Phenolic Compounds From Citrus By-Product: A Way Towards Agriculture of Subsistence and Sustainable Bioeconomy. Waste Biomass Valorization 12:4721–4731. https://doi.org/10.1007/s12649-020-01306-9

Liew SQ, Ngoh GC, Yusoff R, Teoh WH (2016) Sequential ultrasound-microwave assisted acid extraction (UMAE) of pectin from pomelo peels. Int J Biol Macromol 93:426–435. https://doi.org/10.1016/j.ijbiomac.2016.08.065

Liew SS, Ho WY, Yeap SK, bin Sharifudin SA (2018) Phytochemical composition and in vitro antioxidant activities of *Citrus sinensis* peel extracts. PeerJ 6:e5331. https://doi.org/10.7717/peerj.5331

Liew SQ, Teoh WH, Yusoff R, Ngoh GC (2019) Comparisons of process intensifying methods in the extraction of pectin from pomelo peel. Chem Eng Process Process Intensif 143:107586. https://doi.org/10.1016/j.cep.2019.107586

Lin L-Y, Huang B-C, Chen K-C, Peng RY (2020) Integrated anti-hyperlipidemic bioactivity of whole *Citrus grandis* [L.] osbeck fruits—multi-action mechanism evidenced using animal and cell models. Food Funct 11:2978–2996. https://doi.org/10.1039/C9FO02290B

Lipkin M, Reddy B, Newmark H, Lamprecht SA (1999) Dietary factors in human colorectal cancer. Annu Rev Nutr 19:545–586. https://doi.org/10.1146/annurev.nutr.19.1.545

Lopresto CG, Petrillo F, Casazza AA et al (2014) A non-conventional method to extract D-limonene from waste lemon peels and comparison with traditional Soxhlet extraction. Sep Purif Technol 137:13–20. https://doi.org/10.1016/j.seppur.2014.09.015

Lu W, Shi Y, Wang R et al (2021) Antioxidant activity and healthy benefits of natural pigments in fruits: a review. Int J Mol Sci 22. https://doi.org/10.3390/ijms22094945

M'hiri N, Ioannou I, Ghoul M, Mihoubi Boudhrioua N (2017) Phytochemical characteristics of citrus peel and effect of conventional and nonconventional processing on phenolic compounds: a review. Food Rev Intl 33:587–619. https://doi.org/10.1080/87559129.2016.1196489

Mahato N, Sharma K, Sinha M et al (2021) Biotransformation of citrus waste-I: production of biofuel and valuable compounds by fermentation. Processes 9:220. https://doi.org/10.3390/pr9020220

Mamma D, Christakopoulos P (2014) Biotransformation of citrus by-products into value added products. Waste Biomass Valoriz 5:529–549. https://doi.org/10.1007/s12649-013-9250-y

Mannucci C, Calapai F, Cardia L et al (2018) Clinical pharmacology of *Citrus aurantium* and *Citrus sinensis* for the treatment of anxiety. Evid Based Complement Alternat Med 2018:1–18. https://doi.org/10.1155/2018/3624094

Martínez EJ, Rosas JG, Sotres A et al (2018) Codigestion of sludge and citrus peel wastes: Evaluating the effect of biochar addition on microbial communities. Biochem Eng J 137:314–325. https://doi.org/10.1016/j.bej.2018.06.010

Matthaus B, Özcan MM (2012) Chemical evaluation of citrus seeds, an agro-industrial waste, as a new potential source of vegetable oils. Grasas Aceites 63:313–320. https://doi.org/10.3989/gya.118411

Maugeri A, Ferlazzo N, de Luca L et al (2019) The link between the AMPK/SIRT1 axis and a flavonoid-rich extract of *Citrus bergamia* juice: A cell-free, in silico, and in vitro study. Phytother Res 33:1805–1814. https://doi.org/10.1002/ptr.6368

Maurya AK, Mohanty S, Pal A et al (2018) The essential oil from Citrus limetta Risso peels alleviates skin inflammation: In-vitro and in-vivo study. J Ethnopharmacol 212:86–94. https://doi.org/10.1016/j.jep.2017.10.018

Medina-Torres N, Espinosa-Andrews H, Trombotto S et al (2019) Ultrasound-assisted extraction optimization of phenolic compounds from citrus latifolia waste for chitosan bioactive nanoparticles development. Molecules 24:3541. https://doi.org/10.3390/molecules24193541

Mohamed H (2016) Extraction and characterization of pectin from grapefruit peels. MOJ Food Process Technol 2. https://doi.org/10.15406/mojfpt.2016.02.00029

Mohanty S, Maurya AK, Jyotshna et al (2015) Flavonoids Rich fraction of citrus limetta fruit peels reduces proinflammatory cytokine production and attenuates malaria pathogenesis. Curr Pharm Biotechnol 16:544–552. https://doi.org/10.2174/138920101606150407114023

Montero-Calderon A, Cortes C, Zulueta A et al (2019) Green solvents and Ultrasound-Assisted Extraction of bioactive orange (Citrus sinensis) peel compounds. Sci Rep 9:16120. https://doi.org/10.1038/s41598-019-52717-1

Moon S, Lee J, Kim K-T et al (2013) Antimicrobial effect of 7-O-Butylnaringenin, a novel flavonoid, and various natural flavonoids against helicobacter pylori strains. Int J Environ Res Public Health 10:5459–5469. https://doi.org/10.3390/ijerph10115459

Muhammad Aon MK, Zahir ZA, Ahmad R (2015) Low temperature produced citrus peel and green waste biochar improved maize growth and nutrient uptake, and chemical properties of calcareous soil. Pak J Agric Sci 52:627–636

Mulder GJ, Ouwerkerk-Mahadevan S (1997) Modulation of glutathione conjugation in vivo: how to decrease glutathione conjugation in vivo or in intact cellular systems in vitro. Chem Biol Interact 105:17–34. https://doi.org/10.1016/S0009-2797(97)00038-0

Nasiri E, Hesari J, Shekarforoush SS et al (2020) Novel milk-clotting enzyme from sour orange flowers (Citrus aurantium L.) as a coagulant in Iranian white cheese. Eur Food Res Technol 246:139–148. https://doi.org/10.1007/s00217-019-03403-z

Nasser S (2022) The addition of lemon peel powder affects the properties of yogurt. J Food Dairy Sci 13:65–70. https://doi.org/10.21608/jfds.2022.136776.1055

Nauman CM, Johnson JJ (2019) Clinical application of bergamot (Citrus bergamia) for reducing high cholesterol and cardiovascular disease markers. Integr Food Nutr Metab 6. https://doi.org/10.15761/IFNM.1000249

Oboh PA, Abulu EO (1997) The antimicrobial activities of extract of sidiumguajava and Citrus aurantifolia. Niger J Biotechnol 8:25–29

Oikeh EI, Ayevbuomwan M, Irabor F et al (2020) Evaluation of the phenolic content, antioxidant and antimicrobial activities of oil and non-oil extracts of Citrus sinensis (L.) Osbeck Seeds. Prev Nutr Food Sci 25:280–285. https://doi.org/10.3746/pnf.2020.25.3.280

Pacheco MT, Vezza T, Diez-Echave P et al (2018) Anti-inflammatory bowel effect of industrial orange by-products in DSS-treated mice. Food Funct 9:4888–4896. https://doi.org/10.1039/C8FO01060A

Panuccio MR, Attinà E, Basile C et al (2016) Use of recalcitrant agriculture wastes to produce biogas and feasible biofertilizer. Waste Biomass Valoriz 7:267–280. https://doi.org/10.1007/s12649-015-9445-5

Papoutsis K, Pristijono P, Golding JB et al (2018) Screening the effect of four ultrasound-assisted extraction parameters on hesperidin and phenolic acid content of aqueous citrus pomace extracts. Food Biosci 21:20–26. https://doi.org/10.1016/j.fbio.2017.11.001

Patsalou M, Samanides CG, Protopapa E et al (2019) A Citrus Peel Waste Biorefinery for Ethanol and Methane Production. Molecules 24. https://doi.org/10.3390/molecules24132451

Patsalou M, Chrysargyris A, Tzortzakis N, Koutinas M (2020) A biorefinery for conversion of citrus peel waste into essential oils, pectin, fertilizer and succinic acid via different fermentation strategies. Waste Manag 113:469–477. https://doi.org/10.1016/j.wasman.2020.06.020

Pérez-Mendoza JL, García-Hernández F (1983) Fermentation of a waste-product from the industrial processing of the lime (Citrus aurantifolia swingle) for vitamin B12 production. Biotechnol Lett 5:259–264. https://doi.org/10.1007/BF00161126

Pfaltzgraff LA, de Bruyn M, Cooper EC et al (2013) Food waste biomass: a resource for high-value chemicals. Green Chem 15:307. https://doi.org/10.1039/c2gc36978h

Platz EA, Giovannucci E, Rimm EB et al (1997) Dietary fiber and distal colorectal adenoma in men. Cancer Epidemiol Biomark Prev 6:661–670

Rafiq S, Kaul R, Sofi SA et al (2018) Citrus peel as a source of functional ingredient: a review. J Saudi Soc Agric Sci 17:351–358. https://doi.org/10.1016/j.jssas.2016.07.006
Raimondo M, Caracciolo F, Cembalo L et al (2018) Making virtue out of necessity: managing the citrus waste supply chain for bioeconomy applications. Sustainability 10:4821. https://doi.org/10.3390/su10124821
Ramana KVR, Govindarajan VS, Ranganna S, Kefford JF (1981) Citrus fruits — varieties, chemistry, technology, and quality evaluation. Part I: Varieties, production, handling, and storage. CRC Crit Rev Food Sci Nutr 15:353–431. https://doi.org/10.1080/10408398109527321
Risitano R, Currò M, Cirmi S et al (2014) Flavonoid fraction of bergamot juice reduces LPS-induced inflammatory response through SIRT1-Mediated NF-κB inhibition in THP-1 monocytes. PLoS One 9:e107431. https://doi.org/10.1371/journal.pone.0107431
Rosa A, Era B, Masala C et al (2019) Supercritical CO_2 Extraction of Waste Citrus Seeds: Chemical Composition, Nutritional and Biological Properties of Edible Fixed Oils. Eur J Lipid Sci Technol 121:1800502. https://doi.org/10.1002/ejlt.201800502
Russo C, Maugeri A, Lombardo GE et al (2021) The second life of citrus fruit waste: a valuable source of bioactive compounds. Molecules 26:5991. https://doi.org/10.3390/molecules26195991
Šafranko S, Ćorković I, Jerković I et al (2021) Green extraction techniques for obtaining bioactive compounds from mandarin peel (Citrus unshiu var. Kuno): phytochemical analysis and process optimization. Foods 10. https://doi.org/10.3390/foods10051043
Sagar NA, Pareek S, Sharma S et al (2018) Fruit and vegetable waste: bioactive compounds, their extraction, and possible utilization. Compr Rev Food Sci Food Saf 17:512–531. https://doi.org/10.1111/1541-4337.12330
Saini RK, Keum Y-S (2018) Carotenoid extraction methods: a review of recent developments. Food Chem 240:90–103. https://doi.org/10.1016/j.foodchem.2017.07.099
Saini RK, Ranjit A, Sharma K et al (2022) Bioactive compounds of citrus fruits: a review of composition and health benefits of carotenoids, flavonoids, limonoids, and terpenes. Antioxidants 11:239. https://doi.org/10.3390/antiox11020239
Saleem M, Saeed MT (2020) Potential application of waste fruit peels (orange, yellow lemon and banana) as wide range natural antimicrobial agent. J King Saud Univ Sci 32:805–810. https://doi.org/10.1016/j.jksus.2019.02.013
Sanches VL, Cunha TA, Viganó J et al (2022) Comprehensive analysis of phenolics compounds in citrus fruits peels by UPLC-PDA and UPLC-Q/TOF MS using a fused-core column. Food Chem X 14:100262. https://doi.org/10.1016/j.fochx.2022.100262
Sarno M, Ponticorvo E (2020) A new nanohybrid for electrocatalytic biodiesel production from waste Amalfi coast lemon seed oil. Fuel 267:117178. https://doi.org/10.1016/j.fuel.2020.117178
Scordino M, di Mauro A, Passerini A, Maccarone E (2005) Selective recovery of anthocyanins and hydroxycinnamates from a byproduct of citrus processing. J Agric Food Chem 53:651–658. https://doi.org/10.1021/jf048643b
Sharma K, Mahato N, Cho MH, Lee YR (2017) Converting citrus wastes into value-added products: economic and environmently friendly approaches. Nutrition 34:29–46. https://doi.org/10.1016/j.nut.2016.09.006
Sharma M, Usmani Z, Gupta VK, Bhat R (2021) Valorization of fruits and vegetable wastes and by-products to produce natural pigments. Crit Rev Biotechnol 41:535–563. https://doi.org/10.1080/07388551.2021.1873240
Silva AC, Jorge N (2019) Bioactive properties and antioxidant capacity of oils extracted from citrus fruit seeds. Acta Aliment 48:196–203. https://doi.org/10.1556/066.2019.48.2.7
Singh S, Immanuel G (2014) Extraction of antioxidants from fruit peels and its utilization in paneer. J Food Process Technol 05:349. https://doi.org/10.4172/2157-7110.1000349
Singh B, Singh JP, Kaur A, Singh N (2020) Phenolic composition, antioxidant potential and health benefits of citrus peel. Food Res Int 132:109114. https://doi.org/10.1016/j.foodres.2020.109114
Sir Elkhatim KA, Elagib RAA, Hassan AB (2018) Content of phenolic compounds and vitamin C and antioxidant activity in wasted parts of Sudanese citrus fruits. Food Sci Nutr 6:1214–1219. https://doi.org/10.1002/fsn3.660

Slavin JL, Lloyd B (2012) Health benefits of fruits and vegetables. Adv Nutr 3:506–516. https://doi.org/10.3945/an.112.002154

Srimathi N, Subiksha M, Abarna J, Niranjana T (2020) Biological treatment of dairy wastewater using bio enzyme from citrus fruit peels. Int J Recent Technol Eng (IJRTE) 9:292–295. https://doi.org/10.35940/ijrte.A1530.059120

Stella Mary G, Sugumaran P, Niveditha S et al (2016) Production, characterization and evaluation of biochar from pod (Pisum sativum), leaf (Brassica oleracea) and peel (Citrus sinensis) wastes. Int J Recycl Org Waste Agric 5:43–53. https://doi.org/10.1007/s40093-016-0116-8

Suri S, Singh A, Nema PK (2022) Current applications of citrus fruit processing waste: a scientific outlook. Appl Food Res 2:100050. https://doi.org/10.1016/j.afres.2022.100050

Taghizadeh-Alisaraei A, Hosseini SH, Ghobadian B, Motevali A (2017) Biofuel production from citrus wastes: a feasibility study in Iran. Renew Sust Energ Rev 69:1100–1112. https://doi.org/10.1016/j.rser.2016.09.102

Teigiserova DA, Tiruta-Barna L, Ahmadi A et al (2021) A step closer to circular bioeconomy for citrus peel waste: a review of yields and technologies for sustainable management of essential oils. J Environ Manag 280:111832. https://doi.org/10.1016/j.jenvman.2020.111832

Terpstra AHM, Lapré JA, de Vries HT, Beynen AC (2002) The hypocholesterolemic effect of lemon peels, lemon pectin, and the waste stream material of lemon peels in hybrid F1B hamsters. Eur J Nutr 41:19–26. https://doi.org/10.1007/s003940200002

Thakur A, Devi P, Saini S et al (2019) *Citrus limetta* organic waste recycled carbon nanolights: photoelectro catalytic, sensing, and biomedical applications. ACS Sustain Chem Eng 7:502–512. https://doi.org/10.1021/acssuschemeng.8b04025

Tocmo R, Pena-Fronteras J, Calumba KF et al (2020) Valorization of pomelo (*Citrus grandis* Osbeck) peel: a review of current utilization, phytochemistry, bioactivities, and mechanisms of action. Compr Rev Food Sci Food Saf 19:1969–2012. https://doi.org/10.1111/1541-4337.12561

Tuttobene R, Avola G, Gresta F, Abbate V (2009) Industrial orange waste as organic fertilizer in durum wheat. Agron Sustain Dev 29:557–563. https://doi.org/10.1051/agro/2009014

Wedamulla NE, Fan M, Choi Y-J, Kim E-K (2022) Citrus peel as a renewable bioresource: transforming waste to food additives. J Funct Foods 95:105163. https://doi.org/10.1016/j.jff.2022.105163

Widjaja-Adhi MAK, Ramkumar S, Von Lintig J (2018) Protective role of carotenoids in the visual cycle. FASEB J 32:6305–6315. https://doi.org/10.1096/fj.201800467R

Wikandari R, Millati R, Cahyanto MN, Taherzadeh MJ (2014) Biogas production from citrus waste by membrane bioreactor. Membranes (Basel) 4:596–607. https://doi.org/10.3390/membranes4030596

Wilkins MR, Widmer WW, Grohmann K (2007) Simultaneous saccharification and fermentation of citrus peel waste by Saccharomyces cerevisiae to produce ethanol. Process Biochem 42:1614–1619. https://doi.org/10.1016/j.procbio.2007.09.006

Wu P, Tang X, Jian R et al (2021) Chemical composition, antimicrobial and insecticidal activities of essential oils of discarded perfume lemon and leaves (Citrus Limon (L.) Burm. F.) as possible sources of functional botanical agents. Front Chem 9. https://doi.org/10.3389/fchem.2021.679116

Xu G, Ye X, Chen J, Liu D (2007) Effect of heat treatment on the phenolic compounds and antioxidant capacity of citrus peel extract. J Agric Food Chem 55:330–335. https://doi.org/10.1021/jf062517l

Yamamoto M, Suzuki A, Jokura H et al (2008) Glucosyl hesperidin prevents endothelial dysfunction and oxidative stress in spontaneously hypertensive rats. Nutrition 24:470–476. https://doi.org/10.1016/j.nut.2008.01.010

Yang X, Kang S-M, Jeon B-T et al (2011) Isolation and identification of an antioxidant flavonoid compound from citrus-processing by-product. J Sci Food Agric 91:1925–1927. https://doi.org/10.1002/jsfa.4402

Zhang Y, Wang J, Dong J et al (2013) Inhibition of α-toxin production by subinhibitory concentrations of naringenin controls Staphylococcus aureus pneumonia. Fitoterapia 86:92–99. https://doi.org/10.1016/j.fitote.2013.02.001

Chapter 14
Bioactive Compounds in Citrus Fruits: Extraction and Identification

Shweta Suri, Anupama Singh, and Prabhat K. Nema

14.1 Introduction

Citrus fruits are the widely cultivated and consumed fruits all around the globe. Orange (*Citrus sinensis*), Grapefruit (*Citrus paradisi*), Mandarin/Tangerine (*Citrus reticulate*), Lemon (*Citrus limon*), as well as Lime (*Citrus aurantiifolia*) are cultivated worldwide because of their high commercial value (Suri et al. 2022a). The origin and history of the production of citrus fruits are still unknown. It is believed that the tropical, as well as subtropical climates of the Asian Island and the Malaysian Archipelago, started citrus planting at least 4000 years ago, but the exact beginning of citrus farming remains unclear (Berk 2016).

Annual citrus fruit production has ascended rapidly around the world in recent years, increasing from about 51.48 MT in 1975 to 158 MT in 2020. Maximum citrus fruits are produced in Asia (47.7%) then Africa (43.7%), United States of America (8.1%), followed by Europe (0.4%), and Oceania (0.1%). In context to the citrus producing nations, China occupies the first position with 44.63 million tonnes of citrus fruits produced, which is equivalent to 28.16% of the global overall citrus fruit produced in the year 2020. Among other nations, Brazil, India, and Mexico are some important nations, with each nation producing above 5% of the world's total

S. Suri (✉)
Department of Food Engineering, National Institute of Food Technology Entrepreneurship and Management, Sonipat, Haryana, India

Amity Institute of Food Technology, Amity University, Noida, Uttar Pradesh, India

A. Singh
Department of Food Engineering, National Institute of Food Technology Entrepreneurship and Management, Sonipat, Haryana, India

P. K. Nema
Department of Food Engineering, National Institute of Food Technology Entrepreneurship & Management (NIFTEM), Kundli, Haryana, India

S. Singh Purewal et al. (eds.), *Recent Advances in Citrus Fruits*,
https://doi.org/10.1007/978-3-031-37534-7_14

citrus production. Approximately 10.07 million hectares of land worldwide are used to produce citrus fruits. China, India, Brazil, Nigeria, and Mexico are the world's largest producers of citrus fruits (FAOSTAT 2022). Besides, these citrus fruits are highly adaptable to a variety of climates and soil diversities, contributing towards their vast production in tropical and subtropical areas and also in mild temperature areas (Suri et al. 2021).

Citrus fruits are recognized throughout the globe for their pleasant taste and abundance of bioactive substances and nutrients. Citrus fruits provide health benefits by stimulation of the immune system, cardiovascular as well as digestive organs. These fruits also possess anti-bacterial, anti-inflammatory, anti-atherosclerosis, and anti-cancer effects. These effects are partly because of the existence of many bioactive compounds (Czech et al. 2021). These citrus fruits are a rich source of bioactive compounds, particularly polyphenols *viz;* flavonoids & phenolic acids, terpenoids (limonoids & carotenoids), essential oils, dietary fiber, and pectin. Besides, citrus fruits comprise of other nutrients for example vitamins viz.; vitamin B-complex, vitamin C, vitamin E (tocopherols & tocotrienols), and micronutrients (iron, zinc, copper, selenium, and manganese) (Saini et al. 2022). Many of the bioactive components found in citrus fruits possess antioxidative, regulatory, and metabolic-stimulating action that guards the body's tissues as well as fluids from damage related with the occurrence of reactive oxygen species (Czech et al. 2021). The amount as well as the type of bioactive components, and their antioxidant activity, depend broadly on the type of fruit, its cultivar or part of the fruit, and the climate and growing conditions (Bermejo et al. 2011).

Extraction, segregation/isolation, as well as characterization of biologically active components from fruits/vegetables are the important steps for obtaining bioactive compounds from plant matrix. Appropriate measures should be taken to prevent destruction, loss or deterioration of the potent biologically active ingredients during the extraction process. Furthermore, the adequacy of the extraction technique used to recover exact bioactive components should be verified. The use of non-traditional green extraction techniques has become very important due to increased efficiency, little or no use of organic solvents, and reduced consumption of time, energy, and other resources (Anticona et al. 2021). Microwave-assisted extraction, pressurized liquid extraction, enzyme-assisted extraction, ultrasound-assisted extraction, supercritical fluid extraction, and pulsed electric field extraction are among the green extraction methods used by the scientists (Suri et al. 2021).

14.2 Citrus fruits: Taxonomy

Citrus fruits are the member of *Rutaceae* family with taxonomical description:

- Kingdom: *Plantae*
- Order: *Sapindales*
- Family: *Rutaceae*

- Sub-family: *Aurantiodeae*
- Genus: *Citrus L.*
- Species: *Citrus reticulata; Citrus cavaleriei; Citrus maxima; Citrus sinensis; Citrus medica; Citrus micrantha; Citrus japonica; Citrus hystrix*
- Hybrids Species: *Citrus limon; Citrus sinensis; Citrus paradisi; Citrus tangerine; Citrus latifolia; Citrus aurantifolia; etc.*

Taxonomy of citrus fruits is complex and debatable. The citrus family includes 150 genera and 1600 species including mandarin (*Citrus reticulata* L.), lemon (*Citrus limon* L.), sweet orange (*Citrus sinensis* L.), and grapefruit (*Citrus paradisi* L.) is the further most marketable species and is favored for its attractive color, aroma and taste (Karn et al. 2021; Ledesma-Escobarand and de Castro 2014).

Cultivated citrus fruits are derivative of several citrus species present in nature. Some are just assortments from the original wild type of citrus fruits, various other citrus fruits are produced through hybridization amongst two or more original citrus fruit species, while other are backcrosses between the cross and one of the parent species of the hybrid. Citrus plants simply cross amongst species with entirely diverse morphology, and citrus fruits that look alike can have pretty different ancestors (Wu et al. 2014; Velasco and Licciardello 2014). Most of the marketable citrus fruits varieties are derived from one or more "core species" such as mandarins, citrons, as well as pomelos, which contributes a composite floral structure that leads to a more complex fruit. These major core citrus fruit species have led to the development of many hybrid citrus species (Curk et al. 2014).

14.3 Nutritional Characterization of Citrus Fruits

Citrus fruits comprise of excellent nutritional value and have many health-promoting effects, for instances their action on aiding the digestion process, averting cardiovascular ailments and lowering inflammation (Maugeri et al. 2019; Yamada et al. 2011). The nutrients as well as bioactive phytochemical found in citrus fruits contributes towards their excellent health benefits. Major portion of citrus fruits (more than 80%) comprise of water, however citrus fruits are also good source of many other nutrients, for instance simple sugars (glucose, fructose, and sucrose), other carbohydrates (fiber, pectin, cellulose, and hemicellulose), fats & essential oils (D-limonene), vitamins (vitamin B-complex, vitamin C and vitamin E), minerals (potassium, phosphorus and calcium) and plant pigments (carotenoids, and xanthophyll) (Putnik et al. 2017). Citrus fruits also different enzymes like pectinase, pectinestarase, phosphatase etc. The composition, nutritional content and bioactive substances in citrus fruits varies widely on the basis of the variety of citrus fruit, fruit part and stage of growth (Albertini et al. 2006). This might be attributable to variances in hereditary genes, expression of gene, as well as ecological conditions.

The nutritional as well as bioactive components also differed amongst diverse portions of citrus fruit. The pulp is considered to be rich in nutrients *viz;* ascorbic

Table 14.1 Nutritional content found in citrus fruits

Nutrients	Quantity
Macronutrients (g/100 g)	
Protein	0.1–1.3
Lipid	0.07–0.42
Carbohydrate	6.4–13.3
Vitamins (mg/100 g)	
Vitamin A	0–0.058
Vitamin B-complex	0.25–1.38
Vitamin C	19.36–71.00
Vitamin E	0.12–0.22
Minerals (mg/100 g)	
Calcium	04–57
Potassium	102–239
Magnesium	06–14
Phosphorus	07–25
Sodium	0–4 mg/100 g

Source: USDA (United States Department of Agriculture, 2018)

acid and sugars, however majority of biologically active substances (e.g., polyphenols, flavonoids, essential oils and pectin) mainly accrue in the fruit peel and pomace (Tocmo et al. 2020). The seeds of citrus fruits contain high essential oil content in the form of limonoids.

Among nutrients, the carbohydrate present in citrus fruits varies from 6.4 to 13.3 g/100 g. Carbohydrates like simple monosaccharides (glucose, fructose etc.) and disaccharides (sucrose) contribute towards the sweetness of citrus fruits. The protein found in the citrus fruits range from 0.1 to 1.3 g/100 g. Out of all protein found in citrus fruits, most of the proteins consists of enzymes that exhibit wide role in metabolic processes. The lipids present in the citrus fruits vary from 0.07 to 0.42 g/100 g. The lipids content of citrus fruits includes fatty acids, phospholipids and complex fats (USDA 2018). Citrus fruits are rich source of different micronutrients like minerals, water-soluble as well as fat-soluble vitamins. It contains large amounts of minerals like calcium, phosphorus, potassium, sodium, and magnesium. The major water-soluble vitamins found in citrus fruits are ascorbic acid/ vitamin C and vitamin B-complex, whereas the vitamin E and vitamin A are the main fat-soluble vitamins (USDA 2018) (Table 14.1).

14.4 Bioactive Compounds Present in Citrus Fruits

Citrus fruits comprise of beneficial phytochemicals for example polyphenols (phenols and flavonoids- naringin, nobiletin, and hesperidin), carotenoids (α-carotene and β-carotene), essential oils (D-limonene, α-pinene, β-pinene, limonoids, synephrines etc.), pectin, minerals, vitamins *viz;*vitamins A, E, C, coumarins, and other components. Several biological properties of citrus fruits including antioxidants, anti-carcinogen, anti-mutagenic, anti-inflammatory, and anti-aging properties are attributable to these phytochemicals found in them (Rajendran et al. 2014; Zhang et al. 2015; Ke et al. 2015).

14.4.1 Polyphenols

14.4.1.1 Phenols

Citrus fruits contain many naturally occurring bioactive components such as polyphenols, largely phenolic acids as well as flavonoids (Sharma et al. 2017). The phenol content of citrus fruits differs depending on different sections of citrus fruit. Remarkably, citrus by-products encompass greater amounts of polyphenols than the edible portion of citrus (Balasundram et al. 2006). In a study, the maximum amount of phenol was observed in the albedo portion of unripe sweet orange (10,910 mg/kg) and it reported for about half of the total cumulative phenol content found in albedo, flavedo and juice sacs. However, in orange and lemons, major phenol content was observed in flavedo portion while in pummelo, albedo of unripe fruit possesses highest total phenol content (Multari et al. 2020). Further, the polyphenol content of citrus fruit varies depending on different extraction techniques, extraction conditions, solvent used for extraction etc. Phenol contained in citrus fruits can be roughly divided into 2subcategoriesnamely, (1) hydroxybenzoic acid and (2) hydroxycinnamic acid (Ignat et al. 2011). Figures 14.1 and 14.2 presents the chemical diagram showing the basic structure of flavonoids and polyphenols contained in citrus fruits.

14.4.1.2 Flavonoids

Flavonoids belongs to a group of lower molecular mass naturally occurring phenolic components that are broadly dispersed in the plant domain including their existence in fruits, vegetables, flowers, tea leaves etc. This plant based secondary

Fig. 14.1 Basic structure of flavonoids

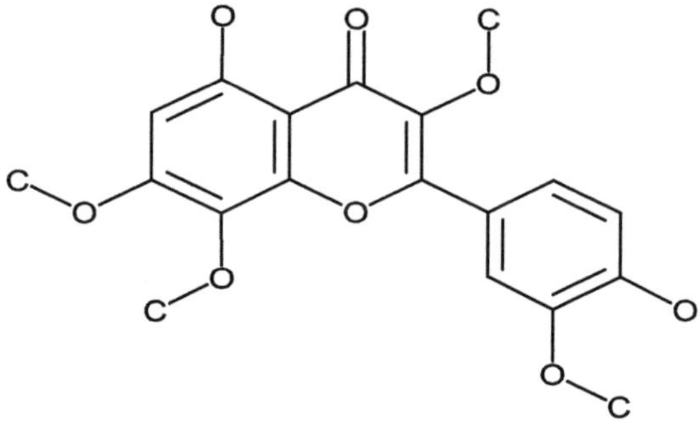

Fig. 14.2 Polyphenols found in citrus fruits

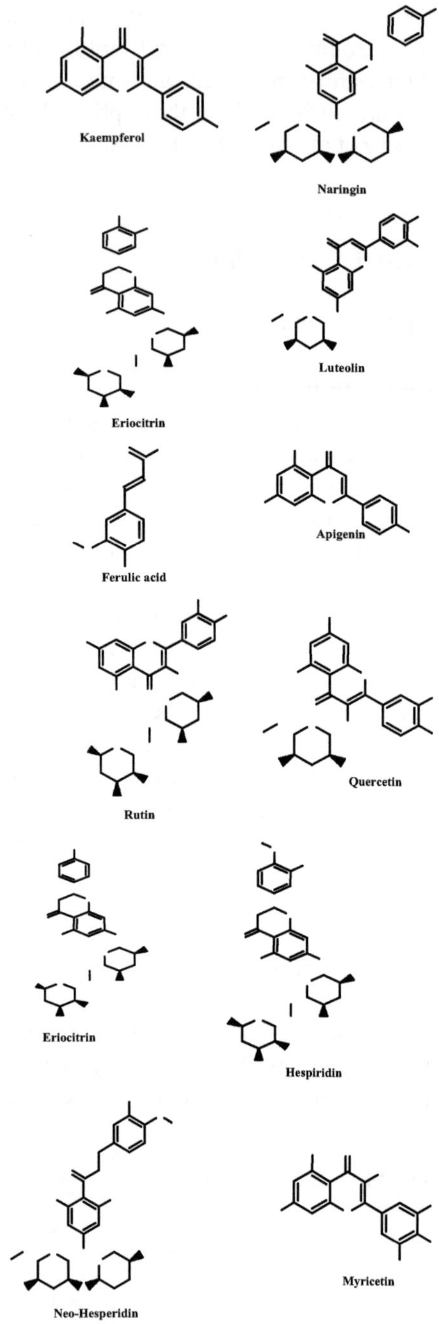

metabolites which are commonly known as flavonoids are related to a variety of health aids and remain vital ingredients in different dietary supplements, nutraceutical, medicine, pharmaceuticals and cosmetics sector. This is owing to their antioxidant, anti-carcinogenic, anti-mutagenic, and anti-inflammatory effects, in addition to their capacity to control main cellular enzyme activity. As per the chemical structure, flavonoids are made up of fifteen carbon body (C6-C3-C6) with two 6-carbon phenyl rings associated with an embedded oxygen-containing heterocycle. Flavonoids may be further subdivided on the basis of the carbon atom of the C-ring to which the B-ring is joined and based on the degree of unsaturation and the oxidation of the C-ring. Those flavonoids whose B-ring is joined at the 3rd position of the C-ring are termed isoflavones, while flavonoids having B-ring attached at fourth position are referred to as neo-flavonoids, and those with the B-ring attached at 2nd position are further categorized into diverse sub-groups based on the structural characteristics of the C-ring. The sub-groups formed are flavonols, flavanones, flavones, flavanonols/catechins, isoflavonoids anthocyanins, and chalcones (Panche et al. 2016). Citrus fruits comprise of significant amounts of flavonols for example; rutin, quercetin, and kaempferol; flavanones like naringin, hesperidin, narirutin, and eriocitrin; flavones for example vitexin, rhoifolin, and diosmin; polymethoxylated flavones commonly abbreviated as PMFs such as tangeritin, nobiletin, and 5-demethyl nobiletin; as well as anthocyanin such as cyanidin and peonidin glucoside (Saini et al. 2022).

Citrus fruits comprise of good amounts of natural flavonoids particularly PMFs which shows a variety of biological as well as physiological functions that are useful not only for plants but also for humans. Citrus PMFs acts as an important barrier to pathogenic attack (Luo et al. 2015). Eighty-one PMFs were noticed in the *Citrus reticulata* leaves, some of which are well-defined by specific names or structures. Thirty-Nine PMFs were recognized in 62 citrus germplasm flavedo, sweet orange and mandarin showed the maximum PMF levels, and lemons and pomelo exhibited lower quantities (Wang et al. 2017). High levels of PMF, for example tangeretin, nobiletin, and sinensetin, were spotted in 4 tissues of 11 citrus genotypes (Durand-Hulak et al. 2015).

Flavonoids show an eminent effect in the removal of reactive oxygen species. Amongst the different flavonoids present in citrus fruits, the antioxidant action of the hesperidin, naringin, and naringenin have been extensively studied (Nakao et al. 2011). Naringin in citrus fruits can greatly improve the immune system's efficiency to prevent organ as well as tissue damage or diseases triggered due to the oxidation by enhancing the action of certain enzymes like glutathione peroxidase, superoxide dismutase, catalase, paraoxonase and further antioxidative enzymes (Mamdouh and Monira 2004). Naringenin is present in major amounts in citrus fruits. Naringenin is usually observed as an aglycone and/or a glycoside (Erlund 2004). Among different flavonoids in citrus fruits, naringin and narirutin are specifically abundant. Naringenin is helpful in inhibiting fatty acid oxidation in the liver by means of regulating fatty acid oxidation enzymes, for example plasma antioxidant enzymes, carnitine palmitoyl transferase, 3-hydroxy3-methylglutaryllCoA reductase, and PON (Jung et al. 2006; Zou et al. 2016).

Hesperidin possess DPPH radical scavenging capacity and may dose-dependently hinder the *in-vitro*Cu^{2+} induced oxidation of low-density lipoprotein thereby promoting pancreatic B-cell rejuvenation and preventing oxidative stress in pregnant diabetic rats' embryo (Toumi et al. 2009). The hesperidin present in candied oranges, lemons and grapefruit were 11 mg/100 g, 22 mg/100 g and 4.37 mg/100 g, correspondingly (Zhou 2016). In a study, 11 major PMFs in fruit flavedo or citrus leaves of 116 citrus plants were assessed in combination with UPLC–DAD–ESI-QTOF-MS/MS and HPLC-DAD examination. All studied citrus plants and their natural or else artificial hybrids have been reported to comprise of enough measurable PMF, particularly in the flavedo portion of wild as well as early grown mandarin which are in the initial periods of fruit development (Peng et al. 2021).

Structural characterization of flavonoid present in pulp of Shatianyu (*Citrus grandis* L. Osbeck) showed that the flavonoids *viz;* naringin and rhoifolin exhibited maximum oxygen radical absorbance ability while melitidin, naringin, and bergamjuicin were main supplier of the ORAC in fruit extract (Deng et al. 2022). Nevertheless, current research has revealed that in citrus fruit albedo (which is inner layer of citrus peel), flavonoids accounted for 89.34% of the polyphenol portions, subjugated by flavanones *viz;* hesperidin and eriocitrin as important components, accounting for 52.81% and 31.31% of total flavonoids (Smeriglio et al. 2019).

Flavonoid concentration in the citrus fruits also vary based on the stage of pollination. Citrus fruit contains higher levels of most of the flavonoids in the intermediate stages of development for example; 60–80 days after pollination. Also, the flavonoid content decreases all through full ripening. This is perhaps because of the higher expression of certain rate limiting enzymes carrying out the biosynthesis of flavonoids *viz;* chalcone synthase-1 and chalcone isomerase (Ledesma-Escobar et al. 2018). In ripened fruit or in the later stages of development, the hesperidin content peaks in the lemon juice sac (*Citrus Akragas*) with a concentration of 2213 mg/kg while in different orange varieties the hesperidin content of 1957 and 1975 mg/kg were observed. However, the flavanone narirutin was abundantly found in grapefruit (292 mg/kg). A substantial quantity of eriocitrin flavanone was noted in lemons (913 mg/kg). Also, larger numbers of flavonoids are found in albedo and flavedo portion of citrus fruit peel in comparison to the citrus juice (Multari et al. 2020).

14.4.2 Essential Oils

Essential oils are those aromatic liquids that have lack of affinity for water and are composed of volatile substances generally present in oil sacs of peels and cuticle of citrus fruit (El Asbahani et al. 2015). Essential oils recovered primarily from the citrus fruit are economically significant with health-promoting action owing to the occurrence of terpenes especially monoterpene i.e., D-limonene and limonoids, sesquiterpenes along with other bioactive ingredients such as flavonoids, carotenoids and coumarins. Citrus fruit essential oil usually has more than 90% volatile content. Citrus essential oil possesses analgesic, antioxidative, anxiolytic, anti-inflammatory, neuroprotective and anti-microbial action. Owing to the properties of citrus essential oil, this could have applicability in food, cosmetics, pharma and perfumery industry (Suri et al. 2021). Because of the intense antimicrobial activity lately, citrus essential oil, has gotten huge consideration as a preservative for natural products, vegetables, meat, and handled food items. Monoterpenes and sesquiterpenes are usually present in the volatile fractions, with limonene as the main component. Limonene is one of most abundant essential oil with a usual concentration ranging from 73.9% to 97% (W/W essential oils). The US Food and Drug Administration (US-FDA) has regarded limonene as a Generally Recognized as Safe (GRAS) material. D-limonene is the main terpenoid accounting for 45–90% of the total terpenoid in orange, mandarin, lemon, tangerine, and grapefruit. The terpenoids mainly γ-terpinene and β-pinene accounts for about 8–20% and 0.3–11% of the total terpenoid, correspondingly in the essential oil present in lemon and mandarin (Raspo et al. 2020). D-Limonene comes from the citrus fruit peels and seeds. Limonene is a renewable organic compound and has many uses as an active ingredient in flavors and fragrances, and functionalized foodstuffs (Ciriminna et al. 2014). (+) Limonene (also R or D-Limonene) has a pleasing orange-like scent, and (−) Limonene (also S or L-Limonene) has a stimulating turpentine-like scent, so chirality of limonene is important aspect in flavor or fragrance (Jongedijk et al. 2016).

(+) Limonene could be simply attained from citrus peels and seeds in a cold pressed extraction procedure and thereafter obtained through the process of centrifugation or else steam distillation (Ciriminna et al. 2014). Another method for recovery of limonene is the conventional solid-liquid/ soxhlet extraction by means of pure and/or varied organic solvents. Amongst the cold pressed essential oil obtained from lemon, tangerine, clementine, sweet orange, bergamot, blood orange, bitter orange and grapefruit, the maximum limonin content (21.2 mg/L) was found in bergamot, whereas the limonin content of the essential oils of Blood Orange and Clementine was the lowermost (0.5–0.9 mg/L). Also, the green varieties of mandarin reported four-times greater limonin (4.5 mg/L) as compared to yellow and red mandarin variety (1.1 mg/L).

Meanwhile, various green extraction methodologies are utilized for retrieval of essential oil from citrus fruits for example, by use of supercritical fluid, microwave assisted extraction, steam explosion, and ultrasound assisted methods (Negro et al. 2016). A larger amount of limonene was recovered from the orange byproducts in a shorter amount of time via microwave extraction technology as compared to conventional heating method. D-limonene extraction consisted of a preliminary extraction phase from the outside of the cell followed by transmembrane diffusion. Microwave extraction technology is reported to exhibit high transmembrane diffusivity thereby resulting in higher yields in comparison to conventional oil extraction (Attard et al. 2014). Apart from microwave extraction, microwave-assisted hydro-distillation is effectively used for obtaining of essential oils from moist citrus peel waste, thereby lowering the costs, evading the use of additives, and enhancing process yields (Bustamante et al. 2016).

14.4.3 Pectin

Citrus fruits and processing by-products like peels contain significant amounts of soluble sugars (glucose, fructose and sucrose) that can be effectively used to produce bio-alcohol. The cellular components of the citrus peel include cellulose, hemicellulose, pectin, galacturonic acid, galactose, and arabinose. The abundance of soluble as well as insoluble sugars indicates their ability to effectively utilize value-added goods by consequential biological procedures (Rivas et al. 2008).

Pectin is one of the complex polysaccharides located in the cellular walls of higher plants. The major components of pectin include D-galacturonic acid that is also termed as sugar acid and is attained from galactose. Pectin acts as an emulsifying agent, thickening agent, stabilizing agent, fat substitute and gelling compound in dairy industry, jams, jellies as well as fruit juices (Suri et al. 2021). A literature review showed that with the advancement in the green chemistry, several scientists use the non-traditional green extraction methods to extract pectin from citrus fruits. For example, pectin yield of 27.81% was achieved upon microwave assisted extraction from grapefruit peel (Bagherian et al. 2011). In another research, microwave-assisted extraction from *Citrus medica* peels yielded 23.83% pectin at power output (660 W) and time (9 min) (Quoc et al. 2015). In addition, a pectin yield of 29.16% from the *Citrus maxima* peel by Bronsted acidic ionic microwave liquid-based extraction (Liu et al. 2017).

14.4.4 Carotenoids

Carotenoids are a ubiquitous assemblage of isoprenoid complexes that take part in photosynthesis as well as signal transduction. As per their chemical configuration, carotenoids can be subcategorized into two chief classes namely,

(a) Hydrocarbon carotenoids involving α-carotene, β-carotene and lycopene.
(b) Oxygenated derivatives of hydrocarbon carotenoids *viz;* β-cryptoxanthin, lutein, xanthophyll-neoxanthin, and violaxanthin.

However, β-carotene, β-cryptoxanthin, lutein, lycopene, as well as zeaxanthin are the most significant carotenoids present in citrus fruits. Carotenoids are those pigments that impart different colors like red, yellow, and orange to flowers plus fruits of different plant classes. Carotenoids are also known to offer the unique colors to birds, fishes and crustaceans (salmon, trout, lobsters, shrimp, pelicans, etc.). Nevertheless, the carotenoid pigment is not synthesized by plants and their color as well as carotenoid content vary depending on their diet (Fraser and Bramley 2004). Besides imparting the color, carotenoids also possess several important biological functions. They are considered as a vital source of vitamin A and can help in preventing the growth of degenerative illnesses *viz;* macular degeneration, metabolic disorder, cardiovascular diseases and cancer (Boukroufa et al. 2017). Carotenoids are involved in photoprotection and light harvesting complex of photosynthetic apparatus (Merchant and Sawaya 2005) They also play a role as a light stabilizer and free radical scavenger (Fraser and Bramley 2004).

Carotenoids present in colored citrus fruits *viz;* yellow and green-colored fruits had gained further importance in food sector as a result of their health promoting action (including richness in provitamin A and anti-cancer effects) (Rao and Rao 2007). Also, the antioxidative action of citrus fruits is mainly attributable to the occurrence of hydrophilic constituents in fruit (Cano et al. 2002). Usually, the carotenoid content of citrus fruits is nearly double that of other fruits. Around 115 different carotenoids are found in citrus fruits and their characteristic color is because of the carotenoid components found in citrus fruits (Tsai et al. 2007). Pink grapefruit contains good amount of β-carotene. Further, citrus fruits also comprise of large amounts of carotenoids for example lutein, zeaxanthin, and β-cryptoxanthin. The red colors of red navel orange and valencia orange are mainly because of high lycopene and cryptoxanthin content, respectively. Pink grapefruit contain high β-carotene content; Other citrus fruits contain high carotenoid content, such as zeaxanthin, lutein, and β-cryptoxanthin (Lee 2001). The carotenoid content in the orange was observed to be 11.25 mg/L when the sample extraction was done using ultrasound technique at power of 208 W/cm^2, temperature of 20 °C and time of 5 min. It was revealed that the carotenoid content of orange was increased by 40% through ultrasound-based extraction in comparison to the conventional extraction (Boukroufa et al. 2017). In another research, carotenoid levels were observed to be significantly greater in tangerine species, such as *Citrus unshiu* and *Citrus reticulata*, as compared to those in the orange variety, *Citrus sinensis* and the hybrid, *Citrus changshanensis* (Abeysinghe et al. 2007). Similarly, cultivar difference was observed among the carotenoid content of different citrus varieties (Fanciullino et al. 2006).

14.5 Extraction & Isolation of Bioactive Compounds

Qualitative and quantitative research on the bioactive components present in plant materials are primarily based on the selection of appropriate extraction methods. The extraction method is sometimes called "sample preparation technique". In most cases, two-thirds of the effort of analytical chemists is spent on sample preparation techniques, but this part of the study is ignored and carried out by untrained personnel. It is true that the advancement of state-of-the-art chromatographical and spectrophotometric procedures makes the study of bioactive components simpler than before, however the accuracy hinge on the type of extraction techniques, input variables, and the specific properties of the plant. Common parameters that affect the extraction process are the matrix features, solvent utilized for extraction, temperature, pressure, and time of the treatment. As a result of these tremendous technological advancements; different food as well as non-food sectors are increasingly interested in bioactive compounds from natural resources (Azmir et al. 2013).

Extraction of plant material could be carried through several extraction methods. Many scientists have been conducting experiments on innovative, inexpensive and effective ways of extraction. The extraction procedure performed to prepare and process the sample is of some interest to the scientific community. Extraction, separation/isolation, as well as description of bioactive ingredients from fruit/vegetables are the basic phases for obtaining bioactive components. Appropriate measures must be taken to prevent the destruction, loss or deterioration of potent bioactive components all through the extraction. Besides, the choice of solvent must be dependent on the properties of the bioactive compound of interest (Suri et al. 2022a). Over the last few years, environmentally friendly, non-conventional green extraction methods have been developed that use less synthetic plus organic chemicals, reduce treatment time, and improve yield and quality of extract. Ultrasonication, pulsed electric fields, supercritical fluids, enzymatic extraction, microwave extraction, extrusion, ohmic heating, high pressure extraction and accelerating solvents are used to enhance the total extraction yield as well as selectivity of bioactive compounds obtained from sample (Lusas and Watkins 1988; Lakkakula et al. 2004; Gaur et al. 2007; Ghafoor et al. 2009; Azmir et al. 2013).

14.5.1 Traditional/Conventional Extraction Methods

Traditional extraction methods *viz;* Soxhlet extraction are taken as a standard/control technique for comparing the achievement of afresh established methods. There are quite a few technical reports and data, in which unconventional methods are critically studied (Smith 2003; Wang and Weller 2006). Existing traditional methods for recovering bioactive components from plants, fruits, vegetables etc. are soxhlet extraction, maceration, and steam distillation/hydro-distillation. To date, numerous traditional and innovative extraction methods for extracting phenol from citrus by-products have been described. The most commonly used method is maceration/immersion in a solvent. It is also considered appropriate at scale-up levels. Nevertheless, this extraction method mainly needs high temperatures (50–150 °C), extended extraction periods (up to some hours), and highly polar solvents such as ethanol (Khan and Dangles 2014). Other traditional methods including soxhlet extraction, hydro-distillation etc. are also used to recover bioactive compounds from citrus fruits.

14.5.1.1 Soxhlet Extraction

A German scientist named Franz Ritter von Soxhlet was the first person who studied and researched soxhlet extractor in 1879. It was developed primarily for the recovery of lipids, yet with advancement it is used for isolation of many different components. Soxhlet extraction method is utilized to extract important bioactive components from a number of natural materials. Besides, it also serves as a model for comparing new/innovative extraction techniques. Usually, for extraction of compounds in soxhlet extractor, a minute quantity of dry sample is positioned in the thimble. The thimbles are thereafter kept in a distillation flask comprising of the appropriate solvent. Later when the solvent reaches to the overflow stage, the thimble holder solution is sucked out through the siphon. The siphon returns the solution to the distillation unit. This solution transports the recovered solute to a bulk liquid. The solute particles persist in the distillation unit and the solvent is returned to the permanent plant bed. This procedure repeats till the complete recovery happens (Azmir et al. 2013).

14.5.1.2 Maceration

Maceration has long been employed in the preparation of home-based tonics. It has become a common and reasonable method to acquire essential oils as well as bioactive components. For small use or laboratory type extractions, maceration usually comprise of different stages. Initially, the plant material is crushed to enhance its surface area for appropriate mixing of solvent. Second, during maceration, a suitable solvent known as menstruum is added to the bolted container. Third, the straining of liquid is done, however the solid filtrate left after this extraction procedure, ex-citrus pomace is pressed for recovery of huge amounts of trapped solution. The sieve obtained and the squeezed liquid are grouped together, and impurities are parted through filtration process. Infrequent shaking during the process of maceration facilitates recovery of bioactive compounds in two ways. (A) It leads to an increase in process of diffusion, (b) It eliminate concentrated solution out of the surface of sample and bring new solvent into menstruum to increase extract yield (Azmir et al. 2013).

14.5.1.3 Steam Distillation/Hydro-Distillation

It is a conventional technique of extracting bioactive components, essential oils etc. from plant sources. It did not contain organic solvents and this procedure can be conducted prior to dehydration of plant material. Hydro-distillation can be done through 3 different methods namely water-based distillation process, direct steam distillation and water coupled steam distillation.

In the hydro-distillation process, the plant material is first sealed in a booth, thereafter water is added, and plant material is boiled. Sequentially, steam is injected directly into the plant material. The major reason behind the release of bioactive components from plant tissues is hot water and steam treatment. The indirect cooling with water leads to condensation of the vapor blend of water plus oil. The mix obtained after condensation moves from the condenser to the separator, at that place the oil as well as bioactive components are involuntary separated from the water. Hydro-distillation process encompasses three major physicochemical processes, firstly hydro-diffusion, secondly hydrolysis and thirdly disintegration through heat. At higher temperatures, few volatile compounds can be destroyed. This limitation of hydro-distillation restricts the use of this method for the recovery of thermo-labile components (Azmir et al. 2013).

14.5.2 Non-traditional/Green Extraction Methods

14.5.2.1 Microwave Assisted Extraction

This type of extraction technology uses electromagnetic waves with frequencies between 0.3 and 300 GHz. This is a frequently used process for recovering bioactive compounds from citrus fruits. Microwave based extraction treatment is said to include 3 consecutive stages namely (a) the disassociation of solutes from the active sites of food material at higher temperature as well as pressure (b) the diffusion of the solvent through the food material (c) the liberation of solutes from the food material into the solvent (Alupului et al. 2012). The main benefits of using microwaves over different traditional used methods are it requires low energy consumption, short time of extraction and low solvent usage (Suri et al. 2021). This recovery technique can extract bioactive components faster than traditional extraction methods and has a higher recovery rate. This is a selective method for recovering the intact organic plus organometallic complexes. It is likewise well-known green extraction technology to lessen the utilization of organic solvents (Alupului et al. 2012). Interestingly, researchers have been working on recovery of pectin, polyphenols, flavonoids, carotenoids, and dietary fibre from citrus fruit peels in a non-conventional way.

Pectin extraction from *Citrullus lanatus* peels using microwaves at different power (160–480 W), irradiation time (60–180 s) and liquid-solid ratio (1: 10–1:30 g/mL). A total yield of pectin (25.79%) was achieved with power (477 W), time

(128 s), and solid-liquid ratio (1:20) (Maran et al. 2014). Besides, the recovery of pectin from the skin of citrus sinensis by surfactant-microwave-assisted extraction exhibited 28% pectin yield at time (7 min), power (400 W), and pH (1.2) (Su et al. 2019). In addition to the recovery of pectin from citrus fruits, polyphenols were also recovered utilizing microwave extraction technique. Phenolic components found in the peel of *Citrus inshiu* was extracted by using microwaves. The study showed hesperidin content of 5860 mg/100 g and narirutin content of 1310 mg/100 g in skin extract of citrus fruit (Inoue et al. 2010). *Citrus limon* peels reported total phenol (1574 mg GAE/100 g) by extraction using microwaves (Dahmoune et al. 2013).

14.5.2.2 Ultrasound Assisted Extraction

Ultrasound is a distinct form of sound wave that goes over and beyond the human hearing. In the field of chemistry, ultrasound usually range from 20 kHz to 100 MHz. Like the other wave types, it penetrates the medium through producing firmness as well as expanding. The ultrasound treatment causes cavitation, which means the formation, development along with collapse of bubbles. By converting kinetic energy into heating of the bladder contents, a large amount of energy can be generated (Azmir et al. 2013). Advantages of the ultrasound method is the reduction in time of extraction, energy and solvent consumption. Ultrasonic energy for recovery of bioactive compounds provides further efficient mixing, quicker transference of energy, condensed temperature gradient and temperature of extraction, extraction of selective compounds, compact equipment size, faster start-up, promote improved production and eliminate processing stages (Chemat et al. 2008). The possible mechanism behind ultrasound assisted extraction is enhanced ultrasonic mass transfer along with accelerated solvent access to cellular components of the plant. The mechanistic action of ultrasonic extraction encompasses two physical phenomena's, (a) diffusion through the cellular walls and (b) releasing of intracellular substances subsequently after rupturing of wall (Mason et al. 1996). Sample moisture, sample particle size, degree of milling, as well as solvent are central features for an effective and efficient extraction process. In addition, the extraction temperature, frequency, pressure, and duration of ultrasonic waves are the determining aspects of ultrasonic wave performance.

Ultrasound extraction has also been utilized with many classical systems as they are believed to improve the effectiveness of a traditional scheme. In the solvent extraction, an ultrasonic device is positioned in a suitable place to improve the effectiveness of extraction (Vinatoru et al. 1998). The ultrasound extraction was lately used to recover hesperidin from peels of Penggan (*Citrus reticulata*) fruit (Ma et al. 2008) polyphenols and flavonone glycosides of *Citrus unshiu* Marc peels and total phenol content of Penggan peels (Ma et al. 2009). Researchers recovered polyphenols from the skin of mandarin and sweet orange by maceration and ultrasound technology, and the efficacy of ultrasonically assisted extraction was greater to that of the previously used techniques. The extraction yields obtained by ultrasonic treatment of mandarin and sweet orange were 5.85% and 12.95%, correspondingly, whereas the extract yields of 5.20% and 12.20% were obtained by the maceration technique (Saini et al. 2019). Recovery of polyphenols from fresh sweet lime (*Citrus sinensis*) peels by utilizing ultrasound treatment at power (200 W), frequency (40 kHz) and treatment time (20 min) exhibited the total phenol of 25.60 mg GAE/g and total flavonoid content of 18.85 mg QE/g (Suri et al. 2022b). Polyphenols obtained from three hybrid varieties of mandarin fruit (*Clemenvilla*, *Ortanique*, and *Nadorcott*) by means of ultrasonic assisted extraction techniques at power (400 W), temperature (40 °C) plus duty cycle (80%) was conducted. Ultrasound treatment for a shorter period of time (in between the first 5 & 15 min) caused the improved physicochemical characteristics, bioactive compounds, along with the antioxidative action of mandarin peel extract (Anticona et al. 2021). This depicts that ultrasonic extraction is an effective as well as sustainable extraction technology.

14.5.2.3 Pulsed Electric Field Extraction

This extraction technique depends on inducing electroporation of cell membranes, ensuing in an improved in extraction yield. In this method, the electric potential passes by the cellular membrane at a short period of time (1–2500 ls) and the particles separate on the basis of the inherent charge. Therefore, repulsive action leads to the establishment of pores and increases their porosity (Azmir et al. 2013). Normally, pulsed electric fields are used to preserve food and extract intracellular components from plant matrices, agro-food waste and their by-products (Suri et al. 2021).

The influence of pulsed electric field on extraction of total phenols as well as flavonoids from orange peels were evaluated. The maximum disintegration index was observed when the time of processing was 60ls. Besides, an increase in the yield of extraction of total phenols by 20%, 129%, 153% and 159% from orange peel was observed following the pulsed electric field treatment at 1, 3, 5 and 7 kV/cm, respectively. Also, in comparison with the untreated material, the pulsed electric field treated orange peels exhibited enhanced flavonoids namely hesperidin and naringin, and total antioxidant property (Luengo et al. 2013).

14.5.2.4 Supercritical Fluid Extraction

All matter has 3 fundamental states *viz;* solid, liquid, and gas. Supercritical state is definite and could only be reached when a material is exposed to temperatures and pressures above the critical point. The critical point is explained as the distinctive temperature as well as pressure beyond which the characteristic gas and liquid phases are absent (Inczedy et al. 1998). In supercritical conditions, certain attributes of gases and liquids are lost. In other words, supercritical fluids cannot be liquefied even if the temperature or pressure is changed. Supercritical fluids have gas-like features such as diffusion, surface tension, viscosity, along with liquid-like density plus solvation (Sihvonen et al. 1999) These features make it appropriate for extracting components in high yields in a quick span of time.

Supercritical extraction technique is considered as an environmental friendly substitution to traditional techniques of extraction. This supercritical method of extraction synthesizes fluids near the critical point at higher temperatures as well as pressures (Diaz-Reinoso et al. 2006). A number of scientists examined the effect of supercritical fluid on extract yield of bioactive components from the citrus fruits. The supercritical fluid process for extracting essential oil from citrus lemon zest led to higher yields of D-limonene (4.5%) (Lopresto et al. 2019). In analogous research, the process of extraction of D-limonene from tangerine peel (*Citrus unshiu* Kuno) using supercritical $CO_2$was carried out, where a high yield of D-limonene (30.65%) at a pressure of 300 bar, however 13.16% yield was reported at a pressure of 100 bar (Safranko et al. 2021). Therefore, although the supercritical extraction technique was known to be advantageous for the extracting oil, the main problems in supercritical fluid extraction were recognized as high maintenance and equipment cost (Suri et al. 2022a).

14.5.2.5 Enzyme Assisted Extraction

Usually some of the bioactive components and plant-based chemicals in the plant matrix are disseminated in the cytoplasm of the cell and few remained in the polysaccharide lignin arrangement by the help of hydrogen bonding or hydrophobicity, that are inaccessible to solvents in the extraction process. Several elements for example; concentration and composition of enzyme, plant particle size, solid-to-liquid ratio, time of hydrolysis, etc. are considered important aspects for the extraction process (Niranjan and Hanmoungjai 2004). Enzymatic pre-treatment step is observed as an innovative and efficient technique for extracting bound compounds thereby improve total yield (Rosenthal et al. 1996). In this process, the utilization of specific enzymes for example cellulase, pectinase, α-amylase etc. in the extraction process improves retrieval by disrupting cell walls and hydrolysing structural plus lipid-soluble polysaccharides. There are 2 methods of enzyme-assisted extraction: (1) enzyme-assisted aqueous based extraction method, (2) enzyme-assisted cold pressed extraction method (Latif and Anwar 2009; Azmir et al. 2013). Traditionally, the enzyme-assisted aqueous extraction functions primarily to extract oil from varied seeds. Enzyme-assisted cold pressing extraction technology uses enzymes to hydrolyze seed cellular walls as polysaccharide-protein colloids are not accessible in this system (Concha et al. 2004).

14.5.2.6 Pressurized Liquid Extraction

This extraction technique is termed through different titles *viz*; pressurized fluid extraction, improved solvent extraction, accelerated fluid extraction, and high-pressure solvent extraction process (Nieto et al. 2010). The idea behind pressurized fluid extraction is to use high pressure to keep the solvent liquid above its usual boiling point. The peculiarity of pressurized liquid extraction is that its elevated pressure carries forward the process of extraction. Automated approach is the major reason for the further expansion of pressure based liquid extraction process, in addition to its benefits like reduced extraction times plus solvent requirements. This extraction technology requires a minimum quantity of solvent due to its high temperature and pressure, which speeds up the recovery of plant materials. Higher extraction temperatures can increase the solubility of the analyte by enhancing both the solvability the mass transfer rate, reducing the viscosity plus surface tension of the solvent, thus increasing the extraction rate (Ibañez et al. 2012). Several studies have explained the applications of pressurized water-based extraction method to obtain bioactive components from citrus fruit peels, pomace etc. High-pressure based extraction of polyphenols from peels of lemon and orange was performed. Testing of pressure-treated citrus peels (300 MPa, 10 min, & 500 MPa, 3 min) showed higher levels of phenol and antioxidants than control. The total phenol present in the extract obtained from the fresh peels of lemon at 300 MPa and 500 MPa was 265.95 mg GAE/100 g and 344.53 mg GAE/100 g, correspondingly, whereas for orange it was 364.57 and 378.90 mg GAE/100 g, correspondingly (Casquete et al. 2014).

14.6 Factors Influencing the Extraction of Bioactive Compounds from Citrus Fruits

14.6.1 Solvent

Solvents utilized for recovering plant bioactive components have a direct influence on the efficiency of phytochemicals like polyphenols, flavonoids, carotenoids etc. With a wide series of solvents from polar to non-polar, different components with respective polarities could be recovered into solvents based on the polarity of the solvent and the target plant material (Dailey and Vuong 2015). In general, coupling of polar as well as less polar solvents is further effective in the recovery of biologically active components from plant matrix (Vuong et al. 2013). Environmental friendly, non-toxic food grade organic solvents such as n-butanol, ethanol, and isopropanol are suggested by the United States Food and Drug Administration for recovery of bioactive constituents (Bartnick et al. 2006). The usage of ethanol in the ultrasound-based process has been known to be more efficient in recovering polyphenols from citrus fruit *viz;* orange peel waste than traditional solvent-based method (Khan et al. 2010). In addition, the non-conventional methods like ultrasound, microwave etc. provided higher yields in natural product extraction than conventional methods, not only on a laboratory scale however also on a pilot plant scale (Boonkird et al. 2008).

14.6.2 *Solid to Solvent Ration*

Solvent volume or the plant material to solvent ratio have been studied to enhance the effectiveness of bioactive components from plant matrix. Hypothetically, the lesser volume of solvent utilized, the lower is the extraction efficiency that occurs due to saturation. Nevertheless, an appropriate plant material (solid)/solvent ratio must be useful for cost-effective purpose as further energy is required to heat the bigger mass. Moreover, greater energy is needed to eliminate the water for more concentrated or powdered production (AL Ubeed et al. 2022).

14.6.3 *Extraction Conditions: Treatment Time, Temperature, Agitation etc.*

Extraction temperature and time greatly influences the efficiency of extraction of bioactive complexes from plant materials (Fig. 14.3). Higher temperature with elongated recovery time generally results in greater extraction efficiency. Yet, the

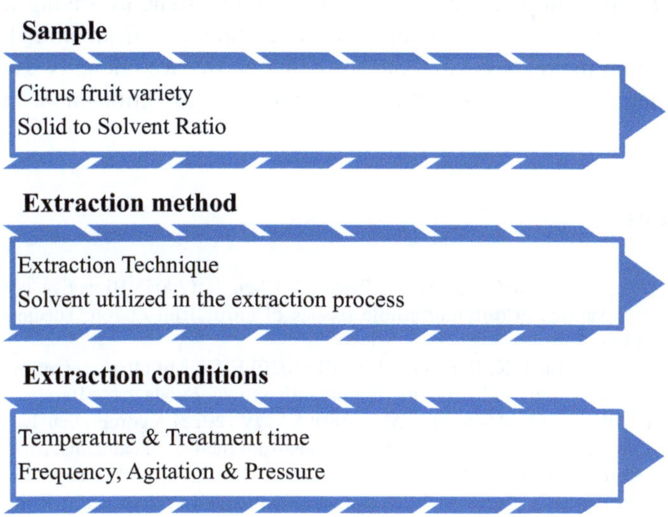

Fig. 14.3 Factors influencing the extraction process

stability of bioactive compounds may decline with prolonged exposure to high temperature since major plant-based compounds are susceptible to heat. Hence, it is vital to regulate the extraction temperature and time to recover high concentrations of bioactive compounds with minimal deterioration.

Process agitation and pressure were observed to impact the proficiency of extraction of phytochemicals compounds from plant matrix. Research has proven that agitation prominently enhances the extraction competence of phytochemicals in comparison to processes where no agitation is used (Ahmed et al. 2020). The treatment time or number of extractions has been shown to influence the effectiveness of extraction process. The larger the extraction time given to the same amount of sample, the greater bioactive compounds can be recovered (Vuong et al. 2011).

14.7 Conclusion

The present chapter gives an in-depth view of different bioactive compounds found in citrus fruits, their methods of extraction and important considerations during extraction processes. It sums up reports on the taxonomy, nutritional, and bioactive content of different varieties of citrus fruits. It can be proposed from the chapter that citrus fruits contain excellent repository of biologically active compounds *viz;* phenols, flavonoids, carotenoids, essential oils, pectin, minerals, vitamins, coumarins, and other compounds that offers several health benefits to mankind. Extraction, isolation and characterization are the basic stages for efficient retrieval of bioactive compounds. Therefore, selection of a potential extraction techniques that can be used to effectively recover these bioactive compounds is utmost important. With the advent of green technology, certain non-conventional energy saving methods are extensively utilized for extraction of bioactive components from citrus fruits. These green extraction methods exhibit various benefits over traditionally used methods. Also, the chapter discusses the influence of extraction condition on the extract yield.

References

Abeysinghe DC, Li X, Sun C, Zhang W, Zhou C, Chen K (2007) Bioactive compounds and antioxidant capacities in different edible tissues of citrus fruit of four species. Food Chem 104:1338–1344

Ahmed M, Ramachandraiah K, Jiang GH, Eun JB (2020) Effects of ultra-sonication and agitation on bioactive compounds and structure of amaranth extract. Foods 9:11–16

Al Ubeed HMS, Bhuyan DJ, Alsherbiny MA, BasuA VQV (2022) A comprehensive review on the techniques for extraction of bioactive compounds from medicinal cannabis. Molecules 27:604

Albertini MV, Carcouet E, Pailly O, Gambotti C, Luro F, Berti L (2006) Changes in organic acids and sugars during early stages of development of acidic and acidless citrus fruit. J Agric Food Chem 54:8335–8339

Alupului A, Calinescu I, Lavric V (2012) Microwave extraction of active principles from medicinal plants. UPB Sci Bul Ser B 74:129–142

Anticona M, Blesa J, Lopez-Malo D, Frigola A, Esteve MJ (2021) Effects of ultrasound-assisted extraction on physicochemical properties, bioactive compounds, and antioxidant capacity for the valorization of hybrid mandarin peels. Food Biosci 42:101185

Attard TM, Watterson B, Budarin VL, Clark JH, Hunt AJ (2014) Microwave assisted extraction as an important technology for valorising orange waste. New J Chem 38:2278–2283

Azmir J, Zaidul ISM, Rahman MM, Sharif KM, Mohamed A, Sahena F et al (2013) Techniques for extraction of bioactive compounds from plant materials: a review. J Food Eng 117:426–436

Bagherian H, Ashtiani FZ, Fouladitajar A, Mohtashamy M (2011) Comparisons between conventional, microwave-and ultrasound-assisted methods for extraction of pectin from grapefruit. Chem Eng Pro Process Intensif 50:1237–1243

Balasundram N, Sundram K, Samman S (2006) Phenolic compounds in plants and agri-industrial by-products: antioxidant activity, occurrence, and potential uses. Food Chem 99:191–203

Bartnick D, Mohler C, Houlihan M (2006) U.S. Patent Application No. 10/972,751

Berk Z (2016) Chapter 1: Introduction: history, production, trade, and utilization. In: Berk Z (ed) Citrus fruit processing. Academic, San Diego, pp 1–8

Bermejo A, Llosa MJ, Cano A (2011) Analysis of bioactive compounds in seven citrus cultivars. Food Sci Tech Int 17:55–62

Boonkird S, Phisalaphong C, Phisalaphong M (2008) Ultrasound-assisted extraction of capsaicinoids from Capsicum frutescens on a lab-and pilot-plant scale. Ultra Sonochem 15:1075–1079

Boukroufa M, Boutekedjiret C, Chemat F (2017) Development of a green procedure of citrus fruits waste processing to recover carotenoids. Resour Eff Technol 3:252–262

Bustamante J, van Stempvoort S, Garcia-Gallarreta M, Houghton JA, Briers HK, Budarin VL, Clark JH (2016) Microwave assisted hydro-distillation of essential oils from wet citrus peel waste. J Clean Prod 137:598–605

Cano A, Alcaraz O, Acosta M, Arnao MB (2002) On-line antioxidant activity determination: comparison of hydrophilic and lipophilic antioxidant activity using the ABTS•+ assay. Redox Rep 7:103–109

Casquete R, Castro SM, Villalobos MC, Serradilla MJ, Queirós RP, Saraiva JA, ... Teixeira P (2014) High pressure extraction of phenolic compounds from citrus peels. High Pres Res 34(4):447–451

Chemat F, Tomao V, Virot M (2008) Ultrasound-assisted extraction in food analysis. In: Handbook of food analysis instruments, pp 85–103

Ciriminna R, Lomeli-Rodriguez M, Cara PD, Lopez-Sanchez JA, Pagliaro M (2014) Limonene: a versatile chemical of the bioeconomy. Chem Comm 50:15288–15296

Concha J, Soto C, Chamy R, Zuniga ME (2004) Enzymatic pretreatment on rose-hip oil extraction: hydrolysis and pressing conditions. J Am Oil Chem Soc 81:549–552

Curk F, Ancillo G, Garcia-Lor A, Luro F, Perrier X, Jacquemoud-Collet JP, Navarro L, Ollitrault P (2014) Next generation haplotyping to decipher nuclear genomic interspecific admixture in citrusspecies: analysis of chromosome 2. BMC Genet 15:1–19

Czech A, Malik A, Sosnowska B, Domaradzki P (2021) Bioactive substances, heavy metals, and antioxidant activity in whole fruit, peel, and pulp of citrus fruits. Int J Food Sci 2021:6662259

Dahmoune F, Boulekbache L, Moussi K, Aoun O, Spigno G, Madani K (2013) Valorization of Citrus limon residues for the recovery of antioxidants: Evaluation and optimization of microwave and ultrasound application to solvent extraction. Indust Crop Prod 50:77–87

Dailey A, Vuong QV (2015) Effect of extraction solvents on recovery of bioactive compounds and antioxidant properties from macadamia (*Macadamia tetraphylla*) skin waste. Cogent Food Agric 1:1115646

Deng M, Jia X, Dong L, Liu L, Huang F, Chi J et al (2022) Structural elucidation of flavonoids from Shatianyu (*Citrus grandis* L. Osbeck) pulp and screening of key antioxidant components. Food Chem 366:130605

Díaz-Reinoso B, Moure A, Domínguez H, Parajó JC (2006) Supercritical CO_2 extraction and purification of compounds with antioxidant activity. J Agril Food Chem 54:2441–2469

Durand-Hulak M, Dugrand A, Duval T, Bidel LP, Jay-Allemand C, Froelicher Y, Bourgaud F, Fanciullino AL (2015) Mapping the genetic and tissular diversity of 64 phenolic compounds in citrus species using a UPLC–MS approach. Ann Bot 115:861–877

El Asbahani A, Miladi K, Badri W, Sala M, Addi EA, Casabianca H et al (2015) Essential oils: from extraction to encapsulation. Int J Pharm 483:220–243
Erlund I (2004) Review of the flavonoids quercetin, hesperetin, and naringenin. Dietary sources, bioactivities, bioavailability, and epidemiology. Nutr Res 24:851–874
Fanciullino AL, Dhuique-Mayer C, Luro F, Casanova J, Morillon R, Ollitrault P (2006) Carotenoid diversity in cultivated citrus is highly influenced by genetic factors. J Agric Food Chem 54:4397–4406
FAOSTAT (2022) Food and agriculture data. Retrieved February 21, 2022, from https://www.fao.org/faostat/en/#data/QC
Fraser PD, Bramley PM (2004) The biosynthesis and nutritional uses of carotenoids. Prog Lipid Res 43(3):228–265
Gaur R, Sharma A, Khare SK, Gupta MN (2007) A novel process for extraction of edible oils: enzyme assisted three phase partitioning (EATPP). Bioresour Technol 98:696–699
Ghafoor K, Choi YH, Jeon JY, Jo IH (2009) Optimization of ultrasound-assisted extraction of phenolic compounds, antioxidants, and anthocyanins from grape (Vitis vinifera) seeds. J Agric Food Chem 57:4988–4994
Ibañez E, Herrero M, Mendiola JA, Castro-Puyana M (2012) Extraction and characterization of bioactive compounds with health benefits from marine resources: macro and micro algae, cyanobacteria, and invertebrates. In: Marine bioactive compounds. Springer, Boston, pp 55–98
Ignat I, Volf I, Popa VI (2011) A critical review of methods for characterisation of polyphenolic compounds in fruits and vegetables. Food Chem 126:1821–1835
Inczédy J, Lengyel T, Ure AM, Gelencsér A, Hulanicki A (1998) Compendium of analytical nomenclature. Blackwell Science, Hoboken
Inoue T, Tsubaki S, Ogawa K, Onishi K, Azuma JI (2010) Isolation of hesperidin from peels of thinned Citrus unshiu fruits by microwave-assisted extraction. Food Chem 123(2):542–547
Jongedijk E, Cankar K, Buchhaupt M, Schrader J, Bouwmeester H, Beekwilder J (2016) Biotechnological production of limonene in microorganisms. Appl Microbiol Biotechnol 100:2927–2938
Jung UJ, Lee MK, Park YB, Kang MA, Choi MS (2006) Effect of citrus flavonoids on lipid metabolism and glucose-regulating enzyme mRNA levels in type-2 diabetic mice. Int J Biochem Cell Biol 38:1134–1145
Karn A, Zhao C, Yang F, Cui J, GaoZ WM et al (2021) *In-vivo* biotransformation of citrus functional components and their effects on health. Crit Rev Food Sci Nutr 61:756–776
Ke Z, Xu X, Nie C, Zhou Z (2015) Citrus flavonoids and human cancers. J Food Nutr Res 3:341–351
Khan MK, Dangles O (2014) A comprehensive review on flavanones, the major citrus polyphenols. J Food Compos Anal 33:85–104
Khan MK, Abert-Vian M, Fabiano-Tixier AS, Dangles O, Chemat F (2010) Ultrasound-assisted extraction of polyphenols (flavanone glycosides) from orange (*Citrus sinensis* L.) peel. Food Chem 119:851–858
Lakkakula NR, Lima M, Walker T (2004) Rice bran stabilization and rice bran oil extraction using ohmic heating. Bioresour Technol 92:157–161
Latif S, Anwar F (2009) Physicochemical studies of hemp (*Cannabis sativa*) seed oil using enzyme-assisted cold-pressing. Eur J Lipid Sci Technol 111:1042–1048
Ledesma-Escobar CA, de Castro MDL (2014) Towards a comprehensive exploitation of citrus. Trends Food Sci Technol 39:63–75
Ledesma-Escobar CA, Priego-Capote F, Robles Olvera VJ, Luque de Castro MD (2018) Targeted analysis of the concentration changes of phenolic compounds in Persian lime (*Citrus latifolia*) during fruit growth. J Agric Food Chem 66:1813–1820
Lee HS (2001) Characterization of carotenoids in juice of red navel orange (Cara Cara). J Agric Food Chem 49:2563–2568
Liu Z, Qiao L, Gu H, Yang F, Yang L (2017) Development of Brönsted acidic ionic liquid based microwave assisted method for simultaneous extraction of pectin and naringin from pomelo peels. Sep Purif Technol 172:326–337
Lopresto CG, Meluso A, Di Sanzo G, Chakraborty S, Calabrò V (2019) Process-intensified waste valorization and environmentally friendly d-limonene extraction. Eur Mediterranean J Env Integr 4:1–12

Luengo E, Álvarez I, Raso J (2013) Improving the pressing extraction of polyphenols of orange peel by pulsed electric fields. Innov Food Sci Emerg Technol 17:79–84
Luo T, Xu K, Luo Y, Chen J, Sheng L, Wang J, Han J, Zeng Y, Xu J, Chen J, Deng X (2015) Distinct carotenoid and flavonoid accumulation in a spontaneous mutant of ponkan (*Citrus reticulata* Blanco) results in yellowish fruit and enhanced postharvest resistance. J Agric Food Chem 63:8601–8614
Lusas EW, Watkins LR (1988) Oilseeds: extrusion for solvent extraction. J Am Oil Chem Soc 65:1109–1114
Ma Y, Ye X, Hao Y, Xu G, Xu G, Liu D (2008) Ultrasound-assisted extraction of hesperidin from Penggan (*Citrus reticulata*) peel. Ultrason Sonochem 15:227–232
Ma YQ, Chen JC, Liu DH, Ye XQ (2009) Simultaneous extraction of phenolic compounds of citrus peel extracts: effect of ultrasound. Ultrason Sonochem 16:57–62
Mamdouh MA, Monira AAEK (2004) The influence of naringin on the oxidative state of rats with streptozotocin-induced acute hyperglycaemia. Zeitschrift für Naturforschung C 59:726–733
Maran JP, Sivakumar V, Thirugnanasambandham K, Sridhar R (2014) Microwave assisted extraction of pectin from waste *Citrullus lanatus* fruit rinds. Carbohydr Polym 101:786–791
Mason TJ, Paniwnyk L, Lorimer JP (1996) The uses of ultrasound in food technology. Ultrason Sonochem 3:S253–S260
Maugeri A, Cirmi S, Minciullo PL, Gangemi S, Calapai G, Mollace V, Navarra M (2019) Citrus fruits and inflammaging: a systematic review. Phytochem Rev 18:1025–1049
Merchant S, Sawaya MR (2005) The light reactions: a guide to recent acquisitions for the picture gallery. Plant Cell 17:648–663
Multari S, Licciardello C, Caruso M, Martens S (2020) Monitoring the changes in phenolic compounds and carotenoids occurring during fruit development in the tissues of four citrus fruits. Food Res Int 134:109228
Nakao K, Murata K, Itoh K, Hanamoto Y, Masuda M, Moriyama K et al (2011) Anti-hyperuricemia effects of extracts of immature *Citrus unshiu* fruit. J Trad Med 28:10–15
Negro V, Mancini G, Ruggeri B, Fino D (2016) Citrus waste as feedstock for bio-based products recovery: review on limonene case study and energy valorization. Bioresource Technol 214:806–815
Nieto A, Borrull F, Pocurull E, Marcé RM (2010) Pressurized liquid extraction: a useful technique to extract pharmaceuticals and personal-care products from sewage sludge. Trends Anal Chem 29:752–764
Niranjan K, Hanmoungjai P (2004) Enzyme-aided aquous extraction. In: Nutritionally enhanced edible oil processing. AOCS Publishing
Panche AN, Diwan AD, Chandra SR (2016) Flavonoids: an overview. J Nutr Sci 5:e47
Peng Z, Zhang H, Li W, Yuan Z, Xie Z, Zhang H et al (2021) Comparative profiling and natural variation of polymethoxylated flavones in various citrus germplasms. Food Chem 354:129499
Putnik P, Barba FJ, Lorenzo JM, Gabrić D, Shpigelman A, Cravotto G, Bursać Kovačević D (2017) An integrated approach to mandarin processing: food safety and nutritional quality, consumer preference, and nutrient bioaccessibility. Compr Rev Food Sci Food Saf 16:1345–1358
Quoc LPT, Huyen VTN, Hue LTN, Hue NTH, Thuan NHD, Tam NTT, Thuan NN, Duy TH (2015) Extraction of pectin from pomelo (Citrus maxima) peels with the assistance of microwave and tartaric acid. Int Food Res J 22:1637
Rajendran P, Nandakumar N, Rengarajan T, Palaniswami R, Gnanadhas EN, Lakshminarasaiah U et al (2014) Antioxidants and human diseases. Clin Chim Acta 436:332–347
Rao AV, Rao LG (2007) Carotenoids and human health. Pharmacol Res 55:207–216
Raspo MA, Vignola MB, Andreatta AE, Juliani HR (2020) Antioxidant and antimicrobial activities of citrus essential oils from Argentina and the United States. Food Biosci 36:100651
Rivas B, Torrado A, Torre P, Converti A, Domínguez JM (2008) Submerged citric acid fermentation on orange peel autohydrolysate. J Agric Food Chem 56:2380–2387
Rosenthal A, Pyle DL, Niranjan K (1996) Aqueous and enzymatic processes for edible oil extraction. Enzyme Microb Technol 19:402–420
Šafranko S, Ćorković I, Jerković I, Jakovljević M, Aladić K, Šubarić D, Jokić S (2021) Green extraction techniques for obtaining bioactive compounds from mandarin peel (*Citrus unshiu* var. Kuno): Phytochemical analysis and process optimization. Foods 10:1043

Saini A, Panesar PS, Bera MB (2019) Comparative study on the extraction and quantification of polyphenols from citrus peels using maceration and ultrasonic technique. Curr Res Nutr Food Sci 7:678

Saini RK, Ranjit A, Sharma K, Prasad P, Shang X, Gowda KGM, Keum YS (2022) Bioactive compounds of citrus fruits: a review of composition and health benefits of carotenoids, flavonoids, limonoids, and terpenes. Antioxidants 11:239

Sharma K, Mahato N, Cho MH, Lee YR (2017) Converting citrus wastes into value-added products: economic and environmently friendly approaches. Nutrition 34:29–46

Sihvonen M, Järvenpää E, Hietaniemi V, Huopalahti R (1999) Advances in supercritical carbon dioxide technologies. Trends Food Sci Technol 10:217–222

Smeriglio A, Cornara L, Denaro M, Barreca D, Burlando B, Xiao J, Trombetta D (2019) Antioxidant and cytoprotective activities of an ancient Mediterranean citrus (*Citrus lumia* Risso) albedo extract: microscopic observations and polyphenol characterization. Food Chem 279:347–355

Smith RM (2003) Before the injection—modern methods of sample preparation for separation techniques. J Chromatogr 1000:3–27

Su DL, Li PJ, Quek SY, Huang ZQ, Yuan YJ, Li GY, Shan Y (2019) Efficient extraction and characterization of pectin from orange peel by a combined surfactant and microwave assisted process. Food Chem 286:1–7

Suri S, Singh A, Nema PK (2021) Recent advances in valorization of citrus fruits processing waste: a way forward towards environmental sustainability. Food Sci Biotechnol 30:1601–1626

Suri S, Singh A, Nema PK (2022a) Current applications of citrus fruit processing waste: a scientific outlook. Appl Food Res 2:100050

Suri S, Singh A, Nema PK, Malakar S, Arora VK (2022b) Sweet lime (*Citrus limetta*) peel waste drying approaches and effect on quality attributes, phytochemical and functional properties. Food Biosci 48:101789

Tocmo R, Pena-Fronteras J, Calumba KF, Mendoza M, Johnson JJ (2020) Valorization of pomelo (*Citrus grandis* Osbeck) peel: a review of current utilization, phytochemistry, bioactivities, and mechanisms of action. Compr Rev Food Sci Food Saf 19:1969–2012

Toumi ML, Merzoug S, Boutefnouchet A, Tahraoui A, Ouali K, Guellati MA (2009) Hesperidin, a natural citrus flavanone, alleviates hyperglycaemic state and attenuates embryopathies in pregnant diabetic mice. J Med Plant Res 3:862–869

Tsai HL, Chang SK, Chang SJ (2007) Antioxidant content and free radical scavenging ability of fresh red pummelo [*Citrus grandis* (L.) Osbeck] juice and freeze-dried products. J Agric Food Chem 55:2867–2872

USDA (2018) National Nutrient Database for Standard Reference (All Nutrients). Nutrient data for 2018, Citrus fruit. https://fdc.nal.usda.gov/fdc-app.html#/. Accessed 29 May 2022

Velasco R, Licciardello C (2014) A genealogy of the citrus family. Nat Biotechnol 32:640–642

Vinatoru M, Toma M, Filip P, Achim T, Stan N, Mason TJ, … Lazurca D (1998) Ultrasonic reactor dedicated to the extraction of active principles from plants. Romanian Patent 98-01014

Vuong QV, Golding JB, Stathopoulos CE, Nguyen MH, Roach PD (2011) Optimizing conditions for the extraction of catechins from green tea using hot water. J Sep Sci 34:3099–3106

Vuong QV, Golding JB, Stathopoulos CE, Roach PD (2013) Effects of aqueous brewing solution pH on the extraction of the major green tea constituents. Food Res Int 53:713–719

Wang L, Weller CL (2006) Recent advances in extraction of nutraceuticals from plants. Trends Food Sci Technol 17:300–312

Wang S, Yang C, Tu H, Zhou J, Liu X, Cheng Y, Luo J, Deng X, Zhang H, Xu J (2017) Characterization and metabolic diversity of flavonoids in citrus species. Sci Rep 7:1–10

Wu GA, Prochnik S, Jenkins J, Salse J, Hellsten U, Murat F, Perrier X, Ruiz M, Scalabrin S, Terol J, Rokhsar D (2014) Complex history of admixture during citrus domestication revealed by genome analysis. Nat Biotechnol 32:LBNL-7054E

Yamada T, Hayasaka S, Shibata Y, Ojima T, Saegusa T, Gotoh T, Ishikawa S, Nakamura Y, Kayaba K (2011) Frequency of citrus fruit intake is associated with the incidence of cardiovascular disease: the Jichi Medical School cohort study. J Epidemiol 21:1103010218

Zhang H, Xi W, Yang Y, Zhou X, Liu X, Yin S et al (2015) An on-line HPLC-FRSD system for rapid evaluation of the total antioxidant capacity of Citrus fruits. Food Chem 172:622–629

Zou Z, Xi W, Hu Y, Nie C, Zhou Z (2016) Antioxidant activity of Citrus fruits. Food Chem 196:885–896

Chapter 15
Potential Benefits of Bioactive Functional Components of Citrus Fruits for Health Promotion and Disease Prevention

Sovan Samanta, Jhimli Banerjee, Rubai Ahmed, and Sandeep Kumar Dash

Abbreviations

ap2	Adipocyte protein 2
cAMP	Cyclic AMP
cGMP	Cyclic GMP
COVID-19	Coronavirus disease 2019
DBP	Diastolic blood pressure
endMT	Endothelial to mesenchymal transition
GI	Gastrointestinal
IBD	Inflammatory bowel disease
LDL	Low-density lipoprotein
LPL	Lipoprotein lipase
LXRβ	Liver X receptor beta
MMP1	Matrix metallopeptidase 1
OA	Osteoarthritis
OECD	Organization for Economic Co-operation and Development
PCD	Programmed cell death
PPARα	Peroxisome proliferator-activated receptor alpha
RMR	Resting metabolic rate
SBP	Systolic blood pressure
SCFAs	Short chain fatty acids
TNF- α	Tumor necrosis factor-α
WHO	World health organization

S. Samanta · J. Banerjee · R. Ahmed · S. K. Dash (✉)
Department of Physiology, University of Gour Banga, Malda, West Bengal, India

S. Singh Purewal et al. (eds.), *Recent Advances in Citrus Fruits*,
https://doi.org/10.1007/978-3-031-37534-7_15

15.1 Introduction

Grown in tropical, subtropical, and temperate climates, citrus is a perennial shrub or small tree in the Rutaceae family that includes lemons, tangerines, grapefruits, mandarins, limes, oranges, and citrons, amid several other hybrids and variations. Citrus genus contains grapefruit (*Citrus paradisi*: 5%), limon and lime (*Citrus limon* and *Citrus aurantifolia*: 12.1%), tangerine (*Citrus reticulata*: 19.9%), sweet orange (*Citrus sinensis*: 61.1% of worldwide citrus percentage), tangerine, lime etc. In the Northern Hemisphere, fruits, especially oranges and grapefruit, achieve their peak maturity between mid-December and April. Among the fruit crops, the yearly production of citrus fruits, which are grown in more than 64 nations worldwide, is 105.4 million tons (Ahmed and Azmat 2019). The ascorbic acid, carotenoids, flavonoids, pectin, calcium, potassium, and other vital nutrients in citrus fruit make it the most significant fruits in the world. Citrus fruits are popular for their valuable origin of both soluble and insoluble fibre, which confers a number of health benefits, one of which is the removal of toxins from the body (Saeid and Ahmed 2021). A healthy bile duct and liver function are two of the many benefits that come from consuming dietary fibre. Additionally, those fibres ensure better gastric adsorption inside the small bowel, which slows the absorption of energy. These inefficacies are essential to human health and are evidenced by these active secondary metabolites, including anti-oxidative, anti-inflammatory, and anti-cancer qualities, as well as cardiovascular and neuroprotective features. Citrus fruits have a long history of use in traditional medicine across several countries in Asia (China, Japan, Korea, and India) (Lv et al. 2015; Sofowora et al. 2013). The Chinese Pharmacopoeia lists nine traditional Chinese medications from six citrus species for proper medical usage which includes *Citrus aurantium L.*, *Citrus reticulata Blanco*, *Citrus medica L.* var. *sarcodactylis Swingle*, *Citrus medica L.*, *Citrus wilsonii Tanaka*, and *Citrus sinensis Osbeck*. The peel or entire (mature or immature) fruits are effective in reducing blood pressure and treating ringworm infections, coughing up blood, dyspepsia, and skin inflammation (Lv et al. 2015; Nambiar and Matela 2012; Asgary and Keshvari 2013). Hesperidin (7-O-rutinoside of hesperetin), a flavanone group flavonoids member, may be extracted in significant quantities from the various citrus species (Roowi and Crozier 2011). Recently, researcher extremely attracted to the bioactivity of hesperidin and its several biological characteristics (Kim et al. 2019; Roohbakhsh et al. 2015; Parhiz et al. 2015; Cheng et al. 2021; Li and Schluesener 2017). Hesperidin is found in significant quantities in several varieties of Citrus sp. like, grapefruit, orange, mandarin, lemon, and lime (Gattuso et al. 2007). Deficiency of it has been linked to nightly soreness in the hands, feet, and legs as well as abnormal capillary leakiness (Hajialyani et al. 2019). Following the worrisome spread of COVID-19 around the planet, human health saw some significant repercussions (Das et al. 2022a, b; Kumar and Rathi 2020). Mental and physical health has been drastically impacted throughout the COVID-19 associated lockdown (Samanta et al. 2022a, b). In this regard, several study groups have found that various citrus species (lemon, orange, limes etc.) significantly able to boost human immunity while also promoting the health of infected individuals (Khanna et al. 2021; Zabetakis et al. 2020; Bhutada et al. 2020). We are attempting to outline the numerous documented potential biological applications of various citrus fruit varieties in this chapter, as well as the effectiveness of these fruits in improving health.

15.2 Overview of Citrus Fruit: Emphasis on Phytonutrients

The tropical flowering plant named as citrus is a member of the Rutaceae family, Aurantioideae subfamily, Citreae tribe, and Citrinae sub-tribe. There are numerous sorts or kinds of species in the genus Citrus, and they differ in terms of their fruits, blooms, leaves, and twigs. Citrus taxonomy is complicated and contentious, owing to sexual diversity within species and genera, as well as polyembryony, which preserves and reproduces maternal genotypes (Inglese and Sortino 2019). Citrus fruit is one of the world's most important horticultural products, and it is also the widely traded commodity (Chavan et al. 2018; Liu et al. 2012). It's unclear where citrus fruits came from exactly, but it's generally accepted that it invented in Southeast Asia and extended throughout the globe. Mediterranean nations, Brazil, China and USA contribute for approximately two-thirds of the world's total citrus fruit harvest (Langgut 2017). Over the past three decades, there has been a consistent rise in the amount of citrus fruits consumed on a per capita basis all over the world (USDA 2022). There is a significant amount of variety in the kinds of citrus fruits that are grown in and consumed by people in the different regions of the world. Mandarins occupy the second largest share (26%) of global citrus production, behind sweet oranges (56%) (Ladaniya et al. 2021). The phytonutrient content of citrus fruits is unusually high compared to that of other fruit types. Phytonutrients are essential for both the elevation of health and the circumvention of disease. Vitamin C is widely known to be abundant in citrus fruits (Mditshwa et al. 2017). In addition, citrus fruits provide a number of other health-promoting bioactive phytochemicals, such as polysaccharides, amino acids, phenolic acids, flavonoids, terpenoids, triterpenes etc. (Ye 2017) (Fig. 15.1). There are around 40 limonoids in citrus fruits, but the most important ones are limonin and nomilin. The bitterness of citrus is partially driven by these compounds that are found in elevated amounts in grapefruit and orange (Okwu 2008).

15.2.1 Phytochemicals in Citrus Fruits

The nutritive and secondary metabolites content of the five most popular citrus species e.g. *Citrus sinensis*, *Citrus reticulate*, *Citrus limonum*, *Citrus aurantifolia*, and *Citrus grandis* was reported in the back-to-back studies of Okwu and Emenike (2006, 2007). These five citrus fruits are high source of crude protein (10.94–17.06%), crude fibre (5.84–7.10%), crude lipid (0.64–1.24%), magnesium (0.49–0.61%),

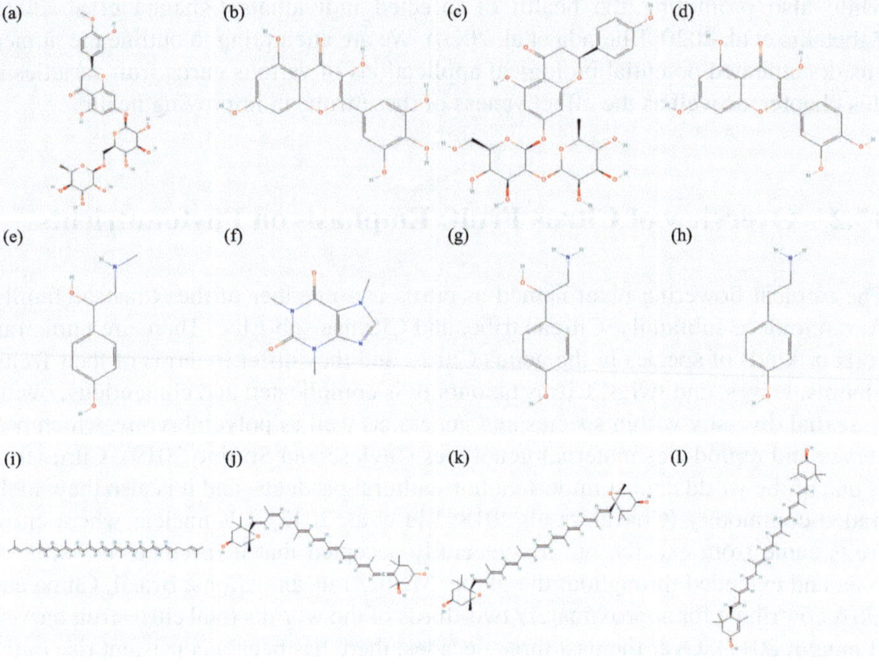

Fig. 15.1 2D Structures of different bioactive compounds present in several of Citrus fruits. All the 2D structures are collected from the PubChem databank (National Library of Medicine; https://pubchem.ncbi.nlm.nih.gov/). Here, (**a**) Hesperidin; PubChem ID: 10621, (**b**) Myricitin; PubChem ID: 5281672, (**c**) Naringin; PubChem ID: 442428, (**d**) Quercetin; PubChem ID: 5280343, (**e**) Synephrine; PubChem ID: 7172, (**f**) Caffeine; PubChem ID: 2519, (**g**) Octopamine; PubChem ID: 4581, (**h**) Tyramine; PubChem ID: 5610, (**i**) Lycopene; PubChem ID: 446925, (**j**) Neoxanthin; PubChem ID: 5282217, (**k**) Violaxanthin; PubChem ID: 448438, (**l**) Lutein; PubChem ID: 5281243

sodium (0.28–0.36%), phosphorus (0.24–0.41%), carbohydrates (70.86–77.10%), potassium (0.28–1.0%), calcium (2.0–3.20%) and food energy (347.04–363 g cal^{-1}). Secondary metabolite concentrations include niacin (0.03–0.43 mg/100 g), alkaloids (0.33–0.04 mg/100 g), phenols (0.01–0.42 mg/100 g), ascorbic acid (19.36–61.60 mg/100 g), flavonoids (0.19–0.57 mg/100 g), thiamine (0.06–0.12 mg/100 g), tannins (0.01–0.04 mg/100 g), and others (Raghavan and Gurunathan 2021).

15.2.1.1 Flavonoids

Flavonoids belong to the category of polyphenolic secondary metabolites that are frequently obtained from plants. They are an essential source of antioxidants in the human nutrition. Most of the colour in flowers and fruits comes from these metabolites (Macheix et al. 2018). Flavonoids are made up of a 15-carbon skeleton (C6-C3-C6), two 6-carbon phenyl rings, and a heterocyclic ring with an oxygen atom in it (Kumar and Pandey 2013). Flavonoids can be split into other segments based on how the heterocyclic ring is changed. These subgroups include flavones, flavonols, flavanones, flavanonols, flavanols (flavan-3-ols), isoflavones, and anthocyanins (Tripoli et al. 2007). Numerous citrus fruits include large concentrations of flavones (serving as rhoifolin, vitexin, and diosmin), flavonols (such as quercetin, rutin, and kaempferol), flavanone-7-O-glycosides (such as naringin, eriocitrin, hesperidin, and narirutin), polymethoxylated flavones (such as nobiletin, tangeritin, and 5-demethyl nobiletin) and anthocyanin (cyanidin and peonidin glucosides) (Tripoli et al. 2007). Most citrus fruits have flavonoids like hesperidin, myricitin, naringin, tangeritin, rutin, and quercetin (Okwu and Emenike 2006).

15.2.1.2 Alkaloids

The indole ring is the primary building block from which alkaloids are derived. Alkaloids can be divided into two categories, non-hetrocyclic (also known as atypical) and heterocyclic (known as typical), according to their structural composition. Synephrine is one of the several alkaloids that are abundant in citrus species (Percy et al. 2010). *Citrus aurantium* (Bitter orange) also yielded various alkaloids including octopamine, tyramine, and n-methyltyramine. *Citrus paradidi* (Grapefruit) and *Citrus maxima* (Pomelo) are the sources from which caffeine, a purine alkaloid, was derived (Kretschmar and Baumann 1999). *Citrus reticulate Blanco* (common mandarin) High amount of acriquinoline A and acriquinoline B are found in *Citrus reticulate Blanco* (common mandarin) (Wansi et al. 2016). Especially in contrast to other Citrus sp., *Citrus aurantium* has a fair amount more active alkaloids, particularly synephrine, which makes up with over 85% of the total protoalkaloid. N-methyltyramine has also been observed in large amount greater tiers as compare to octopamine, tyramine, or hordenine (Lv et al. 2015).

15.2.1.3 Carotenoids and Apocarotenoids

In addition to their role in photosynthesis and signalling, carotenoids are widely dispersed isoprenoid pigments (Saini et al. 2015; Saini and Keum 2018). Based on how they are constructed, carotenoids can be divided into two different categories – (a) hydrocarbon carotenoids like alpha-carotene, beta-carotene and lycopene and (b) Xanthophylls, aerated variants of hydrocarbon carotenoids like neoxanthin, violaxanthin, lutein, and –cryptoxanthin (Saini et al. 2015; Saini and Keum 2018). Since carotenes have a straightforward hydrocarbon structure, they can only be found in their free form. Fatty acids can be esterified with oxygenated functional groups in xanthophylls to produce free or esterified fatty acids. Saturated and unsaturated fatty acids such as caprate (C10:0), laurate (C12:0), myristate (C14:0), palmitate (C16:0), stearate (C18:0), palmitoleate (C16:1) and oleate (C18:1) are commonly acylated to xanthophylls in citrus fruits (Etzbach et al. 2020; Lux et al. 2019). Apocarotenoids are a different class of carotenoids in addition to carotenes and xanthophylls. The fragmentation of carotenoids by the carotenoid cleavage dioxygenases, or (CCDs)/9-cis-epoxycarotenoid dioxygenase (NCED), produces the ecologically and nutritionally significant apocarotenoids (Zheng et al. 2021). Over 115 carotenoids and derivatives are produced by citrus species. Yellow to red hues can be attributed to carotenoids like beta-carotene (Dugo et al. 2006). Wendun (*Citrus grandis Osbeck*), peiyou (*Citrus grandis Osbeck*), and ponkan (*Citrus reticulata Blanco*) all have high concentrations of carotenoids in their peel extracts, which measure 2.04 ± 0.036 0.036 ± 0.0006, and 0.021 ± 0.0004 mg/g dried base (db), respectively (Wang et al. 2008). The most prevalent carotenoids in the pulp of Valencia oranges are esters of violaxanthin, antheraxanthin, beta-cryptoxanthin, and mutatoxanthin esterified mostly with laurate, myristate, and palmitate as monoesters or diesters. A tiny amount of free xanthophylls and carotenes, including alpha and beta carotene, are also present in citrus fruits. Carotenoids are more abundant in citrus fruit peel flavedo than juice sacs, similar to phenolic chemicals. Carotenoid concentrations increase with maturation in contrast to phenolic compounds. Additionally, the albedo only has a tiny quantity of carotenoids compared to phenolic chemicals (Multari et al. 2020; Petry et al. 2019). In a study that compared distinct kinds of oranges (Washington navel and Tarocco), lemons (Akragas), and pummelos (Chandler), the flavedo of perfectly ripe Washington navel oranges contained the highest amount of total carotenoids (159 mg/kg DW), whereas the juice sacs of ripened Tarocco oranges stored the maximum values of total carotenoids (63.7 mg/kg DW). Following the findings of this research, lutein was found to have a significantly higher concentration than any of the other carotenoids in the juice septa of the fruits that were investigated. In fact, lutein was found to account for 83% of the overall carotenoids in the juice cysts of Tarocco oranges. In this contrast, violaxanthin, antheraxanthin, beta-cryptoxanthin, and beta-carotene have been considered as minor carotenoids.

15.2.1.4 Coumarins

Higher plants are the only known sources of the benzo-a-pyrones known as coumarins. Prenylation at C6 or C8, along with the closing of a furan ring, can result from a series of alterations that the abundant coumarins can go through (Bourgaud et al. 2006; Dugrand-Judek et al. 2015). Citrus plants have the ability to create substances that performance as a defensive measure against herbivorous creatures and pathogens. These compounds include coumarins and furanocoumarins. An investigation employing spectrometry and ultra-performance liquid chromatography to determine whether coumarins are present in 61 citrus species is representative of the genetic diversity of the entire citrus family. The coumarin/furanocoumarin diversity and concentrations in citrus peel are generally higher than in the pulp of the same fruits. Based on the chemotypes discovered in the peel, which correspond to the four ancestral taxa (pommelo, mandarin, citron, and papedas) and their respective secondary species offspring, citrus species can be categorized into four categories. Mandarins appear to be lacking in the production of these chemicals, in contrast to the other three ancestral taxa (pummelo, citron, and papeda), which produce considerable amounts of these compounds (Dugrand-Judek et al. 2015). Chromatographic spectra of peel samples from Colombian citrus fruits exhibited both qualitative and quantitative variances. The essential substances contained in citrus fruits have been provisionally identified as coumarins, furanocoumarins, and polymethoxylated flavones. While sweet oranges primarily contain polymethoxylated flavones, Tahitian limes, Key limes, Mandarin limes, and mandarins primarily contain coumarins and furanocoumarins. Limettin, oxypeucedanin hydrate, bergaptene, bergamottin, and isopimpinellin are all examples of coumarins and furanocoumarins that have been identified from Tahitian and Key limes respectively (Ramírez-Pelayo et al. 2019).

15.2.1.5 Ascorbic Acid

Ascorbic acid is a water-soluble micronutrient that plays an important role in the diverse biological procedures occurs inside the human body. Multiple studies have found that consuming citrus fruits can help to reduce the risk of developing cancer and protect a wide range of different types of human cancers (Silalahi 2002; Rekha et al. 2012). It has recently been revealed that the presence of vitamin C is the factor that is accountable for bringing about this property (Du et al. 2012). When associated to the consumption of vitamin C, the ingestion of citrus fruits is strongly attributable to a decrease in total rate of risk. A prior study suggested the concentration of ascorbic acid varies in citrus sp. [Lemon (*Citrus limon*), Bitter Orange (*Citrus aurantium*), Sweet Orange (*Citrus aurantium var. sinensis*), Pomelo (*Citrus maxima*), Grapefruit (*Citrus paradisi*) and Citron (*Citrus medica*)] determined using iodometric titration, dye titration and spectrophotometric methods (Shrestha et al. 2016). Lemon, bitter orange, grapefruit, and sweet orange each had about 34.8 mg/100 ml, 29.89 mg/100 ml, 39.80 mg/100 ml, and 25.11 mg/100 ml of ascorbic acid, respectively. It was found that the average amount of ascorbic acid in pomelo was around 61.29 mg/100 ml, even as the citron's average amount of ascorbic acid was determined 17.4 mg/100 ml (Shrestha et al. 2016). Total amount of ascorbic acid in 99.93 g of pomelo was 25.80 mg, the lowest weight to ascorbic acid ratio (3.87), while the amount in 180 g of citron was 7.2 mg, the greatest weight to ascorbic acid ratio (25.00) (Shrestha et al. 2016).

15.2.1.6 Essential Oils

Citrus' various plant components, including its leaves, blossoms, fruits, and peels, are abundant sources of essential oils (Md Othman et al. 2016). Citrus essential oils have properties that make them anti-inflammatory, anti-diabetic, insecticidal, antifungal, and anti-bacterial. These properties are useful in the pharmaceutical, medical, cosmetic, agricultural, and food industries (Palazzolo et al. 2013; Wolffenbüttel et al. 2018). Aldehydes, ketones, esters, and organic acids are some of the functional groups found in Citrus essential oils. Citrus essential oils in particular are a blend of complex hydrocarbons derived from both terpene and non-terpenoid sources. (Merle et al. 2004). The citrus essential oil's composition varies greatly according on the type, variety, season, and location of the fruit, as well as its degree of maturity (Fisher and Phillips 2008; Droby et al. 2008). These oils include a variety of chemicals, 20 and 60 different molecules among 85–99% of these are made up of volatile molecules, and the remaining 1–15% is made up of nonvolatile compounds (Asadim 2008). A variety of monoterpenes, sesquiterpenes, and sesquiterpenoids are found in the volatile chemicals (Smith et al. 2001). Approximately, 97% of the citrus essential oils are monoterpenes consisting of two isoprene (C_5H_8) units, whereas between 1.8% and 2.2% of the citrus essential oils are alcohols, aldehydes, and esters. In addition, limonene is the most prominent ingredient. Its concentrations range from 32% to 98% depending on the variety, such as in bergamot, lemon, and sweet orange (32–45%, 45–76%, and 68–98% respectively) (Moufida and Marzouk 2003). Citrus essential oils also have a huge quantity of monoterpene hydrocarbons. The peels of extract of *C. microcarpa* (94.0%), *C. grandis* (81.6–96.9%), and *C. aurantifolia* (39.3%) all have limonene as their most important ingredient. On the other hand, the peel of *C. hystrix* is mostly made up of sabinene, which makes up between 36.4% and 48.5% of the peel. Limonene, citronellal, sabinene, and pinene were found to be the most prevalent components in the essential oil from the Kaffir lime (*C. hystrix*), with concentrations of 24.62%, 22.06%, 19.29%, and 10.58%, respectively (Chhikara et al. 2018; Suwannayod et al. 2018; Al-Aamri et al. 2018).

15.3 Putative Activity of Citrus Bioactive Component Towards Various Health Complications

Beneficial biological applications of various Citrus spp. along with their molecular insights are described below. Protective or minimizing activities of some major alarming health threats are mentioned briefly in association with different kinds of citrus bioactive compounds (Figs. 15.1 and 15.2).

15.3.1 Cardiovascular Diseases (CVDs)

An estimated 17.9 million people die each year as a result of cardiovascular diseases (CVDs), making them one of the leading causes of death worldwide. CVDs are a collection of heart and blood vessel disorders that include coronary heart disease, congenital heart disease, cerebro vascular disease, aortic aneurysm, atherosclerosis, and a number of other conditions (Pala et al. 2020). The popularly known medicine can exert various side effects, including myalgia, organ toxicity, and rhabdomyolysis. So, CVDs continue to be a medical and safety concern affecting the majority of people in the world. Recommendations have been made to reduce the social and

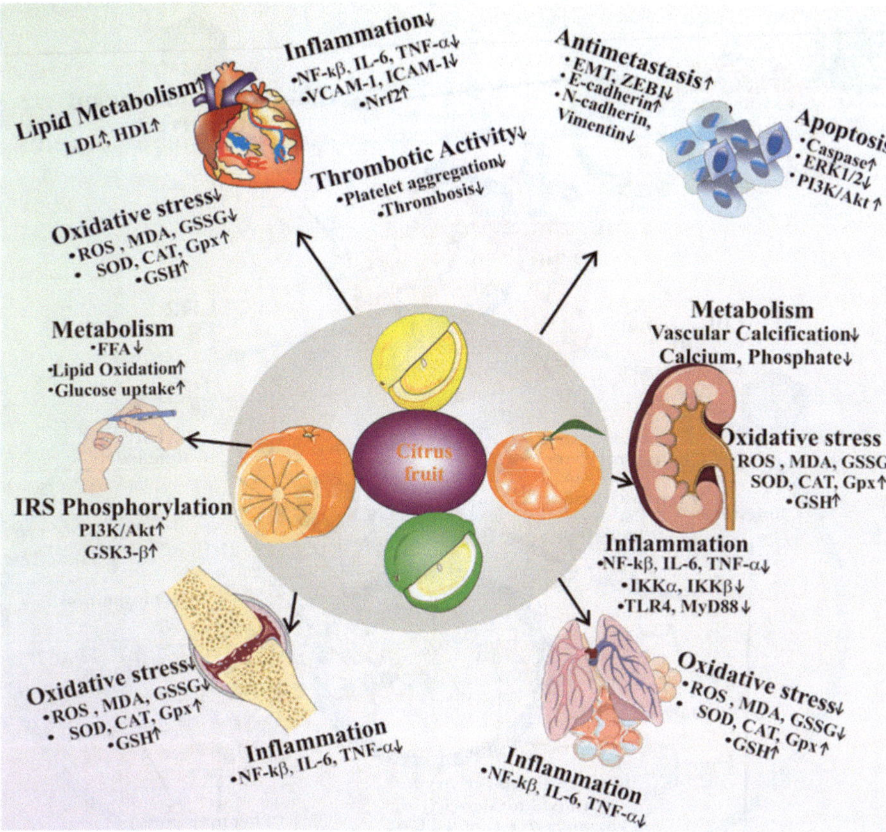

Fig. 15.2 Numerous disease-preventive applications of citrus fruits are depicted in this diagram, along with their health-promoting properties. The beneficial role of citrus fruit against major prevalent diseases was elaborated on in this figure. It mainly acted through the anti-oxidative, anti-inflammatory, metabolic pathways (lipid, glucose etc.), anti-apoptotic, anti-metastasis and many more against cardiovascular diseases, cancer, renal diseases, diabetes, osteoarthritis, respiratory diseases

financial impact of cardiovascular illnesses. Lifestyle and health interventions focusing on numerous risk factors have been suggested. The principle approach to controlling this major health problem is prevention of risk factors by modulating diet and lifestyle. Flavonoids, carotenoids, apocarotenoids, terpenes, and limonoids are the primary bioactive components found in citrus fruits. These components are implicated in the management and prophylaxis of cardiovascular diseases through a variety of different mechanisms. There are various signaling pathways by which these compounds exert protective activity against the CVDs; likewise, inhibition of oxidative stress, anti-inflammatory activity, inhibition of platelet aggregation, improvement in lipid metabolism, regulation of apoptosis and autophagy (Fig. 15.3).

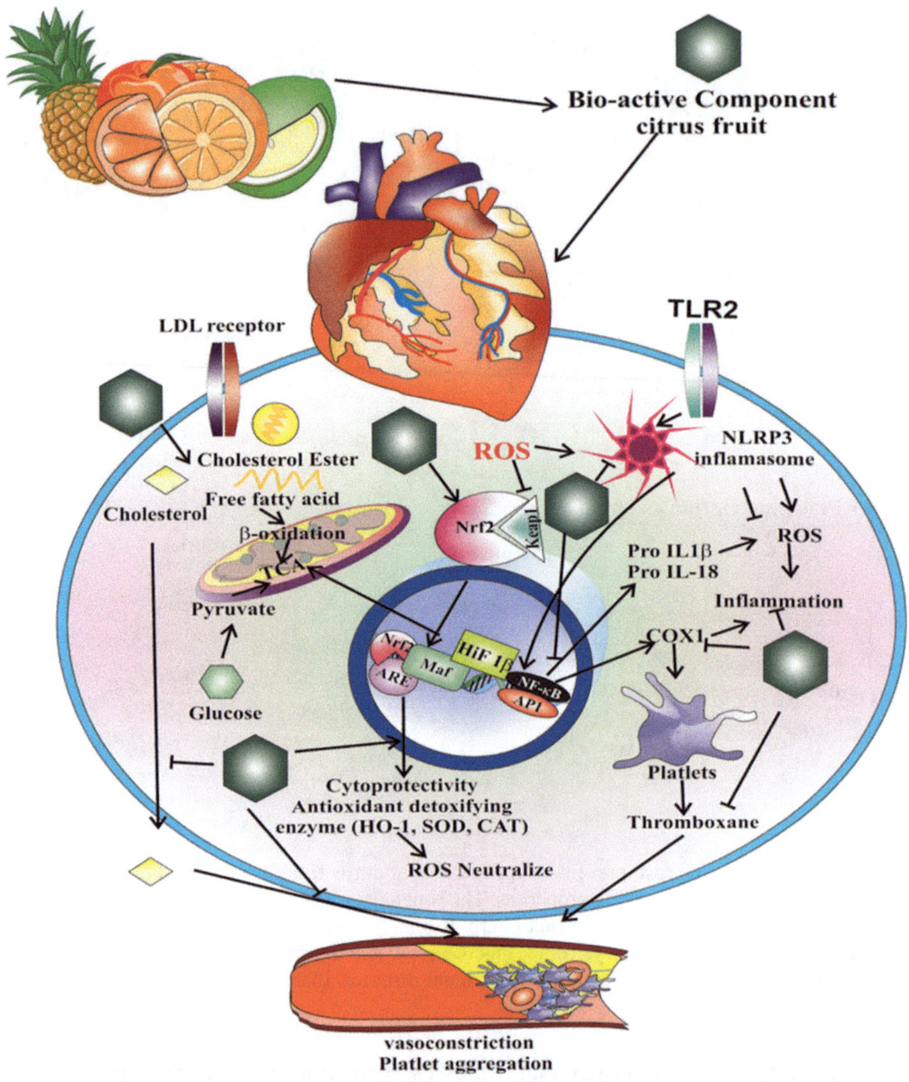

Fig. 15.3 The cardio-protective role of citrus bioactive compounds through various molecular signaling pathways. Here, the inflammatory pathway was altered through TLR-2 cell membrane receptor, which inhibits NLRP3 inflammasome formation and the platelet activation and aggregation also reduced. ROS production is neutralized through the activation of an antioxidant detoxifying enzyme. Cholesterol metabolism was also regulated through the active component of citrus fruit

15.3.1.1 Inhibition of Oxidative Stress

Oxidative stress refers to a state in which the body is unable to maintain a balance between its capacity to detoxify intermediate products and reactive oxygen species (ROS). In general, the human body maintains a homeostasis towards oxidants and antioxidants. Imbalance of the ratio leads to the creation of an enormous cellular ROS level, which induces free radical production and cell damage (Nita and Grzybowski 2016). Although the precise mechanism is unclear, lipid peroxidation damages cell membranes and changes intracellular charge and osmotic pressure, which results in cell death by swelling (Sun et al. 2015). Citrus flavonoids acted as free radical scavengers, such as superoxide anions, hydroxyl radicals, and peroxyl radicals (Cavia-Saiz et al. 2010). In previous studies with catechin, rutin, naringin, quercetin, and hesperidin, the cardiomyocytes' survival was enhanced, antioxidant enzyme levels were increased, and malondialdehyde (MDA) levels were also reduced (Wang et al. 2021a, b). Citrus flavonoids have an antioxidative effect on gene expression that influences cellular cross-communication. Hesperidin increased nuclear factor erythroid-2 related factor 2 (Nrf2), which was linked to the antioxidant defence system in senescent rat hearts (Elavarasan et al. 2012). Carotenoids also efficiently scavenge ROS molecules and play an antioxidant role in preventing CVDs. Atherosclerosis, one of the major CVDs, occurs due to ROS modified LDL oxidation, which leads to coronary heart disease. Carotenoids, especially β-carotene, α-carotene, lycopene, lutein, zeaxanthin, and β-cryptoxanthin had a positive correlation with higher intake of carotenoids and prevention of morbidity and mortality by CVDs (Eggersdorfer and Wyss 2018). Another carotenoid, lycopene, showed its antioxidant role by lowering serum cholesterol in animal model, which magnified its role in CVDs (Müller et al. 2016). Increased intake of this carotenoid may reduce the risk of several heart diseases (Leermakers et al. 2016). In this case, left ventricular function reached its normal activity after intake of Moro and Pera orange juice, either in the case of enhancing antioxidant enzymes or improving metabolic commotion in an animal model. Ribeiro et al. (2021) found that Pera and Moro orange juice reduced the cytotoxic effect of DOX on the energy metabolism of the myocardium.

15.3.1.2 Anti-inflammatory Activity

Inflammation is directly connected to CVDs. Pathogenic, metabolic, and ischemic conditions all induced the release of proinflammatory cytokines such as IL-1, IL-6, IL-18, and TNF-α Previously, the Canakinumab Antiinflammatory Thrombosis Outcomes Study (CANTOS) and the Colchicine Cardiovascular Outcomes Trial clinically proved that anti-inflammatory treatments lowered the rate of CVDs (Ridker et al. 2017, Tardif et al. 2019). Adhesion molecules (ICAM-1, VCAM-1) can also cause inflammation. Arachidonic acid is metabolised by the enzymes cyclooxygenase 2 (COX-2) and lipoxygenase, which are also required for the formation of prostaglandin, leukotriene, and thromboxane. They can all exaggerate the inflammatory response (Wang et al. 2020a, b). The 3′, 4′, or 4′-OH substitutions on ring B of flavonoids functioned as a selective inhibitor of lipoxygenase, but flavones having five or more methoxy groups had stronger phosphodiesterase inhibitory properties (Testai and Calderone 2017). Naringenin efficiently inhibited toll-like receptor 2 (TLR2) in adipose tissue (Yoshida et al. 2013). Furthermore, it reduced

TNF-α and IL-6 protein levels in LPS-induced murine macrophages BMDMs (Liu et al. 2016a, b). Qurectin another important component of citrus fruit exerted anti-inflammatory role in chronic systemic inflammation (Chekalina et al. 2018). In this case, proinflammatory response was decreased along with NF-κB transcriptional activity. Apigenin also improved cardiac dysfunction by lowering NF-κB in cardiomyopathy (Liu et al. 2016a, b). Fisetin was found to protect against isoproterenol-induced myocardial injury by inhibiting the responsible markers, such as the proinflammatory cytokines TNF-α and IL-6 (Garg et al. 2019). In an experiment with LPS-induced rat cardiomyocytes, luteolin reduced TNF-expression while also inhibiting NF-κB (Lv et al. 2011).

15.3.1.3 Antithrombotic Activity

Platelet aggregation has been identified as a major factor in cardiovascular, brain, and peripheral vascular diseases. The intermediate and late phases of atherosclerosis, which encourage and exacerbate CVDs, are often when platelet accumulation and thrombosis occur. Adenosine diphosphate, thromboxane, and other active substances released will also promote the onset of atherosclerosis and have other negative effects. Citrus flavonoids abridged platelet aggregation and thrombosis, which reduced the risk of blood clotting (Faggio et al. 2017). Carotenoids significantly improved endothelial function (Di Pietro et al. 2020). Potential antiplatelet activity of hesperidin was observed in collagen and arachidonic-acid-induced platelet aggregation. Here it intruded on collagen and arachidonic acid signal transduction via phospholipase C-gamma2 (PLCγ) and COX-1 (Jin et al. 2007). Naringin also partially inhibited platelet aggregation through reduction of PLCγ and as a result Thromboxane B2 (TXB2) and secretion of granules decreased (Kim et al. 2015).

15.3.1.4 Improvement in Lipid Metabolism

The lipid profile mainly Fatty acids, triglycerides (TG), and cholesterol play a valid role in the steady state maintained of the cardio-vascular system. A mild alteration of these profiles can be caused by the accumulation of fat and, as a result, atherosclerosis. The main flavonoids, namely hesperidin, hesperetin, naringin, naringenin, and quercetin, have been shown to protect against CVDs via lipid accumulation (Baselga-Escudero et al. 2015; Deng et al. 2021). These naturally occurring bioactive compounds with great safety, can reduce dyslipidemia, treat metabolic conditions brought on by a high-fat diet, and effectively prevent, treat, and even cure CVDs. Hepatic steatosis was prevented and attenuated plasma lipids in animal model by included citrus fruit in the diet (Park et al. 2013). Naringenin and hesperetin blocked translocation of triglycerides with inhibition of acyl-CoA: cholesterol acyltransferase (Lin et al. 2011). Hesperidin enhanced phosphorylation of the PI3K/ERK1 cascade pathway to express the LDLR gene, which in turn lowers LDL levels in plasma (Bawazeer et al. 2016). *Citrus bergamia* was a neglected citrus in comparison to others, but it also significantly exerted lipolysis activity in rats with hyperlipidaemia. These outcomes form the cornerstone of anti-atherogenic activity and are in charge of preventing cardiovascular disease. It also sped up the excretion process of sterols and bile. This reduced lipid levels in plasma and increased lipo-protective profiles in hypercholesterolemic patients (Toth et al. 2016).

15.3.2 Cancer

After cardiovascular diseases, cancer is the second biggest cause of mortality worldwide in 2020, accounting for ten million deaths overall. As a result, research is focusing on two main strategies: avoiding cancer-causing agents or using preventive mechanisms to reduce cancer morbidity. Although several popularly known drugs are available to treat cancer, undesired toxicity is a major burden for this purpose. Draw their attention for the purpose of developing an anticancer drug. Citrus fruit is full of bioactive components which can exhibit various mechanisms to protect and treat cancer (Fig. 15.4).

15.3.2.1 Inhibition of Cancer in Primary Phase

Previous studies have highlighted the anticancer role of bioactive components in citrus fruit against carcinogens and endogenous mutagens. In this context, naringenin and rutin avert DNA damage through free radical scavenging near DNA (Kootstra 1994). Quercetin protects liver cell lines against mercury-induced DNA damage by a mechanism similar to this (Barcelos et al. 2011). The incidence of colon cancer was considerably reduced in rats treated with DMH and hesperetin as a supplement (Aranganathan et al. 2008).

15.3.2.2 Inhibition of Tumor Development via Apoptosis and Antiproliferative Mechanism

Many studies, both *in vivo* and *in vitro*, have shown that citrus extracts inhibit cancer cell proliferation. Human pancreatic cell death is mediated by the apigenin-induced GSK-3/NF-κB signalling cascade (Johnson and De Mejia 2013). Human gastric cancer cell line SGC-7901 proliferation was significantly repressed in a dose-dependent manner by poncirin. Hydroxylated analogues of flavonoids exerted lower cytotoxicity in cancer cells in comparison to methoxylated flavonoids (Wesołowska et al. 2012). Although 5-demethylnobiletin showed an opposite effect on various cancer cells compared with nobiletin (Qiu et al. 2010), Polymethoxyflavones including neohesperidin, nobiletin, tangeretin, and 5-demethylnobiletin exerted an apoptotic effect on three types of gastric cancer:

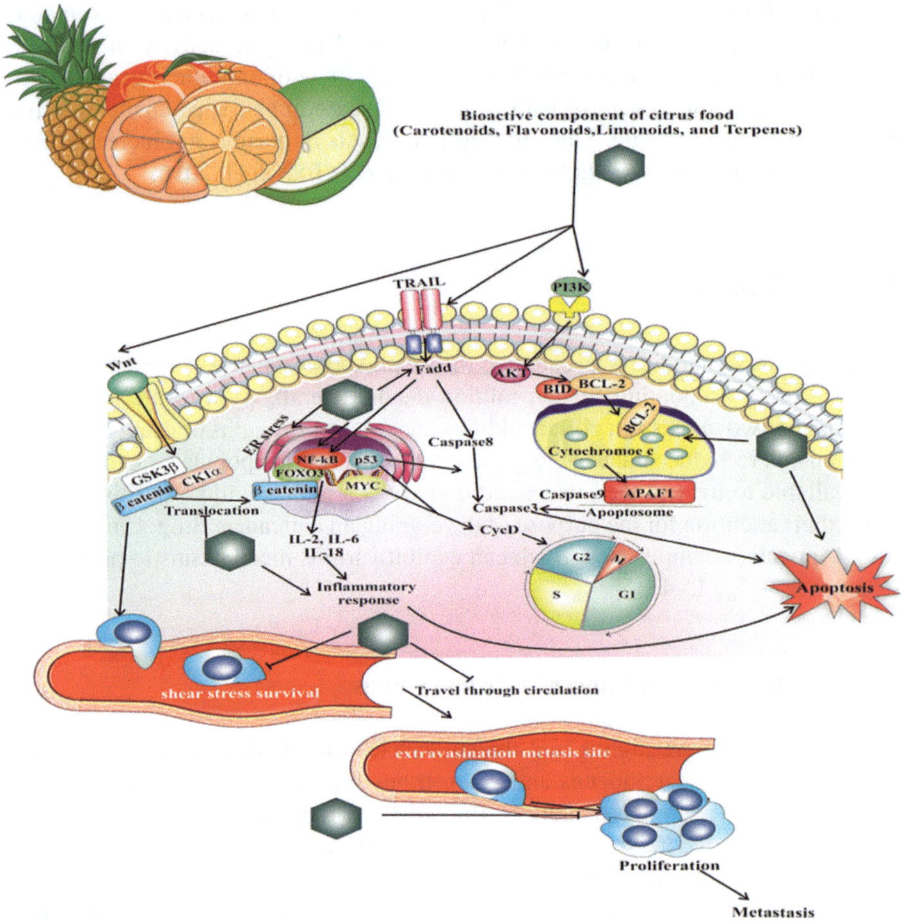

Fig. 15.4 Schematic representation of the anticancer mechanism through regulation of cell proliferation, cell cycle, apoptosis, and metastasis by various bioactive components of citrus fruits. Intrinsic and extrinsic pathways of apoptosis were stimulated by p53, BCL-2, caspase proteins. Inflammatory response is activated through NF-κB and pro-inflammatory cytokines. Cell cycle arrest was also activated through cyclin proteins and anti-metastasis efficacy was found via Wnt pathway

AGS, BGC-823, and SGC-7901. This process increased the expression of the RARB gene, which then activated the caspase 9 and PARP1 proteins (Wang et al. 2020a, b). Naringenin and its derivatized compound naringenin oxime exerted cytotoxic, genotoxic, and apoptotic activity against HT-29, PC-12, MCF-7, and L-929 cells by producing enormous ROS on cancer cells rather than normal cells (Kocyigit et al. 2016). Naringenin decreased the phosphorylation of ERK1/2 in prostate cancer cells, which stopped cell growth and caused apoptosis through the PI3K/Akt pathway (Lim et al. 2017). Limonoids are reported as an antiproliferative component against breast cancer cells, especially in estragon positive breast cancer by caspase 7 dependent pathways (Kim et al. 2013). It showed a synergistic response with curcumin against SW480 by induction of apoptosis (Murthy et al. 2013). Hesperidin induced apoptosis in the NALM-6 leukaemia series by increasing p53 expression, PPAR- and inhibiting NF-κB activation (Ghorbani et al. 2012). Hesperetin stimulated apoptosis in HepG-2 cells through the elevated ROS level and by reducing the Bcl-2/Bax protein ratio, which stimulated the mitochondrial apoptosis pathway (Zhang et al. 2015a, b).

15.3.2.3 Anti-metastasis Activity

Metastasis is a unique mechanism of cancer spreading from its primary neoplasm to a distant site for the formation of another tumour growth known as a secondary tumor, which increases the morbidity rate through cancer. Citrus derived bio elements exhibited antimetastatic activity through the inhibition of cell invasion and metastasis. Naringin reduced ZEB1, one of the key regulators of epithelial-to-mesenchymal transition (EMT) in osteosarcoma cells (Ming et al. 2018). Naringin inhibited matrix metalloprotease activity while also inhibiting ERK phosphorylation, p38, MAPK, and c-Jun. Ultimately, it interrupted and blocked the entire MAPK/ERK/JNK signalling pathway in human glioblastoma cells (Aroui et al. 2016). Nobiletin also inhibited TGF-induced catenin accumulation and induced SLUG protein accumulation in U343 glioma cells (Zhang et al. 2017). Tangeretin increased E-cadherin and decreased N-cadherin and vimentin. This may have stopped radiation from causing SGC7901 gastric cells to go through EMT (Zhang et al. 2015a, b).

15.3.3 Diabetes Mellitus

One of the most important non-communicable diseases receiving attention from all major stakeholders globally is diabetes mellitus which is being treated for its prevalence and accompanying consequences. About 1.6 million people die each year as a result of diabetes mellitus and its complications, making it the third biggest risk factor in early mortality. (Oguntibeju 2019). Citrus fruit has previously been shown to be a regulator of lipid, carbohydrate metabolism, and adipocyte differentiation, as well as a reduction in oxidative stress, insulin sensitivity, and glucose tolerance (Merola et al. 2017; Kou et al. 2019; Gandhi et al. 2020). Previous study reported that flavonoids, including 8-prenylnaringenin, cosmosiin, didymin, diosmin, hesperetin, hesperidin, isosiennsetin, naringenin, naringin and many more citrus flavonoids exerted a potential antidiabetic activity. 8-prenylnaringenin is a prenylflavonoid commonly found in oranges, lemons and grapefruits (Mukai et al. 2012).

15.3.3.1 Lipid and Glucose Metabolism Modulation

Carbohydrate metabolism disorder was one of the most common reasons of DM which directly correlated with hyperglycemia and other common inadequacies like insulin secretion, insulin resistance and activity of insulin. These abnormalities of insulin lead to carbohydrate, protein and fat metabolism disoriented. Lipid alteration profile is responsible for hypertriglyceridemia which developed insulin resistance and β-cell dysfunction. Diosmin a known flavonoid in citrus food was exerted anti-diabetic activity by improving glucose and lipid metabolism instreptozotocin (STZ)-induced animal diabetic model. The biochemical marker was ameliorated included glucose level in plasma, C-reactive protein and glycosylated haemoglobin (HbA1c). On other way it was directly approached on lipid metabolism by declining plasma lipid level of triglycerides (TG), free fatty acids, phospholipids, low-density lipoprotein cholesterol (LDL-C) where asdiminished high-density lipoprotein cholesterol (HDL-C). In cholesterol biosynthesis one of the key enzymes is 3-hydroxy-3-methyl-glutaril-CoA reductase (HMG-CoA reductase) which was elevated in diabetic model and restored after treated with diosmin (Srinivasan and Pari 2013). Hesperidin also had a similar impact on streptozotocin-induced diabetic rats, lowering the plasma glucose levels considerably. Other related parameter such as insulin, glycosylated hemoglobin, hemoglobin alteration occurred in an antidiabetic way. Here the key enzyme expression was observed which involved in glucose metabolism. Experiments exerted elevated hexokinase and glucose-6-phosphate dehydrogenase levels were enhanced and glucose-6-phosphatase and fructose-1,6-bisphosphatase levels were decreased (Sundaram et al. 2019). A synergistic study was shown with limonene and linalool in Whistar diabetic rat model. In this case, the combination of those compounds lowered blood glucose level effectively and also inhibited protein glycation (More et al. 2014). Cosmosiin or apigetrin was exerted antidiabetic role in 3 T3-L1 adipocyte cells via stimulating adiponectin secretion. This secretion induced insulin receptor-β (IR-β) phosphorylation which gave positive signal for translocation of glucose transporter 4 (GLUT4). So, it can be beneficial for glucose uptake in muscles and adipocytes and managed diabetes (Fig. 15.5).

15.3.3.2 Phosphorylation of Insulin Receptor Substrate (IRS)

Insulin is a key regulator for glucose homeostasis, whose reduction or resistance leads to diabetes. Insulin activity is initiated after binding with the receptor present in the plasma membrane of the target cell. Firstly, tyrosine kinase activity is started, which breaks down the insulin receptor substrate (IRS) by phosphorylation and, as a result, the downstream cascade pathway progresses. In this case, either phosphatidylinositol-3 kinase (PI3-K) or the serine/threonine kinase (Akt/PKB) or both act on GLUT4 to move it to the plasma membrane. Translocation of GLUT4 opens the gateway for glucose entry into the target cell. Nobiletin increases glucose uptake in murine 3T3-F442 preadipocytes by the PI3K/Akt and the protein kinase A (PKA) signaling pathway (Onda et al. 2013). Didymin exerted a potent antidiabetic activity

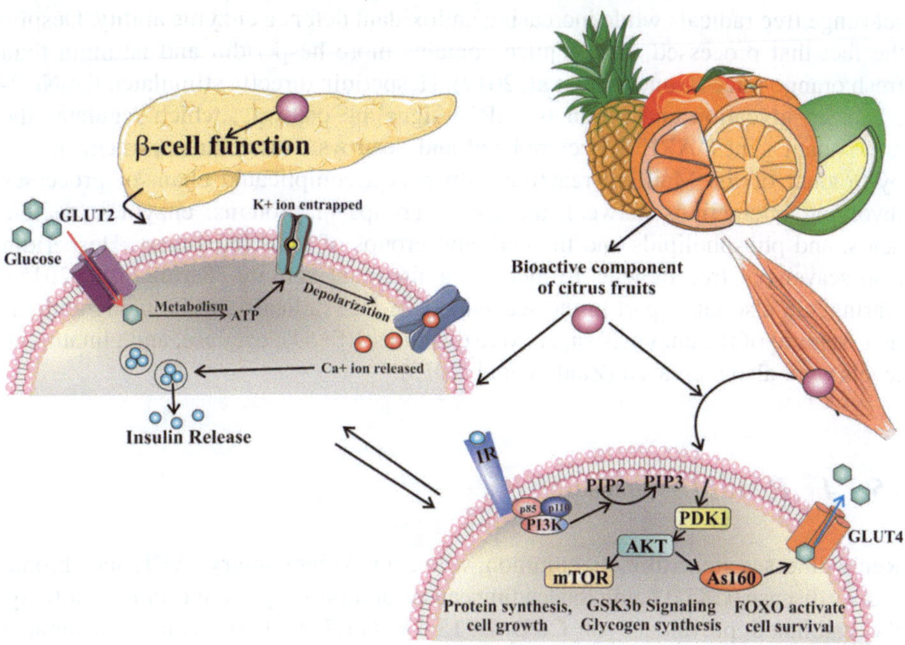

Fig. 15.5 Potential anti-diabetic efficacy of citrus fruits by modulating glucose metabolism pathway. Insulin release and insulin binding ability in muscle cell receptor was enhance and as result cellular homeostasis was maintained by activating PI3k/AKT/mTOR cascade pathway. Protein synthesis, glycogen synthesis and cell survival continue without any disturbance. Glucose channel was also triggered and glucose transport was sustained

by inhibition of human recombinant aldose reductase (HRAR), α-glucosidase, protein tyrosine phosphatase 1B (PTP1B) and advanced glycation end-product (AGE) formation. IRS was activated in insulin-resistant HepG2 cells, a downward cascade mechanism involving PI3K/Akt and glycogen synthasekinase-3 (GSK-3) boosted glucose absorption even in insulin-resistant cells (Ali et al. 2019). Citrus limonoids suppressed adipogenesis through the responsible PPARγ genes and phosphorylation of IRS in 3T3-L1 adipocytes of mice. Here, PI3K/Akt and fork-head transcriptional factor FOXO1 were responsible for the possession of antidiabetic effects by limonoids (Baba et al. 2016) (Fig. 15.5).

15.3.3.3 Antioxidative Property

From previous studies, it was clear that oxidative stress played a big part in how diabetes got worse and spread. As a result of the oxidation of glucose, protein glycation, and oxidative degradation, free radicals are formed quickly in people with diabetes. This caused cellular damage through damage of cell organelles, lipid peroxidation and diabetes development through modulation of insulin secretion and insulin resistance. The primary goal of antioxidants is to inhibit ROS production or scavenge free radicals while increasing antioxidant defence enzyme ability. Despite the fact that processed orange juice contains more hesperidin and naringin than fresh orange juice (Gonçalves et al. 2017). Hesperidin directly stimulated the Nrf2-advanced glycation end products (ARE) signaling pathway, which regulates the antioxidant genes. ARE is accumulated and destroys extracellular protein. It is a by-product of the Maillard reaction, which is a complicated chain of processes involving interactions between the amino groups of proteins, enzymes, nucleic acids, and phospholipids and the carbonyl groups of reducing sugars. Hesperidin also scavenged free radicals to increase antioxidant activity (Parhiz et al. 2015). Naringenin also takes part in the scavenging of free radicals and inducing antioxidant venture of the enzymes superoxide dismutase (SOD), catalase, and glutathione levels were also increased (Zaidun et al. 2018).

15.3.4 Renal Diseases

Renal illnesses are extremely common, with acute kidney injury (AKI) and chronic kidney disease (CKD) being the leading causes of morbidity and mortality. Globally, the estimated prevalence of CKD is 13.4% (11.7–15.1%), with an estimated 4.902–7.083 million individuals requiring renal replacement therapy (Lv and Zhang 2019). CKD initially lessens the functional effects of nephron loss and then ultimately has a negative effect by impairing the size selectivity of the glomerular barrier and causing protein ultrafiltration. Accumulation of excess protein in Bowman space leads to tubulointerstitial damage which activates inflammatory and apoptotic pathways and promotes end-stage renal disease (Klaus et al. 2021).

15.3.4.1 Mineral Metabolism Disorder Modulation

CKD complications were exacerbated by vascular calcification (VC) and mineral metabolism disorders, both of which contribute to atherosclerosis. A decline in the frequency of glomerular filtration rate (GFR) was associated with VC. In VC, the muscular layer of arteries was thickened and lost their elastic nature. Along with VC, mineral metabolism disorder (MMD) also contributes to high mortality in end-stage renal disease (ESRD) patients. MMD was directly linked with phosphate, calcium, and parathyroid hormone metabolism and caused hyperphosphatemia, hypercalcemia, and hyperparathyroidism, respectively (Mehrotra 2006). Adenine-induced animal models were used to study the protective effects of quercetin. Qrectin inhibited VC by modulating the renal functional biomarkers uric acid (UA) and creatinine (CRE) via the p38/MAPK/iNOS pathway (Chang et al. 2017). In a recent study, diosmin reduced blood urea nitrogen (BUN), calcium, and phosphate levels in a CKD animal model. The production of vasodilators and an inhibitory effect on the rennin-angiotensin pathway improved sodium glomerular filtration rate. PTH was one of the minerals that regulated hormones, and it was significantly higher in the diosmin-treated group compared to the untreated group. Diosmin could prevent hypocalcaemia effects by increasing levels of calcitriol (Sharma et al. 2022a, b).

15.3.4.2 Chronic Inflammation Regulation

Inflammation is directly associated with renal disease, which is a complex networking structure formed between the kidney's parenchymal cells and occupant immune cells such as macrophages and dendritic cells, as well as circulate immune cells such as monocytes, lymphocytes, and neutrophils. Toll-like receptor and Nod-like receptor (NLR), nuclear factor κB (NF- κB), and the NLRP3 inflammasome are also members of the inflammation (Andrade-Oliveira et al. 2019). Tangeretin showed protective effects in nephrectomized rats by elevating levels of creatinine, urea, and targeting inflammatory pathways. It regulates inflammatory pathways by inhibiting TNF- and increasing phosphorylation of IB, IKK- α, and IKK- β, which is inhibited by NF- κB activation. This was followed by a decrease in immune cell infiltration in renal tissue as well as a decrease in TNF- α, NO, and the cytokines IL-1 and IL-6 (Wu et al. 2018). In a study, umbelliferae showed an anti-inflammatory effect on streptozotocin (STZ)-induced diabetic nephropathy with a decreased level of uric acid, creatinine. It reduced hyperglycemia by downregulating TLR4, MyD88, p-NF- κB, and p-IB in the kidney tissues of the animal model's treatment group via TNF- α, IL-1, and IL-6 cytokines (Han-Qing et al. 2019). In the same model of diabetic nephropathy induced by STZ or cadmium in the rat, nobiletin showed infiltration of neutrophils diminished in kidney tubules. It hampered the NF- κB downstream pathway and decreased p65 levels with increased Bcl-2 expression (Xu et al. 2021). Citrus-derived flavonoids nobiletin, naringin, and hesperidin inhibited RAW-264 inflammation. TNF- α and IL-6 cytokines were inhibited in 7-macrophages, and NF- κB was also inhibited. LPS was used as a stimulator in this inflammation model. iNOS was suppressed along with the p-65 and MAPK signaling pathway (Kang et al. 2011).

15.3.4.3 Suppressing Oxidative Stress

In renal impairment patients, there was an increased level of oxidant activity and a decreased antioxidant level, which interrupted normal oxidative status and produced oxidative stress. Previous research in Wistar male rats expanded on naringenin's protective role in arsenic trioxide-induced toxicity. There was an enormous amount of free radicals produced and a decreased GSH level after the treatment with arsenic. Naringenin showed antioxidant activity by maintaining elevated antioxidant enzyme levels (SOD, catalase, GPx) and mitigating the arsenic-induced toxicity in an animal model (Mershiba et al. 2013). Taurine (TR) and/or hesperidin (HES) also presented antioxidative activity against carbon tetrachloride (CCl 4)-induced renal toxicity. CCl4 was converted into CCl3, which increased MDA while significantly decreasing CAT and GSH. This transformation directly interacts with DNA, protein, and other biomolecules of cells and causes increased oxidative stress, which may be the principle cause of kidney tissue damage. Taurine restores oxidative status by regulation of cellular ions, free radical scavengers and stabilization of the membrane. Nitrosoglutathione was produced from GSH nitrosylation, which was triggered by taurine. Hesperidin showed its antioxidant property by augmentation in cellular antioxidant resistance.

15.3.5 *Osteoarthritis*

Osteoarthritis (OA) is the most prevalent joint disease in the elderly population known as "wear and tear", which is also a major disease responsible for morbidity and mortality in this age group. Some of the causes that lead to OA include trauma, loss of cartilage structure and function, and an imbalance of pro-inflammatory and anti-inflammatory pathways (Fig. 15.2). OA mostly affects the articular surfaces. Pain and stiffness in the joints are among the signs of OA. The defining feature of OA is the loss of articular cartilage. Articular cartilage cannot restore itself. Although costal chondrocytes are a rich source of donor cells, they have a great propensity for hypertrophy and calcification tendency. There is currently no remedy for OA since it is extremely challenging to replace the cartilage. The only existing treatments concentrate on decreasing symptoms (such as pain and inflammation), preserving joint movement, and preventing the loss of useful abilities. In order to manage OA, it is likely that lowering the production of oxidative stress and inflammation will be beneficial. Recent *in vitro* and preclinical experiments propose the defensive roles of dietary polyphenols in the expansion of OA, in terms of averting chondrocyte inflammation and additional cartilage damage/destruction, through their capacity to interact directly or indirectly with the joint-associated tissues (such as articular cartilage, bone, or synovium), resulting in the reduction of joint pain.

15.3.5.1 Regulating Oxidative Stress

Oxidative stress is directly connected with chondrocyte damage, which induces senescence and secretes inflammatory mediators. Hesperidin showed antioxidant activity against oxidative stress-induced chondrocytes. In this experimental study, hesperidin showed a protective effect by modulating mitochondrial activity, cell proliferation, survival and senescence. Total antioxidant capacity, antioxidant enzymes, and glutathione were restored (Tsai et al. 2019). Another study found that pre-treatment with hesperidin reduced MDA levels, cellular ROS, and inhibited hydrogen peroxide (Gao et al. 2018). Naringenin ameliorates cell damage and apoptosis by the influence of oxidative stress with the decreased expression of MDA (Pan et al. 2022). Induced oxidative stress in animal models was reduced by naringenin by lowering gp91phox, superoxide anion and lipid peroxidation, and the expression of NADPH was elevated (Manchope et al. 2018). Pomegranate peel extract demonstrated antioxidant activity in OA-derived patient erythrocytes, increasing SOD and GPX levels, serum TAC levels, and decreasing lipid peroxidation (Haghighian et al. 2021).

15.3.5.2 Anti-Inflammatory Action

OA is directly related to inflammation. Cytokines and prostaglandin secretion elevated matrix metalloproteinases through chondrocytes and led to inflammation. Tsai et al. (2019) discovered that hesperidin modulated COX-2, IL-1, TNF-α, and unregulated IL-10, TIMP-1, SOX9Foxo1, Foxo3, and Nrf2 cascade pathways. Hesperidin plays an important role in cartilage preservation by lowering IL-1 and TNF-α level while also modulating iNOS and another inflammatory component (Gao et al. 2018). Naringenin had a protective effect on cartilage damage in IOOA mice by lowering matrix metallopeptidase (MMP)3 and MMP13 and increasing NRF2 and hemo oxygenase 1 (HO-1) (Pan et al. 2022). Nobiletin showed anti-inflammatory activity against IL-1β treated chondrocytes of mice. It represented an inhibitory NF-κB pathway in the mouse and lowered the inflammatory protein expression of thrombospondin motifs-5 (ADAMTS5)), cyclooxygenase-2, collagen II, metalloproteinase etc. The protein expression of inducible nitric oxide synthase, matrix metalloproteinase-3, matrix metalloproteinase-13, a disintegrin, and aggrecan was analysed by western blotting (Lin et al. 2019).

15.3.6 Respiratory Diseases

Respiratory diseases, especially those affecting the airways and lungs' related structures, are a key cause of rising mortality and morbidity rates around the globe. The widely emerging respiratory diseases, namely chronic obstructive pulmonary disease (COPD), occupational lung diseases, and asthma, carry a major burden as non-communicable diseases (NCDs). There are various key regulating pathways which are involved in respiratory disease development like various types of growth factors, Notch, Hedgehog, Wingless/Wnt etc., and they further show progressive development by modulating oxidant and antioxidant balance in inflammation. ROS also plays a crucial role in the development of COPD and other respiratory diseases.

15.3.6.1 Regulating Inflammation and Oxidative Stress Against Respiratory Diseases

Inflammation is a defensive response against any pathological or microbial infection to protect vital organs. The main cause was inflammation caused by numerous ROS productions in the body and pathogen invagination in the airways. The primary site of inflammation attracts other immune cells in the infection site, which is the main initiator of oxidative stress in the lungs. This result leads to long-lasting inflammation. In the past decades, natural components have been gaining interest and the individual components of citrus fruit have been studied in detail with respect to treating and combating the rising incidence of respiratory diseases. Citrus-derived bioactive component revealed potential anti-inflammatory activity. In asthma, inflammatory cells were increased and inflammatory mediator alteration, as well as mucus production, was noted. Airway hyperactivity was attenuated by naringenin in the ovalbumin-induced male Wistar model. In this study, IL-4 and IL-13 were reduced by naringenin with the increased expression of glutathione balance and MDA level was lowered (Jasemi et al. 2022). Naringenin lowered ROS levels in the LPS-induced BALF ling injury via regulating H_2O_2 and malondialdehyde (MDA) production (Zhao et al. 2017). An antioxidant enzymatic study was performed in COPD patients in comparison to controls and revealed that SOD, Cat, and GPx were significantly elevated Nrf2 and COX2 also prevent oxidative stress and inflammatory-mediated damage in the lungs and airways (Antus et al. 2018). Flavonoids derived from bergamot and orange juice were tested against H_2O_2-induced oxidative stress in human A549 epithelial cells. The entire cellular antioxidative system was enhanced, including lower levels of ROS and MDA, mitochondrial functional activity developed, and DNA remained intact without any cellular damage (Ferlazzo et al. 2015). Hesperidin also enhanced the antioxidant profile via activating matrix metalloproteinases (MMP) in the nicotine-induced rat model (Balakrishnan and Menon 2007). The mRNA expression was studied after being treated with hesperetin and naringenin for the TTF-1 expression in house dust mites. This phenomenon involved these two reducing lung damage by downregulating TNF- and TGF-1. Citrus alkanie was administered against bleomycin induced fibrosis in the C57BL/6 male rat model. Here, collagen secretion was inhibited, which is the main cause of pulmonary fibrosis and the inflammatory pathway was downregulated through NF-κB and p38 (Wu et al. 2019) (Fig. 15.2).

15.3.7 Neuro-Degenerative Disease

There are significant medical and public health costs associated with degenerative disorders of the nervous system for people all over the world. The three principal neurodegenerative diseases are Alzheimer's disease (AD), Parkinson's disease (PD), and amyotrophic lateral sclerosis (ALS), whose prevalence has been raised with age. In neurodegenerative diseases, the molecular events at the cellular level include deposition of protein, impairment in mitochondria, oxidative stress and apoptosis are induced. Phytochemicals obtained from citrus fruit are capable of modulating this key regulator of these cellular events. Naringin was injected intraperitoneally in a mouse model, which exerted neuroprotective activity through its anti-inflammatory property and stimulated the mTORC1 cascade signaling pathway. The loss of dopamine neurons in the substantia nigra pars compacta (SNc) and the resulting striatal dopamine deficit is a defining feature of Parkinson's disease (Blesa and Przedborski 2014). In a previous experimental study with naringenin against microglial cells, the cells exhibited lower expression of inflammatory markers such as NO, iNOS, COX-2, and stimulated suppressors of cytokine signaling (SOCS)-3. The entire protective activity of naringenin was regulated through modulation of AMPKα and PKCδ pathway (Wu et al. 2016). Memory deficits associated with Aβ deposition and neuroinflammation In addition, Hesperidin significantly reduced ameliorated buildup Aβ deposition in transgenic APP/PS1 *in vivo* studies. Microglial cell activation was also restricted and TGF-1 expression in the cortical cortex and hippocampus was reduced (Li et al. 2015). Hesperidin also exerted its protective efficacy towards apoptosis in the cortical regions of the brain, especially the cerebrum and cerebellum, and also inhibited oxidative stress in the heavy metal (AlCl3) induced AD mice model (Justin Thenmozhi et al. 2017).

15.3.8 Antiviral

In March 2020, coronavirus disease was declared a pandemic by the World Health Organization (WHO). COVID-19 develops severe acute respiratory syndrome, which is responsible for the cause of multiple organ failure, toxemia, and death (Samanta et al. 2022a, b; Das et al. 2022a, b). A strong or well-built immune system is very much needed to provide defence against foreign pathogenic microorganisms, whereas a fragile immune system can enhance sensitivity to infection or disease and make it more severe. Inflammation is the first step of the immune response. Excessive inflammation can cause uncontrolled cell damage and ROS generation. Healthy food is the main source of a well-built immunity. Healthy foods with different dietary nutrients are needed to improve the immune system during this pandemic and have a beneficial effect on health. Foods can have different antiviral activities, which can develop the immune system and also protect cells from ROS, associated with disease or infection. A prior report showed that bio-flavonoids and vitamin C of citrus fruits are very effective for improving the immune boosting power of the human body. Citrus fruits like oranges, limes, mandarins, tangerines, and lemons have crucial nutritional value for improving human health. These fruits are also an abundant source of organic acids, vitamin C, amino acids, multiple minerals (Mg, Ca) and lots of other phytochemicals which are responsible for good health. Vitamin C can minimise viral infection, inflammation, and reduce oxidative stress-mediated cell damage. Regular intake of citrus fruits can reduce the severity of COVID-19 by improving our immune system and acting as powerful antioxidants (Bellavite and Donzelli 2020; Alberca et al. 2020) (Fig. 15.6). Except vitamin C, citrus fruits also contain folate, which can improve the probity of overall immunological barriers and modify the function of several immune cells like NK cells (natural killer cells), B-cells, T-cells, and phagocytes. There are different bioactive compounds or polyphenols present in citrus fruits like naringin, naringenin, hesperidin, narirutin, and narirutin, which all have strong anti-inflammatory effects. In

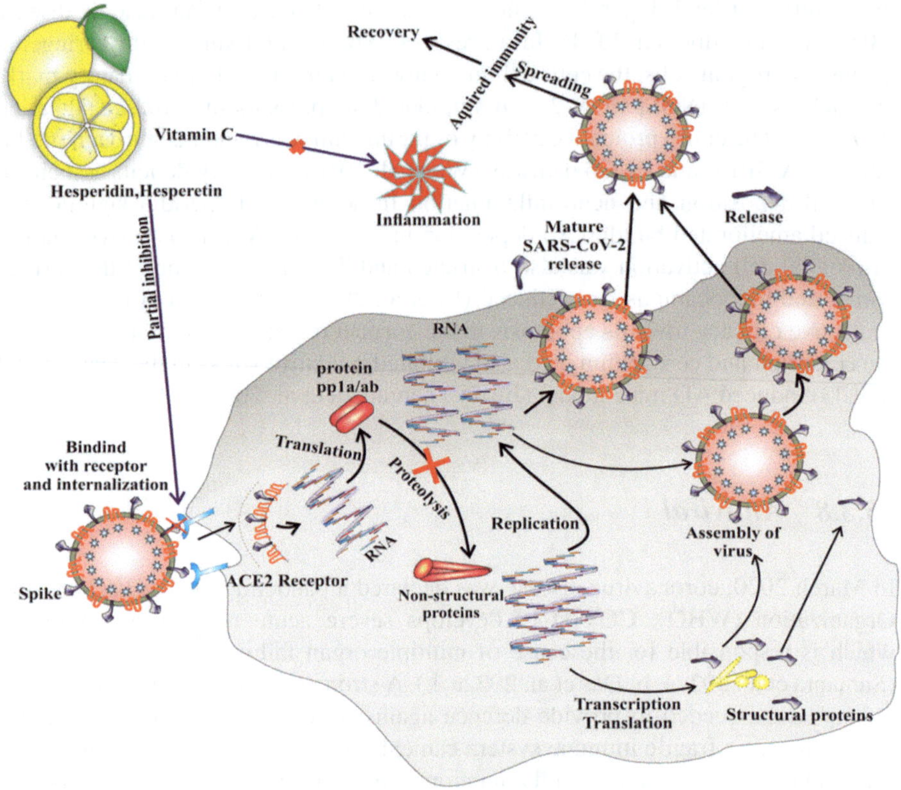

Fig. 15.6 Citrus fruits and their constituent bioactive compounds as treatment for viral infection like SARS-CoV-2. Hesperidin (or Hesperectin) inhibited viral attachment and internalisation into the cell. Citrus-derived vitamin C also prevented the formation of inflammation and restricted viral proliferation

in vivo studies (clinical trials), researchers found that hesperidin reduced inflammatory markers and orange juice limited postprandial inflammation caused by a high fat diet meal and a carbohydrate diet meal in humans. Studies have also shown that drinking orange juice on a regular basis for several weeks can reduce inflammation markers such as C-reactive protein. According to different research studies, bioactive compounds from orange juice can also have antiviral effects (Miles and Calder 2021; Alberca et al. 2020). A prior report suggested that the bioflavonoid of citrus fruit called hesperetin has an important role as an efficient inhibitor of SARS-CoV 3CLpro. Different experiments (*in vivo* and *in vitro*) suggest that naringin has the potency to decrease the expression of several pro-inflammatory cytokines (COX-2, IL-1β, IL-6 and iNOS) induced by lipopolysaccharide (LPS) in a cell line of raw macrophages, and may prevent cytokine by inhibiting the expression of HMGB1 in an *in vivo* experiment (mouse model). The results also showed that naringin has the potency to prevent cytokine storms. Citrus fruits contain several flavonoids which have potential anti-inflammatory activities and can act as effective agents in the treatment of SARS-CoV-2 infection (Liu et al. 2022) (Fig. 15.6). According to several research studies, it was found that bioactive compounds found in citrus fruit named hesperidin contain a shallow binding efficacy with the "spike" protein of the coronavirus and the major protease that can transform the premature or early proteins of the virus (ppa1b and pp1a) into a complex sufficient for viral replication. The binding efficacy or energy of hesperidin to these compounds is lower than indinavir, ritonavir, lopinavir, and indinavir. So, hesperidin could be an effective antiviral agent and also prevent different infectious diseases, including COVID-19 (Bellavite and Donzelli 2020). *Citrus limetta* (Sweet lime), popularly known as Mosambi or Mousambi in India, is very effective in immune boosting during COVID-19. The flavonones of this fruit block pp1b and pp1a (polyproteins), which are responsible for viral replication. Thus, sweet limes have been found to exert inhibitory and a reciprocal therapeutic activity against several RNA viruses, majorly SARS-CoV-2 (Banerjee and Pal 2021). Hesperetin is like an aglycone material or metabolite of hesperidin, having higher bioavailability. A docking simulation study revealed that hesperidin and hesperetin both demonstrated binding potency with cellular proteins, ACE2 (angiotensin-converting enzyme 2) and TMPRSS2 (transmembrane serine protease 2). Both proteins are required for SARS-CoV-2 cellular internalization. Both hesperetin and hesperidin were found to be effective in suppressing contamination of VeroE6 cells with SARS-CoV-2 pseudo-particles (lentiviral-based) variants and wild types of SARS-CoV-2 with spike proteins by blocking or inhibiting the interaction between the ACE2 receptor and the S protein and reducing TMPRSS2 and ACE2 expression (Cheng et al. 2021). According to an analysis, it was identified that plant microRNAs could act against or inhibit the Omicron (B.1.1.529) variant of the coronavirus in several countries in the world. Several dietary plant miRNAs are found in different fruits like oranges (*Citrus sinensis*), peaches (*Prunus persica*), apples (*Malus domestica*), and grapes (*Vitis vinifera*) etc. The strong and effective interaction between the *Citrus sinensis* miRNA cluster (including miR169–3p family) and the CTGCCT conserved site has been noted in the SARS-CoV-2 genome and all Omicron variants from all over the world. This may become a promising therapeutic to induce viral genome silencing (Mangukia et al. 2022). Essential polyphenols (like flavonoids) are present in different portions of citrus fruit, like juice, seed, pulp, skin, and peels. These flavonids have different types of biological activities like antibacterial, antifungal, antiviral etc. (Addi et al. 2021). According to a prior study, essential oil components and effective bioactivity of C. reshni ripe fruit peels (which contain limonene) can have an impact on the influenza virus (H5N1 subtype) (Nagy et al. 2018). Researchers also showed that B Sext (*Citrus bergamia*) and limonin can both inhibit HTLV-1 and HIV-1 expression in affected or infected cells (Balestrieri et al. 2011). Studies also found that a bioflavonoid called quercetin found in citrus fruits can show significant anti-DENV-2 dengue virus inhibitory activity and inhibit DENV-2 virus replication (Zandi et al. 2011). Several research studies have shown that citrus fruit peel can also be used as a green route for synthesis of nanoparticles to fight against viral infection and improve health. Different metallic nanoparticles like Fe, Ag, and Zn are green synthesized with citrus limetta peels. AgNPs, when combined with them, demonstrated the greatest antiviral activity against CHIKV (Choudhary et al. 2020). Citrus fruits, which are a wealthy source of vitamin C, are also effective against or prevent the occurrence of common colds and influenza by activating antiviral pathways in the human body. Flavonoids from citrus fruits, according to the study, may inspire the antiviral pathway mechanism due to their ability to modulate or activate the interferon stimulated response element (ISRE) (Fast et al. 2019). *In vitro* studies found that naringein could be a suitable candidate for ZIKV therapy because it contains efficient antiviral activity even if it is added several hours after infection. This activity of Naringein indicates that it can directly target viral replication. Naringein is effective against ZIKV lineages and can show inhibitory effects in the viral life cycle, probably as a non-competitive inhibitor of protease (NS2B-NS3) (Cataneo et al. 2019). Limonene, the major component found in essential oil of the bitter orange variety (*Citrus x aurantium L*), has also found effective activity against the H1N1 virus (Fadilah et al. 2022).

15.4 Heath Promoting Characteristics of Various Bio-active Compounds of Citrus Species

Health promotion is a progressive strategy of helping human beings to get a better lifestyle to move towards a state of optimal health. A healthy lifestyle always includes various types of effort, like change of behavior, awareness, and good health practises (Viner and Macfarlane 2005). In the 1970s, the concept of health promotion first came out of the discipline of public health (Butler and Friel 2006). Nowadays, health improvement is getting more attention in all educational sectors to maintain the quality of life and literacy during COVID-19 (Merino-Godoy et al. 2022). Good health always depends on health promoting activities (prevention of several chronic diseases like stroke, diabetes, heart disease, cancer etc.) and a healthy lifestyle. Disease prevention and health promotion are very important to educate people about infectious diseases, which are the leading causes of several deaths, population ageing, and the rapidly increasing cost of health care (Rimmer and Braddock 2002). Several types of health-promoting agents are present in the market, like ayurvedic products and different commercial products (Sharma et al. 2007; Cardenas and Fuchs-Tarlovsky 2018). Several vegetables and fruits contain beneficial micronutrients, bio-flavonoids, polyphenols, and anti-oxidants. So, continuous consumption of fruits and veggies can help in the prevention of diabetes, obesity, cardiovascular diseases, even cancers and improve overall health (Hashemian et al. 2022; Atkinson et al. 2005). Among all the fruits Citrus fruits are widely grown, and they are among the most important fruits in the world, as well as among the commercially grown crops that are rich in useful biologically active compounds (biotic and abiotic), antibacterial properties, and phytochemicals with health-promoting properties (Suri et al. 2022; Dutta et al. 2022; Alvi et al. 2021). Citrus fruits are generous sources of different phytochemicals, including flavonoids, carotenoids (antioxidant and pro-vitamin A), ascorbic acid, terpenoids, polyphenols, limonoids, etc., which have a great contribution to health-related improvements (Lado et al. 2018).

15.4.1 Obesity

In the twenty-first century, facing a global threat called obesity; its prevalence has been a multifaceted issue (Haaz et al. 2006; Lu et al. 2013). Adipose tissue serves as an endocrine organ, secreting adipokines, maintain major roles in the body. An excess amount of adipose tissue in the body, called obesity, which is related to a state of chronic pre-infection inflammation and life-threatening disease, affects the quality of life of an individual. According to epidemiology reports to WHO and the Organization for Economic Co-operation and Development (OECD), obesity and overweight are the fifth most common causes of global deaths (Karri et al. 2019; Feng and Wang 2018; Nakajima et al. 2014). The global burden of obesity has been growing rapidly, having a massive impact on overall human health worldwide due to the growing risk of different metabolic diseases (Lai et al. 2015). There are different conventional therapies for obesity like surgical procedures and synthetic moieties, but these have dangerous side effects with chances of severe recurrence (Karri et al. 2019). Several antiobesity drugs or medications are available on the market, but they have no safety assurance and may have threatening side effects (Feng and Wang 2018). In the last few decades, scientists have found some alternative native medicines that have been strongly investigated. In natural therapeutics, citrus fruits (phytochemicals, p-synephrine, etc.) have shown huge potential to fight obesity through different mechanisms (regulating lipid metabolism, regulating energy expenditure and intake, and regulating adipogenesis) (Feng and Wang 2018). *Citrus aurantium* (Bitter orange) contains a variety of compounds like synephrine alkaloids, a safe alternative alkaloid for the treatment of obesity (Haaz et al. 2006). Some bioactive compounds of citrus, like polyphenols and flavonoids, have been assessed for the treatment and prevention of obesity (Nakajima et al. 2014). Citrus fruits have been shown in various scientific studies (*in vivo* and *in vitro* research studies) to be a powerful source of improving anti-obesity therapies (Feng and Wang 2018). Some cell culture assays indicate that the polyphenols of citrus fruits could help in obesity management by reducing lipid content in cells, adipocyte differentiation, and also reducing programmed cell death. But the mechanism is not clear yet. Citrus polyphenols reduce adipose tissue, increase PPAR expression and target genes, and promote oxidation, lipid profile improvement, and glycemia and inflammation improvement by lowering pro-inflammatory cytokines (Nakajima et al. 2014). Citrange (a citrus hybrid of the trifoliate orange and sweet orange) is a kind of citrus fruit that contains neoeriocitrin, poncirin, and naringin as main flavonoid components. A recent *in vivo* (HF induced obese mouse) report suggested that citrange peel extract (CPE) was potently able to minimize the weight gain phenomenon and reduce glucose levels in blood serum along with the decrease of total serum cholesterol. This study also demonstrated that CPE blocked the different mRNA expression of LXRβ, LPL, and ap2 in the mouse liver. Moreover, the citrange flesh and seed extract (CFSE) were found to lower serum glucose, mouse weight, total serum cholesterol level, and serum LDL-c ability. Naringin and poncirin, which are the main bioactive compounds of CPE and CFSE, could be the key bioactive molecules achievable in this procedure (Lu et al. 2013). Previous reports suggested that *C. aurantium* (Bitter orange) contains elements such as tyramine, m-methyltyramine, octopamine, and hordenine responsible for the reduction of body fat. A prior report shows that *C. aurantium* or ephedra, commonly combined with caffeine, also alter thermogenesis and increase RMR in control and obese subjects (Haaz et al. 2006). Some studies demonstrated that nobiletin, a polymethoxylated flavone which is present in citrus fruits, has antiinflammatory, antitumor, insulin resistance effects, and obesity management. Nobiletin increases microRNA expression of PPAR-γ (peroxisome proliferator-activated receptor), sterol regulatory element-binding protein-1c, stearoyl-CoA desaturase-1, synthesis of fatty acid, uncoupling protein-2, carnitine (palmitoyltransferase-1) and adiponectin, and this research also showed that decreasement in microRNA expression of monocyte (chemoattractant protein-1) in WAT and TNF- α (tumor necrosis factor-α). Nobiletin can also up-regulate Akt phosphorylation, glucose transporter-4 protein expression and suppress IκBα degradation in white adipose tissue. Collectively, nobiletin could improve insulin resistance, hyperglycemia, dyslipidemia, and adiposity by regulating lipid metabolism-related expression, regulation of inflammatory maker's expression, the insulin signaling pathway activity, and adipokine genes (Lee et al. 2013). Prior reports showed that if a human being can consume blood orange juice for approximately 2 weeks, then some beneficial effects on endothelial activities are shown in healthy men and women with obesity or over weight. This is the result of the combined action of flavanone and anthocyanin metabolites helping to enhance bioavailability (Li et al. 2020). As per the report, it was suggested that the yuzu peel contains anti-obesity activities through hepatic PPARα and adipocyte PPARγ pathways. Yuzu pomace can be a unique and novel application for anti-obesity (Zang et al. 2014).

15.4.2 Anti-ageing

Aging is an unavoidable and unprompted process in all living beings. Aging is a kind of natural phenomenon with different complexities that manifest a gradual decrease in homeostasis and physiological functions. It leads to age-related injuries and diseases (Parkinson's disease, cardiovascular disease, cognitive declines like Alzheimer's disease, diabetes type II, immune and inflammatory disorders) and ultimately senescence. Numerous studies have been conducted to find ways to delay or even reverse ageing and age-related diseases (Xue et al. 2021; Pan et al. 2012; Howes et al. 2020). Skin is the major protective barrier of the human body and also an overall health indicator. Skin needs proper care and nourishment, which has led to a growing demand for antiaging products day by day. Skin is damaged by different intrinsic and extrinsic factors. Prior research discovered several mechanisms and theories of aging, such as telomere length reduction, decreased proliferative tendency, free radical generation, DNA mutations, and many others. There are different conventional therapeutics using phytocompounds (curcumin, resveratrol, Vitamin E) in cosmetics, liposomes, niosomes, transferosomes, ethosomes, solid lipid nanoparticles, nanostructure lipid carriers, and carbon nanotubes in anti-aging products (Sharma et al. 2022a, b). Continuous exposure to ultraviolet radiation is one of the main reasons for skin ageing (complicated bioprocess) (Amer et al. 2021). Solar radiation and oxidative stress are the probable causes of excess ROS production, which is the primary cause of skin damage (Kim et al. 2016). Although, prevention of ageing is quite impossible, slowing down the premature ageing rate could be possible (Shen et al. 2017). Here, nutraceuticals and different natural dietary interventions played an important role via several mechanisms to stop age-related diseases and received much more attention (Shen et al. 2017; Howes et al. 2020; Pan et al. 2012). Citrus fruits are full of nutrition and different health-promoting properties. Researchers found that it contains lots of photochemicals with several biological functions (Zou et al. 2016). In this research, it was found that CSPE (ethanolic extract of *Citrus sinensis L.* fruit peels) showed a powerful anti-aging effect through down-regulation of mRNA expression of MMP1 along with anti-inflammatory, anti-oxidant effects (Amer et al. 2021). As per the *in vivo* and *in vitro* study reports, liposomal encapsulated citroflavonoid complex (plant derived) is a safe use to reduce age-spots, brighten skin tone, and increase overall skin health. Citroflavonoids have potential anti-inflammatory and antiinflammatory properties, which is the reason for the rapid use of them in anti-ageing cosmetics (Tiedtke et al. 2011). As per the study report, *C. reticulate* (Mandarin orange) peel can be used for an antiwrinkle skin care formula (Apraj and Pandita 2016). Several studies have shown that citrus-juice mixtures with their various bioactive compounds have powerful anti-aging and antioxidant properties in hairless mice and human dermal fibroblasts, acting via the protein kinase pathway (mitogen-activated) and antioxidant enzyme regulation. Citrus juice could reduce intacellular ROS, H_2O_2 and prevent cell damage in human fibroblasts. Pretreatment with citrus-based juice suppressed the elevation of H_2O_2-directed MAPK signaling. Here, different anti oxidant enzymes like MnSOD, glutathione and catalase showed elevated expressions. In an *in vivo* study (UV-B exposed hairless mouse), it was shown that the oral uptake or administration of this juice mixture decreased wrinkle formation, skin thickness, and increased collagen content (Kim et al. 2016). Prior studies found that extended oral intake of nutroxsun (complementary nutrition) reduced the harming effects of sun exposure by inhibition of reactive oxygen species (UVR-induced), cytokines, and lipoperoxides (inflammatory markers) together with their straight action on signaling pathways (intracellular) (Nobile et al. 2016). According to research, hesperidin, which is taken out from orange woodchips, has chelating, antiageing, and antioxidant properties. Hesperidin could reduce dark circles around the eyes, which is a very sensitive and thin skin region. It was also found that the aqueous extraction of hesperidin may be beneficial for human consumption. The nanoformulation of hesperidin was found to be skin-friendly and could be used in several cosmetic products (Stanisic et al. 2020). Studies also found that *C. reticulata* seed waste product is a key beneficial source of skin-improving and is also effective for wound healing at a coetaneous level (Al-Warhi et al. 2022).

15.4.3 Hypertension

Hypertension, diabetes mellitus, obesity, and dyslipidemia are major metabolic disorders among global medical problems. In recent years, this type of illness has been expanding quickly (Siti et al. 2022). Hypertension is a major global public health issue that is influenced by a variety of complex factors such as the external environment, genetics, diet, and gut microbiota dysbiosis (Zhang et al. 2021; Rakib 2021; Ajeigbe et al. 2021). People living in poverty, in particular, who have very poor health education and limited access to healthcare funds, are disproportionately affected by this problem. Management of hypertension is very important to maintain overall health, such as vision impairment, cardiovascular or cerebrovascular events, renal failure, and others. According to different pharmacological therapies, lifestyle modifications through exercise and diet are very effective in the management of hypertension (Feyh et al. 2016). Different kinds of antihypertensive drugs are used for the management of hypertension, such as vasodilators, diuretics, sympatholytic drugs, angiotensin converting enzyme inhibitors and receptor blockers of angiotensin II (Ajeigbe et al. 2021). Research studies have detected that continuous consumption of vegetables and fruit decreases blood pressure (Reshef et al. 2005). Previous reports suggested that citrus fruits (lemons, oranges, tangerines, grapefruits, limes, mandarins, and others) contain lots of vitamin C, calcium, polyphenols, potassium, flavonoids, pectin, and carotenoids, which have lots of health benefits and are used in the treatment of hypertension and heart diseases, so higher intake of several fruits can significantly decrease the risk of hypertension (Tsubota-Utsugi et al. 2011; Audu and Olu 2021; Khan et al. 2021; Mitra et al. 2022a, b). Citrus plant origins carry many advantageous nutrients and different bioactive materials which have anti-inflammatory, antiplatelet aggregation, antidiabetic, antimicrobial, and anticancer activities (Khan et al. 2021). According to the research, it was found that continuous intake of plant-based flavoniods can decrease the risk of hypertension and cardiovascular disease in Australian women (do Rosario et al. 2021). Studies found that *Citrus hystrix* (kaffir lime) was beneficial in metabolic disorders. Different phytochemical compounds found in this plant, including pinene, citronellol, sabinene, and citronellal, which can show potential anti-obesity, antidiabetic, and antihyperlipidemic activity as well as can decrease hypertension development (Siti et al. 2022). Prior studies revealed that *Citrus aurantifolia* (Key lime) may carry hypotensive and antihypertensive effects through multiple pathways or mechanisms in *in vivo* experiments (rats). According to these findings, the major mechanism of the antihypertensive effects could be its vaso-relaxation effect. *Citrus aurantifolia* extract induces vasorelaxation via endothelium-independent mechanisms by increasing cAMP and cGMP levels by inhibiting vascular PDEs (Bukhari et al. 2022). A report suggested that hesperidin reduces blood pressure (systolic) and pulse pressure after prolonged consumption. Orange juice enriched with the bioactive compound hesperidin (Phytochemical) may be a useful co-adjuvant tool for pulse pressure and blood pressure management in early stage hypertensive patients (Valls et al. 2021). Naringin is also effective in treating pulmonary arterial hypertension. It inhibits NF-κB and ERK signaling pathways and EMT in pulmonary arteries (Wu et al. 2021). According to a prior study, high-flavonoid (HF) (narirutin and naringin) sweetie juice is more advantageous than low-flavonoid (LF) sweetie juice in decreasing blood pressure (diastolic) in people with an early stage of hypertension (Reshef et al. 2005). Studies also showed that *Citrus limetta* (Sweet lemon) leaf extract alienated the hypertensive effect caused by angiotensin II (Perez et al. 2010). In a research study, *Citrus depressa Hayata* extract was used in hypertensive rats by spontaneous oral administration for blood pressure regulation. *In vivo* studies revealed that 6 weeks of continuous administration of the extract can lower MBP (average blood pressure), SBP (systolic blood pressure), and DBP (diastolic blood pressure) (Chen et al. 2020). Grapefruit peel, essential oils extracted from orange peels and lemon peels also have a role in maintaining hypertension (Oboh and Ademosun 2011; Oboh et al. 2017). According to recent findings, auraptene also has a standard hypotensive activity. More research is needed to know the proper effects of this phytochemical (Imenshahidi et al. 2013).

15.4.4 Improvement of Gastrointestinal Function

The GI (gastrointestinal) tract plays an important role in our body, including waste disposal, maintaining a proper balance of harmful and helpful microbes, body homeostasis, digestion, and nutrient absorption through intestinal motility and the peristalsis process (Collier et al. 2022; Iswanti 2022; Shahsavari and Parkman 2022). Disruption or abnormalities in GI motility could lead to alteration in GI function; resulting GI disorders (Collier et al. 2022). Several epidemiologic studies have shown that eating vegetables and fruits is good for GI nutrition and lowers the risk of cancers in the digestive tract (upper aero) (Boeing et al. 2006). Citrus flavanoids like, naringin, hesperidins have different beneficial effects like anti-inflammatory, anti-oxidative etc. Research studies shows that naringin and hesperidin effective in the intestinal microbiome. They are metabolized by the bacteria of the intestine, specifically in the proximal colon, and form aglycones of naringenin, hesperetin, and several small phenolic compounds. Researchers also found that citrus flavanonoids and their metabolites are helpful for the gut microbiota composition, have a beneficial effect on gastrointestinal inflammation, improve intestinal barrier function and overall gastrointestinal health (Stevens et al. 2019; Arafah et al. 2020) (Fig. 15.7). *In vivo* research studies found that rats treated with *Fructus aurantii* showed a higher number of VIP-immunopositive cells and 5-HT cells in the duodenum, gastric jejunum, and antrum. So, it was found that *Fructus aurantii* could increase gastrointestinal flexibility by developing VIP and 5-HT expression levels in the gut of rats (Jiang et al. 2014). *Citrus aurantium L. (Au)* and *Citrus reticulate Blanco (Ci)*, local names are bitter orange and mandarin orange, which are commonly used as couplet prokinetics, and *Bupleurum chinense DC (Bup)* is used as an antidepressant in traditional Chinese medicine. Prior studies showed that AuCi (*Citrus aurantium L.* and *Citrus reticulate Blanco*) and AuCiBup together can improve the curative effect on DGE (delayed gastric emptying) related abnormalities (Gong et al. 2022). Excessive oxidative stress and inflammation (physiological) harm the epithelial barrier of the intestine. Nutrients (immunomodulatory) like citrus pectin, fatty acids (omega-3 polyunsaturated), and exosomes (milk-derived) prevented inflammation of the intestinal barrier. These three nutrients helped in the improvement of tight junction protein expression, epithelial proliferation, immunomodulation, bioavailability and bioaccessibility (in the intestinal epithelium) and mucus layer enrichment. Studies have figured out that in the future, with the help of the ideal property of beneficial citrus pectin, it is possible to make a gel-forming matrix with the help of mucin. This could be beneficial for probiotic intestinal adherence and the regeneration of mucosal epithelium cells that are damaged during inflammation (infections), IBD, or antibiotic therapy (Sundaram et al. 2022). Citrus extract with 6.5% naringin and 88.2% hesperidin is very useful for the improvement of gut microbiota (Sost et al. 2021). Studies found that among citrus fruits, bergamot peel contains an essential oil that could be used as antibacterial, antiviral, anti inflammatory, anticancer, analgesic, antioxidant, neuroprotective, antiplatelet, etc. agents. Bergamot is a rich source of flavonoids like neoeriocithrin, neodiosmin, naringin, neoeriocithrin, neohesperidin, eriodictyol, and neodiosmin. According to studies, these flavanones can also affect gene expression, intestinal bacterial development, and the induction of helpful bacteria for the appropriate creation of short-chain fatty acids (SCFAs), which operate as prebiotics. It was also found that bergamot fibre can improve lipid metabolism, increasing the bulk in stool and digestive disorders (Maiuolo et al. 2022). In accordance with *in vivo* research, it was found that ZBVO (the volatile oils of *Citrus aurantium* and *Rhizoma atractylodis Macrocephalae*) administration increased faecal water concentration, promoted intestinal peristalsis, controlled GAS and SP levels (gastrointestinal hormones), decreased the inflammatory response (IL-6 and TNF-α), and regulated brain-intestinal peptides (VIP and 5-HT). It also improved intestinal microbiota by reducing pathogenic bacteria and also helpful for slow transit constipation in patients (Wang et al. 2022). Citrus fruits contain large amount of pectin which strengthen GI immune barrier by direct interaction with several immune cells. The interaction of pectins with the GI immune barrier conducted through pattern recognition receptors (TLR-2/4 or Galectin-3). Prior research suggested that *Citrus aurantium L.* effective in IBD of rats. Oral intake of *Citrus aurantium L.* alleviated weight loss and digestive disorder like, diarrhea, infiltration of colitis inflammatory cells and decrease. Flavonoids of *Citrus aurantium L.* like, naringenin, nobiletin, hesperetin all effective in inflammation and intestine contraction. From this study it was found that *Citrus aurantium L* and its bioflavonoids have different regulatory effects on inflammatory bowel disease (IBD) through intestinal muscle contraction and anti-inflammation (He et al. 2018) (Fig. 15.7).

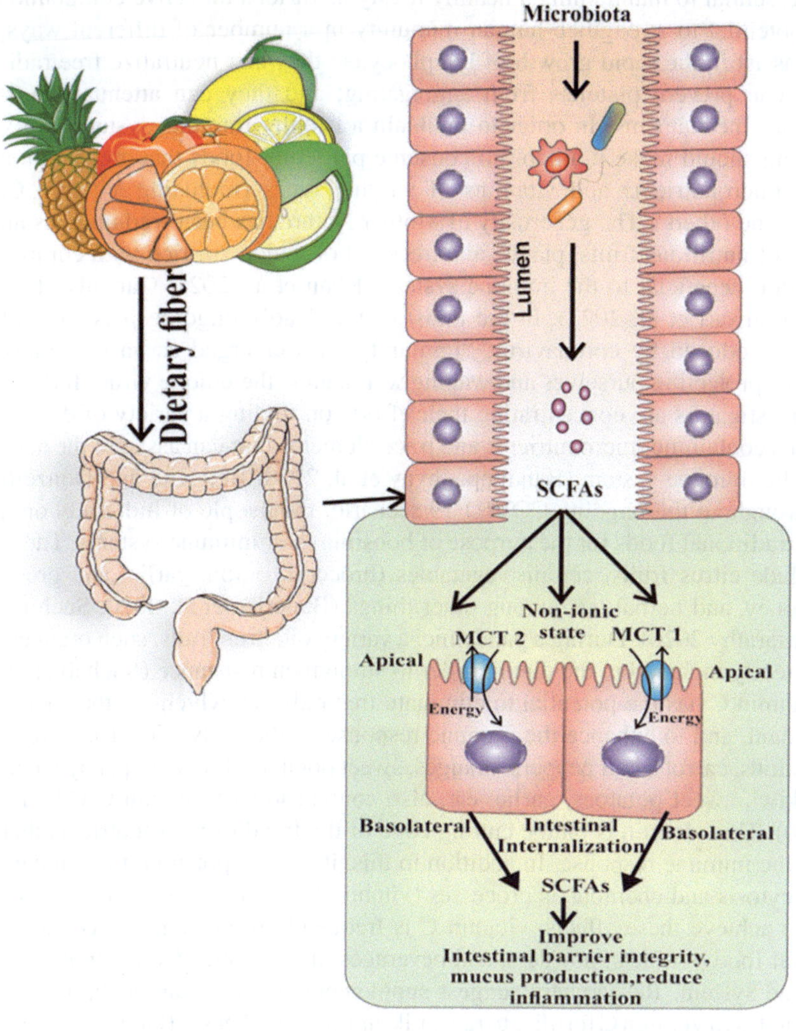

Fig. 15.7 Beneficial role of citrus bioactive compounds on GI function. Citrus fruit dietary fibre stimulates nonpathogenic bacteria in the gut, resulting in improved barrier integrity, mucus production and inhibition of inflammation

15.4.5 Immune Boosting Effect

Having an immune system that is both well-developed and strong is very crucial for maintaining good health. Immunity refers to the capacity of organisms (multicellular or unicellular) to avoid or fight off the harmful effects of pathogens. Our body's immune system hangs on WBC or leucocytes, which build the power to make antibodies to combat several disease-causing foreign pathogens. A robust immune system is essential to maintaining a healthy lifestyle. Various bioactive compounds have the potential to strengthen human immunity in a number of different ways. They can assist in the rapid growth of lymphocytes; they can neutralize free radicals; they can prevent platelets from aggregating; and they can attenuate anti-inflammatory mechanisms. In order to maintain a healthy immune system, every human being should make it a habit to consume particular foods that are rich in a variety of micronutrients and vitamins of varying types (including Vitamin C, Vitamin D, and others). The generosity of Mother Nature has bestowed upon us an abundance of nutritious fruits, plants, and trees, all of which are rich with components that are beneficial to the immune system (Khan et al. 2021; Ganguly et al. 2022; Maheshwari et al. 2022). In the midst of the deadly ongoing pandemic of COVID-19 produced by coronavirus, "immunity" has emerged as an extremely vital tool for protecting ourselves and waging war against the unique virus. Today's modern investigators are concentrating their efforts on finding a variety of dietary components containing micronutrients and trace elements that can assist in the activation of the immune system (Bandyopadhyay et al. 2020; Bellavite and Donzelli 2020). Throughout the ongoing COVID-19 scenario, the people of India rely on a variety of traditional foods for the purpose of boosting their immune systems. These foods include citrus fruits, certain vegetables (broccoli, onion, garlic, and green leaves), honey, and herbal tea, among other things (Banerjee et al. 2020; Suchitra and Parthasarathy 2020). During a pandemic, a variety of citrus fruits, such oranges, are consistently included in the diet in order to strengthen resistance (Shehab et al. 2022). Vitamin C has the potential to eliminate free radicals, which are the source of antioxidant, and to enhance the immune response of the body (Fig. 15.8). Raw apples, lemons, carrots, bell peppers, oranges, sweet potatoes, broccoli, papaya, bell peppers, kiwi, sweet potatoes, garlic, etc. also contain lots of vitamin C (Mishra et al. 2020). This vitamin is on the cutting edge of the trend of micronutrients that stimulate the immune response. In addition to this, it has the potential to stimulate the phagocytosis and chemotaxis processes (vitamin C accumulates in nutrophils). In order to achieve these effects, vitamin C is frequently used as an ingredient in commercial food products, powders, and beverages that are intended to strengthen the immune system. Researchers suggest supplementing this vitamin, liposome-encapsulated, as a vitamin C fortifier to raw milk and cottage cheese (Bandyopadhyay et al. 2020). It was noticed, however, that consistent orange juice consumption over a few weeks may be able to lower a number of inflammation markers, including C-reactive protein. Recent research has shown that the polyphenols present in orange juice may possess antiviral properties (Miles and Calder 2021; Thirumdas et al. 2021). Vitamin C can also be effective in the fight against inflammation by decreasing levels of histamine. Immune cells, including leukocytes and neutrophils, can be protected by the antioxidant effects of vitamin C. Neutrophils are needed for innate immunity and wound healing. It can protect neutrophils from neutrophil paralysis and improve their migration in response to chemotaxis. It also suppresses neutrophil extracellular traps (NETs) and reduces inflammation. It reduced LPS-induced pro-inflammatory cytokine production (TNF-α and IL-6) while increasing anti-inflammatory cytokine production (IL-10) (Basak and Gokhale 2022; Carr and

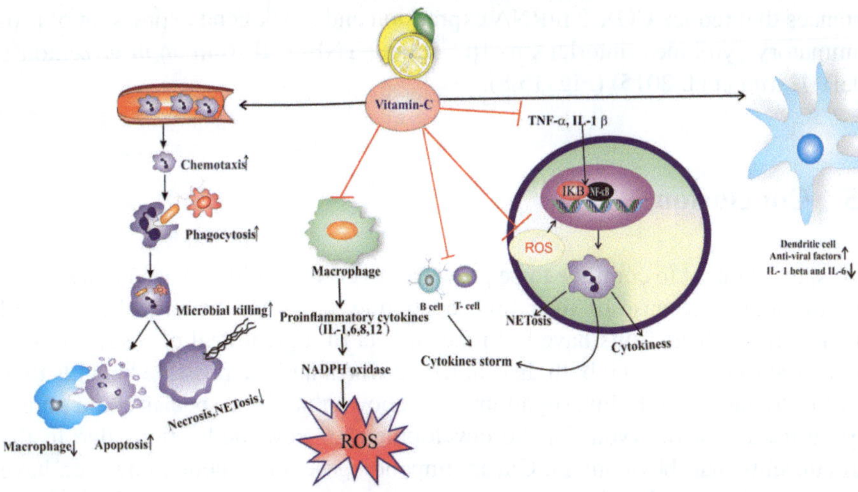

Fig. 15.8 Immune boosting role of vitamin C derive from citrus fruit. Chemotaxis and phagocytosis along with the microbial killing were enhanced through WBC especially neutrophil. ROS expression was lowered which reduced activation of macrophage as well as B and T cells. Cytokines storm and dendritic cell activity also inhibited

Maggini 2017; Bozonet and Carr 2019). Studies have shown that grapefruit has powerful anti-inflammatory and wound healing qualities, making it a popular and healthy fruit. Grapefruit extracellular vesicles can reduce the amount of intracellular reactive oxygen species (ROS) formation in HaCaT cells, which results in increased cell motility and viability. The capacity of treated HUVEC cells to form tubes was improved after treatment with grapefruit extracellular vesicles (GEVs). Savcı et al. (2021) revealed that GEVs have beneficial plant-derived wound healing capacity and can operate as a biotechnological medium or agent in the process of wound healing. However, bioflavonoids found in citrus fruit contribute to the homeostasis of the immune system. Vitamin C has been reported to inhibit the expression of pro-inflammatory mediators such as IL-6 and TNF-α in mature blood cells. It has an effect on gene expression (LPS-induced) in human macrophages (via NF-κB). An *in vitro* research revealed that pretreatment with vitamin C of murine CD40/IgM-activated B cells inhibits apoptosis, whereas low concentrations of vitamin C improve the antioxidant efficacy of activated B cells without affecting cell accretion or the expression of well-defined surface molecules such as CD86 and CD80. Vitamin C also increases IgM and IgA levels and elevated the capacity of NK cells (Mitra et al. 2022a, b) (Fig. 15.8). In accordance with number of scientific findings, both *C. aurantifolia* and *A. galangal* have the potential to be effective anti-inflammatory treatments for asthama. Bioactive compounds from a mixture or extract of *C. aurantifolia* and *A. galangal* paused the expression of CD18/11a by blocking extracellular and intracellular mechanisms, such as those that cause inflammation and that turn on kappa B. Citrus fruits and plants include bioactive substances that reduce COX-2 mRNA expression and block gene expression of pro-inflammatory cytokines (interleukin- 1β, 1α, and TNF-α, IL-6 in an *in vivo* mouse model) (Harun et al. 2015) (Fig. 15.8).

15.5 Conclusion

Citrus species ought to continue to be grown all over the world and used as a significant fruit source because of their high economic value and beneficial biological characteristics. Citrus fruits have been the subject of a great deal of research, with the end goal of establishing both their nutritive worth and the possible benefits they may offer to one's health. Investigations have shown that these species could potentially be used as a reservoir for the development of new medications due to the micronutrients that they contain. Citrus' impending health benefits, however, have not yet been fully explored. From numerous citrus species, a large variety of secondary metabolites and functional metabolites with significant biological value have been identified and classified. Flavonoids, carotenoids, alkaloids, ascorbic acids, essential oils and many more bioactive substances are abundant in various citrus fruits. Together, these secondary metabolic products have the potential to be antioxidants with antibacterial, antiviral, cardioprotective, anticancer etc. applications. If Citrus spp. is successful, additional research is necessary to better understand the mechanisms involved, lower the severity of sickness, and mitigate the negative effects of currently recommended drugs. For the use of bioactive compounds to prevent diseases that pose a threat to human life, there is a need to increase awareness of these substances and their applications. Subsequently, by applying combinational procedures using commercial medications while mitigating the inevitable side effects, the vetting of these phytochemicals can be made significantly more efficacious. Although *in vitro* and *in vivo* investigations are the main techniques used to study the beneficial benefits of citrus fruit bioactive components, their apothecary is still constrained because there is a lack of sufficient clinical data. This chapter describes the vital substances present in citrus that the body needs considerably and whose consumption has a substantial bearing on human health and disease prevention.

Acknowledgement We are grateful to University of Gour Banga for providing the chance to conduct this research at the Department of Physiology.

Authorship Contribution SS, JB and RA: acquisition of information, drafted the article, draw the figures, and final approval of the version to be submitted. SKD: conception and design of the study, revising it critically for important intellectual content, designed the figures, and final approval of the version to be submitted.

Conflicts of Interest/Competing Interests The authors declare that they have no conflict of interest.

Funding No funding was received for conducting this study.

References

Addi M, Elbouzidi A, Abid M, Tungmunnithum D, Elamrani A, Hano C (2021) An overview of bioactive flavonoids from citrus fruits. Appl Sci 12(1):29

Ahmed W, Azmat R (2019) Citrus: an ancient fruits of promise for health benefits. In: Citrus-health benefits and production technology, vol 27, pp 19–30. https://doi.org/10.5772/intechopen.79686

Ajeigbe OF, Ademosun AO, Oboh G (2021) Relieving the tension in hypertension: food–drug interactions and anti-hypertensive mechanisms of food bioactive compounds. J Food Biochem 45(3):e13317

Al-Aamri MS, Al-Abousi NM, Al-Jabri SS, Alam T, Khan SA (2018) Chemical composition and in-vitro antioxidant and antimicrobial activity of the essential oil of Citrus aurantifolia L. leaves grown in Eastern Oman. J Taibah Univ Sci 13(2):108–112

Alberca RW, Teixeira FM, Beserra DR, de Oliveira EA, Andrade MM, Pietrobon AJ, Sato MN (2020) The potential effects of naringenin in COVID-19. Front immunol 11:570919

Ali MY, Zaib S, Rahman MM, Jannat S, Iqbal J, Park SK, Chang MS (2019) Didymin, a dietary citrus flavonoid exhibits anti-diabetic complications and promotes glucose uptake through the activation of PI3K/Akt signaling pathway in insulin-resistant HepG2 cells. Chem Biol Interact 305:180–194

Alvi GB, Iqbal MS, Ghaith MM, Haseeb A, Ahmed B, Qadir MI (2021) Biogenic selenium nanoparticles (SeNPs) from citrus fruit have anti-bacterial activities. Sci Rep 11(1):1–1

Al-Warhi T, Elmaidomy AH, Selim S, Al-Sanea MM, Albqmi M, Mostafa EM, Ibrahim S, Ghoneim MM, Sayed AM, Abdelmohsen UR (2022) Bioactive phytochemicals of Citrus reticulata seeds—an example of waste product rich in healthy skin promoting agents. Antioxidants 11(5):984

Amer RI, Ezzat SM, Aborehab NM, Ragab MF, Mohamed D, Hashad A, Attia D, Salama MM, El Bishbishy MH (2021) Downregulation of MMP1 expression mediates the anti-aging activity of Citrus sinensis peel extract nanoformulation in UV induced photoaging in mice. Biomed Pharmacother 138:111537

Andrade-Oliveira V, Foresto-Neto O, Watanabe IK, Zatz R, Câmara NO (2019) Inflammation in renal diseases: new and old players. Front Pharmacol 10:1192

Antus B, Paska C, Simon B, Barta I (2018) Monitoring antioxidant enzyme activity during exacerbations of chronic obstructive pulmonary disease. COPD 15(5):496–502

Apraj VD, Pandita NS (2016) Evaluation of skin anti-aging potential of Citrus reticulata blanco peel. Pharm Res 8(3):160

Arafah A, Rehman MU, Mir TM, Wali AF, Ali R, Qamar W, Khan R, Ahmad A, Aga SS, Alqahtani S, Almatroudi NM (2020) Multi-therapeutic potential of naringenin (4′, 5, 7-trihydroxyflavonone): experimental evidence and mechanisms. Plan Theory 9(12):1784

Aranganathan S, Selvam JP, Nalini N (2008) Effect of hesperetin, a citrus flavonoid, on bacterial enzymes and carcinogen-induced aberrant crypt foci in colon cancer rats: a dose-dependent study. J Pharm Pharmacol 60(10):1385–1392

Aroui S, Aouey B, Chtourou Y, Meunier AC, Fetoui H, Kenani A (2016) Naringin suppresses cell metastasis and the expression of matrix metalloproteinases (MMP-2 and MMP-9) via the inhibition of ERK-P38-JNK signaling pathway in human glioblastoma. Chem Biol Interact 244:195–203

Asadim BF (2008) Biological effects of essential oils. Food Chem Toxicol 46:446–475

Asgary S, Keshvari M (2013) Effects of citrus sinensis juice on blood pressure. ARYA Atheroscler 9(1):98–101

Atkinson CJ, Nestby R, Ford YY, Dodds PA (2005) Enhancing beneficial antioxidants in fruits: a plant physiological perspective. Biofactors 23(4):229–234

Audu OK, Olu AA (2021) A survey of nutritional benefits and diseases of citrus fruits consumed by humans. J Food Sci Qual Control 7(1):24–28

Baba S, Ueno Y, Kikuchi T, Tanaka R, Fujimori K (2016) A limonoid Kihadanin B from immature Citrus unshiu peels suppresses adipogenesis through repression of the Akt-FOXO1-PPARγ axis in adipocytes. J Agric Food Chem 64(51):9607–9615

Balakrishnan A, Menon VP (2007) Effect of hesperidin on matrix metalloproteinases and antioxidant status during nicotine-induced toxicity. Toxicology 238(2–3):90–98

Balestrieri E, Pizzimenti F, Ferlazzo A, Giofrè SV, Iannazzo D, Piperno A, Romeo R, Chiacchio MA, Mastino A, Macchi B (2011) Antiviral activity of seed extract from Citrus bergamia towards human retroviruses. Bioorg Med Chem 19(6):2084–2089

Banerjee S, Pal SR (2021) Inhibitory and complementary therapeutic effect of sweet lime (Citrus limetta) against RNA-viruses. J Prev Med Holist Health 7(1):37–44

Bandyopadhyay K, Mukherjee M, Deb P, Biswas A, Gope RK, Das S (2020) Study on the fortification techniques of multiple food products using immunity boosting micronutrients and antioxidants: a literature review. Int J Eng Appl Sci Technol 5(6):163–169

Banerjee S, Srivastava S, Giri AK (2020) Possible nutritional approach to cope with COVID-19 in Indian perspective. Adv Res J Med Clin Sci 6(06):207–219

Barcelos GR, Angeli JP, Serpeloni JM, Grotto D, Rocha BA, Bastos JK, Knasmüller S, Júnior FB (2011) Quercetin protects human-derived liver cells against mercury-induced DNA-damage and alterations of the redox status. Mutat Res Genet Toxicol Environ Mutagen 726(2):109–115

Basak S, Gokhale J (2022) Immunity boosting nutraceuticals: current trends and challenges. J Food Biochem 46(3):e13902

Baselga-Escudero L, Pascual-Serrano A, Ribas-Latre A, Casanova E, Salvadó MJ, Arola L, Arola-Arnal A, Bladé C (2015) Long-term supplementation with a low dose of proanthocyanidins normalized liver miR-33a and miR-122 levels in high-fat diet–induced obese rats. Nutr Res 35(4):337–345

Bawazeer NA, Choudhry H, Zamzami MA, Abdulaal WH, Middleton B, Moselhy SS (2016) Role of hesperetin in LDL-receptor expression in hepatoma HepG2 cells. BMC Complement Altern Med 16(1):1–7

Bellavite P, Donzelli A (2020) Hesperidin and SARS-CoV-2: new light on the healthy function of citrus fruits. Antioxidants 9(8):742

Bhutada RS, Rathi R, Dasar D (2020) Immunity boosting diet during Covid 19. Int J Res Pharm Sci 11:832–838

Blesa J, Przedborski S (2014) Parkinson's disease: animal models and dopaminergic cell vulnerability. Front Neuroanat 8(155):1–12

Boeing H, Dietrich T, Hoffmann K, Pischon T, Ferrari P, Lahmann PH, Boutron-Ruault MC, Clavel-Chapelon F, Allen N, Key T, Skeie G (2006) Intake of fruits and vegetables and risk of cancer of the upper aero-digestive tract: the prospective EPIC-study. Cancer Causes Control 17(7):957–969

Bourgaud F, Hehn A, Larbat R, Doerper S, Gontier E, Kellner S, Matern U (2006) Biosynthesis of coumarins in plants: a major pathway still to be unravelled for cytochrome P450 enzymes. Phytochem Rev 5(2):293–308

Bozonet SM, Carr AC (2019) The role of physiological vitamin C concentrations on key functions of neutrophils isolated from healthy individuals. Nutrients 11(6):1363

Bukhari SN, Saeed S, Asim MH, Irfan HM, Ejaz H, Elsherif MA, Junaid K (2022) Antihypertensive and Vasorelaxant effects of Citrus aurantifolia Linn. Fruit: proposed mechanisms. Evid Based Complement Alternat Med 2022:5871424

Butler CD, Friel S (2006) Time to regenerate: ecosystems and health promotion. PLoS Med 3(10):e394

Cardenas D, Fuchs-Tarlovsky V (2018) Is multi-level marketing of nutrition supplements a legal and an ethical practice? Clin Nutr ESPEN 25:133–138

Carr AC, Maggini S (2017) Vitamin C and immune function. Nutrients 9(11):1211

Cataneo AH, Kuczera D, Koishi AC, Zanluca C, Silveira GF, Arruda TB, Suzukawa AA, Bortot LO, Dias-Baruffi M, Verri Jr WA, Robert AW (2019) The citrus flavonoid naringenin impairs the in vitro infection of human cells by Zika virus. Sci Rep 9(1):16348

Cavia-Saiz M, Busto MD, Pilar-Izquierdo MC, Ortega N, Perez-Mateos M, Muñiz P (2010) Antioxidant properties, radical scavenging activity and biomolecule protection capacity of flavonoid naringenin and its glycoside naringin: a comparative study. J Sci Food Agric 90(7):1238–1244

Chang XY, Cui L, Wang XZ, Zhang L, Zhu D, Zhou XR, Hao LR (2017) Quercetin attenuates vascular calcification through suppressed oxidative stress in adenine-induced chronic renal failure rats. Biomed Res Int 2017:5716204

Chavan P, Singh AK, Kaur G (2018) Recent progress in the utilization of industrial waste and by-products of citrus fruits: a review. J Food Process Eng 41(8):e12895

Chekalina N, Burmak Y, Petrov Y, Borisova Z, Manusha Y, Kazakov Y, Kaidashev I (2018) Quercetin reduces the transcriptional activity of NF-kB in stable coronary artery disease. Indian Heart J 70(5):593–597

Chen CC, Liu YT, Chang KC, Chang YX, Lee MH, Lin YS, Chi TY, Hung YC, Lin CY, Chen HY, Huang PM (2020) Extracts of Taiwan tangerine (Citrus depressa Hayata) via supercritical fluid carbon dioxide extraction on the regulation of hypertension in spontaneously hypertensive rats. Adapt Med 12(4):95–101

Cheng FJ, Huynh TK, Yang CS, Hu DW, Shen YC, Tu CY, Wu YC, Tang CH, Huang WC, Chen Y, Ho CY (2021) Hesperidin is a potential inhibitor against SARS-CoV-2 infection. Nutrients 13(8):2800

Chhikara N, Kour R, Jaglan S, Gupta P, Gat Y, Panghal A (2018) Citrus medica: nutritional, phytochemical composition and health benefits–a review. Food Funct 9(4):1978–1992

Choudhary S, Kumar R, Dalal U, Tomar S, Reddy SN (2020) Green synthesis of nanometal impregnated biomass–antiviral potential. Mater Sci Eng C 112:110934

Collier CA, Mendiondo C, Raghavan S (2022) Tissue engineering of the gastrointestinal tract: the historic path to translation. J Biol Eng 16(1):1–2

Das S, Samanta S, Banerjee J, Pal A, Giri B, Kar SS, Dash SK (2022a) Is Omicron the end of pandemic or start of a new innings? Travel Med Infect Dis 23:102332 (1–11)

Das S, Kar SS, Samanta S, Banerjee J, Giri B, Dash SK (2022b) Immunogenic and reactogenic efficacy of Covaxin and Covishield: a comparative review. Immunol Res 22:1–27

Deng Y, Tu Y, Lao S, Wu M, Yin H, Wang L, Liao W (2021) The role and mechanism of citrus flavonoids in cardiovascular diseases prevention and treatment. Crit Rev Food Sci Nutr 10:1–24

Di Pietro N, Baldassarre MP, Cichelli A, Pandolfi A, Formoso G, Pipino C (2020) Role of polyphenols and carotenoids in endothelial dysfunction: an overview from classic to innovative biomarkers. Oxidative Med Cell Longev 2020:6381380

do Rosario VA, Schoenaker DA, Kent K, Weston-Green K, Charlton K (2021) Association between flavonoid intake and risk of hypertension in two cohorts of Australian women: a longitudinal study. Eur J Nutr 60(5):2507–2519

Droby SA, Eick A, Macarisin D, Cohen L, Rafael GI, Stange R, McColum G, Dudai N, Nasser A, Wisniewski M, Shapira R (2008) Role of citrus volatiles in host recognition, germination and growth of Penicillium digitatum and Penicillium italicum. Postharvest Biol Technol 49(3):386–396

Du J, Cullen JJ, Buettner GR (2012) Ascorbic acid: chemistry, biology and the treatment of cancer. Biochim Biophys Acta Rev Cancer 1826(2):443–457

Dugo P, Škeříková V, Kumm T, Trozzi A, Jandera P, Mondello L (2006) Elucidation of carotenoid patterns in citrus products by means of comprehensive normal-phase reversed-phase liquid chromatography. Anal Chem 78(22):7743–7750

Dugrand-Judek A, Olry A, Hehn A, Costantino G, Ollitrault P, Froelicher Y, Bourgaud F (2015) The distribution of coumarins and furanocoumarins in Citrus species closely matches Citrus phylogeny and reflects the organization of biosynthetic pathways. PLoS One 10(11):e0142757

Dutta SK, Gurung G, Yadav A, Laha R, Mishra VK (2022) Factors associated with citrus fruit abscission and management strategies developed so far: a review. N Z J Crop Hortic Sci 22:1–22

Eggersdorfer M, Wyss A (2018) Carotenoids in human nutrition and health. Arch Biochem Biophys 652:18–26

Elavarasan J, Velusamy P, Ganesan T, Ramakrishnan SK, Rajasekaran D, Periandavan K (2012) Hesperidin-mediated expression of Nrf2 and upregulation of antioxidant status in senescent rat heart. J Pharm Pharmacol 64(10):1472–1482

Etzbach L, Stolle R, Anheuser K, Herdegen V, Schieber A, Weber F (2020) Impact of different pasteurization techniques and subsequent ultrasonication on the in vitro bioaccessibility of carotenoids in Valencia Orange (Citrus sinensis (L.) Osbeck) juice. Antioxidants 9(6):534

Faggio C, Sureda A, Morabito S, Sanches-Silva A, Mocan A, Nabavi SF, Nabavi SM (2017) Flavonoids and platelet aggregation: a brief review. Eur J Pharmacol 807:91–101

Fadilah NQ, Jittmittraphap A, Leaungwutiwong P, Pripdeevech P, Dhanushka D, Mahidol C, Ruchirawat S, Kittakoop P (2022) Virucidal activity of essential oils from citrus x aurantium L. against influenza A virus H 1 N 1: Limonene as a potential household disinfectant against virus. Nat Prod Commun17(1):1–7

Fast DJ, Stern NP, Chuang J, Li Y, Scholten JD, Hu C (2019) Flavanones common to citrus fruits activate the interferon-stimulated response element by stimulating expression of IRF7. J Food Bioact 8:58–65

Feng S, Wang Y (2018) Citrus phytochemicals and their potential effects on the prevention and treatment of obesity: review and progress of the past 10 years. J Food Bioact 4:99–106

Ferlazzo N, Visalli G, Smeriglio A, Cirmi S, Lombardo GE, Campiglia P, Di Pietro A, Navarra M (2015) Flavonoid fraction of orange and bergamot juices protect human lung epithelial cells from hydrogen peroxide-induced oxidative stress. Evid Based Complement Altern Med 2015:957031

Feyh A, Bracero L, Lakhani HV, Santhanam P, Shapiro JI, Khitan Z, Sodhi K (2016) Role of dietary components in modulating hypertension. Exp Clin Cardiol 7(4):433

Fisher K, Phillips C (2008) Potential antimicrobial uses of essential oils in food: is citrus the answer? Trends Food Sci Technol 19(3):156–164

Gandhi GR, Vasconcelos AB, Wu DT, Li HB, Antony PJ, Li H, Geng F, Gurgel RQ, Narain N, Gan RY (2020) Citrus flavonoids as promising phytochemicals targeting diabetes and related complications: a systematic review of in vitro and in vivo studies. Nutrients 12(10):2907

Ganguly D, Ghosh S, Banerjee M, Mandal A, Panja S, Choudhury A (2022) A brief review on different plant parts used as immunity boosting agent. J Appl Pharm Sci 10(1):01–06

Gao G, Ding H, Zhuang C, Fan W (2018) Effects of hesperidin on H2O2-treated chondrocytes and cartilage in a rat osteoarthritis model. Med Sci Mon Int Med J Exp Clin Res 24:9177

Garg S, Malhotra RK, Khan SI, Sarkar S, Susrutha PN, Singh V, Goyal S, Nag TC, Ray R, Bhatia J, Arya DS (2019) Fisetin attenuates isoproterenol-induced cardiac ischemic injury in vivo by suppressing RAGE/NF-κB mediated oxidative stress, apoptosis and inflammation. Phytomedicine 56:147–155

Gattuso G, Barreca D, Gargiulli C, Leuzzi U, Caristi C (2007) Flavonoid composition of citrus juices. Molecules 12(8):1641–1673

Ghorbani A, Nazari M, Jeddi-Tehrani M, Zand H (2012) The citrus flavonoid hesperidin induces p53 and inhibits NF-κB activation in order to trigger apoptosis in NALM-6 cells: involvement of PPARγ-dependent mechanism. Eur J Nutr 51(1):39–46

Gonçalves D, Ferreira P, Baldwin E, Cesar T (2017) Health benefits of orange juice and citrus flavonoids. In: Phytochemicals in Citrus: Applications in Functional Foods. CRC Press, Boca Raton, pp 299–324

Gong Y, Liang X, Huang Y, Huang X, Su Q, Ma Y, Chen F (2022) The synergistic effects of Bupleuri radix with Citri Reticulatae Pericarpium and Aurantii Fructus on the improvement of gastrointestinal function. SSRN

Haaz S, Fontaine KR, Cutter G, Limdi N, Perumean-Chaney S, Allison DB (2006) Citrus aurantium and synephrine alkaloids in the treatment of overweight and obesity: an update. Obes Rev 7(1):79–88

Haghighian MK, Rafraf M, Hemmati S, Haghravan S, Asghari-Jafarabadi M (2021) Effects of pomegranate (Punica granatum L.) peel extract supplementation on serum lipid profile and oxidative stress in obese women with knee osteoarthritis: a double blind, randomized, placebo controlled study. Adv Integr Med 8(2):107–113

Hajialyani M, Hosein Farzaei M, Echeverría J, Nabavi SM, Uriarte E, Sobarzo-Sánchez E (2019) Hesperidin as a neuroprotective agent: a review of animal and clinical evidence. Molecules 24(3):648

Han-Qing WA, Sha-Sha WA, Chiufai K, Qi WA, Cheng XL (2019) Umbelliferone ameliorates renal function in diabetic nephropathy rats through regulating inflammation and TLR/NF-κB pathway. Chin J Nat Med 17(5):346–354

Harun NH, Septama AW, Jantan I (2015) Immunomodulatory effects of selected Malaysian plants on the CD18/11a expression and phagocytosis activities of leukocytes. Asian Pac J Trop Biomed 5(1):48–53

Hashemian M, Abdolkarimi M, Nasirzadeh M (2022) Family and school-based educational intervention on fruits and vegetable consumption of female students: application of social cognitive theory. J Educ Health Promot 11:55

He W, Li Y, Liu M, Yu H, Chen Q, Chen Y, Ruan J, Ding Z, Zhang Y, Wang T (2018) Citrus aurantium L. and its flavonoids regulate TNBS-induced inflammatory bowel disease through anti-inflammation and suppressing isolated jejunum contraction. Int J Mol Sci 19(10):3057

Howes MJ, Perry NS, Vásquez-Londoño C, Perry EK (2020) Role of phytochemicals as nutraceuticals for cognitive functions affected in ageing. Br J Pharmacol 177(6):1294–1315

Imenshahidi M, Eghbal M, Sahebkar A, Iranshahi M (2013) Hypotensive activity of auraptene, a monoterpene coumarin from Citrus spp. Pharm Biol 51(5):545–549

Inglese P, Sortino G (2019) Citrus history, taxonomy, breeding, and fruit quality. In: Oxford research Encyclopedia of environmental science. Oxford University Press, Oxford, p 25

Iswanti FC (2022) Gut microbiome diversity as adjuvant marker for immune function. Acta Bio Ina 5(1):80

Jasemi SV, Khazaei H, Fakhri S, Mohammadi-Noori E, Farzaei MH (2022) Naringenin improves ovalbumin-induced allergic asthma in rats through antioxidant and anti-inflammatory effects. Evid Based Complement Alternat Med 2022:9110798

Jiang Y, Bai X, Zhu X, Li J (2014) The effects of Fructus Aurantii extract on the 5-hydroxytryptamine and vasoactive intestinal peptide contents of the rat gastrointestinal tract. Pharm Biol 52(5):581–585

Jin YR, Han XH, Zhang YH, Lee JJ, Lim Y, Chung JH, Yun YP (2007) Antiplatelet activity of hesperetin, a bioflavonoid, is mainly mediated by inhibition of PLC-γ2 phosphorylation and cyclooxygenase-1 activity. Atherosclerosis 194(1):144–152

Johnson JL, De Mejia EG (2013) Flavonoid apigenin modified gene expression associated with inflammation and cancer and induced apoptosis in human pancreatic cancer cells through inhibition of GSK-3β/NF-κ B signaling cascade. Mol Nutr Food Res 57(12):2112–2127

Justin Thenmozhi A, William Raja TR, Manivasagam T, Janakiraman U, Essa MM (2017) Hesperidin ameliorates cognitive dysfunction, oxidative stress and apoptosis against aluminium chloride induced rat model of Alzheimer's disease. Nutr Neurosci 20(6):360–368

Kang SR, Park KI, Park HS, Lee DH, Kim JA, Nagappan A, Kim EH, Lee WS, Shin SC, Park MK, Han DY (2011) Anti-inflammatory effect of flavonoids isolated from Korea Citrus aurantium L. on lipopolysaccharide-induced mouse macrophage RAW 264.7 cells by blocking of nuclear factor-kappa B (NF-κB) and mitogen-activated protein kinase (MAPK) signalling pathways. Food Chem 129(4):1721–1728

Karri S, Sharma S, Hatware K, Patil K (2019) Natural anti-obesity agents and their therapeutic role in management of obesity: a future trend perspective. Biomed Pharmacother 110:224–238

Khan UM, Sameen A, Aadil RM, Shahid M, Sezen S, Zarrabi A, Ozdemir B, Sevindik M, Kaplan DN, Selamoglu Z, Ydyrys A (2021) Citrus genus and its waste utilization: a review on health-promoting activities and industrial application. Evid Based Complement Alternat Med 2021:2488804

Khanna K, Kohli SK, Kaur R, Bhardwaj A, Bhardwaj V, Ohri P, Sharma A, Ahmad A, Bhardwaj R, Ahmad P (2021) Herbal immune-boosters: substantial warriors of pandemic Covid-19 battle. Phytomedicine 85:153361

Kim J, Jayaprakasha GK, Patil BS (2013) Limonoids and their anti-proliferative and anti-aromatase properties in human breast cancer cells. Food Funct 4(2):258–265

Kim TH, Kim HM, Park SW, Jung YS (2015) Inhibitory effects of yuzu and its components on human platelet aggregation. Biomol Ther 23(2):149

Kim DB, Shin GH, Kim JM, Kim YH, Lee JH, Lee JS, Song HJ, Choe SY, Park IJ, Cho JH, Lee OH (2016) Antioxidant and anti-ageing activities of citrus-based juice mixture. Food Chem 194:920–927

Kim J, Wie MB, Ahn M, Tanaka A, Matsuda H, Shin T (2019) Benefits of hesperidin in central nervous system disorders: a review. Anatomy Cell Biol 52(4):369–377

Klaus R, Niyazi M, Lange-Sperandio B (2021) Radiation-induced kidney toxicity: molecular and cellular pathogenesis. Radiat Oncol 16(1):1–1

Kocyigit A, Koyuncu I, Dikilitas M, Bahadori F, Turkkan B (2016) Cytotoxic, genotoxic and apoptotic effects of naringenin-oxime relative to naringenin on normal and cancer cell lines. Asian Pac J Trop Biomed 6(10):872–880

Kootstra A (1994) Protection from UV-B-induced DNA damage by flavonoids. Plant Mol Biol 26(2):771–774

Kou G, Wu J, Liu M, Lin L, Xu X, Zhou Z (2019) Citrus tangeretin reduces the oxidative stress of the myocardium, with the potential for reducing fatigue onset and myocardial damage. J Funct Foods 54:249–253

Kretschmar JA, Baumann TW (1999) Caffeine in citrus flowers. Phytochemistry 52(1):19–23

Kumar S, Pandey AK (2013) Chemistry and biological activities of flavonoids: an overview. Sci World J 2013:162750

Kumar S, Rathi B (2020) Coronavirus disease COVID-19: a new threat to public health. Curr Top Med Chem 20(8):599–600

Ladaniya MS, Marathe RA, Murkute AA, Huchche AD, Das AK, George A, Kolwadkar J (2021) Response of Nagpur mandarin (Citrus reticulata Blanco) to high density planting systems. Sci Rep 11(1):1–1

Lado J, Gambetta G, Zacarias L (2018) Key determinants of citrus fruit quality: metabolites and main changes during maturation. Sci Hortic 233:238–248

Lai CS, Wu JC, Pan MH (2015) Molecular mechanism on functional food bioactives for anti-obesity. Curr Opin Food Sci 2:9–13

Langgut D (2017) The citrus route revealed: from Southeast Asia into the Mediterranean. Hortic Sci 52(6):814–822

Lee YS, Cha BY, Choi SS, Choi BK, Yonezawa T, Teruya T, Nagai K, Woo JT (2013) Nobiletin improves obesity and insulin resistance in high-fat diet-induced obese mice. J Nutr Biochem 24(1):156–162

Leermakers ET, Darweesh SK, Baena CP, Moreira EM, Melo van Lent D, Tielemans MJ, Muka T, Vitezova A, Chowdhury R, Bramer WM, Kiefte-de Jong JC (2016) The effects of lutein on cardiometabolic health across the life course: a systematic review and meta-analysis, 2. Am J Clin Nutr 103(2):481–494

Li C, Schluesener H (2017) Health-promoting effects of the citrus flavanone hesperidin. Crit Rev Food Sci Nutr 57(3):613–631

Li C, Zug C, Qu H, Schluesener H, Zhang Z (2015) Hesperidin ameliorates behavioral impairments and neuropathology of transgenic APP/PS1 mice. Behav Brain Res 281:32–42

Li L, Lyall GK, Martinez-Blazquez JA, Vallejo F, Tomas-Barberan FA, Birch KM, Boesch C (2020) Blood orange juice consumption increases flow-mediated dilation in adults with overweight and obesity: a randomized controlled trial. J Nutr 150(9):2287–2294

Lim W, Park S, Bazer FW, Song G (2017) Naringenin-induced apoptotic cell death in prostate cancer cells is mediated via the PI3K/AKT and MAPK signaling pathways. J Cell Biochem 118(5):1118–1131

Lin Y, Vermeer MA, Bos W, van Buren L, Schuurbiers E, Miret-Catalan S, Trautwein EA (2011) Molecular structures of citrus flavonoids determine their effects on lipid metabolism in HepG2 cells by primarily suppressing apoB secretion. J Agric Food Chem 59(9):4496–4503
Lin Z, Wu D, Huang L, Jiang C, Pan T, Kang X, Pan J (2019) Nobiletin inhibits IL-1β-induced inflammation in chondrocytes via suppression of NF-κB signaling and attenuates osteoarthritis in mice. Front Pharmacol 10:570
Liu Y, Heying E, Tanumihardjo SA (2012) History, global distribution, and nutritional importance of citrus fruits. Compr Rev Food Sci Food Saf 11(6):530–545
Liu H, Yang Z, Deng W, Tang Q (2016a) GW27-e0367 Apigenin attenuates the cardiac remodeling in experimental diabetic cardiomyopathy. J Am Coll Cardiol 68(16S):C14
Liu W, Zheng W, Cheng L, Li M, Huang J, Bao S, Xu Q, Ma Z (2022) Citrus fruits are rich in flavonoids for immunoregulation and potential targeting ACE2. Nat Prod Bioprospect 12(1):4
Liu X, Wang N, Fan S, Zheng X, Yang Y, Zhu Y, Lu Y, Chen Q, Zhou H, Zheng J (2016b) The citrus flavonoid naringenin confers protection in a murine endotoxaemia model through AMPK-ATF3-dependent negative regulation of the TLR4 signalling pathway. Sci Rep 6(1):1–4
Lu Y, Xi W, Ding X, Fan S, Zhang Y, Jiang D, Li Y, Huang C, Zhou Z (2013) Citrange fruit extracts alleviate obesity-associated metabolic disorder in high-fat diet-induced obese C57BL/6 mouse. Int J Mol Sci 14(12):23736–23750
Lux PE, Carle R, Zacarías L, Rodrigo MJ, Schweiggert RM, Steingass CB (2019) Genuine carotenoid profiles in sweet orange [Citrus sinensis (L.) Osbeck cv. Navel] peel and pulp at different maturity stages. J Agric Food Chem 67(47):13164–13175
Lv JC, Zhang LX (2019) Prevalence and disease burden of chronic kidney disease. In: Renal fibrosis: mechanisms and therapies. Springer Nature Singapore, Singapore, pp 3–15
Lv L, Lv L, Zhang Y, Kong Q (2011) Luteolin prevents LPS-induced TNF-α expression in cardiac myocytes through inhibiting NF-κB signaling pathway. Inflammation 34(6):620–629
Lv X, Zhao S, Ning Z, Zeng H, Shu Y, Tao O, Xiao C, Lu C, Liu Y (2015) Citrus fruits as a treasure trove of active natural metabolites that potentially provide benefits for human health. Chem Cent J 9(1):1–4
Macheix JJ, Fleuriet A, Billot J (2018) Fruit phenolics. CRC Press
Maheshwari S, Kumar V, Bhadauria G, Mishra A (2022) Immunomodulatory potential of phytochemicals and other bioactive compounds of fruits: a review. Food Front 3:221–238
Maiuolo J, Musolino V, Gliozzi M, Carresi C, Scarano F, Nucera S, Scicchitano M, Oppedisano F, Bosco F, Macri R, Palma E (2022) Involvement of the intestinal microbiota in the appearance of multiple sclerosis: Aloe vera and Citrus bergamia as potential candidates for intestinal health. Nutrients 14(13):2711
Manchope MF, Artero NA, Fattori V, Mizokami SS, Pitol DL, Issa JP, Fukada SY, Cunha TM, Alves-Filho JC, Cunha FQ, Casagrande R (2018) Naringenin mitigates titanium dioxide (TiO_2)-induced chronic arthritis in mice: role of oxidative stress, cytokines, and NFκB. Inflamm Res 67(11):997–1012
Mangukia N, Rao P, Patel K, Pandya H, Rawal RM (2022) Unveiling the nature's fruit basket to computationally identify Citrus sinensis csi-mir169–3p as a probable plant miRNA against Reference and Omicron SARS-CoV-2 genome. Comput Biol Med 146:105502
Md Othman SN, Hassan MA, Nahar L, Basar N, Jamil S, Sarker SD (2016) Essential oils from the Malaysian citrus (Rutaceae) medicinal plants. Medicines 3(2):13
Mditshwa A, Magwaza LS, Tesfay SZ, Opara UL (2017) Postharvest factors affecting vitamin C content of citrus fruits: a review. Sci Hortic 218:95–104
Mehrotra R (2006) Disordered mineral metabolism and vascular calcification in nondialyzed chronic kidney disease patients. J Ren Nutr 16(2):100–118
Merino-Godoy MD, Moreno-Sánchez E, Gago-Valiente FJ, Costa EI, Sáez-Padilla J (2022) An app about healthy habits as an educational resource during the pandemic. Healthcare 10(1):148
Merle H, Morón M, Blázquez MA, Boira H (2004) Taxonomical contribution of essential oils in mandarins cultivars. Biochem Syst Ecol 32(5):491–497
Merola N, Castillo J, Benavente-García O, Ros G, Nieto G (2017) The effect of consumption of citrus fruit and olive leaf extract on lipid metabolism. Nutrients 9(10):1062

Mershiba SD, Dassprakash MV, Saraswathy SD (2013) Protective effect of naringenin on hepatic and renal dysfunction and oxidative stress in arsenic intoxicated rats. Mol Biol Rep 40(5):3681–3691

Miles EA, Calder PC (2021) Effects of citrus fruit juices and their bioactive components on inflammation and immunity: a narrative review. Front Immunol 12:2558

Ming H, Chuang Q, Jiashi W, Bin L, Guangbin W, Xianglu J (2018) Naringin targets Zeb1 to suppress osteosarcoma cell proliferation and metastasis. Aging 10(12):4141

Mishra P, Mohapatra AK, Maharana S (2020) Role of immunity boosting nutritional foods and COVID-19. Act Sci Nutr Health 4:02–06

Mitra S, Lami MS, Uddin TM, Das R, Islam F, Anjum J, Hossain MJ, Emran TB (2022a) Prospective multifunctional roles and pharmacological potential of dietary flavonoid narirutin. Biomed Pharmacother 150:112932

Mitra S, Paul S, Roy S, Sutradhar H, Bin Emran T, Nainu F, Khandaker MU, Almalki M, Wilairatana P, Mubarak MS (2022b) Exploring the immune-boosting functions of vitamins and minerals as nutritional food bioactive compounds: a comprehensive review. Molecules 27(2):555

More TA, Kulkarni BR, Nalawade ML, Arvindekar AU (2014) Antidiabetic activity of linalool and limonene in streptozotocin-induced diabetic rat: a combinatorial therapy approach. Int J Pharm Pharm Sci 6(8):159–163

Moufida S, Marzouk B (2003) Biochemical characterization of blood orange, sweet orange, lemon, bergamot and bitter orange. Phytochemistry 62(8):1283–1289

Mukai R, Horikawa H, Fujikura Y, Kawamura T, Nemoto H, Nikawa T, Terao J (2012) Prevention of disuse muscle atrophy by dietary ingestion of 8-prenylnaringenin in denervated mice. PLoS One 7(9):e45048

Müller L, Caris-Veyrat C, Lowe G, Böhm V (2016) Lycopene and its antioxidant role in the prevention of cardiovascular diseases—a critical review. Crit Rev Food Sci Nutr 56(11):1868–1879

Multari S, Licciardello C, Caruso M, Martens S (2020) Monitoring the changes in phenolic compounds and carotenoids occurring during fruit development in the tissues of four citrus fruits. Food Res Int 134:109228

Murthy KN, Jayaprakasha GK, Patil BS (2013) Citrus limonoids and curcumin additively inhibit human colon cancer cells. Food Funct 4(5):803–810

Nagy MM, Al-Mahdy DA, Abd El Aziz OM, Kandil AM, Tantawy MA, El Alfy TS (2018) Chemical composition and antiviral activity of essential oils from Citrus reshni hort. ex Tanaka (Cleopatra mandarin) cultivated in Egypt. J Essent Oil-Bear Plants 21(1):264–272

Nakajima VM, Macedo GA, Macedo JA (2014) Citrus bioactive phenolics: role in the obesity treatment. LWT–Food Sci Technol 59(2):1205–1212

Nambiar VS, Matela H (2012) Potential functions of lemon grass (Cymbopogon citratus) in health and disease. Int J Pharm Biol Arch 3(5):1035–1043

Nita M, Grzybowski A (2016) The role of the reactive oxygen species and oxidative stress in the pathomechanism of the age-related ocular diseases and other pathologies of the anterior and posterior eye segments in adults. Oxidative Med Cell Longev 2016:3164734

Nobile V, Michelotti A, Cestone E, Caturla N, Castillo J, Benavente-García O, Pérez-Sánchez A, Micol V (2016) Skin photoprotective and antiageing effects of a combination of rosemary (Rosmarinus officinalis) and grapefruit (Citrus paradisi) polyphenols. Food Nutr Res 60(1):31871

Oboh G, Ademosun AO (2011) Phenolic extracts from grapefruit peels (Citrus paradisi) inhibit key enzymes linked with type 2 diabetes and hypertension. J Food Biochem 35(6):1703–1709

Oboh G, Olasehinde TA, Ademosun AO (2017) Inhibition of enzymes linked to type-2 diabetes and hypertension by essential oils from peels of orange and lemon. Int J Food Prop 20(sup1):S586–S594

Oguntibeju OO (2019) Type 2 diabetes mellitus, oxidative stress and inflammation: examining the links. Int J Physiol Pathophysiol Pharmacol 11(3):45

Okwu DE (2008) Citrus fruits: a rich source of phytochemicals and their roles in human health. Int J Chem Sci 6(2):451–471

Okwu DE, Emenike IN (2006) Evaluation of the phytonutrients and vitamin contents of citrus fruits. Int J Mol Med Adv Sci 2(1):1–6
Okwu DE, Emenike IN (2007) Nutritive value and mineral content of different varieties of citrus fruits. J Food Technol 5(2):105–108
Onda K, Horike N, Suzuki TI, Hirano T (2013) Polymethoxyflavonoidstangeretin and nobiletin increase glucose uptake in murine adipocytes. Phytother Res 27(2):312–316
Pala R, Anju VT, Dyavaiah M, Busi S, Nauli SM (2020) Nanoparticle-mediated drug delivery for the treatment of cardiovascular diseases. Int J Nanomedicine 15:3741
Palazzolo E, Laudicina VA, Germanà MA (2013) Current and potential use of citrus essential oils. Curr Org Chem 17(24):3042–3049
Pan MH, Lai CS, Tsai ML, Wu JC, Ho CT (2012) Molecular mechanisms for anti-aging by natural dietary compounds. Mol Nutr Food Res 56(1):88–115
Pan Z, He Q, Zeng J, Li S, Li M, Chen B, Yang J, Xiao J, Zeng C, Luo H, Wang H (2022) Naringenin protects against iron overload-induced osteoarthritis by suppressing oxidative stress. Phytomedicine 14:154330
Parhiz H, Roohbakhsh A, Soltani F, Rezaee R, Iranshahi M (2015) Antioxidant and anti-inflammatory properties of the citrus flavonoids hesperidin and hesperetin: an updated review of their molecular mechanisms and experimental models. Phytother Res 29(3):323–331
Park HJ, Jung UJ, Cho SJ, Jung HK, Shim S, Choi MS (2013) Citrus unshiu peel extract ameliorates hyperglycemia and hepatic steatosis by altering inflammation and hepatic glucose-and lipid-regulating enzymes in db/db mice. J Nutr Biochem 24(2):419–427
Percy DW, Adcock JL, Conlan XA, Barnett NW, Gange ME, Noonan LK, Henderson LC, Francis PS (2010) Determination of Citrus aurantium protoalkaloids using HPLC with acidic potassium permanganate chemiluminescence detection. Talanta 80(5):2191–2195
Perez YY, Jimenez-Ferrer E, Alonso D, Botello-Amaro CA, Zamilpa A (2010) Citrus limetta leaves extract antagonizes the hypertensive effect of angiotensin II. J Ethnopharmacol 128(3):611–614
Petry FC, de Nadai FB, Cristofani-Yaly M, Latado RR, Mercadante AZ (2019) Carotenoid biosynthesis and quality characteristics of new hybrids between tangor (Citrus reticulata x C. sinensis) cv.'Murcott'and sweet orange (C. sinensis) cv.'Pêra'. Food Res Int 122:461–470
Qiu P, Dong P, Guan H, Li S, Ho CT, Pan MH, McClements DJ, Xiao H (2010) Inhibitory effects of 5-hydroxy polymethoxyflavones on colon cancer cells. Mol Nutr Food Res 54(S2):S244–S252
Raghavan S, Gurunathan J (2021) Citrus species–a golden treasure box of metabolites that is beneficial against disorders. J Herb Med 28:100438
Rakib KH (2021) Hypertension in geriatric population: study in a rural community, Jamalpur, Bangladesh. Sch J Appl Med Sci 1:47–53
Ramírez-Pelayo C, Martínez-Quiñones J, Gil J, Durango D (2019) Coumarins from the peel of citrus grown in Colombia: composition, elicitation and antifungal activity. Heliyon 5(6):e01937
Rekha C, Poornima G, Manasa M, Abhipsa V, Devi JP, Kumar HT, Kekuda TR (2012) Ascorbic acid, total phenol content and antioxidant activity of fresh juices of four ripe and unripe citrus fruits. Chem Sci Trans 1(2):303–310
Reshef N, Hayari Y, Goren C, Boaz M, Madar Z, Knobler H (2005) Antihypertensive effect of sweetie fruit in patients with stage I hypertension. Am J Hypertens 18(10):1360–1363
Ribeiro AP, Pereira AG, Todo MC, Fujimori AS, Dos Santos PP, Dantas D, Fernandes AA, Zanati SG, Hassimotto NM, Zornoff LA, Azevedo PS (2021) Pera orange (Citrus sinensis) and Moro orange (Citrus sinensis (L.) Osbeck) juices attenuate left ventricular dysfunction and oxidative stress and improve myocardial energy metabolism in acute doxorubicin-induced cardiotoxicity in rats. Nutrition 91:111350
Ridker PM, Everett BM, Thuren T, MacFadyen JG, Chang WH, Ballantyne C, Fonseca F, Nicolau J, Koenig W, Anker SD, Kastelein JJ (2017) Antiinflammatory therapy with canakinumab for atherosclerotic disease. N Engl J Med 377(12):1119–1131
Rimmer JH, Braddock D (2002) Health promotion for people with physical, cognitive, and sensory disabilities: an emerging national priority. Am J Health Promot 16(4):220–224

Roohbakhsh A, Parhiz H, Soltani F, Rezaee R, Iranshahi M (2015) Molecular mechanisms behind the biological effects of hesperidin and hesperetin for the prevention of cancer and cardiovascular diseases. Life Sci 124:64–74
Roowi S, Crozier A (2011) Flavonoids in tropical citrus species. J Agric Food Chem 59(22):12217–12225
Saeid A, Ahmed M (2021) Citrus fruits: nutritive value and value-added products. Citrus: Res Dev 3:171
Saini RK, Keum YS (2018) Carotenoid extraction methods: a review of recent developments. Food Chem 240:90–103
Saini RK, Nile SH, Park SW (2015) Carotenoids from fruits and vegetables: chemistry, analysis, occurrence, bioavailability and biological activities. Food Res Int 76:735–750
Samanta S, Banerjee J, Rahaman SN, Ali KM, Ahmed R, Giri B, Pal A, Dash SK (2022a) Alteration of dietary habits and lifestyle pattern during COVID-19 pandemic associated lockdown: an online survey study. Clin Nutr ESPEN 48:234–246
Samanta S, Banerjee J, Kar SS, Ali KM, Giri B, Pal A, Dash SK (2022b) Awareness, knowledge and acceptance of COVID-19 vaccine among the people of West Bengal, India: a web-based survey. Vacunas 23:S46–S55
Savcı Y, Kırbaş OK, Bozkurt BT, Abdik EA, Taşlı PN, Şahin F, Abdik H (2021) Grapefruit-derived extracellular vesicles as a promising cell-free therapeutic tool for wound healing. Food Funct 12(11):5144–5156
Shahsavari D, Parkman HP (2022) Normal gastrointestinal tract physiology. In: Nutrition, weight, and digestive health. Springer, Cham, Springer Nature Switzerland AG, pp 3–28
Sharma H, Chandola HM, Singh G, Basisht G (2007) Utilization of Ayurveda in health care: an approach for prevention, health promotion, and treatment of disease. Part 2—Ayurveda in primary health care. J Altern Complement Med 13(10):1135–1150
Sharma A, Kuhad A, Bhandari R (2022a) Novel nanotechnological approaches for treatment of skin-aging. J Tissue Viability 31(3):374–386
Sharma S, Porwal K, Kulkarni C, Pal S, Sihota P, Kumar S, Tiwari MC, Katekar R, Kumar A, Singh P, Rajput S (2022b) Diosmin, a citrus fruit-derived phlebotonic bioflavonoid protects rats from chronic kidney disease-induced loss of bone mass and strength without deteriorating the renal function. Food Funct 13(4):2184–2199
Shehab NG, Dortaj S, Othman M, Mostafa E, Rezvani P, Alshawi F, Al Ahmad L (2022) Immune-boosting effect of natural remedies and supplements on progress of, and recovery from COVID-19 infection. Trop J Pharm Res 22(2):313–322
Shen CY, Jiang JG, Yang L, Wang DW, Zhu W (2017) Anti-ageing active ingredients from herbs and nutraceuticals used in traditional Chinese medicine: pharmacological mechanisms and implications for drug discovery. Br J Pharmacol 174(11):1395–1425
Shrestha N, Shrestha S, Bhattarai A (2016) Determination of ascorbic acid in different citrus fruits of Kathmandu Valley. J Med Biol Sci Res 2(1):9–14
Silalahi J (2002) Anticancer and health protective properties of citrus fruit components. Asia Pac J Clin Nutr 11(1):79–84
Siti HN, Mohamed S, Kamisah Y (2022) Potential therapeutic effects of Citrus hystrix DC and its bioactive compounds on metabolic disorders. Pharmaceuticals 15(2):167
Smith DC, Forland S, Bachanos E, Matejka M, Barrett V (2001) Qualitative analysis of citrus fruit extracts by GC/MS: an undergraduate experiment. Chem Educ 6(1):28–31
Sofowora A, Ogunbodede E, Onayade A (2013) The role and place of medicinal plants in the strategies for disease prevention. Afr J Tradit Complement Altern Med 10(5):210–229
Sost MM, Ahles S, Verhoeven J, Verbruggen S, Stevens Y, Venema K (2021) A citrus fruit extract high in polyphenols beneficially modulates the gut microbiota of healthy human volunteers in a validated in vitro model of the colon. Nutrients 13(11):3915
Srinivasan S, Pari L (2013) Antihyperlipidemic effect of diosmin: a citrus flavonoid on lipid metabolism in experimental diabetic rats. J Funct Food 5(1):484–492

Stanisic D, Liu LH, Dos Santos RV, Costa AF, Durán N, Tasic L (2020) New sustainable process for hesperidin isolation and anti-ageing effects of hesperidin nanocrystals. Molecules 25(19):4534

Stevens Y, Rymenant EV, Grootaert C, Camp JV, Possemiers S, Masclee A, Jonkers D (2019) The intestinal fate of citrus flavanones and their effects on gastrointestinal health. Nutrients 11(7):1464

Suchitra MR, Parthasarathy S (2020) Nutrition and corona virus: plan a diet in a pandemic. Int J Res Pharm Sci 11:110–114

Sun XY, Gan QZ, Ouyang JM (2015) Calcium oxalate toxicity in renal epithelial cells: the mediation of crystal size on cell death mode. Cell Death Discov 1(1):1–8

Sundaram R, Nandhakumar E, Haseena Banu H (2019) Hesperidin, a citrus flavonoid ameliorates hyperglycemia by regulating key enzymes of carbohydrate metabolism in streptozotocin-induced diabetic rats. Toxicol Mech Methods 29(9):644–653

Sundaram TS, Giromini C, Rebucci R, Pistl J, Bhide M, Baldi A (2022) Role of omega-3 polyunsaturated fatty acids, citrus pectin, and milk-derived exosomes on intestinal barrier integrity and immunity in animals. J Anim Sci Biotechnol 13(1):1–22

Suri S, Singh A, Nema PK (2022) Current applications of citrus fruit processing waste: a scientific outlook. Appl Food Res 27:100050

Suwannayod S, Sukontason KL, Somboon P, Junkum A, Leksomboon R, Chaiwong T, Jones MK, Sripa B, Balthaisong S, Phuyao C, Chareonviriyaphap T (2018) Activity of kaffirlime (Citrus hystrix) essential oil against blow flies and house fly. Southeast Asian J Trop Med Public Health 49(1):32–45

Tardif JC, Kouz S, Waters DD, Bertrand OF, Diaz R, Maggioni AP, Pinto FJ, Ibrahim R, Gamra H, Kiwan GS, Berry C (2019) Efficacy and safety of low-dose colchicine after myocardial infarction. N Engl J Med 381(26):2497–2505

Testai L, Calderone V (2017) Nutraceutical value of citrus flavanones and their implications in cardiovascular disease. Nutrients 9(5):502

Thirumdas R, Kothakota A, Pandiselvam R, Bahrami A, Barba FJ (2021) Role of food nutrients and supplementation in fighting against viral infections and boosting immunity: a review. Trends Food Sci Technol 110:66–77

Tiedtke J, Kiefer S, Weibel M, Smits J, Juch M, Herbst N (2011) Citroflavonoid anti-ageing complex fades age spots and gives skin tone a youthful citrus boost. Cosmetochem International AG, pp 13–17

Toth PP, Patti AM, Nikolic D, Giglio RV, Castellino G, Biancucci T, Geraci F, David S, Montalto G, Rizvi A, Rizzo M (2016) Bergamot reduces plasma lipids, atherogenic small dense LDL, and subclinical atherosclerosis in subjects with moderate hypercholesterolemia: a 6 months prospective study. Front Pharmacol 6:299

Tripoli E, La Guardia M, Giammanco S, Di Majo D, Giammanco M (2007) Citrus flavonoids: molecular structure, biological activity and nutritional properties: a review. Food Chem 104(2):466–479

Tsai YF, Chen YR, Chen JP, Tang Y, Yang KC (2019) Effect of hesperidin on anti-inflammation and cellular antioxidant capacity in hydrogen peroxide-stimulated human articular chondrocytes. Process Biochem 85:175–184

Tsubota-Utsugi M, Ohkubo T, Kikuya M, Metoki H, Kurimoto A, Suzuki K, Fukushima N, Hara A, Asayama K, Satoh H, Tsubono Y (2011) High fruit intake is associated with a lower risk of future hypertension determined by home blood pressure measurement: the OHASAMA study. J Hum Hypertens 25(3):164–171

US Department of Agriculture (2022) Citrus: world markets and trade. https://apps.fas.usda.gov/psdonline/circulars/citrus.pdf

Valls RM, Pedret A, Calderón-Pérez L, Llauradó E, Pla-Pagà L, Companys J, Moragas A, Martín-Luján F, Ortega Y, Giralt M, Romeu M (2021) Effects of hesperidin in orange juice on blood and pulse pressures in mildly hypertensive individuals: a randomized controlled trial (Citrus study). Eur J Nutr 60(3):1277–1288

Viner R, Macfarlane A (2005) Health promotion. BMJ 330(7490):527–529

Wang YC, Chuang YC, Hsu HW (2008) The flavonoid, carotenoid and pectin content in peels of citrus cultivated in Taiwan. Food Chem 106(1):277–284

Wang Y, Chen Y, Zhang H, Chen J, Cao J, Chen Q, Li X, Sun C (2020a) Polymethoxyflavones from citrus inhibited gastric cancer cell proliferation through inducing apoptosis by upregulating RARβ, both in vitro and in vivo. Food Chem Toxicol 146:111811

Wang Y, Liu X, Shi H, Yu Y, Yu Y, Li M, Chen R (2020b) NLRP3 inflammasome, an immune-inflammatory target in pathogenesis and treatment of cardiovascular diseases. Clin Transl Med 10(1):91–106

Wang J, Zhai Y, Ou M, Bian Y, Tang C, Zhang W, Cheng Y, Li G (2021a) Protective effect of lemon peel extract on oxidative stress in H9c2 rat heart cell injury. Drug Des Devel Ther 15:2047

Wang Y, Gallegos JL, Haskell-Ramsay C, Lodge JK (2021b) Effects of chronic consumption of specific fruit (berries, citrus and cherries) on CVD risk factors: a systematic review and meta-analysis of randomised controlled trials. Eur J Nutr 60(2):615–639

Wang L, Wang F, Zhang X, Chen Q, Xu J, Li H, Li F, Yang M (2022) Transdermal Administration of Volatile oil from Citrus aurantium-Rhizoma Atractylodis Macrocephalae alleviates constipation in rats by altering host metabolome and intestinal microbiota composition. Oxidative Med Cell Longev 2022:9965334

Wansi JD, Tcho AT, Toze FA, Nahar L, Martin C, Sarker SD (2016) Cytotoxic acridone and indoloquinazoline alkaloids from Zanthoxylum poggei. Phytochem Lett 17:293–298

Wesołowska O, Wiśniewski J, Środa-Pomianek K, Bielawska-Pohl A, Paprocka M, Duś D, Duarte N, Ferreira MJ, Michalak K (2012) Multidrug resistance reversal and apoptosis induction in human colon cancer cells by some flavonoids present in citrus plants. J Nat Prod 75(11):1896–1902

Wolffenbüttel AN, Zamboni A, Becker G, Dos Santos MK, Borille BT, de Cássia MK, Fagundes AC, de Oliveira Salomón JL, Coelho VR, Ruiz LV, de Moura LV (2018) Citrus essential oils inhalation by mice: Behavioral testing, GCMS plasma analysis, corticosterone, and melatonin levels evaluation. Phytother Res 32(1):160–169

Wu LH, Lin C, Lin HY, Liu YS, Wu CY, Tsai CF, Chang PC, Yeh WL, Lu DY (2016) Naringenin suppresses neuroinflammatory responses through inducing suppressor of cytokine signaling 3 expression. Mol Nurobio 53(2):1080–1091

Wu J, Zhao YM, Deng ZK (2018) Tangeretin ameliorates renal failure via regulating oxidative stress, NF-κB–TNF-α/iNOS signalling and improves memory and cognitive deficits in 5/6 nephrectomized rats. Inflammopharmacology 26(1):119–132

Wu Q, Zhou Y, Zhou XM (2019) Citrus alkaline extract delayed the progression of pulmonary fibrosis by inhibiting p38/NF-B Signaling pathway-induced cell apoptosis. Evid Based Complement Alternat Med 2019:1528586

Wu Y, Cai C, Xiang Y, Zhao H, Lv L, Zeng C (2021) Naringin ameliorates Monocrotaline-induced pulmonary arterial hypertension through endothelial-to-mesenchymal transition inhibition. Front Pharmacol 12:696135

Xu M, Wang R, Fan H, Ni Z (2021) Nobiletin ameliorates streptozotocin-cadmium-induced diabetic nephropathy via NF-κB signalling pathway in rats. Arch Physiol Biochem 3:1–9

Xue F, Li X, Qin L, Liu X, Li C, Adhikari B (2021) Anti-aging properties of phytoconstituents and phyto-nanoemulsions and their application in managing aging-related diseases. Adv Drug Deliv Rev 176:113886

Ye X (2017) Phytochemicals in citrus: applications in functional foods. CRC Press

Yoshida H, Watanabe W, Oomagari H, Tsuruta E, Shida M, Kurokawa M (2013) Citrus flavonoid naringenin inhibits TLR2 expression in adipocytes. J Nutr Biochem 24(7):1276–1284

Zabetakis I, Lordan R, Norton C, Tsoupras A (2020) COVID-19: the inflammation link and the role of nutrition in potential mitigation. Nutrients 12(5):1466

Zaidun NH, Thent ZC, Abd Latiff A (2018) Combating oxidative stress disorders with citrus flavonoid: Naringenin. Life Sci 208:111–122

Zandi K, Teoh BT, Sam SS, Wong PF, Mustafa MR, AbuBakar S (2011) Antiviral activity of four types of bioflavonoid against dengue virus type-2. Virol J 8(1):1–1

Zang L, Shimada Y, Kawajiri J, Tanaka T, Nishimura N (2014) Effects of Yuzu (Citrus junos Siebold ex Tanaka) peel on the diet-induced obesity in a zebrafish model. J Funct Food 10:499–510
Zhang J, Song J, Wu D, Wang J, Dong W (2015a) Hesperetin induces the apoptosis of hepatocellular carcinoma cells via mitochondrial pathway mediated by the increased intracellular reactive oxygen species, ATP and calcium. Med Oncol 32(4):1–1
Zhang X, Zheng L, Sun Y, Wang T, Wang B (2015b) Tangeretin enhances radiosensitivity and inhibits the radiation-induced epithelial-mesenchymal transition of gastric cancer cells. Oncol Rep 34(1):302–310
Zhang X, Zheng K, Li C, Zhao Y, Li H, Liu X, Long Y, Yao J (2017) Nobiletin inhibits invasion via inhibiting AKT/GSK3β/β-catenin signaling pathway in Slug-expressing glioma cells. Oncol Rep 37(5):2847–2856
Zhang GX, Jin L, Jin H, Zheng GS (2021) Influence of dietary components and traditional Chinese medicine on hypertension: a potential role for gut microbiota. Evid Based Complement Alternat Med 2021:5563073
Zhao M, Li C, Shen F, Wang M, Jia N, Wang C (2017) Naringenin ameliorates LPS-induced acute lung injury through its anti-oxidative and anti-inflammatory activity and by inhibition of the PI3K/AKT pathway. Exp Ther Med 14(3):2228–2234
Zheng X, Mi J, Deng X, Al-Babili S (2021) LC–MS-based profiling provides new insights into apocarotenoid biosynthesis and modifications in citrus fruits. J Agric Food Chem 69(6):1842–1851
Zou Z, Xi W, Hu Y, Nie C, Zhou Z (2016) Antioxidant activity of Citrus fruits. Food Chem 196:885–896

Chapter 16
Citrus Diseases and Management

R. Arutselvan, Kalidas Pati, Aria Dolatabadian, and Sudip Kumar Dutta

16.1 Introduction

Citrus is a genus of flowering trees and shrubs belonging to the Rutaceae family. *Citrus* species grow across the globe, particularly in tropical and subtropical regions. Plants in the genus produce highly nutritive fruits containing vitamins A and C, amino acids, sugars, carotenoids, and organic acids. Citrus production is one of the world's largest agricultural industries. It serves as a vital commercial commodity for many businesses in India and holds a standard position in the fruit processing industries. Citrus production contributes USD 6–8 billion to the global economy annually by giving people jobs (harvesting, handling, transportation, storage, and marketing). More than 125 nations in the region around 35° latitude north and south of the equator are known to cultivate citrus (Duncan and Cohn 1990).

Citrus fruits are one of the critical commercial and consumable fruit crops grown worldwide (Tao et al. 2009). *Citrus* originated in the Himalayas in the northeastern region of India. It slowly spread from the east of the Malayan Archipelago to China and Japan and eventually found its way to other parts of the world (Mabberley 2004; Ziegler and Wolfe 1975). The word citrus originated from the Greek word "Kedros", meaning "Fruit of the godly tree", which was Latinized to "Cedrus", and finally "Citrus". Citrus fruits and their products have been associated with human health for their varied aesthetic, medicinal, and consumption purposes since immemorial. Citrus is one of the most economically important fruit crops having extensive

R. Arutselvan · K. Pati (✉)
Regional Centre, ICAR-Central Tuber Crops Research Institute, Bhubaneswar, Odisha, India

A. Dolatabadian
Department of Plant Science, Faculty of Agricultural and Food Sciences,
University of Manitoba, Winnipeg, Canada

S. K. Dutta
ICAR Research Complex for NEH Region, Sikkim Centre, Gangtok, Sikkim, India

S. Singh Purewal et al. (eds.), *Recent Advances in Citrus Fruits*,
https://doi.org/10.1007/978-3-031-37534-7_16

cultivation under tropical and subtropical conditions. With its multifold nutritional and medicinal values, it is one of the leading fruit crops worldwide, including in India.

As per the statistics of 2016, it is commercially cultivated in more than 130 countries, with an annual fruit production of 124.25 million tonnes worldwide (Biswas 2008). India is the third-largest producer of citrus fruits, with an annual production of 12.04 million tonnes. However, it still has one of the lowest yields among the top producers (Horticulture Statistics at a Glance 2018). Due to the varied altitudinal landscape and unique climatic conditions having characteristic high rainfall and humid conditions, the northeastern part of India is often called the "natural home for citrus" and "treasure house for citrus". Out of the 27 *Citrus* species that are found in India, the region is home to (23) species, (1) subspecies, and roughly (65) variations. Citrus, one of India's most essential fruits, is mainly grown in the northeastern part of the country and ranks third in both area and production. Meghalaya is India's most productive and largest state, followed by Manipur, Assam, Tripura, Mizoram, Nagaland and Arunachal Pradesh. Khasi mandarin trees, the largest fruit-growing area in Meghalaya, are grown for trade and export (Das et al. 2009). The principal species of *Citrus* described by (Cooper and Chapot 1977) are as follows (Table 16.1).

Taxonomically, *Citrus* is a name for the genus in the family Rutaceae. It refers to all edible and rootstock species of *Citrus* and a few closely related genera. Common edible fruits in this group are sweet orange, mandarin, grapefruit, lemon, lime, pummelo, *Poncirus* & *Fortunella* spp. *Citrus* spp. are naturally deep-rooted plants (Ford 1954), and for optimum growth, it requires deep alluvial and well-drained soils. Mandarin orange is one of the most valuable and tasty fruit plants commercially grown worldwide.

Table 16.1 List of principal species of *Citrus* genus cultivated worldwide

The common name of species	Species
Citron	*Citrus medica* L.
Sour orange	*C. aurantium* L.
Sweet Orange	*C. sinensis* Osbeck
Pummelo	*C. grandis* Osbeck
Lemon	*C. limon* (L) Burm. f.
Mandarin	*C. reticulate* Blanco
Common lime	*C. aurantifolia* Christm.
Grapefruit	*C. paradise* Macf.
Tachibana	*C. Tachibana* Tan.
Indian wild orange	*C. indica* Tan.
Mauritius papeda	*C. hystrix* CD.
Melanesian Papeda	*C. macrocarpa* Mont.
Celebes papeda	*C. celebica* Koord.
Ichang papeda	*C. ichangensis* Swing
Papeda	*C. microntha* Wester.
Khasi papeda	*C. latipes* Tan.

Citrus spp. possesses greater adaptability, so they are grown to different climatic conditions (tropical, subtropical and even in some favourable parts of the temperate) regions of the world. Citrus grows well under any rainfall regime, providing adequate soil moisture can be maintained. South China and the Himalayan region are places of origin for most citrus fruits. Many *Citrus* species originate in India, and it is considered not less than 78 species of the family Rutaceae as natives of India. India is the leading citrus-growing country in the world and produced over an area of 8,43,000 hectares and contributed to total production and productivity of 75,87,000 MT and 9 MT/ha, respectively (Indian Horticulture Database 2008).

Citrus fruits are a fantastic source of vitamin C and a variety of other vital elements the body needs. Citrus importance is attributed to its diversified use, nutritional value and unique flavour (Goulas and Manganaris 2012). Phytochemicals that own many ecological and physiological roles in citrus fruits are flavonoids, alkaloids, terpenoids, cyanogenic glycosides and phenolic compounds (Chose et al. 2002; Okwu and Morah 2004; Okwu 2005). Citrus plants synthesise and accumulate many phytochemicals in their cells, including low molecular weight compounds like acetophenones, flavonoids, terpenoids, hydroxy benzoic and hydroxycinnamic acids, stilbenes and condensed tannins (Rapisarda et al. 1999; Okwu and Emenike 2006).

An extract of *Citrus* grandis tissue has shown antioxidant activity (Mokbel and Hashinaga 2006). Their extracts can be very useful in preventing oxidation in fruit juices, foods containing essential oils, and dietary supplements. Acetone extracts of *Citrus maxima* root bark revealed nine known coumarins, eleven acridone alkaloids, new coumarins, honyuzicins, new flavones, and honyucitrins (Kim et al. 2013). The antioxidant and antidiabetic effects of grapefruit extract have been confirmed when treated with *Aspergillus saitoi* (Kim et al. 2009). *Citrus maxima* contain high flavanone glycosides such as neohesperidin and naringin (Sawamura and Kuriyama 1988). Citrus fruits efficiently scavenge free radicals of various forms, including DPPH, superoxide anions and H_2O_2 radicals (Chung and Shin 2007) and are rich in antioxidants.

Fungal infections generate considerable challenges in citrus fruit production. Unlike certain systemic infections, these fungi do not cause tree deterioration or death. Some, however, lower yield and fruit size, and their impacts are not always merely ornamental. On the other hand, other fungi create primarily exterior defects on the fruit and must be managed exclusively on fruit going for the fresh market. In citrus-growing regions, heavy rainfall and temperatures lead to more severe fungal infections. Nonetheless, the particular fungus may infect citrus trees using just dew moisture, making them a threat even in semi-arid, winter-rainfall citrus-producing countries. This chapter covers the major citrus diseases to provide readers with essential information and the latest research findings.

16.2 Emerging and Major Fungal Diseases of Citrus

16.2.1 *Foot Rot or Gummosis* (**Phytophthora** *spp.)*

Foot rot is a major disease in Indonesia and other parts of the world (*Phytophthora* spp.). The disease complex of foot rot and Diplodia (*Botryodiplodia theobromae*) has caused damage to about 35% of citrus orchards, mostly in marshes and irrigated fields (Dwiastuti 2020). Infected planting material is the primary cause of the disease. Yellowing of the foliage is the first sign, followed by bark splitting and heavy surface gumming. Citrus trees with severe gumminess eventually develop thoroughly rotten citrus bark, and the girdling action causes the tree to dry out. Blooms profusely and dies at a severe stage before the fruits are fully developed.

Control Effective preventive strategies include cultural and physical treatment (choosing a suitable location with sufficient drainage, using resistant rootstocks, and preventing water contact with the tree stem by using the ring technique of irrigation). Alternately, the diseased areas can be removed and cleaned using a cotton swab dipped in a solution of $HgCl_2$ (0.1%) or $KMnO_4$ solution (1%). Bordeaux is painted onto the portion of the stem 1 m above the ground to help suppress the disease. Ridomil MZ 72@ 2.75 g/l or Aliette (2.5 g/l) are the best treatments for the disease.

16.2.2 *Ganoderma Root-Rot (***Ganoderma lucidum***)*

One or more lateral roots develop the Root-Rot disease in the soil. There are indications of whitish strands of fungus covering the root bark, which eventually turn dark. Later, the fungus penetrates the main trunk's base. Due to water buildup, the afflicted tissues become highly light, bloated, and spongy. Bracket-shaped fungal fructifications emerge at the base of the stem during the wet season.

Control The root rot disease is controlled by removing diseased roots from the field, drenching the base of those roots with a fungicide, and routinely collecting and destroying brackets close to the collar. Isolation of sick trees involves excavating a trench around the tree to keep the diseased roots from coming into contact with the healthy roots and mingling. The disease can be controlled by mixing 0.5 to 1 kg of sulphur powder with the soil in the trench. The disease is effectively controlled by applying Vitavax (500 ppm) and aureofungin solution (750 ppm) to the plant basin.

*16.2.3 Dry Root Rot (*Macrophomina phaseoli, Fusarium *spp.* and Diplodia natalensis*)*

Early symptoms of this disease include moist rot of the citrus bark, while later symptoms include shredded, drying of barks and dead wood beneath. The affected tree loses its leaves and bears an abundant crop of tiny fruits—foul odours from the afflicted roots.

Control The diseased roots should be destroyed or swabbed with Bordeaux paste.

*16.2.4 Pink Disease (*Pellicularia salmonicolour*)*

The pink sickness typically manifests during or immediately following monsoon rains. Branch tips and leaves begin to wilt and perish early on. A silvery-white thin mycelium film has been applied to the damaged branches. The disease is fittingly called the "pink disease"because of the distinctive pink hue the fungus on the branches produces. When the bark becomes seriously affected, it shreds and exposes the timber. Branch breaking and gumming down the length of the branches is another possibility.

Control All the infected areas should be removed, and Bordeaux paste should be applied to the cut ends. The mycelium in the crotches is eliminated when Bordeaux mixture paste combined with crude oil emulsion is sprayed on the infected area.

*16.2.5 Leaf Fall and Fruit-Rot (*Phytophthora palmivora*)*

Leaf shedding begins on the tree's lower branches. Water-soaked areas can be seen on the afflicted leaves. The damaged leaves fall off after the lesions cover the entire leaf. The fruits are also contaminated during various phases of development. The fruit's rind initially becomes covered in wet areas, which later fall off and decay.

Control two sprays of Bordeaux combination at a 1% concentration can effectively control the disease — one in June before the monsoon season begins, and the second in August or September after it has ended.

*16.2.6 Scab (*Elsinoe fawcetti*)*

Initially appearing as tiny semi-translucent spots on the underside of the leaves, the lesions eventually develop into sharply defined pustular elevations. Later phases frequently result in twisted, wrinkled, stunted, and malformed leaves. On the fruit, lesions appear as a corky protrusion that frequently scabs over. There is a circular depression with a pink to crimson centre on the surface opposite the warty growth.

Control Collect and destroy any infected fruit, twigs, or leaves. Spraying Blitox (0.3%) or the Bordeaux mixture is quite effective.

*16.2.7 Powdery Mildew (*Acrosporium tingitaninum*)*

Young leaves and twigs acquire a whitish powdery substance. Those leaves become deformed. The afflicted leaves drop off, and the twigs display symptoms of dieback when the disease is severe. The young fruits' surfaces are also covered in fungus, which causes them to ripen too early, decreasing the yield.

Control Spraying Wettable Sulphur (1.5 kg/200 l of water) will eliminate powdery mildew. Early morning, sulphur dusting (20 kg/hectare) efficiently suppresses the disease. The disease can be controlled more effectively and for a more extended period using systemic fungicides like Bayleton (0.1%), Calaxin (0.04%), or Benomyl (0.05%).

*16.2.8 Anthracnose or Wither Tip (*Colletotrichum gloesporioides and Gloeosporium *spp.)*

Immature leaves, sensitive citrus fruit, and young shoots are frequently targets of the Colletotrichum fungus. Affected leaves display distorted necrotic areas. Dead branches of the twig start to look silvery grey. When impacted, flower buds do not develop into fruits. Fruit loss occurs when fruits are infected.

Control To maintain a healthy orchard, adequate irrigation, manuring, and trimming of diseased twigs are necessary. The disease can be controlled by spraying Bordeaux mixture (1%) or Blitox (2.5%), or Mancozeb (2%).

*16.2.9 Twig Blight (*Diplodia natalensis *and* Fusarium *spp.)*

Citrus leaves with fusarium twig blight dry up and drop to the ground. Small twigs with viscous secretions at the base of the dead twigs die back from the tips. The damaged twigs in the *Diplodia* twig blight case exhibit pycnidial fructifications of the fungus. Unfavourable environmental factors and malnutrition are the disease's primary causes.

Control Twigs die-back disease can be controlled by pruning the dead twigs and applying Benomyl (2.5%) to the affected areas.

*16.2.10 Sooty Mould (***Capnodium citri***)*

Sooty mould is a common problem in citrus orchards where aphids and scale insects cannot be effectively controlled. The disease's main symptom is a velvety black coating on the upper surface of leaves, branches and fruits. The packaging is shallow, and the citrus leaves come off quickly. In warm conditions, the affected leaves curl and drop.

Control Sooty mould can be controlled by pruning and removing infected branches. Disease-causing insects can be repelled with Wettasulphur (0.2%) + Metacid (0.1%) + Acacia Gum (0.3%). When the insect disappears, the sooty mould automatically disappears because there is no suitable environment for the mould to grow.

*16.2.11 Melanose (***Diaporthe citri***)*

Citrus melanosis affects citrus fruits and is found in various citrus-growing regions worldwide. It affects the young leaves and fruits of citrus seeds and cultivars and develops tissue in prolonged warm, damp weather conditions. The signs of this melanotic fungal infection range from small spots to crusty lesions, resulting in a teardrop damage pattern. It is one of the most common citrus diseases worldwide.

Control Prevention is better than cure, so restoration and removal of deadwood are important to eliminate the inoculum of melanin, especially in older citrus orchards. Applying the fungicide pyraclostrobin to citrus trees is much more likely than famoxadone or copper hydroxide to control melanosis, scab, and Alternaria brown spot. In China, mancozeb and fenbuconazole are effective in suppressing citrus melanosis.

*16.2.12 Greasy Spot (***Mycosphaerella citri***)*

Citrus grease spots caused by *Mycosphaerella citri* cause lesions on citrus leaves, leading to premature defoliation and failure of the pericarp, leading to reduced fruit appearance quality and, in turn, reduced marketability. Oil stains are most severe on grapefruit (*Citrus paradisi* Macf.) but also affect most citrus fruits. The primary source of inoculum is sexual ascospores produced from Pseudothesia fruiting bodies formed during the decomposition of leaf litter in orchard soils.

Controls Two sprays of copper fungicide can reduce the appearance, severity and defoliation of citrus greasy stains.

16.3 Citrus Storage Rot

*16.3.1 Green Mould (*Penicillium digitatum*)*

Green mould enters citrus peels through wounds and natural openings. Symptoms show water-soaked areas on the outer surface of the fruit, followed by growth of colourless/white mycelium and sporulation (green).

*16.3.2 Blue Mould (*Penicillium italicum*)*

The blue mould penetrates the citrus peel without breaks and spreads from one fruit to the next in the storage box. Symptoms were similar to green mould, except the fungal spores were blue.

*16.3.3 Alternaria Rot (*Alternaria citri*)*

Alternaria enters the citrus fruit through the joints. Preharvest treatment of citrus fruits with gibberellic acid or post-harvest treatment with 2,4-D can delay bud senescence and subsequent decay by *Alternaria* fungi.

16.3.4 Management

Storage spoilage can be prevented by handling citrus fruits with care at harvest to minimise cuts, abrasions and bruises. Post-harvest fruit loss in citrus can be reduced by treating fruit with Bavistin (carbendazim) 1000 ppm, maintaining optimal temperature range and relative humidity, and eliminating ethylene during transport.

16.4 Citrus Canker

16.4.1 Importance

Among various infectious diseases, fungi, bacteria, algae, viruses, and phytoplasmas play an important role. Citrus bacterial diseases [c.o.: *Xanthomonas axonopodis* (formerly *campestris*) pv. *citri* (Hasse) are among the most severe threats of international economic importance. Citrus cankers mainly cause leaf spots and bark spots. However, conditions favourable for disease development lead to defoliation, shoot death, and fruit drop (Gottwald et al. 2002a). All citrus rootstocks are susceptible to *Xanthomonas axonopodis pv. citri* infection under field conditions. *Xanthomonas axonopodis pv. citri* (Xac) is probably native to India or Southeast Asia and is now found in over thirty countries. Xac is registered in regions of Asia, Africa, Australia, South America, and islands in the Pacific and Indian Oceans and is also found in warm temperate regions of Southwest Asia and the Middle East, including Oman, Saudi Arabia, Iran, Iraq, the United Arab Emirates, and Yemen (Stall and Seymour 1983; Whiteside et al. 1988; Stall and Civerolo 1991; Gottwald et al. 2002a, b, c). Under prevailing climatic conditions, the disease has different severity of canker infections. In India, citrus cankers were first reported in Punjab (Kalita et al. 1995). Between 1954 and 1960, outbreaks were documented in states such as Tamil Nadu, Andhra Pradesh, Karnataka, Madhya Pradesh, Rajasthan, Assam, Uttar Pradesh and West Bengal (Kalita et al. 1996). Among the commercial cultivars, acid lime [*Citrus aurantifolia* (Christm.) Swingle] is the most susceptible cultivar (Das and Dubey 1989), and it is difficult to find an orchard completely free of canker infection. It Is difficult. Yield reductions of up to 50–60% have been observed in citrus fruits in different parts of the world. The disease is mechanically transmitted by infection of the leaves of *Phyllocnistis citrella* Stainton (Lepidoptera: Gracillariidae), which has been reported to exacerbate citrus cankers in Australia, Brazil, Yemen and India (Sohi and Sandhu 1968; Cook 1988; Gottwald et al. 1997; Chagas et al. 2001; Belasque et al. 2005; Christiano et al. 2007; Das et al. 2012).

16.4.2 The Pathogen

Citrus canker is caused by a bacterial plant pathogen known as *Xanthomonas axonopodis* pv. *citri* (*Xanthomonas campestris* pv. *citri*). Another synonym is *Pseudomonas citri, Xanthomonas campestris* pv. *citri* and *Xanthomonas citri.*

The scientific classification of the bacterium has presented in Table 16.2.

Table 16.2 Scientific classification of the bacterium

Kingdom	Bacteria
Phylum	Proteobacteria
Class	Gammaproteobacteria
Order	Xanthomonadales
Family	Xanthomonadaceae
Genus	*Xanthomonas*
Species	*X. axonopodis; Xanthomonas axonopodis*
EPPO Code	XANTCI

The bacterium *Xanthomonas axonopodis* is a polar, rod-shaped Gram-negative bacterium. The bacterium's genome is five megabase pairs long. Under natural circumstances, many pathovars and variations of the bacterium cause various citrus canker disorders (Gottwald et al. 2002a, b, c). Numerous countries that cultivate citrus are subject to strict international phytosanitary regulations about this bacterium, a quarantine pest. Different disease forms are linked to various pathotypes (Gottwald et al. 2002a, b, c). Citrus canker pathotypes can differ in severity, host range, and geographic location. The most severe type of the disease, citrus canker-A (Asiatic canker), which affects most citrus cultivars, is also the most prevalent and economically significant. The disease's most prevalent and harmful variety is called the Asiatic type of canker (canker A), which is brought on by a collection of strains first discovered in Asia. In Argentina, Paraguay, and Uruguay, citrus canker-B (cancrosis-B) severely impacts lemons and little to Mexican lime, sour orange, and pummelo.

Cancrosis B is a disease of lemons, Mexican limes (key limes), bitter oranges, and pomelo brought on by various *X. axonopodis pv. aurantifolii* strains that were first discovered in South America. Mexican limes are linked to citrus canker-C (cancrosis-C) in Brazil. Only key lime and bitter oranges are susceptible to cancrosis C, which is also brought on by strains of *X. axonopodis pv. aurantifolii.* Only key limes are infected by A* strains found in Oman, Saudi Arabia, Iran, and India (Vernière et al. 1998). Réunion and islands around the Indian Ocean have described an atypical form of strain A. It is resistant to some antibiotics to a high degree. D canker strain was reported in 1981 on Mexican lime in Mexico, but its identification remains in discussion (Medina-Urrutia and Stapleton 1985). Pomelos, Mexican lemons and oranges are susceptible to all kinds of citrus diseases. Oranges are moderately early, so they are susceptible to disease; Bitter oranges, lemons and sweet oranges are moderately sensitive; and they have moderate resistance from officials (Schubert et al. 2001).

The same international phytosanitary rules apply to all diseases. There are a variety of techniques that can be used to recognise and distinguish the pathogenic patterns of *Xac*. You can inoculate the leaves of susceptible species, such as grapefruit and Mexican lime, and then monitor the lesions. Regulatory authorities often use serological methods, such as ELISA (enzyme immunoassay), protein electrophoresis, and DNA analysis, to confirm the diagnosis (Gottwald et al. 2002a, b, c). Leaf stomata, other natural openings and wounds are where bacteria are most commonly transmitted. The ability to spread the disease rapidly increases with wind and rain. Cyclones, hurricanes, and tropical storms can spread germs over long distances (Gottwald et al. 2002a, b, c; Polek et al. 2007).

16.4.3 Symptomatology

Broadbent, etc. (1992) Northern, spring-washed Pummeloe (*Citrus maxima*) leaves, twigs, and fruits with 4–5 mm diameter raised, rough, circular, typical citrus cankers. Found cork-like scab-like lesions: Territory, Australia. During an investigation conducted in 1991 as part of the Australian Northern Quarantine Strategy, *Xanthomonas campestris pv citri* (Xcc) is responsible. In inoculated sweet orange, West Indian lime, tart orange, and Ducan grapefruit leaves, all strains produced white tissue lesions that looked like a callus. Lesions were more extensive and frequent than expected on seedlings or removed leaves of West Indian Lime, Sweet Orange, and Ducan Grapefruit.

Wet edges of characteristic citrus canker lesions caused by *Xanthomonas axonopodis pv. citri* (Xac) (Lin et al. 2010). They often congregate in groups on leaf margins, leaf tips, or other places where water likes to collect. The age and type of host tissue at the time of infection influence the size of the lesion (Gottwald et al. 2002a, b, c). As tissues age, they all develop a higher resistance to natural infection (Stall and Seymour 1983). When vulnerable cultivars are severely infected, defoliation may occur. They have a raised, wart-like surface and are the same colour as the branch when they develop on woody tissue. Similar symptoms can be found on fruit and twigs, including elevated, dark brown or black lesions that are corky in texture and have oily or wet edges. As the lesions mature, they appear scabby or corky (Zekri et al. 2011).

Mohammed et al. identified 76 *Xanthomonas*-like strains from various *Citrus* species and separated these strains into two groups based on symptoms produced on grapefruit leaves (Mohammed et al. 2014). Citrus canker pathogens from Taiwan that caused flattened necrotic lesions with a water-soaked edge on the leaves of four *Citrus* species were discovered by Lin-Hsin Cheng et al. in 2008 *viz.*, Mexican lime – *Citrus aurantifolia*, grapefruit – *C. paradisi,* Liucheng - *C. sinensis* and lemon - *C. limon* (Lin et al. 2008). Derso et al. reported citrus canker on lime for the first time in Ethiopia. The bacterium causes defoliation, early fruit drop, and twig dieback by inducing erumpent, callus-like lesions (Derso et al. 2009).

16.4.4 Environmental Parameters Favourable for Canker Disease Development

Citrus canker can occur in dry climes, but it is more common in tropical and subtropical regions with a lot of rainfall and warm temperatures. When damp weather conditions prevail during the stages of young citrus fruit's development and shoot emergence, citrus canker develops into a deadly disease (Schubert et al. 2001). The best conditions for disease development are between 20 °C and 30 °C with uniformly spaced showers (Ramakrishnan 1954; Reedy 1984). According to Palazzo et al. (1987), rain and SE and NW winds at temperatures of 25 °C or higher encouraged the propagation of the citrus canker disease. According to Webb et al., the bacterium *X. campestris pv. citri* could survive in various pH (5.0–8.8) and temperature conditions (29–50 °C) (Webb et al. 1988). Xanthan gum produced during infection protected the bacterial pathogen from an extended range of weather conditions, promoting severe infection.

According to Kalita et al., research conducted in 1993 and 1994 revealed that August was the month with the highest incidence of citrus canker (*X. campestris pv. citri/X. axonopodis pv. citri*) (Kalita et al. 1995). June through September of both years had a consistently high disease incidence. The highest temperatures, rainfall, and relative humidity (R.H.) were observed in these months, indicating a positive association between these factors and the state's high disease incidence. In their study, (Das et al. 2012) found that the disease incidence peaked in July and then declined in September. The month of March had the lowest incidence. They concluded that there is a positive association between the incidence of disease and the weather variables, particularly rainfall, temperature, and relative humidity.

To investigate the combined effects of different temperatures and leaf wetness duration on infection and development of Asiatic citrus canker (*X. c.* subsp. *citri*) on the Tahiti lime plant, (Christiano et al. 2007) used growth chambers. The main component determining disease development was temperature. At the ideal range of temperatures (25–35 °C), disease incidence was 100%. Twelvefold increase in lesion density, a tenfold rise in lesion size, and a sixty-fold increase in disease severity, maximum disease progression was seen at 30–35 °C.

16.4.5 Host Resistance

The trifoliate orange [*Poncirus trifoliata* (L.) Raf.] and its hybrids, which are majorly used as rootstocks, grapefruit, some sweet oranges including Hamlin, Pineapple, and Navel, Mexican (Key) limes, and lemons are the citrus cultivars and rootstocks that are most severely affected by citrus canker (Table 16.3). Citrus canker has made it difficult or impossible for these cultivars to thrive in humid subtropical and tropical climates. Despite varying degrees of susceptibility, all other commercial citrus cultivars are vulnerable enough that when sick or exposed, they must be destroyed as part of an eradication campaign. Other than *Citrus* and *Poncirus*, (Civerolo 1984) provides a list of plants in the Rutaceae family that, in

Table 16.3 Citrus canker resistance or susceptibility in citrus commercial cultivars and species

Rating scale	Citrus commercial cultivars	References
Highly resistant	Kumquats and Calamondin	Goto (1992) and Leite Jr and Mohan (1984)
Resistant	Mandarins – Cleopatra, Ponkan, SunChuSha, Sunki, Satsuma, Tankan	Goto (1992) and Sarkar et al. (2007)
Less susceptible	Tangerines, Tangors, Tangelos (all selections); Sweet oranges and Sour oranges	Zubrzycki and de Zubrzycki (1981)
Susceptible	Sweet-oranges, Pummelo, Limes, Palestine sweet lime, Trifoliate orange, Citranges or Citrumelos	Leite Jr. (2002) and Zubrzycki and de Zubrzycki (1981)
Highly susceptible	Grapefruit, Mexican or Keylime, Lemons and Pointed leaf Hystrix	Gottwald et al. (2002a, b, c)

experimental settings or under severe disease pressure in the wild, can act as Xac hosts. These plants could pose a difficulty as inoculum reservoirs in an eradication or suppression programme, even though they are not anticipated to have any significant impact on citrus canker epidemiology in areas where the disease is endemic (Gottwald et al. 2002a, b, c).

16.4.6 Management

One of the most feared citrus diseases, citrus canker, affects all major *Citrus* species. Citrus crops are severely affected by this disease, and the severity of infection varies by species, variety and local climate. The disease originates in Southeast Asian countries such as India and Japan, and from there has spread to all other continents where citrus fruits are grown, except Europe. However, the disease is prevalent in many places, posing a continuing threat to lemon cultivation, especially in cancer-free areas. Cancer control measures include quarantine or regulatory programs to ban the importation of diseased citrus plant material or fruit and continued rigorous field testing, including rapid eradication of trees. The fundamental goals of various technologies are to prevent, eliminate or destroy pathogens, reduce the amount of inoculum that can become infected, reduce the spread of disease, and protect vulnerable tissues from infection (Civerolo 1981).

A new policy recently enacted in the United States, the "1900 Foot Rule," requires removing and destroying all diseased citrus trees and all healthy citrus plants within 1900 feet of diseased trees (Gottwald et al. 2001). However, such eradication techniques are not considered practical in endemic situations (such as those in India). Most of the year contributes to disease outbreaks in this area. The development of cankers under these circumstances can be prevented by (i) the use of canker-free seedlings, (ii) pre-monsoon pruning and burning of all infected branches, and (iii) regular spraying and appropriate copper-based sterilisation. Pesticides (to control insect damage), and (iv) some precautions (Das and Singh 1999).

16.4.6.1 Cultural and Regulatory Measures

Exclusion is the first and best line of defence against citrus canker. Strict regulations on importing fruit and propagation material from cancer-affected areas to cancer-free areas are essential to eliminate harmful bacteria. Eradication is the second line of defence against citrus canker. Delayed removal of affected trees due to litigation allowed the infection to spread deeper into the country (Gottwald et al. 2002a, b, c). Identifying all potential hosts of a target pest or pathogen is critical for eradication programs. A recent study by Kalita et al. showed that *Xac* inhabits the Indian horny goat weed (*Ageratum conyzoides* L.) (Pabitra et al. 1997). Hygiene plays a vital role in preventing the spread of cancerous bacteria (Timmer 2000). Windshields around citrus orchards can limit or minimise the natural transmission of wind-borne diseases, especially when the wind direction is predominant, but the disease is highly prevalent during stormy weather (Gottwald et al. 2002a, b, c).

16.4.6.2 Biological and Botanical Measures

The early stages of research into the biological treatment of citrus cankers are still underway. Some citrus bacteria, such as *Pseudomonas syringae, Erwinia herbicola, Bacillus subtilis*, and *Pseudomonas fluorescens*, are hostile to canker *in vitro* (Goto et al. 1979; Ota 1983; Unnamalai and Gnanamanickam 1984; Kalita et al. 1996). According to (Mukherjee and Biswas 1984), the disease was reported to be controlled by spraying an emulsion of elasis and hydrocarpus. According to (Takeuchi et al. 1988), bacteria isolated from actinomycete species. It was successfully regulated by benaomycin A and B. They revealed the composition of her body. In addition to their antifungal effects, they also inhibited bacteria that damaged two of her plants, including her *Xanthomonas citri*. (Masroor and Chandra 1989) *Aspergillus clavatus, A. flavus,* and *A. niger* claimed antibiotic antagonist activity against *Xanthomonas campestris pv citri*. (Akhtar et al. 1997) found that most forest trees, plants, and shrubs had an inhibitory effect on bacterial strain XC-100. The potential antimicrobial agent from the diffusion of higher plants could be used in treating citrus canker disease. Using the paper disc agar diffusion method on nutrient glucose agar, (Tongin et al. 2013) Study *Terminalia catappa* Linn. Ethanol leaf extract for its ability to inhibit the growth of *Xanthomonas axonopodis pv. Citri* (Xac) causes calcification cancer. The growth of the calcific bacterial carcinoma strain *Xci*12 was inhibited by *T. catappa* leaf extract (140 mg/ml).

16.4.6.3 Chemical Measures

Numerous chemicals have been sprayed in various locations for a considerable time. According to Beniwal and Chaubey, the most effective antibiotic and fungicide for treating Xanthomonas citri was thiram, which controlled the disease up to 500 ppm and above (Beniwal and Chaubey 1976). Thiram was as effective as agrimycin, streptomycin sulphate, and streptocycline. According to Leite Jr., *Xanthomonas campestris pv. citri* cannot survive for long periods in soil, near nonhost plants, or in plant debris (Leite Jr. 1990). Windbreaks and copper-based fungicides can significantly slow the spread of disease. According to Kale et al. (1994), 100 ppm streptocycline and his 0.1% foliar application of copper oxychloride were used to treat his *Xcc* infection in his 6-year-old Kagji limonene in Maharashtra and India rice fields. Sprays were used at 7, 15, and 21-day intervals, with spraying every week or 2 weeks yielding the least expensive chemical control. In Argentina, in 1990 and 1991, Gottwald and Timmer observed the effects of windbreaks and copper-based fungicides, alone or in combination, on citrus canker incidence and spread (Gottwald and Timmer 1995).

Although not as efficient as windbreaks, copper-based bactericides did lessen the incidence and spread of disease. In trials done by (Chen et al. 2000) in 1997 and 1999, Skaggs Bonanza navel orange trees were exposed to various quantities of chemicals, including 56% cuprous oxide, agro-streptomycin, 77% copper hydroxide (WP), and 50% Shajunwang (WP). According to the findings, the optimum treatment was 50% Shajunwang (WP), which produced a controlled rate of up to 94.5%. For the control of *Xanthomonas citri* (*X. axonopodis pv. citri*), (Zhang et al. 1996) tested copper hydroxide (as Koshad), carbendazim, sulfuric acid, streptomycin, and a bordeaux mixture on Robertson Navel trees that were 6 years old. The best disease control was achieved with copper hydroxide at an 800-times concentration. Copper-based solutions are often used as a regular strategy for treating citrus cankers. According to (Del Campo et al. 2009), copper treatment confers a viable but non-cultivable (VBNC) state in *Xac*, but does not protect vulnerable plants from symptom development. However, copper spraying at 28-day intervals reduced the incidence of citrus canker on leaves and fruits (Behlau et al. 2010).

16.5 Citrus Tristeza Virus

Many viral diseases affect citrus fruit, but 'tristeza' caused by CTV severely impacts worldwide (Moreno et al. 2008). CTV is widespread in India's citrus cultivating geographical zones, with disease incidence ranging from 10% to 90% (Ahlawat 1997; Biswas 2008).

16.5.1 History

The virus is believed to originate in the *Citrus* species. H. Southeast Asian region (Bar-Joseph et al. 1989). Initially, citrus fruits were brought to different regions as fruits or seeds. This suggests that CTV is not seed-transmissible and, therefore, virus-free in citrus growing in importing regions (McClean 1957). Finally, in the nineteenth century, the establishment of maritime trade between Asia and the rest of the West and the growing botanical and commercial interest in citrus fruits led to the migration of numerous exotic *Citrus* species, some of which is an asymptomatic carrier of CTV, leading to its worldwide spread. CTV (Roistacher 1981).

Under different weather conditions, large-scale distribution of CTV favoured interactions with new host cultivars/rootstocks. The disease killed a hundred million sour oranges grafted citrus trees (Moreno et al. 2008). Tristeza 1930 (Argentina), 1937 (Brazil), 1939 (California), 1951 (Florida), 1957 (Spain), 1970 (Israel), 1980 (Venezuela), 1989 (Cyprus), observed in 1992 (Cuba). 1995 (Mexico), 1996 (Dominican Republic), 1997 (India), 2002 (Italy) (Bar-Joseph et al. 1989; Kyriakou et al. 1996; Gottwald et al. 1998; Garnsey et al. 2000; Gottwald et al. 2002a, b, c; Davino et al. 2003). CTV outbreaks have necessitated using CTV-resistant rootstocks to re-establish citrus cultivation in affected countries. The causative agent of Tristeza was unknown worldwide during the epidemic. It was later reported to be of pathogenic origin due to its ability to be transmitted by aphids and transfer infected grafts (Fawcett and Wallace 1946). It was later reported to be associated with filamentous virus particles 2000 nm long and 12 nm in diameter (Kitajima et al. 1964). Inoculation of isolated virions into citrus trees confirmed that these abnormally long particles play a role in causing citrus wilt disease (Garnsey et al. 1977; Garnsey and Muller 1988). Early breeding of sour orange citrus was unsuccessful in South Africa and Australia. As a result, this rootstock did not become widespread and evaded the Tristeza epidemic.

However, another pathogenic interaction was found between citrus bark and CTV in these regions, known as stem attack (SP). This was later observed in several countries where citrus was replanted as Tristeza-resistant rootstocks, followed by Tristeza epidemics (Hughes and Lister 1949; McClean 1956; Muller et al. 1968; Rocha-Peña et al. 1995; Wallace 1978). Regardless of the rootstock, infected trees showed poor fruit yield and quality. CTV SP variants were initially restricted/isolated to Asia, Australia, South Africa, and South America but have since been reported to be present, albeit less frequently, in Florida, California, and the Mediterranean Sea (Ben-Ze'Ev et al. 1989; Kyriakou et al. 1996; Yang et al. 1999; Ruiz-Ruiz et al. 2006).

16.5.2 Symptoms

CTV exhibits three distinct manifestations: (i) quick decline (QD) or tristeza, (ii) stem pitting (SP) and (iii) seedling yellows (SY) in all *Citrus* species and hybrids, viral strains or scion-rootstock combinations. Damages caused by severe QD and SP isolates can be highly severe; however, trees with mild viral isolates show productivity for a certain period (Bar-Joseph et al. 2002).

The infected trees accumulate or produce excessive non-functional phloem and necrosis of the sieve tubes due to CTV infection, leading to scion overgrowth at the bud union and progressive reduction of root mass (Schneider 1957). The scarce supply of water and minerals results in the tree's wilting, leaf chlorosis resembling nitrogen deficiency, unmarketable pale-coloured small fruit, dull greenish-yellow thin foliage, leaf shedding and dieback symptoms in the citrus orchard (Moreno et al. 2008). The Quick decline occurs slowly over several decades and is more available in mature trees. The tristeza syndrome can be reversed or avoided if the tree is grafted onto the decline-tolerant rootstock, upon which it recovers immediately (Atta et al. 2012). The commonly quick decline affected tree species include sweet oranges (*C. sinensis*), mandarins (*C. reticulata*), grapefruits (*C. paradisi*) and limes (*C. aurantifolia*) propagated on sour orange rootstock.

Other syndromes showed by citrus cultivars upon infection by different CTV strains include vein clearing and symptomless expression. Mexican lime is a principal indicator that develops vein clearing and leaf cupping symptoms in response to CTV isolates (Bar-Joseph et al. 1989). But sometimes vein clearing also occurs in sweet orange, Persian lime and other hosts. The intensity of vein clearing is higher in young leaves and decreases with leaf hardening.

16.5.3 Host Range

The genera *Citrus* and *Fortunella* members are natural hosts of CTV, although the virus has been reported to infect other citrus relatives upon experimental inoculation (Moreno et al. 2008). The artificial transmission was achieved using aphid vectors in the non-*Citrus* species *Passiflora gracilis* and *P. coerulea* (Kitajima et al. 1974; Muller et al. 1974), and agroinfiltrated leaves of *N. benthamiana* (Gowda et al. 2005). *Poncirus trifoliate* (L.) Raf., a preferred rootstock, and some trifoliate orange hybrids are resistant to most isolates of CTV, whereas *Fortunella crassifolia* Swing., pummelos and sour orange exhibit resistance to specific CTV strains (Garnsey et al. 1996; Mestre et al. 1997a, b). However, CTV could multiply in *P. trifoliata*, pummelo or sour orange protoplasts by indicating the movement of CTV (Albiach-Marti et al. 2004).

16.5.4 Transmission

CTV is transmitted readily to long distances *via* infected graft and locally with variable efficiency by aphid species in a semi-persistent manner (Bar-Joseph et al. 1989; Moreno et al. 2008). The aphids acquire the phloem-limited virus upon feeding for a minimum of 30 min up to 24 h. The virus cannot replicate or circulate inside the vectors and can infect a healthy plant within 48 h of virus acquisition (D'Onghia et al. 2009). Acquisition of virus through feeding on infected citrus trees by aphid species is confirmed by ELISA (Cambra et al. 1981), RT-PCR (Mehta et al. 1997) and real-time RT-PCR (Bertolini et al. 2008), but only a few species can successfully transmit to a healthy one. Certain aphid species transmit some CTV isolates much more efficiently than others or vice versa (de Mendoza et al. 1984; Raccah et al. 1980). Several factors that influence the spread of CTV in a specific area include the susceptibility of the citrus varieties towards the aphid fauna (Marroqúin et al. 2004), the transmissibility of the CTV viral isolates by the aphid population (Yokomi and Garnsey 1987) and favourable environmental conditions for aphid build up and new flush (Bar-Joseph and Loebenstein 1973).

16.5.5 Diagnosis

The emerging diversity of CTV genotypes led to the quest for more advanced and new differentiation techniques contributing to complicated diagnostic measures. Multiple options/assays are available to detect CTV, and no single technique could be ideal for easy application. These methods should be specific and available with equipment, appropriate raw materials and skilled personnel for handling.

16.5.5.1 Bio-indexing

Bio-indexing is a vital component of control procedures for CTV in citrus. In Israel and California, this method includes indexing thousands of trees annually for CTV infection species or rootstocks to maintain the eradication system (Raccah et al. 1976; Roistacher 1976). The lime test procedure for CTV was significant progress in virus detection, and since then, it has been a critical element in CTV indexing for many years (Wallace 1951). But its limitation is that it takes considerable time (4–6 weeks minimum), requires extensive plant material, insect-free temperature-controlled greenhouse facilities and skilled personnel to grow the host plants on indicator plants and interpret the symptoms critically (Wallace 1951). It does not allow quantitative measurement or detection of CTV from in vitro samples, and it is difficult to detect mild isolates that do not induce symptoms on indicator plants (Wallace 1951).

16.5.5.2 Microscopy

Attempts have been made to find more efficient methods to detect CTV, as the biological index is an unreliable and conventional method for detecting CTV. Negatively stained viral extracts in electron microscopy (EM) were used in an eradication system (Bar-Joseph et al. 1974). It is proved to be of limited application due to the need to concentrate tissue extracts, and he found very few CTV particles in Valencia oranges. Detailed information on the relative titres of his CTV in different tissues at different ages (Bar-Joseph et al. 1979; Garnsey et al. 1979). Debrick developed a technique for detecting plant viruses and named Debrick's serologically specific electron microscopy (SSEM) in 1973 (Debrick 1973). Using SSEM, virus particles are bound to virus antibodies, adsorbed to filmed electron microscope grids, and visualised by electron microscopy. SSEM was reliable because of the development of CTV-specific antisera (Gonsalves et al. 1978).

16.5.5.3 Enzyme-Linked Immunosorbent Assay (ELISA) Based Detection

Successful CTV purification procedure produced a variety of antisera, and it was first shown in 1979 that his CTV could be serologically detected using his ELISA (Bar-Joseph et al. 1979). Several variants of the ELISA method have been developed to improve detection sensitivity. However, DAS-ELISA remained the preferred choice for routine CTV detection and infection screening in large citrus orchards. First monoclonal antibody (mAb) to detect heavy strains of CTV produced by hybrid cells (Vela et al. 1986). Polyclonal antibodies (pAbs) are produced by injecting an antigen into animals such as mice, rabbits, goats, and chickens. Production of pAbs requires large amounts of antigen for subsequent booster administration, including an immunisation schedule (Nolasco and Sequeira 2002). Therefore, bacterially expressed CTV-CP was an excellent source of abundant CTV antigens for producing pAbs.

16.5.5.4 Nucleic Acid-Based Detection

A diagnostic CTV assay that combines the simplicity of ELISA with the advantages of a DNA-based technique called immune-capture or reverses transcription-polymerase chain reaction (IC/RT-PCR) was developed by (Nolasco et al. 1997). Finally, bidirectional (BD)-RT- PCR techniques have been developed (Cevik et al. 1996; Roy and Ramachandran 2012). However, (Yokomi et al. 2010) reported that MCA-13 does not distinguish between isolates that cause regression only and those that cause pitting or mild symptoms and MCA13-reactive isolates. This suggests the need for more selective probes to identify the CTV strains that produce them rapidly. A multiplex RT-PCR technique was developed (Roy et al. 2010) to simplify the identification of CTV genotypes. CTV (175) isolates from 29 citrus-growing countries were successfully analysed using S- and H-RT-PCR (Roy et al. 2010). H-RT-PCR, CTV genotype detection, provided a sensitive, specific, reliable, and rapid method for screening CTV genotypes.

Recently, an immunocapture reverse transcriptase loop-mediated amplification assay (IC-RT-LAMP) was developed to detect heavy strains (Selvaraj et al. 2019). Rapid on-site detection of VT strains was facilitated by modifying RT-LAMP, which uses CTV-IgG to capture his CTV virions from raw citrus leaf juice. The RT-LAMP assay allows real-time fluorescence monitoring with the Genie III fluorometer. Overall, IC-RT-LAMP is reported to be more straightforward, specific, reliable, and economical than RT-qPCR (Selvaraj et al. 2019).

16.5.6 Micro Shoot Tip-Grafting In Vitro

Micrografting is an in vitro transplantation technique involving the aseptic production of seeds or micro propagated cultures by firmly placing meristematic or shoot-tip explants on grown decapitated rootstocks. This method has great potential for fruit tree improvement, large-scale propagation, and successful large-scale production of virus- and viroid-free plants in citrus. (Fifaei et al. 2007) determined various parameters of micro implantation using ELISA and biological indexing and its ability to produce her CTV-free citrus seedlings with shoot sizes of 0.1–0.2 mm. (Hančević et al. 2009) described the effectiveness of CTV removal from infected mandarin oranges by hyperthermia, followed by apical transplantation of citrus fruits with 0.15–0.19 mm buds from the bark into the cambium region of the plant. Make an L-shaped notch. The absence of CTV pathogens in produced seedlings was confirmed by DAS-ELISA and immunocapture (IC/RT-PCR). (Abbas et al. 2008) used his chip micrograft technique to successfully establish basal blocks of tangerine and sweet his orange without CTV infection. Chung and Shin demonstrated the elimination of the Indian citrus ringspot virus (ICRSV) from kinnohis mandarins using shoot sizes ranging from 0.2 to 1.0 mm and micrografted plants using the 0.2 mm technique and verified that no viruses are present (Chung and Shin 2007). Sharma et al. also showed that the highest number of ICRSV-free plants (36.84%) obtained by screening using indirect ELISA and RT was due to hyperthermia followed by micrografting of nodal segments in a 50 °C water bath. It was shown to be the number of ICRSV (Sharma et al. 2007).

Sharma et al. also used chemotherapy in combination with micrografting to eliminate his ICRSV from Kinnow mandarin strains (Sharma et al. 2007). Furthermore, (Vijayakumari et al. 2006) reported on producing virus-free and type-appropriate planting material by micro transplantation. Two hundred three plants from 245 STG transplants were found to be virus-free by biological and serological indexing.

Rapid identification is essential to identify the presence or absence of a virus created by meristematic culture, heat treatment, introduced cultivars, or dubious field sources. (Kapari-Isaia et al. 2008) have typically carried out micrografting utilised for citrus viroids micro indexing and shown that micro transplantation could be successfully diagnosed after 12 days and established that micro indexing in grafted cuttings was more straightforward and more accurate than micro indexing in seedlings or cuttings via injection.

16.6 Conclusion

Citrus fruits offer a wealth of vital nutrients, including a fantastic supply of vitamin C that the body needs. Widespread nutritional advantages result from dietary recommendations that promote the intake of citrus fruit and its by-products. Citrus diseases caused by fungi, bacteria, and viruses endanger citrus production and cause massive economic losses to the growers. Large citrus-growing regions worldwide, including India, are affected by plant disease. The most effective methods for managing the disease are typically integrated approaches that include both general cultural practices and phytosanitary measures, such as eradicating the source of the disease, removing the inoculum, installing suitable windbreaks, and applying protective copper/antibiotic sprays when necessary.

References

Abbas M, Khan MM, Fatima B, Iftikhar Y, Mughal SM, Jaskani MJ, Khan IA, Abbas H (2008) Elimination of citrus tristeza closterovirus (ctv) and production of certified citrus plants through shoot-tip micrografting. Pak J Bot 40:1301–1312

Ahlawat YS (1997) Viruses, greening bacterium and viroids associated with citrus (Citrus species) decline in India. Indian J Agric Sci 67:51–57

Akhtar MA, Rahber-Bhatti MH, Aslam M (1997) Antibacterial activity of plant diffusate against Xanthomonas campestris pv. citri. Int J Pest Manag 43:149–153

Albiach-Marti MR, Grosser JW, Gowda S, Mawassi M, Satyanarayana T, Garnsey SM, Dawson WO (2004) Citrus tristeza virus replicates and forms infectious virions in protoplasts of resistant citrus relatives. Mol Breed 14:117–128

Atta S, Zhou C, Zhou Y, Cao M, Wang X (2012) Distribution and research advances of citrus tristeza virus. J Integr Agric 11:346–358

Bar-Joseph M, Loebenstein G (1973) Effects of strain, source plant, and temperature on the transmissibility of citrus Tristeza virus by the melon aphid. Phytopathology 63:716–720

Bar-Joseph M, Loebenstein G, Oren Y (1974) Use of electron microscopy in eradication of Tristeza sources recently found in Israel. Int Organ Citrus Virol Conf Proc 6:83–85

Bar-Joseph M, Garnsey SM, Gonsalves D, Mira M, Purcifull D, Clark MF, Loebenstein G (1979) The use of enzyme-linked immunosorbent assay for detection of citrus Tristeza virus. Phytopathology 69:190–194

Bar-Joseph M, Che X, Mawassi M, Gowda S, Satyanarayana T, Ayllon MA, Albiach-Martí MR, Garnsey SM, Dawson WO (1989) Citrus tristeza virus. Int Organ Citrus Virol Conf Proc 15:1–7

Bar-Joseph M, Che X, Mawassi M, Gowda S, Satyanarayana T, Ayllón MA, Albiach-Martí MR, Garnsey SM, Dawson WO (2002) The continuous challenge of Citrus tristeza virus molecular research. Int Organ Citrus Virol Conf Proc (1957–2010). 15:1–7

Behlau F, Belasque J, Graham JH, Leite RP (2010) Effect of frequency of copper applications on control of citrus canker and the yield of young bearing sweet orange trees. Crop Prot 29:300–305

Belasque J, Parra-Pedrazzoli AL, Rodrigues Neto J, Yamamoto PT, Chagas MCM, Parra JRP, Vinyard BT, Hartung JS (2005) Adult Citrus Leafminers (Phyllocnistis citrella) are not efficient vectors for Xanthomonas axonopodis pv. citri. Plant Dis 89:590–594

Beniwal SPS, Chaubey SN (1976) Thiram, a fungicide effective against Xanthomonas citri (Hasse) Dowson. Pesticides 10:31–36

Ben-Ze'ev IS, Bar-Joseph M, Nitzan Y, Marcus R (1989) A severe citrus tristeza virus isolate causing the collapse of trees of sour orange before virus is detectable throughout the canopy§. Ann Appl Biol 114:293–300

Bertolini E, Moreno A, Capote N, Olmos A, de Luis A, Vidal E, Pérez-Panadés J, Cambra M (2008) Quantitative detection of Citrus tristeza virus in plant tissues and single aphids by real-time RT-PCR. Eur J Plant Pathol 120:177–188

Biswas KK (2008) Molecular diagnosis of citrus tristeza virus in mandarin (Citrus reticulata) orchards of Darjeeling hills of West Bengal. Indian J Virol 19:26–31

Cambra M, Hermoso de Mendoza A, Moreno P, Navarro L (1981) Use of enzyme-linked immunosorbent assay (ELISA) for detection of citrus tristeza virus (CTV) in different aphid species. Proc Int Soc Citric 1:444–448

Cevik B, Pappu SS, Pappu HR, Benscher D, Irey M, Lee RF, Niblett CL (1996) Application of bi-directional PCR to citrus Tristeza virus: detection and strain differentiation. Int Organ Citrus Virol Conf Proc 13:17–24

Chagas MCM, Parra JRP, Namekata T, Hartung JS, Yamamoto PT (2001) Phyllocnistis citrella Stainton (Lepidoptera: Gracillariidae) and its relationship with the citrus canker bacterium Xanthomonas axonopodis pv citri in Brazil. Neotrop Entomol 30:55–59

Chen YF, Song CQ, Xu HZ, Wu L (2000) Experiment of screening the fungicide for control of citrus canker. S China Fruits 29:18

Chose O, Noël C, Gerbod D, Brenner C, Viscogliosi E, Roseto A (2002) A form of cell death with some features resembling apoptosis in the amitochondrial unicellular organism Trichomonas vaginalis. Exp Cell Res 276:32–39

Christiano RSC, Dalla Pria M, Jesus Junior WC, Parra JRP, Amorim L, Bergamin Filho A (2007) Effect of citrus leaf-miner damage, mechanical damage and inoculum concentration on severity of symptoms of Asiatic citrus canker in Tahiti lime. Crop Prot 26:59–65

Chung HS, Shin JC (2007) Characterization of antioxidant alkaloids and phenolic acids from anthocyanin-pigmented rice (Oryza sativa cv. Heugjinjubyeo). Food Chem 104:1670–1677
Civerolo EL (1981) Citrus bacterial canker disease: an overview. Proc Int Soc Citric 1:390–394
Civerolo EL (1984) Bacterial canker disease of citrus. J Rio Gd Val Hortic Soc 37:127–146
Cook AA (1988) Association of citrus canker pustules with leaf miner tunnels in North Yemen. Plant Dis 72:546
Cooper WC, Chapot H (1977) Fruit production with special emphasis on fruit for processing. In: Nagy S, Shaw PE, Veldhuis MK (eds) Citrus science and technology. AVI Publishing Co., Inc, Westport, pp 1–127
D'Onghia AM, Djelouah K, Roistacher CN (2009) Citrus tristeza virus and Toxoptera citricidus: a serious threat to the Mediterranean citrus industry, Options Méditerranéennes Série B Etudes Recherches. CIHEAM, Bari, p 244
Das BC, Dubey LN (1989) The incidence of citrus canker Xanthomonas citri (Downn) in Assam. J Res Assam Agric Univ 10:80–82
Das AK, Singh S (1999) Management of Bacterial Canker in acid lime. Intensive Agric 36:28–29
Das PT, Tajo L, Goswami J (2009) Assessment of citrus crop condition in Umling block of Ri-Bhoi district using RS and GIS technique. J Indian Soc Remote Sens 37:317
Das R, Mondal B, Mondal P, Khatua DC, Mukherjee N (2012) Disease intensity of citrus canker on acid lime in relation to abiotic and biotic factors. J Agrometeorol 14:107–112
Davino S, Davino M, Sambade A, Guardo M, Caruso A (2003) The first citrus tristeza virus outbreak found in a relevant citrus producing area of Sicily, Italy. Plant Dis 87:314
de Mendoza AH, Ballester-Olmos JF, Pina JA (1984) Transmission of citrus Tristeza virus by aphids (Homoptera, Aphididae) in Spain. Int Organ Citrus Virol Conf Proc (1957–2010). 9:23–27
Debrick KS (1973) Quantitative assay for plant viruses using serologically specific electron microscopy. Virology 56:652–653
Del Campo R, Russi P, Mara P, Mara H, Peyrou M, De León IP, Gaggero C (2009) Xanthomonas axonopodis pv. citri enters the VBNC state after copper treatment and retains its virulence. FEMS Microbiol Lett 298:143–148
Derso E, Vernière C, Pruvost O (2009) First report of Xanthomonas citri pv. citri-A* causing citrus canker on lime in Ethiopia. Plant Dis 93:203
Duncan LW, Cohn E (1990) Nematode parasites of citrus. In: Plant parasitic nematodes in subtropical and tropical agriculture. CAB International, Wallingford, pp 321–346
Dwiastuti ME (2020) Citrus foot rot disease (Phytophthora spp.) control in Indonesia using good agricultural practices efforts green agroindustry. In: IOP conference series: earth and environmental science. IOP Publishing, p 12097
Fawcett HS, Wallace JM (1946) Evidence of the virus nature of citrus quick decline. Calif Citrogr 32:88–89
Fifaei R, Golein B, Taheri H, Tadjvar Y (2007) Elimination of citrus tristeza virus of Washington navel orange (Citrus sinensis [L.] Osbeck) through shoot-tip grafting. Int J Agric Biol 9:27–30
Ford HW (1954) The influence of rootstock and tree age on root distribution of citrus. Proc Amer Soc Hortic Sci 63:4–11
Garnsey SM, Muller GW (1988) Efficiency of mechanical transmission of citrus tristeza virus. Int Organ Citrus Virol Conf Proc
Garnsey SM, Gonsalves D, Purcifull DE (1977) Mechanical transmission of citrus tristeza virus. Phytopathology 67:965–968
Garnsey SM, Gonsalves D, Purcifull DE (1979) Rapid diagnosis of citrus tristeza virus infections by sodium dodecyl sulfate-immunodiffusion procedures. Phytopathology 69:88–95
Garnsey SM, Su HJ, Tsai MC (1996) Differential susceptibility of pummelo and Swingle citrumelo to isolates of citrus tristeza virus. Int Organ Citrus Virol Conf Proc (1957–2010). 13:138–146
Garnsey SM, Gottwald TR, Hilf ME, Matos L, Borbón J (2000) Emergence and spread of severe Strainsof citrus Tristeza virus in The Dominican Republic. Int Organ Citrus Virol Conf Proc (1957–2010). 14:57–68
Gonsalves D, Purcifull DE, Garnsey SM (1978) Purification and serology of citrus tristeza virus. Phytopathology 68:553–559

Goto M (1992) Citrus canker. In: Kumar J, Chaube HS, Singh US, Mukhopadhyay AN (eds) Plant diseases of international importance. Vol. III. Diseases of fruit crops. Prentice Hall, Englewood Cliffs, pp 170–208
Goto M, Tadauchi Y, Okabe N (1979) Interaction between Xanthomonas citri and Erwinia herbicola in vitro and in vivo. Jpn J Phytopathol 45:618–624
Gottwald TR, Timmer LW (1995) The efficacy of windbreaks in reducing the spread of citrus canker caused by Xanthomonas campestris pv. citri. Trop Agric 72(3):194–201
Gottwald TR, Graham JH, Schubert TS (1997) An epidemiological analysis of the spread of citrus canker in urban Miami, Florida, and synergistic interaction with the Asian citrus leafminer. Fruits 6:383–390
Gottwald TR, Garnsey SM, Borbón J (1998) Increase and patterns of spread of citrus tristeza virus infections in Costa Rica and The Dominican Republic in the presence of the brown citrus aphid, Toxoptera citricida. Phytopathology 88:621–636
Gottwald TR, Hughes G, Graham JH, Sun X, Riley T (2001) The citrus canker epidemic in Florida: the scientific basis of regulatory eradication policy for an invasive species. Phytopathology 91:30–34
Gottwald TR, Graham JH, Schubert TS (2002a) Citrus canker: the pathogen and its impact. Plant Heal Prog 3:15
Gottwald TR, Polek M, Riley K (2002b) History, present incidence, and spatial distribution of Citrus tristeza virus in the California Central Valley. Int Organ Citrus Virol Conf Proc
Gottwald TR, Sun X, Riley T, Graham JH, Ferrandino F, Taylor EL (2002c) Geo-referenced spatiotemporal analysis of the urban citrus canker epidemic in Florida. Phytopathology 92:361–377
Goulas V, Manganaris GA (2012) Exploring the phytochemical content and the antioxidant potential of citrus fruits grown in Cyprus. Food Chem 131:39–47
Gowda S, Satyanarayana T, Robertson CJ, Garnsey SM, Dawson WO (2005) Infection of citrus plants with virions generated in Nicotiana benthamiana plants agroinfiltrated with a binary vector based Citrus tristeza virus. Int Organ Citrus Virol Conf Proc
Hančević K, Hartl Musinov D, Černi S, Rošin J, Krajačić M, Gatin Ž, Škorić D (2009) The production of Citrus tristeza virus-free Zorica Rana, a Croatian selection of Satsuma mandarin. J Food Agric Environ 7:254–257
Horticulture Statistics at a Glance (2018) Horticulture Statistics Division, Ministry of Agriculture & Farmers' Welfare, Government of India, Gurgaon. 52–53.
Hughes WA, Lister CA (1949) Lime disease in the Gold Coast. Nature 164:880
Indian Horticulture Database (2008) Ministry of Agriculture, Government of India, Gurgaon. 1–89.
Kale KB, Kolte SO, Peshney NL (1994) Economics of chemical control of citrus canker caused by Xanthomonas campestris pv. citri under field conditions. Indian Phytopathol 47:253–255
Kalita P, Bora LC, Bhagabati KN (1995) Influence of environmental parameters on citrus canker incidence in Assam. J Agric Sci Soc N East India 8:33–35
Kalita P, Bora LC, Bhagabati KN (1996) Phylloplane microflora of citrus and their role in management of citrus canker. Indian Phytopathol 49:234–237
Kapari-Isaia T, Kyriakou A, Papayiannis L, Tsaltas D, Gregoriou S, Psaltis I (2008) Rapid in vitro microindexing of viroids in citrus. Plant Pathol 57:348–353
Kim G-N, Shin J-G, Jang H-D (2009) Antioxidant and antidiabetic activity of Dangyuja (Citrus grandis Osbeck) extract treated with Aspergillus saitoi. Food Chem 117:35–41
Kim SB, Chang BY, Jo YH, Lee SH, Han S-B, Hwang BY, Kim SY, Lee MK (2013) Macrophage activating activity of pyrrole alkaloids from Morus alba fruits. J Ethnopharmacol 145:393–396
Kitajima EW, Silva DM, Oliveira AR, Müller GW, Costa AS (1964) Thread-like particles associated with tristeza disease of citrus. Nature 201:1011–1012
Kitajima EW, Muller GW, Costa AS (1974) Electron microscopy of Tristeza-infected Passiflora gracilis Jacq. Int Organ Citrus Virol Conf Proc
Kyriakou A, Ioannou N, Gavriel J, Bar-Joseph M, Papayiannis C, Kapari T, Savva G (1996) Management of Citrus tristeza virus in Cyprus. Int Organ Citrus Virol Conf Proc (1957–2010). 13:172–178

Leite Jr RP (1990) Citrus canker. Prevention and control in the State of Paraná. Circ Agronômico do Paraná

Leite RP Jr, Mohan SK (1984) Evaluation of citrus cultivars for resistance to canker caused by Xanthomonas campestris pv. citri (Hasse) Dye in the State of Paraná, Brazil. Proc Int Soc Citric 1:385–389

Leite RP Jr (2002) Surviving with citrus canker. Proc Int Soc Citric. (in press)

Lin H-C, Chang H, Tzeng K-C (2008) Characterization of novel strains of citrus canker bacterium from citrus in Taiwan. J Taiwan Agric Res 57:265–278

Lin H-C, Hsu S-T, Chang H, Tzeng K-C (2010) A pectate lyase homologue pel1 from Xanthomonas axonopodis pv. citri is associated with the water-soaked margins of canker lesions. J Plant Pathol 92:149–156

Mabberley DJ (2004) Citrus (Rutaceae): a review of recent advances in etymology, systematics and medical applications. Blumea-Biodivers Evol Biogeogr Plants 49:481–498

Marroquín C, Olmos A, Gorris MT, Bertolini E, Martınez MC, Carbonell EA, de Mendoza AH, Cambra M (2004) Estimation of the number of aphids carrying Citrus tristeza virus that visit adult citrus trees. Virus Res 100:101–108

Masroor MK, Chandra S (1989) Effect of carbon sources on antibiotic production by Aspergillus spp. antagonistic to citrus canker pathogen. Philipp J Sci 118:141–145

McClean APD (1956) Tristeza and stem pitting diseases of citrus in South Africa. FAO Plant Prot Bull 4:88–94

McClean APD (1957) Tristeza virus of citrus: evidence for absence of seed transmission. Plant Dis Rep 41:821

Medina-Urrutia VM, Stapleton JJ (1985) Control of Mexican lime bacteriosis with copper-based products. Proc Florida State Hortic Soc 98:22–24

Mehta P, Brlansky RH, Gowda S, Yokomi RK (1997) Reverse-transcription polymerase chain reaction detection of citrus tristeza virus in aphids. Plant Dis 81:1066–1069

Mestre PF, Asins MJ, Pina JA, Carbonell EA, Navarro L (1997a) Molecular markers flanking citrus tristeza virus resistance gene from Poncirus trifoliata (L.) Raf. Theor Appl Genet 94:458–464

Mestre PF, Asins MJ, Pina JA, Navarro L (1997b) Efficient search for new resistant genotypes to the citrus tristeza closterovirus in the orange subfamily Aurantioideae. Theor Appl Genet 95:1282–1288

Mohammed AA-S, Arya W, Amgad AS, Yasser EI (2014) Distribution and pathotype identification of Xanthomonas citri subsp. citri recovered from south-western region of Saudi Arabia. Afr J Microbiol Res 8:673–679

Mokbel MS, Hashinaga F (2006) Evaluation of the antioxidant activity of extracts from buntan (Citrus grandis Osbeck) fruit tissues. Food Chem 94:529–534

Moreno P, Ambrós S, Albiach-Martí MR, Guerri J, Pena L (2008) Citrus tristeza virus: a pathogen that changed the course of the citrus industry. Mol Plant Pathol 9:251–268

Mukherjee N, Biswas P (1984) Control of plant diseases with plant oils: Their problems and prospects. Indian Agric 14:16–25

Muller GW, Rodriguez O, Costa AS (1968) A tristeza virus complex severe to sweet orange varieties. Int Organ Citrus Virol Conf Proc (1957–2010). 4:64–71

Muller GW, Costa AS, Kitajima EW, Camargo IJB (1974) Additional evidence that Tristeza virus multiplies in Passiflora spp. Int Organ Citrus Virol Conf Proc (1957–2010). 6:75–78

Nolasco G, Sequeira Z (2002) Bacterial expressed coat protein: development of a single antiserum for routine detection of Citrus tristeza virus. Bact expressed coat protein 1000–1008

Nolasco G, Sequeira Z, Bonacalza B, Mendes C, Torres V, Sanchez F, Urgoiti B, Ponz F, Febres VJ, Cevik B (1997) Sensitive CTV diagnosis using immunocapture (IC), reverse transcriptional polymerase chain reaction (RT-PCR) and an exonuclease fluorescent probe assay. Fruits 6:391–396

Okwu DE (2005) Phytochemicals, vitamins and mineral contents of two Nigerian medicinal plants. Int J Mol Med Adv Sci 1:375–381

Okwu DE, Emenike IN (2006) Evaluation of the phytonutrients and vitamin contents of citrus fruits. Int J Mol Med Adv Sci 2:1–6

Okwu DE, Morah FNI (2004) Mineral and nutritive value of Dennettia tripetala fruits. Fruits 59:437–442
Ota T (1983) Interactions in vitro and in vivo between Xanthomonas campestris pv. Citri and antagonistic pseudomonas sp. Jpn J Phytopathol 49:308–315
Pabitra K, Bora LC, Bhagabati KN (1997) Goat weed-a host of citrus canker (Xanthomonas campestris pv. Citri). J Mycol Plant Pathol 27:96–97
Palazzo DH, Nogueiro EMC, Ceravolo LC, Montovanello CM (1987) Epidemiological studies on citrus canker (xcc): progress of the disease in time. Laranja 1:133–140
Polek M, Vidalakis G, Godfrey K (2007) Citrus bacterial canker disease and huanglongbing (citrus greening). UC ANR Publication, Oakland, p 8218
Raccah B, Loebenstein G, Bar-Joseph M, Oren Y (1976) Transmission of tristeza by aphids prevalent on citrus, and operation of the tristeza suppression programme in Israel. Int Organ Citrus Virol Conf Proc (1957–2010). 7:47–49
Raccah B, Loebenstein G, Singer S (1980) Aphid-transmissibility variants of citrus tristeza virus in infected citrus trees. Phytopathology 70:89–93
Ramakrishnan TS (1954) Common diseases of citrus in Madras state. Government of Madras Publication, pp 130–134
Rapisarda P, Tomaino A, Lo Cascio R, Bonina F, De Pasquale A, Saija A (1999) Antioxidant effectiveness as influenced by phenolic content of fresh orange juices. J Agric Food Chem 47:4718–4723
Reedy BC (1984) Incidence of bacterial canker of citrus in relation to weather. Geobios New Rep 3:39–41
Rocha-Peña MA, Lee RF, Lastra R, Niblett CL, Ochoa-Corona FM, Garnsey SM, Yokomi RK (1995) Citrus tristeza virus and its aphid vector Toxoptera citricida: threats to citrus production in the Caribbean and central and North America. Plant Dis 79:437–445
Roistacher CN (1976) Tristeza in the Central Valley: a warning [Citrus, California]. Citrograph
Roistacher CN (1981) Blueprint for disaster-part two: changes in transmissibility of seedling yellows. Citrograph 67:28–32
Roy A, Ramachandran P (2012) Bi-directional PCR-a tool for identifying strains of Citrus tristeza virus. Indian Phytopathol 55:182–186
Roy A, Ananthakrishnan G, Hartung JS, Brlansky RH (2010) Development and application of a multiplex reverse-transcription polymerase chain reaction assay for screening a global collection of citrus tristeza virus isolates. Phytopathology 100:1077–1088
Ruiz-Ruiz S, Moreno P, Guerri J, Ambrós S (2006) The complete nucleotide sequence of a severe stem pitting isolate of Citrus tristeza virus from Spain: comparison with isolates from different origins. Arch Virol 151:387–398
Sarkar J, Saha J, Panja BN, Das A (2007) Evaluation of Citrus species and their relatives of tolerance to bacterial canker disease and prediction of canker incidence based on leaf morphological characters. J Mycopathol Res 45:175–181
Sawamura M, Kuriyama T (1988) Quantitative determination of volatile constituents in the pummelo (Citrus grandis Osbeck forma Tosa-buntan). J Agric Food Chem 36:567–569
Schneider H (1957) The anatomy of tristeza-virus-infected citrus. Int Organ Citrus Virol Conf Proc 1:73–84
Schubert TS, Rizvi SA, Sun X, Gottwald TR, Graham JH, Dixon WN (2001) Meeting the challenge of eradicating citrus canker in Florida—again. Plant Dis 85:340–356
Selvaraj V, Maheshwari Y, Hajeri S, Yokomi R (2019) A rapid detection tool for VT isolates of citrus tristeza virus by immunocapture-reverse transcriptase loop-mediated isothermal amplification assay. PLoS One 14:e0222170
Sharma S, Singh B, Rani G, Zaidi AA, Hallan V, Nagpal A, Virk GS (2007) Production of Indian citrus ringspot virus-free plants of Kinnow mandarin (Citrus nobilis Lour × C. deliciosa Tenora) employing chemotherapy coupled with shoot tip grafting. Vitr Cell Dev Biol 43:254–259
Sohi GS, Sandhu MS (1968) Relationship between citrus leafminer (Phyllocnistis citrella Stainton) injury and citrus canker (Xanthomonas citri (Hasse) Dowson) incidence on citrus leaves. J Res Punjab Agric Univ 5:66–69

Stall RE, Civerolo EL (1991) Research relating to the recent outbreak of citrus canker in Florida. Annu Rev Phytopathol 29:399–420

Stall RE, Seymour CP (1983) Canker, a threat to citrus in the Gulf-Coast states. Plant Dis 67:581–585

Takeuchi T, Hara T, Naganawa H, Okada M, Hamada M, Umezawat H, Gomi S, Sezaki M, Kondo S (1988) New antifungal antibiotics, benanomicins A and B from an actinomycete. J Antibiot (Tokyo) 41:807–811

Tao N, Liu Y, Zhang M (2009) Chemical composition and antimicrobial activities of essential oil from the peel of bingtang sweet orange (Citrus sinensis Osbeck). Int J Food Sci Technol 44:1281–1285

Timmer LW (2000) Inoculum production and epiphytic survival of Xanthomonas campestris pv. citri. Abstr). In: Proceedings of the international citrus canker and research workshop, Ft. Pierce FL, June, 20–22

Tongin Y, Leksomboon C, Jiwajinda S (2013) Efficacy of leaf extract from Terminalia catappa Linn. to inhibit lime canker disease. Proceedings of the 51st Kasetsart university annual conference Bangkok, Thailand, 5–7 Febr 2013, Kasetsart University

Unnamalai N, Gnanamanickam SS (1984) Pseudomonas fluorescens is an antagonist to Xanthomonas citri (Hasse) Dye, the incitant of citrus canker. Curr Sci India 53:703–704

Vela C, Cambra M, Cortés E, Moreno P, Miguet JG, de San Román CP, Sanz A (1986) Production and characterization of monoclonal antibodies specific for citrus tristeza virus and their use for diagnosis. J Gen Virol 67:91–96

Vernière C, Hartung JS, Pruvost OP, Civerolo EL, Alvarez AM, Maestri P, Luisetti J (1998) Characterization of phenotypically distinct strains of Xanthomonas axonopodis pv. citri from Southwest Asia. Eur J Plant Pathol 104:477–487

Vijayakumari N, Ghosh DK, Das AK, Singh A, Singh S (2006) Elimination of citrus tristeza virus and greening pathogens from exotic germplasm through in vitro shoot tip grafting in citrus. Indian J Agric Sci 76:209–210

Wallace JM (1951) Recent developments in studies of quick decline and related diseases. Phytopathology 41:785–793

Wallace JM (1978) Virus and virus like diseases. In: The citrus industry, vol 4. The University of California's Division of Agriculture and Natural Resources, Richmond, pp 67–184

Webb PG, Biggs RH, Gander JE (1988) Citrus canker–the role of enzymes and xanthan gum in infection and spread of the pathogen. In: Proceedings of the annual meeting of the Florida State Horticultural Society

Whiteside J, Garnsey S, Timmer L (1988) Compendium of citrus diseases. APS Press, St. Paul

Yang Z-N, Mathews DM, Dodds JA, Mirkov TE (1999) Molecular characterization of an isolate of citrus tristeza virus that causes severe symptoms in sweet orange. Virus Genes 19:131–142

Yokomi RK, Garnsey SM (1987) Transmission of citrus tristeza virus by Aphis gossypii and Aphis citricola in Florida. Phytophylactica 19:169–172

Yokomi RK, Saponari M, Sieburth PJ (2010) Rapid differentiation and identification of potential severe strains of citrus tristeza virus by real-time reverse transcription-polymerase chain reaction assays. Phytopathology 100:319–327

Zekri M, Dewdney MM, Yates JD, Roberts P (2011) Field identification of citrus canker symptoms and decontamination procedures, Extension service plant pathology series, 214. Florida Cooperative Extension Service, Institute of Food and Agricultural Sciences, University of Florida

Zhang KS, He SZ, Haung Z (1996) Effect of several fungicides for control ofcitrus canker [Xanthomonas campestris pv. citri (Hasse) Dowron]. South China. Fruits 25:20

Ziegler LW, Wolfe HS (1975) Citrus growing in Florida. University of Florida; First Edition edition (1961). University of Florida Press

Zubrzycki HM, de Zubrzycki AD (1981) Resistance to Xanthomonas campestris pv. citri (Hasse) Dowson in oranges (Citrus sinensis (L.) Osbeck). Proc Int Soc Citric 1:405–409

Index

S. Singh Purewal et al. (eds.), *Recent Advances in Citrus Fruits*,
https://doi.org/10.1007/978-3-031-37534-7

MIX
Papier aus verantwortungsvollen Quellen
Paper from responsible sources
FSC® C105338

If you have any concerns about our products, you can contact us on
ProductSafety@springernature.com

In case Publisher is established outside the EU, the EU authorized representative is:
Springer Nature Customer Service Center GmbH
Europaplatz 3, 69115 Heidelberg, Germany

Printed by Libri Plureos GmbH
in Hamburg, Germany